Multivariable Calculus

MATH 120

Eighth Edition

James Stewart

CENGAGE
Learning·

Australia • Brazil • Japan • Korea • Mexico • Singapore • Spain • United Kingdom • United States

Multivariable Calculus: MATH 120, Eighth Edition

Calculus: Early Transcendentals, 8th Edition
James Stewart

For product information and technology assistance, contact us at
Cengage Learning Customer & Sales Support, 1-800-354-9706

For permission to use material from this text or product, submit all requests online at **cengage.com/permissions**
Further permissions questions can be emailed to
permissionrequest@cengage.com

This book contains select works from existing Cengage Learning resources and was produced by Cengage Learning Custom Solutions for collegiate use. As such, those adopting and/or contributing to this work are responsible for editorial content accuracy, continuity and completeness.

Compilation © 2015 Cengage Learning

ISBN: 978-1-305-75208-5

WCN: 01-100-101

Cengage Learning
20 Channel Center Street
Boston, MA 02210
USA

Cengage Learning is a leading provider of customized learning solutions with office locations around the globe, including Singapore, the United Kingdom, Australia, Mexico, Brazil, and Japan. Locate your local office at:
www.international.cengage.com/region.

Cengage Learning products are represented in Canada by Nelson Education, Ltd.

For your lifelong learning solutions, visit **www.cengage.com/custom.**

Visit our corporate website at **www.cengage.com.**

Brief Contents

Preface

> A great discovery solves a great problem but there is a grain of discovery in the solution of any problem. Your problem may be modest; but if it challenges your curiosity and brings into play your inventive faculties, and if you solve it by your own means, you may experience the tension and enjoy the triumph of discovery.
>
> GEORGE POLYA

The art of teaching, Mark Van Doren said, is the art of assisting discovery. I have tried to write a book that assists students in discovering calculus—both for its practical power and its surprising beauty. In this edition, as in the first seven editions, I aim to convey to the student a sense of the utility of calculus and develop technical competence, but I also strive to give some appreciation for the intrinsic beauty of the subject. Newton undoubtedly experienced a sense of triumph when he made his great discoveries. I want students to share some of that excitement.

The emphasis is on understanding concepts. I think that nearly everybody agrees that this should be the primary goal of calculus instruction. In fact, the impetus for the current calculus reform movement came from the Tulane Conference in 1986, which formulated as their first recommendation:

Focus on conceptual understanding.

I have tried to implement this goal through the *Rule of Three:* "Topics should be presented geometrically, numerically, and algebraically." Visualization, numerical and graphical experimentation, and other approaches have changed how we teach conceptual reasoning in fundamental ways. More recently, the Rule of Three has been expanded to become the *Rule of Four* by emphasizing the verbal, or descriptive, point of view as well.

In writing the eighth edition my premise has been that it is possible to achieve conceptual understanding and still retain the best traditions of traditional calculus. The book contains elements of reform, but within the context of a traditional curriculum.

Alternate Versions

I have written several other calculus textbooks that might be preferable for some instructors. Most of them also come in single variable and multivariable versions.

- *Calculus,* Eighth Edition, is similar to the present textbook except that the exponential, logarithmic, and inverse trigonometric functions are covered in the second semester.

- *Essential Calculus*, Second Edition, is a much briefer book (840 pages), though it contains almost all of the topics in *Calculus,* Eighth Edition. The relative brevity is achieved through briefer exposition of some topics and putting some features on the website.

- *Essential Calculus: Early Transcendentals,* Second Edition, resembles *Essential Calculus,* but the exponential, logarithmic, and inverse trigonometric functions are covered in Chapter 3.

- *Calculus: Concepts and Contexts*, Fourth Edition, emphasizes conceptual understanding even more strongly than this book. The coverage of topics is not encyclopedic and the material on transcendental functions and on parametric equations is woven throughout the book instead of being treated in separate chapters.

- *Calculus: Early Vectors* introduces vectors and vector functions in the first semester and integrates them throughout the book. It is suitable for students taking engineering and physics courses concurrently with calculus.

- *Brief Applied Calculus* is intended for students in business, the social sciences, and the life sciences.

- *Biocalculus: Calculus for the Life Sciences* is intended to show students in the life sciences how calculus relates to biology.

- *Biocalculus: Calculus, Probability, and Statistics for the Life Sciences* contains all the content of *Biocalculus: Calculus for the Life Sciences* as well as three additional chapters covering probability and statistics.

What's New in the Eighth Edition?

The changes have resulted from talking with my colleagues and students at the University of Toronto and from reading journals, as well as suggestions from users and reviewers. Here are some of the many improvements that I've incorporated into this edition:

- The data in examples and exercises have been updated to be more timely.

- New examples have been added (see Examples 6.1.5, 11.2.5, and 14.3.3, for instance). And the solutions to some of the existing examples have been amplified.

- Three new projects have been added: The project *Controlling Red Blood Cell Loss During Surgery* (page 244) describes the ANH procedure, in which blood is extracted from the patient before an operation and is replaced by saline solution. This dilutes the patient's blood so that fewer red blood cells are lost during bleeding and the extracted blood is returned to the patient after surgery. The project *Planes and Birds: Minimizing Energy* (page 344) asks how birds can minimize power and energy by flapping their wings versus gliding. In the project *The Speedo LZR Racer* (page 936) it is explained that this suit reduces drag in the water and, as a result, many swimming records were broken. Students are asked why a small decrease in drag can have a big effect on performance.

- I have streamlined Chapter 15 (Multiple Integrals) by combining the first two sections so that iterated integrals are treated earlier.

- More than 20% of the exercises in each chapter are new. Here are some of my favorites: 2.7.61, 2.8.36–38, 3.1.79–80, 3.11.54, 4.1.69, 4.3.34, 4.3.66, 4.4.80, 4.7.39, 4.7.67, 5.1.19–20, 5.2.67–68, 5.4.70, 6.1.51, 8.1.39, 12.5.81, 12.6.29–30, 14.6.65–66. In addition, there are some good new Problems Plus. (See Problems 12–14 on page 272, Problem 13 on page 363, Problems 16–17 on page 426, and Problem 8 on page 986.)

Features

Conceptual Exercises

The most important way to foster conceptual understanding is through the problems that we assign. To that end I have devised various types of problems. Some exercise sets begin with requests to explain the meanings of the basic concepts of the section. (See, for instance, the first few exercises in Sections 2.2, 2.5, 11.2, 14.2, and 14.3.) Similarly, all the review sections begin with a Concept Check and a True-False Quiz. Other exercises test conceptual understanding through graphs or tables (see Exercises 2.7.17, 2.8.35–38, 2.8.47–52, 9.1.11–13, 10.1.24–27, 11.10.2, 13.2.1–2, 13.3.33–39, 14.1.1–2, 14.1.32–38, 14.1.41–44, 14.3.3–10, 14.6.1–2, 14.7.3–4, 15.1.6–8, 16.1.11–18, 16.2.17–18, and 16.3.1–2).

Another type of exercise uses verbal description to test conceptual understanding (see Exercises 2.5.10, 2.8.66, 4.3.69–70, and 7.8.67). I particularly value problems that combine and compare graphical, numerical, and algebraic approaches (see Exercises 2.6.45–46, 3.7.27, and 9.4.4).

Graded Exercise Sets

Each exercise set is carefully graded, progressing from basic conceptual exercises and skill-development problems to more challenging problems involving applications and proofs.

Real-World Data

My assistants and I spent a great deal of time looking in libraries, contacting companies and government agencies, and searching the Internet for interesting real-world data to introduce, motivate, and illustrate the concepts of calculus. As a result, many of the examples and exercises deal with functions defined by such numerical data or graphs. See, for instance, Figure 1 in Section 1.1 (seismograms from the Northridge earthquake), Exercise 2.8.35 (unemployment rates), Exercise 5.1.16 (velocity of the space shuttle *Endeavour*), and Figure 4 in Section 5.4 (San Francisco power consumption). Functions of two variables are illustrated by a table of values of the wind-chill index as a function of air temperature and wind speed (Example 14.1.2). Partial derivatives are introduced in Section 14.3 by examining a column in a table of values of the heat index (perceived air temperature) as a function of the actual temperature and the relative humidity. This example is pursued further in connection with linear approximations (Example 14.4.3). Directional derivatives are introduced in Section 14.6 by using a temperature contour map to estimate the rate of change of temperature at Reno in the direction of Las Vegas. Double integrals are used to estimate the average snowfall in Colorado on December 20–21, 2006 (Example 15.1.9). Vector fields are introduced in Section 16.1 by depictions of actual velocity vector fields showing San Francisco Bay wind patterns.

Projects

One way of involving students and making them active learners is to have them work (perhaps in groups) on extended projects that give a feeling of substantial accomplishment when completed. I have included four kinds of projects: *Applied Projects* involve applications that are designed to appeal to the imagination of students. The project after Section 9.3 asks whether a ball thrown upward takes longer to reach its maximum height or to fall back to its original height. (The answer might surprise you.) The project after Section 14.8 uses Lagrange multipliers to determine the masses of the three stages of a rocket so as to minimize the total mass while enabling the rocket to reach a desired velocity. *Laboratory Projects* involve technology; the one following Section 10.2 shows how to use Bézier curves to design shapes that represent letters for a laser printer. *Writing Projects* ask students to compare present-day methods with those of the founders of calculus—Fermat's method for finding tangents, for instance. Suggested references are supplied. *Discovery Projects* anticipate results to be discussed later or encourage discovery through pattern recognition (see the one following Section 7.6). Others explore aspects of geometry: tetrahedra (after Section 12.4), hyperspheres (after Section 15.6), and intersections of three cylinders (after Section 15.7). Additional projects can be found in the *Instructor's Guide* (see, for instance, Group Exercise 5.1: Position from Samples).

■ Problem Solving

Students usually have difficulties with problems for which there is no single well-defined procedure for obtaining the answer. I think nobody has improved very much on George Polya's four-stage problem-solving strategy and, accordingly, I have included a version of his problem-solving principles following Chapter 1. They are applied, both explicitly and implicitly, throughout the book. After the other chapters I have placed sections called *Problems Plus*, which feature examples of how to tackle challenging calculus problems. In selecting the varied problems for these sections I kept in mind the following advice from David Hilbert: "A mathematical problem should be difficult in order to entice us, yet not inaccessible lest it mock our efforts." When I put these challenging problems on assignments and tests I grade them in a different way. Here I reward a student significantly for ideas toward a solution and for recognizing which problem-solving principles are relevant.

■ Technology

The availability of technology makes it not less important but more important to clearly understand the concepts that underlie the images on the screen. But, when properly used, graphing calculators and computers are powerful tools for discovering and understanding those concepts. This textbook can be used either with or without technology and I use two special symbols to indicate clearly when a particular type of machine is required. The icon ⊞ indicates an exercise that definitely requires the use of such technology, but that is not to say that it can't be used on the other exercises as well. The symbol CAS is reserved for problems in which the full resources of a computer algebra system (like Maple, Mathematica, or the TI-89) are required. But technology doesn't make pencil and paper obsolete. Hand calculation and sketches are often preferable to technology for illustrating and reinforcing some concepts. Both instructors and students need to develop the ability to decide where the hand or the machine is appropriate.

■ Tools for Enriching Calculus

TEC is a companion to the text and is intended to enrich and complement its contents. (It is now accessible in the eBook via CourseMate and Enhanced WebAssign. Selected Visuals and Modules are available at www.stewartcalculus.com.) Developed by Harvey Keynes, Dan Clegg, Hubert Hohn, and myself, TEC uses a discovery and exploratory approach. In sections of the book where technology is particularly appropriate, marginal icons direct students to TEC Modules that provide a laboratory environment in which they can explore the topic in different ways and at different levels. **Visuals are animations of figures in text; Modules are more elaborate activities and include exercises**. Instructors can choose to become involved at several different levels, ranging from simply encouraging students to use the Visuals and Modules for independent exploration, to assigning specific exercises from those included with each Module, or to creating additional exercises, labs, and projects that make use of the Visuals and Modules.

TEC also includes Homework Hints for representative exercises (usually odd-numbered) in every section of the text, indicated by printing the exercise number in red. These hints are usually presented in the form of questions and try to imitate an effective teaching assistant by functioning as a silent tutor. They are constructed so as not to reveal any more of the actual solution than is minimally necessary to make further progress.

■ Enhanced WebAssign

Technology is having an impact on the way homework is assigned to students, particularly in large classes. The use of online homework is growing and its appeal depends on ease of use, grading precision, and reliability. With the Eighth Edition we have been working with the calculus community and WebAssign to develop an online homework system. Up to 70% of the exercises in each section are assignable as online homework, including free response, multiple choice, and multi-part formats.

The system also includes Active Examples, in which students are guided in step-by-step tutorials through text examples, with links to the textbook and to video solutions.

■ Website

Visit CengageBrain.com or stewartcalculus.com for these additional materials:

- Homework Hints
- Algebra Review
- Lies My Calculator and Computer Told Me
- History of Mathematics, with links to the better historical websites
- Additional Topics (complete with exercise sets): Fourier Series, Formulas for the Remainder Term in Taylor Series, Rotation of Axes
- Archived Problems (Drill exercises that appeared in previous editions, together with their solutions)
- Challenge Problems (some from the Problems Plus sections from prior editions)
- Links, for particular topics, to outside Web resources
- Selected Visuals and Modules from Tools for Enriching Calculus (TEC)

Content

Diagnostic Tests The book begins with four diagnostic tests, in Basic Algebra, Analytic Geometry, Functions, and Trigonometry.

A Preview of Calculus This is an overview of the subject and includes a list of questions to motivate the study of calculus.

1 Functions and Models From the beginning, multiple representations of functions are stressed: verbal, numerical, visual, and algebraic. A discussion of mathematical models leads to a review of the standard functions, including exponential and logarithmic functions, from these four points of view.

2 Limits and Derivatives The material on limits is motivated by a prior discussion of the tangent and velocity problems. Limits are treated from descriptive, graphical, numerical, and algebraic points of view. Section 2.4, on the precise definition of a limit, is an optional section. Sections

2.7 and 2.8 deal with derivatives (especially with functions defined graphically and numerically) before the differentiation rules are covered in Chapter 3. Here the examples and exercises explore the meanings of derivatives in various contexts. Higher derivatives are introduced in Section 2.8.

3 Differentiation Rules

All the basic functions, including exponential, logarithmic, and inverse trigonometric functions, are differentiated here. When derivatives are computed in applied situations, students are asked to explain their meanings. Exponential growth and decay are now covered in this chapter.

4 Applications of Differentiation

The basic facts concerning extreme values and shapes of curves are deduced from the Mean Value Theorem. Graphing with technology emphasizes the interaction between calculus and calculators and the analysis of families of curves. Some substantial optimization problems are provided, including an explanation of why you need to raise your head 42° to see the top of a rainbow.

5 Integrals

The area problem and the distance problem serve to motivate the definite integral, with sigma notation introduced as needed. (Full coverage of sigma notation is provided in Appendix E.) Emphasis is placed on explaining the meanings of integrals in various contexts and on estimating their values from graphs and tables.

6 Applications of Integration

Here I present the applications of integration—area, volume, work, average value—that can reasonably be done without specialized techniques of integration. General methods are emphasized. The goal is for students to be able to divide a quantity into small pieces, estimate with Riemann sums, and recognize the limit as an integral.

7 Techniques of Integration

All the standard methods are covered but, of course, the real challenge is to be able to recognize which technique is best used in a given situation. Accordingly, in Section 7.5, I present a strategy for integration. The use of computer algebra systems is discussed in Section 7.6.

8 Further Applications of Integration

Here are the applications of integration—arc length and surface area—for which it is useful to have available all the techniques of integration, as well as applications to biology, economics, and physics (hydrostatic force and centers of mass). I have also included a section on probability. There are more applications here than can realistically be covered in a given course. Instructors should select applications suitable for their students and for which they themselves have enthusiasm.

9 Differential Equations

Modeling is the theme that unifies this introductory treatment of differential equations. Direction fields and Euler's method are studied before separable and linear equations are solved explicitly, so that qualitative, numerical, and analytic approaches are given equal consideration. These methods are applied to the exponential, logistic, and other models for population growth. The first four or five sections of this chapter serve as a good introduction to first-order differential equations. An optional final section uses predator-prey models to illustrate systems of differential equations.

10 Parametric Equations and Polar Coordinates

This chapter introduces parametric and polar curves and applies the methods of calculus to them. Parametric curves are well suited to laboratory projects; the two presented here involve families of curves and Bézier curves. A brief treatment of conic sections in polar coordinates prepares the way for Kepler's Laws in Chapter 13.

11 Infinite Sequences and Series   The convergence tests have intuitive justifications (see page 719) as well as formal proofs. Numerical estimates of sums of series are based on which test was used to prove convergence. The emphasis is on Taylor series and polynomials and their applications to physics. Error estimates include those from graphing devices.

12 Vectors and the Geometry of Space   The material on three-dimensional analytic geometry and vectors is divided into two chapters. Chapter 12 deals with vectors, the dot and cross products, lines, planes, and surfaces.

13 Vector Functions   This chapter covers vector-valued functions, their derivatives and integrals, the length and curvature of space curves, and velocity and acceleration along space curves, culminating in Kepler's laws.

14 Partial Derivatives   Functions of two or more variables are studied from verbal, numerical, visual, and algebraic points of view. In particular, I introduce partial derivatives by looking at a specific column in a table of values of the heat index (perceived air temperature) as a function of the actual temperature and the relative humidity.

15 Multiple Integrals   Contour maps and the Midpoint Rule are used to estimate the average snowfall and average temperature in given regions. Double and triple integrals are used to compute probabilities, surface areas, and (in projects) volumes of hyperspheres and volumes of intersections of three cylinders. Cylindrical and spherical coordinates are introduced in the context of evaluating triple integrals.

16 Vector Calculus   Vector fields are introduced through pictures of velocity fields showing San Francisco Bay wind patterns. The similarities among the Fundamental Theorem for line integrals, Green's Theorem, Stokes' Theorem, and the Divergence Theorem are emphasized.

17 Second-Order Differential Equations   Since first-order differential equations are covered in Chapter 9, this final chapter deals with second-order linear differential equations, their application to vibrating springs and electric circuits, and series solutions.

Ancillaries

Calculus, Early Transcendentals, Eighth Edition, is supported by a complete set of ancillaries developed under my direction. Each piece has been designed to enhance student understanding and to facilitate creative instruction. The tables on pages xxi–xxii describe each of these ancillaries.

Acknowledgments

The preparation of this and previous editions has involved much time spent reading the reasoned (but sometimes contradictory) advice from a large number of astute reviewers. I greatly appreciate the time they spent to understand my motivation for the approach taken. I have learned something from each of them.

■ Eighth Edition Reviewers

Jay Abramson, *Arizona State University*
Adam Bowers, *University of California San Diego*
Neena Chopra, *The Pennsylvania State University*

Edward Dobson, *Mississippi State University*
Isaac Goldbring, *University of Illinois at Chicago*
Lea Jenkins, *Clemson University*
Rebecca Wahl, *Butler University*

■ Technology Reviewers

Maria Andersen, *Muskegon Community College*
Eric Aurand, *Eastfield College*
Joy Becker, *University of Wisconsin–Stout*
Przemyslaw Bogacki, *Old Dominion University*
Amy Elizabeth Bowman, *University of Alabama
   in Huntsville*
Monica Brown, *University of Missouri–St. Louis*
Roxanne Byrne, *University of Colorado at Denver and
   Health Sciences Center*
Teri Christiansen, *University of Missouri–Columbia*
Bobby Dale Daniel, *Lamar University*
Jennifer Daniel, *Lamar University*
Andras Domokos, *California State University, Sacramento*
Timothy Flaherty, *Carnegie Mellon University*
Lee Gibson, *University of Louisville*
Jane Golden, *Hillsborough Community College*
Semion Gutman, *University of Oklahoma*
Diane Hoffoss, *University of San Diego*
Lorraine Hughes, *Mississippi State University*
Jay Jahangiri, *Kent State University*
John Jernigan, *Community College of Philadelphia*

Brian Karasek, *South Mountain Community College*
Jason Kozinski, *University of Florida*
Carole Krueger, *The University of Texas at Arlington*
Ken Kubota, *University of Kentucky*
John Mitchell, *Clark College*
Donald Paul, *Tulsa Community College*
Chad Pierson, *University of Minnesota, Duluth*
Lanita Presson, *University of Alabama in Huntsville*
Karin Reinhold, *State University of New York at Albany*
Thomas Riedel, *University of Louisville*
Christopher Schroeder, *Morehead State University*
Angela Sharp, *University of Minnesota, Duluth*
Patricia Shaw, *Mississippi State University*
Carl Spitznagel, *John Carroll University*
Mohammad Tabanjeh, *Virginia State University*
Capt. Koichi Takagi, *United States Naval Academy*
Lorna TenEyck, *Chemeketa Community College*
Roger Werbylo, *Pima Community College*
David Williams, *Clayton State University*
Zhuan Ye, *Northern Illinois University*

■ Previous Edition Reviewers

B. D. Aggarwala, *University of Calgary*
John Alberghini, *Manchester Community College*
Michael Albert, *Carnegie-Mellon University*
Daniel Anderson, *University of Iowa*
Amy Austin, *Texas A&M University*
Donna J. Bailey, *Northeast Missouri State University*
Wayne Barber, *Chemeketa Community College*
Marilyn Belkin, *Villanova University*
Neil Berger, *University of Illinois, Chicago*
David Berman, *University of New Orleans*
Anthony J. Bevelacqua, *University of North Dakota*
Richard Biggs, *University of Western Ontario*
Robert Blumenthal, *Oglethorpe University*
Martina Bode, *Northwestern University*
Barbara Bohannon, *Hofstra University*
Jay Bourland, *Colorado State University*
Philip L. Bowers, *Florida State University*
Amy Elizabeth Bowman, *University of Alabama in Huntsville*
Stephen W. Brady, *Wichita State University*
Michael Breen, *Tennessee Technological University*
Robert N. Bryan, *University of Western Ontario*

David Buchthal, *University of Akron*
Jenna Carpenter, *Louisiana Tech University*
Jorge Cassio, *Miami-Dade Community College*
Jack Ceder, *University of California, Santa Barbara*
Scott Chapman, *Trinity University*
Zhen-Qing Chen, *University of Washington—Seattle*
James Choike, *Oklahoma State University*
Barbara Cortzen, *DePaul University*
Carl Cowen, *Purdue University*
Philip S. Crooke, *Vanderbilt University*
Charles N. Curtis, *Missouri Southern State College*
Daniel Cyphert, *Armstrong State College*
Robert Dahlin
M. Hilary Davies, *University of Alaska Anchorage*
Gregory J. Davis, *University of Wisconsin–Green Bay*
Elias Deeba, *University of Houston–Downtown*
Daniel DiMaria, *Suffolk Community College*
Seymour Ditor, *University of Western Ontario*
Greg Dresden, *Washington and Lee University*
Daniel Drucker, *Wayne State University*
Kenn Dunn, *Dalhousie University*

Dennis Dunninger, *Michigan State University*
Bruce Edwards, *University of Florida*
David Ellis, *San Francisco State University*
John Ellison, *Grove City College*
Martin Erickson, *Truman State University*
Garret Etgen, *University of Houston*
Theodore G. Faticoni, *Fordham University*
Laurene V. Fausett, *Georgia Southern University*
Norman Feldman, *Sonoma State University*
Le Baron O. Ferguson, *University of California—Riverside*
Newman Fisher, *San Francisco State University*
José D. Flores, *The University of South Dakota*
William Francis, *Michigan Technological University*
James T. Franklin, *Valencia Community College, East*
Stanley Friedlander, *Bronx Community College*
Patrick Gallagher, *Columbia University–New York*
Paul Garrett, *University of Minnesota–Minneapolis*
Frederick Gass, *Miami University of Ohio*
Bruce Gilligan, *University of Regina*
Matthias K. Gobbert, *University of Maryland, Baltimore County*
Gerald Goff, *Oklahoma State University*
Stuart Goldenberg, *California Polytechnic State University*
John A. Graham, *Buckingham Browne & Nichols School*
Richard Grassl, *University of New Mexico*
Michael Gregory, *University of North Dakota*
Charles Groetsch, *University of Cincinnati*
Paul Triantafilos Hadavas, *Armstrong Atlantic State University*
Salim M. Haïdar, *Grand Valley State University*
D. W. Hall, *Michigan State University*
Robert L. Hall, *University of Wisconsin–Milwaukee*
Howard B. Hamilton, *California State University, Sacramento*
Darel Hardy, *Colorado State University*
Shari Harris, *John Wood Community College*
Gary W. Harrison, *College of Charleston*
Melvin Hausner, *New York University/Courant Institute*
Curtis Herink, *Mercer University*
Russell Herman, *University of North Carolina at Wilmington*
Allen Hesse, *Rochester Community College*
Randall R. Holmes, *Auburn University*
James F. Hurley, *University of Connecticut*
Amer Iqbal, *University of Washington—Seattle*
Matthew A. Isom, *Arizona State University*
Gerald Janusz, *University of Illinois at Urbana-Champaign*
John H. Jenkins, *Embry-Riddle Aeronautical University, Prescott Campus*
Clement Jeske, *University of Wisconsin, Platteville*
Carl Jockusch, *University of Illinois at Urbana-Champaign*
Jan E. H. Johansson, *University of Vermont*
Jerry Johnson, *Oklahoma State University*
Zsuzsanna M. Kadas, *St. Michael's College*
Nets Katz, *Indiana University Bloomington*
Matt Kaufman
Matthias Kawski, *Arizona State University*
Frederick W. Keene, *Pasadena City College*
Robert L. Kelley, *University of Miami*
Akhtar Khan, *Rochester Institute of Technology*

Marianne Korten, *Kansas State University*
Virgil Kowalik, *Texas A&I University*
Kevin Kreider, *University of Akron*
Leonard Krop, *DePaul University*
Mark Krusemeyer, *Carleton College*
John C. Lawlor, *University of Vermont*
Christopher C. Leary, *State University of New York at Geneseo*
David Leeming, *University of Victoria*
Sam Lesseig, *Northeast Missouri State University*
Phil Locke, *University of Maine*
Joyce Longman, *Villanova University*
Joan McCarter, *Arizona State University*
Phil McCartney, *Northern Kentucky University*
Igor Malyshev, *San Jose State University*
Larry Mansfield, *Queens College*
Mary Martin, *Colgate University*
Nathaniel F. G. Martin, *University of Virginia*
Gerald Y. Matsumoto, *American River College*
James McKinney, *California State Polytechnic University, Pomona*
Tom Metzger, *University of Pittsburgh*
Richard Millspaugh, *University of North Dakota*
Lon H. Mitchell, *Virginia Commonwealth University*
Michael Montaño, *Riverside Community College*
Teri Jo Murphy, *University of Oklahoma*
Martin Nakashima, *California State Polytechnic University, Pomona*
Ho Kuen Ng, *San Jose State University*
Richard Nowakowski, *Dalhousie University*
Hussain S. Nur, *California State University, Fresno*
Norma Ortiz-Robinson, *Virginia Commonwealth University*
Wayne N. Palmer, *Utica College*
Vincent Panico, *University of the Pacific*
F. J. Papp, *University of Michigan–Dearborn*
Mike Penna, *Indiana University–Purdue University Indianapolis*
Mark Pinsky, *Northwestern University*
Lothar Redlin, *The Pennsylvania State University*
Joel W. Robbin, *University of Wisconsin–Madison*
Lila Roberts, *Georgia College and State University*
E. Arthur Robinson, Jr., *The George Washington University*
Richard Rockwell, *Pacific Union College*
Rob Root, *Lafayette College*
Richard Ruedemann, *Arizona State University*
David Ryeburn, *Simon Fraser University*
Richard St. Andre, *Central Michigan University*
Ricardo Salinas, *San Antonio College*
Robert Schmidt, *South Dakota State University*
Eric Schreiner, *Western Michigan University*
Mihr J. Shah, *Kent State University–Trumbull*
Qin Sheng, *Baylor University*
Theodore Shifrin, *University of Georgia*
Wayne Skrapek, *University of Saskatchewan*
Larry Small, *Los Angeles Pierce College*
Teresa Morgan Smith, *Blinn College*
William Smith, *University of North Carolina*
Donald W. Solomon, *University of Wisconsin–Milwaukee*
Edward Spitznagel, *Washington University*

Joseph Stampfli, *Indiana University*
Kristin Stoley, *Blinn College*
M. B. Tavakoli, *Chaffey College*
Magdalena Toda, *Texas Tech University*
Ruth Trygstad, *Salt Lake Community College*
Paul Xavier Uhlig, *St. Mary's University, San Antonio*
Stan Ver Nooy, *University of Oregon*
Andrei Verona, *California State University–Los Angeles*
Klaus Volpert, *Villanova University*
Russell C. Walker, *Carnegie Mellon University*
William L. Walton, *McCallie School*

Peiyong Wang, *Wayne State University*
Jack Weiner, *University of Guelph*
Alan Weinstein, *University of California, Berkeley*
Theodore W. Wilcox, *Rochester Institute of Technology*
Steven Willard, *University of Alberta*
Robert Wilson, *University of Wisconsin–Madison*
Jerome Wolbert, *University of Michigan–Ann Arbor*
Dennis H. Wortman, *University of Massachusetts, Boston*
Mary Wright, *Southern Illinois University–Carbondale*
Paul M. Wright, *Austin Community College*
Xian Wu, *University of South Carolina*

In addition, I would like to thank R. B. Burckel, Bruce Colletti, David Behrman, John Dersch, Gove Effinger, Bill Emerson, Dan Kalman, Quyan Khan, Alfonso Gracia-Saz, Allan MacIsaac, Tami Martin, Monica Nitsche, Lamia Raffo, Norton Starr, and Jim Trefzger for their suggestions; Al Shenk and Dennis Zill for permission to use exercises from their calculus texts; COMAP for permission to use project material; George Bergman, David Bleecker, Dan Clegg, Victor Kaftal, Anthony Lam, Jamie Lawson, Ira Rosenholtz, Paul Sally, Lowell Smylie, and Larry Wallen for ideas for exercises; Dan Drucker for the roller derby project; Thomas Banchoff, Tom Farmer, Fred Gass, John Ramsay, Larry Riddle, Philip Straffin, and Klaus Volpert for ideas for projects; Dan Anderson, Dan Clegg, Jeff Cole, Dan Drucker, and Barbara Frank for solving the new exercises and suggesting ways to improve them; Marv Riedesel and Mary Johnson for accuracy in proofreading; Andy Bulman-Fleming, Lothar Redlin, Gina Sanders, and Saleem Watson for additional proofreading; and Jeff Cole and Dan Clegg for their careful preparation and proofreading of the answer manuscript.

In addition, I thank those who have contributed to past editions: Ed Barbeau, George Bergman, Fred Brauer, Andy Bulman-Fleming, Bob Burton, David Cusick, Tom DiCiccio, Garret Etgen, Chris Fisher, Leon Gerber, Stuart Goldenberg, Arnold Good, Gene Hecht, Harvey Keynes, E. L. Koh, Zdislav Kovarik, Kevin Kreider, Emile LeBlanc, David Leep, Gerald Leibowitz, Larry Peterson, Mary Pugh, Lothar Redlin, Carl Riehm, John Ringland, Peter Rosenthal, Dusty Sabo, Doug Shaw, Dan Silver, Simon Smith, Saleem Watson, Alan Weinstein, and Gail Wolkowicz.

I also thank Kathi Townes, Stephanie Kuhns, Kristina Elliott, and Kira Abdallah of TECHarts for their production services and the following Cengage Learning staff: Cheryll Linthicum, content project manager; Stacy Green, senior content developer; Samantha Lugtu, associate content developer; Stephanie Kreuz, product assistant; Lynh Pham, media developer; Ryan Ahern, marketing manager; and Vernon Boes, art director. They have all done an outstanding job.

I have been very fortunate to have worked with some of the best mathematics editors in the business over the past three decades: Ron Munro, Harry Campbell, Craig Barth, Jeremy Hayhurst, Gary Ostedt, Bob Pirtle, Richard Stratton, Liz Covello, and now Neha Taleja. All of them have contributed greatly to the success of this book.

JAMES STEWART

Ancillaries for Instructors

Instructor's Guide
by Douglas Shaw
ISBN 978-1-305-39371-4

Each section of the text is discussed from several viewpoints. The Instructor's Guide contains suggested time to allot, points to stress, text discussion topics, core materials for lecture, workshop/discussion suggestions, group work exercises in a form suitable for handout, and suggested homework assignments.

Complete Solutions Manual

Single Variable Early Transcendentals
By Daniel Anderson, Jeffery A. Cole, and Daniel Drucker
ISBN 978-1-305-27239-2

Multivariable
By Dan Clegg and Barbara Frank
ISBN 978-1-305-27611-6

Includes worked-out solutions to all exercises in the text.

Printed Test Bank
By William Steven Harmon
ISBN 978-1-305-38722-5

Contains text-specific multiple-choice and free response test items.

Cengage Learning Testing Powered by Cognero
(login.cengage.com)

This flexible online system allows you to author, edit, and manage test bank content from multiple Cengage Learning solutions; create multiple test versions in an instant; and deliver tests from your LMS, your classroom, or wherever you want.

Ancillaries for Instructors and Students

Stewart Website
www.stewartcalculus.com

Contents: *Homework Hints* ■ *Algebra Review* ■ *Additional Topics* ■ *Drill exercises* ■ *Challenge Problems* ■ *Web Links* ■ *History of Mathematics* ■ *Tools for Enriching Calculus (TEC)*

TEC TOOLS FOR ENRICHING™ CALCULUS
By James Stewart, Harvey Keynes, Dan Clegg, and developer Hubert Hohn

Tools for Enriching Calculus (TEC) functions as both a powerful tool for instructors and as a tutorial environment in which students can explore and review selected topics. The Flash simulation modules in TEC include instructions, written and audio explanations of the concepts, and exercises. TEC is accessible in the eBook via CourseMate and Enhanced WebAssign. Selected Visuals and Modules are available at www.stewartcalculus.com.

ENHANCED WebAssign Enhanced WebAssign®
www.webassign.net
Printed Access Code: ISBN 978-1-285-85826-5
Instant Access Code ISBN: 978-1-285-85825-8

Exclusively from Cengage Learning, Enhanced WebAssign offers an extensive online program for Stewart's Calculus *to encourage the practice that is so critical for concept mastery. The meticulously crafted pedagogy and exercises in our proven texts become even more effective in Enhanced WebAssign, supplemented by multimedia tutorial support and immediate feedback as students complete their assignments. Key features include:*

- *Thousands of homework problems that match your text-book's end-of-section exercises*

- *Opportunities for students to review prerequisite skills and content both at the start of the course and at the beginning of each section*

- Read It *eBook pages,* Watch It *videos,* Master It *tutorials, and* Chat About It *links*

- *A customizable Cengage* YouBook *with highlighting, note-taking, and search features, as well as links to multimedia resources*

- Personal Study Plans *(based on diagnostic quizzing) that identify chapter topics that students will need to master*

- *A WebAssign* Answer Evaluator *that recognizes and accepts equivalent mathematical responses in the same way an instructor grades*

- A Show My Work *feature that gives instructors the option of seeing students' detailed solutions*

- Visualizing Calculus Animations, Lecture Videos, *and more*

■ Electronic items   ■ Printed items

(Table continues on page xxii)

Cengage Customizable YouBook

YouBook *is an eBook that is both interactive and customizable. Containing all the content from Stewart's* Calculus, *YouBook features a text edit tool that allows instructors to modify the textbook narrative as needed. With YouBook, instructors can quickly reorder entire sections and chapters or hide any content they don't teach to create an eBook that perfectly matches their syllabus. Instructors can further customize the text by adding instructor-created or YouTube video links. Additional media assets include animated figures, video clips, highlighting and note-taking features, and more. YouBook is available within Enhanced WebAssign.*

CourseMate

CourseMate is a perfect self-study tool for students, and requires no set up from instructors. CourseMate brings course concepts to life with interactive learning, study, and exam preparation tools that support the printed textbook. Course-Mate for Stewart's Calculus *includes an interactive eBook, Tools for Enriching Calculus, videos, quizzes, flashcards, and more. For instructors, CourseMate includes Engagement Tracker, a first-of-its-kind tool that monitors student engagement.*

CengageBrain.com

To access additional course materials, please visit www.cengagebrain.com. At the CengageBrain.com home page, search for the ISBN of your title (from the back cover of your book) using the search box at the top of the page. This will take you to the product page where these resources can be found.

Ancillaries for Students

Student Solutions Manual

Single Variable Early Transcendentals
By Daniel Anderson, Jeffery A. Cole, and Daniel Drucker
ISBN 978-1-305-27242-2

Multivariable
By Dan Clegg and Barbara Frank
ISBN 978-1-305-27182-1

Provides completely worked-out solutions to all odd-numbered exercises in the text, giving students a chance to

check their answer and ensure they took the correct steps to arrive at the answer. The Student Solutions Manual can be ordered or accessed online as an eBook at www.cengagebrain.com by searching the ISBN.

Study Guide

Single Variable Early Transcendentals
By Richard St. Andre
ISBN 978-1-305-27914-8

Multivariable
By Richard St. Andre
ISBN 978-1-305-27184-5

For each section of the text, the Study Guide provides students with a brief introduction, a short list of concepts to master, and summary and focus questions with explained answers. The Study Guide also contains "Technology Plus" questions and multiple-choice "On Your Own" exam-style questions. The Study Guide can be ordered or accessed online as an eBook at www.cengagebrain.com by searching the ISBN.

A Companion to Calculus
By Dennis Ebersole, Doris Schattschneider, Alicia Sevilla, and Kay Somers
ISBN 978-0-495-01124-8

Written to improve algebra and problem-solving skills of students taking a calculus course, every chapter in this companion is keyed to a calculus topic, providing conceptual background and specific algebra techniques needed to understand and solve calculus problems related to that topic. It is designed for calculus courses that integrate the review of precalculus concepts or for individual use. Order a copy of the text or access the eBook online at www.cengagebrain.com by searching the ISBN.

Linear Algebra for Calculus
by Konrad J. Heuvers, William P. Francis, John H. Kuisti, Deborah F. Lockhart, Daniel S. Moak, and Gene M. Ortner
ISBN 978-0-534-25248-9

This comprehensive book, designed to supplement the calculus course, provides an introduction to and review of the basic ideas of linear algebra. Order a copy of the text or access the eBook online at www.cengagebrain.com by searching the ISBN.

To the Student

Reading a calculus textbook is different from reading a newspaper or a novel, or even a physics book. Don't be discouraged if you have to read a passage more than once in order to understand it. You should have pencil and paper and calculator at hand to sketch a diagram or make a calculation.

Some students start by trying their homework problems and read the text only if they get stuck on an exercise. I suggest that a far better plan is to read and understand a section of the text before attempting the exercises. In particular, you should look at the definitions to see the exact meanings of the terms. And before you read each example, I suggest that you cover up the solution and try solving the problem yourself. You'll get a lot more from looking at the solution if you do so.

Part of the aim of this course is to train you to think logically. Learn to write the solutions of the exercises in a connected, step-by-step fashion with explanatory sentences—not just a string of disconnected equations or formulas.

The answers to the odd-numbered exercises appear at the back of the book, in Appendix I. Some exercises ask for a verbal explanation or interpretation or description. In such cases there is no single correct way of expressing the answer, so don't worry that you haven't found the definitive answer. In addition, there are often several different forms in which to express a numerical or algebraic answer, so if your answer differs from mine, don't immediately assume you're wrong. For example, if the answer given in the back of the book is $\sqrt{2} - 1$ and you obtain $1/(1 + \sqrt{2})$, then you're right and rationalizing the denominator will show that the answers are equivalent.

The icon ⬚ indicates an exercise that definitely requires the use of either a graphing calculator or a computer with graphing software. But that doesn't mean that graphing devices can't be used to check your work on the other exercises as well. The symbol ⬚ is reserved for problems in which the full resources of a computer algebra system (like Maple, Mathematica, or the TI-89) are required.

You will also encounter the symbol ⬚, which warns you against committing an error. I have placed this symbol in the margin in situations where I have observed that a large proportion of my students tend to make the same mistake.

Tools for Enriching Calculus, which is a companion to this text, is referred to by means of the symbol **TEC** and can be accessed in the eBook via Enhanced WebAssign and CourseMate (selected Visuals and Modules are available at www.stewartcalculus.com). It directs you to modules in which you can explore aspects of calculus for which the computer is particularly useful.

You will notice that some exercise numbers are printed in red: 5. This indicates that *Homework Hints* are available for the exercise. These hints can be found on stewartcalculus.com as well as Enhanced WebAssign and CourseMate. The homework hints ask you questions that allow you to make progress toward a solution without actually giving you the answer. You need to pursue each hint in an active manner with pencil and paper to work out the details. If a particular hint doesn't enable you to solve the problem, you can click to reveal the next hint.

I recommend that you keep this book for reference purposes after you finish the course. Because you will likely forget some of the specific details of calculus, the book will serve as a useful reminder when you need to use calculus in subsequent courses. And, because this book contains more material than can be covered in any one course, it can also serve as a valuable resource for a working scientist or engineer.

Calculus is an exciting subject, justly considered to be one of the greatest achievements of the human intellect. I hope you will discover that it is not only useful but also intrinsically beautiful.

JAMES STEWART

Calculators, Computers, and Other Graphing Devices

© Dan Clegg

Advances in technology continue to bring a wider variety of tools for doing mathematics. Handheld calculators are becoming more powerful, as are software programs and Internet resources. In addition, many mathematical applications have been released for smartphones and tablets such as the iPad.

Some exercises in this text are marked with a graphing icon , which indicates that the use of some technology is required. Often this means that we intend for a graphing device to be used in drawing the graph of a function or equation. You might also need technology to find the zeros of a graph or the points of intersection of two graphs. In some cases we will use a calculating device to solve an equation or evaluate a definite integral numerically. Many scientific and graphing calculators have these features built in, such as the Texas Instruments TI-84 or TI-Nspire CX. Similar calculators are made by Hewlett Packard, Casio, and Sharp.

© Dan Clegg

You can also use computer software such as *Graphing Calculator* by Pacific Tech (www.pacifict.com) to perform many of these functions, as well as apps for phones and tablets, like Quick Graph (Colombiamug) or Math-Studio (Pomegranate Apps). Similar functionality is available using a web interface at WolframAlpha.com.

© Dan Clegg

In general, when we use the term "calculator" in this book, we mean the use of any of the resources we have mentioned.

The **CAS** icon is reserved for problems in which the full resources of a *computer algebra system* (CAS) are required. A CAS is capable of doing mathematics (like solving equations, computing derivatives or integrals) *symbolically* rather than just numerically.

Examples of well-established computer algebra systems are the computer software packages Maple and Mathematica. The WolframAlpha website uses the Mathematica engine to provide CAS functionality via the Web.

Many handheld graphing calculators have CAS capabilities, such as the TI-89 and TI-Nspire CX CAS from Texas Instruments. Some tablet and smartphone apps also provide these capabilities, such as the previously mentioned MathStudio.

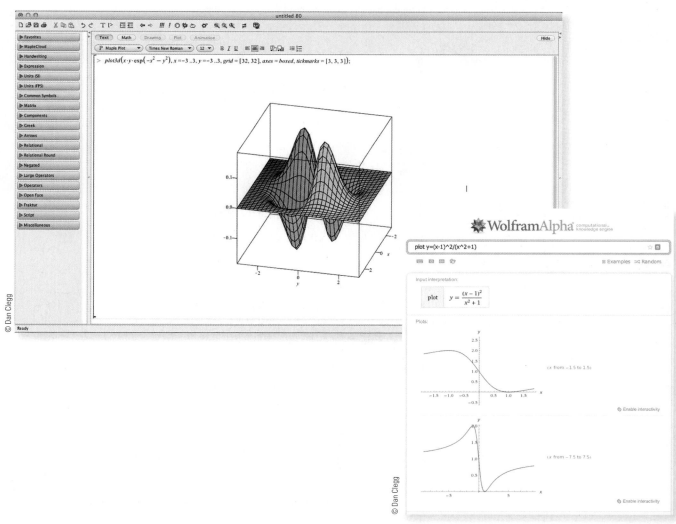

Diagnostic Tests

Success in calculus depends to a large extent on knowledge of the mathematics that precedes calculus: algebra, analytic geometry, functions, and trigonometry. The following tests are intended to diagnose weaknesses that you might have in these areas. After taking each test you can check your answers against the given answers and, if necessary, refresh your skills by referring to the review materials that are provided.

A   Diagnostic Test: Algebra

1. Evaluate each expression without using a calculator.
   (a) $(-3)^4$ (b) -3^4 (c) 3^{-4}

   (d) $\dfrac{5^{23}}{5^{21}}$ (e) $\left(\dfrac{2}{3}\right)^{-2}$ (f) $16^{-3/4}$

2. Simplify each expression. Write your answer without negative exponents.
   (a) $\sqrt{200} - \sqrt{32}$
   (b) $(3a^3b^3)(4ab^2)^2$
   (c) $\left(\dfrac{3x^{3/2}y^3}{x^2y^{-1/2}}\right)^{-2}$

3. Expand and simplify.
   (a) $3(x + 6) + 4(2x - 5)$ (b) $(x + 3)(4x - 5)$
   (c) $\left(\sqrt{a} + \sqrt{b}\right)\left(\sqrt{a} - \sqrt{b}\right)$ (d) $(2x + 3)^2$
   (e) $(x + 2)^3$

4. Factor each expression.
   (a) $4x^2 - 25$ (b) $2x^2 + 5x - 12$
   (c) $x^3 - 3x^2 - 4x + 12$ (d) $x^4 + 27x$
   (e) $3x^{3/2} - 9x^{1/2} + 6x^{-1/2}$ (f) $x^3y - 4xy$

5. Simplify the rational expression.
   (a) $\dfrac{x^2 + 3x + 2}{x^2 - x - 2}$ (b) $\dfrac{2x^2 - x - 1}{x^2 - 9} \cdot \dfrac{x + 3}{2x + 1}$

   (c) $\dfrac{x^2}{x^2 - 4} - \dfrac{x + 1}{x + 2}$ (d) $\dfrac{\dfrac{y}{x} - \dfrac{x}{y}}{\dfrac{1}{y} - \dfrac{1}{x}}$

6. Rationalize the expression and simplify.

(a) $\dfrac{\sqrt{10}}{\sqrt{5} - 2}$

(b) $\dfrac{\sqrt{4 + h} - 2}{h}$

7. Rewrite by completing the square.

(a) $x^2 + x + 1$

(b) $2x^2 - 12x + 11$

8. Solve the equation. (Find only the real solutions.)

(a) $x + 5 = 14 - \frac{1}{2}x$

(b) $\dfrac{2x}{x + 1} = \dfrac{2x - 1}{x}$

(c) $x^2 - x - 12 = 0$

(d) $2x^2 + 4x + 1 = 0$

(e) $x^4 - 3x^2 + 2 = 0$

(f) $3|x - 4| = 10$

(g) $2x(4 - x)^{-1/2} - 3\sqrt{4 - x} = 0$

9. Solve each inequality. Write your answer using interval notation.

(a) $-4 < 5 - 3x \leqslant 17$

(b) $x^2 < 2x + 8$

(c) $x(x - 1)(x + 2) > 0$

(d) $|x - 4| < 3$

(e) $\dfrac{2x - 3}{x + 1} \leqslant 1$

10. State whether each equation is true or false.

(a) $(p + q)^2 = p^2 + q^2$

(b) $\sqrt{ab} = \sqrt{a}\,\sqrt{b}$

(c) $\sqrt{a^2 + b^2} = a + b$

(d) $\dfrac{1 + TC}{C} = 1 + T$

(e) $\dfrac{1}{x - y} = \dfrac{1}{x} - \dfrac{1}{y}$

(f) $\dfrac{1/x}{a/x - b/x} = \dfrac{1}{a - b}$

ANSWERS TO DIAGNOSTIC TEST A: ALGEBRA

1. (a) 81   (b) -81 (c) $\frac{1}{81}$
(d) 25   (e) $\frac{9}{4}$ (f) $\frac{1}{8}$

2. (a) $6\sqrt{2}$ (b) $48a^5b^7$ (c) $\dfrac{x}{9y^7}$

3. (a) $11x - 2$ (b) $4x^2 + 7x - 15$
(c) $a - b$ (d) $4x^2 + 12x + 9$
(e) $x^3 + 6x^2 + 12x + 8$

4. (a) $(2x - 5)(2x + 5)$ (b) $(2x - 3)(x + 4)$
(c) $(x - 3)(x - 2)(x + 2)$ (d) $x(x + 3)(x^2 - 3x + 9)$
(e) $3x^{-1/2}(x - 1)(x - 2)$ (f) $xy(x - 2)(x + 2)$

5. (a) $\dfrac{x + 2}{x - 2}$ (b) $\dfrac{x - 1}{x - 3}$
(c) $\dfrac{1}{x - 2}$ (d) $-(x + y)$

6. (a) $5\sqrt{2} + 2\sqrt{10}$ (b) $\dfrac{1}{\sqrt{4 + h} + 2}$

7. (a) $\left(x + \frac{1}{2}\right)^2 + \frac{3}{4}$ (b) $2(x - 3)^2 - 7$

8. (a) 6   (b) 1   (c) $-3, 4$
(d) $-1 \pm \frac{1}{2}\sqrt{2}$ (e) $\pm 1, \pm\sqrt{2}$ (f) $\frac{2}{3}, \frac{22}{3}$
(g) $\frac{12}{5}$

9. (a) $[-4, 3)$ (b) $(-2, 4)$
(c) $(-2, 0) \cup (1, \infty)$ (d) $(1, 7)$
(e) $(-1, 4]$

10. (a) False   (b) True   (c) False
(d) False   (e) False   (f) True

> If you had difficulty with these problems, you may wish to consult the
> Review of Algebra on the website **www.stewartcalculus.com**.

B   Diagnostic Test: Analytic Geometry

1. Find an equation for the line that passes through the point $(2, -5)$ and
   (a) has slope -3
   (b) is parallel to the x-axis
   (c) is parallel to the y-axis
   (d) is parallel to the line $2x - 4y = 3$

2. Find an equation for the circle that has center $(-1, 4)$ and passes through the point $(3, -2)$.

3. Find the center and radius of the circle with equation $x^2 + y^2 - 6x + 10y + 9 = 0$.

4. Let $A(-7, 4)$ and $B(5, -12)$ be points in the plane.
   (a) Find the slope of the line that contains A and B.
   (b) Find an equation of the line that passes through A and B. What are the intercepts?
   (c) Find the midpoint of the segment AB.
   (d) Find the length of the segment AB.
   (e) Find an equation of the perpendicular bisector of AB.
   (f) Find an equation of the circle for which AB is a diameter.

5. Sketch the region in the xy-plane defined by the equation or inequalities.
   (a) $-1 \leqslant y \leqslant 3$ (b) $|x| < 4$ and $|y| < 2$
   (c) $y < 1 - \frac{1}{2}x$ (d) $y \geqslant x^2 - 1$
   (e) $x^2 + y^2 < 4$ (f) $9x^2 + 16y^2 = 144$

ANSWERS TO DIAGNOSTIC TEST B: ANALYTIC GEOMETRY

1. (a) $y = -3x + 1$ (b) $y = -5$
(c) $x = 2$ (d) $y = \frac{1}{2}x - 6$

2. $(x + 1)^2 + (y - 4)^2 = 52$

3. Center $(3, -5)$, radius 5

4. (a) $-\frac{4}{3}$
(b) $4x + 3y + 16 = 0$; x-intercept -4, y-intercept $-\frac{16}{3}$
(c) $(-1, -4)$
(d) 20
(e) $3x - 4y = 13$
(f) $(x + 1)^2 + (y + 4)^2 = 100$

5.

(a)
(b)
(c)
(d)
(e)
(f)

If you had difficulty with these problems, you may wish to consult the review of analytic geometry in Appendixes B and C.

C Diagnostic Test: Functions

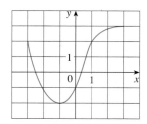

FIGURE FOR PROBLEM 1

1. The graph of a function f is given at the left.
   (a) State the value of $f(-1)$.
   (b) Estimate the value of $f(2)$.
   (c) For what values of x is $f(x) = 2$?
   (d) Estimate the values of x such that $f(x) = 0$.
   (e) State the domain and range of f.

2. If $f(x) = x^3$, evaluate the difference quotient $\dfrac{f(2+h) - f(2)}{h}$ and simplify your answer.

3. Find the domain of the function.

   (a) $f(x) = \dfrac{2x + 1}{x^2 + x - 2}$ (b) $g(x) = \dfrac{\sqrt[3]{x}}{x^2 + 1}$ (c) $h(x) = \sqrt{4 - x} + \sqrt{x^2 - 1}$

4. How are graphs of the functions obtained from the graph of f?
   (a) $y = -f(x)$ (b) $y = 2f(x) - 1$ (c) $y = f(x - 3) + 2$

5. Without using a calculator, make a rough sketch of the graph.
   (a) $y = x^3$ (b) $y = (x + 1)^3$ (c) $y = (x - 2)^3 + 3$
   (d) $y = 4 - x^2$ (e) $y = \sqrt{x}$ (f) $y = 2\sqrt{x}$
   (g) $y = -2^x$ (h) $y = 1 + x^{-1}$

6. Let $f(x) = \begin{cases} 1 - x^2 & \text{if } x \leq 0 \\ 2x + 1 & \text{if } x > 0 \end{cases}$
   (a) Evaluate $f(-2)$ and $f(1)$.   (b) Sketch the graph of f.

7. If $f(x) = x^2 + 2x - 1$ and $g(x) = 2x - 3$, find each of the following functions.
   (a) $f \circ g$ (b) $g \circ f$ (c) $g \circ g \circ g$

ANSWERS TO DIAGNOSTIC TEST C: FUNCTIONS

1. (a) -2 (b) 2.8
   (c) $-3, 1$ (d) $-2.5, 0.3$
   (e) $[-3, 3], [-2, 3]$

2. $12 + 6h + h^2$

3. (a) $(-\infty, -2) \cup (-2, 1) \cup (1, \infty)$
   (b) $(-\infty, \infty)$
   (c) $(-\infty, -1] \cup [1, 4]$

4. (a) Reflect about the x-axis
   (b) Stretch vertically by a factor of 2, then shift 1 unit downward
   (c) Shift 3 units to the right and 2 units upward

5. (a) (b) (c)

   (d) (e) (f)

   (g) (h)

6. (a) $-3, 3$

(b)

7. (a) $(f \circ g)(x) = 4x^2 - 8x + 2$

(b) $(g \circ f)(x) = 2x^2 + 4x - 5$

(c) $(g \circ g \circ g)(x) = 8x - 21$

> If you had difficulty with these problems, you should look at sections 1.1–1.3 of this book.

D   Diagnostic Test: Trigonometry

1. Convert from degrees to radians.

(a) $300°$ (b) $-18°$

2. Convert from radians to degrees.

(a) $5\pi/6$ (b) 2

3. Find the length of an arc of a circle with radius 12 cm if the arc subtends a central angle of $30°$.

4. Find the exact values.

(a) $\tan(\pi/3)$ (b) $\sin(7\pi/6)$ (c) $\sec(5\pi/3)$

5. Express the lengths a and b in the figure in terms of θ.

6. If $\sin x = \frac{1}{3}$ and $\sec y = \frac{5}{4}$, where x and y lie between 0 and $\pi/2$, evaluate $\sin(x + y)$.

7. Prove the identities.

(a) $\tan\theta \sin\theta + \cos\theta = \sec\theta$ (b) $\dfrac{2\tan x}{1 + \tan^2 x} = \sin 2x$

8. Find all values of x such that $\sin 2x = \sin x$ and $0 \le x \le 2\pi$.

9. Sketch the graph of the function $y = 1 + \sin 2x$ without using a calculator.

FIGURE FOR PROBLEM 5

(triangle with hypotenuse 24, side a, base b, angle θ)

ANSWERS TO DIAGNOSTIC TEST D: TRIGONOMETRY

1. (a) $5\pi/3$ (b) $-\pi/10$

2. (a) $150°$ (b) $360°/\pi \approx 114.6°$

3. 2π cm

4. (a) $\sqrt{3}$ (b) $-\frac{1}{2}$ (c) 2

5. (a) $24\sin\theta$ (b) $24\cos\theta$

6. $\frac{1}{15}\left(4 + 6\sqrt{2}\right)$

8. $0, \pi/3, \pi, 5\pi/3, 2\pi$

9.

> If you had difficulty with these problems, you should look at Appendix D of this book.

12 Vectors and the Geometry of Space

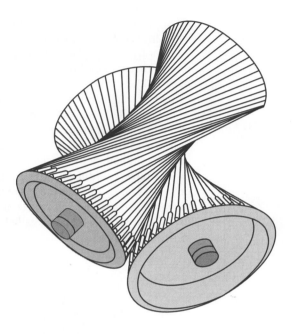

Each of these gears has the shape of a hyperboloid, a type of surface we will study in Section 12.6. The shape allows the gears to transmit motion between skew (neither parallel nor intersecting) axes.

IN THIS CHAPTER WE INTRODUCE vectors and coordinate systems for three-dimensional space. This will be the setting for our study of the calculus of functions of two variables in Chapter 14 because the graph of such a function is a surface in space. In this chapter we will see that vectors provide particularly simple descriptions of lines and planes in space.

12.1 Three-Dimensional Coordinate Systems

FIGURE 1
Coordinate axes

3D Space

To locate a point in a plane, we need two numbers. We know that any point in the plane can be represented as an ordered pair (a, b) of real numbers, where a is the x-coordinate and b is the y-coordinate. For this reason, a plane is called two-dimensional. To locate a point in space, three numbers are required. We represent any point in space by an ordered triple (a, b, c) of real numbers.

In order to represent points in space, we first choose a fixed point O (the origin) and three directed lines through O that are perpendicular to each other, called the **coordinate axes** and labeled the x-axis, y-axis, and z-axis. Usually we think of the x- and y-axes as being horizontal and the z-axis as being vertical, and we draw the orientation of the axes as in Figure 1. The direction of the z-axis is determined by the **right-hand rule** as illustrated in Figure 2: If you curl the fingers of your right hand around the z-axis in the direction of a $90°$ counterclockwise rotation from the positive x-axis to the positive y-axis, then your thumb points in the positive direction of the z-axis.

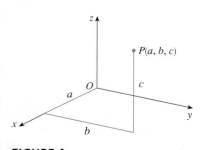

FIGURE 2
Right-hand rule

The three coordinate axes determine the three **coordinate planes** illustrated in Figure 3(a). The xy-plane is the plane that contains the x- and y-axes; the yz-plane contains the y- and z-axes; the xz-plane contains the x- and z-axes. These three coordinate planes divide space into eight parts, called **octants**. The **first octant**, in the foreground, is determined by the positive axes.

FIGURE 3   (a) Coordinate planes   (b)

Because many people have some difficulty visualizing diagrams of three-dimensional figures, you may find it helpful to do the following [see Figure 3(b)]. Look at any bottom corner of a room and call the corner the origin. The wall on your left is in the xz-plane, the wall on your right is in the yz-plane, and the floor is in the xy-plane. The x-axis runs along the intersection of the floor and the left wall. The y-axis runs along the intersection of the floor and the right wall. The z-axis runs up from the floor toward the ceiling along the intersection of the two walls. You are situated in the first octant, and you can now imagine seven other rooms situated in the other seven octants (three on the same floor and four on the floor below), all connected by the common corner point O.

Now if P is any point in space, let a be the (directed) distance from the yz-plane to P, let b be the distance from the xz-plane to P, and let c be the distance from the xy-plane to P. We represent the point P by the ordered triple (a, b, c) of real numbers and we call a, b, and c the **coordinates** of P; a is the x-coordinate, b is the y-coordinate, and c is the z-coordinate. Thus, to locate the point (a, b, c), we can start at the origin O and move a units along the x-axis, then b units parallel to the y-axis, and then c units parallel to the z-axis as in Figure 4.

FIGURE 4

The point $P(a, b, c)$ determines a rectangular box as in Figure 5. If we drop a perpendicular from P to the xy-plane, we get a point Q with coordinates $(a, b, 0)$ called the **projection** of P onto the xy-plane. Similarly, $R(0, b, c)$ and $S(a, 0, c)$ are the projections of P onto the yz-plane and xz-plane, respectively.

As numerical illustrations, the points $(-4, 3, -5)$ and $(3, -2, -6)$ are plotted in Figure 6.

FIGURE 5

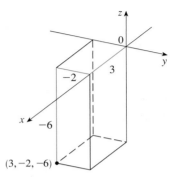

FIGURE 6

The Cartesian product $\mathbb{R} \times \mathbb{R} \times \mathbb{R} = \{(x, y, z) \mid x, y, z \in \mathbb{R}\}$ is the set of all ordered triples of real numbers and is denoted by $\mathbb{R}^3$. We have given a one-to-one correspondence between points P in space and ordered triples (a, b, c) in $\mathbb{R}^3$. It is called a **three-dimensional rectangular coordinate system**. Notice that, in terms of coordinates, the first octant can be described as the set of points whose coordinates are all positive.

◼ Surfaces

In two-dimensional analytic geometry, the graph of an equation involving x and y is a curve in $\mathbb{R}^2$. In three-dimensional analytic geometry, an equation in x, y, and z represents a *surface* in $\mathbb{R}^3$.

EXAMPLE 1 What surfaces in $\mathbb{R}^3$ are represented by the following equations?
(a) $z = 3$ (b) $y = 5$

SOLUTION
(a) The equation $z = 3$ represents the set $\{(x, y, z) \mid z = 3\}$, which is the set of all points in $\mathbb{R}^3$ whose z-coordinate is 3 (x and y can each be any value). This is the horizontal plane that is parallel to the xy-plane and three units above it as in Figure 7(a).

(a) $z = 3$, a plane in $\mathbb{R}^3$

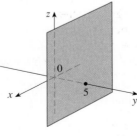

(b) $y = 5$, a plane in $\mathbb{R}^3$

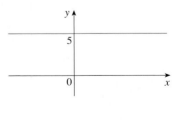

(c) $y = 5$, a line in $\mathbb{R}^2$

FIGURE 7

(b) The equation $y = 5$ represents the set of all points in $\mathbb{R}^3$ whose y-coordinate is 5. This is the vertical plane that is parallel to the xz-plane and five units to the right of it as in Figure 7(b). ◼

NOTE When an equation is given, we must understand from the context whether it represents a curve in $\mathbb{R}^2$ or a surface in $\mathbb{R}^3$. In Example 1, $y = 5$ represents a plane in $\mathbb{R}^3$, but of course $y = 5$ can also represent a line in $\mathbb{R}^2$ if we are dealing with two-dimensional analytic geometry. See Figure 7(b) and (c).

In general, if k is a constant, then $x = k$ represents a plane parallel to the yz-plane, $y = k$ is a plane parallel to the xz-plane, and $z = k$ is a plane parallel to the xy-plane. In Figure 5, the faces of the rectangular box are formed by the three coordinate planes $x = 0$ (the yz-plane), $y = 0$ (the xz-plane), and $z = 0$ (the xy-plane), and the planes $x = a$, $y = b$, and $z = c$.

EXAMPLE 2

(a) Which points (x, y, z) satisfy the equations

$$x^2 + y^2 = 1 \quad \text{and} \quad z = 3$$

(b) What does the equation $x^2 + y^2 = 1$ represent as a surface in $\mathbb{R}^3$?

SOLUTION

(a) Because $z = 3$, the points lie in the horizontal plane $z = 3$ from Example 1(a). Because $x^2 + y^2 = 1$, the points lie on the circle with radius 1 and center on the z-axis. See Figure 8.

(b) Given that $x^2 + y^2 = 1$, with no restrictions on z, we see that the point (x, y, z) could lie on a circle in any horizontal plane $z = k$. So the surface $x^2 + y^2 = 1$ in $\mathbb{R}^3$ consists of all possible horizontal circles $x^2 + y^2 = 1$, $z = k$, and is therefore the circular cylinder with radius 1 whose axis is the z-axis. See Figure 9.

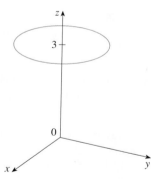

FIGURE 8
The circle $x^2 + y^2 = 1$, $z = 3$

FIGURE 9
The cylinder $x^2 + y^2 = 1$

EXAMPLE 3 Describe and sketch the surface in $\mathbb{R}^3$ represented by the equation $y = x$.

SOLUTION The equation represents the set of all points in $\mathbb{R}^3$ whose x- and y-coordinates are equal, that is, $\{(x, x, z) \mid x \in \mathbb{R}, z \in \mathbb{R}\}$. This is a vertical plane that intersects the xy-plane in the line $y = x$, $z = 0$. The portion of this plane that lies in the first octant is sketched in Figure 10.

■ Distance and Spheres

The familiar formula for the distance between two points in a plane is easily extended to the following three-dimensional formula.

FIGURE 10
The plane $y = x$

4.5d2

> **Distance Formula in Three Dimensions** The distance $|P_1P_2|$ between the points $P_1(x_1, y_1, z_1)$ and $P_2(x_2, y_2, z_2)$ is
>
> $$|P_1P_2| = \sqrt{(x_2 - x_1)^2 + (y_2 - y_1)^2 + (z_2 - z_1)^2}$$

To see why this formula is true, we construct a rectangular box as in Figure 11, where P_1 and P_2 are opposite vertices and the faces of the box are parallel to the coordinate planes. If $A(x_2, y_1, z_1)$ and $B(x_2, y_2, z_1)$ are the vertices of the box indicated in the figure, then

$$|P_1A| = |x_2 - x_1| \qquad |AB| = |y_2 - y_1| \qquad |BP_2| = |z_2 - z_1|$$

Because triangles P_1BP_2 and P_1AB are both right-angled, two applications of the Pythagorean Theorem give

$$|P_1P_2|^2 = |P_1B|^2 + |BP_2|^2$$

and

$$|P_1B|^2 = |P_1A|^2 + |AB|^2$$

Combining these equations, we get

$$|P_1P_2|^2 = |P_1A|^2 + |AB|^2 + |BP_2|^2$$

$$= |x_2 - x_1|^2 + |y_2 - y_1|^2 + |z_2 - z_1|^2$$

$$= (x_2 - x_1)^2 + (y_2 - y_1)^2 + (z_2 - z_1)^2$$

Therefore

$$|P_1P_2| = \sqrt{(x_2 - x_1)^2 + (y_2 - y_1)^2 + (z_2 - z_1)^2}$$

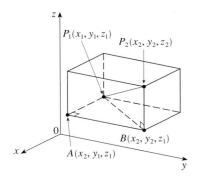

FIGURE 11

EXAMPLE 4 The distance from the point $P(2, -1, 7)$ to the point $Q(1, -3, 5)$ is

$$|PQ| = \sqrt{(1 - 2)^2 + (-3 + 1)^2 + (5 - 7)^2} = \sqrt{1 + 4 + 4} = 3 \qquad \blacksquare$$

EXAMPLE 5 Find an equation of a sphere with radius r and center $C(h, k, l)$.

SOLUTION By definition, a sphere is the set of all points $P(x, y, z)$ whose distance from C is r. (See Figure 12.) Thus P is on the sphere if and only if $|PC| = r$. Squaring both sides, we have $|PC|^2 = r^2$ or

$$(x - h)^2 + (y - k)^2 + (z - l)^2 = r^2 \qquad \blacksquare$$

The result of Example 5 is worth remembering.

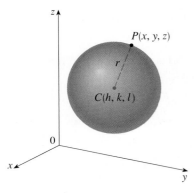

FIGURE 12

> **Equation of a Sphere** An equation of a sphere with center $C(h, k, l)$ and radius r is
>
> $$(x - h)^2 + (y - k)^2 + (z - l)^2 = r^2$$
>
> In particular, if the center is the origin O, then an equation of the sphere is
>
> $$x^2 + y^2 + z^2 = r^2$$

EXAMPLE 6 Show that $x^2 + y^2 + z^2 + 4x - 6y + 2z + 6 = 0$ is the equation of a sphere, and find its center and radius.

SOLUTION We can rewrite the given equation in the form of an equation of a sphere if we complete squares:

$$(x^2 + 4x + 4) + (y^2 - 6y + 9) + (z^2 + 2z + 1) = -6 + 4 + 9 + 1$$

$$(x + 2)^2 + (y - 3)^2 + (z + 1)^2 = 8$$

Comparing this equation with the standard form, we see that it is the equation of a sphere with center $(-2, 3, -1)$ and radius $\sqrt{8} = 2\sqrt{2}$. ∎

EXAMPLE 7 What region in $\mathbb{R}^3$ is represented by the following inequalities?

$$1 \leqslant x^2 + y^2 + z^2 \leqslant 4 \qquad z \leqslant 0$$

SOLUTION The inequalities

$$1 \leqslant x^2 + y^2 + z^2 \leqslant 4$$

can be rewritten as

$$1 \leqslant \sqrt{x^2 + y^2 + z^2} \leqslant 2$$

so they represent the points (x, y, z) whose distance from the origin is at least 1 and at most 2. But we are also given that $z \leqslant 0$, so the points lie on or below the xy-plane. Thus the given inequalities represent the region that lies between (or on) the spheres $x^2 + y^2 + z^2 = 1$ and $x^2 + y^2 + z^2 = 4$ and beneath (or on) the xy-plane. It is sketched in Figure 13. ∎

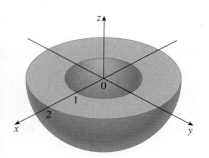

FIGURE 13

12.1 EXERCISES

1. Suppose you start at the origin, move along the x-axis a distance of 4 units in the positive direction, and then move downward a distance of 3 units. What are the coordinates of your position?

2. Sketch the points $(1, 5, 3)$, $(0, 2, -3)$, $(-3, 0, 2)$, and $(2, -2, -1)$ on a single set of coordinate axes.

3. Which of the points $A(-4, 0, -1)$, $B(3, 1, -5)$, and $C(2, 4, 6)$ is closest to the yz-plane? Which point lies in the xz-plane?

4. What are the projections of the point $(2, 3, 5)$ on the xy-, yz-, and xz-planes? Draw a rectangular box with the origin and $(2, 3, 5)$ as opposite vertices and with its faces parallel to the coordinate planes. Label all vertices of the box. Find the length of the diagonal of the box.

5. What does the equation $x = 4$ represent in $\mathbb{R}^2$? What does it represent in $\mathbb{R}^3$? Illustrate with sketches.

6. What does the equation $y = 3$ represent in $\mathbb{R}^3$? What does $z = 5$ represent? What does the pair of equations $y = 3$, $z = 5$ represent? In other words, describe the set of points (x, y, z) such that $y = 3$ and $z = 5$. Illustrate with a sketch.

7. Describe and sketch the surface in $\mathbb{R}^3$ represented by the equation $x + y = 2$.

8. Describe and sketch the surface in $\mathbb{R}^3$ represented by the equation $x^2 + z^2 = 9$.

9–10 Find the lengths of the sides of the triangle PQR. Is it a right triangle? Is it an isosceles triangle?

9. $P(3, -2, -3)$, $Q(7, 0, 1)$, $R(1, 2, 1)$

10. $P(2, -1, 0)$, $Q(4, 1, 1)$, $R(4, -5, 4)$

11. Determine whether the points lie on a straight line.
    (a) $A(2, 4, 2)$, $B(3, 7, -2)$, $C(1, 3, 3)$
    (b) $D(0, -5, 5)$, $E(1, -2, 4)$, $F(3, 4, 2)$

12. Find the distance from $(4, -2, 6)$ to each of the following.
    (a) The xy-plane   (b) The yz-plane
    (c) The xz-plane   (d) The x-axis
    (e) The y-axis   (f) The z-axis

13. Find an equation of the sphere with center $(-3, 2, 5)$ and radius 4. What is the intersection of this sphere with the yz-plane?

14. Find an equation of the sphere with center $(2, -6, 4)$ and radius 5. Describe its intersection with each of the coordinate planes.

15. Find an equation of the sphere that passes through the point $(4, 3, -1)$ and has center $(3, 8, 1)$.

16. Find an equation of the sphere that passes through the origin and whose center is $(1, 2, 3)$.

17–20 Show that the equation represents a sphere, and find its center and radius.

17. $x^2 + y^2 + z^2 - 2x - 4y + 8z = 15$

18. $x^2 + y^2 + z^2 + 8x - 6y + 2z + 17 = 0$

19. $2x^2 + 2y^2 + 2z^2 = 8x - 24z + 1$

20. $3x^2 + 3y^2 + 3z^2 = 10 + 6y + 12z$

21. (a) Prove that the midpoint of the line segment from $P_1(x_1, y_1, z_1)$ to $P_2(x_2, y_2, z_2)$ is

$$\left(\frac{x_1 + x_2}{2}, \frac{y_1 + y_2}{2}, \frac{z_1 + z_2}{2} \right)$$

(b) Find the lengths of the medians of the triangle with vertices $A(1, 2, 3)$, $B(-2, 0, 5)$, and $C(4, 1, 5)$. (A *median* of a triangle is a line segment that joins a vertex to the midpoint of the opposite side.)

22. Find an equation of a sphere if one of its diameters has endpoints $(5, 4, 3)$ and $(1, 6, -9)$.

23. Find equations of the spheres with center $(2, -3, 6)$ that touch (a) the xy-plane, (b) the yz-plane, (c) the xz-plane.

24. Find an equation of the largest sphere with center $(5, 4, 9)$ that is contained in the first octant.

25–38 Describe in words the region of $\mathbb{R}^3$ represented by the equation(s) or inequality.

25. $x = 5$

26. $y = -2$

27. $y < 8$

28. $z \geqslant -1$

29. $0 \leqslant z \leqslant 6$

30. $y^2 = 4$

31. $x^2 + y^2 = 4, \quad z = -1$

32. $x^2 + y^2 = 4$

33. $x^2 + y^2 + z^2 = 4$

34. $x^2 + y^2 + z^2 \leqslant 4$

35. $1 \leqslant x^2 + y^2 + z^2 \leqslant 5$

36. $x = z$

37. $x^2 + z^2 \leqslant 9$

38. $x^2 + y^2 + z^2 > 2z$

39–42 Write inequalities to describe the region.

39. The region between the yz-plane and the vertical plane $x = 5$

40. The solid cylinder that lies on or below the plane $z = 8$ and on or above the disk in the xy-plane with center the origin and radius 2

41. The region consisting of all points between (but not on) the spheres of radius r and R centered at the origin, where $r < R$

42. The solid upper hemisphere of the sphere of radius 2 centered at the origin

43. The figure shows a line L_1 in space and a second line L_2, which is the projection of L_1 onto the xy-plane. (In other words, the points on L_2 are directly beneath, or above, the points on L_1.)
(a) Find the coordinates of the point P on the line L_1.
(b) Locate on the diagram the points A, B, and C, where the line L_1 intersects the xy-plane, the yz-plane, and the xz-plane, respectively.

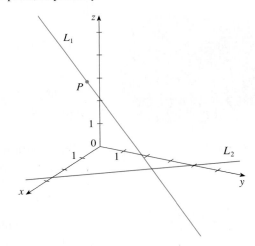

44. Consider the points P such that the distance from P to $A(-1, 5, 3)$ is twice the distance from P to $B(6, 2, -2)$. Show that the set of all such points is a sphere, and find its center and radius.

45. Find an equation of the set of all points equidistant from the points $A(-1, 5, 3)$ and $B(6, 2, -2)$. Describe the set.

46. Find the volume of the solid that lies inside both of the spheres

$$x^2 + y^2 + z^2 + 4x - 2y + 4z + 5 = 0$$

and

$$x^2 + y^2 + z^2 = 4$$

47. Find the distance between the spheres $x^2 + y^2 + z^2 = 4$ and $x^2 + y^2 + z^2 = 4x + 4y + 4z - 11$.

48. Describe and sketch a solid with the following properties. When illuminated by rays parallel to the z-axis, its shadow is a circular disk. If the rays are parallel to the y-axis, its shadow is a square. If the rays are parallel to the x-axis, its shadow is an isosceles triangle.

12.2 Vectors

FIGURE 1
Equivalent vectors

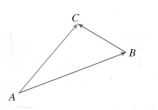

FIGURE 2

The term **vector** is used by scientists to indicate a quantity (such as displacement or velocity or force) that has both magnitude and direction. A vector is often represented by an arrow or a directed line segment. The length of the arrow represents the magnitude of the vector and the arrow points in the direction of the vector. We denote a vector by printing a letter in boldface (**v**) or by putting an arrow above the letter ($\vec{v}$).

For instance, suppose a particle moves along a line segment from point A to point B. The corresponding **displacement vector v**, shown in Figure 1, has **initial point** A (the tail) and **terminal point** B (the tip) and we indicate this by writing $\mathbf{v} = \overrightarrow{AB}$. Notice that the vector $\mathbf{u} = \overrightarrow{CD}$ has the same length and the same direction as **v** even though it is in a different position. We say that **u** and **v** are **equivalent** (or **equal**) and we write $\mathbf{u} = \mathbf{v}$. The **zero vector**, denoted by **0**, has length 0. It is the only vector with no specific direction.

■ Combining Vectors

Suppose a particle moves from A to B, so its displacement vector is $\overrightarrow{AB}$. Then the particle changes direction and moves from B to C, with displacement vector $\overrightarrow{BC}$ as in Figure 2. The combined effect of these displacements is that the particle has moved from A to C. The resulting displacement vector $\overrightarrow{AC}$ is called the *sum* of $\overrightarrow{AB}$ and $\overrightarrow{BC}$ and we write

$$\overrightarrow{AC} = \overrightarrow{AB} + \overrightarrow{BC}$$

In general, if we start with vectors **u** and **v**, we first move **v** so that its tail coincides with the tip of **u** and define the sum of **u** and **v** as follows.

> **Definition of Vector Addition** If **u** and **v** are vectors positioned so the initial point of **v** is at the terminal point of **u**, then the **sum u + v** is the vector from the initial point of **u** to the terminal point of **v**.

The definition of vector addition is illustrated in Figure 3. You can see why this definition is sometimes called the **Triangle Law**.

FIGURE 3
The Triangle Law

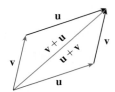

FIGURE 4
The Parallelogram Law

In Figure 4 we start with the same vectors **u** and **v** as in Figure 3 and draw another copy of **v** with the same initial point as **u**. Completing the parallelogram, we see that $\mathbf{u} + \mathbf{v} = \mathbf{v} + \mathbf{u}$. This also gives another way to construct the sum: if we place **u** and **v** so they start at the same point, then $\mathbf{u} + \mathbf{v}$ lies along the diagonal of the parallelogram with **u** and **v** as sides. (This is called the **Parallelogram Law**.)

EXAMPLE 1 Draw the sum of the vectors **a** and **b** shown in Figure 5.

SOLUTION First we move **b** and place its tail at the tip of **a**, being careful to draw a copy of **b** that has the same length and direction. Then we draw the vector $\mathbf{a} + \mathbf{b}$ [see

FIGURE 5

Figure 6(a)] starting at the initial point of **a** and ending at the terminal point of the copy of **b**.

Alternatively, we could place **b** so it starts where **a** starts and construct **a** + **b** by the Parallelogram Law as in Figure 6(b).

TEC Visual 12.2 shows how the Triangle and Parallelogram Laws work for various vectors **a** and **b**.

FIGURE 6   (a)   (b)

It is possible to multiply a vector by a real number c. (In this context we call the real number c a **scalar** to distinguish it from a vector.) For instance, we want $2\mathbf{v}$ to be the same vector as $\mathbf{v} + \mathbf{v}$, which has the same direction as **v** but is twice as long. In general, we multiply a vector by a scalar as follows.

> **Definition of Scalar Multiplication** If c is a scalar and **v** is a vector, then the **scalar multiple** $c\mathbf{v}$ is the vector whose length is $|c|$ times the length of **v** and whose direction is the same as **v** if $c > 0$ and is opposite to **v** if $c < 0$. If $c = 0$ or $\mathbf{v} = \mathbf{0}$, then $c\mathbf{v} = \mathbf{0}$.

This definition is illustrated in Figure 7. We see that real numbers work like scaling factors here; that's why we call them scalars. Notice that two nonzero vectors are **parallel** if they are scalar multiples of one another. In particular, the vector $-\mathbf{v} = (-1)\mathbf{v}$ has the same length as **v** but points in the opposite direction. We call it the **negative** of **v**.

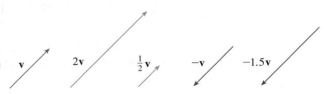

FIGURE 7
Scalar multiples of **v**

By the **difference u** − **v** of two vectors we mean

$$\mathbf{u} - \mathbf{v} = \mathbf{u} + (-\mathbf{v})$$

So we can construct **u** − **v** by first drawing the negative of **v**, −**v**, and then adding it to **u** by the Parallelogram Law as in Figure 8(a). Alternatively, since $\mathbf{v} + (\mathbf{u} - \mathbf{v}) = \mathbf{u}$, the vector **u** − **v**, when added to **v**, gives **u**. So we could construct **u** − **v** as in Figure 8(b) by means of the Triangle Law. Notice that if **u** and **v** both start from the same initial point, then **u** − **v** connects the tip of **v** to the tip of **u**.

FIGURE 8
Drawing **u** − **v**   (a)   (b)

FIGURE 9

FIGURE 10

EXAMPLE 2 If **a** and **b** are the vectors shown in Figure 9, draw **a** − 2**b**.

SOLUTION We first draw the vector −2**b** pointing in the direction opposite to **b** and twice as long. We place it with its tail at the tip of **a** and then use the Triangle Law to draw **a** + (−2**b**) as in Figure 10. ∎

■ Components

For some purposes it's best to introduce a coordinate system and treat vectors algebraically. If we place the initial point of a vector **a** at the origin of a rectangular coordinate system, then the terminal point of **a** has coordinates of the form (a_1, a_2) or (a_1, a_2, a_3), depending on whether our coordinate system is two- or three-dimensional (see Figure 11). These coordinates are called the **components** of **a** and we write

$$\mathbf{a} = \langle a_1, a_2 \rangle \qquad \text{or} \qquad \mathbf{a} = \langle a_1, a_2, a_3 \rangle$$

We use the notation $\langle a_1, a_2 \rangle$ for the ordered pair that refers to a vector so as not to confuse it with the ordered pair (a_1, a_2) that refers to a point in the plane.

FIGURE 11

$$\mathbf{a} = \langle a_1, a_2 \rangle$$

$$\mathbf{a} = \langle a_1, a_2, a_3 \rangle$$

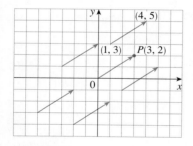

FIGURE 12

Representations of $\mathbf{a} = \langle 3, 2 \rangle$

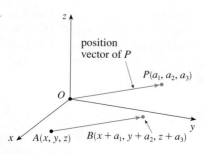

FIGURE 13

Representations of $\mathbf{a} = \langle a_1, a_2, a_3 \rangle$

For instance, the vectors shown in Figure 12 are all equivalent to the vector $\overrightarrow{OP} = \langle 3, 2 \rangle$ whose terminal point is $P(3, 2)$. What they have in common is that the terminal point is reached from the initial point by a displacement of three units to the right and two upward. We can think of all these geometric vectors as **representations** of the algebraic vector $\mathbf{a} = \langle 3, 2 \rangle$. The particular representation $\overrightarrow{OP}$ from the origin to the point $P(3, 2)$ is called the **position vector** of the point P.

In three dimensions, the vector $\mathbf{a} = \overrightarrow{OP} = \langle a_1, a_2, a_3 \rangle$ is the **position vector** of the point $P(a_1, a_2, a_3)$. (See Figure 13.) Let's consider any other representation $\overrightarrow{AB}$ of **a**, where the initial point is $A(x_1, y_1, z_1)$ and the terminal point is $B(x_2, y_2, z_2)$. Then we must have $x_1 + a_1 = x_2$, $y_1 + a_2 = y_2$, and $z_1 + a_3 = z_2$ and so $a_1 = x_2 - x_1$, $a_2 = y_2 - y_1$, and $a_3 = z_2 - z_1$. Thus we have the following result.

> **1** Given the points $A(x_1, y_1, z_1)$ and $B(x_2, y_2, z_2)$, the vector **a** with representation $\overrightarrow{AB}$ is
> $$\mathbf{a} = \langle x_2 - x_1, y_2 - y_1, z_2 - z_1 \rangle$$

EXAMPLE 3 Find the vector represented by the directed line segment with initial point $A(2, -3, 4)$ and terminal point $B(-2, 1, 1)$.

SOLUTION By (1), the vector corresponding to $\overrightarrow{AB}$ is

$$\mathbf{a} = \langle -2 - 2, 1 - (-3), 1 - 4 \rangle = \langle -4, 4, -3 \rangle \qquad ∎$$

The **magnitude** or **length** of the vector $\mathbf{v}$ is the length of any of its representations and is denoted by the symbol $|\mathbf{v}|$ or $\|\mathbf{v}\|$. By using the distance formula to compute the length of a segment OP, we obtain the following formulas.

> The length of the two-dimensional vector $\mathbf{a} = \langle a_1, a_2 \rangle$ is
>
> $$|\mathbf{a}| = \sqrt{a_1^2 + a_2^2}$$
>
> The length of the three-dimensional vector $\mathbf{a} = \langle a_1, a_2, a_3 \rangle$ is
>
> $$|\mathbf{a}| = \sqrt{a_1^2 + a_2^2 + a_3^2}$$

How do we add vectors algebraically? Figure 14 shows that if $\mathbf{a} = \langle a_1, a_2 \rangle$ and $\mathbf{b} = \langle b_1, b_2 \rangle$, then the sum is $\mathbf{a} + \mathbf{b} = \langle a_1 + b_1, a_2 + b_2 \rangle$, at least for the case where the components are positive. In other words, *to add algebraic vectors we add corresponding components.* Similarly, *to subtract vectors we subtract corresponding components.* From the similar triangles in Figure 15 we see that the components of $c\mathbf{a}$ are ca_1 and ca_2. So *to multiply a vector by a scalar we multiply each component by that scalar.*

FIGURE 14

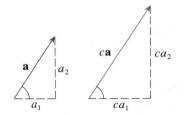

FIGURE 15

> If $\mathbf{a} = \langle a_1, a_2 \rangle$ and $\mathbf{b} = \langle b_1, b_2 \rangle$, then
>
> $$\mathbf{a} + \mathbf{b} = \langle a_1 + b_1, a_2 + b_2 \rangle \qquad \mathbf{a} - \mathbf{b} = \langle a_1 - b_1, a_2 - b_2 \rangle$$
>
> $$c\mathbf{a} = \langle ca_1, ca_2 \rangle$$
>
> Similarly, for three-dimensional vectors,
>
> $$\langle a_1, a_2, a_3 \rangle + \langle b_1, b_2, b_3 \rangle = \langle a_1 + b_1, a_2 + b_2, a_3 + b_3 \rangle$$
>
> $$\langle a_1, a_2, a_3 \rangle - \langle b_1, b_2, b_3 \rangle = \langle a_1 - b_1, a_2 - b_2, a_3 - b_3 \rangle$$
>
> $$c\langle a_1, a_2, a_3 \rangle = \langle ca_1, ca_2, ca_3 \rangle$$

EXAMPLE 4 If $\mathbf{a} = \langle 4, 0, 3 \rangle$ and $\mathbf{b} = \langle -2, 1, 5 \rangle$, find $|\mathbf{a}|$ and the vectors $\mathbf{a} + \mathbf{b}$, $\mathbf{a} - \mathbf{b}$, $3\mathbf{b}$, and $2\mathbf{a} + 5\mathbf{b}$.

SOLUTION
$$|\mathbf{a}| = \sqrt{4^2 + 0^2 + 3^2} = \sqrt{25} = 5$$

$$\mathbf{a} + \mathbf{b} = \langle 4, 0, 3 \rangle + \langle -2, 1, 5 \rangle$$
$$= \langle 4 + (-2), 0 + 1, 3 + 5 \rangle = \langle 2, 1, 8 \rangle$$

$$\mathbf{a} - \mathbf{b} = \langle 4, 0, 3 \rangle - \langle -2, 1, 5 \rangle$$
$$= \langle 4 - (-2), 0 - 1, 3 - 5 \rangle = \langle 6, -1, -2 \rangle$$

$$3\mathbf{b} = 3\langle -2, 1, 5 \rangle = \langle 3(-2), 3(1), 3(5) \rangle = \langle -6, 3, 15 \rangle$$

$$2\mathbf{a} + 5\mathbf{b} = 2\langle 4, 0, 3 \rangle + 5\langle -2, 1, 5 \rangle$$
$$= \langle 8, 0, 6 \rangle + \langle -10, 5, 25 \rangle = \langle -2, 5, 31 \rangle$$

(handwritten annotations:)

$\vec{AB} = \langle -4, 4, -3 \rangle$

$\vec{a+b} = \langle 2, 1, 8 \rangle$

$|a| = \sqrt{4^2 + 3^2} = 5$

$\langle 4, 0, 3 \rangle + \langle 2, -1, -5 \rangle = \langle 6, -1, -2 \rangle$

Vectors in n dimensions are used to list various quantities in an organized way. For instance, the components of a six-dimensional vector

$$\mathbf{p} = \langle p_1, p_2, p_3, p_4, p_5, p_6 \rangle$$

might represent the prices of six different ingredients required to make a particular product. Four-dimensional vectors $\langle x, y, z, t \rangle$ are used in relativity theory, where the first three components specify a position in space and the fourth represents time.

We denote by V_2 the set of all two-dimensional vectors and by V_3 the set of all three-dimensional vectors. More generally, we will later need to consider the set V_n of all n-dimensional vectors. An n-dimensional vector is an ordered n-tuple:

$$\mathbf{a} = \langle a_1, a_2, \ldots, a_n \rangle$$

where $a_1, a_2, \ldots, a_n$ are real numbers that are called the components of $\mathbf{a}$. Addition and scalar multiplication are defined in terms of components just as for the cases $n = 2$ and $n = 3$.

Properties of Vectors   If $\mathbf{a}$, $\mathbf{b}$, and $\mathbf{c}$ are vectors in V_n and c and d are scalars, then

1. $\mathbf{a} + \mathbf{b} = \mathbf{b} + \mathbf{a}$	**2.** $\mathbf{a} + (\mathbf{b} + \mathbf{c}) = (\mathbf{a} + \mathbf{b}) + \mathbf{c}$
3. $\mathbf{a} + \mathbf{0} = \mathbf{a}$	**4.** $\mathbf{a} + (-\mathbf{a}) = \mathbf{0}$
5. $c(\mathbf{a} + \mathbf{b}) = c\mathbf{a} + c\mathbf{b}$	**6.** $(c + d)\mathbf{a} = c\mathbf{a} + d\mathbf{a}$
7. $(cd)\mathbf{a} = c(d\mathbf{a})$	**8.** $1\mathbf{a} = \mathbf{a}$

These eight properties of vectors can be readily verified either geometrically or algebraically. For instance, Property 1 can be seen from Figure 4 (it's equivalent to the Parallelogram Law) or as follows for the case $n = 2$:

$$\mathbf{a} + \mathbf{b} = \langle a_1, a_2 \rangle + \langle b_1, b_2 \rangle = \langle a_1 + b_1, a_2 + b_2 \rangle$$

$$= \langle b_1 + a_1, b_2 + a_2 \rangle = \langle b_1, b_2 \rangle + \langle a_1, a_2 \rangle$$

$$= \mathbf{b} + \mathbf{a}$$

We can see why Property 2 (the associative law) is true by looking at Figure 16 and applying the Triangle Law several times: the vector $\overrightarrow{PQ}$ is obtained either by first constructing $\mathbf{a} + \mathbf{b}$ and then adding $\mathbf{c}$ or by adding $\mathbf{a}$ to the vector $\mathbf{b} + \mathbf{c}$.

Three vectors in V_3 play a special role. Let

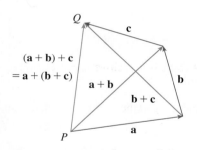

FIGURE 16

$$\mathbf{i} = \langle 1, 0, 0 \rangle \qquad \mathbf{j} = \langle 0, 1, 0 \rangle \qquad \mathbf{k} = \langle 0, 0, 1 \rangle$$

These vectors $\mathbf{i}$, $\mathbf{j}$, and $\mathbf{k}$ are called the **standard basis vectors**. They have length 1 and point in the directions of the positive x-, y-, and z-axes. Similarly, in two dimensions we define $\mathbf{i} = \langle 1, 0 \rangle$ and $\mathbf{j} = \langle 0, 1 \rangle$. (See Figure 17.)

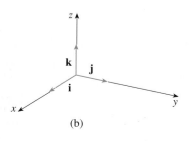

FIGURE 17
Standard basis vectors in V_2 and V_3

(a)   (b)

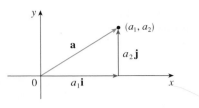

(a) $\mathbf{a} = a_1\mathbf{i} + a_2\mathbf{j}$

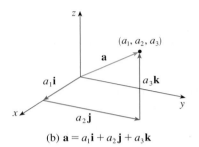

(b) $\mathbf{a} = a_1\mathbf{i} + a_2\mathbf{j} + a_3\mathbf{k}$

FIGURE 18

Gibbs

Josiah Willard Gibbs (1839–1903), a professor of mathematical physics at Yale College, published the first book on vectors, *Vector Analysis*, in 1881. More complicated objects, called quaternions, had earlier been invented by Hamilton as mathematical tools for describing space, but they weren't easy for scientists to use. Quaternions have a scalar part and a vector part. Gibb's idea was to use the vector part separately. Maxwell and Heaviside had similar ideas, but Gibb's approach has proved to be the most convenient way to study space.

$D = \sqrt{4 + 1 + 4} = 9 = 3$

unit vector $= \frac{1}{3} \cdot \langle 2i - j - 2k \rangle$

If $\mathbf{a} = \langle a_1, a_2, a_3 \rangle$, then we can write

$$\mathbf{a} = \langle a_1, a_2, a_3 \rangle = \langle a_1, 0, 0 \rangle + \langle 0, a_2, 0 \rangle + \langle 0, 0, a_3 \rangle$$

$$= a_1\langle 1, 0, 0 \rangle + a_2\langle 0, 1, 0 \rangle + a_3\langle 0, 0, 1 \rangle$$

$\boxed{2}$ $\qquad \mathbf{a} = a_1\mathbf{i} + a_2\mathbf{j} + a_3\mathbf{k}$

Thus any vector in V_3 can be expressed in terms of $\mathbf{i}$, $\mathbf{j}$, and $\mathbf{k}$. For instance,

$$\langle 1, -2, 6 \rangle = \mathbf{i} - 2\mathbf{j} + 6\mathbf{k}$$

Similarly, in two dimensions, we can write

$\boxed{3}$ $\qquad \mathbf{a} = \langle a_1, a_2 \rangle = a_1\mathbf{i} + a_2\mathbf{j}$

See Figure 18 for the geometric interpretation of Equations 3 and 2 and compare with Figure 17.

$2(i + 2j - 3k) + 3(4i + 7k) = \langle 2i + 4j - 6k \rangle + 12i + 21k$

EXAMPLE 5 If $\mathbf{a} = \mathbf{i} + 2\mathbf{j} - 3\mathbf{k}$ and $\mathbf{b} = 4\mathbf{i} + 7\mathbf{k}$, express the vector $2\mathbf{a} + 3\mathbf{b}$ in terms of $\mathbf{i}$, $\mathbf{j}$, and $\mathbf{k}$.

$14i + 4j + 15k$

SOLUTION Using Properties 1, 2, 5, 6, and 7 of vectors, we have

$$2\mathbf{a} + 3\mathbf{b} = 2(\mathbf{i} + 2\mathbf{j} - 3\mathbf{k}) + 3(4\mathbf{i} + 7\mathbf{k})$$

$$= 2\mathbf{i} + 4\mathbf{j} - 6\mathbf{k} + 12\mathbf{i} + 21\mathbf{k} = 14\mathbf{i} + 4\mathbf{j} + 15\mathbf{k} \qquad ∎$$

A **unit vector** is a vector whose length is 1. For instance, $\mathbf{i}$, $\mathbf{j}$, and $\mathbf{k}$ are all unit vectors. In general, if $\mathbf{a} \neq \mathbf{0}$, then the unit vector that has the same direction as $\mathbf{a}$ is

$\boxed{4}$ $\qquad \mathbf{u} = \dfrac{1}{|\mathbf{a}|}\mathbf{a} = \dfrac{\mathbf{a}}{|\mathbf{a}|}$

In order to verify this, we let $c = 1/|\mathbf{a}|$. Then $\mathbf{u} = c\mathbf{a}$ and c is a positive scalar, so $\mathbf{u}$ has the same direction as $\mathbf{a}$. Also

$$|\mathbf{u}| = |c\mathbf{a}| = |c||\mathbf{a}| = \frac{1}{|\mathbf{a}|}|\mathbf{a}| = 1$$

EXAMPLE 6 Find the unit vector in the direction of the vector $2\mathbf{i} - \mathbf{j} - 2\mathbf{k}$.

SOLUTION The given vector has length

$$|2\mathbf{i} - \mathbf{j} - 2\mathbf{k}| = \sqrt{2^2 + (-1)^2 + (-2)^2} = \sqrt{9} = 3$$

so, by Equation 4, the unit vector with the same direction is

$$\tfrac{1}{3}(2\mathbf{i} - \mathbf{j} - 2\mathbf{k}) = \tfrac{2}{3}\mathbf{i} - \tfrac{1}{3}\mathbf{j} - \tfrac{2}{3}\mathbf{k} \qquad ∎$$

■ Applications

Vectors are useful in many aspects of physics and engineering. In Chapter 13 we will see how they describe the velocity and acceleration of objects moving in space. Here we look at forces.

A force is represented by a vector because it has both a magnitude (measured in pounds or newtons) and a direction. If several forces are acting on an object, the **resultant force** experienced by the object is the vector sum of these forces.

EXAMPLE 7 A 100-lb weight hangs from two wires as shown in Figure 19. Find the tensions (forces) T_1 and T_2 in both wires and the magnitudes of the tensions.

FIGURE 19

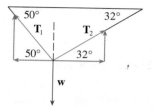

FIGURE 20

SOLUTION We first express T_1 and T_2 in terms of their horizontal and vertical components. From Figure 20 we see that

$$\boxed{5} \qquad T_1 = -|T_1| \cos 50° \, i + |T_1| \sin 50° \, j$$

$$\boxed{6} \qquad T_2 = |T_2| \cos 32° \, i + |T_2| \sin 32° \, j$$

The resultant $T_1 + T_2$ of the tensions counterbalances the weight $w = -100\,j$ and so we must have

$$T_1 + T_2 = -w = 100\,j$$

Thus

$$\left(-|T_1| \cos 50° + |T_2| \cos 32°\right) i + \left(|T_1| \sin 50° + |T_2| \sin 32°\right) j = 100\,j$$

Equating components, we get

$$-|T_1| \cos 50° + |T_2| \cos 32° = 0$$

$$|T_1| \sin 50° + |T_2| \sin 32° = 100$$

Solving the first of these equations for $|T_2|$ and substituting into the second, we get

$$|T_1| \sin 50° + \frac{|T_1| \cos 50°}{\cos 32°} \sin 32° = 100$$

$$|T_1| \left(\sin 50° + \cos 50° \frac{\sin 32°}{\cos 32°} \right) = 100$$

So the magnitudes of the tensions are

$$|T_1| = \frac{100}{\sin 50° + \tan 32° \cos 50°} \approx 85.64 \text{ lb}$$

and

$$|T_2| = \frac{|T_1| \cos 50°}{\cos 32°} \approx 64.91 \text{ lb}$$

Substituting these values in (5) and (6), we obtain the tension vectors

$$T_1 \approx -55.05\,i + 65.60\,j$$

$$T_2 \approx 55.05\,i + 34.40\,j$$

12.2 EXERCISES

1. Are the following quantities vectors or scalars? Explain.
  (a) The cost of a theater ticket
  (b) The current in a river
  (c) The initial flight path from Houston to Dallas
  (d) The population of the world

2. What is the relationship between the point $(4, 7)$ and the vector $\langle 4, 7 \rangle$? Illustrate with a sketch.

3. Name all the equal vectors in the parallelogram shown.

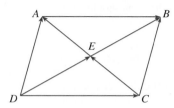

4. Write each combination of vectors as a single vector.
  (a) $\overrightarrow{AB} + \overrightarrow{BC}$
  (b) $\overrightarrow{CD} + \overrightarrow{DB}$
  (c) $\overrightarrow{DB} - \overrightarrow{AB}$
  (d) $\overrightarrow{DC} + \overrightarrow{CA} + \overrightarrow{AB}$

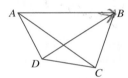

5. Copy the vectors in the figure and use them to draw the following vectors.
  (a) $\mathbf{u} + \mathbf{v}$
  (b) $\mathbf{u} + \mathbf{w}$
  (c) $\mathbf{v} + \mathbf{w}$
  (d) $\mathbf{u} - \mathbf{v}$
  (e) $\mathbf{v} + \mathbf{u} + \mathbf{w}$
  (f) $\mathbf{u} - \mathbf{w} - \mathbf{v}$

6. Copy the vectors in the figure and use them to draw the following vectors.
  (a) $\mathbf{a} + \mathbf{b}$
  (b) $\mathbf{a} - \mathbf{b}$
  (c) $\frac{1}{2}\mathbf{a}$
  (d) $-3\mathbf{b}$
  (e) $\mathbf{a} + 2\mathbf{b}$
  (f) $2\mathbf{b} - \mathbf{a}$

7. In the figure, the tip of $\mathbf{c}$ and the tail of $\mathbf{d}$ are both the midpoint of QR. Express $\mathbf{c}$ and $\mathbf{d}$ in terms of $\mathbf{a}$ and $\mathbf{b}$.

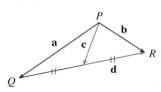

8. If the vectors in the figure satisfy $|\mathbf{u}| = |\mathbf{v}| = 1$ and $\mathbf{u} + \mathbf{v} + \mathbf{w} = \mathbf{0}$, what is $|\mathbf{w}|$?

9–14 Find a vector $\mathbf{a}$ with representation given by the directed line segment $\overrightarrow{AB}$. Draw $\overrightarrow{AB}$ and the equivalent representation starting at the origin.

9. $A(-2, 1)$, $B(1, 2)$
10. $A(-5, -1)$, $B(-3, 3)$
11. $A(3, -1)$, $B(2, 3)$
12. $A(3, 2)$, $B(1, 0)$
13. $A(0, 3, 1)$, $B(2, 3, -1)$
14. $A(0, 6, -1)$, $B(3, 4, 4)$

15–18 Find the sum of the given vectors and illustrate geometrically.

15. $\langle -1, 4 \rangle$, $\langle 6, -2 \rangle$
16. $\langle 3, -1 \rangle$, $\langle -1, 5 \rangle$
17. $\langle 3, 0, 1 \rangle$, $\langle 0, 8, 0 \rangle$
18. $\langle 1, 3, -2 \rangle$, $\langle 0, 0, 6 \rangle$

19–22 Find $\mathbf{a} + \mathbf{b}$, $4\mathbf{a} + 2\mathbf{b}$, $|\mathbf{a}|$, and $|\mathbf{a} - \mathbf{b}|$.

19. $\mathbf{a} = \langle -3, 4 \rangle$, $\mathbf{b} = \langle 9, -1 \rangle$
20. $\mathbf{a} = 5\mathbf{i} + 3\mathbf{j}$, $\mathbf{b} = -\mathbf{i} - 2\mathbf{j}$
21. $\mathbf{a} = 4\mathbf{i} - 3\mathbf{j} + 2\mathbf{k}$, $\mathbf{b} = 2\mathbf{i} - 4\mathbf{k}$
22. $\mathbf{a} = \langle 8, 1, -4 \rangle$, $\mathbf{b} = \langle 5, -2, 1 \rangle$

23–25 Find a unit vector that has the same direction as the given vector.

23. $\langle 6, -2 \rangle$
24. $-5\mathbf{i} + 3\mathbf{j} - \mathbf{k}$
25. $8\mathbf{i} - \mathbf{j} + 4\mathbf{k}$

26. Find the vector that has the same direction as $\langle 6, 2, -3 \rangle$ but has length 4.

27–28 What is the angle between the given vector and the positive direction of the x-axis?

27. $\mathbf{i} + \sqrt{3}\,\mathbf{j}$
28. $8\mathbf{i} + 6\mathbf{j}$

29. If $\mathbf{v}$ lies in the first quadrant and makes an angle $\pi/3$ with the positive x-axis and $|\mathbf{v}| = 4$, find $\mathbf{v}$ in component form.

30. If a child pulls a sled through the snow on a level path with a force of 50 N exerted at an angle of $38°$ above the horizontal, find the horizontal and vertical components of the force.

31. A quarterback throws a football with angle of elevation $40°$ and speed 60 ft/s. Find the horizontal and vertical components of the velocity vector.

32–33 Find the magnitude of the resultant force and the angle it makes with the positive x-axis.

32.

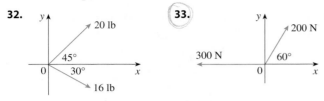

33.

34. The magnitude of a velocity vector is called *speed*. Suppose that a wind is blowing *from* the direction N45°W at a speed of 50 km/h. (This means that the direction from which the wind blows is 45° west of the northerly direction.) A pilot is steering a plane in the direction N60°E at an airspeed (speed in still air) of 250 km/h. The *true course*, or *track*, of the plane is the direction of the resultant of the velocity vectors of the plane and the wind. The *ground speed* of the plane is the magnitude of the resultant. Find the true course and the ground speed of the plane.

35. A woman walks due west on the deck of a ship at 3 mi/h. The ship is moving north at a speed of 22 mi/h. Find the speed and direction of the woman relative to the surface of the water.

36. A crane suspends a 500-lb steel beam horizontally by support cables (with negligible weight) attached from a hook to each end of the beam. The support cables each make an angle of 60° with the beam. Find the tension vector in each support cable and the magnitude of each tension.

37. A block-and-tackle pulley hoist is suspended in a warehouse by ropes of lengths 2 m and 3 m. The hoist weighs 350 N. The ropes, fastened at different heights, make angles of 50° and 38° with the horizontal. Find the tension in each rope and the magnitude of each tension.

38. The tension **T** at each end of a chain has magnitude 25 N (see the figure). What is the weight of the chain?

39. A boatman wants to cross a canal that is 3 km wide and wants to land at a point 2 km upstream from his starting point. The current in the canal flows at 3.5 km/h and the speed of his boat is 13 km/h.
(a) In what direction should he steer?
(b) How long will the trip take?

40. Three forces act on an object. Two of the forces are at an angle of 100° to each other and have magnitudes 25 N and 12 N. The third is perpendicular to the plane of these two forces and has magnitude 4 N. Calculate the magnitude of the force that would exactly counterbalance these three forces.

41. Find the unit vectors that are parallel to the tangent line to the parabola $y = x^2$ at the point $(2, 4)$.

42. (a) Find the unit vectors that are parallel to the tangent line to the curve $y = 2 \sin x$ at the point $(\pi/6, 1)$.
(b) Find the unit vectors that are perpendicular to the tangent line.
(c) Sketch the curve $y = 2 \sin x$ and the vectors in parts (a) and (b), all starting at $(\pi/6, 1)$.

43. If A, B, and C are the vertices of a triangle, find
$$\overrightarrow{AB} + \overrightarrow{BC} + \overrightarrow{CA}$$

44. Let C be the point on the line segment AB that is twice as far from B as it is from A. If $\mathbf{a} = \overrightarrow{OA}$, $\mathbf{b} = \overrightarrow{OB}$, and $\mathbf{c} = \overrightarrow{OC}$, show that $\mathbf{c} = \frac{2}{3}\mathbf{a} + \frac{1}{3}\mathbf{b}$.

45. (a) Draw the vectors $\mathbf{a} = \langle 3, 2 \rangle$, $\mathbf{b} = \langle 2, -1 \rangle$, and $\mathbf{c} = \langle 7, 1 \rangle$.
(b) Show, by means of a sketch, that there are scalars s and t such that $\mathbf{c} = s\mathbf{a} + t\mathbf{b}$.
(c) Use the sketch to estimate the values of s and t.
(d) Find the exact values of s and t.

46. Suppose that $\mathbf{a}$ and $\mathbf{b}$ are nonzero vectors that are not parallel and $\mathbf{c}$ is any vector in the plane determined by $\mathbf{a}$ and $\mathbf{b}$. Give a geometric argument to show that $\mathbf{c}$ can be written as $\mathbf{c} = s\mathbf{a} + t\mathbf{b}$ for suitable scalars s and t. Then give an argument using components.

47. If $\mathbf{r} = \langle x, y, z \rangle$ and $\mathbf{r}_0 = \langle x_0, y_0, z_0 \rangle$, describe the set of all points (x, y, z) such that $|\mathbf{r} - \mathbf{r}_0| = 1$.

48. If $\mathbf{r} = \langle x, y \rangle$, $\mathbf{r}_1 = \langle x_1, y_1 \rangle$, and $\mathbf{r}_2 = \langle x_2, y_2 \rangle$, describe the set of all points (x, y) such that $|\mathbf{r} - \mathbf{r}_1| + |\mathbf{r} - \mathbf{r}_2| = k$, where $k > |\mathbf{r}_1 - \mathbf{r}_2|$.

49. Figure 16 gives a geometric demonstration of Property 2 of vectors. Use components to give an algebraic proof of this fact for the case $n = 2$.

50. Prove Property 5 of vectors algebraically for the case $n = 3$. Then use similar triangles to give a geometric proof.

51. Use vectors to prove that the line joining the midpoints of two sides of a triangle is parallel to the third side and half its length.

52. Suppose the three coordinate planes are all mirrored and a light ray given by the vector $\mathbf{a} = \langle a_1, a_2, a_3 \rangle$ first strikes the xz-plane, as shown in the figure. Use the fact that the angle of incidence equals the angle of reflection to show that the direction of the reflected ray is given by $\mathbf{b} = \langle a_1, -a_2, a_3 \rangle$. Deduce that, after being reflected by all three mutually perpendicular mirrors, the resulting ray is parallel to the initial ray. (American space scientists used this principle, together with laser beams and an array of corner mirrors on the moon, to calculate very precisely the distance from the earth to the moon.)

12.3 The Dot Product

So far we have added two vectors and multiplied a vector by a scalar. The question arises: is it possible to multiply two vectors so that their product is a useful quantity? One such product is the dot product, whose definition follows. Another is the cross product, which is discussed in the next section.

> **1 Definition** If $\mathbf{a} = \langle a_1, a_2, a_3 \rangle$ and $\mathbf{b} = \langle b_1, b_2, b_3 \rangle$, then the **dot product** of $\mathbf{a}$ and $\mathbf{b}$ is the number $\mathbf{a} \cdot \mathbf{b}$ given by
>
> $$\mathbf{a} \cdot \mathbf{b} = a_1 b_1 + a_2 b_2 + a_3 b_3$$

Thus, to find the dot product of $\mathbf{a}$ and $\mathbf{b}$, we multiply corresponding components and add. The result is not a vector. It is a real number, that is, a scalar. For this reason, the dot product is sometimes called the **scalar product** (or **inner product**). Although Definition 1 is given for three-dimensional vectors, the dot product of two-dimensional vectors is defined in a similar fashion:

$$\langle a_1, a_2 \rangle \cdot \langle b_1, b_2 \rangle = a_1 b_1 + a_2 b_2$$

EXAMPLE 1

$$\langle 2, 4 \rangle \cdot \langle 3, -1 \rangle = 2(3) + 4(-1) = 2$$

$$\langle -1, 7, 4 \rangle \cdot \langle 6, 2, -\tfrac{1}{2} \rangle = (-1)(6) + 7(2) + 4\left(-\tfrac{1}{2}\right) = 6$$

$$(\mathbf{i} + 2\mathbf{j} - 3\mathbf{k}) \cdot (2\mathbf{j} - \mathbf{k}) = 1(0) + 2(2) + (-3)(-1) = 7 \qquad \blacksquare$$

The dot product obeys many of the laws that hold for ordinary products of real numbers. These are stated in the following theorem.

> **2 Properties of the Dot Product** If $\mathbf{a}$, $\mathbf{b}$, and $\mathbf{c}$ are vectors in V_3 and c is a scalar, then
>
> **1.** $\mathbf{a} \cdot \mathbf{a} = |\mathbf{a}|^2$ **2.** $\mathbf{a} \cdot \mathbf{b} = \mathbf{b} \cdot \mathbf{a}$
>
> **3.** $\mathbf{a} \cdot (\mathbf{b} + \mathbf{c}) = \mathbf{a} \cdot \mathbf{b} + \mathbf{a} \cdot \mathbf{c}$ **4.** $(c\mathbf{a}) \cdot \mathbf{b} = c(\mathbf{a} \cdot \mathbf{b}) = \mathbf{a} \cdot (c\mathbf{b})$
>
> **5.** $\mathbf{0} \cdot \mathbf{a} = 0$

These properties are easily proved using Definition 1. For instance, here are the proofs of Properties 1 and 3:

1. $\mathbf{a} \cdot \mathbf{a} = a_1^2 + a_2^2 + a_3^2 = |\mathbf{a}|^2$

3. $\mathbf{a} \cdot (\mathbf{b} + \mathbf{c}) = \langle a_1, a_2, a_3 \rangle \cdot \langle b_1 + c_1, b_2 + c_2, b_3 + c_3 \rangle$

$$= a_1(b_1 + c_1) + a_2(b_2 + c_2) + a_3(b_3 + c_3)$$

$$= a_1 b_1 + a_1 c_1 + a_2 b_2 + a_2 c_2 + a_3 b_3 + a_3 c_3$$

$$= (a_1 b_1 + a_2 b_2 + a_3 b_3) + (a_1 c_1 + a_2 c_2 + a_3 c_3)$$

$$= \mathbf{a} \cdot \mathbf{b} + \mathbf{a} \cdot \mathbf{c}$$

The proofs of the remaining properties are left as exercises.   ∎

The dot product $\mathbf{a} \cdot \mathbf{b}$ can be given a geometric interpretation in terms of the **angle θ between $\mathbf{a}$ and $\mathbf{b}$**, which is defined to be the angle between the representations of $\mathbf{a}$ and $\mathbf{b}$ that start at the origin, where $0 \le \theta \le \pi$. In other words, θ is the angle between the line segments $\overrightarrow{OA}$ and $\overrightarrow{OB}$ in Figure 1. Note that if $\mathbf{a}$ and $\mathbf{b}$ are parallel vectors, then $\theta = 0$ or $\theta = \pi$.

The formula in the following theorem is used by physicists as the *definition* of the dot product.

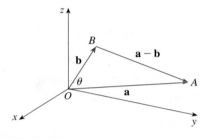

FIGURE 1

---

3 **Theorem** If θ is the angle between the vectors $\mathbf{a}$ and $\mathbf{b}$, then

$$\mathbf{a} \cdot \mathbf{b} = |\mathbf{a}||\mathbf{b}| \cos \theta$$

PROOF If we apply the Law of Cosines to triangle OAB in Figure 1, we get

4 $$|AB|^2 = |OA|^2 + |OB|^2 - 2|OA||OB| \cos \theta$$

(Observe that the Law of Cosines still applies in the limiting cases when $\theta = 0$ or π, or $\mathbf{a} = \mathbf{0}$ or $\mathbf{b} = \mathbf{0}$.) But $|OA| = |\mathbf{a}|$, $|OB| = |\mathbf{b}|$, and $|AB| = |\mathbf{a} - \mathbf{b}|$, so Equation 4 becomes

5 $$|\mathbf{a} - \mathbf{b}|^2 = |\mathbf{a}|^2 + |\mathbf{b}|^2 - 2|\mathbf{a}||\mathbf{b}| \cos \theta$$

Using Properties 1, 2, and 3 of the dot product, we can rewrite the left side of this equation as follows:

$$|\mathbf{a} - \mathbf{b}|^2 = (\mathbf{a} - \mathbf{b}) \cdot (\mathbf{a} - \mathbf{b})$$

$$= \mathbf{a} \cdot \mathbf{a} - \mathbf{a} \cdot \mathbf{b} - \mathbf{b} \cdot \mathbf{a} + \mathbf{b} \cdot \mathbf{b}$$

$$= |\mathbf{a}|^2 - 2\mathbf{a} \cdot \mathbf{b} + |\mathbf{b}|^2$$

Therefore Equation 5 gives

$$|\mathbf{a}|^2 - 2\mathbf{a} \cdot \mathbf{b} + |\mathbf{b}|^2 = |\mathbf{a}|^2 + |\mathbf{b}|^2 - 2|\mathbf{a}||\mathbf{b}| \cos \theta$$

Thus $$-2\mathbf{a} \cdot \mathbf{b} = -2|\mathbf{a}||\mathbf{b}| \cos \theta$$

or $$\mathbf{a} \cdot \mathbf{b} = |\mathbf{a}||\mathbf{b}| \cos \theta$$ ∎

EXAMPLE 2 If the vectors **a** and **b** have lengths 4 and 6, and the angle between them is $\pi/3$, find $\mathbf{a} \cdot \mathbf{b}$.

SOLUTION Using Theorem 3, we have

$$\mathbf{a} \cdot \mathbf{b} = |\mathbf{a}||\mathbf{b}| \cos(\pi/3) = 4 \cdot 6 \cdot \tfrac{1}{2} = 12$$ ■

The formula in Theorem 3 also enables us to find the angle between two vectors.

> **6 Corollary** If θ is the angle between the nonzero vectors **a** and **b**, then
>
> $$\cos \theta = \frac{\mathbf{a} \cdot \mathbf{b}}{|\mathbf{a}||\mathbf{b}|}$$

EXAMPLE 3 Find the angle between the vectors $\mathbf{a} = \langle 2, 2, -1 \rangle$ and $\mathbf{b} = \langle 5, -3, 2 \rangle$.

SOLUTION Since

$$|\mathbf{a}| = \sqrt{2^2 + 2^2 + (-1)^2} = 3 \qquad \text{and} \qquad |\mathbf{b}| = \sqrt{5^2 + (-3)^2 + 2^2} = \sqrt{38}$$

and since

$$\mathbf{a} \cdot \mathbf{b} = 2(5) + 2(-3) + (-1)(2) = 2$$

we have, from Corollary 6,

$$\cos \theta = \frac{\mathbf{a} \cdot \mathbf{b}}{|\mathbf{a}||\mathbf{b}|} = \frac{2}{3\sqrt{38}}$$

So the angle between **a** and **b** is

$$\theta = \cos^{-1}\left(\frac{2}{3\sqrt{38}}\right) \approx 1.46 \quad (\text{or } 84°)$$ ■

Two nonzero vectors **a** and **b** are called **perpendicular** or **orthogonal** if the angle between them is $\theta = \pi/2$. Then Theorem 3 gives

$$\mathbf{a} \cdot \mathbf{b} = |\mathbf{a}||\mathbf{b}| \cos(\pi/2) = 0$$

and conversely if $\mathbf{a} \cdot \mathbf{b} = 0$, then $\cos \theta = 0$, so $\theta = \pi/2$. The zero vector **0** is considered to be perpendicular to all vectors. Therefore we have the following method for determining whether two vectors are orthogonal.

> **7**   Two vectors **a** and **b** are orthogonal if and only if $\mathbf{a} \cdot \mathbf{b} = 0$.

EXAMPLE 4 Show that $2\mathbf{i} + 2\mathbf{j} - \mathbf{k}$ is perpendicular to $5\mathbf{i} - 4\mathbf{j} + 2\mathbf{k}$.

SOLUTION Since

$$(2\mathbf{i} + 2\mathbf{j} - \mathbf{k}) \cdot (5\mathbf{i} - 4\mathbf{j} + 2\mathbf{k}) = 2(5) + 2(-4) + (-1)(2) = 0$$

these vectors are perpendicular by (7).   ■

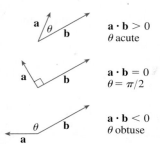

FIGURE 2

TEC Visual 12.3A shows an animation of Figure 2.

FIGURE 3

Because $\cos \theta > 0$ if $0 \leqslant \theta < \pi/2$ and $\cos \theta < 0$ if $\pi/2 < \theta \leqslant \pi$, we see that $\mathbf{a} \cdot \mathbf{b}$ is positive for $\theta < \pi/2$ and negative for $\theta > \pi/2$. We can think of $\mathbf{a} \cdot \mathbf{b}$ as measuring the extent to which $\mathbf{a}$ and $\mathbf{b}$ point in the same direction. The dot product $\mathbf{a} \cdot \mathbf{b}$ is positive if $\mathbf{a}$ and $\mathbf{b}$ point in the same general direction, 0 if they are perpendicular, and negative if they point in generally opposite directions (see Figure 2). In the extreme case where $\mathbf{a}$ and $\mathbf{b}$ point in exactly the same direction, we have $\theta = 0$, so $\cos \theta = 1$ and

$$\mathbf{a} \cdot \mathbf{b} = |\mathbf{a}||\mathbf{b}|$$

If $\mathbf{a}$ and $\mathbf{b}$ point in exactly opposite directions, then we have $\theta = \pi$ and so $\cos \theta = -1$ and $\mathbf{a} \cdot \mathbf{b} = -|\mathbf{a}||\mathbf{b}|$.

■ Direction Angles and Direction Cosines

The **direction angles** of a nonzero vector $\mathbf{a}$ are the angles α, β, and γ (in the interval $[0, \pi]$) that $\mathbf{a}$ makes with the positive x-, y-, and z-axes, respectively. (See Figure 3.)

The cosines of these direction angles, $\cos \alpha$, $\cos \beta$, and $\cos \gamma$, are called the **direction cosines** of the vector $\mathbf{a}$. Using Corollary 6 with $\mathbf{b}$ replaced by $\mathbf{i}$, we obtain

$$\boxed{8} \qquad \cos \alpha = \frac{\mathbf{a} \cdot \mathbf{i}}{|\mathbf{a}||\mathbf{i}|} = \frac{a_1}{|\mathbf{a}|}$$

(This can also be seen directly from Figure 3.)

Similarly, we also have

$$\boxed{9} \qquad \cos \beta = \frac{a_2}{|\mathbf{a}|} \qquad \cos \gamma = \frac{a_3}{|\mathbf{a}|}$$

By squaring the expressions in Equations 8 and 9 and adding, we see that

$$\boxed{10} \qquad \cos^2\alpha + \cos^2\beta + \cos^2\gamma = 1$$

We can also use Equations 8 and 9 to write

$$\mathbf{a} = \langle a_1, a_2, a_3 \rangle = \langle |\mathbf{a}| \cos \alpha, |\mathbf{a}| \cos \beta, |\mathbf{a}| \cos \gamma \rangle$$

$$= |\mathbf{a}|\langle \cos \alpha, \cos \beta, \cos \gamma \rangle$$

Therefore

$$\boxed{11} \qquad \frac{1}{|\mathbf{a}|}\mathbf{a} = \langle \cos \alpha, \cos \beta, \cos \gamma \rangle$$

which says that the direction cosines of $\mathbf{a}$ are the components of the unit vector in the direction of $\mathbf{a}$.

EXAMPLE 5 Find the direction angles of the vector $\mathbf{a} = \langle 1, 2, 3 \rangle$.

SOLUTION Since $|\mathbf{a}| = \sqrt{1^2 + 2^2 + 3^2} = \sqrt{14}$, Equations 8 and 9 give

$$\cos \alpha = \frac{1}{\sqrt{14}} \qquad \cos \beta = \frac{2}{\sqrt{14}} \qquad \cos \gamma = \frac{3}{\sqrt{14}}$$

and so

$$\alpha = \cos^{-1}\left(\frac{1}{\sqrt{14}}\right) \approx 74° \qquad \beta = \cos^{-1}\left(\frac{2}{\sqrt{14}}\right) \approx 58° \qquad \gamma = \cos^{-1}\left(\frac{3}{\sqrt{14}}\right) \approx 37°$$

■

TEC Visual 12.3B shows how Figure 4 changes when we vary **a** and **b**.

FIGURE 4

Vector projections

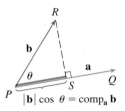

FIGURE 5

Scalar projection

Projections

Figure 4 shows representations $\vec{PQ}$ and $\vec{PR}$ of two vectors **a** and **b** with the same initial point P. If S is the foot of the perpendicular from R to the line containing $\vec{PQ}$, then the vector with representation $\vec{PS}$ is called the **vector projection** of **b** onto **a** and is denoted by $\text{proj}_a \mathbf{b}$. (You can think of it as a shadow of **b**).

The **scalar projection** of **b** onto **a** (also called the **component of b along a**) is defined to be the signed magnitude of the vector projection, which is the number $|\mathbf{b}| \cos \theta$, where θ is the angle between **a** and **b**. (See Figure 5.) This is denoted by $\text{comp}_a \mathbf{b}$. Observe that it is negative if $\pi/2 < \theta \leq \pi$. The equation

$$\mathbf{a} \cdot \mathbf{b} = |\mathbf{a}||\mathbf{b}| \cos \theta = |\mathbf{a}|(|\mathbf{b}| \cos \theta)$$

shows that the dot product of **a** and **b** can be interpreted as the length of **a** times the scalar projection of **b** onto **a**. Since

$$|\mathbf{b}| \cos \theta = \frac{\mathbf{a} \cdot \mathbf{b}}{|\mathbf{a}|} = \frac{\mathbf{a}}{|\mathbf{a}|} \cdot \mathbf{b}$$

the component of **b** along **a** can be computed by taking the dot product of **b** with the unit vector in the direction of **a**. We summarize these ideas as follows.

Scalar projection of **b** onto **a**:	$\text{comp}_a \mathbf{b} = \dfrac{\mathbf{a} \cdot \mathbf{b}}{|\mathbf{a}|}$
Vector projection of **b** onto **a**:	$\text{proj}_a \mathbf{b} = \left(\dfrac{\mathbf{a} \cdot \mathbf{b}}{|\mathbf{a}|}\right) \dfrac{\mathbf{a}}{|\mathbf{a}|} = \dfrac{\mathbf{a} \cdot \mathbf{b}}{|\mathbf{a}|^2} \mathbf{a}$

Notice that the vector projection is the scalar projection times the unit vector in the direction of **a**.

EXAMPLE 6 Find the scalar projection and vector projection of $\mathbf{b} = \langle 1, 1, 2 \rangle$ onto $\mathbf{a} = \langle -2, 3, 1 \rangle$.

SOLUTION Since $|\mathbf{a}| = \sqrt{(-2)^2 + 3^2 + 1^2} = \sqrt{14}$, the scalar projection of **b** onto **a** is

$$\text{comp}_a \mathbf{b} = \frac{\mathbf{a} \cdot \mathbf{b}}{|\mathbf{a}|} = \frac{(-2)(1) + 3(1) + 1(2)}{\sqrt{14}} = \frac{3}{\sqrt{14}}$$

The vector projection is this scalar projection times the unit vector in the direction of **a**:

$$\text{proj}_a \mathbf{b} = \frac{3}{\sqrt{14}} \frac{\mathbf{a}}{|\mathbf{a}|} = \frac{3}{14} \mathbf{a} = \left\langle -\frac{3}{7}, \frac{9}{14}, \frac{3}{14} \right\rangle$$ ∎

One use of projections occurs in physics in calculating work. In Section 6.4 we defined the work done by a constant force F in moving an object through a distance d as $W = Fd$, but this applies only when the force is directed along the line of motion of the object. Suppose, however, that the constant force is a vector $\mathbf{F} = \vec{PR}$ pointing in some other direction, as in Figure 6. If the force moves the object from P to Q, then the **displacement vector** is $\mathbf{D} = \vec{PQ}$. The **work** done by this force is defined to be the product of the component of the force along **D** and the distance moved:

$$W = (|\mathbf{F}| \cos \theta)|\mathbf{D}|$$

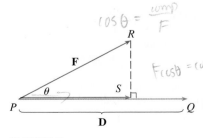

FIGURE 6

But then, from Theorem 3, we have

$$\boxed{12} \qquad W = |\mathbf{F}||\mathbf{D}| \cos \theta = \mathbf{F} \cdot \mathbf{D}$$

Thus the work done by a constant force $\mathbf{F}$ is the dot product $\mathbf{F} \cdot \mathbf{D}$, where $\mathbf{D}$ is the displacement vector.

EXAMPLE 7 A wagon is pulled a distance of 100 m along a horizontal path by a constant force of 70 N. The handle of the wagon is held at an angle of 35° above the horizontal. Find the work done by the force.

SOLUTION If $\mathbf{F}$ and $\mathbf{D}$ are the force and displacement vectors, as pictured in Figure 7, then the work done is

$$W = \mathbf{F} \cdot \mathbf{D} = |\mathbf{F}||\mathbf{D}| \cos 35°$$

$$= (70)(100) \cos 35° \approx 5734 \text{ N} \cdot \text{m} = 5734 \text{ J} \qquad \blacksquare$$

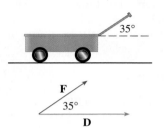

FIGURE 7

EXAMPLE 8 A force is given by a vector $\mathbf{F} = 3\mathbf{i} + 4\mathbf{j} + 5\mathbf{k}$ and moves a particle from the point $P(2, 1, 0)$ to the point $Q(4, 6, 2)$. Find the work done.

SOLUTION The displacement vector is $\mathbf{D} = \overrightarrow{PQ} = \langle 2, 5, 2 \rangle$, so by Equation 12, the work done is

$$W = \mathbf{F} \cdot \mathbf{D} = \langle 3, 4, 5 \rangle \cdot \langle 2, 5, 2 \rangle$$

$$= 6 + 20 + 10 = 36$$

If the unit of length is meters and the magnitude of the force is measured in newtons, then the work done is 36 J. $\qquad \blacksquare$

12.3 EXERCISES

1. Which of the following expressions are meaningful? Which are meaningless? Explain.
  (a) $(\mathbf{a} \cdot \mathbf{b}) \cdot \mathbf{c}$ (b) $(\mathbf{a} \cdot \mathbf{b})\mathbf{c}$
  (c) $|\mathbf{a}|(\mathbf{b} \cdot \mathbf{c})$ (d) $\mathbf{a} \cdot (\mathbf{b} + \mathbf{c})$
  (e) $\mathbf{a} \cdot \mathbf{b} + \mathbf{c}$ (f) $|\mathbf{a}| \cdot (\mathbf{b} + \mathbf{c})$

2–10 Find $\mathbf{a} \cdot \mathbf{b}$.

2. $\mathbf{a} = \langle 5, -2 \rangle$, $\mathbf{b} = \langle 3, 4 \rangle$

3. $\mathbf{a} = \langle 1.5, 0.4 \rangle$, $\mathbf{b} = \langle -4, 6 \rangle$

4. $\mathbf{a} = \langle 6, -2, 3 \rangle$, $\mathbf{b} = \langle 2, 5, -1 \rangle$

5. $\mathbf{a} = \langle 4, 1, \frac{1}{4} \rangle$, $\mathbf{b} = \langle 6, -3, -8 \rangle$

6. $\mathbf{a} = \langle p, -p, 2p \rangle$, $\mathbf{b} = \langle 2q, q, -q \rangle$

7. $\mathbf{a} = 2\mathbf{i} + \mathbf{j}$, $\mathbf{b} = \mathbf{i} - \mathbf{j} + \mathbf{k}$

8. $\mathbf{a} = 3\mathbf{i} + 2\mathbf{j} - \mathbf{k}$, $\mathbf{b} = 4\mathbf{i} + 5\mathbf{k}$

9. $|\mathbf{a}| = 7$, $|\mathbf{b}| = 4$,   the angle between $\mathbf{a}$ and $\mathbf{b}$ is 30°

10. $|\mathbf{a}| = 80$, $|\mathbf{b}| = 50$,   the angle between $\mathbf{a}$ and $\mathbf{b}$ is $3\pi/4$

11–12 If $\mathbf{u}$ is a unit vector, find $\mathbf{u} \cdot \mathbf{v}$ and $\mathbf{u} \cdot \mathbf{w}$.

11.

12.

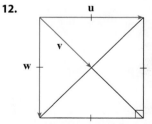

13. (a) Show that $\mathbf{i} \cdot \mathbf{j} = \mathbf{j} \cdot \mathbf{k} = \mathbf{k} \cdot \mathbf{i} = 0$.
  (b) Show that $\mathbf{i} \cdot \mathbf{i} = \mathbf{j} \cdot \mathbf{j} = \mathbf{k} \cdot \mathbf{k} = 1$.

14. A street vendor sells a hamburgers, b hot dogs, and c soft drinks on a given day. He charges \$4 for a hamburger, \$2.50 for a hot dog, and \$1 for a soft drink. If $\mathbf{A} = \langle a, b, c \rangle$ and $\mathbf{P} = \langle 4, 2.5, 1 \rangle$, what is the meaning of the dot product $\mathbf{A} \cdot \mathbf{P}$?

15–20 Find the angle between the vectors. (First find an exact expression and then approximate to the nearest degree.)

15. $\mathbf{a} = \langle 4, 3 \rangle$, $\mathbf{b} = \langle 2, -1 \rangle$

16. $\mathbf{a} = \langle -2, 5 \rangle$, $\mathbf{b} = \langle 5, 12 \rangle$

17. $\mathbf{a} = \langle 1, -4, 1 \rangle$, $\quad \mathbf{b} = \langle 0, 2, -2 \rangle$

18. $\mathbf{a} = \langle -1, 3, 4 \rangle$, $\quad \mathbf{b} = \langle 5, 2, 1 \rangle$

19. $\mathbf{a} = 4\mathbf{i} - 3\mathbf{j} + \mathbf{k}$, $\quad \mathbf{b} = 2\mathbf{i} - \mathbf{k}$

20. $\mathbf{a} = 8\mathbf{i} - \mathbf{j} + 4\mathbf{k}$, $\quad \mathbf{b} = 4\mathbf{j} + 2\mathbf{k}$

21–22 Find, correct to the nearest degree, the three angles of the triangle with the given vertices.

21. $P(2, 0)$, $\quad Q(0, 3)$, $\quad R(3, 4)$

22. $A(1, 0, -1)$, $\quad B(3, -2, 0)$, $\quad C(1, 3, 3)$

23–24 Determine whether the given vectors are orthogonal, parallel, or neither.

23. (a) $\mathbf{a} = \langle 9, 3 \rangle$, $\quad \mathbf{b} = \langle -2, 6 \rangle$
  (b) $\mathbf{a} = \langle 4, 5, -2 \rangle$, $\quad \mathbf{b} = \langle 3, -1, 5 \rangle$
  (c) $\mathbf{a} = -8\mathbf{i} + 12\mathbf{j} + 4\mathbf{k}$, $\quad \mathbf{b} = 6\mathbf{i} - 9\mathbf{j} - 3\mathbf{k}$
  (d) $\mathbf{a} = 3\mathbf{i} - \mathbf{j} + 3\mathbf{k}$, $\quad \mathbf{b} = 5\mathbf{i} + 9\mathbf{j} - 2\mathbf{k}$

24. (a) $\mathbf{u} = \langle -5, 4, -2 \rangle$, $\quad \mathbf{v} = \langle 3, 4, -1 \rangle$
  (b) $\mathbf{u} = 9\mathbf{i} - 6\mathbf{j} + 3\mathbf{k}$, $\quad \mathbf{v} = -6\mathbf{i} + 4\mathbf{j} - 2\mathbf{k}$
  (c) $\mathbf{u} = \langle c, c, c \rangle$, $\quad \mathbf{v} = \langle c, 0, -c \rangle$

---

25. Use vectors to decide whether the triangle with vertices $P(1, -3, -2)$, $Q(2, 0, -4)$, and $R(6, -2, -5)$ is right-angled.

26. Find the values of x such that the angle between the vectors $\langle 2, 1, -1 \rangle$, and $\langle 1, x, 0 \rangle$ is $45°$.

27. Find a unit vector that is orthogonal to both $\mathbf{i} + \mathbf{j}$ and $\mathbf{i} + \mathbf{k}$.

28. Find two unit vectors that make an angle of $60°$ with $\mathbf{v} = \langle 3, 4 \rangle$.

29–30 Find the acute angle between the lines.

29. $2x - y = 3$, $\quad 3x + y = 7$

30. $x + 2y = 7$, $\quad 5x - y = 2$

31–32 Find the acute angles between the curves at their points of intersection. (The angle between two curves is the angle between their tangent lines at the point of intersection.)

31. $y = x^2$, $\quad y = x^3$

32. $y = \sin x$, $\quad y = \cos x$, $\quad 0 \leqslant x \leqslant \pi/2$

33–37 Find the direction cosines and direction angles of the vector. (Give the direction angles correct to the nearest degree.)

33. $\langle 2, 1, 2 \rangle$ **34.** $\langle 6, 3, -2 \rangle$

35. $\mathbf{i} - 2\mathbf{j} - 3\mathbf{k}$ **36.** $\frac{1}{2}\mathbf{i} + \mathbf{j} + \mathbf{k}$

37. $\langle c, c, c \rangle$, $\quad$ where $c > 0$

---

38. If a vector has direction angles $\alpha = \pi/4$ and $\beta = \pi/3$, find the third direction angle γ.

39–44 Find the scalar and vector projections of $\mathbf{b}$ onto $\mathbf{a}$.

39. $\mathbf{a} = \langle -5, 12 \rangle$, $\quad \mathbf{b} = \langle 4, 6 \rangle$

40. $\mathbf{a} = \langle 1, 4 \rangle$, $\quad \mathbf{b} = \langle 2, 3 \rangle$

41. $\mathbf{a} = \langle 4, 7, -4 \rangle$, $\quad \mathbf{b} = \langle 3, -1, 1 \rangle$

42. $\mathbf{a} = \langle -1, 4, 8 \rangle$, $\quad \mathbf{b} = \langle 12, 1, 2 \rangle$

43. $\mathbf{a} = 3\mathbf{i} - 3\mathbf{j} + \mathbf{k}$, $\quad \mathbf{b} = 2\mathbf{i} + 4\mathbf{j} - \mathbf{k}$

44. $\mathbf{a} = \mathbf{i} + 2\mathbf{j} + 3\mathbf{k}$, $\quad \mathbf{b} = 5\mathbf{i} - \mathbf{k}$

---

45. Show that the vector $\text{orth}_a \mathbf{b} = \mathbf{b} - \text{proj}_a \mathbf{b}$ is orthogonal to $\mathbf{a}$. (It is called an **orthogonal projection** of $\mathbf{b}$.)

46. For the vectors in Exercise 40, find $\text{orth}_a \mathbf{b}$ and illustrate by drawing the vectors $\mathbf{a}$, $\mathbf{b}$, $\text{proj}_a \mathbf{b}$, and $\text{orth}_a \mathbf{b}$.

47. If $\mathbf{a} = \langle 3, 0, -1 \rangle$, find a vector $\mathbf{b}$ such that $\text{comp}_a \mathbf{b} = 2$.

48. Suppose that $\mathbf{a}$ and $\mathbf{b}$ are nonzero vectors.
  (a) Under what circumstances is $\text{comp}_a \mathbf{b} = \text{comp}_b \mathbf{a}$?
  (b) Under what circumstances is $\text{proj}_a \mathbf{b} = \text{proj}_b \mathbf{a}$?

49. Find the work done by a force $\mathbf{F} = 8\mathbf{i} - 6\mathbf{j} + 9\mathbf{k}$ that moves an object from the point $(0, 10, 8)$ to the point $(6, 12, 20)$ along a straight line. The distance is measured in meters and the force in newtons.

50. A tow truck drags a stalled car along a road. The chain makes an angle of $30°$ with the road and the tension in the chain is 1500 N. How much work is done by the truck in pulling the car 1 km?

51. A sled is pulled along a level path through snow by a rope. A 30-lb force acting at an angle of $40°$ above the horizontal moves the sled 80 ft. Find the work done by the force.

52. A boat sails south with the help of a wind blowing in the direction S36°E with magnitude 400 lb. Find the work done by the wind as the boat moves 120 ft.

53. Use a scalar projection to show that the distance from a point $P_1(x_1, y_1)$ to the line $ax + by + c = 0$ is

$$\frac{|ax_1 + by_1 + c|}{\sqrt{a^2 + b^2}}$$

Use this formula to find the distance from the point $(-2, 3)$ to the line $3x - 4y + 5 = 0$.

54. If $\mathbf{r} = \langle x, y, z \rangle$, $\mathbf{a} = \langle a_1, a_2, a_3 \rangle$, and $\mathbf{b} = \langle b_1, b_2, b_3 \rangle$, show that the vector equation $(\mathbf{r} - \mathbf{a}) \cdot (\mathbf{r} - \mathbf{b}) = 0$ represents a sphere, and find its center and radius.

55. Find the angle between a diagonal of a cube and one of its edges.

56. Find the angle between a diagonal of a cube and a diagonal of one of its faces.

57. A molecule of methane, CH_4, is structured with the four hydrogen atoms at the vertices of a regular tetrahedron and the carbon atom at the centroid. The *bond angle* is the angle formed by the H—C—H combination; it is the angle between the lines that join the carbon atom to two of the hydrogen atoms. Show that the bond angle is about $109.5°$. $\left[\textit{Hint:}\text{ Take the vertices of the tetrahedron to be the points }(1, 0, 0), (0, 1, 0), (0, 0, 1),\text{ and }(1, 1, 1),\text{ as shown in the figure. Then the centroid is }\left(\frac{1}{2}, \frac{1}{2}, \frac{1}{2}\right).\right]$

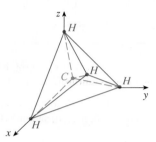

58. If $\mathbf{c} = |\mathbf{a}|\mathbf{b} + |\mathbf{b}|\mathbf{a}$, where $\mathbf{a}$, $\mathbf{b}$, and $\mathbf{c}$ are all nonzero vectors, show that $\mathbf{c}$ bisects the angle between $\mathbf{a}$ and $\mathbf{b}$.

59. Prove Properties 2, 4, and 5 of the dot product (Theorem 2).

60. Suppose that all sides of a quadrilateral are equal in length and opposite sides are parallel. Use vector methods to show that the diagonals are perpendicular.

61. Use Theorem 3 to prove the Cauchy-Schwarz Inequality:
$$|\mathbf{a} \cdot \mathbf{b}| \leq |\mathbf{a}||\mathbf{b}|$$

62. The Triangle Inequality for vectors is
$$|\mathbf{a} + \mathbf{b}| \leq |\mathbf{a}| + |\mathbf{b}|$$

(a) Give a geometric interpretation of the Triangle Inequality.
(b) Use the Cauchy-Schwarz Inequality from Exercise 61 to prove the Triangle Inequality. [*Hint:* Use the fact that $|\mathbf{a} + \mathbf{b}|^2 = (\mathbf{a} + \mathbf{b}) \cdot (\mathbf{a} + \mathbf{b})$ and use Property 3 of the dot product.]

63. The Parallelogram Law states that
$$|\mathbf{a} + \mathbf{b}|^2 + |\mathbf{a} - \mathbf{b}|^2 = 2|\mathbf{a}|^2 + 2|\mathbf{b}|^2$$

(a) Give a geometric interpretation of the Parallelogram Law.
(b) Prove the Parallelogram Law. (See the hint in Exercise 62.)

64. Show that if $\mathbf{u} + \mathbf{v}$ and $\mathbf{u} - \mathbf{v}$ are orthogonal, then the vectors $\mathbf{u}$ and $\mathbf{v}$ must have the same length.

65. If θ is the angle between vectors $\mathbf{a}$ and $\mathbf{b}$, show that
$$\operatorname{proj}_{\mathbf{a}} \mathbf{b} \cdot \operatorname{proj}_{\mathbf{b}} \mathbf{a} = (\mathbf{a} \cdot \mathbf{b}) \cos^2 \theta$$

12.4 The Cross Product

Given two nonzero vectors $\mathbf{a} = \langle a_1, a_2, a_3 \rangle$ and $\mathbf{b} = \langle b_1, b_2, b_3 \rangle$, it is very useful to be able to find a nonzero vector $\mathbf{c}$ that is perpendicular to both $\mathbf{a}$ and $\mathbf{b}$, as we will see in the next section and in Chapters 13 and 14. If $\mathbf{c} = \langle c_1, c_2, c_3 \rangle$ is such a vector, then $\mathbf{a} \cdot \mathbf{c} = 0$ and $\mathbf{b} \cdot \mathbf{c} = 0$ and so

$$\boxed{1} \qquad a_1 c_1 + a_2 c_2 + a_3 c_3 = 0$$

$$\boxed{2} \qquad b_1 c_1 + b_2 c_2 + b_3 c_3 = 0$$

To eliminate c_3 we multiply (1) by b_3 and (2) by a_3 and subtract:

$$\boxed{3} \qquad (a_1 b_3 - a_3 b_1)c_1 + (a_2 b_3 - a_3 b_2)c_2 = 0$$

Equation 3 has the form $pc_1 + qc_2 = 0$, for which an obvious solution is $c_1 = q$ and $c_2 = -p$. So a solution of (3) is

$$c_1 = a_2 b_3 - a_3 b_2 \qquad\qquad c_2 = a_3 b_1 - a_1 b_3$$

Substituting these values into (1) and (2), we then get

$$c_3 = a_1 b_2 - a_2 b_1$$

This means that a vector perpendicular to both $\mathbf{a}$ and $\mathbf{b}$ is

$$\langle c_1, c_2, c_3 \rangle = \langle a_2 b_3 - a_3 b_2, a_3 b_1 - a_1 b_3, a_1 b_2 - a_2 b_1 \rangle$$

The resulting vector is called the *cross product* of $\mathbf{a}$ and $\mathbf{b}$ and is denoted by $\mathbf{a} \times \mathbf{b}$.

4 **Definition** If $\mathbf{a} = \langle a_1, a_2, a_3 \rangle$ and $\mathbf{b} = \langle b_1, b_2, b_3 \rangle$, then the **cross product** of $\mathbf{a}$ and $\mathbf{b}$ is the vector

$$\mathbf{a} \times \mathbf{b} = \langle a_2b_3 - a_3b_2, \; a_3b_1 - a_1b_3, \; a_1b_2 - a_2b_1 \rangle$$

Notice that the **cross product** $\mathbf{a} \times \mathbf{b}$ of two vectors $\mathbf{a}$ and $\mathbf{b}$, unlike the dot product, is a vector. For this reason it is also called the **vector product**. Note that $\mathbf{a} \times \mathbf{b}$ is defined only when $\mathbf{a}$ and $\mathbf{b}$ are *three-dimensional* vectors.

In order to make Definition 4 easier to remember, we use the notation of determinants. A **determinant of order 2** is defined by

$$\begin{vmatrix} a & b \\ c & d \end{vmatrix} = ad - bc$$

(Multiply across the diagonals and subtract.) For example,

$$\begin{vmatrix} 2 & 1 \\ -6 & 4 \end{vmatrix} = 2(4) - 1(-6) = 14$$

A **determinant of order 3** can be defined in terms of second-order determinants as follows:

5
$$\begin{vmatrix} a_1 & a_2 & a_3 \\ b_1 & b_2 & b_3 \\ c_1 & c_2 & c_3 \end{vmatrix} = a_1 \begin{vmatrix} b_2 & b_3 \\ c_2 & c_3 \end{vmatrix} - a_2 \begin{vmatrix} b_1 & b_3 \\ c_1 & c_3 \end{vmatrix} + a_3 \begin{vmatrix} b_1 & b_2 \\ c_1 & c_2 \end{vmatrix}$$

Observe that each term on the right side of Equation 5 involves a number a_i in the first row of the determinant, and a_i is multiplied by the second-order determinant obtained from the left side by deleting the row and column in which a_i appears. Notice also the minus sign in the second term. For example,

$$\begin{vmatrix} 1 & 2 & -1 \\ 3 & 0 & 1 \\ -5 & 4 & 2 \end{vmatrix} = 1 \begin{vmatrix} 0 & 1 \\ 4 & 2 \end{vmatrix} - 2 \begin{vmatrix} 3 & 1 \\ -5 & 2 \end{vmatrix} + (-1) \begin{vmatrix} 3 & 0 \\ -5 & 4 \end{vmatrix}$$

$$= 1(0 - 4) - 2(6 + 5) + (-1)(12 - 0) = -38$$

If we now rewrite Definition 4 using second-order determinants and the standard basis vectors $\mathbf{i}$, $\mathbf{j}$, and $\mathbf{k}$, we see that the cross product of the vectors $\mathbf{a} = a_1\mathbf{i} + a_2\mathbf{j} + a_3\mathbf{k}$ and $\mathbf{b} = b_1\mathbf{i} + b_2\mathbf{j} + b_3\mathbf{k}$ is

6
$$\mathbf{a} \times \mathbf{b} = \begin{vmatrix} a_2 & a_3 \\ b_2 & b_3 \end{vmatrix} \mathbf{i} - \begin{vmatrix} a_1 & a_3 \\ b_1 & b_3 \end{vmatrix} \mathbf{j} + \begin{vmatrix} a_1 & a_2 \\ b_1 & b_2 \end{vmatrix} \mathbf{k}$$

In view of the similarity between Equations 5 and 6, we often write

7
$$\mathbf{a} \times \mathbf{b} = \begin{vmatrix} \mathbf{i} & \mathbf{j} & \mathbf{k} \\ a_1 & a_2 & a_3 \\ b_1 & b_2 & b_3 \end{vmatrix}$$

Although the first row of the symbolic determinant in Equation 7 consists of vectors, if we expand it as if it were an ordinary determinant using the rule in Equation 5, we obtain

Equation 6. The symbolic formula in Equation 7 is probably the easiest way of remembering and computing cross products.

EXAMPLE 1 If $\mathbf{a} = \langle 1, 3, 4 \rangle$ and $\mathbf{b} = \langle 2, 7, -5 \rangle$, then

$$\mathbf{a} \times \mathbf{b} = \begin{vmatrix} \mathbf{i} & \mathbf{j} & \mathbf{k} \\ 1 & 3 & 4 \\ 2 & 7 & -5 \end{vmatrix}$$

$$= \begin{vmatrix} 3 & 4 \\ 7 & -5 \end{vmatrix} \mathbf{i} - \begin{vmatrix} 1 & 4 \\ 2 & -5 \end{vmatrix} \mathbf{j} + \begin{vmatrix} 1 & 3 \\ 2 & 7 \end{vmatrix} \mathbf{k}$$

$$= (-15 - 28) \mathbf{i} - (-5 - 8) \mathbf{j} + (7 - 6) \mathbf{k} = -43\mathbf{i} + 13\mathbf{j} + \mathbf{k} \quad \blacksquare$$

EXAMPLE 2 Show that $\mathbf{a} \times \mathbf{a} = \mathbf{0}$ for any vector $\mathbf{a}$ in V_3.

SOLUTION If $\mathbf{a} = \langle a_1, a_2, a_3 \rangle$, then

$$\mathbf{a} \times \mathbf{a} = \begin{vmatrix} \mathbf{i} & \mathbf{j} & \mathbf{k} \\ a_1 & a_2 & a_3 \\ a_1 & a_2 & a_3 \end{vmatrix}$$

$$= (a_2 a_3 - a_3 a_2) \mathbf{i} - (a_1 a_3 - a_3 a_1) \mathbf{j} + (a_1 a_2 - a_2 a_1) \mathbf{k}$$

$$= 0\mathbf{i} - 0\mathbf{j} + 0\mathbf{k} = \mathbf{0} \quad \blacksquare$$

We constructed the cross product $\mathbf{a} \times \mathbf{b}$ so that it would be perpendicular to both $\mathbf{a}$ and $\mathbf{b}$. This is one of the most important properties of a cross product, so let's emphasize and verify it in the following theorem and give a formal proof.

> **8 Theorem** The vector $\mathbf{a} \times \mathbf{b}$ is orthogonal to both $\mathbf{a}$ and $\mathbf{b}$.

PROOF In order to show that $\mathbf{a} \times \mathbf{b}$ is orthogonal to $\mathbf{a}$, we compute their dot product as follows:

$$(\mathbf{a} \times \mathbf{b}) \cdot \mathbf{a} = \begin{vmatrix} a_2 & a_3 \\ b_2 & b_3 \end{vmatrix} a_1 - \begin{vmatrix} a_1 & a_3 \\ b_1 & b_3 \end{vmatrix} a_2 + \begin{vmatrix} a_1 & a_2 \\ b_1 & b_2 \end{vmatrix} a_3$$

$$= a_1(a_2 b_3 - a_3 b_2) - a_2(a_1 b_3 - a_3 b_1) + a_3(a_1 b_2 - a_2 b_1)$$

$$= a_1 a_2 b_3 - a_1 b_2 a_3 - a_1 a_2 b_3 + b_1 a_2 a_3 + a_1 b_2 a_3 - b_1 a_2 a_3$$

$$= 0$$

A similar computation shows that $(\mathbf{a} \times \mathbf{b}) \cdot \mathbf{b} = 0$. Therefore $\mathbf{a} \times \mathbf{b}$ is orthogonal to both $\mathbf{a}$ and $\mathbf{b}$. $\quad \blacksquare$

If $\mathbf{a}$ and $\mathbf{b}$ are represented by directed line segments with the same initial point (as in Figure 1), then Theorem 8 says that the cross product $\mathbf{a} \times \mathbf{b}$ points in a direction perpendicular to the plane through $\mathbf{a}$ and $\mathbf{b}$. It turns out that the direction of $\mathbf{a} \times \mathbf{b}$ is given by the *right-hand rule:* if the fingers of your right hand curl in the direction of

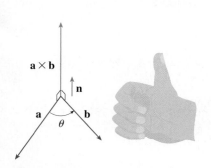

FIGURE 1
The right-hand rule gives the direction
of $\mathbf{a} \times \mathbf{b}$.

a rotation (through an angle less than 180°) from **a** to **b**, then your thumb points in the direction of $\mathbf{a} \times \mathbf{b}$.

Now that we know the direction of the vector $\mathbf{a} \times \mathbf{b}$, the remaining thing we need to complete its geometric description is its length $|\mathbf{a} \times \mathbf{b}|$. This is given by the following theorem.

> **9** **Theorem** If θ is the angle between **a** and **b** (so $0 \leqslant \theta \leqslant \pi$), then
> $$|\mathbf{a} \times \mathbf{b}| = |\mathbf{a}|\,|\mathbf{b}|\sin\theta$$

TEC Visual 12.4 shows how $\mathbf{a} \times \mathbf{b}$ changes as **b** changes.

PROOF From the definitions of the cross product and length of a vector, we have

$$|\mathbf{a} \times \mathbf{b}|^2 = (a_2 b_3 - a_3 b_2)^2 + (a_3 b_1 - a_1 b_3)^2 + (a_1 b_2 - a_2 b_1)^2$$

$$= a_2^2 b_3^2 - 2a_2 a_3 b_2 b_3 + a_3^2 b_2^2 + a_3^2 b_1^2 - 2a_1 a_3 b_1 b_3 + a_1^2 b_3^2$$

$$\quad + a_1^2 b_2^2 - 2a_1 a_2 b_1 b_2 + a_2^2 b_1^2$$

$$= (a_1^2 + a_2^2 + a_3^2)(b_1^2 + b_2^2 + b_3^2) - (a_1 b_1 + a_2 b_2 + a_3 b_3)^2$$

$$= |\mathbf{a}|^2 |\mathbf{b}|^2 - (\mathbf{a} \cdot \mathbf{b})^2$$

$$= |\mathbf{a}|^2 |\mathbf{b}|^2 - |\mathbf{a}|^2 |\mathbf{b}|^2 \cos^2\theta \qquad \text{(by Theorem 12.3.3)}$$

$$= |\mathbf{a}|^2 |\mathbf{b}|^2 (1 - \cos^2\theta)$$

$$= |\mathbf{a}|^2 |\mathbf{b}|^2 \sin^2\theta$$

Taking square roots and observing that $\sqrt{\sin^2\theta} = \sin\theta$ because $\sin\theta \geqslant 0$ when $0 \leqslant \theta \leqslant \pi$, we have

$$|\mathbf{a} \times \mathbf{b}| = |\mathbf{a}|\,|\mathbf{b}|\sin\theta \qquad \blacksquare$$

Geometric characterization of $\mathbf{a} \times \mathbf{b}$

Since a vector is completely determined by its magnitude and direction, we can now say that $\mathbf{a} \times \mathbf{b}$ is the vector that is perpendicular to both **a** and **b**, whose orientation is determined by the right-hand rule, and whose length is $|\mathbf{a}|\,|\mathbf{b}|\sin\theta$. In fact, that is exactly how physicists *define* $\mathbf{a} \times \mathbf{b}$.

> **10** **Corollary** Two nonzero vectors **a** and **b** are parallel if and only if
> $$\mathbf{a} \times \mathbf{b} = \mathbf{0}$$

PROOF Two nonzero vectors **a** and **b** are parallel if and only if $\theta = 0$ or π. In either case $\sin\theta = 0$, so $|\mathbf{a} \times \mathbf{b}| = 0$ and therefore $\mathbf{a} \times \mathbf{b} = \mathbf{0}$. $\qquad \blacksquare$

The geometric interpretation of Theorem 9 can be seen by looking at Figure 2. If **a** and **b** are represented by directed line segments with the same initial point, then they determine a parallelogram with base $|\mathbf{a}|$, altitude $|\mathbf{b}|\sin\theta$, and area

$$A = |\mathbf{a}|\,(|\mathbf{b}|\sin\theta) = |\mathbf{a} \times \mathbf{b}|$$

FIGURE 2

Thus we have the following way of interpreting the magnitude of a cross product.

> The length of the cross product $\mathbf{a} \times \mathbf{b}$ is equal to the area of the parallelogram determined by $\mathbf{a}$ and $\mathbf{b}$.

EXAMPLE 3 Find a vector perpendicular to the plane that passes through the points $P(1, 4, 6)$, $Q(-2, 5, -1)$, and $R(1, -1, 1)$.

SOLUTION The vector $\overrightarrow{PQ} \times \overrightarrow{PR}$ is perpendicular to both $\overrightarrow{PQ}$ and $\overrightarrow{PR}$ and is therefore perpendicular to the plane through P, Q, and R. We know from (12.2.1) that

$$\overrightarrow{PQ} = (-2 - 1)\,\mathbf{i} + (5 - 4)\,\mathbf{j} + (-1 - 6)\,\mathbf{k} = -3\mathbf{i} + \mathbf{j} - 7\mathbf{k}$$

$$\overrightarrow{PR} = (1 - 1)\,\mathbf{i} + (-1 - 4)\,\mathbf{j} + (1 - 6)\,\mathbf{k} = -5\mathbf{j} - 5\mathbf{k}$$

We compute the cross product of these vectors:

$$\overrightarrow{PQ} \times \overrightarrow{PR} = \begin{vmatrix} \mathbf{i} & \mathbf{j} & \mathbf{k} \\ -3 & 1 & -7 \\ 0 & -5 & -5 \end{vmatrix}$$

$$= (-5 - 35)\,\mathbf{i} - (15 - 0)\,\mathbf{j} + (15 - 0)\,\mathbf{k} = -40\mathbf{i} - 15\mathbf{j} + 15\mathbf{k}$$

So the vector $\langle -40, -15, 15 \rangle$ is perpendicular to the given plane. Any nonzero scalar multiple of this vector, such as $\langle -8, -3, 3 \rangle$, is also perpendicular to the plane. ∎

EXAMPLE 4 Find the area of the triangle with vertices $P(1, 4, 6)$, $Q(-2, 5, -1)$, and $R(1, -1, 1)$.

SOLUTION In Example 3 we computed that $\overrightarrow{PQ} \times \overrightarrow{PR} = \langle -40, -15, 15 \rangle$. The area of the parallelogram with adjacent sides PQ and PR is the length of this cross product:

$$\left| \overrightarrow{PQ} \times \overrightarrow{PR} \right| = \sqrt{(-40)^2 + (-15)^2 + 15^2} = 5\sqrt{82}$$

The area A of the triangle PQR is half the area of this parallelogram, that is, $\frac{5}{2}\sqrt{82}$. ∎

If we apply Theorems 8 and 9 to the standard basis vectors $\mathbf{i}$, $\mathbf{j}$, and $\mathbf{k}$ using $\theta = \pi/2$, we obtain

$$\mathbf{i} \times \mathbf{j} = \mathbf{k} \qquad \mathbf{j} \times \mathbf{k} = \mathbf{i} \qquad \mathbf{k} \times \mathbf{i} = \mathbf{j}$$

$$\mathbf{j} \times \mathbf{i} = -\mathbf{k} \qquad \mathbf{k} \times \mathbf{j} = -\mathbf{i} \qquad \mathbf{i} \times \mathbf{k} = -\mathbf{j}$$

Observe that

$$\mathbf{i} \times \mathbf{j} \neq \mathbf{j} \times \mathbf{i}$$

⊘ Thus the cross product is not commutative. Also

$$\mathbf{i} \times (\mathbf{i} \times \mathbf{j}) = \mathbf{i} \times \mathbf{k} = -\mathbf{j}$$

whereas

$$(\mathbf{i} \times \mathbf{i}) \times \mathbf{j} = \mathbf{0} \times \mathbf{j} = \mathbf{0}$$

⊘ So the associative law for multiplication does not usually hold; that is, in general,

$$(\mathbf{a} \times \mathbf{b}) \times \mathbf{c} \neq \mathbf{a} \times (\mathbf{b} \times \mathbf{c})$$

However, some of the usual laws of algebra *do* hold for cross products. The following theorem summarizes the properties of vector products.

---

11 **Properties of the Cross Product** If $\mathbf{a}$, $\mathbf{b}$, and $\mathbf{c}$ are vectors and c is a scalar, then

1. $\mathbf{a} \times \mathbf{b} = -\mathbf{b} \times \mathbf{a}$
2. $(c\mathbf{a}) \times \mathbf{b} = c(\mathbf{a} \times \mathbf{b}) = \mathbf{a} \times (c\mathbf{b})$
3. $\mathbf{a} \times (\mathbf{b} + \mathbf{c}) = \mathbf{a} \times \mathbf{b} + \mathbf{a} \times \mathbf{c}$
4. $(\mathbf{a} + \mathbf{b}) \times \mathbf{c} = \mathbf{a} \times \mathbf{c} + \mathbf{b} \times \mathbf{c}$
5. $\mathbf{a} \cdot (\mathbf{b} \times \mathbf{c}) = (\mathbf{a} \times \mathbf{b}) \cdot \mathbf{c}$
6. $\mathbf{a} \times (\mathbf{b} \times \mathbf{c}) = (\mathbf{a} \cdot \mathbf{c})\mathbf{b} - (\mathbf{a} \cdot \mathbf{b})\mathbf{c}$

These properties can be proved by writing the vectors in terms of their components and using the definition of a cross product. We give the proof of Property 5 and leave the remaining proofs as exercises.

PROOF OF PROPERTY 5 If $\mathbf{a} = \langle a_1, a_2, a_3 \rangle$, $\mathbf{b} = \langle b_1, b_2, b_3 \rangle$, and $\mathbf{c} = \langle c_1, c_2, c_3 \rangle$, then

$$\boxed{12} \quad \mathbf{a} \cdot (\mathbf{b} \times \mathbf{c}) = a_1(b_2c_3 - b_3c_2) + a_2(b_3c_1 - b_1c_3) + a_3(b_1c_2 - b_2c_1)$$

$$= a_1b_2c_3 - a_1b_3c_2 + a_2b_3c_1 - a_2b_1c_3 + a_3b_1c_2 - a_3b_2c_1$$

$$= (a_2b_3 - a_3b_2)c_1 + (a_3b_1 - a_1b_3)c_2 + (a_1b_2 - a_2b_1)c_3$$

$$= (\mathbf{a} \times \mathbf{b}) \cdot \mathbf{c} \qquad ■$$

■ Triple Products

The product $\mathbf{a} \cdot (\mathbf{b} \times \mathbf{c})$ that occurs in Property 5 is called the **scalar triple product** of the vectors $\mathbf{a}$, $\mathbf{b}$, and $\mathbf{c}$. Notice from Equation 12 that we can write the scalar triple product as a determinant:

$$\boxed{13} \qquad \mathbf{a} \cdot (\mathbf{b} \times \mathbf{c}) = \begin{vmatrix} a_1 & a_2 & a_3 \\ b_1 & b_2 & b_3 \\ c_1 & c_2 & c_3 \end{vmatrix}$$

The geometric significance of the scalar triple product can be seen by considering the parallelepiped determined by the vectors $\mathbf{a}$, $\mathbf{b}$, and $\mathbf{c}$. (See Figure 3.) The area of the base parallelogram is $A = |\mathbf{b} \times \mathbf{c}|$. If θ is the angle between $\mathbf{a}$ and $\mathbf{b} \times \mathbf{c}$, then the height h of the parallelepiped is $h = |\mathbf{a}||\cos\theta|$. (We must use $|\cos\theta|$ instead of $\cos\theta$ in case $\theta > \pi/2$.) Therefore the volume of the parallelepiped is

$$V = Ah = |\mathbf{b} \times \mathbf{c}||\mathbf{a}||\cos\theta| = |\mathbf{a} \cdot (\mathbf{b} \times \mathbf{c})|$$

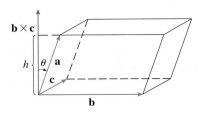

FIGURE 3

Thus we have proved the following formula.

> **14** The volume of the parallelepiped determined by the vectors **a**, **b**, and **c** is the magnitude of their scalar triple product:
>
> $$V = |\mathbf{a} \cdot (\mathbf{b} \times \mathbf{c})|$$

If we use the formula in (14) and discover that the volume of the parallelepiped determined by **a**, **b**, and **c** is 0, then the vectors must lie in the same plane; that is, they are **coplanar**.

EXAMPLE 5 Use the scalar triple product to show that the vectors $\mathbf{a} = \langle 1, 4, -7 \rangle$, $\mathbf{b} = \langle 2, -1, 4 \rangle$, and $\mathbf{c} = \langle 0, -9, 18 \rangle$ are coplanar.

SOLUTION We use Equation 13 to compute their scalar triple product:

$$\mathbf{a} \cdot (\mathbf{b} \times \mathbf{c}) = \begin{vmatrix} 1 & 4 & -7 \\ 2 & -1 & 4 \\ 0 & -9 & 18 \end{vmatrix}$$

$$= 1 \begin{vmatrix} -1 & 4 \\ -9 & 18 \end{vmatrix} - 4 \begin{vmatrix} 2 & 4 \\ 0 & 18 \end{vmatrix} - 7 \begin{vmatrix} 2 & -1 \\ 0 & -9 \end{vmatrix}$$

$$= 1(18) - 4(36) - 7(-18) = 0$$

Therefore, by (14), the volume of the parallelepiped determined by **a**, **b**, and **c** is 0. This means that **a**, **b**, and **c** are coplanar. ∎

The product $\mathbf{a} \times (\mathbf{b} \times \mathbf{c})$ that occurs in Property 6 is called the **vector triple product** of **a**, **b**, and **c**. Property 6 will be used to derive Kepler's First Law of planetary motion in Chapter 13. Its proof is left as Exercise 50.

■ Torque

The idea of a cross product occurs often in physics. In particular, we consider a force **F** acting on a rigid body at a point given by a position vector **r**. (For instance, if we tighten a bolt by applying a force to a wrench as in Figure 4, we produce a turning effect.) The **torque** $\boldsymbol{\tau}$ (relative to the origin) is defined to be the cross product of the position and force vectors

$$\boldsymbol{\tau} = \mathbf{r} \times \mathbf{F}$$

and measures the tendency of the body to rotate about the origin. The direction of the torque vector indicates the axis of rotation. According to Theorem 9, the magnitude of the torque vector is

$$|\boldsymbol{\tau}| = |\mathbf{r} \times \mathbf{F}| = |\mathbf{r}||\mathbf{F}| \sin \theta$$

where θ is the angle between the position and force vectors. Observe that the only component of **F** that can cause a rotation is the one perpendicular to **r**, that is, $|\mathbf{F}| \sin \theta$. The magnitude of the torque is equal to the area of the parallelogram determined by **r** and **F**.

EXAMPLE 6 A bolt is tightened by applying a 40-N force to a 0.25-m wrench as shown in Figure 5. Find the magnitude of the torque about the center of the bolt.

FIGURE 4

FIGURE 5

SOLUTION The magnitude of the torque vector is

$$|\boldsymbol{\tau}| = |\mathbf{r} \times \mathbf{F}| = |\mathbf{r}||\mathbf{F}| \sin 75° = (0.25)(40) \sin 75°$$
$$= 10 \sin 75° \approx 9.66 \text{ N·m}$$

If the bolt is right-threaded, then the torque vector itself is

$$\boldsymbol{\tau} = |\boldsymbol{\tau}| \mathbf{n} \approx 9.66 \, \mathbf{n}$$

where **n** is a unit vector directed down into the page (by the right-hand rule). ∎

12.4 EXERCISES

1–7 Find the cross product $\mathbf{a} \times \mathbf{b}$ and verify that it is orthogonal to both **a** and **b**.

1. $\mathbf{a} = \langle 2, 3, 0 \rangle, \quad \mathbf{b} = \langle 1, 0, 5 \rangle$

2. $\mathbf{a} = \langle 4, 3, -2 \rangle, \quad \mathbf{b} = \langle 2, -1, 1 \rangle$

3. $\mathbf{a} = 2\mathbf{j} - 4\mathbf{k}, \quad \mathbf{b} = -\mathbf{i} + 3\mathbf{j} + \mathbf{k}$

4. $\mathbf{a} = 3\mathbf{i} + 3\mathbf{j} - 3\mathbf{k}, \quad \mathbf{b} = 3\mathbf{i} - 3\mathbf{j} + 3\mathbf{k}$

5. $\mathbf{a} = \frac{1}{2}\mathbf{i} + \frac{1}{3}\mathbf{j} + \frac{1}{4}\mathbf{k}, \quad \mathbf{b} = \mathbf{i} + 2\mathbf{j} - 3\mathbf{k}$

6. $\mathbf{a} = t\mathbf{i} + \cos t\mathbf{j} + \sin t\mathbf{k}, \quad \mathbf{b} = \mathbf{i} - \sin t\mathbf{j} + \cos t\mathbf{k}$

7. $\mathbf{a} = \langle t, 1, 1/t \rangle, \quad \mathbf{b} = \langle t^2, t^2, 1 \rangle$

---

8. If $\mathbf{a} = \mathbf{i} - 2\mathbf{k}$ and $\mathbf{b} = \mathbf{j} + \mathbf{k}$, find $\mathbf{a} \times \mathbf{b}$. Sketch **a**, **b**, and $\mathbf{a} \times \mathbf{b}$ as vectors starting at the origin.

9–12 Find the vector, not with determinants, but by using properties of cross products.

9. $(\mathbf{i} \times \mathbf{j}) \times \mathbf{k}$

10. $\mathbf{k} \times (\mathbf{i} - 2\mathbf{j})$

11. $(\mathbf{j} - \mathbf{k}) \times (\mathbf{k} - \mathbf{i})$

12. $(\mathbf{i} + \mathbf{j}) \times (\mathbf{i} - \mathbf{j})$

---

13. State whether each expression is meaningful. If not, explain why. If so, state whether it is a vector or a scalar.
  (a) $\mathbf{a} \cdot (\mathbf{b} \times \mathbf{c})$
  (b) $\mathbf{a} \times (\mathbf{b} \cdot \mathbf{c})$
  (c) $\mathbf{a} \times (\mathbf{b} \times \mathbf{c})$
  (d) $\mathbf{a} \cdot (\mathbf{b} \cdot \mathbf{c})$
  (e) $(\mathbf{a} \cdot \mathbf{b}) \times (\mathbf{c} \cdot \mathbf{d})$
  (f) $(\mathbf{a} \times \mathbf{b}) \cdot (\mathbf{c} \times \mathbf{d})$

14–15 Find $|\mathbf{u} \times \mathbf{v}|$ and determine whether $\mathbf{u} \times \mathbf{v}$ is directed into the page or out of the page.

14.

15.

16. The figure shows a vector **a** in the xy-plane and a vector **b** in the direction of **k**. Their lengths are $|\mathbf{a}| = 3$ and $|\mathbf{b}| = 2$.
  (a) Find $|\mathbf{a} \times \mathbf{b}|$.

(b) Use the right-hand rule to decide whether the components of $\mathbf{a} \times \mathbf{b}$ are positive, negative, or 0.

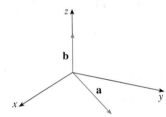

17. If $\mathbf{a} = \langle 2, -1, 3 \rangle$ and $\mathbf{b} = \langle 4, 2, 1 \rangle$, find $\mathbf{a} \times \mathbf{b}$ and $\mathbf{b} \times \mathbf{a}$.

18. If $\mathbf{a} = \langle 1, 0, 1 \rangle$, $\mathbf{b} = \langle 2, 1, -1 \rangle$, and $\mathbf{c} = \langle 0, 1, 3 \rangle$, show that $\mathbf{a} \times (\mathbf{b} \times \mathbf{c}) \neq (\mathbf{a} \times \mathbf{b}) \times \mathbf{c}$.

19. Find two unit vectors orthogonal to both $\langle 3, 2, 1 \rangle$ and $\langle -1, 1, 0 \rangle$.

20. Find two unit vectors orthogonal to both $\mathbf{j} - \mathbf{k}$ and $\mathbf{i} + \mathbf{j}$.

21. Show that $\mathbf{0} \times \mathbf{a} = \mathbf{0} = \mathbf{a} \times \mathbf{0}$ for any vector **a** in V_3.

22. Show that $(\mathbf{a} \times \mathbf{b}) \cdot \mathbf{b} = 0$ for all vectors **a** and **b** in V_3.

23–26 Prove the property of cross products (Theorem 11).

23. Property 1: $\mathbf{a} \times \mathbf{b} = -\mathbf{b} \times \mathbf{a}$

24. Property 2: $(c\mathbf{a}) \times \mathbf{b} = c(\mathbf{a} \times \mathbf{b}) = \mathbf{a} \times (c\mathbf{b})$

25. Property 3: $\mathbf{a} \times (\mathbf{b} + \mathbf{c}) = \mathbf{a} \times \mathbf{b} + \mathbf{a} \times \mathbf{c}$

26. Property 4: $(\mathbf{a} + \mathbf{b}) \times \mathbf{c} = \mathbf{a} \times \mathbf{c} + \mathbf{b} \times \mathbf{c}$

---

27. Find the area of the parallelogram with vertices $A(-3, 0)$, $B(-1, 3)$, $C(5, 2)$, and $D(3, -1)$.

28. Find the area of the parallelogram with vertices $P(1, 0, 2)$, $Q(3, 3, 3)$, $R(7, 5, 8)$, and $S(5, 2, 7)$.

29–32 (a) Find a nonzero vector orthogonal to the plane through the points P, Q, and R, and (b) find the area of triangle PQR.

29. $P(1, 0, 1), \quad Q(-2, 1, 3), \quad R(4, 2, 5)$

30. $P(0, 0, -3), \quad Q(4, 2, 0), \quad R(3, 3, 1)$

31. $P(0, -2, 0), \quad Q(4, 1, -2), \quad R(5, 3, 1)$

32. $P(2, -3, 4)$, $Q(-1, -2, 2)$, $R(3, 1, -3)$

33–34 Find the volume of the parallelepiped determined by the vectors **a**, **b**, and **c**.

33. $\mathbf{a} = \langle 1, 2, 3 \rangle$, $\mathbf{b} = \langle -1, 1, 2 \rangle$, $\mathbf{c} = \langle 2, 1, 4 \rangle$

34. $\mathbf{a} = \mathbf{i} + \mathbf{j}$, $\mathbf{b} = \mathbf{j} + \mathbf{k}$, $\mathbf{c} = \mathbf{i} + \mathbf{j} + \mathbf{k}$

35–36 Find the volume of the parallelepiped with adjacent edges PQ, PR, and PS.

35. $P(-2, 1, 0)$, $Q(2, 3, 2)$, $R(1, 4, -1)$, $S(3, 6, 1)$

36. $P(3, 0, 1)$, $Q(-1, 2, 5)$, $R(5, 1, -1)$, $S(0, 4, 2)$

37. Use the scalar triple product to verify that the vectors $\mathbf{u} = \mathbf{i} + 5\mathbf{j} - 2\mathbf{k}$, $\mathbf{v} = 3\mathbf{i} - \mathbf{j}$, and $\mathbf{w} = 5\mathbf{i} + 9\mathbf{j} - 4\mathbf{k}$ are coplanar.

38. Use the scalar triple product to determine whether the points $A(1, 3, 2)$, $B(3, -1, 6)$, $C(5, 2, 0)$, and $D(3, 6, -4)$ lie in the same plane.

39. A bicycle pedal is pushed by a foot with a 60-N force as shown. The shaft of the pedal is 18 cm long. Find the magnitude of the torque about P.

40. (a) A horizontal force of 20 lb is applied to the handle of a gearshift lever as shown. Find the magnitude of the torque about the pivot point P.
   (b) Find the magnitude of the torque about P if the same force is applied at the elbow Q of the lever.

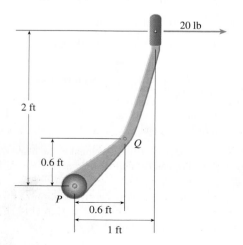

41. A wrench 30 cm long lies along the positive y-axis and grips a bolt at the origin. A force is applied in the direction $\langle 0, 3, -4 \rangle$ at the end of the wrench. Find the magnitude of the force needed to supply $100 \ \mathrm{N \cdot m}$ of torque to the bolt.

42. Let $\mathbf{v} = 5\mathbf{j}$ and let $\mathbf{u}$ be a vector with length 3 that starts at the origin and rotates in the xy-plane. Find the maximum and minimum values of the length of the vector $\mathbf{u} \times \mathbf{v}$. In what direction does $\mathbf{u} \times \mathbf{v}$ point?

43. If $\mathbf{a} \cdot \mathbf{b} = \sqrt{3}$ and $\mathbf{a} \times \mathbf{b} = \langle 1, 2, 2 \rangle$, find the angle between $\mathbf{a}$ and $\mathbf{b}$.

44. (a) Find all vectors $\mathbf{v}$ such that
$$\langle 1, 2, 1 \rangle \times \mathbf{v} = \langle 3, 1, -5 \rangle$$
   (b) Explain why there is no vector $\mathbf{v}$ such that
$$\langle 1, 2, 1 \rangle \times \mathbf{v} = \langle 3, 1, 5 \rangle$$

45. (a) Let P be a point not on the line L that passes through the points Q and R. Show that the distance d from the point P to the line L is
$$d = \frac{|\mathbf{a} \times \mathbf{b}|}{|\mathbf{a}|}$$
   where $\mathbf{a} = \overrightarrow{QR}$ and $\mathbf{b} = \overrightarrow{QP}$.
   (b) Use the formula in part (a) to find the distance from the point $P(1, 1, 1)$ to the line through $Q(0, 6, 8)$ and $R(-1, 4, 7)$.

46. (a) Let P be a point not on the plane that passes through the points Q, R, and S. Show that the distance d from P to the plane is
$$d = \frac{|\mathbf{a} \cdot (\mathbf{b} \times \mathbf{c})|}{|\mathbf{a} \times \mathbf{b}|}$$
   where $\mathbf{a} = \overrightarrow{QR}$, $\mathbf{b} = \overrightarrow{QS}$, and $\mathbf{c} = \overrightarrow{QP}$.
   (b) Use the formula in part (a) to find the distance from the point $P(2, 1, 4)$ to the plane through the points $Q(1, 0, 0)$, $R(0, 2, 0)$, and $S(0, 0, 3)$.

47. Show that $|\mathbf{a} \times \mathbf{b}|^2 = |\mathbf{a}|^2 |\mathbf{b}|^2 - (\mathbf{a} \cdot \mathbf{b})^2$.

48. If $\mathbf{a} + \mathbf{b} + \mathbf{c} = \mathbf{0}$, show that
$$\mathbf{a} \times \mathbf{b} = \mathbf{b} \times \mathbf{c} = \mathbf{c} \times \mathbf{a}$$

49. Prove that $(\mathbf{a} - \mathbf{b}) \times (\mathbf{a} + \mathbf{b}) = 2(\mathbf{a} \times \mathbf{b})$.

50. Prove Property 6 of cross products, that is,
$$\mathbf{a} \times (\mathbf{b} \times \mathbf{c}) = (\mathbf{a} \cdot \mathbf{c})\mathbf{b} - (\mathbf{a} \cdot \mathbf{b})\mathbf{c}$$

51. Use Exercise 50 to prove that
$$\mathbf{a} \times (\mathbf{b} \times \mathbf{c}) + \mathbf{b} \times (\mathbf{c} \times \mathbf{a}) + \mathbf{c} \times (\mathbf{a} \times \mathbf{b}) = \mathbf{0}$$

52. Prove that
$$(\mathbf{a} \times \mathbf{b}) \cdot (\mathbf{c} \times \mathbf{d}) = \begin{vmatrix} \mathbf{a} \cdot \mathbf{c} & \mathbf{b} \cdot \mathbf{c} \\ \mathbf{a} \cdot \mathbf{d} & \mathbf{b} \cdot \mathbf{d} \end{vmatrix}$$

53. Suppose that $\mathbf{a} \neq \mathbf{0}$.
    (a) If $\mathbf{a} \cdot \mathbf{b} = \mathbf{a} \cdot \mathbf{c}$, does it follow that $\mathbf{b} = \mathbf{c}$?
    (b) If $\mathbf{a} \times \mathbf{b} = \mathbf{a} \times \mathbf{c}$, does it follow that $\mathbf{b} = \mathbf{c}$?
    (c) If $\mathbf{a} \cdot \mathbf{b} = \mathbf{a} \cdot \mathbf{c}$ and $\mathbf{a} \times \mathbf{b} = \mathbf{a} \times \mathbf{c}$, does it follow that $\mathbf{b} = \mathbf{c}$?

54. If $\mathbf{v}_1$, $\mathbf{v}_2$, and $\mathbf{v}_3$ are noncoplanar vectors, let

$$\mathbf{k}_1 = \frac{\mathbf{v}_2 \times \mathbf{v}_3}{\mathbf{v}_1 \cdot (\mathbf{v}_2 \times \mathbf{v}_3)} \qquad \mathbf{k}_2 = \frac{\mathbf{v}_3 \times \mathbf{v}_1}{\mathbf{v}_1 \cdot (\mathbf{v}_2 \times \mathbf{v}_3)}$$

$$\mathbf{k}_3 = \frac{\mathbf{v}_1 \times \mathbf{v}_2}{\mathbf{v}_1 \cdot (\mathbf{v}_2 \times \mathbf{v}_3)}$$

(These vectors occur in the study of crystallography. Vectors of the form $n_1 \mathbf{v}_1 + n_2 \mathbf{v}_2 + n_3 \mathbf{v}_3$, where each n_i is an integer, form a *lattice* for a crystal. Vectors written similarly in terms of $\mathbf{k}_1$, $\mathbf{k}_2$, and $\mathbf{k}_3$ form the *reciprocal lattice*.)
    (a) Show that $\mathbf{k}_i$ is perpendicular to $\mathbf{v}_j$ if $i \neq j$.
    (b) Show that $\mathbf{k}_i \cdot \mathbf{v}_i = 1$ for $i = 1, 2, 3$.
    (c) Show that $\mathbf{k}_1 \cdot (\mathbf{k}_2 \times \mathbf{k}_3) = \dfrac{1}{\mathbf{v}_1 \cdot (\mathbf{v}_2 \times \mathbf{v}_3)}$.

DISCOVERY PROJECT   THE GEOMETRY OF A TETRAHEDRON

A tetrahedron is a solid with four vertices, P, Q, R, and S, and four triangular faces, as shown in the figure.

1. Let $\mathbf{v}_1$, $\mathbf{v}_2$, $\mathbf{v}_3$, and $\mathbf{v}_4$ be vectors with lengths equal to the areas of the faces opposite the vertices P, Q, R, and S, respectively, and directions perpendicular to the respective faces and pointing outward. Show that

$$\mathbf{v}_1 + \mathbf{v}_2 + \mathbf{v}_3 + \mathbf{v}_4 = \mathbf{0}$$

2. The volume V of a tetrahedron is one-third the distance from a vertex to the opposite face, times the area of that face.
    (a) Find a formula for the volume of a tetrahedron in terms of the coordinates of its vertices P, Q, R, and S.
    (b) Find the volume of the tetrahedron whose vertices are $P(1, 1, 1)$, $Q(1, 2, 3)$, $R(1, 1, 2)$, and $S(3, -1, 2)$.

3. Suppose the tetrahedron in the figure has a trirectangular vertex S. (This means that the three angles at S are all right angles.) Let A, B, and C be the areas of the three faces that meet at S, and let D be the area of the opposite face PQR. Using the result of Problem 1, or otherwise, show that

$$D^2 = A^2 + B^2 + C^2$$

(This is a three-dimensional version of the Pythagorean Theorem.)

12.5   Equations of Lines and Planes

■ Lines

A line in the xy-plane is determined when a point on the line and the direction of the line (its slope or angle of inclination) are given. The equation of the line can then be written using the point-slope form.

    Likewise, a line L in three-dimensional space is determined when we know a point $P_0(x_0, y_0, z_0)$ on L and the direction of L. In three dimensions the direction of a line is conveniently described by a vector, so we let $\mathbf{v}$ be a vector parallel to L. Let $P(x, y, z)$ be an arbitrary point on L and let $\mathbf{r}_0$ and $\mathbf{r}$ be the position vectors of P_0 and P (that is, they have representations $\overrightarrow{OP_0}$ and $\overrightarrow{OP}$). If $\mathbf{a}$ is the vector with representation $\overrightarrow{P_0P}$, as in Figure 1, then the Triangle Law for vector addition gives $\mathbf{r} = \mathbf{r}_0 + \mathbf{a}$. But, since $\mathbf{a}$ and $\mathbf{v}$ are parallel vectors, there is a scalar t such that $\mathbf{a} = t\mathbf{v}$. Thus

FIGURE 1

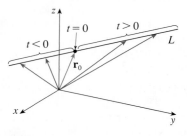

FIGURE 2

$$\boxed{1} \qquad \boxed{\mathbf{r} = \mathbf{r}_0 + t\mathbf{v}}$$

which is a **vector equation** of L. Each value of the **parameter** t gives the position vector $\mathbf{r}$ of a point on L. In other words, as t varies, the line is traced out by the tip of the vector $\mathbf{r}$. As Figure 2 indicates, positive values of t correspond to points on L that lie on one side of P_0, whereas negative values of t correspond to points that lie on the other side of P_0.

If the vector $\mathbf{v}$ that gives the direction of the line L is written in component form as $\mathbf{v} = \langle a, b, c \rangle$, then we have $t\mathbf{v} = \langle ta, tb, tc \rangle$. We can also write $\mathbf{r} = \langle x, y, z \rangle$ and $\mathbf{r}_0 = \langle x_0, y_0, z_0 \rangle$, so the vector equation (1) becomes

$$\langle x, y, z \rangle = \langle x_0 + ta, y_0 + tb, z_0 + tc \rangle$$

Two vectors are equal if and only if corresponding components are equal. Therefore we have the three scalar equations:

$$x = x_0 + at \qquad y = y_0 + bt \qquad z = z_0 + ct$$

where $t \in \mathbb{R}$. These equations are called **parametric equations** of the line L through the point $P_0(x_0, y_0, z_0)$ and parallel to the vector $\mathbf{v} = \langle a, b, c \rangle$. Each value of the parameter t gives a point (x, y, z) on L.

> $\boxed{2}$ Parametric equations for a line through the point (x_0, y_0, z_0) and parallel to the direction vector $\langle a, b, c \rangle$ are
>
> $$x = x_0 + at \qquad y = y_0 + bt \qquad z = z_0 + ct$$

Figure 3 shows the line L in Example 1 and its relation to the given point and to the vector that gives its direction.

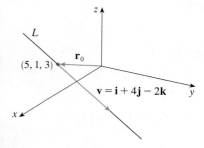

FIGURE 3

EXAMPLE 1

(a) Find a vector equation and parametric equations for the line that passes through the point $(5, 1, 3)$ and is parallel to the vector $\mathbf{i} + 4\mathbf{j} - 2\mathbf{k}$.

(b) Find two other points on the line.

SOLUTION

(a) Here $\mathbf{r}_0 = \langle 5, 1, 3 \rangle = 5\mathbf{i} + \mathbf{j} + 3\mathbf{k}$ and $\mathbf{v} = \mathbf{i} + 4\mathbf{j} - 2\mathbf{k}$, so the vector equation (1) becomes

$$\mathbf{r} = (5\mathbf{i} + \mathbf{j} + 3\mathbf{k}) + t(\mathbf{i} + 4\mathbf{j} - 2\mathbf{k})$$

or

$$\mathbf{r} = (5 + t)\mathbf{i} + (1 + 4t)\mathbf{j} + (3 - 2t)\mathbf{k}$$

Parametric equations are

$$x = 5 + t \qquad y = 1 + 4t \qquad z = 3 - 2t$$

(b) Choosing the parameter value $t = 1$ gives $x = 6$, $y = 5$, and $z = 1$, so $(6, 5, 1)$ is a point on the line. Similarly, $t = -1$ gives the point $(4, -3, 5)$. ∎

The vector equation and parametric equations of a line are not unique. If we change the point or the parameter or choose a different parallel vector, then the equations change. For instance, if, instead of $(5, 1, 3)$, we choose the point $(6, 5, 1)$ in Example 1, then the parametric equations of the line become

$$x = 6 + t \qquad y = 5 + 4t \qquad z = 1 - 2t$$

Or, if we stay with the point $(5, 1, 3)$ but choose the parallel vector $2\mathbf{i} + 8\mathbf{j} - 4\mathbf{k}$, we arrive at the equations

$$x = 5 + 2t \qquad y = 1 + 8t \qquad z = 3 - 4t$$

In general, if a vector $\mathbf{v} = \langle a, b, c \rangle$ is used to describe the direction of a line L, then the numbers a, b, and c are called **direction numbers** of L. Since any vector parallel to $\mathbf{v}$ could also be used, we see that any three numbers proportional to a, b, and c could also be used as a set of direction numbers for L.

Another way of describing a line L is to eliminate the parameter t from Equations 2. If none of a, b, or c is 0, we can solve each of these equations for t:

$$t = \frac{x - x_0}{a} \qquad t = \frac{y - y_0}{b} \qquad t = \frac{z - z_0}{c}$$

Equating the results, we obtain

3
$$\frac{x - x_0}{a} = \frac{y - y_0}{b} = \frac{z - z_0}{c}$$

These equations are called **symmetric equations** of L. Notice that the numbers a, b, and c that appear in the denominators of Equations 3 are direction numbers of L, that is, components of a vector parallel to L. If one of a, b, or c is 0, we can still eliminate t. For instance, if $a = 0$, we could write the equations of L as

$$x = x_0 \qquad \frac{y - y_0}{b} = \frac{z - z_0}{c}$$

This means that L lies in the vertical plane $x = x_0$.

Figure 4 shows the line L in Example 2 and the point P where it intersects the xy-plane.

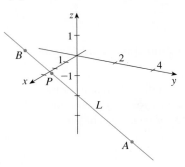

FIGURE 4

EXAMPLE 2
(a) Find parametric equations and symmetric equations of the line that passes through the points $A(2, 4, -3)$ and $B(3, -1, 1)$.
(b) At what point does this line intersect the xy-plane?

SOLUTION

(a) We are not explicitly given a vector parallel to the line, but observe that the vector $\mathbf{v}$ with representation $\overrightarrow{AB}$ is parallel to the line and

$$\mathbf{v} = \langle 3 - 2, -1 - 4, 1 - (-3) \rangle = \langle 1, -5, 4 \rangle$$

Thus direction numbers are $a = 1$, $b = -5$, and $c = 4$. Taking the point $(2, 4, -3)$ as P_0, we see that parametric equations (2) are

$$x = 2 + t \qquad y = 4 - 5t \qquad z = -3 + 4t$$

and symmetric equations (3) are

$$\frac{x - 2}{1} = \frac{y - 4}{-5} = \frac{z + 3}{4}$$

(b) The line intersects the xy-plane when $z = 0$, so we put $z = 0$ in the symmetric equations and obtain

$$\frac{x - 2}{1} = \frac{y - 4}{-5} = \frac{3}{4}$$

This gives $x = \frac{11}{4}$ and $y = \frac{1}{4}$, so the line intersects the xy-plane at the point $\left(\frac{11}{4}, \frac{1}{4}, 0\right)$. ∎

In general, the procedure of Example 2 shows that direction numbers of the line L through the points $P_0(x_0, y_0, z_0)$ and $P_1(x_1, y_1, z_1)$ are $x_1 - x_0$, $y_1 - y_0$, and $z_1 - z_0$ and so symmetric equations of L are

$$\frac{x - x_0}{x_1 - x_0} = \frac{y - y_0}{y_1 - y_0} = \frac{z - z_0}{z_1 - z_0}$$

Often, we need a description, not of an entire line, but of just a line segment. How, for instance, could we describe the line segment AB in Example 2? If we put $t = 0$ in the parametric equations in Example 2(a), we get the point $(2, 4, -3)$ and if we put $t = 1$ we get $(3, -1, 1)$. So the line segment AB is described by the parametric equations

$$x = 2 + t \qquad y = 4 - 5t \qquad z = -3 + 4t \qquad 0 \le t \le 1$$

or by the corresponding vector equation

$$\mathbf{r}(t) = \langle 2 + t, 4 - 5t, -3 + 4t \rangle \qquad 0 \le t \le 1$$

In general, we know from Equation 1 that the vector equation of a line through the (tip of the) vector $\mathbf{r}_0$ in the direction of a vector $\mathbf{v}$ is $\mathbf{r} = \mathbf{r}_0 + t\mathbf{v}$. If the line also passes through (the tip of) $\mathbf{r}_1$, then we can take $\mathbf{v} = \mathbf{r}_1 - \mathbf{r}_0$ and so its vector equation is

$$\mathbf{r} = \mathbf{r}_0 + t(\mathbf{r}_1 - \mathbf{r}_0) = (1 - t)\mathbf{r}_0 + t\mathbf{r}_1$$

The line segment from $\mathbf{r}_0$ to $\mathbf{r}_1$ is given by the parameter interval $0 \le t \le 1$.

> **4**   The line segment from $\mathbf{r}_0$ to $\mathbf{r}_1$ is given by the vector equation
>
> $$\mathbf{r}(t) = (1 - t)\mathbf{r}_0 + t\mathbf{r}_1 \qquad 0 \le t \le 1$$

The lines L_1 and L_2 in Example 3, shown in Figure 5, are skew lines.

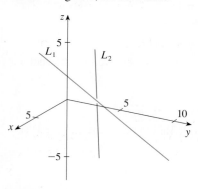

FIGURE 5

EXAMPLE 3 Show that the lines L_1 and L_2 with parametric equations

$$L_1: \quad x = 1 + t \qquad y = -2 + 3t \qquad z = 4 - t$$

$$L_2: \quad x = 2s \qquad\quad y = 3 + s \qquad\quad z = -3 + 4s$$

are **skew lines**; that is, they do not intersect and are not parallel (and therefore do not lie in the same plane).

SOLUTION The lines are not parallel because the corresponding direction vectors $\langle 1, 3, -1 \rangle$ and $\langle 2, 1, 4 \rangle$ are not parallel. (Their components are not proportional.) If L_1 and L_2 had a point of intersection, there would be values of t and s such that

$$1 + t = 2s$$

$$-2 + 3t = 3 + s$$

$$4 - t = -3 + 4s$$

But if we solve the first two equations, we get $t = \frac{11}{5}$ and $s = \frac{8}{5}$, and these values don't satisfy the third equation. Therefore there are no values of t and s that satisfy the three equations, so L_1 and L_2 do not intersect. Thus L_1 and L_2 are skew lines. ∎

Planes

Although a line in space is determined by a point and a direction, a plane in space is more difficult to describe. A single vector parallel to a plane is not enough to convey the

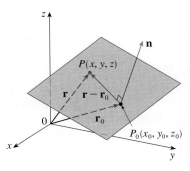

FIGURE 6

"direction" of the plane, but a vector perpendicular to the plane does completely specify its direction. Thus a plane in space is determined by a point $P_0(x_0, y_0, z_0)$ in the plane and a vector $\mathbf{n}$ that is orthogonal to the plane. This orthogonal vector $\mathbf{n}$ is called a **normal vector**. Let $P(x, y, z)$ be an arbitrary point in the plane, and let $\mathbf{r}_0$ and $\mathbf{r}$ be the position vectors of P_0 and P. Then the vector $\mathbf{r} - \mathbf{r}_0$ is represented by $\overrightarrow{P_0P}$. (See Figure 6.) The normal vector $\mathbf{n}$ is orthogonal to every vector in the given plane. In particular, $\mathbf{n}$ is orthogonal to $\mathbf{r} - \mathbf{r}_0$ and so we have

5
$$\mathbf{n} \cdot (\mathbf{r} - \mathbf{r}_0) = 0$$

which can be rewritten as

6
$$\mathbf{n} \cdot \mathbf{r} = \mathbf{n} \cdot \mathbf{r}_0$$

Either Equation 5 or Equation 6 is called a **vector equation of the plane**.

To obtain a scalar equation for the plane, we write $\mathbf{n} = \langle a, b, c \rangle$, $\mathbf{r} = \langle x, y, z \rangle$, and $\mathbf{r}_0 = \langle x_0, y_0, z_0 \rangle$. Then the vector equation (5) becomes

$$\langle a, b, c \rangle \cdot \langle x - x_0, y - y_0, z - z_0 \rangle = 0$$

or

$$a(x - x_0) + b(y - y_0) + c(z - z_0) = 0$$

7 A **scalar equation of the plane** through point $P_0(x_0, y_0, z_0)$ with normal vector $\mathbf{n} = \langle a, b, c \rangle$ is
$$a(x - x_0) + b(y - y_0) + c(z - z_0) = 0$$

EXAMPLE 4 Find an equation of the plane through the point $(2, 4, -1)$ with normal vector $\mathbf{n} = \langle 2, 3, 4 \rangle$. Find the intercepts and sketch the plane.

SOLUTION Putting $a = 2$, $b = 3$, $c = 4$, $x_0 = 2$, $y_0 = 4$, and $z_0 = -1$ in Equation 7, we see that an equation of the plane is

$$2(x - 2) + 3(y - 4) + 4(z + 1) = 0$$

or

$$2x + 3y + 4z = 12$$

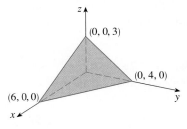

FIGURE 7

To find the x-intercept we set $y = z = 0$ in this equation and obtain $x = 6$. Similarly, the y-intercept is 4 and the z-intercept is 3. This enables us to sketch the portion of the plane that lies in the first octant (see Figure 7). ■

By collecting terms in Equation 7 as we did in Example 4, we can rewrite the equation of a plane as

8
$$ax + by + cz + d = 0$$

where $d = -(ax_0 + by_0 + cz_0)$. Equation 8 is called a **linear equation** in x, y, and z. Conversely, it can be shown that if a, b, and c are not all 0, then the linear equation (8) represents a plane with normal vector $\langle a, b, c \rangle$. (See Exercise 83.)

Figure 8 shows the portion of the plane in Example 5 that is enclosed by triangle PQR.

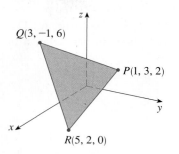

FIGURE 8

EXAMPLE 5 Find an equation of the plane that passes through the points $P(1, 3, 2)$, $Q(3, -1, 6)$, and $R(5, 2, 0)$.

SOLUTION The vectors $\mathbf{a}$ and $\mathbf{b}$ corresponding to $\overrightarrow{PQ}$ and $\overrightarrow{PR}$ are

$$\mathbf{a} = \langle 2, -4, 4 \rangle \qquad \mathbf{b} = \langle 4, -1, -2 \rangle$$

Since both $\mathbf{a}$ and $\mathbf{b}$ lie in the plane, their cross product $\mathbf{a} \times \mathbf{b}$ is orthogonal to the plane and can be taken as the normal vector. Thus

$$\mathbf{n} = \mathbf{a} \times \mathbf{b} = \begin{vmatrix} \mathbf{i} & \mathbf{j} & \mathbf{k} \\ 2 & -4 & 4 \\ 4 & -1 & -2 \end{vmatrix} = 12\mathbf{i} + 20\mathbf{j} + 14\mathbf{k}$$

With the point $P(1, 3, 2)$ and the normal vector $\mathbf{n}$, an equation of the plane is

$$12(x - 1) + 20(y - 3) + 14(z - 2) = 0$$

or

$$6x + 10y + 7z = 50 \qquad\blacksquare$$

EXAMPLE 6 Find the point at which the line with parametric equations $x = 2 + 3t$, $y = -4t$, $z = 5 + t$ intersects the plane $4x + 5y - 2z = 18$.

SOLUTION We substitute the expressions for x, y, and z from the parametric equations into the equation of the plane:

$$4(2 + 3t) + 5(-4t) - 2(5 + t) = 18$$

This simplifies to $-10t = 20$, so $t = -2$. Therefore the point of intersection occurs when the parameter value is $t = -2$. Then $x = 2 + 3(-2) = -4$, $y = -4(-2) = 8$, $z = 5 - 2 = 3$ and so the point of intersection is $(-4, 8, 3)$. $\qquad\blacksquare$

Two planes are **parallel** if their normal vectors are parallel. For instance, the planes $x + 2y - 3z = 4$ and $2x + 4y - 6z = 3$ are parallel because their normal vectors are $\mathbf{n}_1 = \langle 1, 2, -3 \rangle$ and $\mathbf{n}_2 = \langle 2, 4, -6 \rangle$ and $\mathbf{n}_2 = 2\mathbf{n}_1$. If two planes are not parallel, then they intersect in a straight line and the angle between the two planes is defined as the acute angle between their normal vectors (see angle θ in Figure 9).

FIGURE 9

EXAMPLE 7
(a) Find the angle between the planes $x + y + z = 1$ and $x - 2y + 3z = 1$.
(b) Find symmetric equations for the line of intersection L of these two planes.

SOLUTION

(a) The normal vectors of these planes are

$$\mathbf{n}_1 = \langle 1, 1, 1 \rangle \qquad \mathbf{n}_2 = \langle 1, -2, 3 \rangle$$

and so, if θ is the angle between the planes, Corollary 12.3.6 gives

$$\cos \theta = \frac{\mathbf{n}_1 \cdot \mathbf{n}_2}{|\mathbf{n}_1||\mathbf{n}_2|} = \frac{1(1) + 1(-2) + 1(3)}{\sqrt{1 + 1 + 1}\,\sqrt{1 + 4 + 9}} = \frac{2}{\sqrt{42}}$$

$$\theta = \cos^{-1}\left(\frac{2}{\sqrt{42}}\right) \approx 72°$$

Figure 10 shows the planes in Example 7 and their line of intersection L.

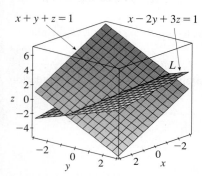

$x + y + z = 1$ $x - 2y + 3z = 1$

FIGURE 10

(b) We first need to find a point on L. For instance, we can find the point where the line intersects the xy-plane by setting $z = 0$ in the equations of both planes. This gives the equations $x + y = 1$ and $x - 2y = 1$, whose solution is $x = 1$, $y = 0$. So the point $(1, 0, 0)$ lies on L.

Now we observe that, since L lies in both planes, it is perpendicular to both of the normal vectors. Thus a vector $\mathbf{v}$ parallel to L is given by the cross product

$$\mathbf{v} = \mathbf{n}_1 \times \mathbf{n}_2 = \begin{vmatrix} \mathbf{i} & \mathbf{j} & \mathbf{k} \\ 1 & 1 & 1 \\ 1 & -2 & 3 \end{vmatrix} = 5\mathbf{i} - 2\mathbf{j} - 3\mathbf{k}$$

Another way to find the line of intersection is to solve the equations of the planes for two of the variables in terms of the third, which can be taken as the parameter.

and so the symmetric equations of L can be written as

$$\frac{x - 1}{5} = \frac{y}{-2} = \frac{z}{-3}$$ ∎

NOTE Since a linear equation in x, y, and z represents a plane and two nonparallel planes intersect in a line, it follows that two linear equations can represent a line. The points (x, y, z) that satisfy both $a_1 x + b_1 y + c_1 z + d_1 = 0$ and $a_2 x + b_2 y + c_2 z + d_2 = 0$ lie on both of these planes, and so the pair of linear equations represents the line of intersection of the planes (if they are not parallel). For instance, in Example 7 the line L was given as the line of intersection of the planes $x + y + z = 1$ and $x - 2y + 3z = 1$. The symmetric equations that we found for L could be written as

$$\frac{x - 1}{5} = \frac{y}{-2} \qquad \text{and} \qquad \frac{y}{-2} = \frac{z}{-3}$$

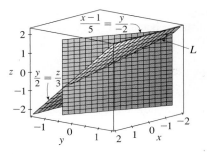

FIGURE 11

Figure 11 shows how the line L in Example 7 can also be regarded as the line of intersection of planes derived from its symmetric equations.

which is again a pair of linear equations. They exhibit L as the line of intersection of the planes $(x - 1)/5 = y/(-2)$ and $y/(-2) = z/(-3)$. (See Figure 11.)

In general, when we write the equations of a line in the symmetric form

$$\frac{x - x_0}{a} = \frac{y - y_0}{b} = \frac{z - z_0}{c}$$

we can regard the line as the line of intersection of the two planes

$$\frac{x - x_0}{a} = \frac{y - y_0}{b} \qquad \text{and} \qquad \frac{y - y_0}{b} = \frac{z - z_0}{c}$$

◼ Distances

EXAMPLE 8 Find a formula for the distance D from a point $P_1(x_1, y_1, z_1)$ to the plane $ax + by + cz + d = 0$.

SOLUTION Let $P_0(x_0, y_0, z_0)$ be any point in the given plane and let $\mathbf{b}$ be the vector corresponding to $\overrightarrow{P_0 P_1}$. Then

$$\mathbf{b} = \langle x_1 - x_0, y_1 - y_0, z_1 - z_0 \rangle$$

From Figure 12 you can see that the distance D from P_1 to the plane is equal to the absolute value of the scalar projection of $\mathbf{b}$ onto the normal vector $\mathbf{n} = \langle a, b, c \rangle$. (See Section 12.3.) Thus

$$D = |\operatorname{comp}_{\mathbf{n}} \mathbf{b}| = \frac{|\mathbf{n} \cdot \mathbf{b}|}{|\mathbf{n}|}$$

$$= \frac{|a(x_1 - x_0) + b(y_1 - y_0) + c(z_1 - z_0)|}{\sqrt{a^2 + b^2 + c^2}}$$

$$= \frac{|(ax_1 + by_1 + cz_1) - (ax_0 + by_0 + cz_0)|}{\sqrt{a^2 + b^2 + c^2}}$$

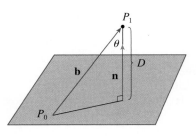

FIGURE 12

Since P_0 lies in the plane, its coordinates satisfy the equation of the plane and so we have $ax_0 + by_0 + cz_0 + d = 0$. Thus the formula for D can be written as

$$\boxed{9} \qquad D = \frac{|ax_1 + by_1 + cz_1 + d|}{\sqrt{a^2 + b^2 + c^2}}$$ ∎

EXAMPLE 9 Find the distance between the parallel planes $10x + 2y - 2z = 5$ and $5x + y - z = 1$.

SOLUTION First we note that the planes are parallel because their normal vectors $\langle 10, 2, -2 \rangle$ and $\langle 5, 1, -1 \rangle$ are parallel. To find the distance D between the planes, we choose any point on one plane and calculate its distance to the other plane. In particular, if we put $y = z = 0$ in the equation of the first plane, we get $10x = 5$ and so $\left(\frac{1}{2}, 0, 0\right)$ is a point in this plane. By Formula 9, the distance between $\left(\frac{1}{2}, 0, 0\right)$ and the plane $5x + y - z - 1 = 0$ is

$$D = \frac{\left|5\left(\frac{1}{2}\right) + 1(0) - 1(0) - 1\right|}{\sqrt{5^2 + 1^2 + (-1)^2}} = \frac{\frac{3}{2}}{3\sqrt{3}} = \frac{\sqrt{3}}{6}$$

So the distance between the planes is $\sqrt{3}/6$. ∎

EXAMPLE 10 In Example 3 we showed that the lines

$$L_1: \quad x = 1 + t \qquad y = -2 + 3t \qquad z = 4 - t$$
$$L_2: \quad x = 2s \qquad\quad y = 3 + s \qquad\quad z = -3 + 4s$$

are skew. Find the distance between them.

SOLUTION Since the two lines L_1 and L_2 are skew, they can be viewed as lying on two parallel planes P_1 and P_2. The distance between L_1 and L_2 is the same as the distance between P_1 and P_2, which can be computed as in Example 9. The common normal vector to both planes must be orthogonal to both $\mathbf{v}_1 = \langle 1, 3, -1 \rangle$ (the direction of L_1) and $\mathbf{v}_2 = \langle 2, 1, 4 \rangle$ (the direction of L_2). So a normal vector is

$$\mathbf{n} = \mathbf{v}_1 \times \mathbf{v}_2 = \begin{vmatrix} \mathbf{i} & \mathbf{j} & \mathbf{k} \\ 1 & 3 & -1 \\ 2 & 1 & 4 \end{vmatrix} = 13\mathbf{i} - 6\mathbf{j} - 5\mathbf{k}$$

If we put $s = 0$ in the equations of L_2, we get the point $(0, 3, -3)$ on L_2 and so an equation for P_2 is

$$13(x - 0) - 6(y - 3) - 5(z + 3) = 0 \qquad \text{or} \qquad 13x - 6y - 5z + 3 = 0$$

If we now set $t = 0$ in the equations for L_1, we get the point $(1, -2, 4)$ on P_1. So the distance between L_1 and L_2 is the same as the distance from $(1, -2, 4)$ to $13x - 6y - 5z + 3 = 0$. By Formula 9, this distance is

$$D = \frac{|13(1) - 6(-2) - 5(4) + 3|}{\sqrt{13^2 + (-6)^2 + (-5)^2}} = \frac{8}{\sqrt{230}} \approx 0.53$$ ∎

FIGURE 13
Skew lines, like those in Example 10, always lie on (nonidentical) parallel planes.

12.5 EXERCISES

1. Determine whether each statement is true or false in $\mathbb{R}^3$.
(a) Two lines parallel to a third line are parallel.
(b) Two lines perpendicular to a third line are parallel.
(c) Two planes parallel to a third plane are parallel.
(d) Two planes perpendicular to a third plane are parallel.
(e) Two lines parallel to a plane are parallel.
(f) Two lines perpendicular to a plane are parallel.
(g) Two planes parallel to a line are parallel.
(h) Two planes perpendicular to a line are parallel.
(i) Two planes either intersect or are parallel.
(j) Two lines either intersect or are parallel.
(k) A plane and a line either intersect or are parallel.

2–5 Find a vector equation and parametric equations for the line.

2. The line through the point $(6, -5, 2)$ and parallel to the vector $\langle 1, 3, -\frac{2}{3} \rangle$

3. The line through the point $(2, 2.4, 3.5)$ and parallel to the vector $3\mathbf{i} + 2\mathbf{j} - \mathbf{k}$

4. The line through the point $(0, 14, -10)$ and parallel to the line $x = -1 + 2t, y = 6 - 3t, z = 3 + 9t$

5. The line through the point $(1, 0, 6)$ and perpendicular to the plane $x + 3y + z = 5$

6–12 Find parametric equations and symmetric equations for the line.

6. The line through the origin and the point $(4, 3, -1)$

7. The line through the points $(0, \frac{1}{2}, 1)$ and $(2, 1, -3)$

8. The line through the points $(1, 2.4, 4.6)$ and $(2.6, 1.2, 0.3)$

9. The line through the points $(-8, 1, 4)$ and $(3, -2, 4)$

10. The line through $(2, 1, 0)$ and perpendicular to both $\mathbf{i} + \mathbf{j}$ and $\mathbf{j} + \mathbf{k}$

11. The line through $(-6, 2, 3)$ and parallel to the line $\frac{1}{2}x = \frac{1}{3}y = z + 1$

12. The line of intersection of the planes $x + 2y + 3z = 1$ and $x - y + z = 1$

13. Is the line through $(-4, -6, 1)$ and $(-2, 0, -3)$ parallel to the line through $(10, 18, 4)$ and $(5, 3, 14)$?

14. Is the line through $(-2, 4, 0)$ and $(1, 1, 1)$ perpendicular to the line through $(2, 3, 4)$ and $(3, -1, -8)$?

15. (a) Find symmetric equations for the line that passes through the point $(1, -5, 6)$ and is parallel to the vector $\langle -1, 2, -3 \rangle$.
(b) Find the points in which the required line in part (a) intersects the coordinate planes.

16. (a) Find parametric equations for the line through $(2, 4, 6)$ that is perpendicular to the plane $x - y + 3z = 7$.

(b) In what points does this line intersect the coordinate planes?

17. Find a vector equation for the line segment from $(6, -1, 9)$ to $(7, 6, 0)$.

18. Find parametric equations for the line segment from $(-2, 18, 31)$ to $(11, -4, 48)$.

19–22 Determine whether the lines L_1 and L_2 are parallel, skew, or intersecting. If they intersect, find the point of intersection.

19. L_1: $x = 3 + 2t, \quad y = 4 - t, \quad z = 1 + 3t$
L_2: $x = 1 + 4s, \quad y = 3 - 2s, \quad z = 4 + 5s$

20. L_1: $x = 5 - 12t, \quad y = 3 + 9t, \quad z = 1 - 3t$
L_2: $x = 3 + 8s, \quad y = -6s, \quad z = 7 + 2s$

21. L_1: $\dfrac{x-2}{1} = \dfrac{y-3}{-2} = \dfrac{z-1}{-3}$
L_2: $\dfrac{x-3}{1} = \dfrac{y+4}{3} = \dfrac{z-2}{-7}$

22. L_1: $\dfrac{x}{1} = \dfrac{y-1}{-1} = \dfrac{z-2}{3}$
L_2: $\dfrac{x-2}{2} = \dfrac{y-3}{-2} = \dfrac{z}{7}$

23–40 Find an equation of the plane.

23. The plane through the origin and perpendicular to the vector $\langle 1, -2, 5 \rangle$

24. The plane through the point $(5, 3, 5)$ and with normal vector $2\mathbf{i} + \mathbf{j} - \mathbf{k}$

25. The plane through the point $(-1, \frac{1}{2}, 3)$ and with normal vector $\mathbf{i} + 4\mathbf{j} + \mathbf{k}$

26. The plane through the point $(2, 0, 1)$ and perpendicular to the line $x = 3t, y = 2 - t, z = 3 + 4t$

27. The plane through the point $(1, -1, -1)$ and parallel to the plane $5x - y - z = 6$

28. The plane through the point $(3, -2, 8)$ and parallel to the plane $z = x + y$

29. The plane through the point $(1, \frac{1}{2}, \frac{1}{3})$ and parallel to the plane $x + y + z = 0$

30. The plane that contains the line $x = 1 + t, y = 2 - t, z = 4 - 3t$ and is parallel to the plane $5x + 2y + z = 1$

31. The plane through the points $(0, 1, 1)$, $(1, 0, 1)$, and $(1, 1, 0)$

32. The plane through the origin and the points $(3, -2, 1)$ and $(1, 1, 1)$

33. The plane through the points $(2, 1, 2)$, $(3, -8, 6)$, and $(-2, -3, 1)$

34. The plane through the points $(3, 0, -1)$, $(-2, -2, 3)$, and $(7, 1, -4)$

35. The plane that passes through the point $(3, 5, -1)$ and contains the line $x = 4 - t$, $y = 2t - 1$, $z = -3t$

36. The plane that passes through the point $(6, -1, 3)$ and contains the line with symmetric equations $x/3 = y + 4 = z/2$

37. The plane that passes through the point $(3, 1, 4)$ and contains the line of intersection of the planes $x + 2y + 3z = 1$ and $2x - y + z = -3$

38. The plane that passes through the points $(0, -2, 5)$ and $(-1, 3, 1)$ and is perpendicular to the plane $2z = 5x + 4y$

39. The plane that passes through the point $(1, 5, 1)$ and is perpendicular to the planes $2x + y - 2z = 2$ and $x + 3z = 4$

40. The plane that passes through the line of intersection of the planes $x - z = 1$ and $y + 2z = 3$ and is perpendicular to the plane $x + y - 2z = 1$

41–44 Use intercepts to help sketch the plane.

41. $2x + 5y + z = 10$ **42.** $3x + y + 2z = 6$

43. $6x - 3y + 4z = 6$ **44.** $6x + 5y - 3z = 15$

45–47 Find the point at which the line intersects the given plane.

45. $x = 2 - 2t$, $y = 3t$, $z = 1 + t$; $x + 2y - z = 7$

46. $x = t - 1$, $y = 1 + 2t$, $z = 3 - t$; $3x - y + 2z = 5$

47. $5x = y/2 = z + 2$; $10x - 7y + 3z + 24 = 0$

48. Where does the line through $(-3, 1, 0)$ and $(-1, 5, 6)$ intersect the plane $2x + y - z = -2$?

49. Find direction numbers for the line of intersection of the planes $x + y + z = 1$ and $x + z = 0$.

50. Find the cosine of the angle between the planes $x + y + z = 0$ and $x + 2y + 3z = 1$.

51–56 Determine whether the planes are parallel, perpendicular, or neither. If neither, find the angle between them. (Round to one decimal place.)

51. $x + 4y - 3z = 1$, $-3x + 6y + 7z = 0$

52. $9x - 3y + 6z = 2$, $2y = 6x + 4z$

53. $x + 2y - z = 2$, $2x - 2y + z = 1$

54. $x - y + 3z = 1$, $3x + y - z = 2$

55. $2x - 3y = z$, $4x = 3 + 6y + 2z$

56. $5x + 2y + 3z = 2$, $y = 4x - 6z$

57–58 (a) Find parametric equations for the line of intersection of the planes and (b) find the angle between the planes.

57. $x + y + z = 1$, $x + 2y + 2z = 1$

58. $3x - 2y + z = 1$, $2x + y - 3z = 3$

59–60 Find symmetric equations for the line of intersection of the planes.

59. $5x - 2y - 2z = 1$, $4x + y + z = 6$

60. $z = 2x - y - 5$, $z = 4x + 3y - 5$

61. Find an equation for the plane consisting of all points that are equidistant from the points $(1, 0, -2)$ and $(3, 4, 0)$.

62. Find an equation for the plane consisting of all points that are equidistant from the points $(2, 5, 5)$ and $(-6, 3, 1)$.

63. Find an equation of the plane with x-intercept a, y-intercept b, and z-intercept c.

64. (a) Find the point at which the given lines intersect:

$$\mathbf{r} = \langle 1, 1, 0 \rangle + t \langle 1, -1, 2 \rangle$$
$$\mathbf{r} = \langle 2, 0, 2 \rangle + s \langle -1, 1, 0 \rangle$$

(b) Find an equation of the plane that contains these lines.

65. Find parametric equations for the line through the point $(0, 1, 2)$ that is parallel to the plane $x + y + z = 2$ and perpendicular to the line $x = 1 + t$, $y = 1 - t$, $z = 2t$.

66. Find parametric equations for the line through the point $(0, 1, 2)$ that is perpendicular to the line $x = 1 + t$, $y = 1 - t$, $z = 2t$ and intersects this line.

67. Which of the following four planes are parallel? Are any of them identical?

P_1: $3x + 6y - 3z = 6$ P_2: $4x - 12y + 8z = 5$

P_3: $9y = 1 + 3x + 6z$ P_4: $z = x + 2y - 2$

68. Which of the following four lines are parallel? Are any of them identical?

L_1: $x = 1 + 6t$, $y = 1 - 3t$, $z = 12t + 5$

L_2: $x = 1 + 2t$, $y = t$, $z = 1 + 4t$

L_3: $2x - 2 = 4 - 4y = z + 1$

L_4: $\mathbf{r} = \langle 3, 1, 5 \rangle + t \langle 4, 2, 8 \rangle$

69–70 Use the formula in Exercise 12.4.45 to find the distance from the point to the given line.

69. $(4, 1, -2)$; $x = 1 + t$, $y = 3 - 2t$, $z = 4 - 3t$

70. $(0, 1, 3)$; $x = 2t$, $y = 6 - 2t$, $z = 3 + t$

71–72 Find the distance from the point to the given plane.

71. $(1, -2, 4)$, $3x + 2y + 6z = 5$

72. $(-6, 3, 5)$, $x - 2y - 4z = 8$

73–74 Find the distance between the given parallel planes.

73. $2x - 3y + z = 4$, $4x - 6y + 2z = 3$

74. $6z = 4y - 2x$, $9z = 1 - 3x + 6y$

---

75. Show that the distance between the parallel planes
$ax + by + cz + d_1 = 0$ and $ax + by + cz + d_2 = 0$ is

$$D = \frac{|d_1 - d_2|}{\sqrt{a^2 + b^2 + c^2}}$$

76. Find equations of the planes that are parallel to the plane
$x + 2y - 2z = 1$ and two units away from it.

77. Show that the lines with symmetric equations $x = y = z$ and
$x + 1 = y/2 = z/3$ are skew, and find the distance between
these lines.

78. Find the distance between the skew lines with parametric
equations $x = 1 + t$, $y = 1 + 6t$, $z = 2t$, and $x = 1 + 2s$,
$y = 5 + 15s$, $z = -2 + 6s$.

79. Let L_1 be the line through the origin and the point $(2, 0, -1)$.
Let L_2 be the line through the points $(1, -1, 1)$ and $(4, 1, 3)$.
Find the distance between L_1 and L_2.

80. Let L_1 be the line through the points $(1, 2, 6)$ and $(2, 4, 8)$.
Let L_2 be the line of intersection of the planes P_1 and P_2,
where P_1 is the plane $x - y + 2z + 1 = 0$ and P_2 is the plane

through the points $(3, 2, -1)$, $(0, 0, 1)$, and $(1, 2, 1)$. Calculate
the distance between L_1 and L_2.

81. Two tanks are participating in a battle simulation. Tank A
is at point $(325, 810, 561)$ and tank B is positioned at point
$(765, 675, 599)$.
  (a) Find parametric equations for the line of sight between the
      tanks.
  (b) If we divide the line of sight into 5 equal segments, the
      elevations of the terrain at the four intermediate points
      from tank A to tank B are 549, 566, 586, and 589. Can the
      tanks see each other?

82. Give a geometric description of each family of planes.
  (a) $x + y + z = c$ (b) $x + y + cz = 1$
  (c) $y \cos \theta + z \sin \theta = 1$

83. If a, b, and c are not all 0, show that the equation
$ax + by + cz + d = 0$ represents a plane and $\langle a, b, c \rangle$ is
a normal vector to the plane.
    Hint: Suppose $a \neq 0$ and rewrite the equation in the form

$$a\left(x + \frac{d}{a}\right) + b(y - 0) + c(z - 0) = 0$$

LABORATORY PROJECT   PUTTING 3D IN PERSPECTIVE

Computer graphics programmers face the same challenge as the great painters of the past: how
to represent a three-dimensional scene as a flat image on a two-dimensional plane (a screen or a
canvas). To create the illusion of perspective, in which closer objects appear larger than those farther away, three-dimensional objects in the computer's memory are projected onto a rectangular
screen window from a viewpoint where the eye, or camera, is located. The viewing volume—the
portion of space that will be visible—is the region contained by the four planes that pass through
the viewpoint and an edge of the screen window. If objects in the scene extend beyond these four
planes, they must be truncated before pixel data are sent to the screen. These planes are therefore
called *clipping planes*.

1. Suppose the screen is represented by a rectangle in the yz-plane with vertices $(0, \pm 400, 0)$
   and $(0, \pm 400, 600)$, and the camera is placed at $(1000, 0, 0)$. A line L in the scene passes
   through the points $(230, -285, 102)$ and $(860, 105, 264)$. At what points should L be clipped
   by the clipping planes?

2. If the clipped line segment is projected onto the screen window, identify the resulting line
   segment.

3. Use parametric equations to plot the edges of the screen window, the clipped line segment, and its projection onto the screen window. Then add sight lines connecting the viewpoint to each end of the clipped segments to verify that the projection is correct.

4. A rectangle with vertices $(621, -147, 206)$, $(563, 31, 242)$, $(657, -111, 86)$, and $(599, 67, 122)$ is added to the scene. The line L intersects this rectangle. To make the rectangle appear opaque, a programmer can use *hidden line rendering,* which removes portions of objects that are behind other objects. Identify the portion of L that should be removed.

12.6   Cylinders and Quadric Surfaces

We have already looked at two special types of surfaces: planes (in Section 12.5) and spheres (in Section 12.1). Here we investigate two other types of surfaces: cylinders and quadric surfaces.

In order to sketch the graph of a surface, it is useful to determine the curves of intersection of the surface with planes parallel to the coordinate planes. These curves are called **traces** (or cross-sections) of the surface.

◼ Cylinders

A **cylinder** is a surface that consists of all lines (called **rulings**) that are parallel to a given line and pass through a given plane curve.

EXAMPLE 1   Sketch the graph of the surface $z = x^2$.

SOLUTION   Notice that the equation of the graph, $z = x^2$, doesn't involve y. This means that any vertical plane with equation $y = k$ (parallel to the xz-plane) intersects the graph in a curve with equation $z = x^2$. So these vertical traces are parabolas. Figure 1 shows how the graph is formed by taking the parabola $z = x^2$ in the xz-plane and moving it in the direction of the y-axis. The graph is a surface, called a **parabolic cylinder**, made up of infinitely many shifted copies of the same parabola. Here the rulings of the cylinder are parallel to the y-axis. ◼

FIGURE 1
The surface $z = x^2$ is a
parabolic cylinder.

We noticed that the variable y is missing from the equation of the cylinder in Example 1. This is typical of a surface whose rulings are parallel to one of the coordinate axes. If one of the variables x, y, or z is missing from the equation of a surface, then the surface is a cylinder.

EXAMPLE 2   Identify and sketch the surfaces.
(a) $x^2 + y^2 = 1$ (b) $y^2 + z^2 = 1$

FIGURE 2
$x^2 + y^2 = 1$

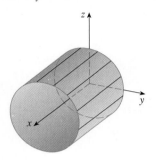

FIGURE 3
$y^2 + z^2 = 1$

SOLUTION

(a) Since z is missing and the equations $x^2 + y^2 = 1$, $z = k$ represent a circle with radius 1 in the plane $z = k$, the surface $x^2 + y^2 = 1$ is a circular cylinder whose axis is the z-axis. (See Figure 2.) Here the rulings are vertical lines.

(b) In this case x is missing and the surface is a circular cylinder whose axis is the x-axis. (See Figure 3.) It is obtained by taking the circle $y^2 + z^2 = 1$, $x = 0$ in the yz-plane and moving it parallel to the x-axis. ∎

NOTE When you are dealing with surfaces, it is important to recognize that an equation like $x^2 + y^2 = 1$ represents a cylinder and not a circle. The trace of the cylinder $x^2 + y^2 = 1$ in the xy-plane is the circle with equations $x^2 + y^2 = 1$, $z = 0$.

Quadric Surfaces

A **quadric surface** is the graph of a second-degree equation in three variables x, y, and z. The most general such equation is

$$Ax^2 + By^2 + Cz^2 + Dxy + Eyz + Fxz + Gx + Hy + Iz + J = 0$$

where $A, B, C, \ldots, J$ are constants, but by translation and rotation it can be brought into one of the two standard forms

$$Ax^2 + By^2 + Cz^2 + J = 0 \qquad \text{or} \qquad Ax^2 + By^2 + Iz = 0$$

Quadric surfaces are the counterparts in three dimensions of the conic sections in the plane. (See Section 10.5 for a review of conic sections.)

EXAMPLE 3 Use traces to sketch the quadric surface with equation

$$x^2 + \frac{y^2}{9} + \frac{z^2}{4} = 1$$

SOLUTION By substituting $z = 0$, we find that the trace in the xy-plane is $x^2 + y^2/9 = 1$, which we recognize as an equation of an ellipse. In general, the horizontal trace in the plane $z = k$ is

$$x^2 + \frac{y^2}{9} = 1 - \frac{k^2}{4} \qquad z = k$$

which is an ellipse, provided that $k^2 < 4$, that is, $-2 < k < 2$.

Similarly, vertical traces parallel to the yz- and xz-planes are also ellipses:

$$\frac{y^2}{9} + \frac{z^2}{4} = 1 - k^2 \qquad x = k \qquad \text{(if } -1 < k < 1\text{)}$$

$$x^2 + \frac{z^2}{4} = 1 - \frac{k^2}{9} \qquad y = k \qquad \text{(if } -3 < k < 3\text{)}$$

Figure 4 shows how drawing some traces indicates the shape of the surface. It's called an **ellipsoid** because all of its traces are ellipses. Notice that it is symmetric with respect to each coordinate plane; this is a reflection of the fact that its equation involves only even powers of x, y, and z. ∎

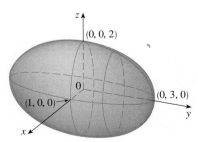

FIGURE 4

The ellipsoid $x^2 + \dfrac{y^2}{9} + \dfrac{z^2}{4} = 1$

EXAMPLE 4 Use traces to sketch the surface $z = 4x^2 + y^2$.

SOLUTION If we put $x = 0$, we get $z = y^2$, so the yz-plane intersects the surface in a parabola. If we put $x = k$ (a constant), we get $z = y^2 + 4k^2$. This means that if we

FIGURE 5

The surface $z = 4x^2 + y^2$ is an elliptic paraboloid. Horizontal traces are ellipses; vertical traces are parabolas.

slice the graph with any plane parallel to the yz-plane, we obtain a parabola that opens upward. Similarly, if $y = k$, the trace is $z = 4x^2 + k^2$, which is again a parabola that opens upward. If we put $z = k$, we get the horizontal traces $4x^2 + y^2 = k$, which we recognize as a family of ellipses. Knowing the shapes of the traces, we can sketch the graph in Figure 5. Because of the elliptical and parabolic traces, the quadric surface $z = 4x^2 + y^2$ is called an **elliptic paraboloid**.

EXAMPLE 5 Sketch the surface $z = y^2 - x^2$.

SOLUTION The traces in the vertical planes $x = k$ are the parabolas $z = y^2 - k^2$, which open upward. The traces in $y = k$ are the parabolas $z = -x^2 + k^2$, which open downward. The horizontal traces are $y^2 - x^2 = k$, a family of hyperbolas. We draw the families of traces in Figure 6, and we show how the traces appear when placed in their correct planes in Figure 7.

FIGURE 6

Vertical traces are parabolas; horizontal traces are hyperbolas. All traces are labeled with the value of k.

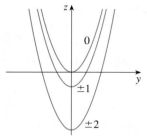

Traces in $x = k$ are $z = y^2 - k^2$.

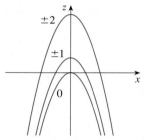

Traces in $y = k$ are $z = -x^2 + k^2$.

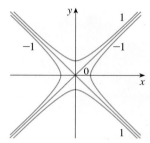

Traces in $z = k$ are $y^2 - x^2 = k$.

FIGURE 7

Traces moved to their correct planes

Traces in $x = k$

Traces in $y = k$

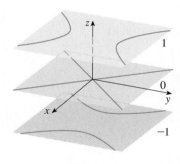

Traces in $z = k$

TEC In Module 12.6A you can investigate how traces determine the shape of a surface.

In Figure 8 we fit together the traces from Figure 7 to form the surface $z = y^2 - x^2$, a **hyperbolic paraboloid**. Notice that the shape of the surface near the origin resembles that of a saddle. This surface will be investigated further in Section 14.7 when we discuss saddle points.

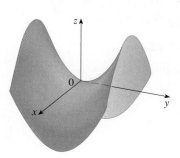

FIGURE 8

Two views of the surface $z = y^2 - x^2$, a hyperbolic paraboloid

FIGURE 9

EXAMPLE 6 Sketch the surface $\dfrac{x^2}{4} + y^2 - \dfrac{z^2}{4} = 1$.

SOLUTION The trace in any horizontal plane $z = k$ is the ellipse

$$\frac{x^2}{4} + y^2 = 1 + \frac{k^2}{4} \qquad z = k$$

but the traces in the xz- and yz-planes are the hyperbolas

$$\frac{x^2}{4} - \frac{z^2}{4} = 1 \qquad y = 0 \qquad \text{and} \qquad y^2 - \frac{z^2}{4} = 1 \qquad x = 0$$

This surface is called a **hyperboloid of one sheet** and is sketched in Figure 9. ∎

The idea of using traces to draw a surface is employed in three-dimensional graphing software. In most such software, traces in the vertical planes $x = k$ and $y = k$ are drawn for equally spaced values of k, and parts of the graph are eliminated using hidden line removal. Table 1 shows computer-drawn graphs of the six basic types of quadric surfaces in standard form. All surfaces are symmetric with respect to the z-axis. If a quadric surface is symmetric about a different axis, its equation changes accordingly.

Table 1 Graphs of Quadric Surfaces

Surface	Equation	Surface	Equation
Ellipsoid	$\dfrac{x^2}{a^2} + \dfrac{y^2}{b^2} + \dfrac{z^2}{c^2} = 1$ All traces are ellipses. If $a = b = c$, the ellipsoid is a sphere.	Cone	$\dfrac{z^2}{c^2} = \dfrac{x^2}{a^2} + \dfrac{y^2}{b^2}$ Horizontal traces are ellipses. Vertical traces in the planes $x = k$ and $y = k$ are hyperbolas if $k \neq 0$ but are pairs of lines if $k = 0$.
Elliptic Paraboloid	$\dfrac{z}{c} = \dfrac{x^2}{a^2} + \dfrac{y^2}{b^2}$ Horizontal traces are ellipses. Vertical traces are parabolas. The variable raised to the first power indicates the axis of the paraboloid.	Hyperboloid of One Sheet	$\dfrac{x^2}{a^2} + \dfrac{y^2}{b^2} - \dfrac{z^2}{c^2} = 1$ Horizontal traces are ellipses. Vertical traces are hyperbolas. The axis of symmetry corresponds to the variable whose coefficient is negative.
Hyperbolic Paraboloid	$\dfrac{z}{c} = \dfrac{x^2}{a^2} - \dfrac{y^2}{b^2}$ Horizontal traces are hyperbolas. Vertical traces are parabolas. The case where $c < 0$ is illustrated.	Hyperboloid of Two Sheets	$-\dfrac{x^2}{a^2} - \dfrac{y^2}{b^2} + \dfrac{z^2}{c^2} = 1$ Horizontal traces in $z = k$ are ellipses if $k > c$ or $k < -c$. Vertical traces are hyperbolas. The two minus signs indicate two sheets.

TEC In Module 12.6B you can see how changing a, b, and c in Table 1 affects the shape of the quadric surface.

EXAMPLE 7 Identify and sketch the surface $4x^2 - y^2 + 2z^2 + 4 = 0$.

SOLUTION Dividing by -4, we first put the equation in standard form:

$$-x^2 + \frac{y^2}{4} - \frac{z^2}{2} = 1$$

Comparing this equation with Table 1, we see that it represents a hyperboloid of two sheets, the only difference being that in this case the axis of the hyperboloid is the y-axis. The traces in the xy- and yz-planes are the hyperbolas

$$-x^2 + \frac{y^2}{4} = 1 \qquad z = 0 \qquad \text{and} \qquad \frac{y^2}{4} - \frac{z^2}{2} = 1 \qquad x = 0$$

The surface has no trace in the xz-plane, but traces in the vertical planes $y = k$ for $|k| > 2$ are the ellipses

$$x^2 + \frac{z^2}{2} = \frac{k^2}{4} - 1 \qquad y = k$$

which can be written as

$$\frac{x^2}{\dfrac{k^2}{4} - 1} + \frac{z^2}{2\left(\dfrac{k^2}{4} - 1\right)} = 1 \qquad y = k$$

These traces are used to make the sketch in Figure 10. ∎

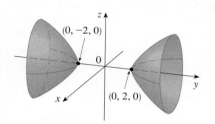

FIGURE 10
$4x^2 - y^2 + 2z^2 + 4 = 0$

EXAMPLE 8 Classify the quadric surface $x^2 + 2z^2 - 6x - y + 10 = 0$.

SOLUTION By completing the square we rewrite the equation as

$$y - 1 = (x - 3)^2 + 2z^2$$

Comparing this equation with Table 1, we see that it represents an elliptic paraboloid. Here, however, the axis of the paraboloid is parallel to the y-axis, and it has been shifted so that its vertex is the point $(3, 1, 0)$. The traces in the plane $y = k$ $(k > 1)$ are the ellipses

$$(x - 3)^2 + 2z^2 = k - 1 \qquad y = k$$

The trace in the xy-plane is the parabola with equation $y = 1 + (x - 3)^2$, $z = 0$. The paraboloid is sketched in Figure 11. ∎

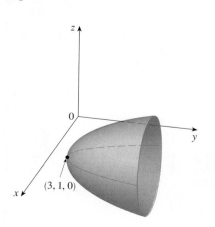

FIGURE 11
$x^2 + 2z^2 - 6x - y + 10 = 0$

◼ Applications of Quadric Surfaces

Examples of quadric surfaces can be found in the world around us. In fact, the world itself is a good example. Although the earth is commonly modeled as a sphere, a more accurate model is an ellipsoid because the earth's rotation has caused a flattening at the poles. (See Exercise 49.)

Circular paraboloids, obtained by rotating a parabola about its axis, are used to collect and reflect light, sound, and radio and television signals. In a radio telescope, for instance, signals from distant stars that strike the bowl are all reflected to the receiver at the focus and are therefore amplified. (The idea is explained in Problem 22 on page 273.) The same principle applies to microphones and satellite dishes in the shape of paraboloids.

Cooling towers for nuclear reactors are usually designed in the shape of hyperboloids of one sheet for reasons of structural stability. Pairs of hyperboloids are used to transmit rotational motion between skew axes. (The cogs of the gears are the generating lines of the hyperboloids. See Exercise 51.)

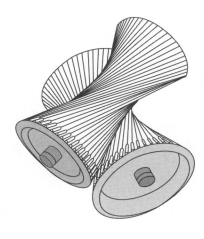

A satellite dish reflects signals to the focus of a paraboloid.

Nuclear reactors have cooling towers in the shape of hyperboloids.

Hperboloids produce gear transmission.

12.6 EXERCISES

1. (a) What does the equation $y = x^2$ represent as a curve in $\mathbb{R}^2$?
  (b) What does it represent as a surface in $\mathbb{R}^3$?
  (c) What does the equation $z = y^2$ represent?

2. (a) Sketch the graph of $y = e^x$ as a curve in $\mathbb{R}^2$.
  (b) Sketch the graph of $y = e^x$ as a surface in $\mathbb{R}^3$.
  (c) Describe and sketch the surface $z = e^y$.

3–8 Describe and sketch the surface.

3. $x^2 + z^2 = 1$ **4.** $4x^2 + y^2 = 4$

5. $z = 1 - y^2$ **6.** $y = z^2$

7. $xy = 1$ **8.** $z = \sin y$

9. (a) Find and identify the traces of the quadric surface $x^2 + y^2 - z^2 = 1$ and explain why the graph looks like the graph of the hyperboloid of one sheet in Table 1.
  (b) If we change the equation in part (a) to $x^2 - y^2 + z^2 = 1$, how is the graph affected?
  (c) What if we change the equation in part (a) to $x^2 + y^2 + 2y - z^2 = 0$?

10. (a) Find and identify the traces of the quadric surface
$-x^2 - y^2 + z^2 = 1$ and explain why the graph looks like
the graph of the hyperboloid of two sheets in Table 1.
(b) If the equation in part (a) is changed to $x^2 - y^2 - z^2 = 1$,
what happens to the graph? Sketch the new graph.

11–20 Use traces to sketch and identify the surface.

11. $x = y^2 + 4z^2$ **12.** $4x^2 + 9y^2 + 9z^2 = 36$

13. $x^2 = 4y^2 + z^2$ **14.** $z^2 - 4x^2 - y^2 = 4$

15. $9y^2 + 4z^2 = x^2 + 36$ **16.** $3x^2 + y + 3z^2 = 0$

17. $\dfrac{x^2}{9} + \dfrac{y^2}{25} + \dfrac{z^2}{4} = 1$ **18.** $3x^2 - y^2 + 3z^2 = 0$

19. $y = z^2 - x^2$ **20.** $x = y^2 - z^2$

21–28 Match the equation with its graph (labeled I–VIII). Give
reasons for your choice.

21. $x^2 + 4y^2 + 9z^2 = 1$ **22.** $9x^2 + 4y^2 + z^2 = 1$

23. $x^2 - y^2 + z^2 = 1$ **24.** $-x^2 + y^2 - z^2 = 1$

25. $y = 2x^2 + z^2$ **26.** $y^2 = x^2 + 2z^2$

27. $x^2 + 2z^2 = 1$ **28.** $y = x^2 - z^2$

I

II

III

IV

V

VI

VII

VIII

29–30 Sketch and identify a quadric surface that could have the
traces shown.

29. Traces in $x = k$ Traces in $y = k$

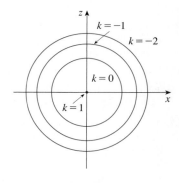

30. Traces in $x = k$ Traces in $z = k$

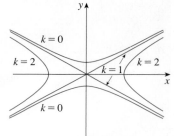

31–38 Reduce the equation to one of the standard forms, classify
the surface, and sketch it.

31. $y^2 = x^2 + \frac{1}{9}z^2$ **32.** $4x^2 - y + 2z^2 = 0$

33. $x^2 + 2y - 2z^2 = 0$ **34.** $y^2 = x^2 + 4z^2 + 4$

35. $x^2 + y^2 - 2x - 6y - z + 10 = 0$

36. $x^2 - y^2 - z^2 - 4x - 2z + 3 = 0$

37. $x^2 - y^2 + z^2 - 4x - 2z = 0$

38. $4x^2 + y^2 + z^2 - 24x - 8y + 4z + 55 = 0$

39–42 Use a computer with three-dimensional graphing software to graph the surface. Experiment with viewpoints and with
domains for the variables until you get a good view of the surface.

39. $-4x^2 - y^2 + z^2 = 1$ **40.** $x^2 - y^2 - z = 0$

41. $-4x^2 - y^2 + z^2 = 0$ **42.** $x^2 - 6x + 4y^2 - z = 0$

---

43. Sketch the region bounded by the surfaces $z = \sqrt{x^2 + y^2}$
and $x^2 + y^2 = 1$ for $1 \le z \le 2$.

44. Sketch the region bounded by the paraboloids $z = x^2 + y^2$
and $z = 2 - x^2 - y^2$.

45. Find an equation for the surface obtained by rotating the curve $y = \sqrt{x}$ about the x-axis.

46. Find an equation for the surface obtained by rotating the line $z = 2y$ about the z-axis.

47. Find an equation for the surface consisting of all points that are equidistant from the point $(-1, 0, 0)$ and the plane $x = 1$. Identify the surface.

48. Find an equation for the surface consisting of all points P for which the distance from P to the x-axis is twice the distance from P to the yz-plane. Identify the surface.

49. Traditionally, the earth's surface has been modeled as a sphere, but the World Geodetic System of 1984 (WGS-84) uses an ellipsoid as a more accurate model. It places the center of the earth at the origin and the north pole on the positive z-axis. The distance from the center to the poles is 6356.523 km and the distance to a point on the equator is 6378.137 km.
  (a) Find an equation of the earth's surface as used by WGS-84.
  (b) Curves of equal latitude are traces in the planes $z = k$. What is the shape of these curves?
  (c) Meridians (curves of equal longitude) are traces in planes of the form $y = mx$. What is the shape of these meridians?

50. A cooling tower for a nuclear reactor is to be constructed in the shape of a hyperboloid of one sheet (see the photo on page 839). The diameter at the base is 280 m and the minimum diameter, 500 m above the base, is 200 m. Find an equation for the tower.

51. Show that if the point (a, b, c) lies on the hyperbolic paraboloid $z = y^2 - x^2$, then the lines with parametric equations $x = a + t, y = b + t, z = c + 2(b - a)t$ and $x = a + t$, $y = b - t, z = c - 2(b + a)t$ both lie entirely on this paraboloid. (This shows that the hyperbolic paraboloid is what is called a **ruled surface**; that is, it can be generated by the motion of a straight line. In fact, this exercise shows that through each point on the hyperbolic paraboloid there are two generating lines. The only other quadric surfaces that are ruled surfaces are cylinders, cones, and hyperboloids of one sheet.)

52. Show that the curve of intersection of the surfaces $x^2 + 2y^2 - z^2 + 3x = 1$ and $2x^2 + 4y^2 - 2z^2 - 5y = 0$ lies in a plane.

53. Graph the surfaces $z = x^2 + y^2$ and $z = 1 - y^2$ on a common screen using the domain $|x| \le 1.2, |y| \le 1.2$ and observe the curve of intersection of these surfaces. Show that the projection of this curve onto the xy-plane is an ellipse.

---

12   REVIEW

CONCEPT CHECK   Answers to the Concept Check can be found on the back endpapers.

1. What is the difference between a vector and a scalar?

2. How do you add two vectors geometrically? How do you add them algebraically?

3. If **a** is a vector and c is a scalar, how is c**a** related to **a** geometrically? How do you find c**a** algebraically?

4. How do you find the vector from one point to another?

5. How do you find the dot product $\mathbf{a} \cdot \mathbf{b}$ of two vectors if you know their lengths and the angle between them? What if you know their components?

6. How are dot products useful?

7. Write expressions for the scalar and vector projections of **b** onto **a**. Illustrate with diagrams.

8. How do you find the cross product $\mathbf{a} \times \mathbf{b}$ of two vectors if you know their lengths and the angle between them? What if you know their components?

9. How are cross products useful?

10. (a) How do you find the area of the parallelogram determined by **a** and **b**?
  (b) How do you find the volume of the parallelepiped determined by **a**, **b**, and **c**?

11. How do you find a vector perpendicular to a plane?

12. How do you find the angle between two intersecting planes?

13. Write a vector equation, parametric equations, and symmetric equations for a line.

14. Write a vector equation and a scalar equation for a plane.

15. (a) How do you tell if two vectors are parallel?
  (b) How do you tell if two vectors are perpendicular?
  (c) How do you tell if two planes are parallel?

16. (a) Describe a method for determining whether three points P, Q, and R lie on the same line.
  (b) Describe a method for determining whether four points P, Q, R, and S lie in the same plane.

17. (a) How do you find the distance from a point to a line?
  (b) How do you find the distance from a point to a plane?
  (c) How do you find the distance between two lines?

18. What are the traces of a surface? How do you find them?

19. Write equations in standard form of the six types of quadric surfaces.

TRUE-FALSE QUIZ

Determine whether the statement is true or false. If it is true, explain why. If it is false, explain why or give an example that disproves the statement.

1. If $\mathbf{u} = \langle u_1, u_2 \rangle$ and $\mathbf{v} = \langle v_1, v_2 \rangle$, then $\mathbf{u} \cdot \mathbf{v} = \langle u_1 v_1, u_2 v_2 \rangle$.

2. For any vectors $\mathbf{u}$ and $\mathbf{v}$ in V_3, $|\mathbf{u} + \mathbf{v}| = |\mathbf{u}| + |\mathbf{v}|$.

3. For any vectors $\mathbf{u}$ and $\mathbf{v}$ in V_3, $|\mathbf{u} \cdot \mathbf{v}| = |\mathbf{u}||\mathbf{v}|$.

4. For any vectors $\mathbf{u}$ and $\mathbf{v}$ in V_3, $|\mathbf{u} \times \mathbf{v}| = |\mathbf{u}||\mathbf{v}|$.

5. For any vectors $\mathbf{u}$ and $\mathbf{v}$ in V_3, $\mathbf{u} \cdot \mathbf{v} = \mathbf{v} \cdot \mathbf{u}$.

6. For any vectors $\mathbf{u}$ and $\mathbf{v}$ in V_3, $\mathbf{u} \times \mathbf{v} = \mathbf{v} \times \mathbf{u}$.

7. For any vectors $\mathbf{u}$ and $\mathbf{v}$ in V_3, $|\mathbf{u} \times \mathbf{v}| = |\mathbf{v} \times \mathbf{u}|$.

8. For any vectors $\mathbf{u}$ and $\mathbf{v}$ in V_3 and any scalar k,
$$k(\mathbf{u} \cdot \mathbf{v}) = (k\mathbf{u}) \cdot \mathbf{v}$$

9. For any vectors $\mathbf{u}$ and $\mathbf{v}$ in V_3 and any scalar k,
$$k(\mathbf{u} \times \mathbf{v}) = (k\mathbf{u}) \times \mathbf{v}$$

10. For any vectors $\mathbf{u}$, $\mathbf{v}$, and $\mathbf{w}$ in V_3,
$$(\mathbf{u} + \mathbf{v}) \times \mathbf{w} = \mathbf{u} \times \mathbf{w} + \mathbf{v} \times \mathbf{w}$$

11. For any vectors $\mathbf{u}$, $\mathbf{v}$, and $\mathbf{w}$ in V_3,
$$\mathbf{u} \cdot (\mathbf{v} \times \mathbf{w}) = (\mathbf{u} \times \mathbf{v}) \cdot \mathbf{w}$$

12. For any vectors $\mathbf{u}$, $\mathbf{v}$, and $\mathbf{w}$ in V_3,
$$\mathbf{u} \times (\mathbf{v} \times \mathbf{w}) = (\mathbf{u} \times \mathbf{v}) \times \mathbf{w}$$

13. For any vectors $\mathbf{u}$ and $\mathbf{v}$ in V_3, $(\mathbf{u} \times \mathbf{v}) \cdot \mathbf{u} = 0$.

14. For any vectors $\mathbf{u}$ and $\mathbf{v}$ in V_3, $(\mathbf{u} + \mathbf{v}) \times \mathbf{v} = \mathbf{u} \times \mathbf{v}$.

15. The vector $\langle 3, -1, 2 \rangle$ is parallel to the plane
$$6x - 2y + 4z = 1$$

16. A linear equation $Ax + By + Cz + D = 0$ represents a line in space.

17. The set of points $\{(x, y, z) \mid x^2 + y^2 = 1\}$ is a circle.

18. In $\mathbb{R}^3$ the graph of $y = x^2$ is a paraboloid.

19. If $\mathbf{u} \cdot \mathbf{v} = 0$, then $\mathbf{u} = \mathbf{0}$ or $\mathbf{v} = \mathbf{0}$.

20. If $\mathbf{u} \times \mathbf{v} = \mathbf{0}$, then $\mathbf{u} = \mathbf{0}$ or $\mathbf{v} = \mathbf{0}$.

21. If $\mathbf{u} \cdot \mathbf{v} = 0$ and $\mathbf{u} \times \mathbf{v} = \mathbf{0}$, then $\mathbf{u} = \mathbf{0}$ or $\mathbf{v} = \mathbf{0}$.

22. If $\mathbf{u}$ and $\mathbf{v}$ are in V_3, then $|\mathbf{u} \cdot \mathbf{v}| \leq |\mathbf{u}||\mathbf{v}|$.

EXERCISES

1. (a) Find an equation of the sphere that passes through the point $(6, -2, 3)$ and has center $(-1, 2, 1)$.
(b) Find the curve in which this sphere intersects the yz-plane.
(c) Find the center and radius of the sphere
$$x^2 + y^2 + z^2 - 8x + 2y + 6z + 1 = 0$$

2. Copy the vectors in the figure and use them to draw each of the following vectors.
(a) $\mathbf{a} + \mathbf{b}$ (b) $\mathbf{a} - \mathbf{b}$ (c) $-\frac{1}{2}\mathbf{a}$ (d) $2\mathbf{a} + \mathbf{b}$

3. If $\mathbf{u}$ and $\mathbf{v}$ are the vectors shown in the figure, find $\mathbf{u} \cdot \mathbf{v}$ and $|\mathbf{u} \times \mathbf{v}|$. Is $\mathbf{u} \times \mathbf{v}$ directed into the page or out of it?

4. Calculate the given quantity if
$$\mathbf{a} = \mathbf{i} + \mathbf{j} - 2\mathbf{k}$$
$$\mathbf{b} = 3\mathbf{i} - 2\mathbf{j} + \mathbf{k}$$
$$\mathbf{c} = \mathbf{j} - 5\mathbf{k}$$

(a) $2\mathbf{a} + 3\mathbf{b}$ (b) $|\mathbf{b}|$
(c) $\mathbf{a} \cdot \mathbf{b}$ (d) $\mathbf{a} \times \mathbf{b}$
(e) $|\mathbf{b} \times \mathbf{c}|$ (f) $\mathbf{a} \cdot (\mathbf{b} \times \mathbf{c})$
(g) $\mathbf{c} \times \mathbf{c}$ (h) $\mathbf{a} \times (\mathbf{b} \times \mathbf{c})$
(i) $\text{comp}_\mathbf{a} \mathbf{b}$ (j) $\text{proj}_\mathbf{a} \mathbf{b}$
(k) The angle between $\mathbf{a}$ and $\mathbf{b}$ (correct to the nearest degree)

5. Find the values of x such that the vectors $\langle 3, 2, x \rangle$ and $\langle 2x, 4, x \rangle$ are orthogonal.

6. Find two unit vectors that are orthogonal to both $\mathbf{j} + 2\mathbf{k}$ and $\mathbf{i} - 2\mathbf{j} + 3\mathbf{k}$.

7. Suppose that $\mathbf{u} \cdot (\mathbf{v} \times \mathbf{w}) = 2$. Find
(a) $(\mathbf{u} \times \mathbf{v}) \cdot \mathbf{w}$ (b) $\mathbf{u} \cdot (\mathbf{w} \times \mathbf{v})$
(c) $\mathbf{v} \cdot (\mathbf{u} \times \mathbf{w})$ (d) $(\mathbf{u} \times \mathbf{v}) \cdot \mathbf{v}$

8. Show that if $\mathbf{a}$, $\mathbf{b}$, and $\mathbf{c}$ are in V_3, then
$$(\mathbf{a} \times \mathbf{b}) \cdot [(\mathbf{b} \times \mathbf{c}) \times (\mathbf{c} \times \mathbf{a})] = [\mathbf{a} \cdot (\mathbf{b} \times \mathbf{c})]^2$$

9. Find the acute angle between two diagonals of a cube.

10. Given the points $A(1, 0, 1)$, $B(2, 3, 0)$, $C(-1, 1, 4)$, and $D(0, 3, 2)$, find the volume of the parallelepiped with adjacent edges AB, AC, and AD.

11. (a) Find a vector perpendicular to the plane through the points $A(1, 0, 0)$, $B(2, 0, -1)$, and $C(1, 4, 3)$.
(b) Find the area of triangle ABC.

12. A constant force $\mathbf{F} = 3\mathbf{i} + 5\mathbf{j} + 10\mathbf{k}$ moves an object along the line segment from $(1, 0, 2)$ to $(5, 3, 8)$. Find the work done if the distance is measured in meters and the force in newtons.

13. A boat is pulled onto shore using two ropes, as shown in the diagram. If a force of 255 N is needed, find the magnitude of the force in each rope.

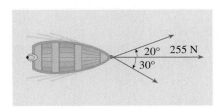

14. Find the magnitude of the torque about P if a 50-N force is applied as shown.

15–17 Find parametric equations for the line.

15. The line through $(4, -1, 2)$ and $(1, 1, 5)$

16. The line through $(1, 0, -1)$ and parallel to the line $\frac{1}{3}(x - 4) = \frac{1}{2}y = z + 2$

17. The line through $(-2, 2, 4)$ and perpendicular to the plane $2x - y + 5z = 12$

18–20 Find an equation of the plane.

18. The plane through $(2, 1, 0)$ and parallel to $x + 4y - 3z = 1$

19. The plane through $(3, -1, 1)$, $(4, 0, 2)$, and $(6, 3, 1)$

20. The plane through $(1, 2, -2)$ that contains the line $x = 2t$, $y = 3 - t$, $z = 1 + 3t$

21. Find the point in which the line with parametric equations $x = 2 - t$, $y = 1 + 3t$, $z = 4t$ intersects the plane $2x - y + z = 2$.

22. Find the distance from the origin to the line $x = 1 + t$, $y = 2 - t$, $z = -1 + 2t$.

23. Determine whether the lines given by the symmetric equations

$$\frac{x - 1}{2} = \frac{y - 2}{3} = \frac{z - 3}{4}$$

and

$$\frac{x + 1}{6} = \frac{y - 3}{-1} = \frac{z + 5}{2}$$

are parallel, skew, or intersecting.

24. (a) Show that the planes $x + y - z = 1$ and $2x - 3y + 4z = 5$ are neither parallel nor perpendicular.
(b) Find, correct to the nearest degree, the angle between these planes.

25. Find an equation of the plane through the line of intersection of the planes $x - z = 1$ and $y + 2z = 3$ and perpendicular to the plane $x + y - 2z = 1$.

26. (a) Find an equation of the plane that passes through the points $A(2, 1, 1)$, $B(-1, -1, 10)$, and $C(1, 3, -4)$.
(b) Find symmetric equations for the line through B that is perpendicular to the plane in part (a).
(c) A second plane passes through $(2, 0, 4)$ and has normal vector $\langle 2, -4, -3 \rangle$. Show that the acute angle between the planes is approximately $43°$.
(d) Find parametric equations for the line of intersection of the two planes.

27. Find the distance between the planes $3x + y - 4z = 2$ and $3x + y - 4z = 24$.

28–36 Identify and sketch the graph of each surface.

28. $x = 3$ **29.** $x = z$

30. $y = z^2$ **31.** $x^2 = y^2 + 4z^2$

32. $4x - y + 2z = 4$ **33.** $-4x^2 + y^2 - 4z^2 = 4$

34. $y^2 + z^2 = 1 + x^2$

35. $4x^2 + 4y^2 - 8y + z^2 = 0$

36. $x = y^2 + z^2 - 2y - 4z + 5$

37. An ellipsoid is created by rotating the ellipse $4x^2 + y^2 = 16$ about the x-axis. Find an equation of the ellipsoid.

38. A surface consists of all points P such that the distance from P to the plane $y = 1$ is twice the distance from P to the point $(0, -1, 0)$. Find an equation for this surface and identify it.

Problems Plus

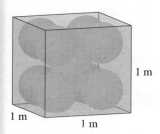

1 m

1 m   1 m

FIGURE FOR PROBLEM 1

1. Each edge of a cubical box has length 1 m. The box contains nine spherical balls with the same radius r. The center of one ball is at the center of the cube and it touches the other eight balls. Each of the other eight balls touches three sides of the box. Thus the balls are tightly packed in the box (see the figure). Find r. (If you have trouble with this problem, read about the problem-solving strategy entitled *Use Analogy* on page 71.)

2. Let B be a solid box with length L, width W, and height H. Let S be the set of all points that are a distance at most 1 from some point of B. Express the volume of S in terms of L, W, and H.

3. Let L be the line of intersection of the planes $cx + y + z = c$ and $x - cy + cz = -1$, where c is a real number.
   (a) Find symmetric equations for L.
   (b) As the number c varies, the line L sweeps out a surface S. Find an equation for the curve of intersection of S with the horizontal plane $z = t$ (the trace of S in the plane $z = t$).
   (c) Find the volume of the solid bounded by S and the planes $z = 0$ and $z = 1$.

4. A plane is capable of flying at a speed of 180 km/h in still air. The pilot takes off from an airfield and heads due north according to the plane's compass. After 30 minutes of flight time, the pilot notices that, due to the wind, the plane has actually traveled 80 km at an angle 5° east of north.
   (a) What is the wind velocity?
   (b) In what direction should the pilot have headed to reach the intended destination?

5. Suppose $\mathbf{v}_1$ and $\mathbf{v}_2$ are vectors with $|\mathbf{v}_1| = 2$, $|\mathbf{v}_2| = 3$, and $\mathbf{v}_1 \cdot \mathbf{v}_2 = 5$. Let $\mathbf{v}_3 = \text{proj}_{\mathbf{v}_1} \mathbf{v}_2$, $\mathbf{v}_4 = \text{proj}_{\mathbf{v}_2} \mathbf{v}_3$, $\mathbf{v}_5 = \text{proj}_{\mathbf{v}_3} \mathbf{v}_4$, and so on. Compute $\sum_{n=1}^{\infty} |\mathbf{v}_n|$.

6. Find an equation of the largest sphere that passes through the point $(-1, 1, 4)$ and is such that each of the points (x, y, z) inside the sphere satisfies the condition

$$x^2 + y^2 + z^2 < 136 + 2(x + 2y + 3z)$$

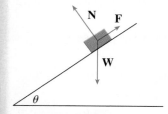

N   F

W

θ

FIGURE FOR PROBLEM 7

7. Suppose a block of mass m is placed on an inclined plane, as shown in the figure. The block's descent down the plane is slowed by friction; if θ is not too large, friction will prevent the block from moving at all. The forces acting on the block are the weight $\mathbf{W}$, where $|\mathbf{W}| = mg$ (g is the acceleration due to gravity); the normal force $\mathbf{N}$ (the normal component of the reactionary force of the plane on the block), where $|\mathbf{N}| = n$; and the force $\mathbf{F}$ due to friction, which acts parallel to the inclined plane, opposing the direction of motion. If the block is at rest and θ is increased, $|\mathbf{F}|$ must also increase until ultimately $|\mathbf{F}|$ reaches its maximum, beyond which the block begins to slide. At this angle θ_s, it has been observed that $|\mathbf{F}|$ is proportional to n. Thus, when $|\mathbf{F}|$ is maximal, we can say that $|\mathbf{F}| = \mu_s n$, where μ_s is called the *coefficient of static friction* and depends on the materials that are in contact.
   (a) Observe that $\mathbf{N} + \mathbf{F} + \mathbf{W} = \mathbf{0}$ and deduce that $\mu_s = \tan(\theta_s)$.
   (b) Suppose that, for $\theta > \theta_s$, an additional outside force $\mathbf{H}$ is applied to the block, horizontally from the left, and let $|\mathbf{H}| = h$. If h is small, the block may still slide down the plane; if h is large enough, the block will move up the plane. Let h_{min} be the smallest value of h that allows the block to remain motionless (so that $|\mathbf{F}|$ is maximal).
       By choosing the coordinate axes so that $\mathbf{F}$ lies along the x-axis, resolve each force into components parallel and perpendicular to the inclined plane and show that

$$h_{min} \sin\theta + mg \cos\theta = n \qquad \text{and} \qquad h_{min} \cos\theta + \mu_s n = mg \sin\theta$$

   (c) Show that

$$h_{min} = mg \tan(\theta - \theta_s)$$

   Does this equation seem reasonable? Does it make sense for $\theta = \theta_s$? Does it make sense as $\theta \to 90°$? Explain.

(d) Let h_{max} be the largest value of h that allows the block to remain motionless. (In which direction is **F** heading?) Show that

$$h_{max} = mg \tan(\theta + \theta_s)$$

Does this equation seem reasonable? Explain.

8. A solid has the following properties. When illuminated by rays parallel to the z-axis, its shadow is a circular disk. If the rays are parallel to the y-axis, its shadow is a square. If the rays are parallel to the x-axis, its shadow is an isosceles triangle. (In Exercise 12.1.48 you were asked to describe and sketch an example of such a solid, but there are many such solids.) Assume that the projection onto the xz-plane is a square whose sides have length 1.
(a) What is the volume of the largest such solid?
(b) Is there a smallest volume?

13 Vector Functions

© Natalia Davydenko / Shutterstock.com

The paths of objects moving through space like the jet planes pictured here can be described by vector functions. In Section 13.1 we will see how to use these vector functions to determine whether or not two such objects will collide.

THE FUNCTIONS THAT WE HAVE been using so far have been real-valued functions. We now study functions whose values are vectors because such functions are needed to describe curves and surfaces in space. We will also use vector-valued functions to describe the motion of objects through space. In particular, we will use them to derive Kepler's laws of planetary motion.

13.1 Vector Functions and Space Curves

In general, a function is a rule that assigns to each element in the domain an element in the range. A **vector-valued function**, or **vector function**, is simply a function whose domain is a set of real numbers and whose range is a set of vectors. We are most interested in vector functions **r** whose values are three-dimensional vectors. This means that for every number t in the domain of **r** there is a unique vector in V_3 denoted by $\mathbf{r}(t)$. If $f(t)$, $g(t)$, and $h(t)$ are the components of the vector $\mathbf{r}(t)$, then f, g, and h are real-valued functions called the **component functions** of **r** and we can write

$$\mathbf{r}(t) = \langle f(t), g(t), h(t) \rangle = f(t)\,\mathbf{i} + g(t)\,\mathbf{j} + h(t)\,\mathbf{k}$$

We use the letter t to denote the independent variable because it represents time in most applications of vector functions.

EXAMPLE 1 If

$$\mathbf{r}(t) = \left\langle t^3, \ln(3 - t), \sqrt{t} \right\rangle$$

then the component functions are

$$f(t) = t^3 \qquad g(t) = \ln(3 - t) \qquad h(t) = \sqrt{t}$$

By our usual convention, the domain of **r** consists of all values of t for which the expression for $\mathbf{r}(t)$ is defined. The expressions t^3, $\ln(3 - t)$, and $\sqrt{t}$ are all defined when $3 - t > 0$ and $t \geqslant 0$. Therefore the domain of **r** is the interval $[0, 3)$. ∎

■ Limits and Continuity

The **limit** of a vector function **r** is defined by taking the limits of its component functions as follows.

If $\lim_{t \to a} \mathbf{r}(t) = \mathbf{L}$, this definition is equivalent to saying that the length and direction of the vector $\mathbf{r}(t)$ approach the length and direction of the vector **L**.

> **1** If $\mathbf{r}(t) = \langle f(t), g(t), h(t) \rangle$, then
>
> $$\lim_{t \to a} \mathbf{r}(t) = \left\langle \lim_{t \to a} f(t), \lim_{t \to a} g(t), \lim_{t \to a} h(t) \right\rangle$$
>
> provided the limits of the component functions exist.

Equivalently, we could have used an ε-δ definition (see Exercise 54). Limits of vector functions obey the same rules as limits of real-valued functions (see Exercise 53).

EXAMPLE 2 Find $\lim_{t \to 0} \mathbf{r}(t)$, where $\mathbf{r}(t) = (1 + t^3)\,\mathbf{i} + te^{-t}\,\mathbf{j} + \dfrac{\sin t}{t}\,\mathbf{k}$.

SOLUTION According to Definition 1, the limit of **r** is the vector whose components are the limits of the component functions of **r**:

$$\lim_{t \to 0} \mathbf{r}(t) = \left[\lim_{t \to 0} (1 + t^3) \right] \mathbf{i} + \left[\lim_{t \to 0} te^{-t} \right] \mathbf{j} + \left[\lim_{t \to 0} \frac{\sin t}{t} \right] \mathbf{k}$$

$$= \mathbf{i} + \mathbf{k} \qquad \text{(by Equation 3.3.2)}$$

∎

A vector function $\mathbf{r}$ is **continuous at** a if

$$\lim_{t \to a} \mathbf{r}(t) = \mathbf{r}(a)$$

In view of Definition 1, we see that $\mathbf{r}$ is continuous at a if and only if its component functions f, g, and h are continuous at a.

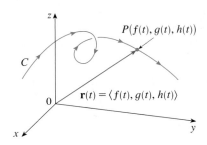

FIGURE 1

C is traced out by the tip of a moving position vector $\mathbf{r}(t)$.

◼ Space Curves

There is a close connection between continuous vector functions and space curves. Suppose that f, g, and h are continuous real-valued functions on an interval I. Then the set C of all points (x, y, z) in space, where

$$\boxed{2} \qquad\qquad x = f(t) \qquad y = g(t) \qquad z = h(t)$$

and t varies throughout the interval I, is called a **space curve**. The equations in (2) are called **parametric equations of C** and t is called a **parameter**. We can think of C as being traced out by a moving particle whose position at time t is $(f(t), g(t), h(t))$. If we now consider the vector function $\mathbf{r}(t) = \langle f(t), g(t), h(t) \rangle$, then $\mathbf{r}(t)$ is the position vector of the point $P(f(t), g(t), h(t))$ on C. Thus any continuous vector function $\mathbf{r}$ defines a space curve C that is traced out by the tip of the moving vector $\mathbf{r}(t)$, as shown in Figure 1.

EXAMPLE 3 Describe the curve defined by the vector function

$$\mathbf{r}(t) = \langle 1 + t, 2 + 5t, -1 + 6t \rangle$$

SOLUTION The corresponding parametric equations are

$$x = 1 + t \qquad y = 2 + 5t \qquad z = -1 + 6t$$

which we recognize from Equations 12.5.2 as parametric equations of a line passing through the point $(1, 2, -1)$ and parallel to the vector $\langle 1, 5, 6 \rangle$. Alternatively, we could observe that the function can be written as $\mathbf{r} = \mathbf{r}_0 + t\mathbf{v}$, where $\mathbf{r}_0 = \langle 1, 2, -1 \rangle$ and $\mathbf{v} = \langle 1, 5, 6 \rangle$, and this is the vector equation of a line as given by Equation 12.5.1. ◼

TEC Visual 13.1A shows several curves being traced out by position vectors, including those in Figures 1 and 2.

Plane curves can also be represented in vector notation. For instance, the curve given by the parametric equations $x = t^2 - 2t$ and $y = t + 1$ (see Example 10.1.1) could also be described by the vector equation

$$\mathbf{r}(t) = \langle t^2 - 2t, t + 1 \rangle = (t^2 - 2t)\,\mathbf{i} + (t + 1)\,\mathbf{j}$$

where $\mathbf{i} = \langle 1, 0 \rangle$ and $\mathbf{j} = \langle 0, 1 \rangle$.

EXAMPLE 4 Sketch the curve whose vector equation is

$$\mathbf{r}(t) = \cos t\,\mathbf{i} + \sin t\,\mathbf{j} + t\,\mathbf{k}$$

SOLUTION The parametric equations for this curve are

$$x = \cos t \qquad y = \sin t \qquad z = t$$

Since $x^2 + y^2 = \cos^2 t + \sin^2 t = 1$ for all values of t, the curve must lie on the circular cylinder $x^2 + y^2 = 1$. The point (x, y, z) lies directly above the point $(x, y, 0)$, which moves counterclockwise around the circle $x^2 + y^2 = 1$ in the xy-plane. (The projection of the curve onto the xy-plane has vector equation $\mathbf{r}(t) = \langle \cos t, \sin t, 0 \rangle$. See Example 10.1.2.) Since $z = t$, the curve spirals upward around the cylinder as t increases. The curve, shown in Figure 2, is called a **helix**. ◼

FIGURE 2

FIGURE 3
A double helix

The corkscrew shape of the helix in Example 4 is familiar from its occurrence in coiled springs. It also occurs in the model of DNA (deoxyribonucleic acid, the genetic material of living cells). In 1953 James Watson and Francis Crick showed that the structure of the DNA molecule is that of two linked, parallel helixes that are intertwined as in Figure 3.

In Examples 3 and 4 we were given vector equations of curves and asked for a geometric description or sketch. In the next two examples we are given a geometric description of a curve and are asked to find parametric equations for the curve.

EXAMPLE 5 Find a vector equation and parametric equations for the line segment that joins the point $P(1, 3, -2)$ to the point $Q(2, -1, 3)$.

SOLUTION In Section 12.5 we found a vector equation for the line segment that joins the tip of the vector $\mathbf{r}_0$ to the tip of the vector $\mathbf{r}_1$:

$$\mathbf{r}(t) = (1 - t)\mathbf{r}_0 + t\mathbf{r}_1 \qquad 0 \leq t \leq 1$$

Figure 4 shows the line segment PQ in Example 5.

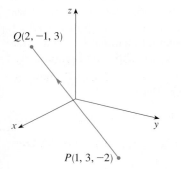

FIGURE 4

(See Equation 12.5.4.) Here we take $\mathbf{r}_0 = \langle 1, 3, -2 \rangle$ and $\mathbf{r}_1 = \langle 2, -1, 3 \rangle$ to obtain a vector equation of the line segment from P to Q:

$$\mathbf{r}(t) = (1 - t)\langle 1, 3, -2 \rangle + t\langle 2, -1, 3 \rangle \qquad 0 \leq t \leq 1$$

or

$$\mathbf{r}(t) = \langle 1 + t, 3 - 4t, -2 + 5t \rangle \qquad 0 \leq t \leq 1$$

The corresponding parametric equations are

$$x = 1 + t \qquad y = 3 - 4t \qquad z = -2 + 5t \qquad 0 \leq t \leq 1 \qquad \blacksquare$$

EXAMPLE 6 Find a vector function that represents the curve of intersection of the cylinder $x^2 + y^2 = 1$ and the plane $y + z = 2$.

SOLUTION Figure 5 shows how the plane and the cylinder intersect, and Figure 6 shows the curve of intersection C, which is an ellipse.

FIGURE 5

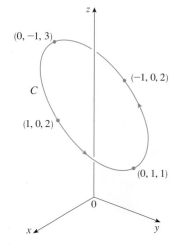

FIGURE 6

The projection of C onto the xy-plane is the circle $x^2 + y^2 = 1$, $z = 0$. So we know from Example 10.1.2 that we can write

$$x = \cos t \qquad y = \sin t \qquad 0 \le t \le 2\pi$$

From the equation of the plane, we have

$$z = 2 - y = 2 - \sin t$$

So we can write parametric equations for C as

$$x = \cos t \qquad y = \sin t \qquad z = 2 - \sin t \qquad 0 \le t \le 2\pi$$

The corresponding vector equation is

$$\mathbf{r}(t) = \cos t \, \mathbf{i} + \sin t \, \mathbf{j} + (2 - \sin t) \, \mathbf{k} \qquad 0 \le t \le 2\pi$$

This equation is called a *parametrization* of the curve C. The arrows in Figure 6 indicate the direction in which C is traced as the parameter t increases. ∎

■ Using Computers to Draw Space Curves

Space curves are inherently more difficult to draw by hand than plane curves; for an accurate representation we need to use technology. For instance, Figure 7 shows a computer-generated graph of the curve with parametric equations

$$x = (4 + \sin 20t) \cos t \qquad y = (4 + \sin 20t) \sin t \qquad z = \cos 20t$$

It's called a **toroidal spiral** because it lies on a torus. Another interesting curve, the **trefoil knot**, with equations

$$x = (2 + \cos 1.5t) \cos t \qquad y = (2 + \cos 1.5t) \sin t \qquad z = \sin 1.5t$$

is graphed in Figure 8. It wouldn't be easy to plot either of these curves by hand.

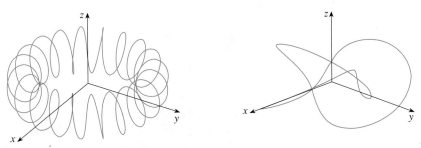

FIGURE 7
A toroidal spiral

FIGURE 8
A trefoil knot

Even when a computer is used to draw a space curve, optical illusions make it difficult to get a good impression of what the curve really looks like. (This is especially true in Figure 8. See Exercise 52.) The next example shows how to cope with this problem.

EXAMPLE 7 Use a computer to draw the curve with vector equation $\mathbf{r}(t) = \langle t, t^2, t^3 \rangle$. This curve is called a **twisted cubic**.

SOLUTION We start by using the computer to plot the curve with parametric equations $x = t$, $y = t^2$, $z = t^3$ for $-2 \le t \le 2$. The result is shown in Figure 9(a), but it's hard to see the true nature of the curve from that graph alone. Most three-dimensional computer graphing programs allow the user to enclose a curve or surface in a box instead of displaying the coordinate axes. When we look at the same curve in a box in Figure 9(b), we have a much clearer picture of the curve. We can see that it climbs from a lower corner of the box to the upper corner nearest us, and it twists as it climbs.

(a)

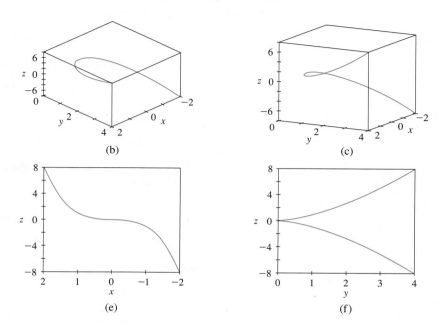

(b)

(c)

(d)

(e)

(f)

FIGURE 9 Views of the twisted cubic

TEC In Visual 13.1B you can rotate the box in Figure 9 to see the curve from any viewpoint.

We get an even better idea of the curve when we view it from different vantage points. Part (c) shows the result of rotating the box to give another viewpoint. Parts (d), (e), and (f) show the views we get when we look directly at a face of the box. In particular, part (d) shows the view from directly above the box. It is the projection of the curve onto the xy-plane, namely, the parabola $y = x^2$. Part (e) shows the projection onto the xz-plane, the cubic curve $z = x^3$. It's now obvious why the given curve is called a twisted cubic.

Another method of visualizing a space curve is to draw it on a surface. For instance, the twisted cubic in Example 7 lies on the parabolic cylinder $y = x^2$. (Eliminate the parameter from the first two parametric equations, $x = t$ and $y = t^2$.) Figure 10 shows both the cylinder and the twisted cubic, and we see that the curve moves upward from the origin along the surface of the cylinder. We also used this method in Example 4 to visualize the helix lying on the circular cylinder (see Figure 2).

FIGURE 10

A third method for visualizing the twisted cubic is to realize that it also lies on the cylinder $z = x^3$. So it can be viewed as the curve of intersection of the cylinders $y = x^2$ and $z = x^3$. (See Figure 11.)

TEC Visual 13.1C shows how curves arise as intersections of surfaces.

FIGURE 11

Some computer algebra systems provide us with a clearer picture of a space curve by enclosing it in a tube. Such a plot enables us to see whether one part of a curve passes in front of or behind another part of the curve. For example, Figure 13 shows the curve of Figure 12(b) as rendered by the `tubeplot` command in Maple.

We have seen that an interesting space curve, the helix, occurs in the model of DNA. Another notable example of a space curve in science is the trajectory of a positively charged particle in orthogonally oriented electric and magnetic fields $\mathbf{E}$ and $\mathbf{B}$. Depending on the initial velocity given the particle at the origin, the path of the particle is either a space curve whose projection onto the horizontal plane is the cycloid we studied in Section 10.1 [Figure 12(a)] or a curve whose projection is the trochoid investigated in Exercise 10.1.40 [Figure 12(b)].

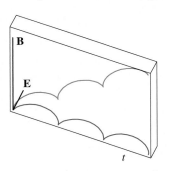

(a) $\mathbf{r}(t) = \langle t - \sin t, 1 - \cos t, t \rangle$

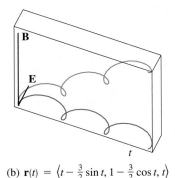

(b) $\mathbf{r}(t) = \langle t - \frac{3}{2}\sin t, 1 - \frac{3}{2}\cos t, t \rangle$

FIGURE 12
Motion of a charged particle in orthogonally oriented electric and magnetic fields

FIGURE 13

For further details concerning the physics involved and animations of the trajectories of the particles, see the following websites:

- www.physics.ucla.edu/plasma-exp/Beam/
- www.phy.ntnu.edu.tw/ntnujava/index.php?topic=36

13.1 EXERCISES

1–2 Find the domain of the vector function.

1. $\mathbf{r}(t) = \left\langle \ln(t+1), \dfrac{t}{\sqrt{9-t^2}}, 2^t \right\rangle$

2. $\mathbf{r}(t) = \cos t\,\mathbf{i} + \ln t\,\mathbf{j} + \dfrac{1}{t-2}\,\mathbf{k}$

3–6 Find the limit.

3. $\displaystyle\lim_{t \to 0} \left(e^{-3t}\,\mathbf{i} + \dfrac{t^2}{\sin^2 t}\,\mathbf{j} + \cos 2t\,\mathbf{k} \right)$

4. $\displaystyle\lim_{t \to 1} \left(\dfrac{t^2 - t}{t - 1}\,\mathbf{i} + \sqrt{t+8}\,\mathbf{j} + \dfrac{\sin \pi t}{\ln t}\,\mathbf{k} \right)$

5. $\displaystyle\lim_{t\to\infty}\left\langle \dfrac{1+t^2}{1-t^2},\ \tan^{-1}t,\ \dfrac{1-e^{-2t}}{t}\right\rangle$

6. $\displaystyle\lim_{t\to\infty}\left\langle te^{-t},\ \dfrac{t^3+t}{2t^3-1},\ t\sin\dfrac{1}{t}\right\rangle$

7–14 Sketch the curve with the given vector equation. Indicate with an arrow the direction in which t increases.

7. $\mathbf{r}(t)=\langle \sin t, t\rangle$ **8.** $\mathbf{r}(t)=\langle t^2-1, t\rangle$

9. $\mathbf{r}(t)=\langle t, 2-t, 2t\rangle$ **10.** $\mathbf{r}(t)=\langle \sin \pi t, t, \cos \pi t\rangle$

11. $\mathbf{r}(t)=\langle 3, t, 2-t^2\rangle$

12. $\mathbf{r}(t)=2\cos t\,\mathbf{i}+2\sin t\,\mathbf{j}+\mathbf{k}$

13. $\mathbf{r}(t)=t^2\mathbf{i}+t^4\mathbf{j}+t^6\mathbf{k}$

14. $\mathbf{r}(t)=\cos t\,\mathbf{i}-\cos t\,\mathbf{j}+\sin t\,\mathbf{k}$

15–16 Draw the projections of the curve on the three coordinate planes. Use these projections to help sketch the curve.

15. $\mathbf{r}(t)=\langle t, \sin t, 2\cos t\rangle$ **16.** $\mathbf{r}(t)=\langle t, t, t^2\rangle$

17–20 Find a vector equation and parametric equations for the line segment that joins P to Q.

17. $P(2, 0, 0)$, $Q(6, 2, -2)$ **18.** $P(-1, 2, -2)$, $Q(-3, 5, 1)$

19. $P(0, -1, 1)$, $Q\!\left(\tfrac{1}{2}, \tfrac{1}{3}, \tfrac{1}{4}\right)$ **20.** $P(a, b, c)$, $Q(u, v, w)$

21–26 Match the parametric equations with the graphs (labeled I–VI). Give reasons for your choices.

I

II

III

IV

V

VI

21. $x=t\cos t$, $y=t$, $z=t\sin t$, $t\geq 0$

22. $x=\cos t$, $y=\sin t$, $z=1/(1+t^2)$

23. $x=t$, $y=1/(1+t^2)$, $z=t^2$

24. $x=\cos t$, $y=\sin t$, $z=\cos 2t$

25. $x=\cos 8t$, $y=\sin 8t$, $z=e^{0.8t}$, $t\geq 0$

26. $x=\cos^2 t$, $y=\sin^2 t$, $z=t$

---

27. Show that the curve with parametric equations $x=t\cos t$, $y=t\sin t$, $z=t$ lies on the cone $z^2=x^2+y^2$, and use this fact to help sketch the curve.

28. Show that the curve with parametric equations $x=\sin t$, $y=\cos t$, $z=\sin^2 t$ is the curve of intersection of the surfaces $z=x^2$ and $x^2+y^2=1$. Use this fact to help sketch the curve.

29. Find three different surfaces that contain the curve $\mathbf{r}(t)=2t\mathbf{i}+e^t\mathbf{j}+e^{2t}\mathbf{k}$.

30. Find three different surfaces that contain the curve $\mathbf{r}(t)=t^2\mathbf{i}+\ln t\,\mathbf{j}+(1/t)\,\mathbf{k}$.

31. At what points does the curve $\mathbf{r}(t)=t\,\mathbf{i}+(2t-t^2)\,\mathbf{k}$ intersect the paraboloid $z=x^2+y^2$?

32. At what points does the helix $\mathbf{r}(t)=\langle \sin t, \cos t, t\rangle$ intersect the sphere $x^2+y^2+z^2=5$?

33–37 Use a computer to graph the curve with the given vector equation. Make sure you choose a parameter domain and viewpoints that reveal the true nature of the curve.

33. $\mathbf{r}(t)=\langle \cos t\sin 2t, \sin t\sin 2t, \cos 2t\rangle$

34. $\mathbf{r}(t)=\langle te^t, e^{-t}, t\rangle$

35. $\mathbf{r}(t)=\left\langle \sin 3t\cos t, \tfrac{1}{4}t, \sin 3t\sin t\right\rangle$

36. $\mathbf{r}(t)=\langle \cos(8\cos t)\sin t, \sin(8\cos t)\sin t, \cos t\rangle$

37. $\mathbf{r}(t)=\langle \cos 2t, \cos 3t, \cos 4t\rangle$

---

38. Graph the curve with parametric equations $x=\sin t$, $y=\sin 2t$, $z=\cos 4t$. Explain its shape by graphing its projections onto the three coordinate planes.

39. Graph the curve with parametric equations

$$x=(1+\cos 16t)\cos t$$

$$y=(1+\cos 16t)\sin t$$

$$z=1+\cos 16t$$

Explain the appearance of the graph by showing that it lies on a cone.

40. Graph the curve with parametric equations

$$x=\sqrt{1-0.25\cos^2 10t}\,\cos t$$

$$y=\sqrt{1-0.25\cos^2 10t}\,\sin t$$

$$z=0.5\cos 10t$$

Explain the appearance of the graph by showing that it lies on a sphere.

41. Show that the curve with parametric equations $x = t^2$, $y = 1 - 3t$, $z = 1 + t^3$ passes through the points $(1, 4, 0)$ and $(9, -8, 28)$ but not through the point $(4, 7, -6)$.

42–46 Find a vector function that represents the curve of intersection of the two surfaces.

42. The cylinder $x^2 + y^2 = 4$ and the surface $z = xy$

43. The cone $z = \sqrt{x^2 + y^2}$ and the plane $z = 1 + y$

44. The paraboloid $z = 4x^2 + y^2$ and the parabolic cylinder $y = x^2$

45. The hyperboloid $z = x^2 - y^2$ and the cylinder $x^2 + y^2 = 1$

46. The semiellipsoid $x^2 + y^2 + 4z^2 = 4$, $y \geq 0$, and the cylinder $x^2 + z^2 = 1$

47. Try to sketch by hand the curve of intersection of the circular cylinder $x^2 + y^2 = 4$ and the parabolic cylinder $z = x^2$. Then find parametric equations for this curve and use these equations and a computer to graph the curve.

48. Try to sketch by hand the curve of intersection of the parabolic cylinder $y = x^2$ and the top half of the ellipsoid $x^2 + 4y^2 + 4z^2 = 16$. Then find parametric equations for this curve and use these equations and a computer to graph the curve.

49. If two objects travel through space along two different curves, it's often important to know whether they will collide. (Will a missile hit its moving target? Will two aircraft collide?) The curves might intersect, but we need to know whether the objects are in the same position *at the same time*. Suppose the trajectories of two particles are given by the vector functions

$$\mathbf{r}_1(t) = \langle t^2, 7t - 12, t^2 \rangle \qquad \mathbf{r}_2(t) = \langle 4t - 3, t^2, 5t - 6 \rangle$$

for $t \geq 0$. Do the particles collide?

50. Two particles travel along the space curves

$$\mathbf{r}_1(t) = \langle t, t^2, t^3 \rangle \qquad \mathbf{r}_2(t) = \langle 1 + 2t, 1 + 6t, 1 + 14t \rangle$$

Do the particles collide? Do their paths intersect?

51. (a) Graph the curve with parametric equations

$$x = \tfrac{27}{26} \sin 8t - \tfrac{8}{39} \sin 18t$$

$$y = -\tfrac{27}{26} \cos 8t + \tfrac{8}{39} \cos 18t$$

$$z = \tfrac{144}{65} \sin 5t$$

(b) Show that the curve lies on the hyperboloid of one sheet $144x^2 + 144y^2 - 25z^2 = 100$.

52. The view of the trefoil knot shown in Figure 8 is accurate, but it doesn't reveal the whole story. Use the parametric equations

$$x = (2 + \cos 1.5t) \cos t$$

$$y = (2 + \cos 1.5t) \sin t$$

$$z = \sin 1.5t$$

to sketch the curve by hand as viewed from above, with gaps indicating where the curve passes over itself. Start by showing that the projection of the curve onto the xy-plane has polar coordinates $r = 2 + \cos 1.5t$ and $\theta = t$, so r varies between 1 and 3. Then show that z has maximum and minimum values when the projection is halfway between $r = 1$ and $r = 3$.

When you have finished your sketch, use a computer to draw the curve with viewpoint directly above and compare with your sketch. Then use the computer to draw the curve from several other viewpoints. You can get a better impression of the curve if you plot a tube with radius 0.2 around the curve. (Use the `tubeplot` command in Maple or the `tubecurve` or `Tube` command in Mathematica.)

53. Suppose $\mathbf{u}$ and $\mathbf{v}$ are vector functions that possess limits as $t \to a$ and let c be a constant. Prove the following properties of limits.

(a) $\lim\limits_{t \to a} [\mathbf{u}(t) + \mathbf{v}(t)] = \lim\limits_{t \to a} \mathbf{u}(t) + \lim\limits_{t \to a} \mathbf{v}(t)$

(b) $\lim\limits_{t \to a} c\mathbf{u}(t) = c \lim\limits_{t \to a} \mathbf{u}(t)$

(c) $\lim\limits_{t \to a} [\mathbf{u}(t) \cdot \mathbf{v}(t)] = \lim\limits_{t \to a} \mathbf{u}(t) \cdot \lim\limits_{t \to a} \mathbf{v}(t)$

(d) $\lim\limits_{t \to a} [\mathbf{u}(t) \times \mathbf{v}(t)] = \lim\limits_{t \to a} \mathbf{u}(t) \times \lim\limits_{t \to a} \mathbf{v}(t)$

54. Show that $\lim_{t \to a} \mathbf{r}(t) = \mathbf{b}$ if and only if for every $\varepsilon > 0$ there is a number $\delta > 0$ such that

$$\text{if } 0 < |t - a| < \delta \quad \text{then} \quad |\mathbf{r}(t) - \mathbf{b}| < \varepsilon$$

13.2 Derivatives and Integrals of Vector Functions

Later in this chapter we are going to use vector functions to describe the motion of planets and other objects through space. Here we prepare the way by developing the calculus of vector functions.

■ Derivatives

The derivative $\mathbf{r}'$ of a vector function $\mathbf{r}$ is defined in much the same way as for real-valued functions:

$$\boxed{1} \qquad \frac{d\mathbf{r}}{dt} = \mathbf{r}'(t) = \lim_{h \to 0} \frac{\mathbf{r}(t+h) - \mathbf{r}(t)}{h}$$

if this limit exists. The geometric significance of this definition is shown in Figure 1. If the points P and Q have position vectors $\mathbf{r}(t)$ and $\mathbf{r}(t+h)$, then $\overrightarrow{PQ}$ represents the vector $\mathbf{r}(t+h) - \mathbf{r}(t)$, which can therefore be regarded as a secant vector. If $h > 0$, the scalar multiple $(1/h)(\mathbf{r}(t+h) - \mathbf{r}(t))$ has the same direction as $\mathbf{r}(t+h) - \mathbf{r}(t)$. As $h \to 0$, it appears that this vector approaches a vector that lies on the tangent line. For this reason, the vector $\mathbf{r}'(t)$ is called the **tangent vector** to the curve defined by $\mathbf{r}$ at the point P, provided that $\mathbf{r}'(t)$ exists and $\mathbf{r}'(t) \neq \mathbf{0}$. The **tangent line** to C at P is defined to be the line through P parallel to the tangent vector $\mathbf{r}'(t)$. We will also have occasion to consider the **unit tangent vector**, which is

Notice that when $0 < h < 1$, multiplying the secant vector by $1/h$ *stretches* the vector, as shown in Figure 1(b).

$$\mathbf{T}(t) = \frac{\mathbf{r}'(t)}{|\mathbf{r}'(t)|}$$

TEC Visual 13.2 shows an animation of Figure 1.

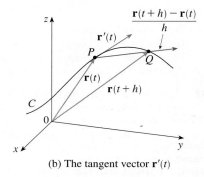

FIGURE 1

(a) The secant vector $\overrightarrow{PQ}$

(b) The tangent vector $\mathbf{r}'(t)$

The following theorem gives us a convenient method for computing the derivative of a vector function $\mathbf{r}$: just differentiate each component of $\mathbf{r}$.

$\boxed{2}$ **Theorem** If $\mathbf{r}(t) = \langle f(t), g(t), h(t) \rangle = f(t)\,\mathbf{i} + g(t)\,\mathbf{j} + h(t)\,\mathbf{k}$, where f, g, and h are differentiable functions, then

$$\mathbf{r}'(t) = \langle f'(t), g'(t), h'(t) \rangle = f'(t)\,\mathbf{i} + g'(t)\,\mathbf{j} + h'(t)\,\mathbf{k}$$

PROOF

$$\mathbf{r}'(t) = \lim_{\Delta t \to 0} \frac{1}{\Delta t} [\mathbf{r}(t + \Delta t) - \mathbf{r}(t)]$$

$$= \lim_{\Delta t \to 0} \frac{1}{\Delta t} [\langle f(t + \Delta t), g(t + \Delta t), h(t + \Delta t) \rangle - \langle f(t), g(t), h(t) \rangle]$$

$$= \lim_{\Delta t \to 0} \left\langle \frac{f(t + \Delta t) - f(t)}{\Delta t}, \frac{g(t + \Delta t) - g(t)}{\Delta t}, \frac{h(t + \Delta t) - h(t)}{\Delta t} \right\rangle$$

$$= \left\langle \lim_{\Delta t \to 0} \frac{f(t + \Delta t) - f(t)}{\Delta t}, \lim_{\Delta t \to 0} \frac{g(t + \Delta t) - g(t)}{\Delta t}, \lim_{\Delta t \to 0} \frac{h(t + \Delta t) - h(t)}{\Delta t} \right\rangle$$

$$= \langle f'(t), g'(t), h'(t) \rangle \qquad \blacksquare$$

EXAMPLE 1

(a) Find the derivative of $\mathbf{r}(t) = (1 + t^3)\mathbf{i} + te^{-t}\mathbf{j} + \sin 2t\,\mathbf{k}$.

(b) Find the unit tangent vector at the point where $t = 0$.

SOLUTION

(a) According to Theorem 2, we differentiate each component of $\mathbf{r}$:

$$\mathbf{r}'(t) = 3t^2\mathbf{i} + (1 - t)e^{-t}\mathbf{j} + 2\cos 2t\,\mathbf{k}$$

(b) Since $\mathbf{r}(0) = \mathbf{i}$ and $\mathbf{r}'(0) = \mathbf{j} + 2\mathbf{k}$, the unit tangent vector at the point $(1, 0, 0)$ is

$$\mathbf{T}(0) = \frac{\mathbf{r}'(0)}{|\mathbf{r}'(0)|} = \frac{\mathbf{j} + 2\mathbf{k}}{\sqrt{1 + 4}} = \frac{1}{\sqrt{5}}\mathbf{j} + \frac{2}{\sqrt{5}}\mathbf{k} \qquad \blacksquare$$

EXAMPLE 2 For the curve $\mathbf{r}(t) = \sqrt{t}\,\mathbf{i} + (2 - t)\,\mathbf{j}$, find $\mathbf{r}'(t)$ and sketch the position vector $\mathbf{r}(1)$ and the tangent vector $\mathbf{r}'(1)$.

SOLUTION We have

$$\mathbf{r}'(t) = \frac{1}{2\sqrt{t}}\mathbf{i} - \mathbf{j} \qquad \text{and} \qquad \mathbf{r}'(1) = \frac{1}{2}\mathbf{i} - \mathbf{j}$$

The curve is a plane curve and elimination of the parameter from the equations $x = \sqrt{t}$, $y = 2 - t$ gives $y = 2 - x^2$, $x \geqslant 0$. In Figure 2 we draw the position vector $\mathbf{r}(1) = \mathbf{i} + \mathbf{j}$ starting at the origin and the tangent vector $\mathbf{r}'(1)$ starting at the corresponding point $(1, 1)$. $\qquad \blacksquare$

EXAMPLE 3 Find parametric equations for the tangent line to the helix with parametric equations

$$x = 2\cos t \qquad y = \sin t \qquad z = t$$

at the point $(0, 1, \pi/2)$.

SOLUTION The vector equation of the helix is $\mathbf{r}(t) = \langle 2\cos t, \sin t, t \rangle$, so

$$\mathbf{r}'(t) = \langle -2\sin t, \cos t, 1 \rangle$$

The parameter value corresponding to the point $(0, 1, \pi/2)$ is $t = \pi/2$, so the tangent vector there is $\mathbf{r}'(\pi/2) = \langle -2, 0, 1 \rangle$. The tangent line is the line through $(0, 1, \pi/2)$ parallel to the vector $\langle -2, 0, 1 \rangle$, so by Equations 12.5.2 its parametric equations are

$$x = -2t \qquad y = 1 \qquad z = \frac{\pi}{2} + t \qquad \blacksquare$$

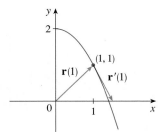

FIGURE 2

Notice from Figure 2 that the tangent vector points in the direction of increasing t. (See Exercise 58.)

The helix and the tangent line in Example 3 are shown in Figure 3.

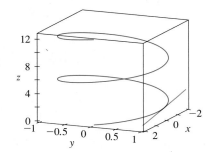

FIGURE 3

In Section 13.4 we will see how $\mathbf{r}'(t)$ and $\mathbf{r}''(t)$ can be interpreted as the velocity and acceleration vectors of a particle moving through space with position vector $\mathbf{r}(t)$ at time t.

Just as for real-valued functions, the **second derivative** of a vector function $\mathbf{r}$ is the derivative of $\mathbf{r}'$, that is, $\mathbf{r}'' = (\mathbf{r}')'$. For instance, the second derivative of the function in Example 3 is

$$\mathbf{r}''(t) = \langle -2\cos t, -\sin t, 0 \rangle$$

■ Differentiation Rules

The next theorem shows that the differentiation formulas for real-valued functions have their counterparts for vector-valued functions.

> **3** **Theorem** Suppose $\mathbf{u}$ and $\mathbf{v}$ are differentiable vector functions, c is a scalar, and f is a real-valued function. Then
>
> 1. $\dfrac{d}{dt}[\mathbf{u}(t) + \mathbf{v}(t)] = \mathbf{u}'(t) + \mathbf{v}'(t)$
>
> 2. $\dfrac{d}{dt}[c\mathbf{u}(t)] = c\mathbf{u}'(t)$
>
> 3. $\dfrac{d}{dt}[f(t)\,\mathbf{u}(t)] = f'(t)\,\mathbf{u}(t) + f(t)\,\mathbf{u}'(t)$
>
> 4. $\dfrac{d}{dt}[\mathbf{u}(t) \cdot \mathbf{v}(t)] = \mathbf{u}'(t) \cdot \mathbf{v}(t) + \mathbf{u}(t) \cdot \mathbf{v}'(t)$
>
> 5. $\dfrac{d}{dt}[\mathbf{u}(t) \times \mathbf{v}(t)] = \mathbf{u}'(t) \times \mathbf{v}(t) + \mathbf{u}(t) \times \mathbf{v}'(t)$
>
> 6. $\dfrac{d}{dt}[\mathbf{u}(f(t))] = f'(t)\,\mathbf{u}'(f(t))$ (Chain Rule)

This theorem can be proved either directly from Definition 1 or by using Theorem 2 and the corresponding differentiation formulas for real-valued functions. The proof of Formula 4 follows; the remaining formulas are left as exercises.

PROOF OF FORMULA 4 Let

$$\mathbf{u}(t) = \langle f_1(t), f_2(t), f_3(t) \rangle \qquad \mathbf{v}(t) = \langle g_1(t), g_2(t), g_3(t) \rangle$$

Then
$$\mathbf{u}(t) \cdot \mathbf{v}(t) = f_1(t)\,g_1(t) + f_2(t)\,g_2(t) + f_3(t)\,g_3(t) = \sum_{i=1}^{3} f_i(t)\,g_i(t)$$

so the ordinary Product Rule gives

$$\frac{d}{dt}[\mathbf{u}(t) \cdot \mathbf{v}(t)] = \frac{d}{dt}\sum_{i=1}^{3} f_i(t)\,g_i(t) = \sum_{i=1}^{3}\frac{d}{dt}[f_i(t)\,g_i(t)]$$

$$= \sum_{i=1}^{3}[f_i'(t)\,g_i(t) + f_i(t)\,g_i'(t)]$$

$$= \sum_{i=1}^{3} f_i'(t)\,g_i(t) + \sum_{i=1}^{3} f_i(t)\,g_i'(t)$$

$$= \mathbf{u}'(t) \cdot \mathbf{v}(t) + \mathbf{u}(t) \cdot \mathbf{v}'(t) \qquad ■$$

EXAMPLE 4 Show that if $|\mathbf{r}(t)| = c$ (a constant), then $\mathbf{r}'(t)$ is orthogonal to $\mathbf{r}(t)$ for all t.

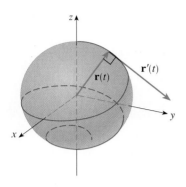

FIGURE 4

SOLUTION Since

$$\mathbf{r}(t) \cdot \mathbf{r}(t) = |\mathbf{r}(t)|^2 = c^2$$

and c^2 is a constant, Formula 4 of Theorem 3 gives

$$0 = \frac{d}{dt}[\mathbf{r}(t) \cdot \mathbf{r}(t)] = \mathbf{r}'(t) \cdot \mathbf{r}(t) + \mathbf{r}(t) \cdot \mathbf{r}'(t) = 2\mathbf{r}'(t) \cdot \mathbf{r}(t)$$

Thus $\mathbf{r}'(t) \cdot \mathbf{r}(t) = 0$, which says that $\mathbf{r}'(t)$ is orthogonal to $\mathbf{r}(t)$.

Geometrically, this result says that if a curve lies on a sphere with center the origin, then the tangent vector $\mathbf{r}'(t)$ is always perpendicular to the position vector $\mathbf{r}(t)$. (See Figure 4.) ∎

■ Integrals

The **definite integral** of a continuous vector function $\mathbf{r}(t)$ can be defined in much the same way as for real-valued functions except that the integral is a vector. But then we can express the integral of $\mathbf{r}$ in terms of the integrals of its component functions f, g, and h as follows. (We use the notation of Chapter 5.)

$$\int_a^b \mathbf{r}(t)\, dt = \lim_{n \to \infty} \sum_{i=1}^n \mathbf{r}(t_i^*)\, \Delta t$$

$$= \lim_{n \to \infty} \left[\left(\sum_{i=1}^n f(t_i^*)\, \Delta t \right) \mathbf{i} + \left(\sum_{i=1}^n g(t_i^*)\, \Delta t \right) \mathbf{j} + \left(\sum_{i=1}^n h(t_i^*)\, \Delta t \right) \mathbf{k} \right]$$

and so

$$\int_a^b \mathbf{r}(t)\, dt = \left(\int_a^b f(t)\, dt \right) \mathbf{i} + \left(\int_a^b g(t)\, dt \right) \mathbf{j} + \left(\int_a^b h(t)\, dt \right) \mathbf{k}$$

This means that we can evaluate an integral of a vector function by integrating each component function.

We can extend the Fundamental Theorem of Calculus to continuous vector functions as follows:

$$\int_a^b \mathbf{r}(t)\, dt = \mathbf{R}(t) \Big]_a^b = \mathbf{R}(b) - \mathbf{R}(a)$$

where $\mathbf{R}$ is an antiderivative of $\mathbf{r}$, that is, $\mathbf{R}'(t) = \mathbf{r}(t)$. We use the notation $\int \mathbf{r}(t)\, dt$ for indefinite integrals (antiderivatives).

EXAMPLE 5 If $\mathbf{r}(t) = 2\cos t\, \mathbf{i} + \sin t\, \mathbf{j} + 2t\, \mathbf{k}$, then

$$\int \mathbf{r}(t)\, dt = \left(\int 2\cos t\, dt \right) \mathbf{i} + \left(\int \sin t\, dt \right) \mathbf{j} + \left(\int 2t\, dt \right) \mathbf{k}$$

$$= 2\sin t\, \mathbf{i} - \cos t\, \mathbf{j} + t^2\, \mathbf{k} + \mathbf{C}$$

where $\mathbf{C}$ is a vector constant of integration, and

$$\int_0^{\pi/2} \mathbf{r}(t)\, dt = \left[2\sin t\, \mathbf{i} - \cos t\, \mathbf{j} + t^2\, \mathbf{k} \right]_0^{\pi/2} = 2\mathbf{i} + \mathbf{j} + \frac{\pi^2}{4}\mathbf{k} \qquad ∎$$

13.2 EXERCISES

1. The figure shows a curve C given by a vector function $\mathbf{r}(t)$.
   (a) Draw the vectors $\mathbf{r}(4.5) - \mathbf{r}(4)$ and $\mathbf{r}(4.2) - \mathbf{r}(4)$.
   (b) Draw the vectors

$$\frac{\mathbf{r}(4.5) - \mathbf{r}(4)}{0.5} \quad \text{and} \quad \frac{\mathbf{r}(4.2) - \mathbf{r}(4)}{0.2}$$

   (c) Write expressions for $\mathbf{r}'(4)$ and the unit tangent vector $\mathbf{T}(4)$.
   (d) Draw the vector $\mathbf{T}(4)$.

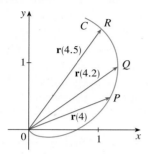

2. (a) Make a large sketch of the curve described by the vector function $\mathbf{r}(t) = \langle t^2, t \rangle$, $0 \le t \le 2$, and draw the vectors $\mathbf{r}(1)$, $\mathbf{r}(1.1)$, and $\mathbf{r}(1.1) - \mathbf{r}(1)$.
   (b) Draw the vector $\mathbf{r}'(1)$ starting at $(1, 1)$, and compare it with the vector

$$\frac{\mathbf{r}(1.1) - \mathbf{r}(1)}{0.1}$$

   Explain why these vectors are so close to each other in length and direction.

3–8
(a) Sketch the plane curve with the given vector equation.
(b) Find $\mathbf{r}'(t)$.
(c) Sketch the position vector $\mathbf{r}(t)$ and the tangent vector $\mathbf{r}'(t)$ for the given value of t.

3. $\mathbf{r}(t) = \langle t - 2, t^2 + 1 \rangle$, $t = -1$

4. $\mathbf{r}(t) = \langle t^2, t^3 \rangle$, $t = 1$

5. $\mathbf{r}(t) = e^{2t}\,\mathbf{i} + e^t\,\mathbf{j}$, $t = 0$

6. $\mathbf{r}(t) = e^t\,\mathbf{i} + 2t\,\mathbf{j}$, $t = 0$

7. $\mathbf{r}(t) = 4 \sin t\,\mathbf{i} - 2 \cos t\,\mathbf{j}$, $t = 3\pi/4$

8. $\mathbf{r}(t) = (\cos t + 1)\,\mathbf{i} + (\sin t - 1)\,\mathbf{j}$, $t = -\pi/3$

9–16 Find the derivative of the vector function.

9. $\mathbf{r}(t) = \langle \sqrt{t-2}, 3, 1/t^2 \rangle$

10. $\mathbf{r}(t) = \langle e^{-t}, t - t^3, \ln t \rangle$

11. $\mathbf{r}(t) = t^2\,\mathbf{i} + \cos(t^2)\,\mathbf{j} + \sin^2 t\,\mathbf{k}$

12. $\mathbf{r}(t) = \dfrac{1}{1+t}\,\mathbf{i} + \dfrac{t}{1+t}\,\mathbf{j} + \dfrac{t^2}{1+t}\,\mathbf{k}$

13. $\mathbf{r}(t) = t \sin t\,\mathbf{i} + e^t \cos t\,\mathbf{j} + \sin t \cos t\,\mathbf{k}$

14. $\mathbf{r}(t) = \sin^2 at\,\mathbf{i} + te^{bt}\,\mathbf{j} + \cos^2 ct\,\mathbf{k}$

15. $\mathbf{r}(t) = \mathbf{a} + t\,\mathbf{b} + t^2\,\mathbf{c}$

16. $\mathbf{r}(t) = t\,\mathbf{a} \times (\mathbf{b} + t\,\mathbf{c})$

17–20 Find the unit tangent vector $\mathbf{T}(t)$ at the point with the given value of the parameter t.

17. $\mathbf{r}(t) = \langle t^2 - 2t, 1 + 3t, \frac{1}{3}t^3 + \frac{1}{2}t^2 \rangle$, $t = 2$

18. $\mathbf{r}(t) = \langle \tan^{-1} t, 2e^{2t}, 8te^t \rangle$, $t = 0$

19. $\mathbf{r}(t) = \cos t\,\mathbf{i} + 3t\,\mathbf{j} + 2 \sin 2t\,\mathbf{k}$, $t = 0$

20. $\mathbf{r}(t) = \sin^2 t\,\mathbf{i} + \cos^2 t\,\mathbf{j} + \tan^2 t\,\mathbf{k}$, $t = \pi/4$

21. If $\mathbf{r}(t) = \langle t, t^2, t^3 \rangle$, find $\mathbf{r}'(t)$, $\mathbf{T}(1)$, $\mathbf{r}''(t)$, and $\mathbf{r}'(t) \times \mathbf{r}''(t)$.

22. If $\mathbf{r}(t) = \langle e^{2t}, e^{-2t}, te^{2t} \rangle$, find $\mathbf{T}(0)$, $\mathbf{r}''(0)$, and $\mathbf{r}'(t) \cdot \mathbf{r}''(t)$.

23–26 Find parametric equations for the tangent line to the curve with the given parametric equations at the specified point.

23. $x = t^2 + 1$, $y = 4\sqrt{t}$, $z = e^{t^2 - t}$; $(2, 4, 1)$

24. $x = \ln(t + 1)$, $y = t \cos 2t$, $z = 2^t$; $(0, 0, 1)$

25. $x = e^{-t} \cos t$, $y = e^{-t} \sin t$, $z = e^{-t}$; $(1, 0, 1)$

26. $x = \sqrt{t^2 + 3}$, $y = \ln(t^2 + 3)$, $z = t$; $(2, \ln 4, 1)$

27. Find a vector equation for the tangent line to the curve of intersection of the cylinders $x^2 + y^2 = 25$ and $y^2 + z^2 = 20$ at the point $(3, 4, 2)$.

28. Find the point on the curve $\mathbf{r}(t) = \langle 2 \cos t, 2 \sin t, e^t \rangle$, $0 \le t \le \pi$, where the tangent line is parallel to the plane $\sqrt{3}\,x + y = 1$.

CAS 29–31 Find parametric equations for the tangent line to the curve with the given parametric equations at the specified point. Illustrate by graphing both the curve and the tangent line on a common screen.

29. $x = t$, $y = e^{-t}$, $z = 2t - t^2$; $(0, 1, 0)$

30. $x = 2 \cos t$, $y = 2 \sin t$, $z = 4 \cos 2t$; $(\sqrt{3}, 1, 2)$

31. $x = t \cos t$, $y = t$, $z = t \sin t$; $(-\pi, \pi, 0)$

32. (a) Find the point of intersection of the tangent lines to the curve $\mathbf{r}(t) = \langle \sin \pi t, 2 \sin \pi t, \cos \pi t \rangle$ at the points where $t = 0$ and $t = 0.5$.
   (b) Illustrate by graphing the curve and both tangent lines.

33. The curves $\mathbf{r}_1(t) = \langle t, t^2, t^3 \rangle$ and $\mathbf{r}_2(t) = \langle \sin t, \sin 2t, t \rangle$ intersect at the origin. Find their angle of intersection correct to the nearest degree.

34. At what point do the curves $r_1(t) = \langle t, 1 - t, 3 + t^2 \rangle$ and $r_2(s) = \langle 3 - s, s - 2, s^2 \rangle$ intersect? Find their angle of intersection correct to the nearest degree.

35–40 Evaluate the integral.

35. $\int_0^2 (t\,\mathbf{i} - t^3\,\mathbf{j} + 3t^5\,\mathbf{k})\,dt$

36. $\int_1^4 \left(2t^{3/2}\,\mathbf{i} + (t + 1)\sqrt{t}\,\mathbf{k}\right)dt$

37. $\int_0^1 \left(\frac{1}{t + 1}\,\mathbf{i} + \frac{1}{t^2 + 1}\,\mathbf{j} + \frac{t}{t^2 + 1}\,\mathbf{k}\right)dt$

38. $\int_0^{\pi/4} (\sec t \tan t\,\mathbf{i} + t \cos 2t\,\mathbf{j} + \sin^2 2t \cos 2t\,\mathbf{k})\,dt$

39. $\int (\sec^2 t\,\mathbf{i} + t(t^2 + 1)^3\,\mathbf{j} + t^2 \ln t\,\mathbf{k})\,dt$

40. $\int \left(te^{2t}\,\mathbf{i} + \frac{t}{1 - t}\,\mathbf{j} + \frac{1}{\sqrt{1 - t^2}}\,\mathbf{k}\right)dt$

41. Find $r(t)$ if $r'(t) = 2t\,\mathbf{i} + 3t^2\,\mathbf{j} + \sqrt{t}\,\mathbf{k}$ and $r(1) = \mathbf{i} + \mathbf{j}$.

42. Find $r(t)$ if $r'(t) = t\,\mathbf{i} + e^t\,\mathbf{j} + te^t\,\mathbf{k}$ and $r(0) = \mathbf{i} + \mathbf{j} + \mathbf{k}$.

43. Prove Formula 1 of Theorem 3.

44. Prove Formula 3 of Theorem 3.

45. Prove Formula 5 of Theorem 3.

46. Prove Formula 6 of Theorem 3.

47. If $u(t) = \langle \sin t, \cos t, t \rangle$ and $v(t) = \langle t, \cos t, \sin t \rangle$, use Formula 4 of Theorem 3 to find
$$\frac{d}{dt}[u(t) \cdot v(t)]$$

48. If u and v are the vector functions in Exercise 47, use Formula 5 of Theorem 3 to find
$$\frac{d}{dt}[u(t) \times v(t)]$$

49. Find $f'(2)$, where $f(t) = u(t) \cdot v(t)$, $u(2) = \langle 1, 2, -1 \rangle$, $u'(2) = \langle 3, 0, 4 \rangle$, and $v(t) = \langle t, t^2, t^3 \rangle$.

50. If $r(t) = u(t) \times v(t)$, where u and v are the vector functions in Exercise 49, find $r'(2)$.

51. If $r(t) = a \cos \omega t + b \sin \omega t$, where a and b are constant vectors, show that $r(t) \times r'(t) = \omega a \times b$.

52. If r is the vector function in Exercise 51, show that $r''(t) + \omega^2 r(t) = 0$.

53. Show that if r is a vector function such that r'' exists, then
$$\frac{d}{dt}[r(t) \times r'(t)] = r(t) \times r''(t)$$

54. Find an expression for $\dfrac{d}{dt}[u(t) \cdot (v(t) \times w(t))]$.

55. If $r(t) \neq 0$, show that $\dfrac{d}{dt}|r(t)| = \dfrac{1}{|r(t)|}\,r(t) \cdot r'(t)$.

[*Hint:* $|r(t)|^2 = r(t) \cdot r(t)$]

56. If a curve has the property that the position vector $r(t)$ is always perpendicular to the tangent vector $r'(t)$, show that the curve lies on a sphere with center the origin.

57. If $u(t) = r(t) \cdot [r'(t) \times r''(t)]$, show that
$$u'(t) = r(t) \cdot [r'(t) \times r'''(t)]$$

58. Show that the tangent vector to a curve defined by a vector function $r(t)$ points in the direction of increasing t. [*Hint:* Refer to Figure 1 and consider the cases $h > 0$ and $h < 0$ separately.]

13.3 Arc Length and Curvature

FIGURE 1
The length of a space curve is the limit of lengths of inscribed polygons.

■ Length of a Curve

In Section 10.2 we defined the length of a plane curve with parametric equations $x = f(t)$, $y = g(t)$, $a \leqslant t \leqslant b$, as the limit of lengths of inscribed polygons and, for the case where f' and g' are continuous, we arrived at the formula

$$\boxed{1} \qquad L = \int_a^b \sqrt{[f'(t)]^2 + [g'(t)]^2}\,dt = \int_a^b \sqrt{\left(\frac{dx}{dt}\right)^2 + \left(\frac{dy}{dt}\right)^2}\,dt$$

The length of a space curve is defined in exactly the same way (see Figure 1). Suppose that the curve has the vector equation $r(t) = \langle f(t), g(t), h(t) \rangle$, $a \leqslant t \leqslant b$, or, equivalently, the parametric equations $x = f(t)$, $y = g(t)$, $z = h(t)$, where f', g', and h' are continuous. If the curve is traversed exactly once as t increases from a to b, then it can be shown that its length is

$$\boxed{2} \qquad \begin{aligned} L &= \int_a^b \sqrt{[f'(t)]^2 + [g'(t)]^2 + [h'(t)]^2}\; dt \\ &= \int_a^b \sqrt{\left(\frac{dx}{dt}\right)^2 + \left(\frac{dy}{dt}\right)^2 + \left(\frac{dz}{dt}\right)^2}\; dt \end{aligned}$$

Notice that both of the arc length formulas (1) and (2) can be put into the more compact form

In the next section we will see that if $\mathbf{r}(t)$ is the position vector of a moving object at time t, then $\mathbf{r}'(t)$ is the velocity vector and $|\mathbf{r}'(t)|$ is the speed. Thus Equation 3 says that to compute distance traveled, we integrate speed.

$$\boxed{3} \qquad L = \int_a^b |\mathbf{r}'(t)|\; dt$$

because, for plane curves $\mathbf{r}(t) = f(t)\,\mathbf{i} + g(t)\,\mathbf{j}$,

$$|\mathbf{r}'(t)| = |f'(t)\,\mathbf{i} + g'(t)\,\mathbf{j}| = \sqrt{[f'(t)]^2 + [g'(t)]^2}$$

and for space curves $\mathbf{r}(t) = f(t)\,\mathbf{i} + g(t)\,\mathbf{j} + h(t)\,\mathbf{k}$,

$$|\mathbf{r}'(t)| = |f'(t)\,\mathbf{i} + g'(t)\,\mathbf{j} + h'(t)\,\mathbf{k}| = \sqrt{[f'(t)]^2 + [g'(t)]^2 + [h'(t)]^2}$$

Figure 2 shows the arc of the helix whose length is computed in Example 1.

EXAMPLE 1 Find the length of the arc of the circular helix with vector equation $\mathbf{r}(t) = \cos t\,\mathbf{i} + \sin t\,\mathbf{j} + t\,\mathbf{k}$ from the point $(1, 0, 0)$ to the point $(1, 0, 2\pi)$.

SOLUTION Since $\mathbf{r}'(t) = -\sin t\,\mathbf{i} + \cos t\,\mathbf{j} + \mathbf{k}$, we have

$$|\mathbf{r}'(t)| = \sqrt{(-\sin t)^2 + \cos^2 t + 1} = \sqrt{2}$$

The arc from $(1, 0, 0)$ to $(1, 0, 2\pi)$ is described by the parameter interval $0 \le t \le 2\pi$ and so, from Formula 3, we have

$$L = \int_0^{2\pi} |\mathbf{r}'(t)|\; dt = \int_0^{2\pi} \sqrt{2}\; dt = 2\sqrt{2}\,\pi \qquad \blacksquare$$

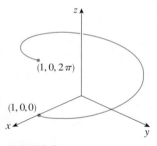

FIGURE 2

A single curve C can be represented by more than one vector function. For instance, the twisted cubic

$$\boxed{4} \qquad \mathbf{r}_1(t) = \langle t, t^2, t^3 \rangle \qquad 1 \le t \le 2$$

could also be represented by the function

$$\boxed{5} \qquad \mathbf{r}_2(u) = \langle e^u, e^{2u}, e^{3u} \rangle \qquad 0 \le u \le \ln 2$$

where the connection between the parameters t and u is given by $t = e^u$. We say that Equations 4 and 5 are **parametrizations** of the curve C. If we were to use Equation 3 to compute the length of C using Equations 4 and 5, we would get the same answer. In general, it can be shown that when Equation 3 is used to compute arc length, the answer is independent of the parametrization that is used.

▮ The Arc Length Function

Now we suppose that C is a curve given by a vector function

$$\mathbf{r}(t) = f(t)\,\mathbf{i} + g(t)\,\mathbf{j} + h(t)\,\mathbf{k} \qquad a \le t \le b$$

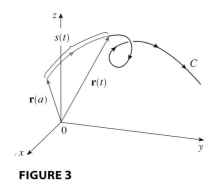

FIGURE 3

where $\mathbf{r}'$ is continuous and C is traversed exactly once as t increases from a to b. We define its **arc length function** s by

$$\boxed{6} \qquad s(t) = \int_a^t |\mathbf{r}'(u)|\, du = \int_a^t \sqrt{\left(\frac{dx}{du}\right)^2 + \left(\frac{dy}{du}\right)^2 + \left(\frac{dz}{du}\right)^2}\; du$$

Thus $s(t)$ is the length of the part of C between $\mathbf{r}(a)$ and $\mathbf{r}(t)$. (See Figure 3.) If we differentiate both sides of Equation 6 using Part 1 of the Fundamental Theorem of Calculus, we obtain

$$\boxed{7} \qquad \frac{ds}{dt} = |\mathbf{r}'(t)|$$

It is often useful to **parametrize a curve with respect to arc length** because arc length arises naturally from the shape of the curve and does not depend on a particular coordinate system. If a curve $\mathbf{r}(t)$ is already given in terms of a parameter t and $s(t)$ is the arc length function given by Equation 6, then we may be able to solve for t as a function of s: $t = t(s)$. Then the curve can be reparametrized in terms of s by substituting for t: $\mathbf{r} = \mathbf{r}(t(s))$. Thus, if $s = 3$ for instance, $\mathbf{r}(t(3))$ is the position vector of the point 3 units of length along the curve from its starting point.

EXAMPLE 2 Reparametrize the helix $\mathbf{r}(t) = \cos t\, \mathbf{i} + \sin t\, \mathbf{j} + t\, \mathbf{k}$ with respect to arc length measured from $(1, 0, 0)$ in the direction of increasing t.

SOLUTION The initial point $(1, 0, 0)$ corresponds to the parameter value $t = 0$. From Example 1 we have

$$\frac{ds}{dt} = |\mathbf{r}'(t)| = \sqrt{2}$$

and so

$$s = s(t) = \int_0^t |\mathbf{r}'(u)|\, du = \int_0^t \sqrt{2}\; du = \sqrt{2}\, t$$

Therefore $t = s/\sqrt{2}$ and the required reparametrization is obtained by substituting for t:

$$\mathbf{r}(t(s)) = \cos\!\left(s/\sqrt{2}\right) \mathbf{i} + \sin\!\left(s/\sqrt{2}\right) \mathbf{j} + \left(s/\sqrt{2}\right) \mathbf{k} \qquad \blacksquare$$

▮ Curvature

TEC Visual 13.3A shows animated unit tangent vectors, like those in Figure 4, for a variety of plane curves and space curves.

FIGURE 4
Unit tangent vectors at equally spaced points on C

A parametrization $\mathbf{r}(t)$ is called **smooth** on an interval I if $\mathbf{r}'$ is continuous and $\mathbf{r}'(t) \neq \mathbf{0}$ on I. A curve is called **smooth** if it has a smooth parametrization. A smooth curve has no sharp corners or cusps; when the tangent vector turns, it does so continuously.

If C is a smooth curve defined by the vector function $\mathbf{r}$, recall that the unit tangent vector $\mathbf{T}(t)$ is given by

$$\mathbf{T}(t) = \frac{\mathbf{r}'(t)}{|\mathbf{r}'(t)|}$$

and indicates the direction of the curve. From Figure 4 you can see that $\mathbf{T}(t)$ changes direction very slowly when C is fairly straight, but it changes direction more quickly when C bends or twists more sharply.

The curvature of C at a given point is a measure of how quickly the curve changes direction at that point. Specifically, we define it to be the magnitude of the rate of change of the unit tangent vector with respect to arc length. (We use arc length so that the curvature will be independent of the parametrization.) Because the unit tangent vector has constant length, only changes in direction contribute to the rate of change of $\mathbf{T}$.

> **8** **Definition** The **curvature** of a curve is
> $$\kappa = \left| \frac{d\mathbf{T}}{ds} \right|$$
> where $\mathbf{T}$ is the unit tangent vector.

The curvature is easier to compute if it is expressed in terms of the parameter t instead of s, so we use the Chain Rule (Theorem 13.2.3, Formula 6) to write

$$\frac{d\mathbf{T}}{dt} = \frac{d\mathbf{T}}{ds}\frac{ds}{dt} \quad \text{and} \quad \kappa = \left| \frac{d\mathbf{T}}{ds} \right| = \left| \frac{d\mathbf{T}/dt}{ds/dt} \right|$$

But $ds/dt = |\mathbf{r}'(t)|$ from Equation 7, so

> **9**
> $$\kappa(t) = \frac{|\mathbf{T}'(t)|}{|\mathbf{r}'(t)|}$$

EXAMPLE 3 Show that the curvature of a circle of radius a is $1/a$.

SOLUTION We can take the circle to have center the origin, and then a parametrization is

$$\mathbf{r}(t) = a\cos t\,\mathbf{i} + a\sin t\,\mathbf{j}$$

Therefore $\mathbf{r}'(t) = -a\sin t\,\mathbf{i} + a\cos t\,\mathbf{j}$ and $|\mathbf{r}'(t)| = a$

so $$\mathbf{T}(t) = \frac{\mathbf{r}'(t)}{|\mathbf{r}'(t)|} = -\sin t\,\mathbf{i} + \cos t\,\mathbf{j}$$

and $$\mathbf{T}'(t) = -\cos t\,\mathbf{i} - \sin t\,\mathbf{j}$$

This gives $|\mathbf{T}'(t)| = 1$, so using Formula 9, we have

$$\kappa(t) = \frac{|\mathbf{T}'(t)|}{|\mathbf{r}'(t)|} = \frac{1}{a}$$ ∎

The result of Example 3 shows that small circles have large curvature and large circles have small curvature, in accordance with our intuition. We can see directly from the definition of curvature that the curvature of a straight line is always 0 because the tangent vector is constant.

Although Formula 9 can be used in all cases to compute the curvature, the formula given by the following theorem is often more convenient to apply.

> **10** **Theorem** The curvature of the curve given by the vector function $\mathbf{r}$ is
> $$\kappa(t) = \frac{|\mathbf{r}'(t) \times \mathbf{r}''(t)|}{|\mathbf{r}'(t)|^3}$$

PROOF Since $\mathbf{T} = \mathbf{r}'/|\mathbf{r}'|$ and $|\mathbf{r}'| = ds/dt$, we have

$$\mathbf{r}' = |\mathbf{r}'|\mathbf{T} = \frac{ds}{dt}\mathbf{T}$$

so the Product Rule (Theorem 13.2.3, Formula 3) gives

$$\mathbf{r}'' = \frac{d^2s}{dt^2}\mathbf{T} + \frac{ds}{dt}\mathbf{T}'$$

Using the fact that $\mathbf{T} \times \mathbf{T} = \mathbf{0}$ (see Example 12.4.2), we have

$$\mathbf{r}' \times \mathbf{r}'' = \left(\frac{ds}{dt}\right)^2 (\mathbf{T} \times \mathbf{T}')$$

Now $|\mathbf{T}(t)| = 1$ for all t, so $\mathbf{T}$ and $\mathbf{T}'$ are orthogonal by Example 13.2.4. Therefore, by Theorem 12.4.9,

$$|\mathbf{r}' \times \mathbf{r}''| = \left(\frac{ds}{dt}\right)^2 |\mathbf{T} \times \mathbf{T}'| = \left(\frac{ds}{dt}\right)^2 |\mathbf{T}||\mathbf{T}'| = \left(\frac{ds}{dt}\right)^2 |\mathbf{T}'|$$

Thus

$$|\mathbf{T}'| = \frac{|\mathbf{r}' \times \mathbf{r}''|}{(ds/dt)^2} = \frac{|\mathbf{r}' \times \mathbf{r}''|}{|\mathbf{r}'|^2}$$

and

$$\kappa = \frac{|\mathbf{T}'|}{|\mathbf{r}'|} = \frac{|\mathbf{r}' \times \mathbf{r}''|}{|\mathbf{r}'|^3} \qquad \blacksquare$$

EXAMPLE 4 Find the curvature of the twisted cubic $\mathbf{r}(t) = \langle t, t^2, t^3 \rangle$ at a general point and at $(0, 0, 0)$.

SOLUTION We first compute the required ingredients:

$$\mathbf{r}'(t) = \langle 1, 2t, 3t^2 \rangle \qquad \mathbf{r}''(t) = \langle 0, 2, 6t \rangle$$

$$|\mathbf{r}'(t)| = \sqrt{1 + 4t^2 + 9t^4}$$

$$\mathbf{r}'(t) \times \mathbf{r}''(t) = \begin{vmatrix} \mathbf{i} & \mathbf{j} & \mathbf{k} \\ 1 & 2t & 3t^2 \\ 0 & 2 & 6t \end{vmatrix} = 6t^2\,\mathbf{i} - 6t\,\mathbf{j} + 2\,\mathbf{k}$$

$$|\mathbf{r}'(t) \times \mathbf{r}''(t)| = \sqrt{36t^4 + 36t^2 + 4} = 2\sqrt{9t^4 + 9t^2 + 1}$$

Theorem 10 then gives

$$\kappa(t) = \frac{|\mathbf{r}'(t) \times \mathbf{r}''(t)|}{|\mathbf{r}'(t)|^3} = \frac{2\sqrt{1 + 9t^2 + 9t^4}}{(1 + 4t^2 + 9t^4)^{3/2}}$$

At the origin, where $t = 0$, the curvature is $\kappa(0) = 2$. $\qquad \blacksquare$

For the special case of a plane curve with equation $y = f(x)$, we choose x as the parameter and write $\mathbf{r}(x) = x\,\mathbf{i} + f(x)\,\mathbf{j}$. Then $\mathbf{r}'(x) = \mathbf{i} + f'(x)\,\mathbf{j}$ and $\mathbf{r}''(x) = f''(x)\,\mathbf{j}$. Since $\mathbf{i} \times \mathbf{j} = \mathbf{k}$ and $\mathbf{j} \times \mathbf{j} = \mathbf{0}$, it follows that $\mathbf{r}'(x) \times \mathbf{r}''(x) = f''(x)\,\mathbf{k}$. We also have $|\mathbf{r}'(x)| = \sqrt{1 + [f'(x)]^2}$ and so, by Theorem 10,

11

$$\kappa(x) = \frac{|f''(x)|}{[1 + (f'(x))^2]^{3/2}}$$

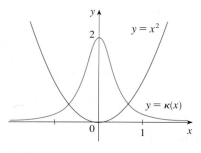

FIGURE 5
The parabola $y = x^2$ and its curvature function

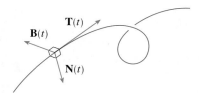

FIGURE 6

Figure 7 illustrates Example 6 by showing the vectors **T**, **N**, and **B** at two locations on the helix. In general, the vectors **T**, **N**, and **B**, starting at the various points on a curve, form a set of orthogonal vectors, called the **TNB frame**, that moves along the curve as t varies. This **TNB** frame plays an important role in the branch of mathematics known as differential geometry and in its applications to the motion of spacecraft.

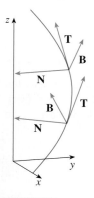

FIGURE 7

EXAMPLE 5 Find the curvature of the parabola $y = x^2$ at the points $(0, 0)$, $(1, 1)$, and $(2, 4)$.

SOLUTION Since $y' = 2x$ and $y'' = 2$, Formula 11 gives

$$\kappa(x) = \frac{|y''|}{[1 + (y')^2]^{3/2}} = \frac{2}{(1 + 4x^2)^{3/2}}$$

The curvature at $(0, 0)$ is $\kappa(0) = 2$. At $(1, 1)$ it is $\kappa(1) = 2/5^{3/2} \approx 0.18$. At $(2, 4)$ it is $\kappa(2) = 2/17^{3/2} \approx 0.03$. Observe from the expression for $\kappa(x)$ or the graph of κ in Figure 5 that $\kappa(x) \to 0$ as $x \to \pm\infty$. This corresponds to the fact that the parabola appears to become flatter as $x \to \pm\infty$. ∎

◼ The Normal and Binormal Vectors

At a given point on a smooth space curve $\mathbf{r}(t)$, there are many vectors that are orthogonal to the unit tangent vector $\mathbf{T}(t)$. We single out one by observing that, because $|\mathbf{T}(t)| = 1$ for all t, we have $\mathbf{T}(t) \cdot \mathbf{T}'(t) = 0$ by Example 13.2.4, so $\mathbf{T}'(t)$ is orthogonal to $\mathbf{T}(t)$. Note that, typically, $\mathbf{T}'(t)$ is itself not a unit vector. But at any point where $\kappa \neq 0$ we can define the **principal unit normal vector** $\mathbf{N}(t)$ (or simply **unit normal**) as

$$\mathbf{N}(t) = \frac{\mathbf{T}'(t)}{|\mathbf{T}'(t)|}$$

We can think of the unit normal vector as indicating the direction in which the curve is turning at each point. The vector $\mathbf{B}(t) = \mathbf{T}(t) \times \mathbf{N}(t)$ is called the **binormal vector**. It is perpendicular to both $\mathbf{T}$ and $\mathbf{N}$ and is also a unit vector. (See Figure 6.)

EXAMPLE 6 Find the unit normal and binormal vectors for the circular helix

$$\mathbf{r}(t) = \cos t\,\mathbf{i} + \sin t\,\mathbf{j} + t\,\mathbf{k}$$

SOLUTION We first compute the ingredients needed for the unit normal vector:

$$\mathbf{r}'(t) = -\sin t\,\mathbf{i} + \cos t\,\mathbf{j} + \mathbf{k} \qquad |\mathbf{r}'(t)| = \sqrt{2}$$

$$\mathbf{T}(t) = \frac{\mathbf{r}'(t)}{|\mathbf{r}'(t)|} = \frac{1}{\sqrt{2}}(-\sin t\,\mathbf{i} + \cos t\,\mathbf{j} + \mathbf{k})$$

$$\mathbf{T}'(t) = \frac{1}{\sqrt{2}}(-\cos t\,\mathbf{i} - \sin t\,\mathbf{j}) \qquad |\mathbf{T}'(t)| = \frac{1}{\sqrt{2}}$$

$$\mathbf{N}(t) = \frac{\mathbf{T}'(t)}{|\mathbf{T}'(t)|} = -\cos t\,\mathbf{i} - \sin t\,\mathbf{j} = \langle -\cos t, -\sin t, 0 \rangle$$

This shows that the normal vector at any point on the helix is horizontal and points toward the z-axis. The binormal vector is

$$\mathbf{B}(t) = \mathbf{T}(t) \times \mathbf{N}(t) = \frac{1}{\sqrt{2}} \begin{bmatrix} \mathbf{i} & \mathbf{j} & \mathbf{k} \\ -\sin t & \cos t & 1 \\ -\cos t & -\sin t & 0 \end{bmatrix}$$

$$= \frac{1}{\sqrt{2}} \langle \sin t, -\cos t, 1 \rangle \qquad ∎$$

TEC Visual 13.3B shows how the **TNB** frame moves along several curves.

The plane determined by the normal and binormal vectors **N** and **B** at a point P on a curve C is called the **normal plane** of C at P. It consists of all lines that are orthogonal to the tangent vector **T**. The plane determined by the vectors **T** and **N** is called the **osculating plane** of C at P. The name comes from the Latin *osculum*, meaning "kiss." It is the plane that comes closest to containing the part of the curve near P. (For a plane curve, the osculating plane is simply the plane that contains the curve.)

The circle that lies in the osculating plane of C at P, has the same tangent as C at P, lies on the concave side of C (toward which **N** points), and has radius $\rho = 1/\kappa$ (the reciprocal of the curvature) is called the **osculating circle** (or the **circle of curvature**) of C at P. It is the circle that best describes how C behaves near P; it shares the same tangent, normal, and curvature at P.

EXAMPLE 7 Find equations of the normal plane and osculating plane of the helix in Example 6 at the point $P(0, 1, \pi/2)$.

Figure 8 shows the helix and the osculating plane in Example 7.

SOLUTION The point P corresponds to $t = \pi/2$ and the normal plane there has normal vector $\mathbf{r}'(\pi/2) = \langle -1, 0, 1 \rangle$, so an equation is

$$-1(x - 0) + 0(y - 1) + 1\left(z - \frac{\pi}{2}\right) = 0 \quad \text{or} \quad z = x + \frac{\pi}{2}$$

The osculating plane at P contains the vectors **T** and **N**, so its normal vector is $\mathbf{T} \times \mathbf{N} = \mathbf{B}$. From Example 6 we have

$$\mathbf{B}(t) = \frac{1}{\sqrt{2}}\langle \sin t, -\cos t, 1 \rangle \qquad \mathbf{B}\left(\frac{\pi}{2}\right) = \left\langle \frac{1}{\sqrt{2}}, 0, \frac{1}{\sqrt{2}} \right\rangle$$

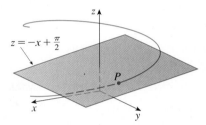

FIGURE 8

A simpler normal vector is $\langle 1, 0, 1 \rangle$, so an equation of the osculating plane is

$$1(x - 0) + 0(y - 1) + 1\left(z - \frac{\pi}{2}\right) = 0 \quad \text{or} \quad z = -x + \frac{\pi}{2} \qquad ■$$

EXAMPLE 8 Find and graph the osculating circle of the parabola $y = x^2$ at the origin.

SOLUTION From Example 5, the curvature of the parabola at the origin is $\kappa(0) = 2$. So the radius of the osculating circle at the origin is $1/\kappa = \frac{1}{2}$ and its center is $\left(0, \frac{1}{2}\right)$. Its equation is therefore

$$x^2 + \left(y - \tfrac{1}{2}\right)^2 = \tfrac{1}{4}$$

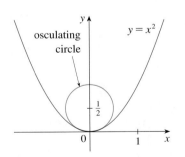

FIGURE 9

Notice that the circle and the parabola appear to bend similarly at the origin.

For the graph in Figure 9 we use parametric equations of this circle:

$$x = \tfrac{1}{2}\cos t \qquad y = \tfrac{1}{2} + \tfrac{1}{2}\sin t \qquad ■$$

TEC Visual 13.3C shows how the osculating circle changes as a point moves along a curve.

We summarize here the formulas for unit tangent, unit normal and binormal vectors, and curvature.

$$\mathbf{T}(t) = \frac{\mathbf{r}'(t)}{|\mathbf{r}'(t)|} \qquad \mathbf{N}(t) = \frac{\mathbf{T}'(t)}{|\mathbf{T}'(t)|} \qquad \mathbf{B}(t) = \mathbf{T}(t) \times \mathbf{N}(t)$$

$$\kappa = \left|\frac{d\mathbf{T}}{ds}\right| = \frac{|\mathbf{T}'(t)|}{|\mathbf{r}'(t)|} = \frac{|\mathbf{r}'(t) \times \mathbf{r}''(t)|}{|\mathbf{r}'(t)|^3}$$

13.3 EXERCISES

1–6 Find the length of the curve.

1. $\mathbf{r}(t) = \langle t, 3\cos t, 3\sin t \rangle, \quad -5 \leqslant t \leqslant 5$

2. $\mathbf{r}(t) = \langle 2t, t^2, \frac{1}{3}t^3 \rangle, \quad 0 \leqslant t \leqslant 1$

3. $\mathbf{r}(t) = \sqrt{2}\,t\,\mathbf{i} + e^t\,\mathbf{j} + e^{-t}\,\mathbf{k}, \quad 0 \leqslant t \leqslant 1$

4. $\mathbf{r}(t) = \cos t\,\mathbf{i} + \sin t\,\mathbf{j} + \ln\cos t\,\mathbf{k}, \quad 0 \leqslant t \leqslant \pi/4$

5. $\mathbf{r}(t) = \mathbf{i} + t^2\,\mathbf{j} + t^3\,\mathbf{k}, \quad 0 \leqslant t \leqslant 1$

6. $\mathbf{r}(t) = t^2\,\mathbf{i} + 9t\,\mathbf{j} + 4t^{3/2}\,\mathbf{k}, \quad 1 \leqslant t \leqslant 4$

7–9 Find the length of the curve correct to four decimal places. (Use a calculator to approximate the integral.)

7. $\mathbf{r}(t) = \langle t^2, t^3, t^4 \rangle, \quad 0 \leqslant t \leqslant 2$

8. $\mathbf{r}(t) = \langle t, e^{-t}, te^{-t} \rangle, \quad 1 \leqslant t \leqslant 3$

9. $\mathbf{r}(t) = \langle \cos \pi t, 2t, \sin 2\pi t \rangle, \quad$ from $(1, 0, 0)$ to $(1, 4, 0)$

10. Graph the curve with parametric equations $x = \sin t$, $y = \sin 2t$, $z = \sin 3t$. Find the total length of this curve correct to four decimal places.

11. Let C be the curve of intersection of the parabolic cylinder $x^2 = 2y$ and the surface $3z = xy$. Find the exact length of C from the origin to the point $(6, 18, 36)$.

12. Find, correct to four decimal places, the length of the curve of intersection of the cylinder $4x^2 + y^2 = 4$ and the plane $x + y + z = 2$.

13–14 (a) Find the arc length function for the curve measured from the point P in the direction of increasing t and then reparametrize the curve with respect to arc length starting from P, and (b) find the point 4 units along the curve (in the direction of increasing t) from P.

13. $\mathbf{r}(t) = (5 - t)\,\mathbf{i} + (4t - 3)\,\mathbf{j} + 3t\,\mathbf{k}, \quad P(4, 1, 3)$

14. $\mathbf{r}(t) = e^t \sin t\,\mathbf{i} + e^t \cos t\,\mathbf{j} + \sqrt{2}\,e^t\,\mathbf{k}, \quad P(0, 1, \sqrt{2})$

15. Suppose you start at the point $(0, 0, 3)$ and move 5 units along the curve $x = 3\sin t$, $y = 4t$, $z = 3\cos t$ in the positive direction. Where are you now?

16. Reparametrize the curve

$$\mathbf{r}(t) = \left(\frac{2}{t^2 + 1} - 1 \right) \mathbf{i} + \frac{2t}{t^2 + 1}\,\mathbf{j}$$

with respect to arc length measured from the point $(1, 0)$ in the direction of increasing t. Express the reparametrization in its simplest form. What can you conclude about the curve?

17–20
(a) Find the unit tangent and unit normal vectors $\mathbf{T}(t)$ and $\mathbf{N}(t)$.
(b) Use Formula 9 to find the curvature.

17. $\mathbf{r}(t) = \langle t, 3\cos t, 3\sin t \rangle$

18. $\mathbf{r}(t) = \langle t^2, \sin t - t\cos t, \cos t + t\sin t \rangle, \quad t > 0$

19. $\mathbf{r}(t) = \langle \sqrt{2}\,t, e^t, e^{-t} \rangle$

20. $\mathbf{r}(t) = \langle t, \frac{1}{2}t^2, t^2 \rangle$

21–23 Use Theorem 10 to find the curvature.

21. $\mathbf{r}(t) = t^3\,\mathbf{j} + t^2\,\mathbf{k}$

22. $\mathbf{r}(t) = t\,\mathbf{i} + t^2\,\mathbf{j} + e^t\,\mathbf{k}$

23. $\mathbf{r}(t) = \sqrt{6}\,t^2\,\mathbf{i} + 2t\,\mathbf{j} + 2t^3\,\mathbf{k}$

24. Find the curvature of $\mathbf{r}(t) = \langle t^2, \ln t, t\ln t \rangle$ at the point $(1, 0, 0)$.

25. Find the curvature of $\mathbf{r}(t) = \langle t, t^2, t^3 \rangle$ at the point $(1, 1, 1)$.

26. Graph the curve with parametric equations $x = \cos t$, $y = \sin t$, $z = \sin 5t$ and find the curvature at the point $(1, 0, 0)$.

27–29 Use Formula 11 to find the curvature.

27. $y = x^4$ **28.** $y = \tan x$ **29.** $y = xe^x$

30–31 At what point does the curve have maximum curvature? What happens to the curvature as $x \to \infty$?

30. $y = \ln x$ **31.** $y = e^x$

32. Find an equation of a parabola that has curvature 4 at the origin.

33. (a) Is the curvature of the curve C shown in the figure greater at P or at Q? Explain.
(b) Estimate the curvature at P and at Q by sketching the osculating circles at those points.

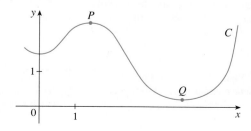

34–35 Use a graphing calculator or computer to graph both the curve and its curvature function $\kappa(x)$ on the same screen. Is the graph of κ what you would expect?

34. $y = x^4 - 2x^2$ **35.** $y = x^{-2}$

CAS 36–37 Plot the space curve and its curvature function $\kappa(t)$. Comment on how the curvature reflects the shape of the curve.

36. $\mathbf{r}(t) = \langle t - \sin t, 1 - \cos t, 4 \cos(t/2) \rangle$, $0 \le t \le 8\pi$

37. $\mathbf{r}(t) = \left\langle te^t, e^{-t}, \sqrt{2}\,t \right\rangle$, $-5 \le t \le 5$

38–39 Two graphs, a and b, are shown. One is a curve $y = f(x)$ and the other is the graph of its curvature function $y = \kappa(x)$. Identify each curve and explain your choices.

38.

39.

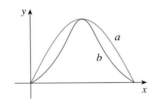

CAS 40. (a) Graph the curve $\mathbf{r}(t) = \langle \sin 3t, \sin 2t, \sin 3t \rangle$. At how many points on the curve does it appear that the curvature has a local or absolute maximum?
    (b) Use a CAS to find and graph the curvature function. Does this graph confirm your conclusion from part (a)?

CAS 41. The graph of $\mathbf{r}(t) = \left\langle t - \frac{3}{2} \sin t, 1 - \frac{3}{2} \cos t, t \right\rangle$ is shown in Figure 13.1.12(b). Where do you think the curvature is largest? Use a CAS to find and graph the curvature function. For which values of t is the curvature largest?

42. Use Theorem 10 to show that the curvature of a plane parametric curve $x = f(t)$, $y = g(t)$ is

$$\kappa = \frac{|\dot{x}\ddot{y} - \dot{y}\ddot{x}|}{[\dot{x}^2 + \dot{y}^2]^{3/2}}$$

where the dots indicate derivatives with respect to t.

43–45 Use the formula in Exercise 42 to find the curvature.

43. $x = t^2$, $y = t^3$

44. $x = a \cos \omega t$, $y = b \sin \omega t$

45. $x = e^t \cos t$, $y = e^t \sin t$

---

46. Consider the curvature at $x = 0$ for each member of the family of functions $f(x) = e^{cx}$. For which members is $\kappa(0)$ largest?

47–48 Find the vectors $\mathbf{T}$, $\mathbf{N}$, and $\mathbf{B}$ at the given point.

47. $\mathbf{r}(t) = \left\langle t^2, \frac{2}{3}t^3, t \right\rangle$, $\left(1, \frac{2}{3}, 1 \right)$

48. $\mathbf{r}(t) = \langle \cos t, \sin t, \ln \cos t \rangle$, $(1, 0, 0)$

49–50 Find equations of the normal plane and osculating plane of the curve at the given point.

49. $x = \sin 2t$, $y = -\cos 2t$, $z = 4t$; $(0, 1, 2\pi)$

50. $x = \ln t$, $y = 2t$, $z = t^2$; $(0, 2, 1)$

---

51. Find equations of the osculating circles of the ellipse $9x^2 + 4y^2 = 36$ at the points $(2, 0)$ and $(0, 3)$. Use a graphing calculator or computer to graph the ellipse and both osculating circles on the same screen.

52. Find equations of the osculating circles of the parabola $y = \frac{1}{2}x^2$ at the points $(0, 0)$ and $\left(1, \frac{1}{2} \right)$. Graph both osculating circles and the parabola on the same screen.

53. At what point on the curve $x = t^3$, $y = 3t$, $z = t^4$ is the normal plane parallel to the plane $6x + 6y - 8z = 1$?

CAS 54. Is there a point on the curve in Exercise 53 where the osculating plane is parallel to the plane $x + y + z = 1$? [*Note:* You will need a CAS for differentiating, for simplifying, and for computing a cross product.]

55. Find equations of the normal and osculating planes of the curve of intersection of the parabolic cylinders $x = y^2$ and $z = x^2$ at the point $(1, 1, 1)$.

56. Show that the osculating plane at every point on the curve $\mathbf{r}(t) = \left\langle t + 2, 1 - t, \frac{1}{2}t^2 \right\rangle$ is the same plane. What can you conclude about the curve?

57. Show that at every point on the curve

$$\mathbf{r}(t) = \langle e^t \cos t, e^t \sin t, e^t \rangle$$

the angle between the unit tangent vector and the z-axis is the same. Then show that the same result holds true for the unit normal and binormal vectors.

58. The *rectifying plane* of a curve at a point is the plane that contains the vectors $\mathbf{T}$ and $\mathbf{B}$ at that point. Find the rectifying plane of the curve $\mathbf{r}(t) = \sin t\,\mathbf{i} + \cos t\,\mathbf{j} + \tan t\,\mathbf{k}$ at the point $\left(\sqrt{2}/2, \sqrt{2}/2, 1 \right)$.

59. Show that the curvature κ is related to the tangent and normal vectors by the equation

$$\frac{d\mathbf{T}}{ds} = \kappa \mathbf{N}$$

60. Show that the curvature of a plane curve is $\kappa = |d\phi/ds|$, where ϕ is the angle between $\mathbf{T}$ and $\mathbf{i}$; that is, ϕ is the angle of inclination of the tangent line. (This shows that the definition of curvature is consistent with the definition for plane curves given in Exercise 10.2.69.)

61. (a) Show that $d\mathbf{B}/ds$ is perpendicular to $\mathbf{B}$.
  (b) Show that $d\mathbf{B}/ds$ is perpendicular to $\mathbf{T}$.
  (c) Deduce from parts (a) and (b) that $d\mathbf{B}/ds = -\tau(s)\mathbf{N}$ for some number $\tau(s)$ called the **torsion** of the curve. (The torsion measures the degree of twisting of a curve.)
  (d) Show that for a plane curve the torsion is $\tau(s) = 0$.

62. The following formulas, called the **Frenet-Serret formulas**, are of fundamental importance in differential geometry:

  1. $d\mathbf{T}/ds = \kappa\mathbf{N}$

  2. $d\mathbf{N}/ds = -\kappa\mathbf{T} + \tau\mathbf{B}$

  3. $d\mathbf{B}/ds = -\tau\mathbf{N}$

(Formula 1 comes from Exercise 59 and Formula 3 comes from Exercise 61.) Use the fact that $\mathbf{N} = \mathbf{B} \times \mathbf{T}$ to deduce Formula 2 from Formulas 1 and 3.

63. Use the Frenet-Serret formulas to prove each of the following. (Primes denote derivatives with respect to t. Start as in the proof of Theorem 10.)

  (a) $\mathbf{r}'' = s''\mathbf{T} + \kappa(s')^2\mathbf{N}$

  (b) $\mathbf{r}' \times \mathbf{r}'' = \kappa(s')^3\mathbf{B}$

  (c) $\mathbf{r}''' = [s''' - \kappa^2(s')^3]\mathbf{T} + [3\kappa s' s'' + \kappa'(s')^2]\mathbf{N} + \kappa\tau(s')^3\mathbf{B}$

  (d) $\tau = \dfrac{(\mathbf{r}' \times \mathbf{r}'') \cdot \mathbf{r}'''}{|\mathbf{r}' \times \mathbf{r}''|^2}$

64. Show that the circular helix $\mathbf{r}(t) = \langle a\cos t, a\sin t, bt\rangle$, where a and b are positive constants, has constant curvature and constant torsion. [Use the result of Exercise 63(d).]

65. Use the formula in Exercise 63(d) to find the torsion of the curve $\mathbf{r}(t) = \langle t, \frac{1}{2}t^2, \frac{1}{3}t^3\rangle$.

66. Find the curvature and torsion of the curve $x = \sinh t$, $y = \cosh t$, $z = t$ at the point $(0, 1, 0)$.

67. The DNA molecule has the shape of a double helix (see Figure 3 on page 850). The radius of each helix is about 10 angstroms ($1\ \text{Å} = 10^{-8}$ cm). Each helix rises about 34 Å during each complete turn, and there are about 2.9×10^8 complete turns. Estimate the length of each helix.

68. Let's consider the problem of designing a railroad track to make a smooth transition between sections of straight track. Existing track along the negative x-axis is to be joined smoothly to a track along the line $y = 1$ for $x \geqslant 1$.
  (a) Find a polynomial $P = P(x)$ of degree 5 such that the function F defined by

$$F(x) = \begin{cases} 0 & \text{if } x \leqslant 0 \\ P(x) & \text{if } 0 < x < 1 \\ 1 & \text{if } x \geqslant 1 \end{cases}$$

  is continuous and has continuous slope and continuous curvature.
  (b) Graph F.

13.4 Motion in Space: Velocity and Acceleration

In this section we show how the ideas of tangent and normal vectors and curvature can be used in physics to study the motion of an object, including its velocity and acceleration, along a space curve. In particular, we follow in the footsteps of Newton by using these methods to derive Kepler's First Law of planetary motion.

Suppose a particle moves through space so that its position vector at time t is $\mathbf{r}(t)$. Notice from Figure 1 that, for small values of h, the vector

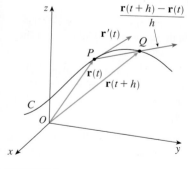

FIGURE 1

$$\boxed{1} \qquad \frac{\mathbf{r}(t + h) - \mathbf{r}(t)}{h}$$

approximates the direction of the particle moving along the curve $\mathbf{r}(t)$. Its magnitude measures the size of the displacement vector per unit time. The vector (1) gives the average velocity over a time interval of length h and its limit is the **velocity vector** $\mathbf{v}(t)$ at time t:

$$\boxed{2} \qquad \mathbf{v}(t) = \lim_{h \to 0} \frac{\mathbf{r}(t + h) - \mathbf{r}(t)}{h} = \mathbf{r}'(t)$$

Thus the velocity vector is also the tangent vector and points in the direction of the tangent line.

The **speed** of the particle at time t is the magnitude of the velocity vector, that is, $|\mathbf{v}(t)|$. This is appropriate because, from (2) and from Equation 13.3.7, we have

$$|\mathbf{v}(t)| = |\mathbf{r}'(t)| = \frac{ds}{dt} = \text{rate of change of distance with respect to time}$$

As in the case of one-dimensional motion, the **acceleration** of the particle is defined as the derivative of the velocity:

$$\mathbf{a}(t) = \mathbf{v}'(t) = \mathbf{r}''(t)$$

EXAMPLE 1 The position vector of an object moving in a plane is given by $\mathbf{r}(t) = t^3\,\mathbf{i} + t^2\,\mathbf{j}$. Find its velocity, speed, and acceleration when $t = 1$ and illustrate geometrically.

SOLUTION The velocity and acceleration at time t are

$$\mathbf{v}(t) = \mathbf{r}'(t) = 3t^2\,\mathbf{i} + 2t\,\mathbf{j}$$

$$\mathbf{a}(t) = \mathbf{r}''(t) = 6t\,\mathbf{i} + 2\,\mathbf{j}$$

and the speed is

$$|\mathbf{v}(t)| = \sqrt{(3t^2)^2 + (2t)^2} = \sqrt{9t^4 + 4t^2}$$

When $t = 1$, we have

$$\mathbf{v}(1) = 3\,\mathbf{i} + 2\,\mathbf{j} \qquad \mathbf{a}(1) = 6\,\mathbf{i} + 2\,\mathbf{j} \qquad |\mathbf{v}(1)| = \sqrt{13}$$

These velocity and acceleration vectors are shown in Figure 2. ∎

FIGURE 2

TEC Visual 13.4 shows animated velocity and acceleration vectors for objects moving along various curves.

EXAMPLE 2 Find the velocity, acceleration, and speed of a particle with position vector $\mathbf{r}(t) = \langle t^2, e^t, te^t \rangle$.

SOLUTION

$$\mathbf{v}(t) = \mathbf{r}'(t) = \langle 2t, e^t, (1 + t)e^t \rangle$$

$$\mathbf{a}(t) = \mathbf{v}'(t) = \langle 2, e^t, (2 + t)e^t \rangle$$

$$|\mathbf{v}(t)| = \sqrt{4t^2 + e^{2t} + (1 + t)^2 e^{2t}}$$

∎

Figure 3 shows the path of the particle in Example 2 with the velocity and acceleration vectors when $t = 1$.

FIGURE 3

The vector integrals that were introduced in Section 13.2 can be used to find position vectors when velocity or acceleration vectors are known, as in the next example.

EXAMPLE 3 A moving particle starts at an initial position $\mathbf{r}(0) = \langle 1, 0, 0 \rangle$ with initial velocity $\mathbf{v}(0) = \mathbf{i} - \mathbf{j} + \mathbf{k}$. Its acceleration is $\mathbf{a}(t) = 4t\,\mathbf{i} + 6t\,\mathbf{j} + \mathbf{k}$. Find its velocity and position at time t.

SOLUTION Since $\mathbf{a}(t) = \mathbf{v}'(t)$, we have

$$\mathbf{v}(t) = \int \mathbf{a}(t)\,dt = \int (4t\,\mathbf{i} + 6t\,\mathbf{j} + \mathbf{k})\,dt$$

$$= 2t^2\,\mathbf{i} + 3t^2\,\mathbf{j} + t\,\mathbf{k} + \mathbf{C}$$

To determine the value of the constant vector $\mathbf{C}$, we use the fact that $\mathbf{v}(0) = \mathbf{i} - \mathbf{j} + \mathbf{k}$. The preceding equation gives $\mathbf{v}(0) = \mathbf{C}$, so $\mathbf{C} = \mathbf{i} - \mathbf{j} + \mathbf{k}$ and

$$\mathbf{v}(t) = 2t^2\,\mathbf{i} + 3t^2\,\mathbf{j} + t\,\mathbf{k} + \mathbf{i} - \mathbf{j} + \mathbf{k}$$

$$= (2t^2 + 1)\,\mathbf{i} + (3t^2 - 1)\,\mathbf{j} + (t + 1)\,\mathbf{k}$$

The expression for $\mathbf{r}(t)$ that we obtained in Example 3 was used to plot the path of the particle in Figure 4 for $0 \leq t \leq 3$.

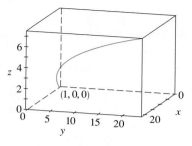

FIGURE 4

Since $\mathbf{v}(t) = \mathbf{r}'(t)$, we have

$$\mathbf{r}(t) = \int \mathbf{v}(t)\, dt$$

$$= \int \left[(2t^2 + 1)\,\mathbf{i} + (3t^2 - 1)\,\mathbf{j} + (t + 1)\,\mathbf{k} \right] dt$$

$$= \left(\tfrac{2}{3}t^3 + t \right)\mathbf{i} + (t^3 - t)\,\mathbf{j} + \left(\tfrac{1}{2}t^2 + t \right)\mathbf{k} + \mathbf{D}$$

Putting $t = 0$, we find that $\mathbf{D} = \mathbf{r}(0) = \mathbf{i}$, so the position at time t is given by

$$\mathbf{r}(t) = \left(\tfrac{2}{3}t^3 + t + 1 \right)\mathbf{i} + (t^3 - t)\,\mathbf{j} + \left(\tfrac{1}{2}t^2 + t \right)\mathbf{k} \qquad \blacksquare$$

In general, vector integrals allow us to recover velocity when acceleration is known and position when velocity is known:

$$\mathbf{v}(t) = \mathbf{v}(t_0) + \int_{t_0}^{t} \mathbf{a}(u)\, du \qquad\qquad \mathbf{r}(t) = \mathbf{r}(t_0) + \int_{t_0}^{t} \mathbf{v}(u)\, du$$

If the force that acts on a particle is known, then the acceleration can be found from **Newton's Second Law of Motion**. The vector version of this law states that if, at any time t, a force $\mathbf{F}(t)$ acts on an object of mass m producing an acceleration $\mathbf{a}(t)$, then

$$\mathbf{F}(t) = m\mathbf{a}(t)$$

EXAMPLE 4 An object with mass m that moves in a circular path with constant angular speed ω has position vector $\mathbf{r}(t) = a \cos \omega t\, \mathbf{i} + a \sin \omega t\, \mathbf{j}$. Find the force acting on the object and show that it is directed toward the origin.

The object moving with position P has angular speed $\omega = d\theta/dt$, where θ is the angle shown in Figure 5.

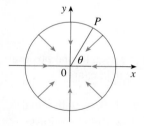

FIGURE 5

SOLUTION To find the force, we first need to know the acceleration:

$$\mathbf{v}(t) = \mathbf{r}'(t) = -a\omega \sin \omega t\, \mathbf{i} + a\omega \cos \omega t\, \mathbf{j}$$

$$\mathbf{a}(t) = \mathbf{v}'(t) = -a\omega^2 \cos \omega t\, \mathbf{i} - a\omega^2 \sin \omega t\, \mathbf{j}$$

Therefore Newton's Second Law gives the force as

$$\mathbf{F}(t) = m\mathbf{a}(t) = -m\omega^2 (a \cos \omega t\, \mathbf{i} + a \sin \omega t\, \mathbf{j})$$

Notice that $\mathbf{F}(t) = -m\omega^2 \mathbf{r}(t)$. This shows that the force acts in the direction opposite to the radius vector $\mathbf{r}(t)$ and therefore points toward the origin (see Figure 5). Such a force is called a *centripetal* (center-seeking) force. $\qquad \blacksquare$

■ Projectile Motion

EXAMPLE 5 A projectile is fired with angle of elevation α and initial velocity $\mathbf{v}_0$. (See Figure 6.) Assuming that air resistance is negligible and the only external force is due to gravity, find the position function $\mathbf{r}(t)$ of the projectile. What value of α maximizes the range (the horizontal distance traveled)?

SOLUTION We set up the axes so that the projectile starts at the origin. Since the force due to gravity acts downward, we have

$$\mathbf{F} = m\mathbf{a} = -mg\,\mathbf{j}$$

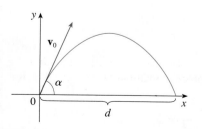

FIGURE 6

where $g = |\mathbf{a}| \approx 9.8$ m/s^2. Thus

$$\mathbf{a} = -g\,\mathbf{j}$$

Since $\mathbf{v}'(t) = \mathbf{a}$, we have

$$\mathbf{v}(t) = -gt\,\mathbf{j} + \mathbf{C}$$

where $\mathbf{C} = \mathbf{v}(0) = \mathbf{v}_0$. Therefore

$$\mathbf{r}'(t) = \mathbf{v}(t) = -gt\,\mathbf{j} + \mathbf{v}_0$$

Integrating again, we obtain

$$\mathbf{r}(t) = -\tfrac{1}{2}gt^2\,\mathbf{j} + t\,\mathbf{v}_0 + \mathbf{D}$$

But $\mathbf{D} = \mathbf{r}(0) = \mathbf{0}$, so the position vector of the projectile is given by

$$\boxed{3} \qquad \mathbf{r}(t) = -\tfrac{1}{2}gt^2\,\mathbf{j} + t\,\mathbf{v}_0$$

If we write $|\mathbf{v}_0| = v_0$ (the initial speed of the projectile), then

$$\mathbf{v}_0 = v_0 \cos \alpha \,\mathbf{i} + v_0 \sin \alpha \,\mathbf{j}$$

and Equation 3 becomes

$$\mathbf{r}(t) = (v_0 \cos \alpha)t\,\mathbf{i} + \left[(v_0 \sin \alpha)t - \tfrac{1}{2}gt^2\right]\mathbf{j}$$

The parametric equations of the trajectory are therefore

If you eliminate t from Equations 4, you will see that y is a quadratic function of x. So the path of the projectile is part of a parabola.

$$\boxed{4} \qquad x = (v_0 \cos \alpha)t \qquad y = (v_0 \sin \alpha)t - \tfrac{1}{2}gt^2$$

The horizontal distance d is the value of x when $y = 0$. Setting $y = 0$, we obtain $t = 0$ or $t = (2v_0 \sin \alpha)/g$. This second value of t then gives

$$d = x = (v_0 \cos \alpha)\frac{2v_0 \sin \alpha}{g} = \frac{v_0^2(2 \sin \alpha \cos \alpha)}{g} = \frac{v_0^2 \sin 2\alpha}{g}$$

Clearly, d has its maximum value when $\sin 2\alpha = 1$, that is, $\alpha = 45°$. ∎

EXAMPLE 6 A projectile is fired with muzzle speed 150 m/s and angle of elevation 45° from a position 10 m above ground level. Where does the projectile hit the ground, and with what speed?

SOLUTION If we place the origin at ground level, then the initial position of the projectile is $(0, 10)$ and so we need to adjust Equations 4 by adding 10 to the expression for y. With $v_0 = 150$ m/s, $\alpha = 45°$, and $g = 9.8$ m/s^2, we have

$$x = 150 \cos(45°)t = 75\sqrt{2}\,t$$

$$y = 10 + 150 \sin(45°)t - \tfrac{1}{2}(9.8)t^2 = 10 + 75\sqrt{2}\,t - 4.9t^2$$

Impact occurs when $y = 0$, that is, $4.9t^2 - 75\sqrt{2}\,t - 10 = 0$. Using the quadratic formula to solve this equation (and taking only the positive value of t), we get

$$t = \frac{75\sqrt{2} + \sqrt{11{,}250 + 196}}{9.8} \approx 21.74$$

Then $x \approx 75\sqrt{2}\,(21.74) \approx 2306$, so the projectile hits the ground about 2306 m away.

The velocity of the projectile is

$$\mathbf{v}(t) = \mathbf{r}'(t) = 75\sqrt{2}\,\mathbf{i} + \left(75\sqrt{2} - 9.8t\right)\mathbf{j}$$

So its speed at impact is

$$|\mathbf{v}(21.74)| = \sqrt{(75\sqrt{2})^2 + (75\sqrt{2} - 9.8 \cdot 21.74)^2} \approx 151 \text{ m/s} \qquad \blacksquare$$

■ Tangential and Normal Components of Acceleration

When we study the motion of a particle, it is often useful to resolve the acceleration into two components, one in the direction of the tangent and the other in the direction of the normal. If we write $v = |\mathbf{v}|$ for the speed of the particle, then

$$\mathbf{T}(t) = \frac{\mathbf{r}'(t)}{|\mathbf{r}'(t)|} = \frac{\mathbf{v}(t)}{|\mathbf{v}(t)|} = \frac{\mathbf{v}}{v}$$

and so

$$\mathbf{v} = v\mathbf{T}$$

If we differentiate both sides of this equation with respect to t, we get

$$\boxed{5} \qquad \mathbf{a} = \mathbf{v}' = v'\mathbf{T} + v\mathbf{T}'$$

If we use the expression for the curvature given by Equation 13.3.9, then we have

$$\boxed{6} \qquad \kappa = \frac{|\mathbf{T}'|}{|\mathbf{r}'|} = \frac{|\mathbf{T}'|}{v} \qquad \text{so} \qquad |\mathbf{T}'| = \kappa v$$

The unit normal vector was defined in the preceding section as $\mathbf{N} = \mathbf{T}'/|\mathbf{T}'|$, so (6) gives

$$\mathbf{T}' = |\mathbf{T}'|\mathbf{N} = \kappa v\mathbf{N}$$

and Equation 5 becomes

$$\boxed{7} \qquad \boxed{\mathbf{a} = v'\mathbf{T} + \kappa v^2\mathbf{N}}$$

Writing a_T and a_N for the tangential and normal components of acceleration, we have

$$\mathbf{a} = a_T\mathbf{T} + a_N\mathbf{N}$$

where

$$\boxed{8} \qquad a_T = v' \qquad \text{and} \qquad a_N = \kappa v^2$$

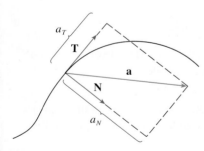

FIGURE 7

This resolution is illustrated in Figure 7.

Let's look at what Formula 7 says. The first thing to notice is that the binormal vector **B** is absent. No matter how an object moves through space, its acceleration always lies in the plane of **T** and **N** (the osculating plane). (Recall that **T** gives the direction of motion and **N** points in the direction the curve is turning.) Next we notice that the tangential component of acceleration is v', the rate of change of speed, and the normal component of acceleration is κv^2, the curvature times the square of the speed. This makes sense if we think of a passenger in a car—a sharp turn in a road means a large value of the curvature κ, so the component of the acceleration perpendicular to the motion is large and the passenger is thrown against a car door. High speed around the turn has the same effect; in fact, if you double your speed, a_N is increased by a factor of 4.

Although we have expressions for the tangential and normal components of acceleration in Equations 8, it's desirable to have expressions that depend only on $\mathbf{r}$, $\mathbf{r}'$, and $\mathbf{r}''$. To this end we take the dot product of $\mathbf{v} = v\mathbf{T}$ with $\mathbf{a}$ as given by Equation 7:

$$\mathbf{v} \cdot \mathbf{a} = v\mathbf{T} \cdot (v'\mathbf{T} + \kappa v^2 \mathbf{N})$$

$$= vv'\mathbf{T} \cdot \mathbf{T} + \kappa v^3 \mathbf{T} \cdot \mathbf{N}$$

$$= vv' \quad (\text{since } \mathbf{T} \cdot \mathbf{T} = 1 \text{ and } \mathbf{T} \cdot \mathbf{N} = 0)$$

Therefore

$$\boxed{9} \qquad a_T = v' = \frac{\mathbf{v} \cdot \mathbf{a}}{v} = \frac{\mathbf{r}'(t) \cdot \mathbf{r}''(t)}{|\mathbf{r}'(t)|}$$

Using the formula for curvature given by Theorem 13.3.10, we have

$$\boxed{10} \qquad a_N = \kappa v^2 = \frac{|\mathbf{r}'(t) \times \mathbf{r}''(t)|}{|\mathbf{r}'(t)|^3} |\mathbf{r}'(t)|^2 = \frac{|\mathbf{r}'(t) \times \mathbf{r}''(t)|}{|\mathbf{r}'(t)|}$$

EXAMPLE 7 A particle moves with position function $\mathbf{r}(t) = \langle t^2, t^2, t^3 \rangle$. Find the tangential and normal components of acceleration.

SOLUTION
$$\mathbf{r}(t) = t^2\,\mathbf{i} + t^2\,\mathbf{j} + t^3\,\mathbf{k}$$

$$\mathbf{r}'(t) = 2t\,\mathbf{i} + 2t\,\mathbf{j} + 3t^2\,\mathbf{k}$$

$$\mathbf{r}''(t) = 2\,\mathbf{i} + 2\,\mathbf{j} + 6t\,\mathbf{k}$$

$$|\mathbf{r}'(t)| = \sqrt{8t^2 + 9t^4}$$

Therefore Equation 9 gives the tangential component as

$$a_T = \frac{\mathbf{r}'(t) \cdot \mathbf{r}''(t)}{|\mathbf{r}'(t)|} = \frac{8t + 18t^3}{\sqrt{8t^2 + 9t^4}}$$

Since
$$\mathbf{r}'(t) \times \mathbf{r}''(t) = \begin{vmatrix} \mathbf{i} & \mathbf{j} & \mathbf{k} \\ 2t & 2t & 3t^2 \\ 2 & 2 & 6t \end{vmatrix} = 6t^2\,\mathbf{i} - 6t^2\,\mathbf{j}$$

Equation 10 gives the normal component as

$$a_N = \frac{|\mathbf{r}'(t) \times \mathbf{r}''(t)|}{|\mathbf{r}'(t)|} = \frac{6\sqrt{2}\,t^2}{\sqrt{8t^2 + 9t^4}} \qquad \blacksquare$$

■ Kepler's Laws of Planetary Motion

We now describe one of the great accomplishments of calculus by showing how the material of this chapter can be used to prove Kepler's laws of planetary motion. After 20 years of studying the astronomical observations of the Danish astronomer Tycho Brahe, the German mathematician and astronomer Johannes Kepler (1571–1630) formulated the following three laws.

> **Kepler's Laws**
>
> **1.** A planet revolves around the sun in an elliptical orbit with the sun at one focus.
>
> **2.** The line joining the sun to a planet sweeps out equal areas in equal times.
>
> **3.** The square of the period of revolution of a planet is proportional to the cube of the length of the major axis of its orbit.

In his book *Principia Mathematica* of 1687, Sir Isaac Newton was able to show that these three laws are consequences of two of his own laws, the Second Law of Motion and the Law of Universal Gravitation. In what follows we prove Kepler's First Law. The remaining laws are left as exercises (with hints).

Since the gravitational force of the sun on a planet is so much larger than the forces exerted by other celestial bodies, we can safely ignore all bodies in the universe except the sun and one planet revolving about it. We use a coordinate system with the sun at the origin and we let $\mathbf{r} = \mathbf{r}(t)$ be the position vector of the planet. (Equally well, $\mathbf{r}$ could be the position vector of the moon or a satellite moving around the earth or a comet moving around a star.) The velocity vector is $\mathbf{v} = \mathbf{r}'$ and the acceleration vector is $\mathbf{a} = \mathbf{r}''$. We use the following laws of Newton:

$$\text{Second Law of Motion:} \quad \mathbf{F} = m\mathbf{a}$$

$$\text{Law of Gravitation:} \quad \mathbf{F} = -\frac{GMm}{r^3}\mathbf{r} = -\frac{GMm}{r^2}\mathbf{u}$$

where $\mathbf{F}$ is the gravitational force on the planet, m and M are the masses of the planet and the sun, G is the gravitational constant, $r = |\mathbf{r}|$, and $\mathbf{u} = (1/r)\mathbf{r}$ is the unit vector in the direction of $\mathbf{r}$.

We first show that the planet moves in one plane. By equating the expressions for $\mathbf{F}$ in Newton's two laws, we find that

$$\mathbf{a} = -\frac{GM}{r^3}\mathbf{r}$$

and so $\mathbf{a}$ is parallel to $\mathbf{r}$. It follows that $\mathbf{r} \times \mathbf{a} = \mathbf{0}$. We use Formula 5 in Theorem 13.2.3 to write

$$\frac{d}{dt}(\mathbf{r} \times \mathbf{v}) = \mathbf{r}' \times \mathbf{v} + \mathbf{r} \times \mathbf{v}'$$

$$= \mathbf{v} \times \mathbf{v} + \mathbf{r} \times \mathbf{a} = \mathbf{0} + \mathbf{0} = \mathbf{0}$$

Therefore $$\mathbf{r} \times \mathbf{v} = \mathbf{h}$$

where $\mathbf{h}$ is a constant vector. (We may assume that $\mathbf{h} \neq \mathbf{0}$; that is, $\mathbf{r}$ and $\mathbf{v}$ are not parallel.) This means that the vector $\mathbf{r} = \mathbf{r}(t)$ is perpendicular to $\mathbf{h}$ for all values of t, so the planet always lies in the plane through the origin perpendicular to $\mathbf{h}$. Thus the orbit of the planet is a plane curve.

To prove Kepler's First Law we rewrite the vector $\mathbf{h}$ as follows:

$$\mathbf{h} = \mathbf{r} \times \mathbf{v} = \mathbf{r} \times \mathbf{r}' = r\mathbf{u} \times (r\mathbf{u})'$$

$$= r\mathbf{u} \times (r\mathbf{u}' + r'\mathbf{u}) = r^2(\mathbf{u} \times \mathbf{u}') + rr'(\mathbf{u} \times \mathbf{u})$$

$$= r^2(\mathbf{u} \times \mathbf{u}')$$

Then

$$\mathbf{a} \times \mathbf{h} = \frac{-GM}{r^2} \mathbf{u} \times (r^2 \mathbf{u} \times \mathbf{u}') = -GM \mathbf{u} \times (\mathbf{u} \times \mathbf{u}')$$

$$= -GM[(\mathbf{u} \cdot \mathbf{u}')\mathbf{u} - (\mathbf{u} \cdot \mathbf{u})\mathbf{u}'] \quad \text{(by Theorem 12.4.11, Property 6)}$$

But $\mathbf{u} \cdot \mathbf{u} = |\mathbf{u}|^2 = 1$ and, since $|\mathbf{u}(t)| = 1$, it follows from Example 13.2.4 that

$$\mathbf{u} \cdot \mathbf{u}' = 0$$

Therefore

$$\mathbf{a} \times \mathbf{h} = GM \mathbf{u}'$$

and so

$$(\mathbf{v} \times \mathbf{h})' = \mathbf{v}' \times \mathbf{h} = \mathbf{a} \times \mathbf{h} = GM \mathbf{u}'$$

Integrating both sides of this equation, we get

$$\boxed{11} \qquad \mathbf{v} \times \mathbf{h} = GM \mathbf{u} + \mathbf{c}$$

where $\mathbf{c}$ is a constant vector.

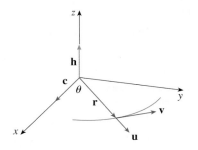

FIGURE 8

At this point it is convenient to choose the coordinate axes so that the standard basis vector $\mathbf{k}$ points in the direction of the vector $\mathbf{h}$. Then the planet moves in the xy-plane. Since both $\mathbf{v} \times \mathbf{h}$ and $\mathbf{u}$ are perpendicular to $\mathbf{h}$, Equation 11 shows that $\mathbf{c}$ lies in the xy-plane. This means that we can choose the x- and y-axes so that the vector $\mathbf{i}$ lies in the direction of $\mathbf{c}$, as shown in Figure 8.

If θ is the angle between $\mathbf{c}$ and $\mathbf{r}$, then (r, θ) are polar coordinates of the planet. From Equation 11 we have

$$\mathbf{r} \cdot (\mathbf{v} \times \mathbf{h}) = \mathbf{r} \cdot (GM \mathbf{u} + \mathbf{c}) = GM \mathbf{r} \cdot \mathbf{u} + \mathbf{r} \cdot \mathbf{c}$$

$$= GMr \mathbf{u} \cdot \mathbf{u} + |\mathbf{r}||\mathbf{c}| \cos \theta = GMr + rc \cos \theta$$

where $c = |\mathbf{c}|$. Then

$$r = \frac{\mathbf{r} \cdot (\mathbf{v} \times \mathbf{h})}{GM + c \cos \theta} = \frac{1}{GM} \frac{\mathbf{r} \cdot (\mathbf{v} \times \mathbf{h})}{1 + e \cos \theta}$$

where $e = c/(GM)$. But

$$\mathbf{r} \cdot (\mathbf{v} \times \mathbf{h}) = (\mathbf{r} \times \mathbf{v}) \cdot \mathbf{h} = \mathbf{h} \cdot \mathbf{h} = |\mathbf{h}|^2 = h^2$$

where $h = |\mathbf{h}|$. So

$$r = \frac{h^2/(GM)}{1 + e \cos \theta} = \frac{eh^2/c}{1 + e \cos \theta}$$

Writing $d = h^2/c$, we obtain the equation

$$\boxed{12} \qquad r = \frac{ed}{1 + e \cos \theta}$$

Comparing with Theorem 10.6.6, we see that Equation 12 is the polar equation of a conic section with focus at the origin and eccentricity e. We know that the orbit of a planet is a closed curve and so the conic must be an ellipse.

This completes the derivation of Kepler's First Law. We will guide you through the derivation of the Second and Third Laws in the Applied Project on page 880. The proofs of these three laws show that the methods of this chapter provide a powerful tool for describing some of the laws of nature.

13.4   EXERCISES

1. The table gives coordinates of a particle moving through space along a smooth curve.
(a) Find the average velocities over the time intervals [0, 1], [0.5, 1], [1, 2], and [1, 1.5].
(b) Estimate the velocity and speed of the particle at $t = 1$.

t	x	y	z
0	2.7	9.8	3.7
0.5	3.5	7.2	3.3
1.0	4.5	6.0	3.0
1.5	5.9	6.4	2.8
2.0	7.3	7.8	2.7

2. The figure shows the path of a particle that moves with position vector $\mathbf{r}(t)$ at time t.
(a) Draw a vector that represents the average velocity of the particle over the time interval $2 \le t \le 2.4$.
(b) Draw a vector that represents the average velocity over the time interval $1.5 \le t \le 2$.
(c) Write an expression for the velocity vector $\mathbf{v}(2)$.
(d) Draw an approximation to the vector $\mathbf{v}(2)$ and estimate the speed of the particle at $t = 2$.

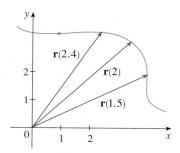

3–8 Find the velocity, acceleration, and speed of a particle with the given position function. Sketch the path of the particle and draw the velocity and acceleration vectors for the specified value of t.

3. $\mathbf{r}(t) = \left\langle -\frac{1}{2}t^2, t \right\rangle$, $t = 2$

4. $\mathbf{r}(t) = \langle t^2, 1/t^2 \rangle$, $t = 1$

5. $\mathbf{r}(t) = 3 \cos t \, \mathbf{i} + 2 \sin t \, \mathbf{j}$, $t = \pi/3$

6. $\mathbf{r}(t) = e^t \mathbf{i} + e^{2t} \mathbf{j}$, $t = 0$

7. $\mathbf{r}(t) = t \, \mathbf{i} + t^2 \mathbf{j} + 2 \, \mathbf{k}$, $t = 1$

8. $\mathbf{r}(t) = t \, \mathbf{i} + 2 \cos t \, \mathbf{j} + \sin t \, \mathbf{k}$, $t = 0$

9–14 Find the velocity, acceleration, and speed of a particle with the given position function.

9. $\mathbf{r}(t) = \langle t^2 + t, t^2 - t, t^3 \rangle$

10. $\mathbf{r}(t) = \langle 2 \cos t, 3t, 2 \sin t \rangle$

11. $\mathbf{r}(t) = \sqrt{2}\, t \, \mathbf{i} + e^t \mathbf{j} + e^{-t} \mathbf{k}$

12. $\mathbf{r}(t) = t^2 \mathbf{i} + 2t \, \mathbf{j} + \ln t \, \mathbf{k}$

13. $\mathbf{r}(t) = e^t (\cos t \, \mathbf{i} + \sin t \, \mathbf{j} + t \, \mathbf{k})$

14. $\mathbf{r}(t) = \langle t^2, \sin t - t \cos t, \cos t + t \sin t \rangle$, $t \ge 0$

15–16 Find the velocity and position vectors of a particle that has the given acceleration and the given initial velocity and position.

15. $\mathbf{a}(t) = 2 \, \mathbf{i} + 2t \, \mathbf{k}$, $\mathbf{v}(0) = 3 \, \mathbf{i} - \mathbf{j}$, $\mathbf{r}(0) = \mathbf{j} + \mathbf{k}$

16. $\mathbf{a}(t) = \sin t \, \mathbf{i} + 2 \cos t \, \mathbf{j} + 6t \, \mathbf{k}$,
$\mathbf{v}(0) = -\mathbf{k}$, $\mathbf{r}(0) = \mathbf{j} - 4 \, \mathbf{k}$

17–18
(a) Find the position vector of a particle that has the given acceleration and the specified initial velocity and position.
(b) Use a computer to graph the path of the particle.

17. $\mathbf{a}(t) = 2t \, \mathbf{i} + \sin t \, \mathbf{j} + \cos 2t \, \mathbf{k}$, $\mathbf{v}(0) = \mathbf{i}$, $\mathbf{r}(0) = \mathbf{j}$

18. $\mathbf{a}(t) = t \, \mathbf{i} + e^t \mathbf{j} + e^{-t} \mathbf{k}$, $\mathbf{v}(0) = \mathbf{k}$, $\mathbf{r}(0) = \mathbf{j} + \mathbf{k}$

19. The position function of a particle is given by $\mathbf{r}(t) = \langle t^2, 5t, t^2 - 16t \rangle$. When is the speed a minimum?

20. What force is required so that a particle of mass m has the position function $\mathbf{r}(t) = t^3 \mathbf{i} + t^2 \mathbf{j} + t^3 \mathbf{k}$?

21. A force with magnitude 20 N acts directly upward from the xy-plane on an object with mass 4 kg. The object starts at the origin with initial velocity $\mathbf{v}(0) = \mathbf{i} - \mathbf{j}$. Find its position function and its speed at time t.

22. Show that if a particle moves with constant speed, then the velocity and acceleration vectors are orthogonal.

23. A projectile is fired with an initial speed of 200 m/s and angle of elevation 60°. Find (a) the range of the projectile, (b) the maximum height reached, and (c) the speed at impact.

24. Rework Exercise 23 if the projectile is fired from a position 100 m above the ground.

25. A ball is thrown at an angle of 45° to the ground. If the ball lands 90 m away, what was the initial speed of the ball?

26. A projectile is fired from a tank with initial speed 400 m/s. Find two angles of elevation that can be used to hit a target 3000 m away.

27. A rifle is fired with angle of elevation 36°. What is the muzzle speed if the maximum height of the bullet is 1600 ft?

28. A batter hits a baseball 3 ft above the ground toward the center field fence, which is 10 ft high and 400 ft from home plate. The ball leaves the bat with speed 115 ft/s at an angle 50° above the horizontal. Is it a home run? (In other words, does the ball clear the fence?)

29. A medieval city has the shape of a square and is protected by walls with length 500 m and height 15 m. You are the commander of an attacking army and the closest you can get to the wall is 100 m. Your plan is to set fire to the city by catapulting heated rocks over the wall (with an initial speed of 80 m/s). At what range of angles should you tell your men to set the catapult? (Assume the path of the rocks is perpendicular to the wall.)

30. Show that a projectile reaches three-quarters of its maximum height in half the time needed to reach its maximum height.

31. A ball is thrown eastward into the air from the origin (in the direction of the positive x-axis). The initial velocity is $50\,\mathbf{i} + 80\,\mathbf{k}$, with speed measured in feet per second. The spin of the ball results in a southward acceleration of 4 ft/s^2, so the acceleration vector is $\mathbf{a} = -4\,\mathbf{j} - 32\,\mathbf{k}$. Where does the ball land and with what speed?

32. A ball with mass 0.8 kg is thrown southward into the air with a speed of 30 m/s at an angle of 30° to the ground. A west wind applies a steady force of 4 N to the ball in an easterly direction. Where does the ball land and with what speed?

33. Water traveling along a straight portion of a river normally flows fastest in the middle, and the speed slows to almost zero at the banks. Consider a long straight stretch of river flowing north, with parallel banks 40 m apart. If the maximum water speed is 3 m/s, we can use a quadratic function as a basic model for the rate of water flow x units from the west bank: $f(x) = \frac{3}{400} x(40 - x)$.
   (a) A boat proceeds at a constant speed of 5 m/s from a point A on the west bank while maintaining a heading perpendicular to the bank. How far down the river on the opposite bank will the boat touch shore? Graph the path of the boat.
   (b) Suppose we would like to pilot the boat to land at the point B on the east bank directly opposite A. If we maintain a constant speed of 5 m/s and a constant heading, find the angle at which the boat should head. Then graph the actual path the boat follows. Does the path seem realistic?

34. Another reasonable model for the water speed of the river in Exercise 33 is a sine function: $f(x) = 3\sin(\pi x/40)$. If a boater would like to cross the river from A to B with constant heading and a constant speed of 5 m/s, determine the angle at which the boat should head.

35. A particle has position function $\mathbf{r}(t)$. If $\mathbf{r}'(t) = \mathbf{c} \times \mathbf{r}(t)$, where $\mathbf{c}$ is a constant vector, describe the path of the particle.

36. (a) If a particle moves along a straight line, what can you say about its acceleration vector?
   (b) If a particle moves with constant speed along a curve, what can you say about its acceleration vector?

37–40 Find the tangential and normal components of the acceleration vector.

37. $\mathbf{r}(t) = (t^2 + 1)\,\mathbf{i} + t^3\,\mathbf{j}, \quad t \geq 0$

38. $\mathbf{r}(t) = 2t^2\,\mathbf{i} + \left(\frac{2}{3}t^3 - 2t\right)\mathbf{j}$

39. $\mathbf{r}(t) = \cos t\,\mathbf{i} + \sin t\,\mathbf{j} + t\,\mathbf{k}$

40. $\mathbf{r}(t) = t\,\mathbf{i} + 2e^t\,\mathbf{j} + e^{2t}\,\mathbf{k}$

41–42 Find the tangential and normal components of the acceleration vector at the given point.

41. $\mathbf{r}(t) = \ln t\,\mathbf{i} + (t^2 + 3t)\,\mathbf{j} + 4\sqrt{t}\,\mathbf{k}, \quad (0, 4, 4)$

42. $\mathbf{r}(t) = \dfrac{1}{t}\,\mathbf{i} + \dfrac{1}{t^2}\,\mathbf{j} + \dfrac{1}{t^3}\,\mathbf{k}, \quad (1, 1, 1)$

43. The magnitude of the acceleration vector $\mathbf{a}$ is 10 cm/s^2. Use the figure to estimate the tangential and normal components of $\mathbf{a}$.

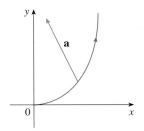

44. If a particle with mass m moves with position vector $\mathbf{r}(t)$, then its **angular momentum** is defined as $\mathbf{L}(t) = m\mathbf{r}(t) \times \mathbf{v}(t)$ and its **torque** as $\boldsymbol{\tau}(t) = m\mathbf{r}(t) \times \mathbf{a}(t)$. Show that $\mathbf{L}'(t) = \boldsymbol{\tau}(t)$. Deduce that if $\boldsymbol{\tau}(t) = \mathbf{0}$ for all t, then $\mathbf{L}(t)$ is constant. (This is the *law of conservation of angular momentum*.)

45. The position function of a spaceship is
$$\mathbf{r}(t) = (3 + t)\,\mathbf{i} + (2 + \ln t)\,\mathbf{j} + \left(7 - \frac{4}{t^2 + 1}\right)\mathbf{k}$$
and the coordinates of a space station are $(6, 4, 9)$. The captain wants the spaceship to coast into the space station. When should the engines be turned off?

46. A rocket burning its onboard fuel while moving through space has velocity $\mathbf{v}(t)$ and mass $m(t)$ at time t. If the exhaust gases escape with velocity $\mathbf{v}_e$ relative to the rocket, it can be deduced from Newton's Second Law of Motion that
$$m\frac{d\mathbf{v}}{dt} = \frac{dm}{dt}\mathbf{v}_e$$
   (a) Show that $\mathbf{v}(t) = \mathbf{v}(0) - \ln\dfrac{m(0)}{m(t)}\,\mathbf{v}_e$.
   (b) For the rocket to accelerate in a straight line from rest to twice the speed of its own exhaust gases, what fraction of its initial mass would the rocket have to burn as fuel?

APPLIED PROJECT   **KEPLER'S LAWS**

Johannes Kepler stated the following three laws of planetary motion on the basis of massive amounts of data on the positions of the planets at various times.

> **Kepler's Laws**
>
> **1.** A planet revolves around the sun in an elliptical orbit with the sun at one focus.
>
> **2.** The line joining the sun to a planet sweeps out equal areas in equal times.
>
> **3.** The square of the period of revolution of a planet is proportional to the cube of the length of the major axis of its orbit.

Kepler formulated these laws because they fitted the astronomical data. He wasn't able to see why they were true or how they related to each other. But Sir Isaac Newton, in his *Principia Mathematica* of 1687, showed how to deduce Kepler's three laws from two of Newton's own laws, the Second Law of Motion and the Law of Universal Gravitation. In Section 13.4 we proved Kepler's First Law using the calculus of vector functions. In this project we guide you through the proofs of Kepler's Second and Third Laws and explore some of their consequences.

1. Use the following steps to prove Kepler's Second Law. The notation is the same as in the proof of the First Law in Section 13.4. In particular, use polar coordinates so that $\mathbf{r} = (r\cos\theta)\,\mathbf{i} + (r\sin\theta)\,\mathbf{j}$.

(a) Show that $\mathbf{h} = r^2 \dfrac{d\theta}{dt}\,\mathbf{k}$.

(b) Deduce that $r^2 \dfrac{d\theta}{dt} = h$.

(c) If $A = A(t)$ is the area swept out by the radius vector $\mathbf{r} = \mathbf{r}(t)$ in the time interval $[t_0, t]$ as in the figure, show that

$$\frac{dA}{dt} = \tfrac{1}{2} r^2 \frac{d\theta}{dt}$$

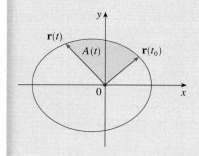

(d) Deduce that

$$\frac{dA}{dt} = \tfrac{1}{2} h = \text{constant}$$

This says that the rate at which A is swept out is constant and proves Kepler's Second Law.

2. Let T be the period of a planet about the sun; that is, T is the time required for it to travel once around its elliptical orbit. Suppose that the lengths of the major and minor axes of the ellipse are $2a$ and $2b$.

(a) Use part (d) of Problem 1 to show that $T = 2\pi ab/h$.

(b) Show that $\dfrac{h^2}{GM} = ed = \dfrac{b^2}{a}$.

(c) Use parts (a) and (b) to show that $T^2 = \dfrac{4\pi^2}{GM}\,a^3$.

This proves Kepler's Third Law. [Notice that the proportionality constant $4\pi^2/(GM)$ is independent of the planet.]

3. The period of the earth's orbit is approximately 365.25 days. Use this fact and Kepler's Third Law to find the length of the major axis of the earth's orbit. You will need the mass of the sun, $M = 1.99 \times 10^{30}$ kg, and the gravitational constant, $G = 6.67 \times 10^{-11}$ N·m²/kg².

4. It's possible to place a satellite into orbit about the earth so that it remains fixed above a given location on the equator. Compute the altitude that is needed for such a satellite. The earth's mass is 5.98×10^{24} kg; its radius is 6.37×10^6 m. (This orbit is called the Clarke Geosynchronous Orbit after Arthur C. Clarke, who first proposed the idea in 1945. The first such satellite, *Syncom II,* was launched in July 1963.)

13   REVIEW

CONCEPT CHECK

Answers to the Concept Check can be found on the back endpapers.

1. What is a vector function? How do you find its derivative and its integral?

2. What is the connection between vector functions and space curves?

3. How do you find the tangent vector to a smooth curve at a point? How do you find the tangent line? The unit tangent vector?

4. If **u** and **v** are differentiable vector functions, c is a scalar, and f is a real-valued function, write the rules for differentiating the following vector functions.
(a) $\mathbf{u}(t) + \mathbf{v}(t)$ (b) $c\mathbf{u}(t)$ (c) $f(t)\mathbf{u}(t)$
(d) $\mathbf{u}(t) \cdot \mathbf{v}(t)$ (e) $\mathbf{u}(t) \times \mathbf{v}(t)$ (f) $\mathbf{u}(f(t))$

5. How do you find the length of a space curve given by a vector function $\mathbf{r}(t)$?

6. (a) What is the definition of curvature?
(b) Write a formula for curvature in terms of $\mathbf{r}'(t)$ and $\mathbf{T}'(t)$.
(c) Write a formula for curvature in terms of $\mathbf{r}'(t)$ and $\mathbf{r}''(t)$.
(d) Write a formula for the curvature of a plane curve with equation $y = f(x)$.

7. (a) Write formulas for the unit normal and binormal vectors of a smooth space curve $\mathbf{r}(t)$.
(b) What is the normal plane of a curve at a point? What is the osculating plane? What is the osculating circle?

8. (a) How do you find the velocity, speed, and acceleration of a particle that moves along a space curve?
(b) Write the acceleration in terms of its tangential and normal components.

9. State Kepler's Laws.

TRUE-FALSE QUIZ

Determine whether the statement is true or false. If it is true, explain why. If it is false, explain why or give an example that disproves the statement.

1. The curve with vector equation $\mathbf{r}(t) = t^3\mathbf{i} + 2t^3\mathbf{j} + 3t^3\mathbf{k}$ is a line.

2. The curve $\mathbf{r}(t) = \langle 0, t^2, 4t \rangle$ is a parabola.

3. The curve $\mathbf{r}(t) = \langle 2t, 3 - t, 0 \rangle$ is a line that passes through the origin.

4. The derivative of a vector function is obtained by differentiating each component function.

5. If $\mathbf{u}(t)$ and $\mathbf{v}(t)$ are differentiable vector functions, then
$$\frac{d}{dt}[\mathbf{u}(t) \times \mathbf{v}(t)] = \mathbf{u}'(t) \times \mathbf{v}'(t)$$

6. If $\mathbf{r}(t)$ is a differentiable vector function, then
$$\frac{d}{dt}|\mathbf{r}(t)| = |\mathbf{r}'(t)|$$

7. If $\mathbf{T}(t)$ is the unit tangent vector of a smooth curve, then the curvature is $\kappa = |d\mathbf{T}/dt|$.

8. The binormal vector is $\mathbf{B}(t) = \mathbf{N}(t) \times \mathbf{T}(t)$.

9. Suppose f is twice continuously differentiable. At an inflection point of the curve $y = f(x)$, the curvature is 0.

10. If $\kappa(t) = 0$ for all t, the curve is a straight line.

11. If $|\mathbf{r}(t)| = 1$ for all t, then $|\mathbf{r}'(t)|$ is a constant.

12. If $|\mathbf{r}(t)| = 1$ for all t, then $\mathbf{r}'(t)$ is orthogonal to $\mathbf{r}(t)$ for all t.

13. The osculating circle of a curve C at a point has the same tangent vector, normal vector, and curvature as C at that point.

14. Different parametrizations of the same curve result in identical tangent vectors at a given point on the curve.

EXERCISES

1. (a) Sketch the curve with vector function

$$\mathbf{r}(t) = t\,\mathbf{i} + \cos \pi t\,\mathbf{j} + \sin \pi t\,\mathbf{k} \qquad t \geqslant 0$$

  (b) Find $\mathbf{r}'(t)$ and $\mathbf{r}''(t)$.

2. Let $\mathbf{r}(t) = \left\langle \sqrt{2 - t},\, (e^t - 1)/t,\, \ln(t + 1) \right\rangle$.
  (a) Find the domain of $\mathbf{r}$.
  (b) Find $\lim_{t \to 0} \mathbf{r}(t)$.
  (c) Find $\mathbf{r}'(t)$.

3. Find a vector function that represents the curve of intersection of the cylinder $x^2 + y^2 = 16$ and the plane $x + z = 5$.

4. Find parametric equations for the tangent line to the curve $x = 2 \sin t$, $y = 2 \sin 2t$, $z = 2 \sin 3t$ at the point $\left(1, \sqrt{3}, 2\right)$. Graph the curve and the tangent line on a common screen.

5. If $\mathbf{r}(t) = t^2\,\mathbf{i} + t \cos \pi t\,\mathbf{j} + \sin \pi t\,\mathbf{k}$, evaluate $\int_0^1 \mathbf{r}(t)\,dt$.

6. Let C be the curve with equations $x = 2 - t^3$, $y = 2t - 1$, $z = \ln t$. Find (a) the point where C intersects the xz-plane, (b) parametric equations of the tangent line at $(1, 1, 0)$, and (c) an equation of the normal plane to C at $(1, 1, 0)$.

7. Use Simpson's Rule with $n = 6$ to estimate the length of the arc of the curve with equations $x = t^2$, $y = t^3$, $z = t^4$, $0 \leqslant t \leqslant 3$.

8. Find the length of the curve $\mathbf{r}(t) = \langle 2t^{3/2}, \cos 2t, \sin 2t \rangle$, $0 \leqslant t \leqslant 1$.

9. The helix $\mathbf{r}_1(t) = \cos t\,\mathbf{i} + \sin t\,\mathbf{j} + t\,\mathbf{k}$ intersects the curve $\mathbf{r}_2(t) = (1 + t)\,\mathbf{i} + t^2\,\mathbf{j} + t^3\,\mathbf{k}$ at the point $(1, 0, 0)$. Find the angle of intersection of these curves.

10. Reparametrize the curve $\mathbf{r}(t) = e^t\,\mathbf{i} + e^t \sin t\,\mathbf{j} + e^t \cos t\,\mathbf{k}$ with respect to arc length measured from the point $(1, 0, 1)$ in the direction of increasing t.

11. For the curve given by $\mathbf{r}(t) = \langle \sin^3 t, \cos^3 t, \sin^2 t \rangle$, $0 \leqslant t \leqslant \pi/2$, find
  (a) the unit tangent vector,
  (b) the unit normal vector,
  (c) the unit binormal vector, and
  (d) the curvature.

12. Find the curvature of the ellipse $x = 3 \cos t$, $y = 4 \sin t$ at the points $(3, 0)$ and $(0, 4)$.

13. Find the curvature of the curve $y = x^4$ at the point $(1, 1)$.

14. Find an equation of the osculating circle of the curve $y = x^4 - x^2$ at the origin. Graph both the curve and its osculating circle.

15. Find an equation of the osculating plane of the curve $x = \sin 2t$, $y = t$, $z = \cos 2t$ at the point $(0, \pi, 1)$.

16. The figure shows the curve C traced by a particle with position vector $\mathbf{r}(t)$ at time t.
  (a) Draw a vector that represents the average velocity of the particle over the time interval $3 \leqslant t \leqslant 3.2$.
  (b) Write an expression for the velocity $\mathbf{v}(3)$.
  (c) Write an expression for the unit tangent vector $\mathbf{T}(3)$ and draw it.

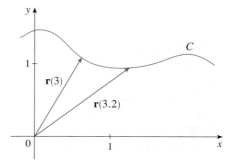

17. A particle moves with position function $\mathbf{r}(t) = t \ln t\,\mathbf{i} + t\,\mathbf{j} + e^{-t}\,\mathbf{k}$. Find the velocity, speed, and acceleration of the particle.

18. Find the velocity, speed, and acceleration of a particle moving with position function $\mathbf{r}(t) = (2t^2 - 3)\,\mathbf{i} + 2t\,\mathbf{j}$. Sketch the path of the particle and draw the position, velocity, and acceleration vectors for $t = 1$.

19. A particle starts at the origin with initial velocity $\mathbf{i} - \mathbf{j} + 3\,\mathbf{k}$. Its acceleration is $\mathbf{a}(t) = 6t\,\mathbf{i} + 12t^2\,\mathbf{j} - 6t\,\mathbf{k}$. Find its position function.

20. An athlete throws a shot at an angle of $45°$ to the horizontal at an initial speed of 43 ft/s. It leaves his hand 7 ft above the ground.
  (a) Where is the shot 2 seconds later?
  (b) How high does the shot go?
  (c) Where does the shot land?

21. A projectile is launched with an initial speed of 40 m/s from the floor of a tunnel whose height is 30 m. What angle of elevation should be used to achieve the maximum possible horizontal range of the projectile? What is the maximum range?

22. Find the tangential and normal components of the acceleration vector of a particle with position function

$$\mathbf{r}(t) = t\,\mathbf{i} + 2t\,\mathbf{j} + t^2\,\mathbf{k}$$

23. A disk of radius 1 is rotating in the counterclockwise direction at a constant angular speed ω. A particle starts at the center of the disk and moves toward the edge along a fixed radius so that its position at time t, $t \geqslant 0$, is given by

$\mathbf{r}(t) = t\mathbf{R}(t)$, where

$$\mathbf{R}(t) = \cos \omega t\, \mathbf{i} + \sin \omega t\, \mathbf{j}$$

(a) Show that the velocity $\mathbf{v}$ of the particle is

$$\mathbf{v} = \cos \omega t\, \mathbf{i} + \sin \omega t\, \mathbf{j} + t\mathbf{v}_d$$

where $\mathbf{v}_d = \mathbf{R}'(t)$ is the velocity of a point on the edge of the disk.
(b) Show that the acceleration $\mathbf{a}$ of the particle is

$$\mathbf{a} = 2\mathbf{v}_d + t\mathbf{a}_d$$

where $\mathbf{a}_d = \mathbf{R}''(t)$ is the acceleration of a point on the edge of the disk. The extra term $2\mathbf{v}_d$ is called the *Coriolis acceleration*; it is the result of the interaction of the rotation of the disk and the motion of the particle. One can obtain a physical demonstration of this acceleration by walking toward the edge of a moving merry-go-round.
(c) Determine the Coriolis acceleration of a particle that moves on a rotating disk according to the equation

$$\mathbf{r}(t) = e^{-t} \cos \omega t\, \mathbf{i} + e^{-t} \sin \omega t\, \mathbf{j}$$

24. In designing *transfer curves* to connect sections of straight railroad tracks, it's important to realize that the acceleration of the train should be continuous so that the reactive force exerted by the train on the track is also continuous. Because of the formulas for the components of acceleration in Section 13.4, this will be the case if the curvature varies continuously.
(a) A logical candidate for a transfer curve to join existing tracks given by $y = 1$ for $x \le 0$ and $y = \sqrt{2} - x$ for $x \ge 1/\sqrt{2}$ might be the function $f(x) = \sqrt{1 - x^2}$, $0 < x < 1/\sqrt{2}$, whose graph is the arc of the circle shown in the figure. It looks reasonable at first glance. Show that the function

$$F(x) = \begin{cases} 1 & \text{if } x \le 0 \\ \sqrt{1 - x^2} & \text{if } 0 < x < 1/\sqrt{2} \\ \sqrt{2} - x & \text{if } x \ge 1/\sqrt{2} \end{cases}$$

is continuous and has continuous slope, but does not have continuous curvature. Therefore f is not an appropriate transfer curve.

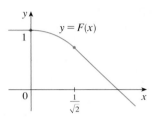

(b) Find a fifth-degree polynomial to serve as a transfer curve between the following straight line segments: $y = 0$ for $x \le 0$ and $y = x$ for $x \ge 1$. Could this be done with a fourth-degree polynomial? Use a graphing calculator or computer to sketch the graph of the "connected" function and check to see that it looks like the one in the figure.

Problems Plus

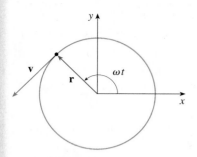

FIGURE FOR PROBLEM 1

1. A particle P moves with constant angular speed ω around a circle whose center is at the origin and whose radius is R. The particle is said to be in *uniform circular motion*. Assume that the motion is counterclockwise and that the particle is at the point $(R, 0)$ when $t = 0$. The position vector at time $t \geq 0$ is $\mathbf{r}(t) = R \cos \omega t \, \mathbf{i} + R \sin \omega t \, \mathbf{j}$.

(a) Find the velocity vector $\mathbf{v}$ and show that $\mathbf{v} \cdot \mathbf{r} = 0$. Conclude that $\mathbf{v}$ is tangent to the circle and points in the direction of the motion.

(b) Show that the speed $|\mathbf{v}|$ of the particle is the constant ωR. The *period* T of the particle is the time required for one complete revolution. Conclude that

$$T = \frac{2\pi R}{|\mathbf{v}|} = \frac{2\pi}{\omega}$$

(c) Find the acceleration vector $\mathbf{a}$. Show that it is proportional to $\mathbf{r}$ and that it points toward the origin. An acceleration with this property is called a *centripetal acceleration*. Show that the magnitude of the acceleration vector is $|\mathbf{a}| = R\omega^2$.

(d) Suppose that the particle has mass m. Show that the magnitude of the force $\mathbf{F}$ that is required to produce this motion, called a *centripetal force*, is

$$|\mathbf{F}| = \frac{m|\mathbf{v}|^2}{R}$$

FIGURE FOR PROBLEM 2

2. A circular curve of radius R on a highway is banked at an angle θ so that a car can safely traverse the curve without skidding when there is no friction between the road and the tires. The loss of friction could occur, for example, if the road is covered with a film of water or ice. The rated speed v_R of the curve is the maximum speed that a car can attain without skidding. Suppose a car of mass m is traversing the curve at the rated speed v_R. Two forces are acting on the car: the vertical force, mg, due to the weight of the car, and a force $\mathbf{F}$ exerted by, and normal to, the road (see the figure).

The vertical component of $\mathbf{F}$ balances the weight of the car, so that $|\mathbf{F}| \cos \theta = mg$. The horizontal component of $\mathbf{F}$ produces a centripetal force on the car so that, by Newton's Second Law and part (d) of Problem 1,

$$|\mathbf{F}| \sin \theta = \frac{m v_R^2}{R}$$

(a) Show that $v_R^2 = Rg \tan \theta$.

(b) Find the rated speed of a circular curve with radius 400 ft that is banked at an angle of $12°$.

(c) Suppose the design engineers want to keep the banking at $12°$, but wish to increase the rated speed by 50%. What should the radius of the curve be?

3. A projectile is fired from the origin with angle of elevation α and initial speed v_0. Assuming that air resistance is negligible and that the only force acting on the projectile is gravity, g, we showed in Example 13.4.5 that the position vector of the projectile is

$$\mathbf{r}(t) = (v_0 \cos \alpha)t \, \mathbf{i} + \left[(v_0 \sin \alpha)t - \tfrac{1}{2}gt^2 \right] \mathbf{j}$$

We also showed that the maximum horizontal distance of the projectile is achieved when $\alpha = 45°$ and in this case the range is $R = v_0^2/g$.

(a) At what angle should the projectile be fired to achieve maximum height and what is the maximum height?

(b) Fix the initial speed v_0 and consider the parabola $x^2 + 2Ry - R^2 = 0$, whose graph is shown in the figure at the left. Show that the projectile can hit any target inside or on the boundary of the region bounded by the parabola and the x-axis, and that it can't hit any target outside this region.

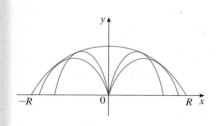

FIGURE FOR PROBLEM 3

(c) Suppose that the gun is elevated to an angle of inclination α in order to aim at a target that is suspended at a height h directly over a point D units downrange (see the figure below). The target is released at the instant the gun is fired. Show that the projectile always hits the target, regardless of the value v_0, provided the projectile does not hit the ground "before" D.

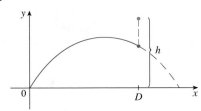

4. (a) A projectile is fired from the origin down an inclined plane that makes an angle θ with the horizontal. The angle of elevation of the gun and the initial speed of the projectile are α and v_0, respectively. Find the position vector of the projectile and the parametric equations of the path of the projectile as functions of the time t. (Ignore air resistance.)

   (b) Show that the angle of elevation α that will maximize the downhill range is the angle halfway between the plane and the vertical.

   (c) Suppose the projectile is fired up an inclined plane whose angle of inclination is θ. Show that, in order to maximize the (uphill) range, the projectile should be fired in the direction halfway between the plane and the vertical.

   (d) In a paper presented in 1686, Edmond Halley summarized the laws of gravity and projectile motion and applied them to gunnery. One problem he posed involved firing a projectile to hit a target a distance R up an inclined plane. Show that the angle at which the projectile should be fired to hit the target but use the least amount of energy is the same as the angle in part (c). (Use the fact that the energy needed to fire the projectile is proportional to the square of the initial speed, so minimizing the energy is equivalent to minimizing the initial speed.)

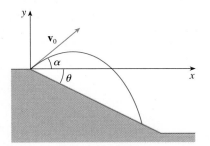

FIGURE FOR PROBLEM 4

5. A ball rolls off a table with a speed of 2 ft/s. The table is 3.5 ft high.

   (a) Determine the point at which the ball hits the floor and find its speed at the instant of impact.

   (b) Find the angle θ between the path of the ball and the vertical line drawn through the point of impact (see the figure).

   (c) Suppose the ball rebounds from the floor at the same angle with which it hits the floor, but loses 20% of its speed due to energy absorbed by the ball on impact. Where does the ball strike the floor on the second bounce?

FIGURE FOR PROBLEM 5

6. Find the curvature of the curve with parametric equations

$$x = \int_0^t \sin\left(\tfrac{1}{2}\pi\theta^2\right) d\theta \qquad y = \int_0^t \cos\left(\tfrac{1}{2}\pi\theta^2\right) d\theta$$

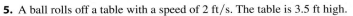

7. If a projectile is fired with angle of elevation α and initial speed v, then parametric equations for its trajectory are

$$x = (v \cos \alpha)t \qquad y = (v \sin \alpha)t - \tfrac{1}{2}gt^2$$

(See Example 13.4.5.) We know that the range (horizontal distance traveled) is maximized when $\alpha = 45°$. What value of α maximizes the total distance traveled by the projectile? (State your answer correct to the nearest degree.)

8. A cable has radius r and length L and is wound around a spool with radius R without overlapping. What is the shortest length along the spool that is covered by the cable?

9. Show that the curve with vector equation

$$\mathbf{r}(t) = \langle a_1 t^2 + b_1 t + c_1, a_2 t^2 + b_2 t + c_2, a_3 t^2 + b_3 t + c_3 \rangle$$

lies in a plane and find an equation of the plane.

14 Partial Derivatives

Courtesy Speedo and ANSYS, Inc.

In 2008 Speedo introduced the LZR Racer and, because it reduced drag in the water, many swimming records were broken. In the project on page 936 you are asked to use partial derivatives to explain why a small decrease in drag can have a big effect on performance.

SO FAR WE HAVE DEALT with the calculus of functions of a single variable. But, in the real world, physical quantities often depend on two or more variables, so in this chapter we turn our attention to functions of several variables and extend the basic ideas of differential calculus to such functions.

14.1   Functions of Several Variables

In this section we study functions of two or more variables from four points of view:

- verbally   (by a description in words)
- numerically   (by a table of values)
- algebraically   (by an explicit formula)
- visually   (by a graph or level curves)

■ Functions of Two Variables

The temperature T at a point on the surface of the earth at any given time depends on the longitude x and latitude y of the point. We can think of T as being a function of the two variables x and y, or as a function of the pair (x, y). We indicate this functional dependence by writing $T = f(x, y)$.

The volume V of a circular cylinder depends on its radius r and its height h. In fact, we know that $V = \pi r^2 h$. We say that V is a function of r and h, and we write $V(r, h) = \pi r^2 h$.

> **Definition** A **function f of two variables** is a rule that assigns to each ordered pair of real numbers (x, y) in a set D a unique real number denoted by $f(x, y)$. The set D is the **domain** of f and its **range** is the set of values that f takes on, that is, $\{f(x, y) \mid (x, y) \in D\}$.

We often write $z = f(x, y)$ to make explicit the value taken on by f at the general point (x, y). The variables x and y are **independent variables** and z is the **dependent variable**. [Compare this with the notation $y = f(x)$ for functions of a single variable.]

A function of two variables is just a function whose domain is a subset of $\mathbb{R}^2$ and whose range is a subset of $\mathbb{R}$. One way of visualizing such a function is by means of an arrow diagram (see Figure 1), where the domain D is represented as a subset of the xy-plane and the range is a set of numbers on a real line, shown as a z-axis. For instance, if $f(x, y)$ represents the temperature at a point (x, y) in a flat metal plate with the shape of D, we can think of the z-axis as a thermometer displaying the recorded temperatures.

If a function f is given by a formula and no domain is specified, then the domain of f is understood to be the set of all pairs (x, y) for which the given expression is a well-defined real number.

FIGURE 1

EXAMPLE 1   For each of the following functions, evaluate $f(3, 2)$ and find and sketch the domain.

(a) $f(x, y) = \dfrac{\sqrt{x + y + 1}}{x - 1}$

(b) $f(x, y) = x \ln(y^2 - x)$

SOLUTION

(a) $f(3, 2) = \dfrac{\sqrt{3 + 2 + 1}}{3 - 1} = \dfrac{\sqrt{6}}{2}$

The expression for f makes sense if the denominator is not 0 and the quantity under the square root sign is nonnegative. So the domain of f is

$$D = \{(x, y) \mid x + y + 1 \geqslant 0, \ x \neq 1\}$$

The inequality $x + y + 1 \geqslant 0$, or $y \geqslant -x - 1$, describes the points that lie on or

$x + y + 1 = 0$

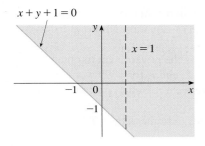

FIGURE 2

Domain of $f(x, y) = \dfrac{\sqrt{x + y + 1}}{x - 1}$

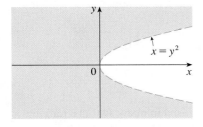

FIGURE 3

Domain of $f(x, y) = x \ln(y^2 - x)$

The Wind-Chill Index

The wind-chill index measures how cold it feels when it's windy. It is based on a model of how fast a human face loses heat. It was developed through clinical trials in which volunteers were exposed to a variety of temperatures and wind speeds in a refrigerated wind tunnel.

above the line $y = -x - 1$, while $x \neq 1$ means that the points on the line $x = 1$ must be excluded from the domain. (See Figure 2.)

(b) $$f(3, 2) = 3 \ln(2^2 - 3) = 3 \ln 1 = 0$$

Since $\ln(y^2 - x)$ is defined only when $y^2 - x > 0$, that is, $x < y^2$, the domain of f is $D = \{(x, y) \mid x < y^2\}$. This is the set of points to the left of the parabola $x = y^2$. (See Figure 3.) ∎

Not all functions can be represented by explicit formulas. The function in the next example is described verbally and by numerical estimates of its values.

EXAMPLE 2 In regions with severe winter weather, the *wind-chill index* is often used to describe the apparent severity of the cold. This index W is a subjective temperature that depends on the actual temperature T and the wind speed v. So W is a function of T and v, and we can write $W = f(T, v)$. Table 1 records values of W compiled by the US National Weather Service and the Meteorological Service of Canada.

Table 1 Wind-chill index as a function of air temperature and wind speed

Wind speed (km/h)

T \ v	5	10	15	20	25	30	40	50	60	70	80
5	4	3	2	1	1	0	−1	−1	−2	−2	−3
0	−2	−3	−4	−5	−6	−6	−7	−8	−9	−9	−10
−5	−7	−9	−11	−12	−12	−13	−14	−15	−16	−16	−17
−10	−13	−15	−17	−18	−19	−20	−21	−22	−23	−23	−24
−15	−19	−21	−23	−24	−25	−26	−27	−29	−30	−30	−31
−20	−24	−27	−29	−30	−32	−33	−34	−35	−36	−37	−38
−25	−30	−33	−35	−37	−38	−39	−41	−42	−43	−44	−45
−30	−36	−39	−41	−43	−44	−46	−48	−49	−50	−51	−52
−35	−41	−45	−48	−49	−51	−52	−54	−56	−57	−58	−60
−40	−47	−51	−54	−56	−57	−59	−61	−63	−64	−65	−67

Actual temperature (°C)

For instance, the table shows that if the temperature is $-5°C$ and the wind speed is 50 km/h, then subjectively it would feel as cold as a temperature of about $-15°C$ with no wind. So

$$f(-5, 50) = -15$$ ∎

EXAMPLE 3 In 1928 Charles Cobb and Paul Douglas published a study in which they modeled the growth of the American economy during the period 1899–1922. They considered a simplified view of the economy in which production output is determined by the amount of labor involved and the amount of capital invested. While there are many other factors affecting economic performance, their model proved to be remarkably accurate. The function they used to model production was of the form

$$\boxed{1} \qquad P(L, K) = bL^\alpha K^{1-\alpha}$$

where P is the total production (the monetary value of all goods produced in a year), L is the amount of labor (the total number of person-hours worked in a year), and K is

Table 2

Year	P	L	K
1899	100	100	100
1900	101	105	107
1901	112	110	114
1902	122	117	122
1903	124	122	131
1904	122	121	138
1905	143	125	149
1906	152	134	163
1907	151	140	176
1908	126	123	185
1909	155	143	198
1910	159	147	208
1911	153	148	216
1912	177	155	226
1913	184	156	236
1914	169	152	244
1915	189	156	266
1916	225	183	298
1917	227	198	335
1918	223	201	366
1919	218	196	387
1920	231	194	407
1921	179	146	417
1922	240	161	431

the amount of capital invested (the monetary worth of all machinery, equipment, and buildings). In Section 14.3 we will show how the form of Equation 1 follows from certain economic assumptions.

Cobb and Douglas used economic data published by the government to obtain Table 2. They took the year 1899 as a baseline and P, L, and K for 1899 were each assigned the value 100. The values for other years were expressed as percentages of the 1899 figures.

Cobb and Douglas used the method of least squares to fit the data of Table 2 to the function

$$\boxed{2} \qquad P(L, K) = 1.01L^{0.75}K^{0.25}$$

(See Exercise 81 for the details.)

If we use the model given by the function in Equation 2 to compute the production in the years 1910 and 1920, we get the values

$$P(147, 208) = 1.01(147)^{0.75}(208)^{0.25} \approx 161.9$$

$$P(194, 407) = 1.01(194)^{0.75}(407)^{0.25} \approx 235.8$$

which are quite close to the actual values, 159 and 231.

The production function (1) has subsequently been used in many settings, ranging from individual firms to global economics. It has become known as the **Cobb-Douglas production function**. Its domain is $\{(L, K) \mid L \geqslant 0, K \geqslant 0\}$ because L and K represent labor and capital and are therefore never negative. ∎

EXAMPLE 4 Find the domain and range of $g(x, y) = \sqrt{9 - x^2 - y^2}$.

SOLUTION The domain of g is

$$D = \{(x, y) \mid 9 - x^2 - y^2 \geqslant 0\} = \{(x, y) \mid x^2 + y^2 \leqslant 9\}$$

which is the disk with center $(0, 0)$ and radius 3. (See Figure 4.) The range of g is

$$\left\{z \mid z = \sqrt{9 - x^2 - y^2}, (x, y) \in D\right\}$$

Since z is a positive square root, $z \geqslant 0$. Also, because $9 - x^2 - y^2 \leqslant 9$, we have

$$\sqrt{9 - x^2 - y^2} \leqslant 3$$

So the range is

$$\{z \mid 0 \leqslant z \leqslant 3\} = [0, 3]$$ ∎

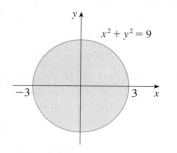

FIGURE 4
Domain of $g(x, y) = \sqrt{9 - x^2 - y^2}$

■ Graphs

Another way of visualizing the behavior of a function of two variables is to consider its graph.

> **Definition** If f is a function of two variables with domain D, then the **graph** of f is the set of all points (x, y, z) in $\mathbb{R}^3$ such that $z = f(x, y)$ and (x, y) is in D.

Just as the graph of a function f of one variable is a curve C with equation $y = f(x)$, so the graph of a function f of two variables is a surface S with equation $z = f(x, y)$. We can visualize the graph S of f as lying directly above or below its domain D in the xy-plane (see Figure 5).

FIGURE 5

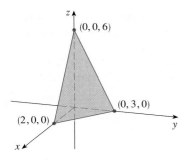

FIGURE 6

EXAMPLE 5 Sketch the graph of the function $f(x, y) = 6 - 3x - 2y$.

SOLUTION The graph of f has the equation $z = 6 - 3x - 2y$, or $3x + 2y + z = 6$, which represents a plane. To graph the plane we first find the intercepts. Putting $y = z = 0$ in the equation, we get $x = 2$ as the x-intercept. Similarly, the y-intercept is 3 and the z-intercept is 6. This helps us sketch the portion of the graph that lies in the first octant in Figure 6. ∎

The function in Example 5 is a special case of the function

$$f(x, y) = ax + by + c$$

which is called a **linear function**. The graph of such a function has the equation

$$z = ax + by + c \qquad \text{or} \qquad ax + by - z + c = 0$$

so it is a plane. In much the same way that linear functions of one variable are important in single-variable calculus, we will see that linear functions of two variables play a central role in multivariable calculus.

EXAMPLE 6 Sketch the graph of $g(x, y) = \sqrt{9 - x^2 - y^2}$.

SOLUTION The graph has equation $z = \sqrt{9 - x^2 - y^2}$. We square both sides of this equation to obtain $z^2 = 9 - x^2 - y^2$, or $x^2 + y^2 + z^2 = 9$, which we recognize as an equation of the sphere with center the origin and radius 3. But, since $z \geqslant 0$, the graph of g is just the top half of this sphere (see Figure 7). ∎

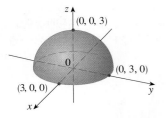

FIGURE 7
Graph of $g(x, y) = \sqrt{9 - x^2 - y^2}$

NOTE An entire sphere can't be represented by a single function of x and y. As we saw in Example 6, the upper hemisphere of the sphere $x^2 + y^2 + z^2 = 9$ is represented by the function $g(x, y) = \sqrt{9 - x^2 - y^2}$. The lower hemisphere is represented by the function $h(x, y) = -\sqrt{9 - x^2 - y^2}$.

EXAMPLE 7 Use a computer to draw the graph of the Cobb-Douglas production function $P(L, K) = 1.01 L^{0.75} K^{0.25}$.

SOLUTION Figure 8 shows the graph of P for values of the labor L and capital K that lie between 0 and 300. The computer has drawn the surface by plotting vertical traces. We see from these traces that the value of the production P increases as either L or K increases, as is to be expected.

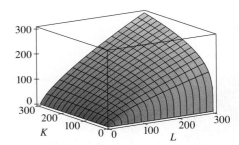

FIGURE 8

∎

EXAMPLE 8 Find the domain and range and sketch the graph of $h(x, y) = 4x^2 + y^2$.

SOLUTION Notice that $h(x, y)$ is defined for all possible ordered pairs of real numbers (x, y), so the domain is $\mathbb{R}^2$, the entire xy-plane. The range of h is the set $[0, \infty)$ of all nonnegative real numbers. [Notice that $x^2 \geqslant 0$ and $y^2 \geqslant 0$, so $h(x, y) \geqslant 0$ for all x

and y.] The graph of h has the equation $z = 4x^2 + y^2$, which is the elliptic paraboloid that we sketched in Example 12.6.4. Horizontal traces are ellipses and vertical traces are parabolas (see Figure 9).

FIGURE 9
Graph of $h(x, y) = 4x^2 + y^2$

        Computer programs are readily available for graphing functions of two variables. In most such programs, traces in the vertical planes $x = k$ and $y = k$ are drawn for equally spaced values of k and parts of the graph are eliminated using hidden line removal.

        Figure 10 shows computer-generated graphs of several functions. Notice that we get an especially good picture of a function when rotation is used to give views from dif-

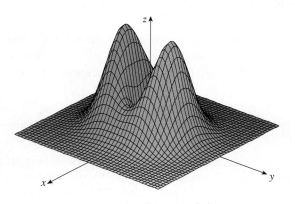

(a) $f(x, y) = (x^2 + 3y^2)e^{-x^2-y^2}$

(b) $f(x, y) = (x^2 + 3y^2)e^{-x^2-y^2}$

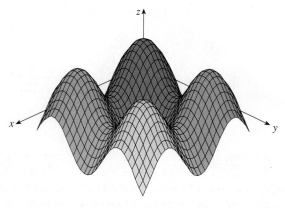

(c) $f(x, y) = \sin x + \sin y$

(d) $f(x, y) = \dfrac{\sin x \, \sin y}{xy}$

FIGURE 10

ferent vantage points. In parts (a) and (b) the graph of f is very flat and close to the xy-plane except near the origin; this is because $e^{-x^2-y^2}$ is very small when x or y is large.

◼ Level Curves

So far we have two methods for visualizing functions: arrow diagrams and graphs. A third method, borrowed from mapmakers, is a contour map on which points of constant elevation are joined to form *contour curves*, or *level curves*.

> **Definition** The **level curves** of a function f of two variables are the curves with equations $f(x, y) = k$, where k is a constant (in the range of f).

A level curve $f(x, y) = k$ is the set of all points in the domain of f at which f takes on a given value k. In other words, it shows where the graph of f has height k.

You can see from Figure 11 the relation between level curves and horizontal traces. The level curves $f(x, y) = k$ are just the traces of the graph of f in the horizontal plane $z = k$ projected down to the xy-plane. So if you draw the level curves of a function and visualize them being lifted up to the surface at the indicated height, then you can mentally piece together a picture of the graph. The surface is steep where the level curves are close together. It is somewhat flatter where they are farther apart.

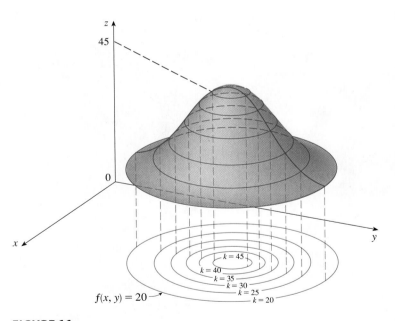

$f(x, y) = 20$
$k = 45$
$k = 40$
$k = 35$
$k = 30$
$k = 25$
$k = 20$

FIGURE 11

4000
4500
5000
LONESOME MTN. • A
5500
B • 5500
5000
4500
Lonesome Creek

FIGURE 12

TEC Visual 14.1A animates Figure 11 by showing level curves being lifted up to graphs of functions.

One common example of level curves occurs in topographic maps of mountainous regions, such as the map in Figure 12. The level curves are curves of constant elevation above sea level. If you walk along one of these contour lines, you neither ascend nor descend. Another common example is the temperature function introduced in the opening paragraph of this section. Here the level curves are called **isothermals** and join loca-

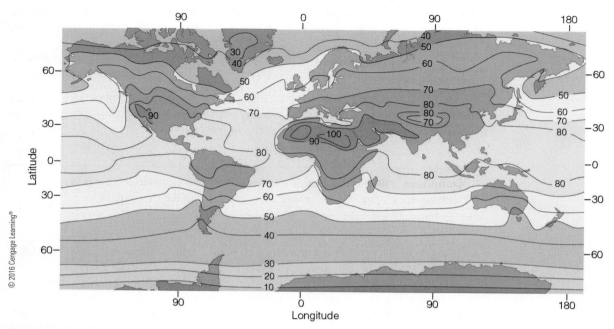

© 2016 Cengage Learning®

FIGURE 13 Average air temperature near sea level in July (°F)

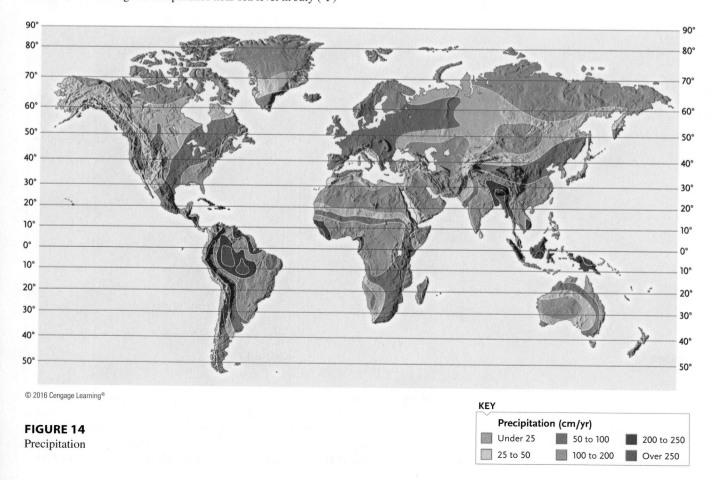

© 2016 Cengage Learning®

FIGURE 14
Precipitation

KEY

Precipitation (cm/yr)

Under 25	50 to 100	200 to 250
25 to 50	100 to 200	Over 250

tions with the same temperature. Figure 13 shows a weather map of the world indicating the average July temperatures. The isothermals are the curves that separate the colored bands.

In weather maps of atmospheric pressure at a given time as a function of longitude and latitude, the level curves are called **isobars** and join locations with the same pressure. (See Exercise 34.) Surface winds tend to flow from areas of high pressure across the isobars toward areas of low pressure, and are strongest where the isobars are tightly packed.

A contour map of world-wide precipitation is shown in Figure 14. Here the level curves are not labeled but they separate the colored regions and the amount of precipitation in each region is indicated in the color key.

FIGURE 15

EXAMPLE 9 A contour map for a function f is shown in Figure 15. Use it to estimate the values of $f(1, 3)$ and $f(4, 5)$.

SOLUTION The point $(1, 3)$ lies partway between the level curves with z-values 70 and 80. We estimate that

$$f(1, 3) \approx 73$$

Similarly, we estimate that $\qquad f(4, 5) \approx 56$ ∎

EXAMPLE 10 Sketch the level curves of the function $f(x, y) = 6 - 3x - 2y$ for the values $k = -6, 0, 6, 12$.

SOLUTION The level curves are

$$6 - 3x - 2y = k \qquad \text{or} \qquad 3x + 2y + (k - 6) = 0$$

This is a family of lines with slope $-\frac{3}{2}$. The four particular level curves with $k = -6, 0, 6$, and 12 are $3x + 2y - 12 = 0$, $3x + 2y - 6 = 0$, $3x + 2y = 0$, and $3x + 2y + 6 = 0$. They are sketched in Figure 16. The level curves are equally spaced parallel lines because the graph of f is a plane (see Figure 6).

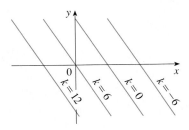

FIGURE 16
Contour map of
$f(x, y) = 6 - 3x - 2y$

∎

EXAMPLE 11 Sketch the level curves of the function

$$g(x, y) = \sqrt{9 - x^2 - y^2} \qquad \text{for} \quad k = 0, 1, 2, 3$$

SOLUTION The level curves are

$$\sqrt{9 - x^2 - y^2} = k \qquad \text{or} \qquad x^2 + y^2 = 9 - k^2$$

This is a family of concentric circles with center $(0, 0)$ and radius $\sqrt{9 - k^2}$. The cases

$k = 0, 1, 2, 3$ are shown in Figure 17. Try to visualize these level curves lifted up to form a surface and compare with the graph of g (a hemisphere) in Figure 7. (See TEC Visual 14.1A.)

FIGURE 17
Contour map of
$g(x, y) = \sqrt{9 - x^2 - y^2}$

EXAMPLE 12 Sketch some level curves of the function $h(x, y) = 4x^2 + y^2 + 1$.

SOLUTION The level curves are

$$4x^2 + y^2 + 1 = k \qquad \text{or} \qquad \frac{x^2}{\frac{1}{4}(k - 1)} + \frac{y^2}{k - 1} = 1$$

which, for $k > 1$, describes a family of ellipses with semiaxes $\frac{1}{2}\sqrt{k - 1}$ and $\sqrt{k - 1}$. Figure 18(a) shows a contour map of h drawn by a computer. Figure 18(b) shows these level curves lifted up to the graph of h (an elliptic paraboloid) where they become horizontal traces. We see from Figure 18 how the graph of h is put together from the level curves.

TEC Visual 14.1B demonstrates the connection between surfaces and their contour maps.

FIGURE 18
The graph of $h(x, y) = 4x^2 + y^2 + 1$
is formed by lifting the level curves.

(a) Contour map   (b) Horizontal traces are raised level curves

EXAMPLE 13 Plot level curves for the Cobb-Douglas production function of Example 3.

SOLUTION In Figure 19 we use a computer to draw a contour plot for the Cobb-Douglas production function

$$P(L, K) = 1.01L^{0.75}K^{0.25}$$

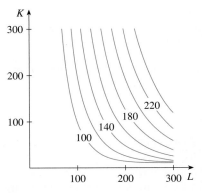

FIGURE 19

Level curves are labeled with the value of the production P. For instance, the level curve labeled 140 shows all values of the labor L and capital investment K that result in a production of $P = 140$. We see that, for a fixed value of P, as L increases K decreases, and vice versa.

For some purposes, a contour map is more useful than a graph. That is certainly true in Example 13. (Compare Figure 19 with Figure 8.) It is also true in estimating function values, as in Example 9.

Figure 20 shows some computer-generated level curves together with the corresponding computer-generated graphs. Notice that the level curves in part (c) crowd together near the origin. That corresponds to the fact that the graph in part (d) is very steep near the origin.

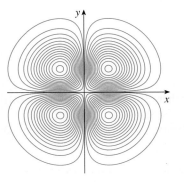

(a) Level curves of $f(x, y) = -xye^{-x^2-y^2}$

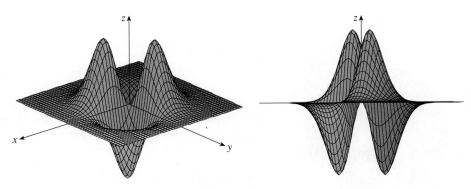

(b) Two views of $f(x, y) = -xye^{-x^2-y^2}$

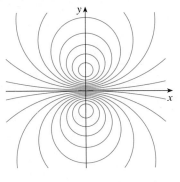

(c) Level curves of $f(x, y) = \dfrac{-3y}{x^2 + y^2 + 1}$

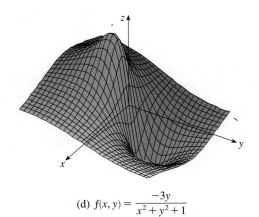

(d) $f(x, y) = \dfrac{-3y}{x^2 + y^2 + 1}$

FIGURE 20

■ Functions of Three or More Variables

A **function of three variables**, f, is a rule that assigns to each ordered triple (x, y, z) in a domain $D \subset \mathbb{R}^3$ a unique real number denoted by $f(x, y, z)$. For instance, the temperature T at a point on the surface of the earth depends on the longitude x and latitude y of the point and on the time t, so we could write $T = f(x, y, t)$.

EXAMPLE 14 Find the domain of f if

$$f(x, y, z) = \ln(z - y) + xy \sin z$$

SOLUTION The expression for $f(x, y, z)$ is defined as long as $z - y > 0$, so the domain of f is

$$D = \{(x, y, z) \in \mathbb{R}^3 \mid z > y\}$$

This is a **half-space** consisting of all points that lie above the plane $z = y$. ∎

It's very difficult to visualize a function f of three variables by its graph, since that would lie in a four-dimensional space. However, we do gain some insight into f by examining its **level surfaces**, which are the surfaces with equations $f(x, y, z) = k$, where k is a constant. If the point (x, y, z) moves along a level surface, the value of $f(x, y, z)$ remains fixed.

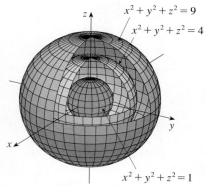

$x^2 + y^2 + z^2 = 9$
$x^2 + y^2 + z^2 = 4$
$x^2 + y^2 + z^2 = 1$

FIGURE 21

EXAMPLE 15 Find the level surfaces of the function

$$f(x, y, z) = x^2 + y^2 + z^2$$

SOLUTION The level surfaces are $x^2 + y^2 + z^2 = k$, where $k \geqslant 0$. These form a family of concentric spheres with radius $\sqrt{k}$. (See Figure 21.) Thus, as (x, y, z) varies over any sphere with center O, the value of $f(x, y, z)$ remains fixed. ∎

Functions of any number of variables can be considered. A **function of n variables** is a rule that assigns a number $z = f(x_1, x_2, \ldots, x_n)$ to an n-tuple $(x_1, x_2, \ldots, x_n)$ of real numbers. We denote by $\mathbb{R}^n$ the set of all such n-tuples. For example, if a company uses n different ingredients in making a food product, c_i is the cost per unit of the ith ingredient, and x_i units of the ith ingredient are used, then the total cost C of the ingredients is a function of the n variables $x_1, x_2, \ldots, x_n$:

$$\boxed{3} \qquad C = f(x_1, x_2, \ldots, x_n) = c_1 x_1 + c_2 x_2 + \cdots + c_n x_n$$

The function f is a real-valued function whose domain is a subset of $\mathbb{R}^n$. Sometimes we will use vector notation to write such functions more compactly: If $\mathbf{x} = \langle x_1, x_2, \ldots, x_n \rangle$, we often write $f(\mathbf{x})$ in place of $f(x_1, x_2, \ldots, x_n)$. With this notation we can rewrite the function defined in Equation 3 as

$$f(\mathbf{x}) = \mathbf{c} \cdot \mathbf{x}$$

where $\mathbf{c} = \langle c_1, c_2, \ldots, c_n \rangle$ and $\mathbf{c} \cdot \mathbf{x}$ denotes the dot product of the vectors $\mathbf{c}$ and $\mathbf{x}$ in V_n.

In view of the one-to-one correspondence between points $(x_1, x_2, \ldots, x_n)$ in $\mathbb{R}^n$ and their position vectors $\mathbf{x} = \langle x_1, x_2, \ldots, x_n \rangle$ in V_n, we have three ways of looking at a function f defined on a subset of $\mathbb{R}^n$:

1. As a function of n real variables $x_1, x_2, \ldots, x_n$

2. As a function of a single point variable $(x_1, x_2, \ldots, x_n)$

3. As a function of a single vector variable $\mathbf{x} = \langle x_1, x_2, \ldots, x_n \rangle$

We will see that all three points of view are useful.

14.1 EXERCISES

1. In Example 2 we considered the function $W = f(T, v)$, where W is the wind-chill index, T is the actual temperature, and v is the wind speed. A numerical representation is given in Table 1 on page 889.
  (a) What is the value of $f(-15, 40)$? What is its meaning?
  (b) Describe in words the meaning of the question "For what value of v is $f(-20, v) = -30$?" Then answer the question.
  (c) Describe in words the meaning of the question "For what value of T is $f(T, 20) = -49$?" Then answer the question.
  (d) What is the meaning of the function $W = f(-5, v)$? Describe the behavior of this function.
  (e) What is the meaning of the function $W = f(T, 50)$? Describe the behavior of this function.

2. The *temperature-humidity index* I (or humidex, for short) is the perceived air temperature when the actual temperature is T and the relative humidity is h, so we can write $I = f(T, h)$. The following table of values of I is an excerpt from a table compiled by the National Oceanic & Atmospheric Administration.

Table 3   Apparent temperature as a function of temperature and humidity

Relative humidity (%)

T \ h	20	30	40	50	60	70
80	77	78	79	81	82	83
85	82	84	86	88	90	93
90	87	90	93	96	100	106
95	93	96	101	107	114	124
100	99	104	110	120	132	144

Actual temperature (°F)

  (a) What is the value of $f(95, 70)$? What is its meaning?
  (b) For what value of h is $f(90, h) = 100$?
  (c) For what value of T is $f(T, 50) = 88$?
  (d) What are the meanings of the functions $I = f(80, h)$ and $I = f(100, h)$? Compare the behavior of these two functions of h.

3. A manufacturer has modeled its yearly production function P (the monetary value of its entire production in millions of dollars) as a Cobb-Douglas function

$$P(L, K) = 1.47L^{0.65}K^{0.35}$$

where L is the number of labor hours (in thousands) and K is the invested capital (in millions of dollars). Find $P(120, 20)$ and interpret it.

4. Verify for the Cobb-Douglas production function

$$P(L, K) = 1.01L^{0.75}K^{0.25}$$

discussed in Example 3 that the production will be doubled if both the amount of labor and the amount of capital are doubled. Determine whether this is also true for the general production function

$$P(L, K) = bL^{\alpha}K^{1-\alpha}$$

5. A model for the surface area of a human body is given by the function

$$S = f(w, h) = 0.1091w^{0.425}h^{0.725}$$

where w is the weight (in pounds), h is the height (in inches), and S is measured in square feet.
  (a) Find $f(160, 70)$ and interpret it.
  (b) What is your own surface area?

6. The wind-chill index W discussed in Example 2 has been modeled by the following function:

$$W(T, v) = 13.12 + 0.6215T - 11.37v^{0.16} + 0.3965Tv^{0.16}$$

Check to see how closely this model agrees with the values in Table 1 for a few values of T and v.

7. The wave heights h in the open sea depend on the speed v of the wind and the length of time t that the wind has been blowing at that speed. Values of the function $h = f(v, t)$ are recorded in feet in Table 4.
  (a) What is the value of $f(40, 15)$? What is its meaning?
  (b) What is the meaning of the function $h = f(30, t)$? Describe the behavior of this function.
  (c) What is the meaning of the function $h = f(v, 30)$? Describe the behavior of this function.

Table 4

Duration (hours)

v \ t	5	10	15	20	30	40	50
10	2	2	2	2	2	2	2
15	4	4	5	5	5	5	5
20	5	7	8	8	9	9	9
30	9	13	16	17	18	19	19
40	14	21	25	28	31	33	33
50	19	29	36	40	45	48	50
60	24	37	47	54	62	67	69

Wind speed (knots)

8. A company makes three sizes of cardboard boxes: small, medium, and large. It costs $2.50 to make a small box,

$4.00 for a medium box, and $4.50 for a large box. Fixed costs are $8000.

(a) Express the cost of making x small boxes, y medium boxes, and z large boxes as a function of three variables: $C = f(x, y, z)$.

(b) Find $f(3000, 5000, 4000)$ and interpret it.

(c) What is the domain of f?

9. Let $g(x, y) = \cos(x + 2y)$.
  (a) Evaluate $g(2, -1)$.
  (b) Find the domain of g.
  (c) Find the range of g.

10. Let $F(x, y) = 1 + \sqrt{4 - y^2}$.
  (a) Evaluate $F(3, 1)$.
  (b) Find and sketch the domain of F.
  (c) Find the range of F.

11. Let $f(x, y, z) = \sqrt{x} + \sqrt{y} + \sqrt{z} + \ln(4 - x^2 - y^2 - z^2)$.
  (a) Evaluate $f(1, 1, 1)$.
  (b) Find and describe the domain of f.

12. Let $g(x, y, z) = x^3 y^2 z \sqrt{10 - x - y - z}$.
  (a) Evaluate $g(1, 2, 3)$.
  (b) Find and describe the domain of g.

13–22 Find and sketch the domain of the function.

13. $f(x, y) = \sqrt{x - 2} + \sqrt{y - 1}$

14. $f(x, y) = \sqrt[4]{x - 3y}$

15. $f(x, y) = \ln(9 - x^2 - 9y^2)$ **16.** $f(x, y) = \sqrt{x^2 + y^2 - 4}$

17. $g(x, y) = \dfrac{x - y}{x + y}$ **18.** $g(x, y) = \dfrac{\ln(2 - x)}{1 - x^2 - y^2}$

19. $f(x, y) = \dfrac{\sqrt{y - x^2}}{1 - x^2}$

20. $f(x, y) = \sin^{-1}(x + y)$

21. $f(x, y, z) = \sqrt{4 - x^2} + \sqrt{9 - y^2} + \sqrt{1 - z^2}$

22. $f(x, y, z) = \ln(16 - 4x^2 - 4y^2 - z^2)$

23–31 Sketch the graph of the function.

23. $f(x, y) = y$ **24.** $f(x, y) = x^2$

25. $f(x, y) = 10 - 4x - 5y$ **26.** $f(x, y) = \cos y$

27. $f(x, y) = \sin x$ **28.** $f(x, y) = 2 - x^2 - y^2$

29. $f(x, y) = x^2 + 4y^2 + 1$ **30.** $f(x, y) = \sqrt{4x^2 + y^2}$

31. $f(x, y) = \sqrt{4 - 4x^2 - y^2}$

---

32. Match the function with its graph (labeled I–VI). Give reasons for your choices.

(a) $f(x, y) = \dfrac{1}{1 + x^2 + y^2}$ (b) $f(x, y) = \dfrac{1}{1 + x^2 y^2}$

(c) $f(x, y) = \ln(x^2 + y^2)$ (d) $f(x, y) = \cos \sqrt{x^2 + y^2}$

(e) $f(x, y) = |xy|$ (f) $f(x, y) = \cos(xy)$

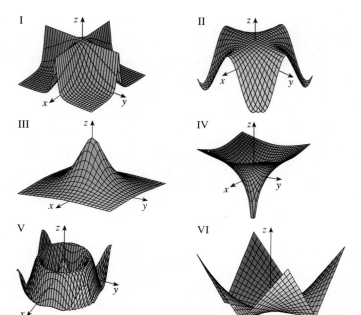

33. A contour map for a function f is shown. Use it to estimate the values of $f(-3, 3)$ and $f(3, -2)$. What can you say about the shape of the graph?

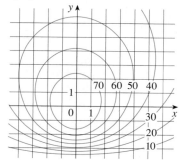

34. Shown is a contour map of atmospheric pressure in North America on August 12, 2008. On the level curves (called isobars) the pressure is indicated in millibars (mb).
(a) Estimate the pressure at C (Chicago), N (Nashville), S (San Francisco), and V (Vancouver).
(b) At which of these locations were the winds strongest?

35. Level curves (isothermals) are shown for the typical water temperature (in °C) in Long Lake (Minnesota) as a function of depth and time of year. Estimate the temperature in the lake on June 9 (day 160) at a depth of 10 m and on June 29 (day 180) at a depth of 5 m.

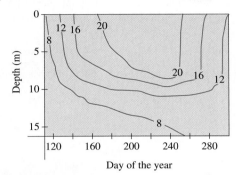

36. Two contour maps are shown. One is for a function f whose graph is a cone. The other is for a function g whose graph is a paraboloid. Which is which, and why?

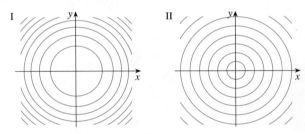

37. Locate the points A and B on the map of Lonesome Mountain (Figure 12). How would you describe the terrain near A? Near B?

38. Make a rough sketch of a contour map for the function whose graph is shown.

39. The *body mass index* (BMI) of a person is defined by

$$B(m, h) = \frac{m}{h^2}$$

where m is the person's mass (in kilograms) and h is the height (in meters). Draw the level curves $B(m, h) = 18.5$, $B(m, h) = 25$, $B(m, h) = 30$, and $B(m, h) = 40$. A rough guideline is that a person is underweight if the BMI is less than

18.5; optimal if the BMI lies between 18.5 and 25; overweight if the BMI lies between 25 and 30; and obese if the BMI exceeds 30. Shade the region corresponding to optimal BMI. Does someone who weighs 62 kg and is 152 cm tall fall into this category?

40. The body mass index is defined in Exercise 39. Draw the level curve of this function corresponding to someone who is 200 cm tall and weighs 80 kg. Find the weights and heights of two other people with that same level curve.

41–44 A contour map of a function is shown. Use it to make a rough sketch of the graph of f.

45–52 Draw a contour map of the function showing several level curves.

45. $f(x, y) = x^2 - y^2$ **46.** $f(x, y) = xy$

47. $f(x, y) = \sqrt{x} + y$ **48.** $f(x, y) = \ln(x^2 + 4y^2)$

49. $f(x, y) = ye^x$ **50.** $f(x, y) = y - \arctan x$

51. $f(x, y) = \sqrt[3]{x^2 + y^2}$ **52.** $f(x, y) = y/(x^2 + y^2)$

53–54 Sketch both a contour map and a graph of the function and compare them.

53. $f(x, y) = x^2 + 9y^2$ **54.** $f(x, y) = \sqrt{36 - 9x^2 - 4y^2}$

55. A thin metal plate, located in the xy-plane, has temperature $T(x, y)$ at the point (x, y). Sketch some level curves (isothermals) if the temperature function is given by

$$T(x, y) = \frac{100}{1 + x^2 + 2y^2}$$

56. If $V(x, y)$ is the electric potential at a point (x, y) in the xy-plane, then the level curves of V are called *equipotential curves* because at all points on such a curve the electric potential is the same. Sketch some equipotential curves if $V(x, y) = c/\sqrt{r^2 - x^2 - y^2}$, where c is a positive constant.

57–60 Use a computer to graph the function using various domains and viewpoints. Get a printout of one that, in your opinion, gives a good view. If your software also produces level curves, then plot some contour lines of the same function and compare with the graph.

57. $f(x, y) = xy^2 - x^3$ (monkey saddle)

58. $f(x, y) = xy^3 - yx^3$ (dog saddle)

59. $f(x, y) = e^{-(x^2+y^2)/3}(\sin(x^2) + \cos(y^2))$

60. $f(x, y) = \cos x \cos y$

61–66 Match the function (a) with its graph (labeled A–F below) and (b) with its contour map (labeled I–VI). Give reasons for your choices.

61. $z = \sin(xy)$

62. $z = e^x \cos y$

63. $z = \sin(x - y)$

64. $z = \sin x - \sin y$

65. $z = (1 - x^2)(1 - y^2)$

66. $z = \dfrac{x - y}{1 + x^2 + y^2}$

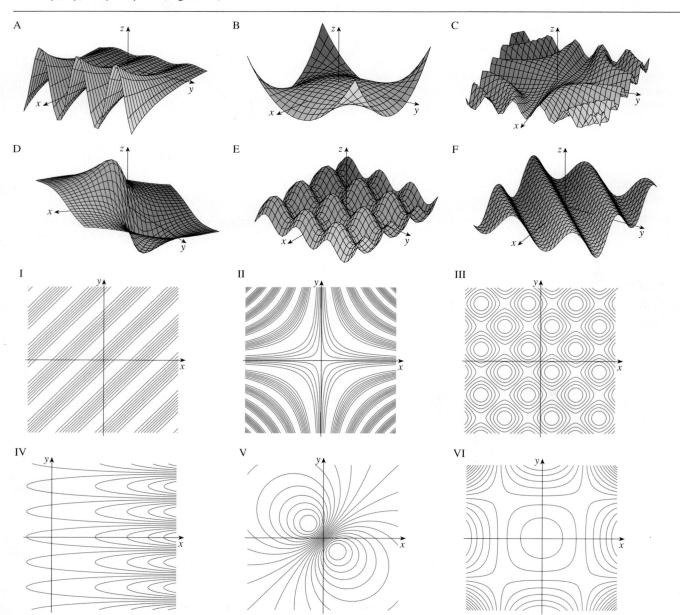

67–70 Describe the level surfaces of the function.

67. $f(x, y, z) = x + 3y + 5z$

68. $f(x, y, z) = x^2 + 3y^2 + 5z^2$

69. $f(x, y, z) = y^2 + z^2$

70. $f(x, y, z) = x^2 - y^2 - z^2$

71–72 Describe how the graph of g is obtained from the graph of f.

71. (a) $g(x, y) = f(x, y) + 2$
(b) $g(x, y) = 2f(x, y)$
(c) $g(x, y) = -f(x, y)$
(d) $g(x, y) = 2 - f(x, y)$

72. (a) $g(x, y) = f(x - 2, y)$
(b) $g(x, y) = f(x, y + 2)$
(c) $g(x, y) = f(x + 3, y - 4)$

73–74 Use a computer to graph the function using various domains and viewpoints. Get a printout that gives a good view of the "peaks and valleys." Would you say the function has a maximum value? Can you identify any points on the graph that you might consider to be "local maximum points"? What about "local minimum points"?

73. $f(x, y) = 3x - x^4 - 4y^2 - 10xy$

74. $f(x, y) = xye^{-x^2-y^2}$

75–76 Graph the function using various domains and viewpoints. Comment on the limiting behavior of the function. What happens as both x and y become large? What happens as (x, y) approaches the origin?

75. $f(x, y) = \dfrac{x + y}{x^2 + y^2}$ **76.** $f(x, y) = \dfrac{xy}{x^2 + y^2}$

77. Investigate the family of functions $f(x, y) = e^{cx^2+y^2}$. How does the shape of the graph depend on c?

78. Use a computer to investigate the family of surfaces

$$z = (ax^2 + by^2)e^{-x^2-y^2}$$

How does the shape of the graph depend on the numbers a and b?

79. Use a computer to investigate the family of surfaces $z = x^2 + y^2 + cxy$. In particular, you should determine the transitional values of c for which the surface changes from one type of quadric surface to another.

80. Graph the functions

$$f(x, y) = \sqrt{x^2 + y^2}$$

$$f(x, y) = e^{\sqrt{x^2+y^2}}$$

$$f(x, y) = \ln\sqrt{x^2 + y^2}$$

$$f(x, y) = \sin\left(\sqrt{x^2 + y^2}\right)$$

and $$f(x, y) = \frac{1}{\sqrt{x^2 + y^2}}$$

In general, if g is a function of one variable, how is the graph of

$$f(x, y) = g\left(\sqrt{x^2 + y^2}\right)$$

obtained from the graph of g?

81. (a) Show that, by taking logarithms, the general Cobb-Douglas function $P = bL^\alpha K^{1-\alpha}$ can be expressed as

$$\ln \frac{P}{K} = \ln b + \alpha \ln \frac{L}{K}$$

(b) If we let $x = \ln(L/K)$ and $y = \ln(P/K)$, the equation in part (a) becomes the linear equation $y = \alpha x + \ln b$. Use Table 2 (in Example 3) to make a table of values of $\ln(L/K)$ and $\ln(P/K)$ for the years 1899–1922. Then use a graphing calculator or computer to find the least squares regression line through the points $(\ln(L/K), \ln(P/K))$.

(c) Deduce that the Cobb-Douglas production function is $P = 1.01L^{0.75}K^{0.25}$.

14.2 Limits and Continuity

Let's compare the behavior of the functions

$$f(x, y) = \frac{\sin(x^2 + y^2)}{x^2 + y^2} \quad \text{and} \quad g(x, y) = \frac{x^2 - y^2}{x^2 + y^2}$$

as x and y both approach 0 [and therefore the point (x, y) approaches the origin].

Tables 1 and 2 show values of $f(x, y)$ and $g(x, y)$, correct to three decimal places, for points (x, y) near the origin. (Notice that neither function is defined at the origin.)

Table 1 Values of $f(x, y)$

x \ y	−1.0	−0.5	−0.2	0	0.2	0.5	1.0
−1.0	0.455	0.759	0.829	0.841	0.829	0.759	0.455
−0.5	0.759	0.959	0.986	0.990	0.986	0.959	0.759
−0.2	0.829	0.986	0.999	1.000	0.999	0.986	0.829
0	0.841	0.990	1.000		1.000	0.990	0.841
0.2	0.829	0.986	0.999	1.000	0.999	0.986	0.829
0.5	0.759	0.959	0.986	0.990	0.986	0.959	0.759
1.0	0.455	0.759	0.829	0.841	0.829	0.759	0.455

Table 2 Values of $g(x, y)$

x \ y	−1.0	−0.5	−0.2	0	0.2	0.5	1.0
−1.0	0.000	0.600	0.923	1.000	0.923	0.600	0.000
−0.5	−0.600	0.000	0.724	1.000	0.724	0.000	−0.600
−0.2	−0.923	−0.724	0.000	1.000	0.000	−0.724	−0.923
0	−1.000	−1.000	−1.000		−1.000	−1.000	−1.000
0.2	−0.923	−0.724	0.000	1.000	0.000	−0.724	−0.923
0.5	−0.600	0.000	0.724	1.000	0.724	0.000	−0.600
1.0	0.000	0.600	0.923	1.000	0.923	0.600	0.000

It appears that as (x, y) approaches $(0, 0)$, the values of $f(x, y)$ are approaching 1 whereas the values of $g(x, y)$ aren't approaching any number. It turns out that these guesses based on numerical evidence are correct, and we write

$$\lim_{(x, y) \to (0, 0)} \frac{\sin(x^2 + y^2)}{x^2 + y^2} = 1 \quad \text{and} \quad \lim_{(x, y) \to (0, 0)} \frac{x^2 - y^2}{x^2 + y^2} \quad \text{does not exist}$$

In general, we use the notation

$$\lim_{(x, y) \to (a, b)} f(x, y) = L$$

to indicate that the values of $f(x, y)$ approach the number L as the point (x, y) approaches the point (a, b) along any path that stays within the domain of f. In other words, we can make the values of $f(x, y)$ as close to L as we like by taking the point (x, y) sufficiently close to the point (a, b), but not equal to (a, b). A more precise definition follows.

> **1 Definition** Let f be a function of two variables whose domain D includes points arbitrarily close to (a, b). Then we say that the **limit of $f(x, y)$ as (x, y) approaches (a, b)** is L and we write
>
> $$\lim_{(x, y) \to (a, b)} f(x, y) = L$$
>
> if for every number $\varepsilon > 0$ there is a corresponding number $\delta > 0$ such that
>
> if $(x, y) \in D$ and $0 < \sqrt{(x - a)^2 + (y - b)^2} < \delta$ then $|f(x, y) - L| < \varepsilon$

Other notations for the limit in Definition 1 are

$$\lim_{\substack{x \to a \\ y \to b}} f(x, y) = L \quad \text{and} \quad f(x, y) \to L \text{ as } (x, y) \to (a, b)$$

Notice that $|f(x, y) - L|$ is the distance between the numbers $f(x, y)$ and L, and $\sqrt{(x - a)^2 + (y - b)^2}$ is the distance between the point (x, y) and the point (a, b). Thus Definition 1 says that the distance between $f(x, y)$ and L can be made arbitrarily small by making the distance from (x, y) to (a, b) sufficiently small (but not 0). Figure 1 illustrates Definition 1 by means of an arrow diagram. If any small interval $(L - \varepsilon, L + \varepsilon)$ is given around L, then we can find a disk D_δ with center (a, b) and radius $\delta > 0$ such that f maps all the points in D_δ [except possibly (a, b)] into the interval $(L - \varepsilon, L + \varepsilon)$.

FIGURE 1

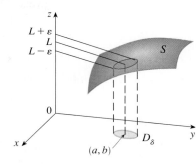

FIGURE 2

Another illustration of Definition 1 is given in Figure 2 where the surface S is the graph of f. If $\varepsilon > 0$ is given, we can find $\delta > 0$ such that if (x, y) is restricted to lie in the disk D_δ and $(x, y) \neq (a, b)$, then the corresponding part of S lies between the horizontal planes $z = L - \varepsilon$ and $z = L + \varepsilon$.

For functions of a single variable, when we let x approach a, there are only two possible directions of approach, from the left or from the right. We recall from Chapter 2 that if $\lim_{x \to a^-} f(x) \neq \lim_{x \to a^+} f(x)$, then $\lim_{x \to a} f(x)$ does not exist.

For functions of two variables the situation is not as simple because we can let (x, y) approach (a, b) from an infinite number of directions in any manner whatsoever (see Figure 3) as long as (x, y) stays within the domain of f.

FIGURE 3

Definition 1 says that the distance between $f(x, y)$ and L can be made arbitrarily small by making the distance from (x, y) to (a, b) sufficiently small (but not 0). The definition refers only to the *distance* between (x, y) and (a, b). It does not refer to the direction of approach. Therefore, if the limit exists, then $f(x, y)$ must approach the same limit no matter how (x, y) approaches (a, b). Thus, if we can find two different paths of approach along which the function $f(x, y)$ has different limits, then it follows that $\lim_{(x, y) \to (a, b)} f(x, y)$ does not exist.

> If $f(x, y) \to L_1$ as $(x, y) \to (a, b)$ along a path C_1 and $f(x, y) \to L_2$ as $(x, y) \to (a, b)$ along a path C_2, where $L_1 \neq L_2$, then $\lim_{(x, y) \to (a, b)} f(x, y)$ does not exist.

EXAMPLE 1   Show that $\displaystyle \lim_{(x, y) \to (0, 0)} \frac{x^2 - y^2}{x^2 + y^2}$ does not exist.

SOLUTION   Let $f(x, y) = (x^2 - y^2)/(x^2 + y^2)$. First let's approach $(0, 0)$ along the x-axis. Then $y = 0$ gives $f(x, 0) = x^2/x^2 = 1$ for all $x \neq 0$, so

$$f(x, y) \to 1 \qquad \text{as} \qquad (x, y) \to (0, 0) \text{ along the } x\text{-axis}$$

We now approach along the y-axis by putting $x = 0$. Then $f(0, y) = \dfrac{-y^2}{y^2} = -1$ for all $y \neq 0$, so

$$f(x, y) \to -1 \qquad \text{as} \qquad (x, y) \to (0, 0) \text{ along the } y\text{-axis}$$

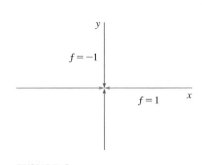

FIGURE 4

(See Figure 4.) Since f has two different limits along two different lines, the given limit does not exist. (This confirms the conjecture we made on the basis of numerical evidence at the beginning of this section.) ∎

EXAMPLE 2 If $f(x, y) = xy/(x^2 + y^2)$, does $\lim\limits_{(x, y) \to (0, 0)} f(x, y)$ exist?

SOLUTION If $y = 0$, then $f(x, 0) = 0/x^2 = 0$. Therefore

$$f(x, y) \to 0 \qquad \text{as} \qquad (x, y) \to (0, 0) \text{ along the } x\text{-axis}$$

If $x = 0$, then $f(0, y) = 0/y^2 = 0$, so

$$f(x, y) \to 0 \qquad \text{as} \qquad (x, y) \to (0, 0) \text{ along the } y\text{-axis}$$

Although we have obtained identical limits along the axes, that does not show that the given limit is 0. Let's now approach $(0, 0)$ along another line, say $y = x$. For all $x \neq 0$,

$$f(x, x) = \frac{x^2}{x^2 + x^2} = \frac{1}{2}$$

Therefore $f(x, y) \to \frac{1}{2} \qquad$ as $\qquad (x, y) \to (0, 0)$ along $y = x$

(See Figure 5.) Since we have obtained different limits along different paths, the given limit does not exist. ∎

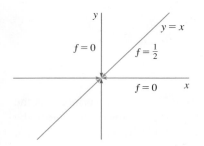

FIGURE 5

Figure 6 sheds some light on Example 2. The ridge that occurs above the line $y = x$ corresponds to the fact that $f(x, y) = \frac{1}{2}$ for all points (x, y) on that line except the origin.

TEC In Visual 14.2 a rotating line on the surface in Figure 6 shows different limits at the origin from different directions.

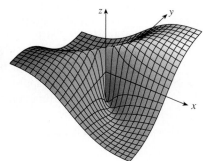

FIGURE 6

$$f(x, y) = \frac{xy}{x^2 + y^2}$$

EXAMPLE 3 If $f(x, y) = \dfrac{xy^2}{x^2 + y^4}$, does $\lim\limits_{(x, y) \to (0, 0)} f(x, y)$ exist?

SOLUTION With the solution of Example 2 in mind, let's try to save time by letting $(x, y) \to (0, 0)$ along any line through the origin. If the line is not the y-axis, then $y = mx$, where m is the slope, and

$$f(x, y) = f(x, mx) = \frac{x(mx)^2}{x^2 + (mx)^4} = \frac{m^2 x^3}{x^2 + m^4 x^4} = \frac{m^2 x}{1 + m^4 x^2}$$

So $f(x, y) \to 0 \qquad$ as $\qquad (x, y) \to (0, 0)$ along $y = mx$

We get the same result as $(x, y) \to (0, 0)$ along the line $x = 0$. Thus f has the same limiting value along every line through the origin. But that does not show that the given

Figure 7 shows the graph of the function in Example 3. Notice the ridge above the parabola $x = y^2$.

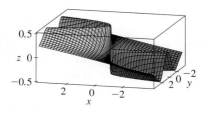

FIGURE 7

limit is 0, for if we now let $(x, y) \to (0, 0)$ along the parabola $x = y^2$, we have

$$f(x, y) = f(y^2, y) = \frac{y^2 \cdot y^2}{(y^2)^2 + y^4} = \frac{y^4}{2y^4} = \frac{1}{2}$$

so $f(x, y) \to \frac{1}{2}$ as $(x, y) \to (0, 0)$ along $x = y^2$

Since different paths lead to different limiting values, the given limit does not exist. ∎

Now let's look at limits that *do* exist. Just as for functions of one variable, the calculation of limits for functions of two variables can be greatly simplified by the use of properties of limits. The Limit Laws listed in Section 2.3 can be extended to functions of two variables: the limit of a sum is the sum of the limits, the limit of a product is the product of the limits, and so on. In particular, the following equations are true.

$$\boxed{2} \qquad \lim_{(x, y) \to (a, b)} x = a \qquad \lim_{(x, y) \to (a, b)} y = b \qquad \lim_{(x, y) \to (a, b)} c = c$$

The Squeeze Theorem also holds.

EXAMPLE 4 Find $\displaystyle\lim_{(x, y) \to (0, 0)} \frac{3x^2 y}{x^2 + y^2}$ if it exists.

SOLUTION As in Example 3, we could show that the limit along any line through the origin is 0. This doesn't prove that the given limit is 0, but the limits along the parabolas $y = x^2$ and $x = y^2$ also turn out to be 0, so we begin to suspect that the limit does exist and is equal to 0.

Let $\varepsilon > 0$. We want to find $\delta > 0$ such that

$$\text{if} \quad 0 < \sqrt{x^2 + y^2} < \delta \quad \text{then} \quad \left| \frac{3x^2 y}{x^2 + y^2} - 0 \right| < \varepsilon$$

that is,   if $0 < \sqrt{x^2 + y^2} < \delta$ then $\dfrac{3x^2 |y|}{x^2 + y^2} < \varepsilon$

But $x^2 \le x^2 + y^2$ since $y^2 \ge 0$, so $x^2/(x^2 + y^2) \le 1$ and therefore

$$\boxed{3} \qquad \frac{3x^2 |y|}{x^2 + y^2} \le 3 |y| = 3 \sqrt{y^2} \le 3 \sqrt{x^2 + y^2}$$

Thus if we choose $\delta = \varepsilon/3$ and let $0 < \sqrt{x^2 + y^2} < \delta$, then

$$\left| \frac{3x^2 y}{x^2 + y^2} - 0 \right| \le 3 \sqrt{x^2 + y^2} < 3\delta = 3\left(\frac{\varepsilon}{3} \right) = \varepsilon$$

Hence, by Definition 1,

$$\lim_{(x, y) \to (0, 0)} \frac{3x^2 y}{x^2 + y^2} = 0$$

∎

Another way to do Example 4 is to use the Squeeze Theorem instead of Definition 1. From (2) it follows that

$$\lim_{(x, y) \to (0, 0)} 3 |y| = 0$$

and so the first inequality in (3) shows that the given limit is 0.

■ Continuity

Recall that evaluating limits of *continuous* functions of a single variable is easy. It can be accomplished by direct substitution because the defining property of a continuous function is $\lim_{x \to a} f(x) = f(a)$. Continuous functions of two variables are also defined by the direct substitution property.

---

4 **Definition** A function f of two variables is called **continuous at** (a, b) if

$$\lim_{(x, y) \to (a, b)} f(x, y) = f(a, b)$$

We say f is **continuous on** D if f is continuous at every point (a, b) in D.

The intuitive meaning of continuity is that if the point (x, y) changes by a small amount, then the value of $f(x, y)$ changes by a small amount. This means that a surface that is the graph of a continuous function has no hole or break.

Using the properties of limits, you can see that sums, differences, products, and quotients of continuous functions are continuous on their domains. Let's use this fact to give examples of continuous functions.

A **polynomial function of two variables** (or polynomial, for short) is a sum of terms of the form $cx^m y^n$, where c is a constant and m and n are nonnegative integers. A **rational function** is a ratio of polynomials. For instance,

$$f(x, y) = x^4 + 5x^3 y^2 + 6xy^4 - 7y + 6$$

is a polynomial, whereas

$$g(x, y) = \frac{2xy + 1}{x^2 + y^2}$$

is a rational function.

The limits in (2) show that the functions $f(x, y) = x$, $g(x, y) = y$, and $h(x, y) = c$ are continuous. Since any polynomial can be built up out of the simple functions f, g, and h by multiplication and addition, it follows that *all polynomials are continuous on* $\mathbb{R}^2$. Likewise, any rational function is continuous on its domain because it is a quotient of continuous functions.

EXAMPLE 5 Evaluate $\lim_{(x, y) \to (1, 2)} (x^2 y^3 - x^3 y^2 + 3x + 2y)$.

SOLUTION Since $f(x, y) = x^2 y^3 - x^3 y^2 + 3x + 2y$ is a polynomial, it is continuous everywhere, so we can find the limit by direct substitution:

$$\lim_{(x, y) \to (1, 2)} (x^2 y^3 - x^3 y^2 + 3x + 2y) = 1^2 \cdot 2^3 - 1^3 \cdot 2^2 + 3 \cdot 1 + 2 \cdot 2 = 11 \quad \blacksquare$$

EXAMPLE 6 Where is the function $f(x, y) = \dfrac{x^2 - y^2}{x^2 + y^2}$ continuous?

SOLUTION The function f is discontinuous at $(0, 0)$ because it is not defined there. Since f is a rational function, it is continuous on its domain, which is the set $D = \{(x, y) \mid (x, y) \neq (0, 0)\}$. $\quad \blacksquare$

EXAMPLE 7 Let

$$g(x, y) = \begin{cases} \dfrac{x^2 - y^2}{x^2 + y^2} & \text{if } (x, y) \neq (0, 0) \\ 0 & \text{if } (x, y) = (0, 0) \end{cases}$$

Here g is defined at $(0, 0)$ but g is still discontinuous there because $\lim_{(x, y) \to (0, 0)} g(x, y)$ does not exist (see Example 1).

Figure 8 shows the graph of the continuous function in Example 8.

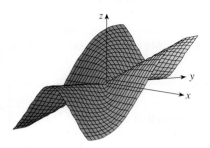

FIGURE 8

EXAMPLE 8 Let

$$f(x, y) = \begin{cases} \dfrac{3x^2 y}{x^2 + y^2} & \text{if } (x, y) \neq (0, 0) \\ 0 & \text{if } (x, y) = (0, 0) \end{cases}$$

We know f is continuous for $(x, y) \neq (0, 0)$ since it is equal to a rational function there. Also, from Example 4, we have

$$\lim_{(x, y) \to (0, 0)} f(x, y) = \lim_{(x, y) \to (0, 0)} \frac{3x^2 y}{x^2 + y^2} = 0 = f(0, 0)$$

Therefore f is continuous at $(0, 0)$, and so it is continuous on $\mathbb{R}^2$.

Just as for functions of one variable, composition is another way of combining two continuous functions to get a third. In fact, it can be shown that if f is a continuous function of two variables and g is a continuous function of a single variable that is defined on the range of f, then the composite function $h = g \circ f$ defined by $h(x, y) = g(f(x, y))$ is also a continuous function.

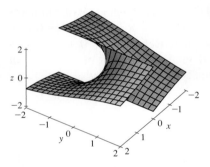

FIGURE 9
The function $h(x, y) = \arctan(y/x)$ is discontinuous where $x = 0$.

EXAMPLE 9 Where is the function $h(x, y) = \arctan(y/x)$ continuous?

SOLUTION The function $f(x, y) = y/x$ is a rational function and therefore continuous except on the line $x = 0$. The function $g(t) = \arctan t$ is continuous everywhere. So the composite function

$$g(f(x, y)) = \arctan(y/x) = h(x, y)$$

is continuous except where $x = 0$. The graph in Figure 9 shows the break in the graph of h above the y-axis.

Functions of Three or More Variables

Everything that we have done in this section can be extended to functions of three or more variables. The notation

$$\lim_{(x, y, z) \to (a, b, c)} f(x, y, z) = L$$

means that the values of $f(x, y, z)$ approach the number L as the point (x, y, z) approaches the point (a, b, c) along any path in the domain of f. Because the distance between two points (x, y, z) and (a, b, c) in $\mathbb{R}^3$ is given by $\sqrt{(x - a)^2 + (y - b)^2 + (z - c)^2}$, we can write the precise definition as follows: for every number $\varepsilon > 0$ there is a corresponding number $\delta > 0$ such that

$$\text{if } (x, y, z) \text{ is in the domain of } f \text{ and } 0 < \sqrt{(x - a)^2 + (y - b)^2 + (z - c)^2} < \delta$$

$$\text{then } |f(x, y, z) - L| < \varepsilon$$

The function f is **continuous** at (a, b, c) if

$$\lim_{(x, y, z) \to (a, b, c)} f(x, y, z) = f(a, b, c)$$

For instance, the function

$$f(x, y, z) = \frac{1}{x^2 + y^2 + z^2 - 1}$$

is a rational function of three variables and so is continuous at every point in $\mathbb{R}^3$ except where $x^2 + y^2 + z^2 = 1$. In other words, it is discontinuous on the sphere with center the origin and radius 1.

If we use the vector notation introduced at the end of Section 14.1, then we can write the definitions of a limit for functions of two or three variables in a single compact form as follows.

> **5** If f is defined on a subset D of $\mathbb{R}^n$, then $\lim_{\mathbf{x} \to \mathbf{a}} f(\mathbf{x}) = L$ means that for every number $\varepsilon > 0$ there is a corresponding number $\delta > 0$ such that
>
> $$\text{if} \quad \mathbf{x} \in D \quad \text{and} \quad 0 < |\mathbf{x} - \mathbf{a}| < \delta \quad \text{then} \quad |f(\mathbf{x}) - L| < \varepsilon$$

Notice that if $n = 1$, then $\mathbf{x} = x$ and $\mathbf{a} = a$, and (5) is just the definition of a limit for functions of a single variable. For the case $n = 2$, we have $\mathbf{x} = \langle x, y \rangle$, $\mathbf{a} = \langle a, b \rangle$, and $|\mathbf{x} - \mathbf{a}| = \sqrt{(x - a)^2 + (y - b)^2}$, so (5) becomes Definition 1. If $n = 3$, then $\mathbf{x} = \langle x, y, z \rangle$, $\mathbf{a} = \langle a, b, c \rangle$, and (5) becomes the definition of a limit of a function of three variables. In each case the definition of continuity can be written as

$$\lim_{\mathbf{x} \to \mathbf{a}} f(\mathbf{x}) = f(\mathbf{a})$$

14.2 EXERCISES

1. Suppose that $\lim_{(x, y) \to (3, 1)} f(x, y) = 6$. What can you say about the value of $f(3, 1)$? What if f is continuous?

2. Explain why each function is continuous or discontinuous.
   (a) The outdoor temperature as a function of longitude, latitude, and time
   (b) Elevation (height above sea level) as a function of longitude, latitude, and time
   (c) The cost of a taxi ride as a function of distance traveled and time

3–4 Use a table of numerical values of $f(x, y)$ for (x, y) near the origin to make a conjecture about the value of the limit of $f(x, y)$ as $(x, y) \to (0, 0)$. Then explain why your guess is correct.

3. $f(x, y) = \dfrac{x^2 y^3 + x^3 y^2 - 5}{2 - xy}$ **4.** $f(x, y) = \dfrac{2xy}{x^2 + 2y^2}$

5–22 Find the limit, if it exists, or show that the limit does not exist.

5. $\lim\limits_{(x, y) \to (3, 2)} (x^2 y^3 - 4y^2)$ **6.** $\lim\limits_{(x, y) \to (2, -1)} \dfrac{x^2 y + xy^2}{x^2 - y^2}$

7. $\lim\limits_{(x, y) \to (\pi, \pi/2)} y \sin(x - y)$

8. $\lim\limits_{(x, y) \to (3, 2)} e^{\sqrt{2x - y}}$

9. $\lim\limits_{(x, y) \to (0, 0)} \dfrac{x^4 - 4y^2}{x^2 + 2y^2}$

10. $\lim\limits_{(x, y) \to (0, 0)} \dfrac{5y^4 \cos^2 x}{x^4 + y^4}$

11. $\lim\limits_{(x, y) \to (0, 0)} \dfrac{y^2 \sin^2 x}{x^4 + y^4}$

12. $\lim\limits_{(x, y) \to (1, 0)} \dfrac{xy - y}{(x - 1)^2 + y^2}$

13. $\lim\limits_{(x, y) \to (0, 0)} \dfrac{xy}{\sqrt{x^2 + y^2}}$

14. $\lim\limits_{(x, y) \to (0, 0)} \dfrac{x^3 - y^3}{x^2 + xy + y^2}$

15. $\lim\limits_{(x, y) \to (0, 0)} \dfrac{xy^2 \cos y}{x^2 + y^4}$

16. $\lim\limits_{(x, y) \to (0, 0)} \dfrac{xy^4}{x^4 + y^4}$

17. $\lim\limits_{(x, y) \to (0, 0)} \dfrac{x^2 + y^2}{\sqrt{x^2 + y^2 + 1} - 1}$

18. $\lim\limits_{(x, y) \to (0, 0)} \dfrac{xy^4}{x^2 + y^8}$

19. $\lim\limits_{(x, y, z) \to (\pi, 0, 1/3)} e^{y^2} \tan(xz)$

20. $\lim\limits_{(x, y, z) \to (0, 0, 0)} \dfrac{xy + yz}{x^2 + y^2 + z^2}$

21. $\lim\limits_{(x, y, z)\to(0, 0, 0)} \dfrac{xy + yz^2 + xz^2}{x^2 + y^2 + z^4}$

22. $\lim\limits_{(x, y, z)\to(0, 0, 0)} \dfrac{x^2y^2z^2}{x^2 + y^2 + z^2}$

23–24 Use a computer graph of the function to explain why the limit does not exist.

23. $\lim\limits_{(x, y)\to(0, 0)} \dfrac{2x^2 + 3xy + 4y^2}{3x^2 + 5y^2}$ **24.** $\lim\limits_{(x, y)\to(0, 0)} \dfrac{xy^3}{x^2 + y^6}$

25–26 Find $h(x, y) = g(f(x, y))$ and the set of points at which h is continuous.

25. $g(t) = t^2 + \sqrt{t}, \quad f(x, y) = 2x + 3y - 6$

26. $g(t) = t + \ln t, \quad f(x, y) = \dfrac{1 - xy}{1 + x^2y^2}$

27–28 Graph the function and observe where it is discontinuous. Then use the formula to explain what you have observed.

27. $f(x, y) = e^{1/(x-y)}$ **28.** $f(x, y) = \dfrac{1}{1 - x^2 - y^2}$

29–38 Determine the set of points at which the function is continuous.

29. $F(x, y) = \dfrac{xy}{1 + e^{x-y}}$ **30.** $F(x, y) = \cos\sqrt{1 + x - y}$

31. $F(x, y) = \dfrac{1 + x^2 + y^2}{1 - x^2 - y^2}$ **32.** $H(x, y) = \dfrac{e^x + e^y}{e^{xy} - 1}$

33. $G(x, y) = \sqrt{x} + \sqrt{1 - x^2 - y^2}$

34. $G(x, y) = \ln(1 + x - y)$

35. $f(x, y, z) = \arcsin(x^2 + y^2 + z^2)$

36. $f(x, y, z) = \sqrt{y - x^2}\, \ln z$

37. $f(x, y) = \begin{cases} \dfrac{x^2y^3}{2x^2 + y^2} & \text{if } (x, y) \neq (0, 0) \\ 1 & \text{if } (x, y) = (0, 0) \end{cases}$

38. $f(x, y) = \begin{cases} \dfrac{xy}{x^2 + xy + y^2} & \text{if } (x, y) \neq (0, 0) \\ 0 & \text{if } (x, y) = (0, 0) \end{cases}$

39–41 Use polar coordinates to find the limit. [If (r, θ) are polar coordinates of the point (x, y) with $r \geq 0$, note that $r \to 0^+$ as $(x, y) \to (0, 0)$.]

39. $\lim\limits_{(x, y)\to(0, 0)} \dfrac{x^3 + y^3}{x^2 + y^2}$

40. $\lim\limits_{(x, y)\to(0, 0)} (x^2 + y^2) \ln(x^2 + y^2)$

41. $\lim\limits_{(x, y)\to(0, 0)} \dfrac{e^{-x^2-y^2} - 1}{x^2 + y^2}$

---

42. At the beginning of this section we considered the function

$$f(x, y) = \frac{\sin(x^2 + y^2)}{x^2 + y^2}$$

and guessed on the basis of numerical evidence that $f(x, y) \to 1$ as $(x, y) \to (0, 0)$. Use polar coordinates to confirm the value of the limit. Then graph the function.

43. Graph and discuss the continuity of the function

$$f(x, y) = \begin{cases} \dfrac{\sin xy}{xy} & \text{if } xy \neq 0 \\ 1 & \text{if } xy = 0 \end{cases}$$

44. Let

$$f(x, y) = \begin{cases} 0 & \text{if } y \leq 0 \quad \text{or} \quad y \geq x^4 \\ 1 & \text{if } 0 < y < x^4 \end{cases}$$

(a) Show that $f(x, y) \to 0$ as $(x, y) \to (0, 0)$ along any path through $(0, 0)$ of the form $y = mx^a$ with $0 < a < 4$.
(b) Despite part (a), show that f is discontinuous at $(0, 0)$.
(c) Show that f is discontinuous on two entire curves.

45. Show that the function f given by $f(\mathbf{x}) = |\mathbf{x}|$ is continuous on $\mathbb{R}^n$.   [*Hint:* Consider $|\mathbf{x} - \mathbf{a}|^2 = (\mathbf{x} - \mathbf{a}) \cdot (\mathbf{x} - \mathbf{a})$.]

46. If $\mathbf{c} \in V_n$, show that the function f given by $f(\mathbf{x}) = \mathbf{c} \cdot \mathbf{x}$ is continuous on $\mathbb{R}^n$.

14.3 Partial Derivatives

On a hot day, extreme humidity makes us think the temperature is higher than it really is, whereas in very dry air we perceive the temperature to be lower than the thermometer indicates. The National Weather Service has devised the *heat index* (also called the temperature-humidity index, or humidex, in some countries) to describe the combined

effects of temperature and humidity. The heat index I is the perceived air temperature when the actual temperature is T and the relative humidity is H. So I is a function of
T and H and we can write $I = f(T, H)$. The following table of values of I is an excerpt
from a table compiled by the National Weather Service.

Table 1 Heat index I as a function of temperature and humidity

Relative humidity (%)

T \ H	50	55	60	65	70	75	80	85	90
90	96	98	100	103	106	109	112	115	119
92	100	103	105	108	112	115	119	123	128
94	104	107	111	114	118	122	127	132	137
96	109	113	116	121	125	130	135	141	146
98	114	118	123	127	133	138	144	150	157
100	119	124	129	135	141	147	154	161	168

Actual temperature (°F)

If we concentrate on the highlighted column of the table, which corresponds to a relative humidity of $H = 70\%$, we are considering the heat index as a function of the single
variable T for a fixed value of H. Let's write $g(T) = f(T, 70)$. Then $g(T)$ describes how
the heat index I increases as the actual temperature T increases when the relative humidity is 70%. The derivative of g when $T = 96°F$ is the rate of change of I with respect to
T when $T = 96°F$:

$$g'(96) = \lim_{h \to 0} \frac{g(96 + h) - g(96)}{h} = \lim_{h \to 0} \frac{f(96 + h, 70) - f(96, 70)}{h}$$

We can approximate $g'(96)$ using the values in Table 1 by taking $h = 2$ and -2:

$$g'(96) \approx \frac{g(98) - g(96)}{2} = \frac{f(98, 70) - f(96, 70)}{2} = \frac{133 - 125}{2} = 4$$

$$g'(96) \approx \frac{g(94) - g(96)}{-2} = \frac{f(94, 70) - f(96, 70)}{-2} = \frac{118 - 125}{-2} = 3.5$$

Averaging these values, we can say that the derivative $g'(96)$ is approximately 3.75. This
means that, when the actual temperature is $96°F$ and the relative humidity is 70%, the
apparent temperature (heat index) rises by about $3.75°F$ for every degree that the actual
temperature rises!

Now let's look at the highlighted row in Table 1, which corresponds to a fixed temperature of $T = 96°F$. The numbers in this row are values of the function $G(H) = f(96, H)$,
which describes how the heat index increases as the relative humidity H increases when
the actual temperature is $T = 96°F$. The derivative of this function when $H = 70\%$ is
the rate of change of I with respect to H when $H = 70\%$:

$$G'(70) = \lim_{h \to 0} \frac{G(70 + h) - G(70)}{h} = \lim_{h \to 0} \frac{f(96, 70 + h) - f(96, 70)}{h}$$

By taking $h = 5$ and -5, we approximate $G'(70)$ using the tabular values:

$$G'(70) \approx \frac{G(75) - G(70)}{5} = \frac{f(96, 75) - f(96, 70)}{5} = \frac{130 - 125}{5} = 1$$

$$G'(70) \approx \frac{G(65) - G(70)}{-5} = \frac{f(96, 65) - f(96, 70)}{-5} = \frac{121 - 125}{-5} = 0.8$$

By averaging these values we get the estimate $G'(70) \approx 0.9$. This says that, when the temperature is 96°F and the relative humidity is 70%, the heat index rises about 0.9°F for every percent that the relative humidity rises.

In general, if f is a function of two variables x and y, suppose we let only x vary while keeping y fixed, say $y = b$, where b is a constant. Then we are really considering a function of a single variable x, namely, $g(x) = f(x, b)$. If g has a derivative at a, then we call it the **partial derivative of f with respect to x at** (a, b) and denote it by $f_x(a, b)$. Thus

1
$$f_x(a, b) = g'(a) \qquad \text{where} \qquad g(x) = f(x, b)$$

By the definition of a derivative, we have

$$g'(a) = \lim_{h \to 0} \frac{g(a + h) - g(a)}{h}$$

and so Equation 1 becomes

2
$$f_x(a, b) = \lim_{h \to 0} \frac{f(a + h, b) - f(a, b)}{h}$$

Similarly, the **partial derivative of f with respect to y at** (a, b), denoted by $f_y(a, b)$, is obtained by keeping x fixed ($x = a$) and finding the ordinary derivative at b of the function $G(y) = f(a, y)$:

3
$$f_y(a, b) = \lim_{h \to 0} \frac{f(a, b + h) - f(a, b)}{h}$$

With this notation for partial derivatives, we can write the rates of change of the heat index I with respect to the actual temperature T and relative humidity H when $T = 96°F$ and $H = 70\%$ as follows:

$$f_T(96, 70) \approx 3.75 \qquad f_H(96, 70) \approx 0.9$$

If we now let the point (a, b) vary in Equations 2 and 3, f_x and f_y become functions of two variables.

4 If f is a function of two variables, its **partial derivatives** are the functions f_x and f_y defined by

$$f_x(x, y) = \lim_{h \to 0} \frac{f(x + h, y) - f(x, y)}{h}$$

$$f_y(x, y) = \lim_{h \to 0} \frac{f(x, y + h) - f(x, y)}{h}$$

There are many alternative notations for partial derivatives. For instance, instead of f_x we can write f_1 or $D_1 f$ (to indicate differentiation with respect to the *first* variable) or $\partial f / \partial x$. But here $\partial f / \partial x$ can't be interpreted as a ratio of differentials.

Notations for Partial Derivatives If $z = f(x, y)$, we write

$$f_x(x, y) = f_x = \frac{\partial f}{\partial x} = \frac{\partial}{\partial x} f(x, y) = \frac{\partial z}{\partial x} = f_1 = D_1 f = D_x f$$

$$f_y(x, y) = f_y = \frac{\partial f}{\partial y} = \frac{\partial}{\partial y} f(x, y) = \frac{\partial z}{\partial y} = f_2 = D_2 f = D_y f$$

To compute partial derivatives, all we have to do is remember from Equation 1 that the partial derivative with respect to x is just the *ordinary* derivative of the function g of a single variable that we get by keeping y fixed. Thus we have the following rule.

Rule for Finding Partial Derivatives of $z = f(x, y)$

1. To find f_x, regard y as a constant and differentiate $f(x, y)$ with respect to x.

2. To find f_y, regard x as a constant and differentiate $f(x, y)$ with respect to y.

EXAMPLE 1 If $f(x, y) = x^3 + x^2 y^3 - 2y^2$, find $f_x(2, 1)$ and $f_y(2, 1)$.

SOLUTION Holding y constant and differentiating with respect to x, we get

$$f_x(x, y) = 3x^2 + 2xy^3$$

and so

$$f_x(2, 1) = 3 \cdot 2^2 + 2 \cdot 2 \cdot 1^3 = 16$$

Holding x constant and differentiating with respect to y, we get

$$f_y(x, y) = 3x^2 y^2 - 4y$$

$$f_y(2, 1) = 3 \cdot 2^2 \cdot 1^2 - 4 \cdot 1 = 8$$

∎

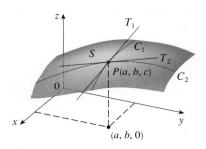

FIGURE 1
The partial derivatives of f at (a, b) are the slopes of the tangents to C_1 and C_2.

■ Interpretations of Partial Derivatives

To give a geometric interpretation of partial derivatives, we recall that the equation $z = f(x, y)$ represents a surface S (the graph of f). If $f(a, b) = c$, then the point $P(a, b, c)$ lies on S. By fixing $y = b$, we are restricting our attention to the curve C_1 in which the vertical plane $y = b$ intersects S. (In other words, C_1 is the trace of S in the plane $y = b$.) Likewise, the vertical plane $x = a$ intersects S in a curve C_2. Both of the curves C_1 and C_2 pass through the point P. (See Figure 1.)

Note that the curve C_1 is the graph of the function $g(x) = f(x, b)$, so the slope of its tangent T_1 at P is $g'(a) = f_x(a, b)$. The curve C_2 is the graph of the function $G(y) = f(a, y)$, so the slope of its tangent T_2 at P is $G'(b) = f_y(a, b)$.

Thus the partial derivatives $f_x(a, b)$ and $f_y(a, b)$ can be interpreted geometrically as the slopes of the tangent lines at $P(a, b, c)$ to the traces C_1 and C_2 of S in the planes $y = b$ and $x = a$.

As we have seen in the case of the heat index function, partial derivatives can also be interpreted as *rates of change*. If $z = f(x, y)$, then $\partial z / \partial x$ represents the rate of change of z with respect to x when y is fixed. Similarly, $\partial z / \partial y$ represents the rate of change of z with respect to y when x is fixed.

EXAMPLE 2 If $f(x, y) = 4 - x^2 - 2y^2$, find $f_x(1, 1)$ and $f_y(1, 1)$ and interpret these numbers as slopes.

SOLUTION We have

$$f_x(x, y) = -2x \qquad f_y(x, y) = -4y$$

$$f_x(1, 1) = -2 \qquad f_y(1, 1) = -4$$

The graph of f is the paraboloid $z = 4 - x^2 - 2y^2$ and the vertical plane $y = 1$ intersects it in the parabola $z = 2 - x^2$, $y = 1$. (As in the preceding discussion, we label it C_1 in Figure 2.) The slope of the tangent line to this parabola at the point $(1, 1, 1)$ is $f_x(1, 1) = -2$. Similarly, the curve C_2 in which the plane $x = 1$ intersects the paraboloid is the parabola $z = 3 - 2y^2$, $x = 1$, and the slope of the tangent line at $(1, 1, 1)$ is $f_y(1, 1) = -4$. (See Figure 3.)

FIGURE 2

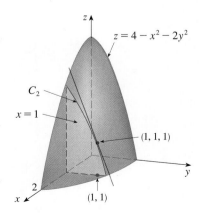

FIGURE 3

Figure 4 is a computer-drawn counterpart to Figure 2. Part (a) shows the plane $y = 1$ intersecting the surface to form the curve C_1 and part (b) shows C_1 and T_1. [We have used the vector equations $\mathbf{r}(t) = \langle t, 1, 2 - t^2 \rangle$ for C_1 and $\mathbf{r}(t) = \langle 1 + t, 1, 1 - 2t \rangle$ for T_1.] Similarly, Figure 5 corresponds to Figure 3.

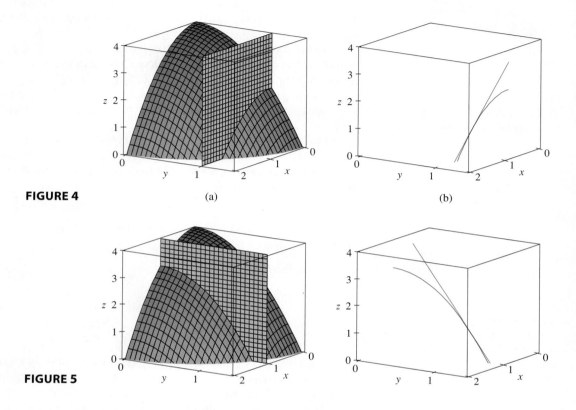

FIGURE 4   (a)   (b)

FIGURE 5

EXAMPLE 3 In Exercise 14.1.39 we defined the body mass index of a person as

$$B(m, h) = \frac{m}{h^2}$$

Calculate the partial derivatives of B for a young man with $m = 64$ kg and $h = 1.68$ m and interpret them.

SOLUTION Regarding h as a constant, we see that the partial derivative with respect to m is

$$\frac{\partial B}{\partial m}(m, h) = \frac{\partial}{\partial m}\left(\frac{m}{h^2}\right) = \frac{1}{h^2}$$

so

$$\frac{\partial B}{\partial m}(64, 1.68) = \frac{1}{(1.68)^2} \approx 0.35 \ (\text{kg/m}^2)/\text{kg}$$

This is the rate at which the man's BMI increases with respect to his weight when he weighs 64 kg and his height is 1.68 m. So if his weight increases by a small amount, one kilogram for instance, and his height remains unchanged, then his BMI will increase by about 0.35.

Now we regard m as a constant. The partial derivative with respect to h is

$$\frac{\partial B}{\partial h}(m, h) = \frac{\partial}{\partial h}\left(\frac{m}{h^2}\right) = m\left(-\frac{2}{h^3}\right) = -\frac{2m}{h^3}$$

so

$$\frac{\partial B}{\partial h}(64, 1.68) = -\frac{2 \cdot 64}{(1.68)^3} \approx -27 \ (\text{kg/m}^2)/\text{m}$$

This is the rate at which the man's BMI increases with respect to his height when he weighs 64 kg and his height is 1.68 m. So if the man is still growing and his weight stays unchanged while his height increases by a small amount, say 1 cm, then his BMI will *decrease* by about $27(0.01) = 0.27$. ∎

EXAMPLE 4 If $f(x, y) = \sin\left(\dfrac{x}{1 + y}\right)$, calculate $\dfrac{\partial f}{\partial x}$ and $\dfrac{\partial f}{\partial y}$.

SOLUTION Using the Chain Rule for functions of one variable, we have

$$\frac{\partial f}{\partial x} = \cos\left(\frac{x}{1 + y}\right) \cdot \frac{\partial}{\partial x}\left(\frac{x}{1 + y}\right) = \cos\left(\frac{x}{1 + y}\right) \cdot \frac{1}{1 + y}$$

$$\frac{\partial f}{\partial y} = \cos\left(\frac{x}{1 + y}\right) \cdot \frac{\partial}{\partial y}\left(\frac{x}{1 + y}\right) = -\cos\left(\frac{x}{1 + y}\right) \cdot \frac{x}{(1 + y)^2} \quad ∎$$

EXAMPLE 5 Find $\partial z/\partial x$ and $\partial z/\partial y$ if z is defined implicitly as a function of x and y by the equation

$$x^3 + y^3 + z^3 + 6xyz = 1$$

SOLUTION To find $\partial z/\partial x$, we differentiate implicitly with respect to x, being careful to treat y as a constant:

$$3x^2 + 3z^2\frac{\partial z}{\partial x} + 6yz + 6xy\frac{\partial z}{\partial x} = 0$$

Solving this equation for $\partial z/\partial x$, we obtain

$$\frac{\partial z}{\partial x} = -\frac{x^2 + 2yz}{z^2 + 2xy}$$

Similarly, implicit differentiation with respect to y gives

$$\frac{\partial z}{\partial y} = -\frac{y^2 + 2xz}{z^2 + 2xy} \quad ∎$$

Some computer software can plot surfaces defined by implicit equations in three variables. Figure 6 shows such a plot of the surface defined by the equation in Example 5.

FIGURE 6

▪ Functions of More Than Two Variables

Partial derivatives can also be defined for functions of three or more variables. For example, if f is a function of three variables x, y, and z, then its partial derivative with respect to x is defined as

$$f_x(x, y, z) = \lim_{h \to 0} \frac{f(x + h, y, z) - f(x, y, z)}{h}$$

and it is found by regarding y and z as constants and differentiating $f(x, y, z)$ with respect to x. If $w = f(x, y, z)$, then $f_x = \partial w/\partial x$ can be interpreted as the rate of change of w with respect to x when y and z are held fixed. But we can't interpret it geometrically because the graph of f lies in four-dimensional space.

In general, if u is a function of n variables, $u = f(x_1, x_2, \ldots, x_n)$, its partial derivative with respect to the ith variable x_i is

$$\frac{\partial u}{\partial x_i} = \lim_{h \to 0} \frac{f(x_1, \ldots, x_{i-1}, x_i + h, x_{i+1}, \ldots, x_n) - f(x_1, \ldots, x_i, \ldots, x_n)}{h}$$

and we also write

$$\frac{\partial u}{\partial x_i} = \frac{\partial f}{\partial x_i} = f_{x_i} = f_i = D_i f$$

EXAMPLE 6 Find f_x, f_y, and f_z if $f(x, y, z) = e^{xy} \ln z$.

SOLUTION Holding y and z constant and differentiating with respect to x, we have

$$f_x = y e^{xy} \ln z$$

Similarly, $$f_y = x e^{xy} \ln z \quad \text{and} \quad f_z = \frac{e^{xy}}{z} \qquad ■$$

▪ Higher Derivatives

If f is a function of two variables, then its partial derivatives f_x and f_y are also functions of two variables, so we can consider their partial derivatives $(f_x)_x$, $(f_x)_y$, $(f_y)_x$, and $(f_y)_y$, which are called the **second partial derivatives** of f. If $z = f(x, y)$, we use the following notation:

$$(f_x)_x = f_{xx} = f_{11} = \frac{\partial}{\partial x}\left(\frac{\partial f}{\partial x}\right) = \frac{\partial^2 f}{\partial x^2} = \frac{\partial^2 z}{\partial x^2}$$

$$(f_x)_y = f_{xy} = f_{12} = \frac{\partial}{\partial y}\left(\frac{\partial f}{\partial x}\right) = \frac{\partial^2 f}{\partial y \, \partial x} = \frac{\partial^2 z}{\partial y \, \partial x}$$

$$(f_y)_x = f_{yx} = f_{21} = \frac{\partial}{\partial x}\left(\frac{\partial f}{\partial y}\right) = \frac{\partial^2 f}{\partial x \, \partial y} = \frac{\partial^2 z}{\partial x \, \partial y}$$

$$(f_y)_y = f_{yy} = f_{22} = \frac{\partial}{\partial y}\left(\frac{\partial f}{\partial y}\right) = \frac{\partial^2 f}{\partial y^2} = \frac{\partial^2 z}{\partial y^2}$$

Thus the notation f_{xy} (or $\partial^2 f / \partial y \, \partial x$) means that we first differentiate with respect to x and then with respect to y, whereas in computing f_{yx} the order is reversed.

EXAMPLE 7 Find the second partial derivatives of

$$f(x, y) = x^3 + x^2 y^3 - 2y^2$$

SOLUTION In Example 1 we found that

$$f_x(x, y) = 3x^2 + 2xy^3 \qquad f_y(x, y) = 3x^2 y^2 - 4y$$

Therefore

$$f_{xx} = \frac{\partial}{\partial x}(3x^2 + 2xy^3) = 6x + 2y^3 \qquad f_{xy} = \frac{\partial}{\partial y}(3x^2 + 2xy^3) = 6xy^2$$

$$f_{yx} = \frac{\partial}{\partial x}(3x^2 y^2 - 4y) = 6xy^2 \qquad f_{yy} = \frac{\partial}{\partial y}(3x^2 y^2 - 4y) = 6x^2 y - 4 \quad ■$$

Figure 7 shows the graph of the function f in Example 7 and the graphs of its first- and second-order partial derivatives for $-2 \leqslant x \leqslant 2$, $-2 \leqslant y \leqslant 2$. Notice that these graphs are consistent with our interpretations of f_x and f_y as slopes of tangent lines to traces of the graph of f. For instance, the graph of f decreases if we start at $(0, -2)$ and move in the positive x-direction. This is reflected in the negative values of f_x. You should compare the graphs of f_{yx} and f_{yy} with the graph of f_y to see the relationships.

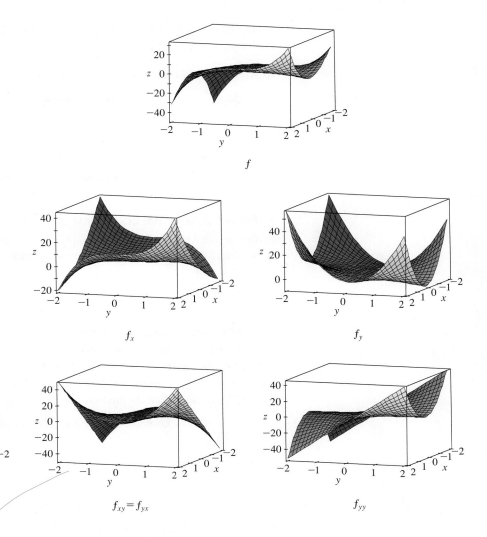

f

f_x

f_y

f_{xx}

$f_{xy} = f_{yx}$

f_{yy}

FIGURE 7

Notice that $f_{xy} = f_{yx}$ in Example 7. This is not just a coincidence. It turns out that the mixed partial derivatives f_{xy} and f_{yx} are equal for most functions that one meets in practice. The following theorem, which was discovered by the French mathematician Alexis Clairaut (1713–1765), gives conditions under which we can assert that $f_{xy} = f_{yx}$. The proof is given in Appendix F.

Clairaut

Alexis Clairaut was a child prodigy in mathematics: he read l'Hospital's textbook on calculus when he was ten and presented a paper on geometry to the French Academy of Sciences when he was 13. At the age of 18, Clairaut published *Recherches sur les courbes à double courbure,* which was the first systematic treatise on three-dimensional analytic geometry and included the calculus of space curves.

Clairaut's Theorem Suppose f is defined on a disk D that contains the point (a, b). If the functions f_{xy} and f_{yx} are both continuous on D, then

$$f_{xy}(a, b) = f_{yx}(a, b)$$

Partial derivatives of order 3 or higher can also be defined. For instance,

$$f_{xyy} = (f_{xy})_y = \frac{\partial}{\partial y}\left(\frac{\partial^2 f}{\partial y \, \partial x}\right) = \frac{\partial^3 f}{\partial y^2 \, \partial x}$$

and using Clairaut's Theorem it can be shown that $f_{xyy} = f_{yxy} = f_{yyx}$ if these functions are continuous.

EXAMPLE 8　Calculate f_{xxyz} if $f(x, y, z) = \sin(3x + yz)$.

SOLUTION

$$f_x = 3\cos(3x + yz)$$

$$f_{xx} = -9\sin(3x + yz)$$

$$f_{xxy} = -9z\cos(3x + yz)$$

$$f_{xxyz} = -9\cos(3x + yz) + 9yz\sin(3x + yz)$$ ∎

■ Partial Differential Equations

Partial derivatives occur in *partial differential equations* that express certain physical laws. For instance, the partial differential equation

$$\frac{\partial^2 u}{\partial x^2} + \frac{\partial^2 u}{\partial y^2} = 0$$

is called **Laplace's equation** after Pierre Laplace (1749–1827). Solutions of this equation are called **harmonic functions**; they play a role in problems of heat conduction, fluid flow, and electric potential.

EXAMPLE 9　Show that the function $u(x, y) = e^x \sin y$ is a solution of Laplace's equation.

SOLUTION　We first compute the needed second-order partial derivatives:

$$u_x = e^x \sin y \qquad u_y = e^x \cos y$$

$$u_{xx} = e^x \sin y \qquad u_{yy} = -e^x \sin y$$

So

$$u_{xx} + u_{yy} = e^x \sin y - e^x \sin y = 0$$

Therefore u satisfies Laplace's equation. ∎

The **wave equation**

$$\frac{\partial^2 u}{\partial t^2} = a^2 \frac{\partial^2 u}{\partial x^2}$$

FIGURE 8

describes the motion of a waveform, which could be an ocean wave, a sound wave, a light wave, or a wave traveling along a vibrating string. For instance, if $u(x, t)$ represents the displacement of a vibrating violin string at time t and at a distance x from one end of the string (as in Figure 8), then $u(x, t)$ satisfies the wave equation. Here the constant a depends on the density of the string and on the tension in the string.

EXAMPLE 10　Verify that the function $u(x, t) = \sin(x - at)$ satisfies the wave equation.

SOLUTION　　$u_x = \cos(x - at)$　　　　$u_t = -a\cos(x - at)$

$$u_{xx} = -\sin(x - at) \qquad u_{tt} = -a^2 \sin(x - at) = a^2 u_{xx}$$

So u satisfies the wave equation. ∎

Partial differential equations involving functions of three variables are also very important in science and engineering. The three-dimensional Laplace equation is

$$\boxed{5} \qquad \frac{\partial^2 u}{\partial x^2} + \frac{\partial^2 u}{\partial y^2} + \frac{\partial^2 u}{\partial z^2} = 0$$

and one place it occurs is in geophysics. If $u(x, y, z)$ represents magnetic field strength at position (x, y, z), then it satisfies Equation 5. The strength of the magnetic field indicates the distribution of iron-rich minerals and reflects different rock types and the location of faults. Figure 9 shows a contour map of the earth's magnetic field as recorded from an aircraft carrying a magnetometer and flying 200 m above the surface of the ground. The contour map is enhanced by color-coding of the regions between the level curves.

FIGURE 9
Magnetic field strength of the earth

Figure 10 shows a contour map for the second-order partial derivative of u in the vertical direction, that is, u_{zz}. It turns out that the values of the partial derivatives u_{xx} and u_{yy} are relatively easily measured from a map of the magnetic field. Then values of u_{zz} can be calculated from Laplace's equation (5).

FIGURE 10
Second vertical derivative
of the magnetic field

■ The Cobb-Douglas Production Function

In Example 14.1.3 we described the work of Cobb and Douglas in modeling the total production P of an economic system as a function of the amount of labor L and the capital investment K. Here we use partial derivatives to show how the particular form of their model follows from certain assumptions they made about the economy.

If the production function is denoted by $P = P(L, K)$, then the partial derivative $\partial P/\partial L$ is the rate at which production changes with respect to the amount of labor. Economists call it the marginal production with respect to labor or the **marginal productivity of labor**. Likewise, the partial derivative $\partial P/\partial K$ is the rate of change of production with respect to capital and is called the **marginal productivity of capital**. In these terms, the assumptions made by Cobb and Douglas can be stated as follows.

  (i) If either labor or capital vanishes, then so will production.

  (ii) The marginal productivity of labor is proportional to the amount of production per unit of labor.

  (iii) The marginal productivity of capital is proportional to the amount of production per unit of capital.

Because the production per unit of labor is P/L, assumption (ii) says that

$$\frac{\partial P}{\partial L} = \alpha \, \frac{P}{L}$$

for some constant α. If we keep K constant ($K = K_0$), then this partial differential equation becomes an ordinary differential equation:

$$\boxed{6} \qquad\qquad \frac{dP}{dL} = \alpha \, \frac{P}{L}$$

If we solve this separable differential equation by the methods of Section 9.3 (see also Exercise 85), we get

$$\boxed{7} \qquad\qquad P(L, K_0) = C_1(K_0)L^{\alpha}$$

Notice that we have written the constant C_1 as a function of K_0 because it could depend on the value of K_0.

Similarly, assumption (iii) says that

$$\frac{\partial P}{\partial K} = \beta \, \frac{P}{K}$$

and we can solve this differential equation to get

$$\boxed{8} \qquad\qquad P(L_0, K) = C_2(L_0)K^{\beta}$$

Comparing Equations 7 and 8, we have

$$\boxed{9} \qquad\qquad P(L, K) = bL^{\alpha}K^{\beta}$$

where b is a constant that is independent of both L and K. Assumption (i) shows that $\alpha > 0$ and $\beta > 0$.

Notice from Equation 9 that if labor and capital are both increased by a factor m, then

$$P(mL, mK) = b(mL)^{\alpha}(mK)^{\beta} = m^{\alpha+\beta}bL^{\alpha}K^{\beta} = m^{\alpha+\beta}P(L, K)$$

If $\alpha + \beta = 1$, then $P(mL, mK) = mP(L, K)$, which means that production is also increased by a factor of m. That is why Cobb and Douglas assumed that $\alpha + \beta = 1$ and therefore

$$P(L, K) = bL^{\alpha}K^{1-\alpha}$$

This is the Cobb-Douglas production function that we discussed in Section 14.1.

14.3 EXERCISES

1. The temperature T (in °C) at a location in the Northern Hemisphere depends on the longitude x, latitude y, and time t, so we can write $T = f(x, y, t)$. Let's measure time in hours from the beginning of January.
   (a) What are the meanings of the partial derivatives $\partial T/\partial x$, $\partial T/\partial y$, and $\partial T/\partial t$?
   (b) Honolulu has longitude 158°W and latitude 21°N. Suppose that at 9:00 AM on January 1 the wind is blowing hot air to the northeast, so the air to the west and south is warm and the air to the north and east is cooler. Would you expect $f_x(158, 21, 9)$, $f_y(158, 21, 9)$, and $f_t(158, 21, 9)$ to be positive or negative? Explain.

2. At the beginning of this section we discussed the function $I = f(T, H)$, where I is the heat index, T is the temperature, and H is the relative humidity. Use Table 1 to estimate $f_T(92, 60)$ and $f_H(92, 60)$. What are the practical interpretations of these values?

3. The wind-chill index W is the perceived temperature when the actual temperature is T and the wind speed is v, so we can write $W = f(T, v)$. The following table of values is an excerpt from Table 1 in Section 14.1.

Wind speed (km/h)

T \ v	20	30	40	50	60	70
−10	−18	−20	−21	−22	−23	−23
−15	−24	−26	−27	−29	−30	−30
−20	−30	−33	−34	−35	−36	−37
−25	−37	−39	−41	−42	−43	−44

Actual temperature (°C)

(a) Estimate the values of $f_T(-15, 30)$ and $f_v(-15, 30)$. What are the practical interpretations of these values?

(b) In general, what can you say about the signs of $\partial W/\partial T$ and $\partial W/\partial v$?
(c) What appears to be the value of the following limit?

$$\lim_{v \to \infty} \frac{\partial W}{\partial v}$$

4. The wave heights h in the open sea depend on the speed v of the wind and the length of time t that the wind has been blowing at that speed. Values of the function $h = f(v, t)$ are recorded in feet in the following table.

Duration (hours)

v \ t	5	10	15	20	30	40	50
10	2	2	2	2	2	2	2
15	4	4	5	5	5	5	5
20	5	7	8	8	9	9	9
30	9	13	16	17	18	19	19
40	14	21	25	28	31	33	33
50	19	29	36	40	45	48	50
60	24	37	47	54	62	67	69

Wind speed (knots)

(a) What are the meanings of the partial derivatives $\partial h/\partial v$ and $\partial h/\partial t$?
(b) Estimate the values of $f_v(40, 15)$ and $f_t(40, 15)$. What are the practical interpretations of these values?
(c) What appears to be the value of the following limit?

$$\lim_{t \to \infty} \frac{\partial h}{\partial t}$$

5–8 Determine the signs of the partial derivatives for the function f whose graph is shown.

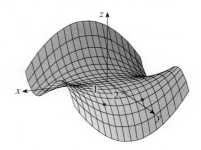

5. (a) $f_x(1, 2)$ (b) $f_y(1, 2)$

6. (a) $f_x(-1, 2)$ (b) $f_y(-1, 2)$

7. (a) $f_{xx}(-1, 2)$ (b) $f_{yy}(-1, 2)$

8. (a) $f_{xy}(1, 2)$ (b) $f_{xy}(-1, 2)$

9. The following surfaces, labeled a, b, and c, are graphs of a function f and its partial derivatives f_x and f_y. Identify each surface and give reasons for your choices.

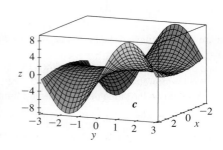

10. A contour map is given for a function f. Use it to estimate $f_x(2, 1)$ and $f_y(2, 1)$.

11. If $f(x, y) = 16 - 4x^2 - y^2$, find $f_x(1, 2)$ and $f_y(1, 2)$ and interpret these numbers as slopes. Illustrate with either hand-drawn sketches or computer plots.

12. If $f(x, y) = \sqrt{4 - x^2 - 4y^2}$, find $f_x(1, 0)$ and $f_y(1, 0)$ and interpret these numbers as slopes. Illustrate with either hand-drawn sketches or computer plots.

13–14 Find f_x and f_y and graph f, f_x, and f_y with domains and viewpoints that enable you to see the relationships between them.

13. $f(x, y) = x^2y^3$

14. $f(x, y) = \dfrac{y}{1 + x^2y^2}$

15–40 Find the first partial derivatives of the function.

15. $f(x, y) = x^4 + 5xy^3$

16. $f(x, y) = x^2y - 3y^4$

17. $f(x, t) = t^2e^{-x}$

18. $f(x, t) = \sqrt{3x + 4t}$

19. $z = \ln(x + t^2)$

20. $z = x\sin(xy)$

21. $f(x, y) = \dfrac{x}{y}$

22. $f(x, y) = \dfrac{x}{(x + y)^2}$

23. $f(x, y) = \dfrac{ax + by}{cx + dy}$

24. $w = \dfrac{e^v}{u + v^2}$

25. $g(u, v) = (u^2v - v^3)^5$

26. $u(r, \theta) = \sin(r\cos\theta)$

27. $R(p, q) = \tan^{-1}(pq^2)$

28. $f(x, y) = x^y$

29. $F(x, y) = \displaystyle\int_y^x \cos(e^t)\,dt$

30. $F(\alpha, \beta) = \displaystyle\int_\alpha^\beta \sqrt{t^3 + 1}\,dt$

31. $f(x, y, z) = x^3yz^2 + 2yz$

32. $f(x, y, z) = xy^2e^{-xz}$

33. $w = \ln(x + 2y + 3z)$

34. $w = y\tan(x + 2z)$

35. $p = \sqrt{t^4 + u^2\cos v}$

36. $u = x^{y/z}$

37. $h(x, y, z, t) = x^2y\cos(z/t)$

38. $\phi(x, y, z, t) = \dfrac{\alpha x + \beta y^2}{\gamma z + \delta t^2}$

39. $u = \sqrt{x_1^2 + x_2^2 + \cdots + x_n^2}$

40. $u = \sin(x_1 + 2x_2 + \cdots + nx_n)$

41–44 Find the indicated partial derivative.

41. $R(s, t) = te^{s/t}$; $R_t(0, 1)$

42. $f(x, y) = y \sin^{-1}(xy)$; $f_y\left(1, \frac{1}{2}\right)$

43. $f(x, y, z) = \ln \dfrac{1 - \sqrt{x^2 + y^2 + z^2}}{1 + \sqrt{x^2 + y^2 + z^2}}$; $f_y(1, 2, 2)$

44. $f(x, y, z) = x^{yz}$; $f_z(e, 1, 0)$

45–46 Use the definition of partial derivatives as limits (4) to find $f_x(x, y)$ and $f_y(x, y)$.

45. $f(x, y) = xy^2 - x^3y$

46. $f(x, y) = \dfrac{x}{x + y^2}$

47–50 Use implicit differentiation to find $\partial z / \partial x$ and $\partial z / \partial y$.

47. $x^2 + 2y^2 + 3z^2 = 1$

48. $x^2 - y^2 + z^2 - 2z = 4$

49. $e^z = xyz$

50. $yz + x \ln y = z^2$

51–52 Find $\partial z / \partial x$ and $\partial z / \partial y$.

51. (a) $z = f(x) + g(y)$ (b) $z = f(x + y)$

52. (a) $z = f(x)g(y)$ (b) $z = f(xy)$

(c) $z = f(x/y)$

53–58 Find all the second partial derivatives.

53. $f(x, y) = x^4y - 2x^3y^2$

54. $f(x, y) = \ln(ax + by)$

55. $z = \dfrac{y}{2x + 3y}$

56. $T = e^{-2r} \cos \theta$

57. $v = \sin(s^2 - t^2)$

58. $w = \sqrt{1 + uv^2}$

59–62 Verify that the conclusion of Clairaut's Theorem holds, that is, $u_{xy} = u_{yx}$.

59. $u = x^4y^3 - y^4$

60. $u = e^{xy} \sin y$

61. $u = \cos(x^2y)$

62. $u = \ln(x + 2y)$

63–70 Find the indicated partial derivative(s).

63. $f(x, y) = x^4y^2 - x^3y$; f_{xxx}, f_{xyx}

64. $f(x, y) = \sin(2x + 5y)$; f_{yxy}

65. $f(x, y, z) = e^{xyz^2}$; f_{xyz}

66. $g(r, s, t) = e^r \sin(st)$; g_{rst}

67. $W = \sqrt{u + v^2}$; $\dfrac{\partial^3 W}{\partial u^2 \, \partial v}$

68. $V = \ln(r + s^2 + t^3)$; $\dfrac{\partial^3 V}{\partial r \, \partial s \, \partial t}$

69. $w = \dfrac{x}{y + 2z}$; $\dfrac{\partial^3 w}{\partial z \, \partial y \, \partial x}$, $\dfrac{\partial^3 w}{\partial x^2 \, \partial y}$

70. $u = x^a y^b z^c$; $\dfrac{\partial^6 u}{\partial x \, \partial y^2 \, \partial z^3}$

71. If $f(x, y, z) = xy^2z^3 + \arcsin(x\sqrt{z})$, find f_{xzy}.
[*Hint:* Which order of differentiation is easiest?]

72. If $g(x, y, z) = \sqrt{1 + xz} + \sqrt{1 - xy}$, find g_{xyz}. [*Hint:* Use a different order of differentiation for each term.]

73. Use the table of values of $f(x, y)$ to estimate the values of $f_x(3, 2)$, $f_x(3, 2.2)$, and $f_{xy}(3, 2)$.

x \ y	1.8	2.0	2.2
2.5	12.5	10.2	9.3
3.0	18.1	17.5	15.9
3.5	20.0	22.4	26.1

74. Level curves are shown for a function f. Determine whether the following partial derivatives are positive or negative at the point P.

(a) f_x (b) f_y (c) f_{xx}

(d) f_{xy} (e) f_{yy}

75. Verify that the function $u = e^{-\alpha^2 k^2 t} \sin kx$ is a solution of the *heat conduction equation* $u_t = \alpha^2 u_{xx}$.

76. Determine whether each of the following functions is a solution of Laplace's equation $u_{xx} + u_{yy} = 0$.

(a) $u = x^2 + y^2$ (b) $u = x^2 - y^2$

(c) $u = x^3 + 3xy^2$ (d) $u = \ln \sqrt{x^2 + y^2}$

(e) $u = \sin x \cosh y + \cos x \sinh y$

(f) $u = e^{-x} \cos y - e^{-y} \cos x$

77. Verify that the function $u = 1/\sqrt{x^2 + y^2 + z^2}$ is a solution of the three-dimensional Laplace equation $u_{xx} + u_{yy} + u_{zz} = 0$.

78. Show that each of the following functions is a solution of the wave equation $u_{tt} = a^2 u_{xx}$.

(a) $u = \sin(kx) \sin(akt)$ (b) $u = t/(a^2t^2 - x^2)$

(c) $u = (x - at)^6 + (x + at)^6$

(d) $u = \sin(x - at) + \ln(x + at)$

79. If f and g are twice differentiable functions of a single variable, show that the function

$$u(x, t) = f(x + at) + g(x - at)$$

is a solution of the wave equation given in Exercise 78.

80. If $u = e^{a_1 x_1 + a_2 x_2 + \cdots + a_n x_n}$, where $a_1^2 + a_2^2 + \cdots + a_n^2 = 1$, show that

$$\frac{\partial^2 u}{\partial x_1^2} + \frac{\partial^2 u}{\partial x_2^2} + \cdots + \frac{\partial^2 u}{\partial x_n^2} = u$$

81. The *diffusion equation*

$$\frac{\partial c}{\partial t} = D \frac{\partial^2 c}{\partial x^2}$$

where D is a positive constant, describes the diffusion of heat through a solid, or the concentration of a pollutant at time t at a distance x from the source of the pollution, or the invasion of alien species into a new habitat. Verify that the function

$$c(x, t) = \frac{1}{\sqrt{4\pi Dt}} e^{-x^2/(4Dt)}$$

is a solution of the diffusion equation.

82. The temperature at a point (x, y) on a flat metal plate is given by $T(x, y) = 60/(1 + x^2 + y^2)$, where T is measured in °C and x, y in meters. Find the rate of change of temperature with respect to distance at the point $(2, 1)$ in (a) the x-direction and (b) the y-direction.

83. The total resistance R produced by three conductors with resistances R_1, R_2, R_3 connected in a parallel electrical circuit is given by the formula

$$\frac{1}{R} = \frac{1}{R_1} + \frac{1}{R_2} + \frac{1}{R_3}$$

Find $\partial R/\partial R_1$.

84. Show that the Cobb-Douglas production function $P = bL^\alpha K^\beta$ satisfies the equation

$$L\frac{\partial P}{\partial L} + K\frac{\partial P}{\partial K} = (\alpha + \beta)P$$

85. Show that the Cobb-Douglas production function satisfies $P(L, K_0) = C_1(K_0)L^\alpha$ by solving the differential equation

$$\frac{dP}{dL} = \alpha \frac{P}{L}$$

(See Equation 6.)

86. Cobb and Douglas used the equation $P(L, K) = 1.01L^{0.75}K^{0.25}$ to model the American economy from 1899 to 1922, where L is the amount of labor and K is the amount of capital. (See Example 14.1.3.)
   (a) Calculate P_L and P_K.
   (b) Find the marginal productivity of labor and the marginal productivity of capital in the year 1920, when $L = 194$ and $K = 407$ (compared with the assigned values $L = 100$ and $K = 100$ in 1899). Interpret the results.
   (c) In the year 1920, which would have benefited production more, an increase in capital investment or an increase in spending on labor?

87. The *van der Waals equation* for n moles of a gas is

$$\left(P + \frac{n^2 a}{V^2}\right)(V - nb) = nRT$$

where P is the pressure, V is the volume, and T is the temperature of the gas. The constant R is the universal gas constant and a and b are positive constants that are characteristic of a particular gas. Calculate $\partial T/\partial P$ and $\partial P/\partial V$.

88. The gas law for a fixed mass m of an ideal gas at absolute temperature T, pressure P, and volume V is $PV = mRT$, where R is the gas constant. Show that

$$\frac{\partial P}{\partial V}\frac{\partial V}{\partial T}\frac{\partial T}{\partial P} = -1$$

89. For the ideal gas of Exercise 88, show that

$$T\frac{\partial P}{\partial T}\frac{\partial V}{\partial T} = mR$$

90. The wind-chill index is modeled by the function

$$W = 13.12 + 0.6215T - 11.37v^{0.16} + 0.3965Tv^{0.16}$$

where T is the temperature (°C) and v is the wind speed (km/h). When $T = -15$°C and $v = 30$ km/h, by how much would you expect the apparent temperature W to drop if the actual temperature decreases by 1°C? What if the wind speed increases by 1 km/h?

91. A model for the surface area of a human body is given by the function

$$S = f(w, h) = 0.1091w^{0.425}h^{0.725}$$

where w is the weight (in pounds), h is the height (in inches), and S is measured in square feet. Calculate and interpret the partial derivatives.

(a) $\dfrac{\partial S}{\partial w}(160, 70)$ (b) $\dfrac{\partial S}{\partial h}(160, 70)$

92. One of Poiseuille's laws states that the resistance of blood flowing through an artery is

$$R = C\frac{L}{r^4}$$

where L and r are the length and radius of the artery and C is a positive constant determined by the viscosity of the blood. Calculate $\partial R/\partial L$ and $\partial R/\partial r$ and interpret them.

93. In the project on page 344 we expressed the power needed by a bird during its flapping mode as

$$P(v, x, m) = Av^3 + \frac{B(mg/x)^2}{v}$$

where A and B are constants specific to a species of bird, v is the velocity of the bird, m is the mass of the bird, and x is the fraction of the flying time spent in flapping mode. Calculate $\partial P/\partial v$, $\partial P/\partial x$, and $\partial P/\partial m$ and interpret them.

94. The average energy E (in kcal) needed for a lizard to walk or run a distance of 1 km has been modeled by the equation

$$E(m, v) = 2.65m^{0.66} + \frac{3.5m^{0.75}}{v}$$

where m is the body mass of the lizard (in grams) and v is its speed (in km/h). Calculate $E_m(400, 8)$ and $E_v(400, 8)$ and interpret your answers.

Source: C. Robbins, *Wildlife Feeding and Nutrition*, 2d ed. (San Diego: Academic Press, 1993).

95. The kinetic energy of a body with mass m and velocity v is $K = \frac{1}{2}mv^2$. Show that

$$\frac{\partial K}{\partial m}\frac{\partial^2 K}{\partial v^2} = K$$

96. If a, b, c are the sides of a triangle and A, B, C are the opposite angles, find $\partial A/\partial a$, $\partial A/\partial b$, $\partial A/\partial c$ by implicit differentiation of the Law of Cosines.

97. You are told that there is a function f whose partial derivatives are $f_x(x, y) = x + 4y$ and $f_y(x, y) = 3x - y$. Should you believe it?

98. The paraboloid $z = 6 - x - x^2 - 2y^2$ intersects the plane $x = 1$ in a parabola. Find parametric equations for the tangent line to this parabola at the point $(1, 2, -4)$. Use a computer to graph the paraboloid, the parabola, and the tangent line on the same screen.

99. The ellipsoid $4x^2 + 2y^2 + z^2 = 16$ intersects the plane $y = 2$ in an ellipse. Find parametric equations for the tangent line to this ellipse at the point $(1, 2, 2)$.

100. In a study of frost penetration it was found that the temperature T at time t (measured in days) at a depth x (measured in feet) can be modeled by the function

$$T(x, t) = T_0 + T_1e^{-\lambda x}\sin(\omega t - \lambda x)$$

where $\omega = 2\pi/365$ and λ is a positive constant.
(a) Find $\partial T/\partial x$. What is its physical significance?

(b) Find $\partial T/\partial t$. What is its physical significance?
(c) Show that T satisfies the heat equation $T_t = kT_{xx}$ for a certain constant k.
(d) If $\lambda = 0.2$, $T_0 = 0$, and $T_1 = 10$, use a computer to graph $T(x, t)$.
(e) What is the physical significance of the term $-\lambda x$ in the expression $\sin(\omega t - \lambda x)$?

101. Use Clairaut's Theorem to show that if the third-order partial derivatives of f are continuous, then

$$f_{xyy} = f_{yxy} = f_{yyx}$$

102. (a) How many nth-order partial derivatives does a function of two variables have?
(b) If these partial derivatives are all continuous, how many of them can be distinct?
(c) Answer the question in part (a) for a function of three variables.

103. If

$$f(x, y) = x(x^2 + y^2)^{-3/2}e^{\sin(x^2 y)}$$

find $f_x(1, 0)$. [*Hint:* Instead of finding $f_x(x, y)$ first, note that it's easier to use Equation 1 or Equation 2.]

104. If $f(x, y) = \sqrt[3]{x^3 + y^3}$, find $f_x(0, 0)$.

105. Let

$$f(x, y) = \begin{cases} \dfrac{x^3y - xy^3}{x^2 + y^2} & \text{if } (x, y) \neq (0, 0) \\ 0 & \text{if } (x, y) = (0, 0) \end{cases}$$

(a) Use a computer to graph f.
(b) Find $f_x(x, y)$ and $f_y(x, y)$ when $(x, y) \neq (0, 0)$.
(c) Find $f_x(0, 0)$ and $f_y(0, 0)$ using Equations 2 and 3.
(d) Show that $f_{xy}(0, 0) = -1$ and $f_{yx}(0, 0) = 1$.
(e) Does the result of part (d) contradict Clairaut's Theorem? Use graphs of f_{xy} and f_{yx} to illustrate your answer.

14.4 Tangent Planes and Linear Approximations

One of the most important ideas in single-variable calculus is that as we zoom in toward a point on the graph of a differentiable function, the graph becomes indistinguishable from its tangent line and we can approximate the function by a linear function. (See Section 3.10.) Here we develop similar ideas in three dimensions. As we zoom in toward a point on a surface that is the graph of a differentiable function of two variables, the surface looks more and more like a plane (its tangent plane) and we can approximate the function by a linear function of two variables. We also extend the idea of a differential to functions of two or more variables.

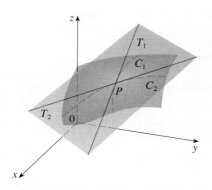

FIGURE 1
The tangent plane contains the tangent lines T_1 and T_2.

■ Tangent Planes

Suppose a surface S has equation $z = f(x, y)$, where f has continuous first partial derivatives, and let $P(x_0, y_0, z_0)$ be a point on S. As in the preceding section, let C_1 and C_2 be the curves obtained by intersecting the vertical planes $y = y_0$ and $x = x_0$ with the surface S. Then the point P lies on both C_1 and C_2. Let T_1 and T_2 be the tangent lines to the curves C_1 and C_2 at the point P. Then the **tangent plane** to the surface S at the point P is defined to be the plane that contains both tangent lines T_1 and T_2. (See Figure 1.)

We will see in Section 14.6 that if C is any other curve that lies on the surface S and passes through P, then its tangent line at P also lies in the tangent plane. Therefore you can think of the tangent plane to S at P as consisting of all possible tangent lines at P to curves that lie on S and pass through P. The tangent plane at P is the plane that most closely approximates the surface S near the point P.

We know from Equation 12.5.7 that any plane passing through the point $P(x_0, y_0, z_0)$ has an equation of the form

$$A(x - x_0) + B(y - y_0) + C(z - z_0) = 0$$

By dividing this equation by C and letting $a = -A/C$ and $b = -B/C$, we can write it in the form

$$\boxed{1} \qquad z - z_0 = a(x - x_0) + b(y - y_0)$$

If Equation 1 represents the tangent plane at P, then its intersection with the plane $y = y_0$ must be the tangent line T_1. Setting $y = y_0$ in Equation 1 gives

$$z - z_0 = a(x - x_0) \qquad \text{where } y = y_0$$

and we recognize this as the equation (in point-slope form) of a line with slope a. But from Section 14.3 we know that the slope of the tangent T_1 is $f_x(x_0, y_0)$. Therefore $a = f_x(x_0, y_0)$.

Similarly, putting $x = x_0$ in Equation 1, we get $z - z_0 = b(y - y_0)$, which must represent the tangent line T_2, so $b = f_y(x_0, y_0)$.

Note the similarity between the equation of a tangent plane and the equation of a tangent line:

$$y - y_0 = f'(x_0)(x - x_0)$$

> $\boxed{2}$ Suppose f has continuous partial derivatives. An equation of the tangent plane to the surface $z = f(x, y)$ at the point $P(x_0, y_0, z_0)$ is
>
> $$z - z_0 = f_x(x_0, y_0)(x - x_0) + f_y(x_0, y_0)(y - y_0)$$

EXAMPLE 1 Find the tangent plane to the elliptic paraboloid $z = 2x^2 + y^2$ at the point $(1, 1, 3)$.

SOLUTION Let $f(x, y) = 2x^2 + y^2$. Then

$$f_x(x, y) = 4x \qquad f_y(x, y) = 2y$$
$$f_x(1, 1) = 4 \qquad f_y(1, 1) = 2$$

Then (2) gives the equation of the tangent plane at $(1, 1, 3)$ as

$$z - 3 = 4(x - 1) + 2(y - 1)$$

or

$$z = 4x + 2y - 3 \qquad \blacksquare$$

Figure 2(a) shows the elliptic paraboloid and its tangent plane at $(1, 1, 3)$ that we found in Example 1. In parts (b) and (c) we zoom in toward the point $(1, 1, 3)$ by restrict-

TEC Visual 14.4 shows an animation of Figures 2 and 3.

ing the domain of the function $f(x, y) = 2x^2 + y^2$. Notice that the more we zoom in, the flatter the graph appears and the more it resembles its tangent plane.

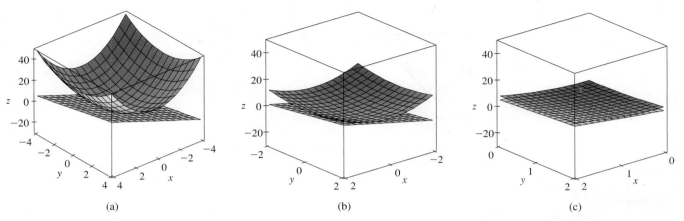

(a)   (b)   (c)

FIGURE 2 The elliptic paraboloid $z = 2x^2 + y^2$ appears to coincide with its tangent plane as we zoom in toward $(1, 1, 3)$.

In Figure 3 we corroborate this impression by zooming in toward the point $(1, 1)$ on a contour map of the function $f(x, y) = 2x^2 + y^2$. Notice that the more we zoom in, the more the level curves look like equally spaced parallel lines, which is characteristic of a plane.

FIGURE 3
Zooming in toward $(1, 1)$
on a contour map of
$f(x, y) = 2x^2 + y^2$

  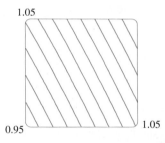

■ Linear Approximations

In Example 1 we found that an equation of the tangent plane to the graph of the function $f(x, y) = 2x^2 + y^2$ at the point $(1, 1, 3)$ is $z = 4x + 2y - 3$. Therefore, in view of the visual evidence in Figures 2 and 3, the linear function of two variables

$$L(x, y) = 4x + 2y - 3$$

is a good approximation to $f(x, y)$ when (x, y) is near $(1, 1)$. The function L is called the *linearization* of f at $(1, 1)$ and the approximation

$$f(x, y) \approx 4x + 2y - 3$$

is called the *linear approximation* or *tangent plane approximation* of f at $(1, 1)$.

For instance, at the point $(1.1, 0.95)$ the linear approximation gives

$$f(1.1, 0.95) \approx 4(1.1) + 2(0.95) - 3 = 3.3$$

which is quite close to the true value of $f(1.1, 0.95) = 2(1.1)^2 + (0.95)^2 = 3.3225$. But if we take a point farther away from $(1, 1)$, such as $(2, 3)$, we no longer get a good approximation. In fact, $L(2, 3) = 11$ whereas $f(2, 3) = 17$.

In general, we know from (2) that an equation of the tangent plane to the graph of a function f of two variables at the point $(a, b, f(a, b))$ is

$$z = f(a, b) + f_x(a, b)(x - a) + f_y(a, b)(y - b)$$

The linear function whose graph is this tangent plane, namely

$$\boxed{3} \qquad L(x, y) = f(a, b) + f_x(a, b)(x - a) + f_y(a, b)(y - b)$$

is called the **linearization** of f at (a, b) and the approximation

$$\boxed{4} \qquad f(x, y) \approx f(a, b) + f_x(a, b)(x - a) + f_y(a, b)(y - b)$$

is called the **linear approximation** or the **tangent plane approximation** of f at (a, b).

We have defined tangent planes for surfaces $z = f(x, y)$, where f has continuous first partial derivatives. What happens if f_x and f_y are not continuous? Figure 4 pictures such a function; its equation is

$$f(x, y) = \begin{cases} \dfrac{xy}{x^2 + y^2} & \text{if } (x, y) \neq (0, 0) \\ 0 & \text{if } (x, y) = (0, 0) \end{cases}$$

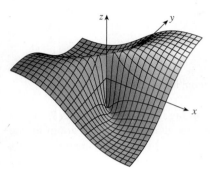

FIGURE 4
$f(x, y) = \dfrac{xy}{x^2 + y^2}$ if $(x, y) \neq (0, 0)$,
$f(0, 0) = 0$

You can verify (see Exercise 46) that its partial derivatives exist at the origin and, in fact, $f_x(0, 0) = 0$ and $f_y(0, 0) = 0$, but f_x and f_y are not continuous. The linear approximation would be $f(x, y) \approx 0$, but $f(x, y) = \frac{1}{2}$ at all points on the line $y = x$. So a function of two variables can behave badly even though both of its partial derivatives exist. To rule out such behavior, we formulate the idea of a differentiable function of two variables.

Recall that for a function of one variable, $y = f(x)$, if x changes from a to $a + \Delta x$, we defined the increment of y as

$$\Delta y = f(a + \Delta x) - f(a)$$

In Chapter 3 we showed that if f is differentiable at a, then

This is Equation 3.4.7.

$$\boxed{5} \qquad \Delta y = f'(a)\, \Delta x + \varepsilon\, \Delta x \qquad \text{where } \varepsilon \to 0 \text{ as } \Delta x \to 0$$

Now consider a function of two variables, $z = f(x, y)$, and suppose x changes from a to $a + \Delta x$ and y changes from b to $b + \Delta y$. Then the corresponding **increment** of z is

$$\boxed{6} \qquad \Delta z = f(a + \Delta x, b + \Delta y) - f(a, b)$$

Thus the increment Δz represents the change in the value of f when (x, y) changes from (a, b) to $(a + \Delta x, b + \Delta y)$. By analogy with (5) we define the differentiability of a function of two variables as follows.

> **7** **Definition** If $z = f(x, y)$, then f is **differentiable** at (a, b) if Δz can be expressed in the form
>
> $$\Delta z = f_x(a, b)\, \Delta x + f_y(a, b)\, \Delta y + \varepsilon_1\, \Delta x + \varepsilon_2\, \Delta y$$
>
> where ε_1 and $\varepsilon_2 \to 0$ as $(\Delta x,\ \Delta y) \to (0, 0)$.

Definition 7 says that a differentiable function is one for which the linear approximation (4) is a good approximation when (x, y) is near (a, b). In other words, the tangent plane approximates the graph of f well near the point of tangency.

It's sometimes hard to use Definition 7 directly to check the differentiability of a function, but the next theorem provides a convenient sufficient condition for differentiability.

Theorem 8 is proved in Appendix F.

> **8 Theorem** If the partial derivatives f_x and f_y exist near (a, b) and are continuous at (a, b), then f is differentiable at (a, b).

EXAMPLE 2 Show that $f(x, y) = xe^{xy}$ is differentiable at $(1, 0)$ and find its linearization there. Then use it to approximate $f(1.1, -0.1)$.

Figure 5 shows the graphs of the function f and its linearization L in Example 2.

SOLUTION The partial derivatives are

$$f_x(x, y) = e^{xy} + xye^{xy} \qquad f_y(x, y) = x^2 e^{xy}$$
$$f_x(1, 0) = 1 \qquad\qquad f_y(1, 0) = 1$$

Both f_x and f_y are continuous functions, so f is differentiable by Theorem 8. The linearization is

$$L(x, y) = f(1, 0) + f_x(1, 0)(x - 1) + f_y(1, 0)(y - 0)$$
$$= 1 + 1(x - 1) + 1 \cdot y = x + y$$

The corresponding linear approximation is

$$xe^{xy} \approx x + y$$

so

$$f(1.1, -0.1) \approx 1.1 - 0.1 = 1$$

Compare this with the actual value of $f(1.1, -0.1) = 1.1e^{-0.11} \approx 0.98542$. ∎

FIGURE 5

EXAMPLE 3 At the beginning of Section 14.3 we discussed the heat index (perceived temperature) I as a function of the actual temperature T and the relative humidity H and gave the following table of values from the National Weather Service.

Relative humidity (%)

T \ H	50	55	60	65	70	75	80	85	90
90	96	98	100	103	106	109	112	115	119
92	100	103	105	108	112	115	119	123	128
94	104	107	111	114	118	122	127	132	137
96	109	113	116	121	125	130	135	141	146
98	114	118	123	127	133	138	144	150	157
100	119	124	129	135	141	147	154	161	168

Actual temperature (°F)

Find a linear approximation for the heat index $I = f(T, H)$ when T is near 96°F and H is near 70%. Use it to estimate the heat index when the temperature is 97°F and the relative humidity is 72%.

SOLUTION We read from the table that $f(96, 70) = 125$. In Section 14.3 we used the tabular values to estimate that $f_T(96, 70) \approx 3.75$ and $f_H(96, 70) \approx 0.9$. (See pages 912–13.) So the linear approximation is

$$f(T, H) \approx f(96, 70) + f_T(96, 70)(T - 96) + f_H(96, 70)(H - 70)$$
$$\approx 125 + 3.75(T - 96) + 0.9(H - 70)$$

In particular,

$$f(97, 72) \approx 125 + 3.75(1) + 0.9(2) = 130.55$$

Therefore, when $T = 97°F$ and $H = 72\%$, the heat index is

$$I \approx 131°F$$

■ Differentials

For a differentiable function of one variable, $y = f(x)$, we define the differential dx to be an independent variable; that is, dx can be given the value of any real number. The differential of y is then defined as

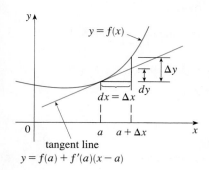

y = f(x)

$$\boxed{9} \qquad dy = f'(x) \, dx$$

(See Section 3.10.) Figure 6 shows the relationship between the increment Δy and the differential dy: Δy represents the change in height of the curve $y = f(x)$ and dy represents the change in height of the tangent line when x changes by an amount $dx = \Delta x$.

For a differentiable function of two variables, $z = f(x, y)$, we define the **differentials** dx and dy to be independent variables; that is, they can be given any values. Then the **differential** dz, also called the **total differential**, is defined by

$$\boxed{10} \qquad dz = f_x(x, y) \, dx + f_y(x, y) \, dy = \frac{\partial z}{\partial x} \, dx + \frac{\partial z}{\partial y} \, dy$$

tangent line
$y = f(a) + f'(a)(x - a)$

FIGURE 6

(Compare with Equation 9.) Sometimes the notation df is used in place of dz.

If we take $dx = \Delta x = x - a$ and $dy = \Delta y = y - b$ in Equation 10, then the differential of z is

$$dz = f_x(a, b)(x - a) + f_y(a, b)(y - b)$$

So, in the notation of differentials, the linear approximation (4) can be written as

$$f(x, y) \approx f(a, b) + dz$$

Figure 7 is the three-dimensional counterpart of Figure 6 and shows the geometric interpretation of the differential dz and the increment Δz: dz represents the change in height of the tangent plane, whereas Δz represents the change in height of the surface $z = f(x, y)$ when (x, y) changes from (a, b) to $(a + \Delta x, b + \Delta y)$.

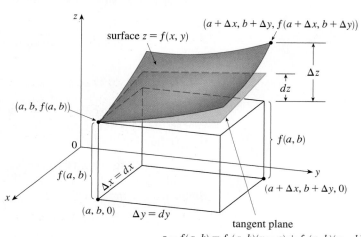

FIGURE 7

EXAMPLE 4

(a) If $z = f(x, y) = x^2 + 3xy - y^2$, find the differential dz.

(b) If x changes from 2 to 2.05 and y changes from 3 to 2.96, compare the values of Δz and dz.

SOLUTION

(a) Definition 10 gives

$$dz = \frac{\partial z}{\partial x}\, dx + \frac{\partial z}{\partial y}\, dy = (2x + 3y)\, dx + (3x - 2y)\, dy$$

(b) Putting $x = 2$, $dx = \Delta x = 0.05$, $y = 3$, and $dy = \Delta y = -0.04$, we get

$$dz = [2(2) + 3(3)]0.05 + [3(2) - 2(3)](-0.04) = 0.65$$

The increment of z is

$$\Delta z = f(2.05, 2.96) - f(2, 3)$$
$$= [(2.05)^2 + 3(2.05)(2.96) - (2.96)^2] - [2^2 + 3(2)(3) - 3^2]$$
$$= 0.6449$$

In Example 4, dz is close to Δz because the tangent plane is a good approximation to the surface $z = x^2 + 3xy - y^2$ near $(2, 3, 13)$. (See Figure 8.)

FIGURE 8

Notice that $\Delta z \approx dz$ but dz is easier to compute. ■

EXAMPLE 5 The base radius and height of a right circular cone are measured as 10 cm and 25 cm, respectively, with a possible error in measurement of as much as 0.1 cm in each. Use differentials to estimate the maximum error in the calculated volume of the cone.

SOLUTION The volume V of a cone with base radius r and height h is $V = \pi r^2 h/3$. So the differential of V is

$$dV = \frac{\partial V}{\partial r}\, dr + \frac{\partial V}{\partial h}\, dh = \frac{2\pi r h}{3}\, dr + \frac{\pi r^2}{3}\, dh$$

Since each error is at most 0.1 cm, we have $|\Delta r| \leq 0.1$, $|\Delta h| \leq 0.1$. To estimate the largest error in the volume we take the largest error in the measurement of r and of h. Therefore we take $dr = 0.1$ and $dh = 0.1$ along with $r = 10$, $h = 25$. This gives

$$dV = \frac{500\pi}{3}(0.1) + \frac{100\pi}{3}(0.1) = 20\pi$$

Thus the maximum error in the calculated volume is about 20π cm$^3 \approx 63$ cm^3. ■

■ **Functions of Three or More Variables**

Linear approximations, differentiability, and differentials can be defined in a similar manner for functions of more than two variables. A differentiable function is defined by an expression similar to the one in Definition 7. For such functions the **linear approximation** is

$$f(x, y, z) \approx f(a, b, c) + f_x(a, b, c)(x - a) + f_y(a, b, c)(y - b) + f_z(a, b, c)(z - c)$$

and the linearization $L(x, y, z)$ is the right side of this expression.

If $w = f(x, y, z)$, then the **increment** of w is

$$\Delta w = f(x + \Delta x, y + \Delta y, z + \Delta z) - f(x, y, z)$$

The **differential** dw is defined in terms of the differentials dx, dy, and dz of the independent variables by

$$dw = \frac{\partial w}{\partial x}\, dx + \frac{\partial w}{\partial y}\, dy + \frac{\partial w}{\partial z}\, dz$$

EXAMPLE 6 The dimensions of a rectangular box are measured to be 75 cm, 60 cm, and 40 cm, and each measurement is correct to within 0.2 cm. Use differentials to estimate the largest possible error when the volume of the box is calculated from these measurements.

SOLUTION If the dimensions of the box are x, y, and z, its volume is $V = xyz$ and so

$$dV = \frac{\partial V}{\partial x}\, dx + \frac{\partial V}{\partial y}\, dy + \frac{\partial V}{\partial z}\, dz = yz\, dx + xz\, dy + xy\, dz$$

We are given that $|\Delta x| \le 0.2$, $|\Delta y| \le 0.2$, and $|\Delta z| \le 0.2$. To estimate the largest error in the volume, we therefore use $dx = 0.2$, $dy = 0.2$, and $dz = 0.2$ together with $x = 75$, $y = 60$, and $z = 40$:

$$\Delta V \approx dV = (60)(40)(0.2) + (75)(40)(0.2) + (75)(60)(0.2) = 1980$$

Thus an error of only 0.2 cm in measuring each dimension could lead to an error of approximately 1980 cm³ in the calculated volume! This may seem like a large error, but it's only about 1% of the volume of the box. ∎

14.4 EXERCISES

1–6 Find an equation of the tangent plane to the given surface at the specified point.

1. $z = 2x^2 + y^2 - 5y$, $(1, 2, -4)$

2. $z = (x + 2)^2 - 2(y - 1)^2 - 5$, $(2, 3, 3)$

3. $z = e^{x-y}$, $(2, 2, 1)$

4. $z = x/y^2$, $(-4, 2, -1)$

5. $z = x \sin(x + y)$, $(-1, 1, 0)$

6. $z = \ln(x - 2y)$, $(3, 1, 0)$

7–8 Graph the surface and the tangent plane at the given point. (Choose the domain and viewpoint so that you get a good view of both the surface and the tangent plane.) Then zoom in until the surface and the tangent plane become indistinguishable.

7. $z = x^2 + xy + 3y^2$, $(1, 1, 5)$

8. $z = \sqrt{9 + x^2y^2}$, $(2, 2, 5)$

9–10 Draw the graph of f and its tangent plane at the given point. (Use your computer algebra system both to compute the partial derivatives and to graph the surface and its tangent plane.)

Then zoom in until the surface and the tangent plane become indistinguishable.

9. $f(x, y) = \dfrac{1 + \cos^2(x - y)}{1 + \cos^2(x + y)}$, $\left(\dfrac{\pi}{3}, \dfrac{\pi}{6}, \dfrac{7}{4}\right)$

10. $f(x, y) = e^{-xy/10}\left(\sqrt{x} + \sqrt{y} + \sqrt{xy}\right)$, $(1, 1, 3e^{-0.1})$

11–16 Explain why the function is differentiable at the given point. Then find the linearization $L(x, y)$ of the function at that point.

11. $f(x, y) = 1 + x \ln(xy - 5)$, $(2, 3)$

12. $f(x, y) = \sqrt{xy}$, $(1, 4)$

13. $f(x, y) = x^2 e^y$, $(1, 0)$

14. $f(x, y) = \dfrac{1 + y}{1 + x}$, $(1, 3)$

15. $f(x, y) = 4 \arctan(xy)$, $(1, 1)$

16. $f(x, y) = y + \sin(x/y)$, $(0, 3)$

17–18 Verify the linear approximation at $(0, 0)$.

17. $e^x \cos(xy) \approx x + 1$

18. $\dfrac{y - 1}{x + 1} \approx x + y - 1$

19. Given that f is a differentiable function with $f(2, 5) = 6$, $f_x(2, 5) = 1$, and $f_y(2, 5) = -1$, use a linear approximation to estimate $f(2.2, 4.9)$.

20. Find the linear approximation of the function $f(x, y) = 1 - xy \cos \pi y$ at $(1, 1)$ and use it to approximate $f(1.02, 0.97)$. Illustrate by graphing f and the tangent plane.

21. Find the linear approximation of the function $f(x, y, z) = \sqrt{x^2 + y^2 + z^2}$ at $(3, 2, 6)$ and use it to approximate the number $\sqrt{(3.02)^2 + (1.97)^2 + (5.99)^2}$.

22. The wave heights h in the open sea depend on the speed v of the wind and the length of time t that the wind has been blowing at that speed. Values of the function $h = f(v, t)$ are recorded in feet in the following table. Use the table to find a linear approximation to the wave height function when v is near 40 knots and t is near 20 hours. Then estimate the wave heights when the wind has been blowing for 24 hours at 43 knots.

Duration (hours)

v \ t	5	10	15	20	30	40	50
20	5	7	8	8	9	9	9
30	9	13	16	17	18	19	19
40	14	21	25	28	31	33	33
50	19	29	36	40	45	48	50
60	24	37	47	54	62	67	69

Wind speed (knots)

23. Use the table in Example 3 to find a linear approximation to the heat index function when the temperature is near 94°F and the relative humidity is near 80%. Then estimate the heat index when the temperature is 95°F and the relative humidity is 78%.

24. The wind-chill index W is the perceived temperature when the actual temperature is T and the wind speed is v, so we can write $W = f(T, v)$. The following table of values is an excerpt from Table 1 in Section 14.1. Use the table to find a linear approximation to the wind-chill index function when T is near -15°C and v is near 50 km/h. Then estimate the wind-chill index when the temperature is -17°C and the wind speed is 55 km/h.

Wind speed (km/h)

T \ v	20	30	40	50	60	70
-10	-18	-20	-21	-22	-23	-23
-15	-24	-26	-27	-29	-30	-30
-20	-30	-33	-34	-35	-36	-37
-25	-37	-39	-41	-42	-43	-44

Actual temperature (°C)

25–30 Find the differential of the function.

25. $z = e^{-2x} \cos 2\pi t$

26. $u = \sqrt{x^2 + 3y^2}$

27. $m = p^5 q^3$

28. $T = \dfrac{v}{1 + uvw}$

29. $R = \alpha \beta^2 \cos \gamma$

30. $L = xze^{-y^2 - z^2}$

31. If $z = 5x^2 + y^2$ and (x, y) changes from $(1, 2)$ to $(1.05, 2.1)$, compare the values of Δz and dz.

32. If $z = x^2 - xy + 3y^2$ and (x, y) changes from $(3, -1)$ to $(2.96, -0.95)$, compare the values of Δz and dz.

33. The length and width of a rectangle are measured as 30 cm and 24 cm, respectively, with an error in measurement of at most 0.1 cm in each. Use differentials to estimate the maximum error in the calculated area of the rectangle.

34. Use differentials to estimate the amount of metal in a closed cylindrical can that is 10 cm high and 4 cm in diameter if the metal in the top and bottom is 0.1 cm thick and the metal in the sides is 0.05 cm thick.

35. Use differentials to estimate the amount of tin in a closed tin can with diameter 8 cm and height 12 cm if the tin is 0.04 cm thick.

36. The wind-chill index is modeled by the function

$$W = 13.12 + 0.6215T - 11.37v^{0.16} + 0.3965Tv^{0.16}$$

where T is the temperature (in °C) and v is the wind speed (in km/h). The wind speed is measured as 26 km/h, with a possible error of ± 2 km/h, and the temperature is measured as -11°C, with a possible error of ± 1°C. Use differentials to estimate the maximum error in the calculated value of W due to the measurement errors in T and v.

37. The tension T in the string of the yo-yo in the figure is

$$T = \frac{mgR}{2r^2 + R^2}$$

where m is the mass of the yo-yo and g is acceleration due to gravity. Use differentials to estimate the change in the tension if R is increased from 3 cm to 3.1 cm and r is increased from 0.7 cm to 0.8 cm. Does the tension increase or decrease?

38. The pressure, volume, and temperature of a mole of an ideal gas are related by the equation $PV = 8.31T$, where P is measured in kilopascals, V in liters, and T in kelvins. Use differentials to find the approximate change in the pressure if the volume increases from 12 L to 12.3 L and the temperature decreases from 310 K to 305 K.

39. If R is the total resistance of three resistors, connected in parallel, with resistances R_1, R_2, R_3, then

$$\frac{1}{R} = \frac{1}{R_1} + \frac{1}{R_2} + \frac{1}{R_3}$$

If the resistances are measured in ohms as $R_1 = 25 \ \Omega$, $R_2 = 40 \ \Omega$, and $R_3 = 50 \ \Omega$, with a possible error of 0.5% in each case, estimate the maximum error in the calculated value of R.

40. A model for the surface area of a human body is given by $S = 0.1091 w^{0.425} h^{0.725}$, where w is the weight (in pounds), h is the height (in inches), and S is measured in square feet. If the errors in measurement of w and h are at most 2%, use differentials to estimate the maximum percentage error in the calculated surface area.

41. In Exercise 14.1.39 and Example 14.3.3, the body mass index of a person was defined as $B(m, h) = m/h^2$, where m is the mass in kilograms and h is the height in meters.
(a) What is the linear approximation of $B(m, h)$ for a child with mass 23 kg and height 1.10 m?
(b) If the child's mass increases by 1 kg and height by 3 cm, use the linear approximation to estimate the new BMI. Compare with the actual new BMI.

42. Suppose you need to know an equation of the tangent plane to a surface S at the point $P(2, 1, 3)$. You don't have an equation

for S but you know that the curves

$$\mathbf{r}_1(t) = \langle 2 + 3t, 1 - t^2, 3 - 4t + t^2 \rangle$$
$$\mathbf{r}_2(u) = \langle 1 + u^2, 2u^3 - 1, 2u + 1 \rangle$$

both lie on S. Find an equation of the tangent plane at P.

43–44 Show that the function is differentiable by finding values of ε_1 and ε_2 that satisfy Definition 7.

43. $f(x, y) = x^2 + y^2$ **44.** $f(x, y) = xy - 5y^2$

45. Prove that if f is a function of two variables that is differentiable at (a, b), then f is continuous at (a, b).
Hint: Show that

$$\lim_{(\Delta x, \Delta y) \to (0, 0)} f(a + \Delta x, b + \Delta y) = f(a, b)$$

46. (a) The function

$$f(x, y) = \begin{cases} \dfrac{xy}{x^2 + y^2} & \text{if } (x, y) \neq (0, 0) \\ 0 & \text{if } (x, y) = (0, 0) \end{cases}$$

was graphed in Figure 4. Show that $f_x(0, 0)$ and $f_y(0, 0)$ both exist but f is not differentiable at $(0, 0)$. [*Hint:* Use the result of Exercise 45.]
(b) Explain why f_x and f_y are not continuous at $(0, 0)$.

APPLIED PROJECT   THE SPEEDO LZR RACER

Many technological advances have occurred in sports that have contributed to increased athletic performance. One of the best known is the introduction, in 2008, of the Speedo LZR racer. It was claimed that this full-body swimsuit reduced a swimmer's drag in the water. Figure 1 shows the number of world records broken in men's and women's long-course freestyle swimming events from 1990 to 2011.[1] The dramatic increase in 2008 when the suit was introduced led people to claim that such suits are a form of technological doping. As a result all full-body suits were banned from competition starting in 2010.

FIGURE 1 Number of world records set in long-course men's and women's freestyle swimming event 1990–2011

It might be surprising that a simple reduction in drag could have such a big effect on performance. We can gain some insight into this using a simple mathematical model.[2]

---

1. L. Foster et al., "Influence of Full Body Swimsuits on Competitive Performance," *Procedia Engineering* 34 (2012): 712–17.
2. Adapted from http://plus.maths.org/content/swimming.

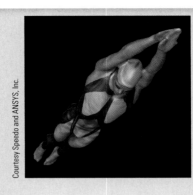

Courtesy Speedo and ANSYS, Inc.

The speed v of an object being propelled through water is given by

$$v(P, C) = \left(\frac{2P}{kC}\right)^{1/3}$$

where P is the power being used to propel the object, C is the drag coefficient, and k is a positive constant. Athletes can therefore increase their swimming speeds by increasing their power or reducing their drag coefficients. But how effective is each of these?

To compare the effect of increasing power versus reducing drag, we need to somehow compare the two in common units. The most common approach is to determine the percentage change in speed that results from a given percentage change in power and in drag.

If we work with percentages as fractions, then when power is changed by a fraction x (with x corresponding to $100x$ percent), P changes from P to $P + xP$. Likewise, if the drag coefficient is changed by a fraction y, this means that it has changed from C to $C + yC$. Finally, the fractional change in speed resulting from both effects is

$$\boxed{1} \qquad \frac{v(P + xP, C + yC) - v(P, C)}{v(P, C)}$$

1. Expression 1 gives the fractional change in speed that results from a change x in power and a change y in drag. Show that this reduces to the function

$$f(x, y) = \left(\frac{1 + x}{1 + y}\right)^{1/3} - 1$$

Given the context, what is the domain of f?

2. Suppose that the possible changes in power x and drag y are small. Find the linear approximation to the function $f(x, y)$. What does this approximation tell you about the effect of a small increase in power versus a small decrease in drag?

3. Calculate $f_{xx}(x, y)$ and $f_{yy}(x, y)$. Based on the signs of these derivatives, does the linear approximation in Problem 2 result in an overestimate or an underestimate for an increase in power? What about for a decrease in drag? Use your answer to explain why, for changes in power or drag that are not very small, a decrease in drag is more effective.

4. Graph the level curves of $f(x, y)$. Explain how the shapes of these curves relate to your answers to Problems 2 and 3.

14.5 The Chain Rule

Recall that the Chain Rule for functions of a single variable gives the rule for differentiating a composite function: If $y = f(x)$ and $x = g(t)$, where f and g are differentiable functions, then y is indirectly a differentiable function of t and

$$\boxed{1} \qquad \frac{dy}{dt} = \frac{dy}{dx}\frac{dx}{dt}$$

For functions of more than one variable, the Chain Rule has several versions, each of them giving a rule for differentiating a composite function. The first version (Theorem 2) deals with the case where $z = f(x, y)$ and each of the variables x and y is, in turn, a function of a variable t. This means that z is indirectly a function of t, $z = f(g(t), h(t))$, and the Chain Rule gives a formula for differentiating z as a function of t. We assume that f

is differentiable (Definition 14.4.7). Recall that this is the case when f_x and f_y are continuous (Theorem 14.4.8).

---

2 **The Chain Rule (Case 1)** Suppose that $z = f(x, y)$ is a differentiable function of x and y, where $x = g(t)$ and $y = h(t)$ are both differentiable functions of t. Then z is a differentiable function of t and

$$\frac{dz}{dt} = \frac{\partial f}{\partial x}\frac{dx}{dt} + \frac{\partial f}{\partial y}\frac{dy}{dt}$$

PROOF A change of Δt in t produces changes of Δx in x and Δy in y. These, in turn, produce a change of Δz in z, and from Definition 14.4.7 we have

$$\Delta z = \frac{\partial f}{\partial x}\Delta x + \frac{\partial f}{\partial y}\Delta y + \varepsilon_1\, \Delta x + \varepsilon_2\, \Delta y$$

where $\varepsilon_1 \to 0$ and $\varepsilon_2 \to 0$ as $(\Delta x, \Delta y) \to (0, 0)$. [If the functions ε_1 and ε_2 are not defined at $(0, 0)$, we can define them to be 0 there.] Dividing both sides of this equation by Δt, we have

$$\frac{\Delta z}{\Delta t} = \frac{\partial f}{\partial x}\frac{\Delta x}{\Delta t} + \frac{\partial f}{\partial y}\frac{\Delta y}{\Delta t} + \varepsilon_1\frac{\Delta x}{\Delta t} + \varepsilon_2\frac{\Delta y}{\Delta t}$$

If we now let $\Delta t \to 0$, then $\Delta x = g(t + \Delta t) - g(t) \to 0$ because g is differentiable and therefore continuous. Similarly, $\Delta y \to 0$. This, in turn, means that $\varepsilon_1 \to 0$ and $\varepsilon_2 \to 0$, so

$$\frac{dz}{dt} = \lim_{\Delta t \to 0}\frac{\Delta z}{\Delta t}$$

$$= \frac{\partial f}{\partial x}\lim_{\Delta t \to 0}\frac{\Delta x}{\Delta t} + \frac{\partial f}{\partial y}\lim_{\Delta t \to 0}\frac{\Delta y}{\Delta t} + \left(\lim_{\Delta t \to 0}\varepsilon_1\right)\lim_{\Delta t \to 0}\frac{\Delta x}{\Delta t} + \left(\lim_{\Delta t \to 0}\varepsilon_2\right)\lim_{\Delta t \to 0}\frac{\Delta y}{\Delta t}$$

$$= \frac{\partial f}{\partial x}\frac{dx}{dt} + \frac{\partial f}{\partial y}\frac{dy}{dt} + 0 \cdot \frac{dx}{dt} + 0 \cdot \frac{dy}{dt}$$

$$= \frac{\partial f}{\partial x}\frac{dx}{dt} + \frac{\partial f}{\partial y}\frac{dy}{dt} \qquad\blacksquare$$

Since we often write $\partial z / \partial x$ in place of $\partial f / \partial x$, we can rewrite the Chain Rule in the form

Notice the similarity to the definition of the differential:

$$dz = \frac{\partial z}{\partial x}dx + \frac{\partial z}{\partial y}dy$$

$$\frac{dz}{dt} = \frac{\partial z}{\partial x}\frac{dx}{dt} + \frac{\partial z}{\partial y}\frac{dy}{dt}$$

EXAMPLE 1 If $z = x^2y + 3xy^4$, where $x = \sin 2t$ and $y = \cos t$, find dz/dt when $t = 0$.

SOLUTION The Chain Rule gives

$$\frac{dz}{dt} = \frac{\partial z}{\partial x}\frac{dx}{dt} + \frac{\partial z}{\partial y}\frac{dy}{dt}$$

$$= (2xy + 3y^4)(2\cos 2t) + (x^2 + 12xy^3)(-\sin t)$$

It's not necessary to substitute the expressions for x and y in terms of t. We simply observe that when $t = 0$, we have $x = \sin 0 = 0$ and $y = \cos 0 = 1$. Therefore

$$\left.\frac{dz}{dt}\right|_{t=0} = (0 + 3)(2 \cos 0) + (0 + 0)(-\sin 0) = 6 \qquad \blacksquare$$

FIGURE 1
The curve $x = \sin 2t,\ y = \cos t$

The derivative in Example 1 can be interpreted as the rate of change of z with respect to t as the point (x, y) moves along the curve C with parametric equations $x = \sin 2t$, $y = \cos t$. (See Figure 1.) In particular, when $t = 0$, the point (x, y) is $(0, 1)$ and $dz/dt = 6$ is the rate of increase as we move along the curve C through $(0, 1)$. If, for instance, $z = T(x, y) = x^2 y + 3xy^4$ represents the temperature at the point (x, y), then the composite function $z = T(\sin 2t, \cos t)$ represents the temperature at points on C and the derivative dz/dt represents the rate at which the temperature changes along C.

EXAMPLE 2 The pressure P (in kilopascals), volume V (in liters), and temperature T (in kelvins) of a mole of an ideal gas are related by the equation $PV = 8.31T$. Find the rate at which the pressure is changing when the temperature is 300 K and increasing at a rate of 0.1 K/s and the volume is 100 L and increasing at a rate of 0.2 L/s.

SOLUTION If t represents the time elapsed in seconds, then at the given instant we have $T = 300$, $dT/dt = 0.1$, $V = 100$, $dV/dt = 0.2$. Since

$$P = 8.31\frac{T}{V}$$

the Chain Rule gives

$$\frac{dP}{dt} = \frac{\partial P}{\partial T}\frac{dT}{dt} + \frac{\partial P}{\partial V}\frac{dV}{dt} = \frac{8.31}{V}\frac{dT}{dt} - \frac{8.31T}{V^2}\frac{dV}{dt}$$

$$= \frac{8.31}{100}(0.1) - \frac{8.31(300)}{100^2}(0.2) = -0.04155$$

The pressure is decreasing at a rate of about 0.042 kPa/s. $\qquad \blacksquare$

We now consider the situation where $z = f(x, y)$ but each of x and y is a function of two variables s and t: $x = g(s, t)$, $y = h(s, t)$. Then z is indirectly a function of s and t and we wish to find $\partial z/\partial s$ and $\partial z/\partial t$. Recall that in computing $\partial z/\partial t$ we hold s fixed and compute the ordinary derivative of z with respect to t. Therefore we can apply Theorem 2 to obtain

$$\frac{\partial z}{\partial t} = \frac{\partial z}{\partial x}\frac{\partial x}{\partial t} + \frac{\partial z}{\partial y}\frac{\partial y}{\partial t}$$

A similar argument holds for $\partial z/\partial s$ and so we have proved the following version of the Chain Rule.

---

3 **The Chain Rule (Case 2)** Suppose that $z = f(x, y)$ is a differentiable function of x and y, where $x = g(s, t)$ and $y = h(s, t)$ are differentiable functions of s and t. Then

$$\frac{\partial z}{\partial s} = \frac{\partial z}{\partial x}\frac{\partial x}{\partial s} + \frac{\partial z}{\partial y}\frac{\partial y}{\partial s} \qquad\qquad \frac{\partial z}{\partial t} = \frac{\partial z}{\partial x}\frac{\partial x}{\partial t} + \frac{\partial z}{\partial y}\frac{\partial y}{\partial t}$$

EXAMPLE 3 If $z = e^x \sin y$, where $x = st^2$ and $y = s^2 t$, find $\partial z / \partial s$ and $\partial z / \partial t$.

SOLUTION Applying Case 2 of the Chain Rule, we get

$$\frac{\partial z}{\partial s} = \frac{\partial z}{\partial x}\frac{\partial x}{\partial s} + \frac{\partial z}{\partial y}\frac{\partial y}{\partial s} = (e^x \sin y)(t^2) + (e^x \cos y)(2st)$$

$$= t^2 e^{st^2} \sin(s^2 t) + 2st e^{st^2} \cos(s^2 t)$$

$$\frac{\partial z}{\partial t} = \frac{\partial z}{\partial x}\frac{\partial x}{\partial t} + \frac{\partial z}{\partial y}\frac{\partial y}{\partial t} = (e^x \sin y)(2st) + (e^x \cos y)(s^2)$$

$$= 2st e^{st^2} \sin(s^2 t) + s^2 e^{st^2} \cos(s^2 t) \qquad \blacksquare$$

Case 2 of the Chain Rule contains three types of variables: s and t are **independent** variables, x and y are called **intermediate** variables, and z is the **dependent** variable. Notice that Theorem 3 has one term for each intermediate variable and each of these terms resembles the one-dimensional Chain Rule in Equation 1.

To remember the Chain Rule, it's helpful to draw the **tree diagram** in Figure 2. We draw branches from the dependent variable z to the intermediate variables x and y to indicate that z is a function of x and y. Then we draw branches from x and y to the independent variables s and t. On each branch we write the corresponding partial derivative. To find $\partial z / \partial s$, we find the product of the partial derivatives along each path from z to s and then add these products:

$$\frac{\partial z}{\partial s} = \frac{\partial z}{\partial x}\frac{\partial x}{\partial s} + \frac{\partial z}{\partial y}\frac{\partial y}{\partial s}$$

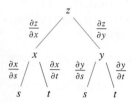

FIGURE 2

Similarly, we find $\partial z / \partial t$ by using the paths from z to t.

Now we consider the general situation in which a dependent variable u is a function of n intermediate variables $x_1, \ldots, x_n$, each of which is, in turn, a function of m independent variables $t_1, \ldots, t_m$. Notice that there are n terms, one for each intermediate variable. The proof is similar to that of Case 1.

---

4 **The Chain Rule (General Version)** Suppose that u is a differentiable function of the n variables $x_1, x_2, \ldots, x_n$ and each x_j is a differentiable function of the m variables $t_1, t_2, \ldots, t_m$. Then u is a function of $t_1, t_2, \ldots, t_m$ and

$$\frac{\partial u}{\partial t_i} = \frac{\partial u}{\partial x_1}\frac{\partial x_1}{\partial t_i} + \frac{\partial u}{\partial x_2}\frac{\partial x_2}{\partial t_i} + \cdots + \frac{\partial u}{\partial x_n}\frac{\partial x_n}{\partial t_i}$$

for each $i = 1, 2, \ldots, m$.

EXAMPLE 4 Write out the Chain Rule for the case where $w = f(x, y, z, t)$ and $x = x(u, v)$, $y = y(u, v)$, $z = z(u, v)$, and $t = t(u, v)$.

SOLUTION We apply Theorem 4 with $n = 4$ and $m = 2$. Figure 3 shows the tree diagram. Although we haven't written the derivatives on the branches, it's understood that if a branch leads from y to u, then the partial derivative for that branch is $\partial y / \partial u$. With the aid of the tree diagram, we can now write the required expressions:

FIGURE 3

$$\frac{\partial w}{\partial u} = \frac{\partial w}{\partial x}\frac{\partial x}{\partial u} + \frac{\partial w}{\partial y}\frac{\partial y}{\partial u} + \frac{\partial w}{\partial z}\frac{\partial z}{\partial u} + \frac{\partial w}{\partial t}\frac{\partial t}{\partial u}$$

$$\frac{\partial w}{\partial v} = \frac{\partial w}{\partial x}\frac{\partial x}{\partial v} + \frac{\partial w}{\partial y}\frac{\partial y}{\partial v} + \frac{\partial w}{\partial z}\frac{\partial z}{\partial v} + \frac{\partial w}{\partial t}\frac{\partial t}{\partial v} \qquad \blacksquare$$

EXAMPLE 5 If $u = x^4 y + y^2 z^3$, where $x = rse^t$, $y = rs^2 e^{-t}$, and $z = r^2 s \sin t$, find the value of $\partial u / \partial s$ when $r = 2$, $s = 1$, $t = 0$.

SOLUTION With the help of the tree diagram in Figure 4, we have

$$\frac{\partial u}{\partial s} = \frac{\partial u}{\partial x}\frac{\partial x}{\partial s} + \frac{\partial u}{\partial y}\frac{\partial y}{\partial s} + \frac{\partial u}{\partial z}\frac{\partial z}{\partial s}$$

$$= (4x^3 y)(re^t) + (x^4 + 2yz^3)(2rse^{-t}) + (3y^2 z^2)(r^2 \sin t)$$

FIGURE 4

When $r = 2$, $s = 1$, and $t = 0$, we have $x = 2$, $y = 2$, and $z = 0$, so

$$\frac{\partial u}{\partial s} = (64)(2) + (16)(4) + (0)(0) = 192 \qquad \blacksquare$$

EXAMPLE 6 If $g(s, t) = f(s^2 - t^2, t^2 - s^2)$ and f is differentiable, show that g satisfies the equation

$$t\frac{\partial g}{\partial s} + s\frac{\partial g}{\partial t} = 0$$

SOLUTION Let $x = s^2 - t^2$ and $y = t^2 - s^2$. Then $g(s, t) = f(x, y)$ and the Chain Rule gives

$$\frac{\partial g}{\partial s} = \frac{\partial f}{\partial x}\frac{\partial x}{\partial s} + \frac{\partial f}{\partial y}\frac{\partial y}{\partial s} = \frac{\partial f}{\partial x}(2s) + \frac{\partial f}{\partial y}(-2s)$$

$$\frac{\partial g}{\partial t} = \frac{\partial f}{\partial x}\frac{\partial x}{\partial t} + \frac{\partial f}{\partial y}\frac{\partial y}{\partial t} = \frac{\partial f}{\partial x}(-2t) + \frac{\partial f}{\partial y}(2t)$$

Therefore

$$t\frac{\partial g}{\partial s} + s\frac{\partial g}{\partial t} = \left(2st\frac{\partial f}{\partial x} - 2st\frac{\partial f}{\partial y}\right) + \left(-2st\frac{\partial f}{\partial x} + 2st\frac{\partial f}{\partial y}\right) = 0 \qquad \blacksquare$$

EXAMPLE 7 If $z = f(x, y)$ has continuous second-order partial derivatives and $x = r^2 + s^2$ and $y = 2rs$, find (a) $\partial z / \partial r$ and (b) $\partial^2 z / \partial r^2$.

SOLUTION
(a) The Chain Rule gives

$$\frac{\partial z}{\partial r} = \frac{\partial z}{\partial x}\frac{\partial x}{\partial r} + \frac{\partial z}{\partial y}\frac{\partial y}{\partial r} = \frac{\partial z}{\partial x}(2r) + \frac{\partial z}{\partial y}(2s)$$

(b) Applying the Product Rule to the expression in part (a), we get

$$\frac{\partial^2 z}{\partial r^2} = \frac{\partial}{\partial r}\left(2r\frac{\partial z}{\partial x} + 2s\frac{\partial z}{\partial y}\right)$$

[5]

$$= 2\frac{\partial z}{\partial x} + 2r\frac{\partial}{\partial r}\left(\frac{\partial z}{\partial x}\right) + 2s\frac{\partial}{\partial r}\left(\frac{\partial z}{\partial y}\right)$$

$\dfrac{\partial z}{\partial x}$

FIGURE 5

But, using the Chain Rule again (see Figure 5), we have

$$\frac{\partial}{\partial r}\left(\frac{\partial z}{\partial x}\right) = \frac{\partial}{\partial x}\left(\frac{\partial z}{\partial x}\right)\frac{\partial x}{\partial r} + \frac{\partial}{\partial y}\left(\frac{\partial z}{\partial x}\right)\frac{\partial y}{\partial r} = \frac{\partial^2 z}{\partial x^2}(2r) + \frac{\partial^2 z}{\partial y\,\partial x}(2s)$$

$$\frac{\partial}{\partial r}\left(\frac{\partial z}{\partial y}\right) = \frac{\partial}{\partial x}\left(\frac{\partial z}{\partial y}\right)\frac{\partial x}{\partial r} + \frac{\partial}{\partial y}\left(\frac{\partial z}{\partial y}\right)\frac{\partial y}{\partial r} = \frac{\partial^2 z}{\partial x\,\partial y}(2r) + \frac{\partial^2 z}{\partial y^2}(2s)$$

Putting these expressions into Equation 5 and using the equality of the mixed second-order derivatives, we obtain

$$\frac{\partial^2 z}{\partial r^2} = 2\,\frac{\partial z}{\partial x} + 2r\left(2r\,\frac{\partial^2 z}{\partial x^2} + 2s\,\frac{\partial^2 z}{\partial y\,\partial x}\right) + 2s\left(2r\,\frac{\partial^2 z}{\partial x\,\partial y} + 2s\,\frac{\partial^2 z}{\partial y^2}\right)$$

$$= 2\,\frac{\partial z}{\partial x} + 4r^2\,\frac{\partial^2 z}{\partial x^2} + 8rs\,\frac{\partial^2 z}{\partial x\,\partial y} + 4s^2\,\frac{\partial^2 z}{\partial y^2}.$$

◼

◼ Implicit Differentiation

The Chain Rule can be used to give a more complete description of the process of implicit differentiation that was introduced in Sections 3.5 and 14.3. We suppose that an equation of the form $F(x, y) = 0$ defines y implicitly as a differentiable function of x, that is, $y = f(x)$, where $F(x, f(x)) = 0$ for all x in the domain of f. If F is differentiable, we can apply Case 1 of the Chain Rule to differentiate both sides of the equation $F(x, y) = 0$ with respect to x. Since both x and y are functions of x, we obtain

$$\frac{\partial F}{\partial x}\frac{dx}{dx} + \frac{\partial F}{\partial y}\frac{dy}{dx} = 0$$

But $dx/dx = 1$, so if $\partial F/\partial y \neq 0$ we solve for dy/dx and obtain

$$\boxed{6} \qquad \frac{dy}{dx} = -\frac{\dfrac{\partial F}{\partial x}}{\dfrac{\partial F}{\partial y}} = -\frac{F_x}{F_y}$$

To derive this equation we assumed that $F(x, y) = 0$ defines y implicitly as a function of x. The **Implicit Function Theorem**, proved in advanced calculus, gives conditions under which this assumption is valid: it states that if F is defined on a disk containing (a, b), where $F(a, b) = 0$, $F_y(a, b) \neq 0$, and F_x and F_y are continuous on the disk, then the equation $F(x, y) = 0$ defines y as a function of x near the point (a, b) and the derivative of this function is given by Equation 6.

EXAMPLE 8 Find y' if $x^3 + y^3 = 6xy$.

SOLUTION The given equation can be written as

$$F(x, y) = x^3 + y^3 - 6xy = 0$$

so Equation 6 gives

The solution to Example 8 should be compared to the one in Example 3.5.2.

$$\frac{dy}{dx} = -\frac{F_x}{F_y} = -\frac{3x^2 - 6y}{3y^2 - 6x} = -\frac{x^2 - 2y}{y^2 - 2x}$$

◼

Now we suppose that z is given implicitly as a function $z = f(x, y)$ by an equation of the form $F(x, y, z) = 0$. This means that $F(x, y, f(x, y)) = 0$ for all (x, y) in the domain of f. If F and f are differentiable, then we can use the Chain Rule to differentiate the equation $F(x, y, z) = 0$ as follows:

$$\frac{\partial F}{\partial x}\frac{\partial x}{\partial x} + \frac{\partial F}{\partial y}\frac{\partial y}{\partial x} + \frac{\partial F}{\partial z}\frac{\partial z}{\partial x} = 0$$

But

$$\frac{\partial}{\partial x}(x) = 1 \quad \text{and} \quad \frac{\partial}{\partial x}(y) = 0$$

so this equation becomes

$$\frac{\partial F}{\partial x} + \frac{\partial F}{\partial z}\frac{\partial z}{\partial x} = 0$$

If $\partial F/\partial z \neq 0$, we solve for $\partial z/\partial x$ and obtain the first formula in Equations 7. The formula for $\partial z/\partial y$ is obtained in a similar manner.

$$\boxed{7} \qquad \frac{\partial z}{\partial x} = -\frac{\dfrac{\partial F}{\partial x}}{\dfrac{\partial F}{\partial z}} \qquad \frac{\partial z}{\partial y} = -\frac{\dfrac{\partial F}{\partial y}}{\dfrac{\partial F}{\partial z}}$$

Again, a version of the **Implicit Function Theorem** stipulates conditions under which our assumption is valid: if F is defined within a sphere containing (a, b, c), where $F(a, b, c) = 0$, $F_z(a, b, c) \neq 0$, and F_x, F_y, and F_z are continuous inside the sphere, then the equation $F(x, y, z) = 0$ defines z as a function of x and y near the point (a, b, c) and this function is differentiable, with partial derivatives given by (7).

EXAMPLE 9 Find $\dfrac{\partial z}{\partial x}$ and $\dfrac{\partial z}{\partial y}$ if $x^3 + y^3 + z^3 + 6xyz = 1$.

SOLUTION Let $F(x, y, z) = x^3 + y^3 + z^3 + 6xyz - 1$. Then, from Equations 7, we have

$$\frac{\partial z}{\partial x} = -\frac{F_x}{F_z} = -\frac{3x^2 + 6yz}{3z^2 + 6xy} = -\frac{x^2 + 2yz}{z^2 + 2xy}$$

The solution to Example 9 should be compared to the one in Example 14.3.5.

$$\frac{\partial z}{\partial y} = -\frac{F_y}{F_z} = -\frac{3y^2 + 6xz}{3z^2 + 6xy} = -\frac{y^2 + 2xz}{z^2 + 2xy}$$

14.5 EXERCISES

1–6 Use the Chain Rule to find dz/dt or dw/dt.

1. $z = xy^3 - x^2y$, $\quad x = t^2 + 1$, $\quad y = t^2 - 1$

2. $z = \dfrac{x - y}{x + 2y}$, $\quad x = e^{\pi t}$, $\quad y = e^{-\pi t}$

3. $z = \sin x \cos y$, $\quad x = \sqrt{t}$, $\quad y = 1/t$

4. $z = \sqrt{1 + xy}$, $\quad x = \tan t$, $\quad y = \arctan t$

5. $w = xe^{y/z}$, $\quad x = t^2$, $\quad y = 1 - t$, $\quad z = 1 + 2t$

6. $w = \ln\sqrt{x^2 + y^2 + z^2}$, $\quad x = \sin t$, $\quad y = \cos t$, $\quad z = \tan t$

7–12 Use the Chain Rule to find $\partial z/\partial s$ and $\partial z/\partial t$.

7. $z = (x - y)^5$, $\quad x = s^2t$, $\quad y = st^2$

8. $z = \tan^{-1}(x^2 + y^2)$, $\quad x = s\ln t$, $\quad y = te^s$

9. $z = \ln(3x + 2y)$, $x = s \sin t$, $y = t \cos s$

10. $z = \sqrt{x}\, e^{xy}$, $x = 1 + st$, $y = s^2 - t^2$

11. $z = e^r \cos\theta$, $r = st$, $\theta = \sqrt{s^2 + t^2}$

12. $z = \tan(u/v)$, $u = 2s + 3t$, $v = 3s - 2t$

---

13. Let $p(t) = f(g(t), h(t))$, where f is differentiable, $g(2) = 4$, $g'(2) = -3$, $h(2) = 5$, $h'(2) = 6$, $f_x(4, 5) = 2$, $f_y(4, 5) = 8$. Find $p'(2)$.

14. Let $R(s, t) = G(u(s, t), v(s, t))$, where G, u, and v are differentiable, $u(1, 2) = 5$, $u_s(1, 2) = 4$, $u_t(1, 2) = -3$, $v(1, 2) = 7$, $v_s(1, 2) = 2$, $v_t(1, 2) = 6$, $G_u(5, 7) = 9$, $G_v(5, 7) = -2$. Find $R_s(1, 2)$ and $R_t(1, 2)$.

15. Suppose f is a differentiable function of x and y, and $g(u, v) = f(e^u + \sin v, e^u + \cos v)$. Use the table of values to calculate $g_u(0, 0)$ and $g_v(0, 0)$.

	f	g	f_x	f_y
(0, 0)	3	6	4	8
(1, 2)	6	3	2	5

16. Suppose f is a differentiable function of x and y, and $g(r, s) = f(2r - s, s^2 - 4r)$. Use the table of values in Exercise 15 to calculate $g_r(1, 2)$ and $g_s(1, 2)$.

17–20 Use a tree diagram to write out the Chain Rule for the given case. Assume all functions are differentiable.

17. $u = f(x, y)$,   where $x = x(r, s, t)$, $y = y(r, s, t)$

18. $w = f(x, y, z)$,   where $x = x(u, v)$, $y = y(u, v)$, $z = z(u, v)$

19. $T = F(p, q, r)$,   where $p = p(x, y, z)$, $q = q(x, y, z)$, $r = r(x, y, z)$

20. $R = F(t, u)$ where $t = t(w, x, y, z)$, $u = u(w, x, y, z)$

21–26 Use the Chain Rule to find the indicated partial derivatives.

21. $z = x^4 + x^2 y$, $x = s + 2t - u$, $y = stu^2$;

$\dfrac{\partial z}{\partial s}$, $\dfrac{\partial z}{\partial t}$, $\dfrac{\partial z}{\partial u}$ when $s = 4, t = 2, u = 1$

22. $T = \dfrac{v}{2u + v}$, $u = pq\sqrt{r}$, $v = p\sqrt{q}\, r$;

$\dfrac{\partial T}{\partial p}$, $\dfrac{\partial T}{\partial q}$, $\dfrac{\partial T}{\partial r}$ when $p = 2, q = 1, r = 4$

23. $w = xy + yz + zx$, $x = r \cos\theta$, $y = r \sin\theta$, $z = r\theta$;

$\dfrac{\partial w}{\partial r}$, $\dfrac{\partial w}{\partial \theta}$ when $r = 2, \theta = \pi/2$

24. $P = \sqrt{u^2 + v^2 + w^2}$, $u = xe^y$, $v = ye^x$, $w = e^{xy}$;

$\dfrac{\partial P}{\partial x}$, $\dfrac{\partial P}{\partial y}$ when $x = 0, y = 2$

25. $N = \dfrac{p + q}{p + r}$, $p = u + vw$, $q = v + uw$, $r = w + uv$;

$\dfrac{\partial N}{\partial u}$, $\dfrac{\partial N}{\partial v}$, $\dfrac{\partial N}{\partial w}$ when $u = 2, v = 3, w = 4$

26. $u = xe^{ty}$, $x = \alpha^2\beta$, $y = \beta^2\gamma$, $t = \gamma^2\alpha$;

$\dfrac{\partial u}{\partial \alpha}$, $\dfrac{\partial u}{\partial \beta}$, $\dfrac{\partial u}{\partial \gamma}$ when $\alpha = -1, \beta = 2, \gamma = 1$

27–30 Use Equation 6 to find dy/dx.

27. $y \cos x = x^2 + y^2$

28. $\cos(xy) = 1 + \sin y$

29. $\tan^{-1}(x^2 y) = x + xy^2$

30. $e^y \sin x = x + xy$

31–34 Use Equations 7 to find $\partial z/\partial x$ and $\partial z/\partial y$.

31. $x^2 + 2y^2 + 3z^2 = 1$

32. $x^2 - y^2 + z^2 - 2z = 4$

33. $e^z = xyz$

34. $yz + x \ln y = z^2$

35. The temperature at a point (x, y) is $T(x, y)$, measured in degrees Celsius. A bug crawls so that its position after t seconds is given by $x = \sqrt{1 + t}$, $y = 2 + \frac{1}{3}t$, where x and y are measured in centimeters. The temperature function satisfies $T_x(2, 3) = 4$ and $T_y(2, 3) = 3$. How fast is the temperature rising on the bug's path after 3 seconds?

36. Wheat production W in a given year depends on the average temperature T and the annual rainfall R. Scientists estimate that the average temperature is rising at a rate of $0.15°\text{C/year}$ and rainfall is decreasing at a rate of 0.1 cm/year. They also estimate that at current production levels, $\partial W/\partial T = -2$ and $\partial W/\partial R = 8$.
(a) What is the significance of the signs of these partial derivatives?
(b) Estimate the current rate of change of wheat production, dW/dt.

37. The speed of sound traveling through ocean water with salinity 35 parts per thousand has been modeled by the equation

$$C = 1449.2 + 4.6T - 0.055T^2 + 0.00029T^3 + 0.016D$$

where C is the speed of sound (in meters per second), T is the temperature (in degrees Celsius), and D is the depth below the ocean surface (in meters). A scuba diver began a leisurely dive into the ocean water; the diver's depth and the surrounding water temperature over time are recorded in the following graphs. Estimate the rate of change (with respect to time) of the speed of sound through the ocean water experienced by the diver 20 minutes into the dive. What are the units?

38. The radius of a right circular cone is increasing at a rate of 1.8 in/s while its height is decreasing at a rate of 2.5 in/s. At what rate is the volume of the cone changing when the radius is 120 in. and the height is 140 in.?

39. The length ℓ, width w, and height h of a box change with time. At a certain instant the dimensions are $\ell = 1$ m and $w = h = 2$ m, and ℓ and w are increasing at a rate of 2 m/s while h is decreasing at a rate of 3 m/s. At that instant find the rates at which the following quantities are changing.
(a) The volume
(b) The surface area
(c) The length of a diagonal

40. The voltage V in a simple electrical circuit is slowly decreasing as the battery wears out. The resistance R is slowly increasing as the resistor heats up. Use Ohm's Law, $V = IR$, to find how the current I is changing at the moment when $R = 400\ \Omega$, $I = 0.08$ A, $dV/dt = -0.01$ V/s, and $dR/dt = 0.03\ \Omega$/s.

41. The pressure of 1 mole of an ideal gas is increasing at a rate of 0.05 kPa/s and the temperature is increasing at a rate of 0.15 K/s. Use the equation $PV = 8.31T$ in Example 2 to find the rate of change of the volume when the pressure is 20 kPa and the temperature is 320 K.

42. A manufacturer has modeled its yearly production function P (the value of its entire production, in millions of dollars) as a Cobb-Douglas function

$$P(L, K) = 1.47L^{0.65}K^{0.35}$$

where L is the number of labor hours (in thousands) and K is the invested capital (in millions of dollars). Suppose that when $L = 30$ and $K = 8$, the labor force is decreasing at a rate of 2000 labor hours per year and capital is increasing at a rate of $500,000 per year. Find the rate of change of production.

43. One side of a triangle is increasing at a rate of 3 cm/s and a second side is decreasing at a rate of 2 cm/s. If the area of the triangle remains constant, at what rate does the angle between the sides change when the first side is 20 cm long, the second side is 30 cm, and the angle is $\pi/6$?

44. A sound with frequency f_s is produced by a source traveling along a line with speed v_s. If an observer is traveling with speed v_o along the same line from the opposite direction toward the source, then the frequency of the sound heard by the observer is

$$f_o = \left(\frac{c + v_o}{c - v_s}\right)f_s$$

where c is the speed of sound, about 332 m/s. (This is the **Doppler effect**.) Suppose that, at a particular moment, you are in a train traveling at 34 m/s and accelerating at 1.2 m/s². A train is approaching you from the opposite direction on the other track at 40 m/s, accelerating at 1.4 m/s², and sounds its whistle, which has a frequency of 460 Hz. At that instant, what is the perceived frequency that you hear and how fast is it changing?

45–48 Assume that all the given functions are differentiable.

45. If $z = f(x, y)$, where $x = r\cos\theta$ and $y = r\sin\theta$, (a) find $\partial z/\partial r$ and $\partial z/\partial\theta$ and (b) show that

$$\left(\frac{\partial z}{\partial x}\right)^2 + \left(\frac{\partial z}{\partial y}\right)^2 = \left(\frac{\partial z}{\partial r}\right)^2 + \frac{1}{r^2}\left(\frac{\partial z}{\partial\theta}\right)^2$$

46. If $u = f(x, y)$, where $x = e^s\cos t$ and $y = e^s\sin t$, show that

$$\left(\frac{\partial u}{\partial x}\right)^2 + \left(\frac{\partial u}{\partial y}\right)^2 = e^{-2s}\left[\left(\frac{\partial u}{\partial s}\right)^2 + \left(\frac{\partial u}{\partial t}\right)^2\right]$$

47. If $z = \dfrac{1}{x}[f(x - y) + g(x + y)]$, show that

$$\frac{\partial}{\partial x}\left(x^2\frac{\partial z}{\partial x}\right) = x^2\frac{\partial^2 z}{\partial y^2}$$

48. If $z = \dfrac{1}{y}[f(ax + y) + g(ax - y)]$, show that

$$\frac{\partial^2 z}{\partial x^2} = \frac{a^2}{y^2}\frac{\partial}{\partial y}\left(y^2\frac{\partial z}{\partial y}\right)$$

49–54 Assume that all the given functions have continuous second-order partial derivatives.

49. Show that any function of the form

$$z = f(x + at) + g(x - at)$$

is a solution of the wave equation

$$\frac{\partial^2 z}{\partial t^2} = a^2\frac{\partial^2 z}{\partial x^2}$$

[*Hint:* Let $u = x + at$, $v = x - at$.]

50. If $u = f(x, y)$, where $x = e^s\cos t$ and $y = e^s\sin t$, show that

$$\frac{\partial^2 u}{\partial x^2} + \frac{\partial^2 u}{\partial y^2} = e^{-2s}\left[\frac{\partial^2 u}{\partial s^2} + \frac{\partial^2 u}{\partial t^2}\right]$$

51. If $z = f(x, y)$, where $x = r^2 + s^2$ and $y = 2rs$, find $\partial^2 z/\partial r\,\partial s$. (Compare with Example 7.)

52. If $z = f(x, y)$, where $x = r\cos\theta$ and $y = r\sin\theta$, find (a) $\partial z/\partial r$, (b) $\partial z/\partial\theta$, and (c) $\partial^2 z/\partial r\,\partial\theta$.

53. If $z = f(x, y)$, where $x = r\cos\theta$ and $y = r\sin\theta$, show that

$$\frac{\partial^2 z}{\partial x^2} + \frac{\partial^2 z}{\partial y^2} = \frac{\partial^2 z}{\partial r^2} + \frac{1}{r^2}\frac{\partial^2 z}{\partial\theta^2} + \frac{1}{r}\frac{\partial z}{\partial r}$$

54. Suppose $z = f(x, y)$, where $x = g(s, t)$ and $y = h(s, t)$.
(a) Show that

$$\frac{\partial^2 z}{\partial t^2} = \frac{\partial^2 z}{\partial x^2}\left(\frac{\partial x}{\partial t}\right)^2 + 2\frac{\partial^2 z}{\partial x\,\partial y}\frac{\partial x}{\partial t}\frac{\partial y}{\partial t} + \frac{\partial^2 z}{\partial y^2}\left(\frac{\partial y}{\partial t}\right)^2$$

$$+ \frac{\partial z}{\partial x}\frac{\partial^2 x}{\partial t^2} + \frac{\partial z}{\partial y}\frac{\partial^2 y}{\partial t^2}$$

(b) Find a similar formula for $\partial^2 z/\partial s\,\partial t$.

55. A function f is called **homogeneous of degree n** if it satisfies the equation

$$f(tx, ty) = t^n f(x, y)$$

for all t, where n is a positive integer and f has continuous second-order partial derivatives.

(a) Verify that $f(x, y) = x^2 y + 2xy^2 + 5y^3$ is homogeneous of degree 3.

(b) Show that if f is homogeneous of degree n, then

$$x \frac{\partial f}{\partial x} + y \frac{\partial f}{\partial y} = nf(x, y)$$

[*Hint:* Use the Chain Rule to differentiate $f(tx, ty)$ with respect to t.]

56. If f is homogeneous of degree n, show that

$$x^2 \frac{\partial^2 f}{\partial x^2} + 2xy \frac{\partial^2 f}{\partial x\, \partial y} + y^2 \frac{\partial^2 f}{\partial y^2} = n(n-1)f(x, y)$$

57. If f is homogeneous of degree n, show that

$$f_x(tx, ty) = t^{n-1} f_x(x, y)$$

58. Suppose that the equation $F(x, y, z) = 0$ implicitly defines each of the three variables x, y, and z as functions of the other two: $z = f(x, y)$, $y = g(x, z)$, $x = h(y, z)$. If F is differentiable and F_x, F_y, and F_z are all nonzero, show that

$$\frac{\partial z}{\partial x} \frac{\partial x}{\partial y} \frac{\partial y}{\partial z} = -1$$

59. Equation 6 is a formula for the derivative dy/dx of a function defined implicitly by an equation $F(x, y) = 0$, provided that F is differentiable and $F_y \neq 0$. Prove that if F has continuous second derivatives, then a formula for the second derivative of y is

$$\frac{d^2 y}{dx^2} = -\frac{F_{xx} F_y^2 - 2F_{xy} F_x F_y + F_{yy} F_x^2}{F_y^3}$$

14.6 Directional Derivatives and the Gradient Vector

FIGURE 1

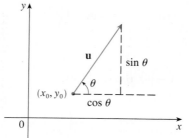

FIGURE 2
A unit vector
$\mathbf{u} = \langle a, b \rangle = \langle \cos u, \sin u \rangle$

The weather map in Figure 1 shows a contour map of the temperature function $T(x, y)$ for the states of California and Nevada at 3:00 PM on a day in October. The level curves, or isothermals, join locations with the same temperature. The partial derivative T_x at a location such as Reno is the rate of change of temperature with respect to distance if we travel east from Reno; T_y is the rate of change of temperature if we travel north. But what if we want to know the rate of change of temperature when we travel southeast (toward Las Vegas), or in some other direction? In this section we introduce a type of derivative, called a *directional derivative,* that enables us to find the rate of change of a function of two or more variables in any direction.

■ Directional Derivatives

Recall that if $z = f(x, y)$, then the partial derivatives f_x and f_y are defined as

$$\boxed{1} \qquad f_x(x_0, y_0) = \lim_{h \to 0} \frac{f(x_0 + h, y_0) - f(x_0, y_0)}{h}$$

$$f_y(x_0, y_0) = \lim_{h \to 0} \frac{f(x_0, y_0 + h) - f(x_0, y_0)}{h}$$

and represent the rates of change of z in the x- and y-directions, that is, in the directions of the unit vectors $\mathbf{i}$ and $\mathbf{j}$.

Suppose that we now wish to find the rate of change of z at (x_0, y_0) in the direction of an arbitrary unit vector $\mathbf{u} = \langle a, b \rangle$. (See Figure 2.) To do this we consider the surface S with the equation $z = f(x, y)$ (the graph of f) and we let $z_0 = f(x_0, y_0)$. Then the point $P(x_0, y_0, z_0)$ lies on S. The vertical plane that passes through P in the direction of $\mathbf{u}$ intersects S in a curve C. (See Figure 3.) The slope of the tangent line T to C at the point P is the rate of change of z in the direction of **u**.

TEC Visual 14.6A animates Figure 3 by rotating **u** and therefore T.

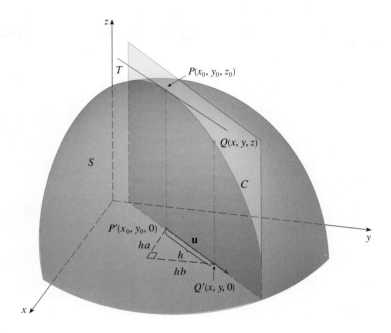

FIGURE 3

If $Q(x, y, z)$ is another point on C and P', Q' are the projections of P, Q onto the xy-plane, then the vector $\overrightarrow{P'Q'}$ is parallel to **u** and so

$$\overrightarrow{P'Q'} = h\mathbf{u} = \langle ha, hb \rangle$$

for some scalar h. Therefore $x - x_0 = ha$, $y - y_0 = hb$, so $x = x_0 + ha$, $y = y_0 + hb$, and

$$\frac{\Delta z}{h} = \frac{z - z_0}{h} = \frac{f(x_0 + ha, y_0 + hb) - f(x_0, y_0)}{h}$$

If we take the limit as $h \to 0$, we obtain the rate of change of z (with respect to distance) in the direction of **u**, which is called the directional derivative of f in the direction of **u**.

---

2 Definition The **directional derivative** of f at (x_0, y_0) in the direction of a unit vector $\mathbf{u} = \langle a, b \rangle$ is

$$D_{\mathbf{u}} f(x_0, y_0) = \lim_{h \to 0} \frac{f(x_0 + ha, y_0 + hb) - f(x_0, y_0)}{h}$$

if this limit exists.

By comparing Definition 2 with Equations 1, we see that if $\mathbf{u} = \mathbf{i} = \langle 1, 0 \rangle$, then $D_{\mathbf{i}} f = f_x$ and if $\mathbf{u} = \mathbf{j} = \langle 0, 1 \rangle$, then $D_{\mathbf{j}} f = f_y$. In other words, the partial derivatives of f with respect to x and y are just special cases of the directional derivative.

EXAMPLE 1 Use the weather map in Figure 1 to estimate the value of the directional derivative of the temperature function at Reno in the southeasterly direction.

SOLUTION The unit vector directed toward the southeast is $\mathbf{u} = (\mathbf{i} - \mathbf{j})/\sqrt{2}$, but we won't need to use this expression. We start by drawing a line through Reno toward the southeast (see Figure 4).

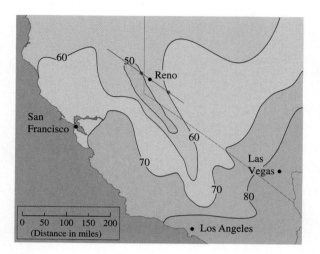

FIGURE 4

We approximate the directional derivative $D_\mathbf{u} T$ by the average rate of change of the temperature between the points where this line intersects the isothermals $T = 50$ and $T = 60$. The temperature at the point southeast of Reno is $T = 60°F$ and the temperature at the point northwest of Reno is $T = 50°F$. The distance between these points looks to be about 75 miles. So the rate of change of the temperature in the southeasterly direction is

$$D_\mathbf{u} T \approx \frac{60 - 50}{75} = \frac{10}{75} \approx 0.13°F/mi \qquad \blacksquare$$

When we compute the directional derivative of a function defined by a formula, we generally use the following theorem.

$\boxed{3}$ **Theorem** If f is a differentiable function of x and y, then f has a directional derivative in the direction of any unit vector $\mathbf{u} = \langle a, b \rangle$ and

$$D_\mathbf{u} f(x, y) = f_x(x, y)\, a + f_y(x, y)\, b$$

PROOF If we define a function g of the single variable h by

$$g(h) = f(x_0 + ha, y_0 + hb)$$

then, by the definition of a derivative, we have

$$\boxed{4} \quad g'(0) = \lim_{h \to 0} \frac{g(h) - g(0)}{h} = \lim_{h \to 0} \frac{f(x_0 + ha, y_0 + hb) - f(x_0, y_0)}{h}$$

$$= D_\mathbf{u} f(x_0, y_0)$$

On the other hand, we can write $g(h) = f(x, y)$, where $x = x_0 + ha$, $y = y_0 + hb$, so the Chain Rule (Theorem 14.5.2) gives

$$g'(h) = \frac{\partial f}{\partial x}\frac{dx}{dh} + \frac{\partial f}{\partial y}\frac{dy}{dh} = f_x(x, y)\,a + f_y(x, y)\,b$$

If we now put $h = 0$, then $x = x_0$, $y = y_0$, and

$$\boxed{5} \qquad\qquad g'(0) = f_x(x_0, y_0)\,a + f_y(x_0, y_0)\,b$$

Comparing Equations 4 and 5, we see that

$$D_{\mathbf{u}}f(x_0, y_0) = f_x(x_0, y_0)\,a + f_y(x_0, y_0)\,b \qquad\qquad \blacksquare$$

If the unit vector $\mathbf{u}$ makes an angle θ with the positive x-axis (as in Figure 2), then we can write $\mathbf{u} = \langle \cos\theta, \sin\theta \rangle$ and the formula in Theorem 3 becomes

$$\boxed{6} \qquad\qquad D_{\mathbf{u}}f(x, y) = f_x(x, y)\,\cos\theta + f_y(x, y)\,\sin\theta$$

The directional derivative $D_{\mathbf{u}} f(1, 2)$ in Example 2 represents the rate of change of z in the direction of $\mathbf{u}$. This is the slope of the tangent line to the curve of intersection of the surface $z = x^3 - 3xy + 4y^2$ and the vertical plane through $(1, 2, 0)$ in the direction of $\mathbf{u}$ shown in Figure 5.

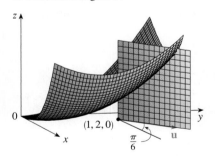

FIGURE 5

EXAMPLE 2 Find the directional derivative $D_{\mathbf{u}} f(x, y)$ if

$$f(x, y) = x^3 - 3xy + 4y^2$$

and $\mathbf{u}$ is the unit vector given by angle $\theta = \pi/6$. What is $D_{\mathbf{u}} f(1, 2)$?

SOLUTION Formula 6 gives

$$D_{\mathbf{u}}f(x, y) = f_x(x, y)\cos\frac{\pi}{6} + f_y(x, y)\sin\frac{\pi}{6}$$

$$= (3x^2 - 3y)\frac{\sqrt{3}}{2} + (-3x + 8y)\tfrac{1}{2}$$

$$= \tfrac{1}{2}\left[3\sqrt{3}\,x^2 - 3x + \left(8 - 3\sqrt{3}\right)y\right]$$

Therefore

$$D_{\mathbf{u}}f(1, 2) = \tfrac{1}{2}\left[3\sqrt{3}\,(1)^2 - 3(1) + \left(8 - 3\sqrt{3}\right)(2)\right] = \frac{13 - 3\sqrt{3}}{2} \qquad \blacksquare$$

■ The Gradient Vector

Notice from Theorem 3 that the directional derivative of a differentiable function can be written as the dot product of two vectors:

$$\boxed{7} \qquad\qquad D_{\mathbf{u}}f(x, y) = f_x(x, y)\,a + f_y(x, y)\,b$$

$$= \langle f_x(x, y), f_y(x, y) \rangle \cdot \langle a, b \rangle$$

$$= \langle f_x(x, y), f_y(x, y) \rangle \cdot \mathbf{u}$$

The first vector in this dot product occurs not only in computing directional derivatives but in many other contexts as well. So we give it a special name (the *gradient* of f) and a special notation (**grad** f or ∇f, which is read "del f").

> **8**   **Definition** If f is a function of two variables x and y, then the **gradient** of f is the vector function ∇f defined by
>
> $$\nabla f(x, y) = \langle f_x(x, y), f_y(x, y) \rangle = \frac{\partial f}{\partial x}\,\mathbf{i} + \frac{\partial f}{\partial y}\,\mathbf{j}$$

EXAMPLE 3 If $f(x, y) = \sin x + e^{xy}$, then

$$\nabla f(x, y) = \langle f_x, f_y \rangle = \langle \cos x + ye^{xy},\ xe^{xy} \rangle$$

and

$$\nabla f(0, 1) = \langle 2, 0 \rangle$$

      With this notation for the gradient vector, we can rewrite Equation 7 for the directional derivative of a differentiable function as

> **9** $$D_{\mathbf{u}} f(x, y) = \nabla f(x, y) \cdot \mathbf{u}$$

This expresses the directional derivative in the direction of a unit vector $\mathbf{u}$ as the scalar projection of the gradient vector onto $\mathbf{u}$.

The gradient vector $\nabla f(2, -1)$ in Example 4 is shown in Figure 6 with initial point $(2, -1)$. Also shown is the vector $\mathbf{v}$ that gives the direction of the directional derivative. Both of these vectors are superimposed on a contour plot of the graph of f.

EXAMPLE 4 Find the directional derivative of the function $f(x, y) = x^2 y^3 - 4y$ at the point $(2, -1)$ in the direction of the vector $\mathbf{v} = 2\mathbf{i} + 5\mathbf{j}$.

SOLUTION We first compute the gradient vector at $(2, -1)$:

$$\nabla f(x, y) = 2xy^3\,\mathbf{i} + (3x^2 y^2 - 4)\,\mathbf{j}$$

$$\nabla f(2, -1) = -4\mathbf{i} + 8\mathbf{j}$$

Note that $\mathbf{v}$ is not a unit vector, but since $|\mathbf{v}| = \sqrt{29}$, the unit vector in the direction of $\mathbf{v}$ is

$$\mathbf{u} = \frac{\mathbf{v}}{|\mathbf{v}|} = \frac{2}{\sqrt{29}}\,\mathbf{i} + \frac{5}{\sqrt{29}}\,\mathbf{j}$$

Therefore, by Equation 9, we have

$$D_{\mathbf{u}} f(2, -1) = \nabla f(2, -1) \cdot \mathbf{u} = (-4\mathbf{i} + 8\mathbf{j}) \cdot \left(\frac{2}{\sqrt{29}}\,\mathbf{i} + \frac{5}{\sqrt{29}}\,\mathbf{j} \right)$$

$$= \frac{-4 \cdot 2 + 8 \cdot 5}{\sqrt{29}} = \frac{32}{\sqrt{29}}$$

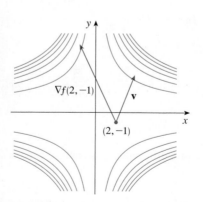

FIGURE 6

Functions of Three Variables

For functions of three variables we can define directional derivatives in a similar manner. Again $D_{\mathbf{u}} f(x, y, z)$ can be interpreted as the rate of change of the function in the direction of a unit vector $\mathbf{u}$.

---

10 **Definition** The **directional derivative** of f at (x_0, y_0, z_0) in the direction of a unit vector $\mathbf{u} = \langle a, b, c \rangle$ is

$$D_{\mathbf{u}} f(x_0, y_0, z_0) = \lim_{h \to 0} \frac{f(x_0 + ha, y_0 + hb, z_0 + hc) - f(x_0, y_0, z_0)}{h}$$

if this limit exists.

If we use vector notation, then we can write both definitions (2 and 10) of the directional derivative in the compact form

11
$$D_{\mathbf{u}} f(\mathbf{x}_0) = \lim_{h \to 0} \frac{f(\mathbf{x}_0 + h\mathbf{u}) - f(\mathbf{x}_0)}{h}$$

where $\mathbf{x}_0 = \langle x_0, y_0 \rangle$ if $n = 2$ and $\mathbf{x}_0 = \langle x_0, y_0, z_0 \rangle$ if $n = 3$. This is reasonable because the vector equation of the line through $\mathbf{x}_0$ in the direction of the vector $\mathbf{u}$ is given by $\mathbf{x} = \mathbf{x}_0 + t\mathbf{u}$ (Equation 12.5.1) and so $f(\mathbf{x}_0 + h\mathbf{u})$ represents the value of f at a point on this line.

If $f(x, y, z)$ is differentiable and $\mathbf{u} = \langle a, b, c \rangle$, then the same method that was used to prove Theorem 3 can be used to show that

12
$$D_{\mathbf{u}} f(x, y, z) = f_x(x, y, z)\, a + f_y(x, y, z)\, b + f_z(x, y, z)\, c$$

For a function f of three variables, the **gradient vector**, denoted by ∇f or **grad** f, is

$$\nabla f(x, y, z) = \langle f_x(x, y, z), f_y(x, y, z), f_z(x, y, z) \rangle$$

or, for short,

13
$$\nabla f = \langle f_x, f_y, f_z \rangle = \frac{\partial f}{\partial x}\mathbf{i} + \frac{\partial f}{\partial y}\mathbf{j} + \frac{\partial f}{\partial z}\mathbf{k}$$

Then, just as with functions of two variables, Formula 12 for the directional derivative can be rewritten as

14
$$D_{\mathbf{u}} f(x, y, z) = \nabla f(x, y, z) \cdot \mathbf{u}$$

EXAMPLE 5 If $f(x, y, z) = x \sin yz$, (a) find the gradient of f and (b) find the directional derivative of f at $(1, 3, 0)$ in the direction of $\mathbf{v} = \mathbf{i} + 2\mathbf{j} - \mathbf{k}$.

SOLUTION
(a) The gradient of f is

$$\nabla f(x, y, z) = \langle f_x(x, y, z), f_y(x, y, z), f_z(x, y, z) \rangle$$

$$= \langle \sin yz, xz \cos yz, xy \cos yz \rangle$$

(b) At $(1, 3, 0)$ we have $\nabla f(1, 3, 0) = \langle 0, 0, 3 \rangle$. The unit vector in the direction of $\mathbf{v} = \mathbf{i} + 2\mathbf{j} - \mathbf{k}$ is

$$\mathbf{u} = \frac{1}{\sqrt{6}}\mathbf{i} + \frac{2}{\sqrt{6}}\mathbf{j} - \frac{1}{\sqrt{6}}\mathbf{k}$$

Therefore Equation 14 gives

$$D_{\mathbf{u}} f(1, 3, 0) = \nabla f(1, 3, 0) \cdot \mathbf{u}$$

$$= 3\mathbf{k} \cdot \left(\frac{1}{\sqrt{6}}\mathbf{i} + \frac{2}{\sqrt{6}}\mathbf{j} - \frac{1}{\sqrt{6}}\mathbf{k} \right)$$

$$= 3 \left(-\frac{1}{\sqrt{6}} \right) = -\sqrt{\frac{3}{2}}$$

■ Maximizing the Directional Derivative

Suppose we have a function f of two or three variables and we consider all possible directional derivatives of f at a given point. These give the rates of change of f in all possible directions. We can then ask the questions: in which of these directions does f change fastest and what is the maximum rate of change? The answers are provided by the following theorem.

TEC Visual 14.6B provides visual confirmation of Theorem 15.

> **15 Theorem** Suppose f is a differentiable function of two or three variables. The maximum value of the directional derivative $D_{\mathbf{u}} f(\mathbf{x})$ is $|\nabla f(\mathbf{x})|$ and it occurs when $\mathbf{u}$ has the same direction as the gradient vector $\nabla f(\mathbf{x})$.

PROOF From Equation 9 or 14 we have

$$D_{\mathbf{u}} f = \nabla f \cdot \mathbf{u} = |\nabla f| |\mathbf{u}| \cos \theta = |\nabla f| \cos \theta$$

where θ is the angle between ∇f and $\mathbf{u}$. The maximum value of $\cos \theta$ is 1 and this occurs when $\theta = 0$. Therefore the maximum value of $D_{\mathbf{u}} f$ is $|\nabla f|$ and it occurs when $\theta = 0$, that is, when $\mathbf{u}$ has the same direction as ∇f. ■

EXAMPLE 6

(a) If $f(x, y) = xe^y$, find the rate of change of f at the point $P(2, 0)$ in the direction from P to $Q(\frac{1}{2}, 2)$.

(b) In what direction does f have the maximum rate of change? What is this maximum rate of change?

SOLUTION

(a) We first compute the gradient vector:

$$\nabla f(x, y) = \langle f_x, f_y \rangle = \langle e^y, xe^y \rangle$$

$$\nabla f(2, 0) = \langle 1, 2 \rangle$$

The unit vector in the direction of $\overrightarrow{PQ} = \langle -\frac{3}{2}, 2 \rangle$ is $\mathbf{u} = \langle -\frac{3}{5}, \frac{4}{5} \rangle$, so the rate of change of f in the direction from P to Q is

$$D_{\mathbf{u}} f(2, 0) = \nabla f(2, 0) \cdot \mathbf{u} = \langle 1, 2 \rangle \cdot \langle -\tfrac{3}{5}, \tfrac{4}{5} \rangle$$

$$= 1(-\tfrac{3}{5}) + 2(\tfrac{4}{5}) = 1$$

(b) According to Theorem 15, f increases fastest in the direction of the gradient vector $\nabla f(2, 0) = \langle 1, 2 \rangle$. The maximum rate of change is

$$|\nabla f(2, 0)| = |\langle 1, 2 \rangle| = \sqrt{5} \qquad \blacksquare$$

At $(2, 0)$ the function in Example 6 increases fastest in the direction of the gradient vector $\nabla f(2, 0) = \langle 1, 2 \rangle$. Notice from Figure 7 that this vector appears to be perpendicular to the level curve through $(2, 0)$. Figure 8 shows the graph of f and the gradient vector.

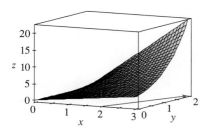

FIGURE 7   **FIGURE 8**

EXAMPLE 7 Suppose that the temperature at a point (x, y, z) in space is given by $T(x, y, z) = 80/(1 + x^2 + 2y^2 + 3z^2)$, where T is measured in degrees Celsius and x, y, z in meters. In which direction does the temperature increase fastest at the point $(1, 1, -2)$? What is the maximum rate of increase?

SOLUTION The gradient of T is

$$\nabla T = \frac{\partial T}{\partial x}\mathbf{i} + \frac{\partial T}{\partial y}\mathbf{j} + \frac{\partial T}{\partial z}\mathbf{k}$$

$$= -\frac{160x}{(1 + x^2 + 2y^2 + 3z^2)^2}\mathbf{i} - \frac{320y}{(1 + x^2 + 2y^2 + 3z^2)^2}\mathbf{j} - \frac{480z}{(1 + x^2 + 2y^2 + 3z^2)^2}\mathbf{k}$$

$$= \frac{160}{(1 + x^2 + 2y^2 + 3z^2)^2}(-x\mathbf{i} - 2y\mathbf{j} - 3z\mathbf{k})$$

At the point $(1, 1, -2)$ the gradient vector is

$$\nabla T(1, 1, -2) = \tfrac{160}{256}(-\mathbf{i} - 2\mathbf{j} + 6\mathbf{k}) = \tfrac{5}{8}(-\mathbf{i} - 2\mathbf{j} + 6\mathbf{k})$$

By Theorem 15 the temperature increases fastest in the direction of the gradient vector $\nabla T(1, 1, -2) = \tfrac{5}{8}(-\mathbf{i} - 2\mathbf{j} + 6\mathbf{k})$ or, equivalently, in the direction of $-\mathbf{i} - 2\mathbf{j} + 6\mathbf{k}$ or the unit vector $(-\mathbf{i} - 2\mathbf{j} + 6\mathbf{k})/\sqrt{41}$. The maximum rate of increase is the length of the gradient vector:

$$|\nabla T(1, 1, -2)| = \tfrac{5}{8}|-\mathbf{i} - 2\mathbf{j} + 6\mathbf{k}| = \tfrac{5}{8}\sqrt{41}$$

Therefore the maximum rate of increase of temperature is $\tfrac{5}{8}\sqrt{41} \approx 4°C/m$. $\qquad \blacksquare$

Tangent Planes to Level Surfaces

Suppose S is a surface with equation $F(x, y, z) = k$, that is, it is a level surface of a function F of three variables, and let $P(x_0, y_0, z_0)$ be a point on S. Let C be any curve that lies on the surface S and passes through the point P. Recall from Section 13.1 that the curve C is described by a continuous vector function $\mathbf{r}(t) = \langle x(t), y(t), z(t) \rangle$. Let t_0 be the parameter value corresponding to P; that is, $\mathbf{r}(t_0) = \langle x_0, y_0, z_0 \rangle$. Since C lies on S, any point $(x(t), y(t), z(t))$ must satisfy the equation of S, that is,

$$\boxed{16} \qquad F(x(t), y(t), z(t)) = k$$

If x, y, and z are differentiable functions of t and F is also differentiable, then we can use the Chain Rule to differentiate both sides of Equation 16 as follows:

$$\boxed{17} \qquad \frac{\partial F}{\partial x}\frac{dx}{dt} + \frac{\partial F}{\partial y}\frac{dy}{dt} + \frac{\partial F}{\partial z}\frac{dz}{dt} = 0$$

But, since $\nabla F = \langle F_x, F_y, F_z \rangle$ and $\mathbf{r}'(t) = \langle x'(t), y'(t), z'(t) \rangle$, Equation 17 can be written in terms of a dot product as

$$\nabla F \cdot \mathbf{r}'(t) = 0$$

In particular, when $t = t_0$ we have $\mathbf{r}(t_0) = \langle x_0, y_0, z_0 \rangle$, so

$$\boxed{18} \qquad \nabla F(x_0, y_0, z_0) \cdot \mathbf{r}'(t_0) = 0$$

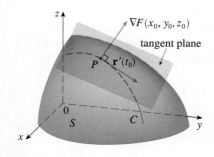

FIGURE 9

Equation 18 says that *the gradient vector at P*, $\nabla F(x_0, y_0, z_0)$, *is perpendicular to the tangent vector* $\mathbf{r}'(t_0)$ *to any curve C on S that passes through P.* (See Figure 9.) If $\nabla F(x_0, y_0, z_0) \neq \mathbf{0}$, it is therefore natural to define the **tangent plane to the level surface** $F(x, y, z) = k$ **at** $P(x_0, y_0, z_0)$ as the plane that passes through P and has normal vector $\nabla F(x_0, y_0, z_0)$. Using the standard equation of a plane (Equation 12.5.7), we can write the equation of this tangent plane as

$$\boxed{19} \qquad F_x(x_0, y_0, z_0)(x - x_0) + F_y(x_0, y_0, z_0)(y - y_0) + F_z(x_0, y_0, z_0)(z - z_0) = 0$$

The **normal line** to S at P is the line passing through P and perpendicular to the tangent plane. The direction of the normal line is therefore given by the gradient vector $\nabla F(x_0, y_0, z_0)$ and so, by Equation 12.5.3, its symmetric equations are

$$\boxed{20} \qquad \frac{x - x_0}{F_x(x_0, y_0, z_0)} = \frac{y - y_0}{F_y(x_0, y_0, z_0)} = \frac{z - z_0}{F_z(x_0, y_0, z_0)}$$

In the special case in which the equation of a surface S is of the form $z = f(x, y)$ (that is, S is the graph of a function f of two variables), we can rewrite the equation as

$$F(x, y, z) = f(x, y) - z = 0$$

and regard S as a level surface (with $k = 0$) of F. Then

$$F_x(x_0, y_0, z_0) = f_x(x_0, y_0)$$

$$F_y(x_0, y_0, z_0) = f_y(x_0, y_0)$$

$$F_z(x_0, y_0, z_0) = -1$$

so Equation 19 becomes

$$f_x(x_0, y_0)(x - x_0) + f_y(x_0, y_0)(y - y_0) - (z - z_0) = 0$$

which is equivalent to Equation 14.4.2. Thus our new, more general, definition of a tangent plane is consistent with the definition that was given for the special case of Section 14.4.

EXAMPLE 8   Find the equations of the tangent plane and normal line at the point $(-2, 1, -3)$ to the ellipsoid

$$\frac{x^2}{4} + y^2 + \frac{z^2}{9} = 3$$

SOLUTION   The ellipsoid is the level surface (with $k = 3$) of the function

$$F(x, y, z) = \frac{x^2}{4} + y^2 + \frac{z^2}{9}$$

Therefore we have

$$F_x(x, y, z) = \frac{x}{2} \qquad F_y(x, y, z) = 2y \qquad F_z(x, y, z) = \frac{2z}{9}$$

$$F_x(-2, 1, -3) = -1 \qquad F_y(-2, 1, -3) = 2 \qquad F_z(-2, 1, -3) = -\frac{2}{3}$$

Then Equation 19 gives the equation of the tangent plane at $(-2, 1, -3)$ as

$$-1(x + 2) + 2(y - 1) - \tfrac{2}{3}(z + 3) = 0$$

which simplifies to $3x - 6y + 2z + 18 = 0$.

By Equation 20, symmetric equations of the normal line are

$$\frac{x + 2}{-1} = \frac{y - 1}{2} = \frac{z + 3}{-\frac{2}{3}}$$

Figure 10 shows the ellipsoid, tangent plane, and normal line in Example 8.

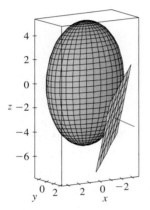

FIGURE 10

■ Significance of the Gradient Vector

We now summarize the ways in which the gradient vector is significant. We first consider a function f of three variables and a point $P(x_0, y_0, z_0)$ in its domain. On the one hand, we know from Theorem 15 that the gradient vector $\nabla f(x_0, y_0, z_0)$ gives the direction of fastest increase of f. On the other hand, we know that $\nabla f(x_0, y_0, z_0)$ is orthogonal to the level surface S of f through P. (Refer to Figure 9.) These two properties are quite compatible intuitively because as we move away from P on the level surface S, the value of f does not change at all. So it seems reasonable that if we move in the perpendicular direction, we get the maximum increase.

In like manner we consider a function f of two variables and a point $P(x_0, y_0)$ in its domain. Again the gradient vector $\nabla f(x_0, y_0)$ gives the direction of fastest increase of f. Also, by considerations similar to our discussion of tangent planes, it can be shown that $\nabla f(x_0, y_0)$ is perpendicular to the level curve $f(x, y) = k$ that passes through P. Again this is intuitively plausible because the values of f remain constant as we move along the curve. (See Figure 11.)

If we consider a topographical map of a hill and let $f(x, y)$ represent the height above sea level at a point with coordinates (x, y), then a curve of steepest ascent can be drawn

FIGURE 11

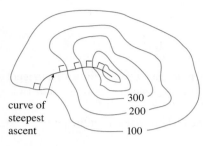

curve of
steepest
ascent

FIGURE 12

as in Figure 12 by making it perpendicular to all of the contour lines. This phenomenon can also be noticed in Figure 14.1.12, where Lonesome Creek follows a curve of steepest descent.

Computer algebra systems have commands that plot sample gradient vectors. Each gradient vector $\nabla f(a, b)$ is plotted starting at the point (a, b). Figure 13 shows such a plot (called a *gradient vector field*) for the function $f(x, y) = x^2 - y^2$ superimposed on a contour map of f. As expected, the gradient vectors point "uphill" and are perpendicular to the level curves.

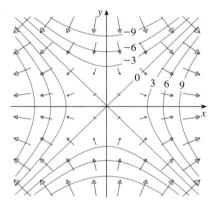

FIGURE 13

14.6 EXERCISES

1. Level curves for barometric pressure (in millibars) are shown for 6:00 AM on a day in November. A deep low with pressure 972 mb is moving over northeast Iowa. The distance along the red line from K (Kearney, Nebraska) to S (Sioux City, Iowa) is 300 km. Estimate the value of the directional derivative of the pressure function at Kearney in the direction of Sioux City. What are the units of the directional derivative?

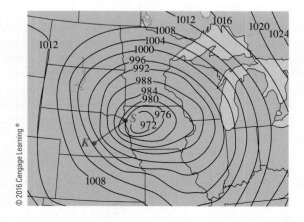

2. The contour map shows the average maximum temperature for November 2004 (in °C). Estimate the value of the directional

derivative of this temperature function at Dubbo, New South Wales, in the direction of Sydney. What are the units?

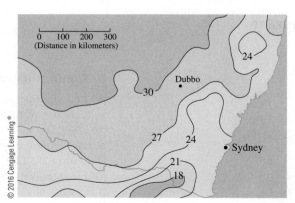

3. A table of values for the wind-chill index $W = f(T, v)$ is given in Exercise 14.3.3 on page 923. Use the table to estimate the value of $D_{\mathbf{u}} f(-20, 30)$, where $\mathbf{u} = (\mathbf{i} + \mathbf{j})/\sqrt{2}$.

4–6 Find the directional derivative of f at the given point in the direction indicated by the angle θ.

4. $f(x, y) = xy^3 - x^2$, $(1, 2)$, $\theta = \pi/3$

5. $f(x, y) = y \cos(xy)$, $(0, 1)$, $\theta = \pi/4$

6. $f(x, y) = \sqrt{2x + 3y}$, $(3, 1)$, $\theta = -\pi/6$

7–10
(a) Find the gradient of f.
(b) Evaluate the gradient at the point P.
(c) Find the rate of change of f at P in the direction of the vector $\mathbf{u}$.

7. $f(x, y) = x/y$, $P(2, 1)$, $\mathbf{u} = \frac{3}{5}\mathbf{i} + \frac{4}{5}\mathbf{j}$

8. $f(x, y) = x^2 \ln y$, $P(3, 1)$, $\mathbf{u} = -\frac{5}{13}\mathbf{i} + \frac{12}{13}\mathbf{j}$

9. $f(x, y, z) = x^2 yz - xyz^3$, $P(2, -1, 1)$, $\mathbf{u} = \left\langle 0, \frac{4}{5}, -\frac{3}{5} \right\rangle$

10. $f(x, y, z) = y^2 e^{xyz}$, $P(0, 1, -1)$, $\mathbf{u} = \left\langle \frac{3}{13}, \frac{4}{13}, \frac{12}{13} \right\rangle$

11–17 Find the directional derivative of the function at the given point in the direction of the vector $\mathbf{v}$.

11. $f(x, y) = e^x \sin y$, $(0, \pi/3)$, $\mathbf{v} = \langle -6, 8 \rangle$

12. $f(x, y) = \dfrac{x}{x^2 + y^2}$, $(1, 2)$, $\mathbf{v} = \langle 3, 5 \rangle$

13. $g(s, t) = s \sqrt{t}$, $(2, 4)$, $\mathbf{v} = 2\mathbf{i} - \mathbf{j}$

14. $g(u, v) = u^2 e^{-v}$, $(3, 0)$, $\mathbf{v} = 3\mathbf{i} + 4\mathbf{j}$

15. $f(x, y, z) = x^2 y + y^2 z$, $(1, 2, 3)$, $\mathbf{v} = \langle 2, -1, 2 \rangle$

16. $f(x, y, z) = xy^2 \tan^{-1} z$, $(2, 1, 1)$, $\mathbf{v} = \langle 1, 1, 1 \rangle$

17. $h(r, s, t) = \ln(3r + 6s + 9t)$, $(1, 1, 1)$,
$\mathbf{v} = 4\mathbf{i} + 12\mathbf{j} + 6\mathbf{k}$

---

18. Use the figure to estimate $D_{\mathbf{u}} f(2, 2)$.

19. Find the directional derivative of $f(x, y) = \sqrt{xy}$ at $P(2, 8)$ in the direction of $Q(5, 4)$.

20. Find the directional derivative of $f(x, y, z) = xy^2 z^3$ at $P(2, 1, 1)$ in the direction of $Q(0, -3, 5)$.

21–26 Find the maximum rate of change of f at the given point and the direction in which it occurs.

21. $f(x, y) = 4y\sqrt{x}$, $(4, 1)$

22. $f(s, t) = te^{st}$, $(0, 2)$

23. $f(x, y) = \sin(xy)$, $(1, 0)$

24. $f(x, y, z) = x \ln(yz)$, $\left(1, 2, \frac{1}{2}\right)$

25. $f(x, y, z) = x/(y + z)$, $(8, 1, 3)$

26. $f(p, q, r) = \arctan(pqr)$, $(1, 2, 1)$

---

27. (a) Show that a differentiable function f decreases most rapidly at $\mathbf{x}$ in the direction opposite to the gradient vector, that is, in the direction of $-\nabla f(\mathbf{x})$.
(b) Use the result of part (a) to find the direction in which the function $f(x, y) = x^4 y - x^2 y^3$ decreases fastest at the point $(2, -3)$.

28. Find the directions in which the directional derivative of $f(x, y) = x^2 + xy^3$ at the point $(2, 1)$ has the value 2.

29. Find all points at which the direction of fastest change of the function $f(x, y) = x^2 + y^2 - 2x - 4y$ is $\mathbf{i} + \mathbf{j}$.

30. Near a buoy, the depth of a lake at the point with coordinates (x, y) is $z = 200 + 0.02x^2 - 0.001y^3$, where x, y, and z are measured in meters. A fisherman in a small boat starts at the point $(80, 60)$ and moves toward the buoy, which is located at $(0, 0)$. Is the water under the boat getting deeper or shallower when he departs? Explain.

31. The temperature T in a metal ball is inversely proportional to the distance from the center of the ball, which we take to be the origin. The temperature at the point $(1, 2, 2)$ is $120°$.
(a) Find the rate of change of T at $(1, 2, 2)$ in the direction toward the point $(2, 1, 3)$.
(b) Show that at any point in the ball the direction of greatest increase in temperature is given by a vector that points toward the origin.

32. The temperature at a point (x, y, z) is given by

$$T(x, y, z) = 200e^{-x^2 - 3y^2 - 9z^2}$$

where T is measured in °C and x, y, z in meters.
(a) Find the rate of change of temperature at the point $P(2, -1, 2)$ in the direction toward the point $(3, -3, 3)$.
(b) In which direction does the temperature increase fastest at P?
(c) Find the maximum rate of increase at P.

33. Suppose that over a certain region of space the electrical potential V is given by $V(x, y, z) = 5x^2 - 3xy + xyz$.
(a) Find the rate of change of the potential at $P(3, 4, 5)$ in the direction of the vector $\mathbf{v} = \mathbf{i} + \mathbf{j} - \mathbf{k}$.
(b) In which direction does V change most rapidly at P?
(c) What is the maximum rate of change at P?

34. Suppose you are climbing a hill whose shape is given by the equation $z = 1000 - 0.005x^2 - 0.01y^2$, where x, y, and z are measured in meters, and you are standing at a point with coordinates $(60, 40, 966)$. The positive x-axis points east and the positive y-axis points north.
(a) If you walk due south, will you start to ascend or descend? At what rate?

(b) If you walk northwest, will you start to ascend or descend? At what rate?

(c) In which direction is the slope largest? What is the rate of ascent in that direction? At what angle above the horizontal does the path in that direction begin?

35. Let f be a function of two variables that has continuous partial derivatives and consider the points $A(1, 3)$, $B(3, 3)$, $C(1, 7)$, and $D(6, 15)$. The directional derivative of f at A in the direction of the vector $\overrightarrow{AB}$ is 3 and the directional derivative at A in the direction of $\overrightarrow{AC}$ is 26. Find the directional derivative of f at A in the direction of the vector $\overrightarrow{AD}$.

36. Shown is a topographic map of Blue River Pine Provincial Park in British Columbia. Draw curves of steepest descent from point A (descending to Mud Lake) and from point B.

37. Show that the operation of taking the gradient of a function has the given property. Assume that u and v are differentiable functions of x and y and that a, b are constants.

(a) $\nabla(au + bv) = a\,\nabla u + b\,\nabla v$

(b) $\nabla(uv) = u\,\nabla v + v\,\nabla u$

(c) $\nabla\left(\dfrac{u}{v}\right) = \dfrac{v\,\nabla u - u\,\nabla v}{v^2}$ (d) $\nabla u^n = nu^{n-1}\,\nabla u$

38. Sketch the gradient vector $\nabla f(4, 6)$ for the function f whose level curves are shown. Explain how you chose the direction and length of this vector.

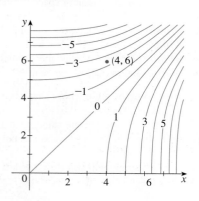

39. The **second directional derivative** of $f(x, y)$ is

$$D_{\mathbf{u}}^2 f(x, y) = D_{\mathbf{u}}[D_{\mathbf{u}} f(x, y)]$$

If $f(x, y) = x^3 + 5x^2 y + y^3$ and $\mathbf{u} = \left\langle \frac{3}{5}, \frac{4}{5} \right\rangle$, calculate $D_{\mathbf{u}}^2 f(2, 1)$.

40. (a) If $\mathbf{u} = \langle a, b \rangle$ is a unit vector and f has continuous second partial derivatives, show that

$$D_{\mathbf{u}}^2 f = f_{xx} a^2 + 2f_{xy} ab + f_{yy} b^2$$

(b) Find the second directional derivative of $f(x, y) = xe^{2y}$ in the direction of $\mathbf{v} = \langle 4, 6 \rangle$.

41–46 Find equations of (a) the tangent plane and (b) the normal line to the given surface at the specified point.

41. $2(x - 2)^2 + (y - 1)^2 + (z - 3)^2 = 10$, $(3, 3, 5)$

42. $x = y^2 + z^2 + 1$, $(3, 1, -1)$

43. $xy^2 z^3 = 8$, $(2, 2, 1)$

44. $xy + yz + zx = 5$, $(1, 2, 1)$

45. $x + y + z = e^{xyz}$, $(0, 0, 1)$

46. $x^4 + y^4 + z^4 = 3x^2 y^2 z^2$, $(1, 1, 1)$

47–48 Use a computer to graph the surface, the tangent plane, and the normal line on the same screen. Choose the domain carefully so that you avoid extraneous vertical planes. Choose the viewpoint so that you get a good view of all three objects.

47. $xy + yz + zx = 3$, $(1, 1, 1)$ **48.** $xyz = 6$, $(1, 2, 3)$

49. If $f(x, y) = xy$, find the gradient vector $\nabla f(3, 2)$ and use it to find the tangent line to the level curve $f(x, y) = 6$ at the point $(3, 2)$. Sketch the level curve, the tangent line, and the gradient vector.

50. If $g(x, y) = x^2 + y^2 - 4x$, find the gradient vector $\nabla g(1, 2)$ and use it to find the tangent line to the level curve $g(x, y) = 1$ at the point $(1, 2)$. Sketch the level curve, the tangent line, and the gradient vector.

51. Show that the equation of the tangent plane to the ellipsoid $x^2/a^2 + y^2/b^2 + z^2/c^2 = 1$ at the point (x_0, y_0, z_0) can be written as

$$\frac{xx_0}{a^2} + \frac{yy_0}{b^2} + \frac{zz_0}{c^2} = 1$$

52. Find the equation of the tangent plane to the hyperboloid $x^2/a^2 + y^2/b^2 - z^2/c^2 = 1$ at (x_0, y_0, z_0) and express it in a form similar to the one in Exercise 51.

53. Show that the equation of the tangent plane to the elliptic paraboloid $z/c = x^2/a^2 + y^2/b^2$ at the point (x_0, y_0, z_0) can be written as

$$\frac{2xx_0}{a^2} + \frac{2yy_0}{b^2} = \frac{z + z_0}{c}$$

54. At what point on the ellipsoid $x^2 + y^2 + 2z^2 = 1$ is the tangent plane parallel to the plane $x + 2y + z = 1$?

55. Are there any points on the hyperboloid $x^2 - y^2 - z^2 = 1$ where the tangent plane is parallel to the plane $z = x + y$?

56. Show that the ellipsoid $3x^2 + 2y^2 + z^2 = 9$ and the sphere $x^2 + y^2 + z^2 - 8x - 6y - 8z + 24 = 0$ are tangent to each other at the point $(1, 1, 2)$. (This means that they have a common tangent plane at the point.)

57. Show that every plane that is tangent to the cone $x^2 + y^2 = z^2$ passes through the origin.

58. Show that every normal line to the sphere $x^2 + y^2 + z^2 = r^2$ passes through the center of the sphere.

59. Where does the normal line to the paraboloid $z = x^2 + y^2$ at the point $(1, 1, 2)$ intersect the paraboloid a second time?

60. At what points does the normal line through the point $(1, 2, 1)$ on the ellipsoid $4x^2 + y^2 + 4z^2 = 12$ intersect the sphere $x^2 + y^2 + z^2 = 102$?

61. Show that the sum of the x-, y-, and z-intercepts of any tangent plane to the surface $\sqrt{x} + \sqrt{y} + \sqrt{z} = \sqrt{c}$ is a constant.

62. Show that the pyramids cut off from the first octant by any tangent planes to the surface $xyz = 1$ at points in the first octant must all have the same volume.

63. Find parametric equations for the tangent line to the curve of intersection of the paraboloid $z = x^2 + y^2$ and the ellipsoid $4x^2 + y^2 + z^2 = 9$ at the point $(-1, 1, 2)$.

64. (a) The plane $y + z = 3$ intersects the cylinder $x^2 + y^2 = 5$ in an ellipse. Find parametric equations for the tangent line to this ellipse at the point $(1, 2, 1)$.
　　(b) Graph the cylinder, the plane, and the tangent line on the same screen.

65. Where does the helix $\mathbf{r}(t) = \langle \cos \pi t, \sin \pi t, t \rangle$ intersect the paraboloid $z = x^2 + y^2$? What is the angle of intersection between the helix and the paraboloid? (This is the angle between the tangent vector to the curve and the tangent plane to the paraboloid.)

66. The helix $\mathbf{r}(t) = \langle \cos(\pi t/2), \sin(\pi t/2), t \rangle$ intersects the sphere $x^2 + y^2 + z^2 = 2$ in two points. Find the angle of intersection at each point.

67. (a) Two surfaces are called **orthogonal** at a point of intersection if their normal lines are perpendicular at that point. Show that surfaces with equations $F(x, y, z) = 0$ and $G(x, y, z) = 0$ are orthogonal at a point P where $\nabla F \neq \mathbf{0}$ and $\nabla G \neq \mathbf{0}$ if and only if
$$F_x G_x + F_y G_y + F_z G_z = 0 \quad \text{at } P$$
　　(b) Use part (a) to show that the surfaces $z^2 = x^2 + y^2$ and $x^2 + y^2 + z^2 = r^2$ are orthogonal at every point of intersection. Can you see why this is true without using calculus?

68. (a) Show that the function $f(x, y) = \sqrt[3]{xy}$ is continuous and the partial derivatives f_x and f_y exist at the origin but the directional derivatives in all other directions do not exist.
　　(b) Graph f near the origin and comment on how the graph confirms part (a).

69. Suppose that the directional derivatives of $f(x, y)$ are known at a given point in two nonparallel directions given by unit vectors $\mathbf{u}$ and $\mathbf{v}$. Is it possible to find ∇f at this point? If so, how would you do it?

70. Show that if $z = f(x, y)$ is differentiable at $\mathbf{x}_0 = \langle x_0, y_0 \rangle$, then
$$\lim_{\mathbf{x} \to \mathbf{x}_0} \frac{f(\mathbf{x}) - f(\mathbf{x}_0) - \nabla f(\mathbf{x}_0) \cdot (\mathbf{x} - \mathbf{x}_0)}{|\mathbf{x} - \mathbf{x}_0|} = 0$$
[*Hint:* Use Definition 14.4.7 directly.]

14.7   Maximum and Minimum Values

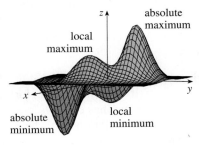

FIGURE 1

As we saw in Chapter 4, one of the main uses of ordinary derivatives is in finding maximum and minimum values (extreme values). In this section we see how to use partial derivatives to locate maxima and minima of functions of two variables. In particular, in Example 6 we will see how to maximize the volume of a box without a lid if we have a fixed amount of cardboard to work with.

Look at the hills and valleys in the graph of f shown in Figure 1. There are two points (a, b) where f has a *local maximum*, that is, where $f(a, b)$ is larger than nearby values of $f(x, y)$. The larger of these two values is the *absolute maximum*. Likewise, f has two *local minima*, where $f(a, b)$ is smaller than nearby values. The smaller of these two values is the *absolute minimum*.

> **1** **Definition** A function of two variables has a **local maximum** at (a, b) if $f(x, y) \leq f(a, b)$ when (x, y) is near (a, b). [This means that $f(x, y) \leq f(a, b)$ for all points (x, y) in some disk with center (a, b).] The number $f(a, b)$ is called a **local maximum value**. If $f(x, y) \geq f(a, b)$ when (x, y) is near (a, b), then f has a **local minimum** at (a, b) and $f(a, b)$ is a **local minimum value**.

If the inequalities in Definition 1 hold for *all* points (x, y) in the domain of f, then f has an **absolute maximum** (or **absolute minimum**) at (a, b).

Notice that the conclusion of Theorem 2 can be stated in the notation of gradient vectors as $\nabla f(a, b) = \mathbf{0}$.

> **2** **Theorem** If f has a local maximum or minimum at (a, b) and the first-order partial derivatives of f exist there, then $f_x(a, b) = 0$ and $f_y(a, b) = 0$.

PROOF Let $g(x) = f(x, b)$. If f has a local maximum (or minimum) at (a, b), then g has a local maximum (or minimum) at a, so $g'(a) = 0$ by Fermat's Theorem (see Theorem 4.1.4). But $g'(a) = f_x(a, b)$ (see Equation 14.3.1) and so $f_x(a, b) = 0$. Similarly, by applying Fermat's Theorem to the function $G(y) = f(a, y)$, we obtain $f_y(a, b) = 0$. ∎

If we put $f_x(a, b) = 0$ and $f_y(a, b) = 0$ in the equation of a tangent plane (Equation 14.4.2), we get $z = z_0$. Thus the geometric interpretation of Theorem 2 is that if the graph of f has a tangent plane at a local maximum or minimum, then the tangent plane must be horizontal.

A point (a, b) is called a **critical point** (or *stationary point*) of f if $f_x(a, b) = 0$ and $f_y(a, b) = 0$, or if one of these partial derivatives does not exist. Theorem 2 says that if f has a local maximum or minimum at (a, b), then (a, b) is a critical point of f. However, as in single-variable calculus, not all critical points give rise to maxima or minima. At a critical point, a function could have a local maximum or a local minimum or neither.

EXAMPLE 1 Let $f(x, y) = x^2 + y^2 - 2x - 6y + 14$. Then

$$f_x(x, y) = 2x - 2 \qquad f_y(x, y) = 2y - 6$$

These partial derivatives are equal to 0 when $x = 1$ and $y = 3$, so the only critical point is $(1, 3)$. By completing the square, we find that

$$f(x, y) = 4 + (x - 1)^2 + (y - 3)^2$$

Since $(x - 1)^2 \geq 0$ and $(y - 3)^2 \geq 0$, we have $f(x, y) \geq 4$ for all values of x and y. Therefore $f(1, 3) = 4$ is a local minimum, and in fact it is the absolute minimum of f. This can be confirmed geometrically from the graph of f, which is the elliptic paraboloid with vertex $(1, 3, 4)$ shown in Figure 2. ∎

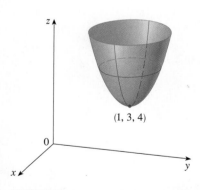

FIGURE 2
$z = x^2 + y^2 - 2x - 6y + 14$

EXAMPLE 2 Find the extreme values of $f(x, y) = y^2 - x^2$.

SOLUTION Since $f_x = -2x$ and $f_y = 2y$, the only critical point is $(0, 0)$. Notice that for points on the x-axis we have $y = 0$, so $f(x, y) = -x^2 < 0$ (if $x \neq 0$). However, for points on the y-axis we have $x = 0$, so $f(x, y) = y^2 > 0$ (if $y \neq 0$). Thus every disk with center $(0, 0)$ contains points where f takes positive values as well as points where f takes negative values. Therefore $f(0, 0) = 0$ can't be an extreme value for f, so f has no extreme value. ∎

FIGURE 3
$z = y^2 - x^2$

Example 2 illustrates the fact that a function need not have a maximum or minimum value at a critical point. Figure 3 shows how this is possible. The graph of f is the hyperbolic paraboloid $z = y^2 - x^2$, which has a horizontal tangent plane ($z = 0$) at the origin. You can see that $f(0, 0) = 0$ is a maximum in the direction of the x-axis but a minimum in the direction of the y-axis. Near the origin the graph has the shape of a saddle and so $(0, 0)$ is called a *saddle point* of f.

A mountain pass also has the shape of a saddle. As the photograph of the geological formation illustrates, for people hiking in one direction the saddle point is the lowest point on their route, while for those traveling in a different direction the saddle point is the highest point.

We need to be able to determine whether or not a function has an extreme value at a critical point. The following test, which is proved at the end of this section, is analogous to the Second Derivative Test for functions of one variable.

---

3 **Second Derivatives Test** Suppose the second partial derivatives of f are continuous on a disk with center (a, b), and suppose that $f_x(a, b) = 0$ and $f_y(a, b) = 0$ [that is, (a, b) is a critical point of f]. Let

$$D = D(a, b) = f_{xx}(a, b)\,f_{yy}(a, b) - [\,f_{xy}(a, b)]^2$$

(a) If $D > 0$ and $f_{xx}(a, b) > 0$, then $f(a, b)$ is a local minimum.

(b) If $D > 0$ and $f_{xx}(a, b) < 0$, then $f(a, b)$ is a local maximum.

(c) If $D < 0$, then $f(a, b)$ is not a local maximum or minimum.

NOTE 1 In case (c) the point (a, b) is called a **saddle point** of f and the graph of f crosses its tangent plane at (a, b).

NOTE 2 If $D = 0$, the test gives no information: f could have a local maximum or local minimum at (a, b), or (a, b) could be a saddle point of f.

NOTE 3 To remember the formula for D, it's helpful to write it as a determinant:

$$D = \begin{vmatrix} f_{xx} & f_{xy} \\ f_{yx} & f_{yy} \end{vmatrix} = f_{xx}f_{yy} - (f_{xy})^2$$

EXAMPLE 3 Find the local maximum and minimum values and saddle points of $f(x, y) = x^4 + y^4 - 4xy + 1$.

SOLUTION We first locate the critical points:

$$f_x = 4x^3 - 4y \qquad f_y = 4y^3 - 4x$$

Setting these partial derivatives equal to 0, we obtain the equations

$$x^3 - y = 0 \qquad \text{and} \qquad y^3 - x = 0$$

To solve these equations we substitute $y = x^3$ from the first equation into the second one. This gives

$$0 = x^9 - x = x(x^8 - 1) = x(x^4 - 1)(x^4 + 1) = x(x^2 - 1)(x^2 + 1)(x^4 + 1)$$

so there are three real roots: $x = 0, 1, -1$. The three critical points are $(0, 0)$, $(1, 1)$, and $(-1, -1)$.

FIGURE 4
$z = x^4 + y^4 - 4xy + 1$

Next we calculate the second partial derivatives and $D(x, y)$:

$$f_{xx} = 12x^2 \qquad f_{xy} = -4 \qquad f_{yy} = 12y^2$$

$$D(x, y) = f_{xx}f_{yy} - (f_{xy})^2 = 144x^2y^2 - 16$$

Since $D(0, 0) = -16 < 0$, it follows from case (c) of the Second Derivatives Test that the origin is a saddle point; that is, f has no local maximum or minimum at $(0, 0)$. Since $D(1, 1) = 128 > 0$ and $f_{xx}(1, 1) = 12 > 0$, we see from case (a) of the test that $f(1, 1) = -1$ is a local minimum. Similarly, we have $D(-1, -1) = 128 > 0$ and $f_{xx}(-1, -1) = 12 > 0$, so $f(-1, -1) = -1$ is also a local minimum.

The graph of f is shown in Figure 4.   ∎

A contour map of the function f in Example 3 is shown in Figure 5. The level curves near $(1, 1)$ and $(-1, -1)$ are oval in shape and indicate that as we move away from $(1, 1)$ or $(-1, -1)$ in any direction the values of f are increasing. The level curves near $(0, 0)$, on the other hand, resemble hyperbolas. They reveal that as we move away from the origin (where the value of f is 1), the values of f decrease in some directions but increase in other directions. Thus the contour map suggests the presence of the minima and saddle point that we found in Example 3.

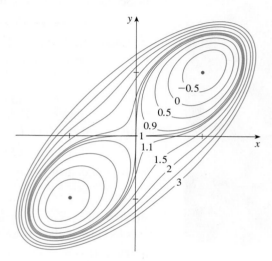

FIGURE 5

TEC In Module 14.7 you can use contour maps to estimate the locations of critical points.

EXAMPLE 4 Find and classify the critical points of the function

$$f(x, y) = 10x^2y - 5x^2 - 4y^2 - x^4 - 2y^4$$

Also find the highest point on the graph of f.

SOLUTION The first-order partial derivatives are

$$f_x = 20xy - 10x - 4x^3 \qquad f_y = 10x^2 - 8y - 8y^3$$

So to find the critical points we need to solve the equations

4                $$2x(10y - 5 - 2x^2) = 0$$

5                $$5x^2 - 4y - 4y^3 = 0$$

From Equation 4 we see that either

$$x = 0 \quad \text{or} \quad 10y - 5 - 2x^2 = 0$$

In the first case $(x = 0)$, Equation 5 becomes $-4y(1 + y^2) = 0$, so $y = 0$ and we have the critical point $(0, 0)$.

In the second case ($10y - 5 - 2x^2 = 0$), we get

$$[6] \qquad\qquad x^2 = 5y - 2.5$$

and, putting this in Equation 5, we have $25y - 12.5 - 4y - 4y^3 = 0$. So we have to solve the cubic equation

$$[7] \qquad\qquad 4y^3 - 21y + 12.5 = 0$$

Using a graphing calculator or computer to graph the function

$$g(y) = 4y^3 - 21y + 12.5$$

FIGURE 6

as in Figure 6, we see that Equation 7 has three real roots. By zooming in, we can find the roots to four decimal places:

$$y \approx -2.5452 \qquad y \approx 0.6468 \qquad y \approx 1.8984$$

(Alternatively, we could have used Newton's method or solved numerically using a calculator or computer to locate these roots.) From Equation 6, the corresponding x-values are given by

$$x = \pm\sqrt{5y - 2.5}$$

If $y \approx -2.5452$, then x has no corresponding real values. If $y \approx 0.6468$, then $x \approx \pm 0.8567$. If $y \approx 1.8984$, then $x \approx \pm 2.6442$. So we have a total of five critical points, which are analyzed in the following chart. All quantities are rounded to two decimal places.

Critical point	Value of f	f_{xx}	D	Conclusion
$(0, 0)$	0.00	-10.00	80.00	local maximum
$(\pm 2.64, 1.90)$	8.50	-55.93	2488.72	local maximum
$(\pm 0.86, 0.65)$	-1.48	-5.87	-187.64	saddle point

Figures 7 and 8 give two views of the graph of f and we see that the surface opens downward. [This can also be seen from the expression for $f(x, y)$: the dominant terms are $-x^4 - 2y^4$ when $|x|$ and $|y|$ are large.] Comparing the values of f at its local maximum points, we see that the absolute maximum value of f is $f(\pm 2.64, 1.90) \approx 8.50$. In other words, the highest points on the graph of f are $(\pm 2.64, 1.90, 8.50)$.

TEC Visual 14.7 shows several families of surfaces. The surface in Figures 7 and 8 is a member of one of these families.

FIGURE 7   **FIGURE 8**

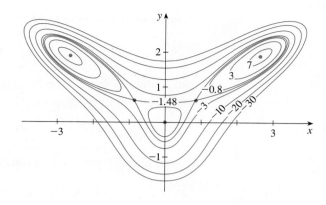

The five critical points of the function f in Example 4 are shown in red in the contour map of f in Figure 9.

FIGURE 9

EXAMPLE 5 Find the shortest distance from the point $(1, 0, -2)$ to the plane $x + 2y + z = 4$.

SOLUTION The distance from any point (x, y, z) to the point $(1, 0, -2)$ is

$$d = \sqrt{(x-1)^2 + y^2 + (z+2)^2}$$

but if (x, y, z) lies on the plane $x + 2y + z = 4$, then $z = 4 - x - 2y$ and so we have $d = \sqrt{(x-1)^2 + y^2 + (6-x-2y)^2}$. We can minimize d by minimizing the simpler expression

$$d^2 = f(x, y) = (x-1)^2 + y^2 + (6-x-2y)^2$$

By solving the equations

$$f_x = 2(x-1) - 2(6-x-2y) = 4x + 4y - 14 = 0$$

$$f_y = 2y - 4(6-x-2y) = 4x + 10y - 24 = 0$$

we find that the only critical point is $\left(\frac{11}{6}, \frac{5}{3}\right)$. Since $f_{xx} = 4$, $f_{xy} = 4$, and $f_{yy} = 10$, we have $D(x, y) = f_{xx}f_{yy} - (f_{xy})^2 = 24 > 0$ and $f_{xx} > 0$, so by the Second Derivatives Test f has a local minimum at $\left(\frac{11}{6}, \frac{5}{3}\right)$. Intuitively, we can see that this local minimum is actually an absolute minimum because there must be a point on the given plane that is closest to $(1, 0, -2)$. If $x = \frac{11}{6}$ and $y = \frac{5}{3}$, then

$$d = \sqrt{(x-1)^2 + y^2 + (6-x-2y)^2} = \sqrt{\left(\tfrac{5}{6}\right)^2 + \left(\tfrac{5}{3}\right)^2 + \left(\tfrac{5}{6}\right)^2} = \tfrac{5}{6}\sqrt{6}$$

Example 5 could also be solved using vectors. Compare with the methods of Section 12.5.

The shortest distance from $(1, 0, -2)$ to the plane $x + 2y + z = 4$ is $\frac{5}{6}\sqrt{6}$. ∎

EXAMPLE 6 A rectangular box without a lid is to be made from 12 m² of cardboard. Find the maximum volume of such a box.

SOLUTION Let the length, width, and height of the box (in meters) be x, y, and z, as shown in Figure 10. Then the volume of the box is

$$V = xyz$$

We can express V as a function of just two variables x and y by using the fact that the area of the four sides and the bottom of the box is

FIGURE 10

$$2xz + 2yz + xy = 12$$

Solving this equation for z, we get $z = (12 - xy)/[2(x + y)]$, so the expression for V becomes

$$V = xy \frac{12 - xy}{2(x + y)} = \frac{12xy - x^2 y^2}{2(x + y)}$$

We compute the partial derivatives:

$$\frac{\partial V}{\partial x} = \frac{y^2(12 - 2xy - x^2)}{2(x + y)^2} \qquad \frac{\partial V}{\partial y} = \frac{x^2(12 - 2xy - y^2)}{2(x + y)^2}$$

If V is a maximum, then $\partial V/\partial x = \partial V/\partial y = 0$, but $x = 0$ or $y = 0$ gives $V = 0$, so we must solve the equations

$$12 - 2xy - x^2 = 0 \qquad 12 - 2xy - y^2 = 0$$

These imply that $x^2 = y^2$ and so $x = y$. (Note that x and y must both be positive in this problem.) If we put $x = y$ in either equation we get $12 - 3x^2 = 0$, which gives $x = 2$, $y = 2$, and $z = (12 - 2 \cdot 2)/[2(2 + 2)] = 1$.

We could use the Second Derivatives Test to show that this gives a local maximum of V, or we could simply argue from the physical nature of this problem that there must be an absolute maximum volume, which has to occur at a critical point of V, so it must occur when $x = 2$, $y = 2$, $z = 1$. Then $V = 2 \cdot 2 \cdot 1 = 4$, so the maximum volume of the box is 4 m^3. ∎

■ Absolute Maximum and Minimum Values

For a function f of one variable, the Extreme Value Theorem says that if f is continuous on a closed interval $[a, b]$, then f has an absolute minimum value and an absolute maximum value. According to the Closed Interval Method in Section 4.1, we found these by evaluating f not only at the critical numbers but also at the endpoints a and b.

There is a similar situation for functions of two variables. Just as a closed interval contains its endpoints, a **closed set** in $\mathbb{R}^2$ is one that contains all its boundary points. [A boundary point of D is a point (a, b) such that every disk with center (a, b) contains points in D and also points not in D.] For instance, the disk

$$D = \{(x, y) \mid x^2 + y^2 \leqslant 1\}$$

which consists of all points on or inside the circle $x^2 + y^2 = 1$, is a closed set because it contains all of its boundary points (which are the points on the circle $x^2 + y^2 = 1$). But if even one point on the boundary curve were omitted, the set would not be closed. (See Figure 11.)

A **bounded set** in $\mathbb{R}^2$ is one that is contained within some disk. In other words, it is finite in extent. Then, in terms of closed and bounded sets, we can state the following counterpart of the Extreme Value Theorem in two dimensions.

(a) Closed sets

(b) Sets that are not closed

FIGURE 11

---

8 **Extreme Value Theorem for Functions of Two Variables** If f is continuous on a closed, bounded set D in $\mathbb{R}^2$, then f attains an absolute maximum value $f(x_1, y_1)$ and an absolute minimum value $f(x_2, y_2)$ at some points (x_1, y_1) and (x_2, y_2) in D.

To find the extreme values guaranteed by Theorem 8, we note that, by Theorem 2, if f has an extreme value at (x_1, y_1), then (x_1, y_1) is either a critical point of f or a boundary point of D. Thus we have the following extension of the Closed Interval Method.

> **9** To find the absolute maximum and minimum values of a continuous function f on a closed, bounded set D:
>
> **1.** Find the values of f at the critical points of f in D.
>
> **2.** Find the extreme values of f on the boundary of D.
>
> **3.** The largest of the values from steps 1 and 2 is the absolute maximum value; the smallest of these values is the absolute minimum value.

EXAMPLE 7 Find the absolute maximum and minimum values of the function $f(x, y) = x^2 - 2xy + 2y$ on the rectangle $D = \{(x, y) \mid 0 \leqslant x \leqslant 3, 0 \leqslant y \leqslant 2\}$.

SOLUTION Since f is a polynomial, it is continuous on the closed, bounded rectangle D, so Theorem 8 tells us there is both an absolute maximum and an absolute minimum. According to step 1 in (9), we first find the critical points. These occur when

$$f_x = 2x - 2y = 0 \qquad f_y = -2x + 2 = 0$$

so the only critical point is $(1, 1)$, and the value of f there is $f(1, 1) = 1$.

In step 2 we look at the values of f on the boundary of D, which consists of the four line segments L_1, L_2, L_3, L_4 shown in Figure 12. On L_1 we have $y = 0$ and

$$f(x, 0) = x^2 \qquad 0 \leqslant x \leqslant 3$$

This is an increasing function of x, so its minimum value is $f(0, 0) = 0$ and its maximum value is $f(3, 0) = 9$. On L_2 we have $x = 3$ and

$$f(3, y) = 9 - 4y \qquad 0 \leqslant y \leqslant 2$$

This is a decreasing function of y, so its maximum value is $f(3, 0) = 9$ and its minimum value is $f(3, 2) = 1$. On L_3 we have $y = 2$ and

$$f(x, 2) = x^2 - 4x + 4 \qquad 0 \leqslant x \leqslant 3$$

By the methods of Chapter 4, or simply by observing that $f(x, 2) = (x - 2)^2$, we see that the minimum value of this function is $f(2, 2) = 0$ and the maximum value is $f(0, 2) = 4$. Finally, on L_4 we have $x = 0$ and

$$f(0, y) = 2y \qquad 0 \leqslant y \leqslant 2$$

with maximum value $f(0, 2) = 4$ and minimum value $f(0, 0) = 0$. Thus, on the boundary, the minimum value of f is 0 and the maximum is 9.

In step 3 we compare these values with the value $f(1, 1) = 1$ at the critical point and conclude that the absolute maximum value of f on D is $f(3, 0) = 9$ and the absolute minimum value is $f(0, 0) = f(2, 2) = 0$. Figure 13 shows the graph of f.

FIGURE 12

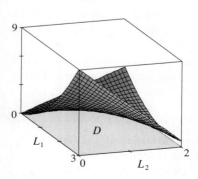

FIGURE 13
$f(x, y) = x^2 - 2xy + 2y$

We close this section by giving a proof of the first part of the Second Derivatives Test. Part (b) has a similar proof.

PROOF OF THEOREM 3, PART (a) We compute the second-order directional derivative of f in the direction of $\mathbf{u} = \langle h, k \rangle$. The first-order derivative is given by Theorem 14.6.3:

$$D_{\mathbf{u}} f = f_x h + f_y k$$

Applying this theorem a second time, we have

$$D_{\mathbf{u}}^2 f = D_{\mathbf{u}}(D_{\mathbf{u}} f) = \frac{\partial}{\partial x} (D_{\mathbf{u}} f) h + \frac{\partial}{\partial y} (D_{\mathbf{u}} f) k$$

$$= (f_{xx} h + f_{yx} k) h + (f_{xy} h + f_{yy} k) k$$

$$= f_{xx} h^2 + 2 f_{xy} hk + f_{yy} k^2 \qquad \text{(by Clairaut's Theorem)}$$

If we complete the square in this expression, we obtain

$$\boxed{10} \qquad D_{\mathbf{u}}^2 f = f_{xx} \left(h + \frac{f_{xy}}{f_{xx}} k \right)^2 + \frac{k^2}{f_{xx}} (f_{xx} f_{yy} - f_{xy}^2)$$

We are given that $f_{xx}(a, b) > 0$ and $D(a, b) > 0$. But f_{xx} and $D = f_{xx} f_{yy} - f_{xy}^2$ are continuous functions, so there is a disk B with center (a, b) and radius $\delta > 0$ such that $f_{xx}(x, y) > 0$ and $D(x, y) > 0$ whenever (x, y) is in B. Therefore, by looking at Equation 10, we see that $D_{\mathbf{u}}^2 f(x, y) > 0$ whenever (x, y) is in B. This means that if C is the curve obtained by intersecting the graph of f with the vertical plane through $P(a, b, f(a, b))$ in the direction of $\mathbf{u}$, then C is concave upward on an interval of length 2δ. This is true in the direction of every vector $\mathbf{u}$, so if we restrict (x, y) to lie in B, the graph of f lies above its horizontal tangent plane at P. Thus $f(x, y) \geq f(a, b)$ whenever (x, y) is in B. This shows that $f(a, b)$ is a local minimum. ∎

14.7 EXERCISES

1. Suppose $(1, 1)$ is a critical point of a function f with continuous second derivatives. In each case, what can you say about f?

(a) $f_{xx}(1, 1) = 4$, $f_{xy}(1, 1) = 1$, $f_{yy}(1, 1) = 2$

(b) $f_{xx}(1, 1) = 4$, $f_{xy}(1, 1) = 3$, $f_{yy}(1, 1) = 2$

2. Suppose $(0, 2)$ is a critical point of a function g with continuous second derivatives. In each case, what can you say about g?

(a) $g_{xx}(0, 2) = -1$, $g_{xy}(0, 2) = 6$, $g_{yy}(0, 2) = 1$

(b) $g_{xx}(0, 2) = -1$, $g_{xy}(0, 2) = 2$, $g_{yy}(0, 2) = -8$

(c) $g_{xx}(0, 2) = 4$, $g_{xy}(0, 2) = 6$, $g_{yy}(0, 2) = 9$

3–4 Use the level curves in the figure to predict the location of the critical points of f and whether f has a saddle point or a local maximum or minimum at each critical point. Explain your reasoning. Then use the Second Derivatives Test to confirm your predictions.

3. $f(x, y) = 4 + x^3 + y^3 - 3xy$

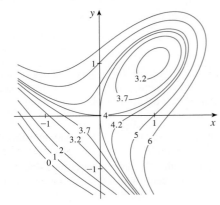

4. $f(x, y) = 3x - x^3 - 2y^2 + y^4$

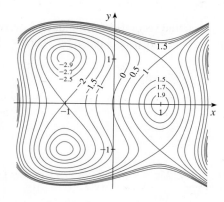

5–20 Find the local maximum and minimum values and saddle point(s) of the function. If you have three-dimensional graphing software, graph the function with a domain and viewpoint that reveal all the important aspects of the function.

5. $f(x, y) = x^2 + xy + y^2 + y$

6. $f(x, y) = xy - 2x - 2y - x^2 - y^2$

7. $f(x, y) = (x - y)(1 - xy)$

8. $f(x, y) = y(e^x - 1)$

9. $f(x, y) = x^2 + y^4 + 2xy$

10. $f(x, y) = 2 - x^4 + 2x^2 - y^2$

11. $f(x, y) = x^3 - 3x + 3xy^2$

12. $f(x, y) = x^3 + y^3 - 3x^2 - 3y^2 - 9x$

13. $f(x, y) = x^4 - 2x^2 + y^3 - 3y$

14. $f(x, y) = y \cos x$

15. $f(x, y) = e^x \cos y$

16. $f(x, y) = xye^{-(x^2+y^2)/2}$

17. $f(x, y) = xy + e^{-xy}$

18. $f(x, y) = (x^2 + y^2)e^{-x}$

19. $f(x, y) = y^2 - 2y \cos x, \quad -1 \le x \le 7$

20. $f(x, y) = \sin x \sin y, \quad -\pi < x < \pi, \quad -\pi < y < \pi$

21. Show that $f(x, y) = x^2 + 4y^2 - 4xy + 2$ has an infinite number of critical points and that $D = 0$ at each one. Then show that f has a local (and absolute) minimum at each critical point.

22. Show that $f(x, y) = x^2ye^{-x^2-y^2}$ has maximum values at $(\pm 1, 1/\sqrt{2})$ and minimum values at $(\pm 1, -1/\sqrt{2})$. Show also that f has infinitely many other critical points and $D = 0$ at each of them. Which of them give rise to maximum values? Minimum values? Saddle points?

23–26 Use a graph or level curves or both to estimate the local maximum and minimum values and saddle point(s) of the function. Then use calculus to find these values precisely.

23. $f(x, y) = x^2 + y^2 + x^{-2}y^{-2}$

24. $f(x, y) = (x - y)e^{-x^2-y^2}$

25. $f(x, y) = \sin x + \sin y + \sin(x + y),$
$0 \le x \le 2\pi, \ 0 \le y \le 2\pi$

26. $f(x, y) = \sin x + \sin y + \cos(x + y),$
$0 \le x \le \pi/4, \ 0 \le y \le \pi/4$

27–30 Use a graphing device as in Example 4 (or Newton's method or solve numerically using a calculator or computer) to find the critical points of f correct to three decimal places. Then classify the critical points and find the highest or lowest points on the graph, if any.

27. $f(x, y) = x^4 + y^4 - 4x^2y + 2y$

28. $f(x, y) = y^6 - 2y^4 + x^2 - y^2 + y$

29. $f(x, y) = x^4 + y^3 - 3x^2 + y^2 + x - 2y + 1$

30. $f(x, y) = 20e^{-x^2-y^2}\sin 3x \cos 3y, \quad |x| \le 1, \quad |y| \le 1$

31–38 Find the absolute maximum and minimum values of f on the set D.

31. $f(x, y) = x^2 + y^2 - 2x,$ D is the closed triangular region with vertices $(2, 0)$, $(0, 2)$, and $(0, -2)$

32. $f(x, y) = x + y - xy,$ D is the closed triangular region with vertices $(0, 0)$, $(0, 2)$, and $(4, 0)$

33. $f(x, y) = x^2 + y^2 + x^2y + 4,$
$D = \{(x, y) \mid |x| \le 1, |y| \le 1\}$

34. $f(x, y) = x^2 + xy + y^2 - 6y,$
$D = \{(x, y) \mid -3 \le x \le 3, 0 \le y \le 5\}$

35. $f(x, y) = x^2 + 2y^2 - 2x - 4y + 1,$
$D = \{(x, y) \mid 0 \le x \le 2, 0 \le y \le 3\}$

36. $f(x, y) = xy^2,$ $D = \{(x, y) \mid x \ge 0, y \ge 0, x^2 + y^2 \le 3\}$

37. $f(x, y) = 2x^3 + y^4,$ $D = \{(x, y) \mid x^2 + y^2 \le 1\}$

38. $f(x, y) = x^3 - 3x - y^3 + 12y,$ D is the quadrilateral whose vertices are $(-2, 3)$, $(2, 3)$, $(2, 2)$, and $(-2, -2)$

39. For functions of one variable it is impossible for a continuous function to have two local maxima and no local minimum. But for functions of two variables such functions exist. Show that the function

$$f(x, y) = -(x^2 - 1)^2 - (x^2y - x - 1)^2$$

has only two critical points, but has local maxima at both of them. Then use a computer to produce a graph with a carefully chosen domain and viewpoint to see how this is possible.

40. If a function of one variable is continuous on an interval and has only one critical number, then a local maximum has to be an absolute maximum. But this is not true for functions of two variables. Show that the function

$$f(x, y) = 3xe^y - x^3 - e^{3y}$$

has exactly one critical point, and that f has a local maximum there that is not an absolute maximum. Then use a computer to produce a graph with a carefully chosen domain and viewpoint to see how this is possible.

41. Find the shortest distance from the point $(2, 0, -3)$ to the plane $x + y + z = 1$.

42. Find the point on the plane $x - 2y + 3z = 6$ that is closest to the point $(0, 1, 1)$.

43. Find the points on the cone $z^2 = x^2 + y^2$ that are closest to the point $(4, 2, 0)$.

44. Find the points on the surface $y^2 = 9 + xz$ that are closest to the origin.

45. Find three positive numbers whose sum is 100 and whose product is a maximum.

46. Find three positive numbers whose sum is 12 and the sum of whose squares is as small as possible.

47. Find the maximum volume of a rectangular box that is inscribed in a sphere of radius r.

48. Find the dimensions of the box with volume 1000 cm^3 that has minimal surface area.

49. Find the volume of the largest rectangular box in the first octant with three faces in the coordinate planes and one vertex in the plane $x + 2y + 3z = 6$.

50. Find the dimensions of the rectangular box with largest volume if the total surface area is given as 64 cm^2.

51. Find the dimensions of a rectangular box of maximum volume such that the sum of the lengths of its 12 edges is a constant c.

52. The base of an aquarium with given volume V is made of slate and the sides are made of glass. If slate costs five times as much (per unit area) as glass, find the dimensions of the aquarium that minimize the cost of the materials.

53. A cardboard box without a lid is to have a volume of $32{,}000 \text{ cm}^3$. Find the dimensions that minimize the amount of cardboard used.

54. A rectangular building is being designed to minimize heat loss. The east and west walls lose heat at a rate of 10 units/m² per day, the north and south walls at a rate of 8 units/m² per day, the floor at a rate of 1 unit/m² per day, and the roof at a rate of 5 units/m² per day. Each wall must be at least 30 m long, the height must be at least 4 m, and the volume must be exactly 4000 m^3.
   (a) Find and sketch the domain of the heat loss as a function of the lengths of the sides.

(b) Find the dimensions that minimize heat loss. (Check both the critical points and the points on the boundary of the domain.)
   (c) Could you design a building with even less heat loss if the restrictions on the lengths of the walls were removed?

55. If the length of the diagonal of a rectangular box must be L, what is the largest possible volume?

56. A model for the yield Y of an agricultural crop as a function of the nitrogen level N and phosphorus level P in the soil (measured in appropriate units) is

$$Y(N, P) = kNPe^{-N-P}$$

where k is a positive constant. What levels of nitrogen and phosphorus result in the best yield?

57. The Shannon index (sometimes called the Shannon-Wiener index or Shannon-Weaver index) is a measure of diversity in an ecosystem. For the case of three species, it is defined as

$$H = -p_1 \ln p_1 - p_2 \ln p_2 - p_3 \ln p_3$$

where p_i is the proportion of species i in the ecosystem.
   (a) Express H as a function of two variables using the fact that $p_1 + p_2 + p_3 = 1$.
   (b) What is the domain of H?
   (c) Find the maximum value of H. For what values of p_1, p_2, p_3 does it occur?

58. Three alleles (alternative versions of a gene) A, B, and O determine the four blood types A (AA or AO), B (BB or BO), O (OO), and AB. The Hardy-Weinberg Law states that the proportion of individuals in a population who carry two different alleles is

$$P = 2pq + 2pr + 2rq$$

where p, q, and r are the proportions of A, B, and O in the population. Use the fact that $p + q + r = 1$ to show that P is at most $\frac{2}{3}$.

59. Suppose that a scientist has reason to believe that two quantities x and y are related linearly, that is, $y = mx + b$, at least approximately, for some values of m and b. The scientist performs an experiment and collects data in the form of points $(x_1, y_1), (x_2, y_2), \ldots, (x_n, y_n)$, and then plots these points. The points don't lie exactly on a straight line, so the scientist wants to find constants m and b so that the line $y = mx + b$ "fits" the points as well as possible (see the figure).

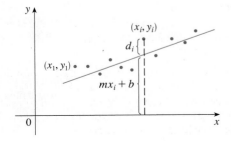

Let $d_i = y_i - (mx_i + b)$ be the vertical deviation of the point (x_i, y_i) from the line. The **method of least squares** determines m and b so as to minimize $\sum_{i=1}^{n} d_i^2$, the sum of the squares of these deviations. Show that, according to this method, the line of best fit is obtained when

$$m \sum_{i=1}^{n} x_i + bn = \sum_{i=1}^{n} y_i$$

and

$$m \sum_{i=1}^{n} x_i^2 + b \sum_{i=1}^{n} x_i = \sum_{i=1}^{n} x_i y_i$$

Thus the line is found by solving these two equations in the two unknowns m and b. (See Section 1.2 for a further discussion and applications of the method of least squares.)

60. Find an equation of the plane that passes through the point $(1, 2, 3)$ and cuts off the smallest volume in the first octant.

APPLIED PROJECT   DESIGNING A DUMPSTER

For this project we locate a rectangular trash Dumpster in order to study its shape and construction. We then attempt to determine the dimensions of a container of similar design that minimize construction cost.

1. First locate a trash Dumpster in your area. Carefully study and describe all details of its construction, and determine its volume. Include a sketch of the container.

2. While maintaining the general shape and method of construction, determine the dimensions such a container of the same volume should have in order to minimize the cost of construction. Use the following assumptions in your analysis:

   • The sides, back, and front are to be made from 12-gauge (0.1046 inch thick) steel sheets, which cost $0.70 per square foot (including any required cuts or bends).

   • The base is to be made from a 10-gauge (0.1345 inch thick) steel sheet, which costs $0.90 per square foot.

   • Lids cost approximately $50.00 each, regardless of dimensions.

   • Welding costs approximately $0.18 per foot for material and labor combined.

   Give justification of any further assumptions or simplifications made of the details of construction.

3. Describe how any of your assumptions or simplifications may affect the final result.

4. If you were hired as a consultant on this investigation, what would your conclusions be? Would you recommend altering the design of the Dumpster? If so, describe the savings that would result.

DISCOVERY PROJECT   QUADRATIC APPROXIMATIONS AND CRITICAL POINTS

The Taylor polynomial approximation to functions of one variable that we discussed in Chapter 11 can be extended to functions of two or more variables. Here we investigate quadratic approximations to functions of two variables and use them to give insight into the Second Derivatives Test for classifying critical points.

In Section 14.4 we discussed the linearization of a function f of two variables at a point (a, b):

$$L(x, y) = f(a, b) + f_x(a, b)(x - a) + f_y(a, b)(y - b)$$

Recall that the graph of L is the tangent plane to the surface $z = f(x, y)$ at $(a, b, f(a, b))$ and the corresponding linear approximation is $f(x, y) \approx L(x, y)$. The linearization L is also called the **first-degree Taylor polynomial** of f at (a, b).

1. If f has continuous second-order partial derivatives at (a, b), then the **second-degree Taylor polynomial** of f at (a, b) is

$$Q(x, y) = f(a, b) + f_x(a, b)(x - a) + f_y(a, b)(y - b)$$
$$+ \tfrac{1}{2} f_{xx}(a, b)(x - a)^2 + f_{xy}(a, b)(x - a)(y - b) + \tfrac{1}{2} f_{yy}(a, b)(y - b)^2$$

and the approximation $f(x, y) \approx Q(x, y)$ is called the **quadratic approximation** to f at (a, b). Verify that Q has the same first- and second-order partial derivatives as f at (a, b).

2. (a) Find the first- and second-degree Taylor polynomials L and Q of $f(x, y) = e^{-x^2 - y^2}$ at $(0, 0)$.

(b) Graph f, L, and Q. Comment on how well L and Q approximate f.

3. (a) Find the first- and second-degree Taylor polynomials L and Q for $f(x, y) = xe^y$ at $(1, 0)$.

(b) Compare the values of L, Q, and f at $(0.9, 0.1)$.

(c) Graph f, L, and Q. Comment on how well L and Q approximate f.

4. In this problem we analyze the behavior of the polynomial $f(x, y) = ax^2 + bxy + cy^2$ (without using the Second Derivatives Test) by identifying the graph as a paraboloid.

(a) By completing the square, show that if $a \neq 0$, then

$$f(x, y) = ax^2 + bxy + cy^2 = a\left[\left(x + \frac{b}{2a} y \right)^2 + \left(\frac{4ac - b^2}{4a^2} \right) y^2 \right]$$

(b) Let $D = 4ac - b^2$. Show that if $D > 0$ and $a > 0$, then f has a local minimum at $(0, 0)$.

(c) Show that if $D > 0$ and $a < 0$, then f has a local maximum at $(0, 0)$.

(d) Show that if $D < 0$, then $(0, 0)$ is a saddle point.

5. (a) Suppose f is any function with continuous second-order partial derivatives such that $f(0, 0) = 0$ and $(0, 0)$ is a critical point of f. Write an expression for the second-degree Taylor polynomial, Q, of f at $(0, 0)$.

(b) What can you conclude about Q from Problem 4?

(c) In view of the quadratic approximation $f(x, y) \approx Q(x, y)$, what does part (b) suggest about f?

14.8   Lagrange Multipliers

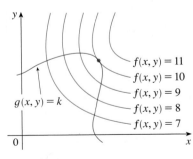

FIGURE 1

TEC Visual 14.8 animates Figure 1 for both level curves and level surfaces.

In Example 14.7.6 we maximized a volume function $V = xyz$ subject to the constraint $2xz + 2yz + xy = 12$, which expressed the side condition that the surface area was 12 m². In this section we present Lagrange's method for maximizing or minimizing a general function $f(x, y, z)$ subject to a constraint (or side condition) of the form $g(x, y, z) = k$.

It's easier to explain the geometric basis of Lagrange's method for functions of two variables. So we start by trying to find the extreme values of $f(x, y)$ subject to a constraint of the form $g(x, y) = k$. In other words, we seek the extreme values of $f(x, y)$ when the point (x, y) is restricted to lie on the level curve $g(x, y) = k$. Figure 1 shows this curve together with several level curves of f. These have the equations $f(x, y) = c$, where $c = 7, 8, 9, 10, 11$. To maximize $f(x, y)$ subject to $g(x, y) = k$ is to find the largest value of c such that the level curve $f(x, y) = c$ intersects $g(x, y) = k$. It appears from Figure 1 that this happens when these curves just touch each other, that is, when they have a common tangent line. (Otherwise, the value of c could be increased further.) This means that the normal lines at the point (x_0, y_0) where they touch are identical. So the gradient vectors are parallel; that is, $\nabla f(x_0, y_0) = \lambda \nabla g(x_0, y_0)$ for some scalar λ.

This kind of argument also applies to the problem of finding the extreme values of $f(x, y, z)$ subject to the constraint $g(x, y, z) = k$. Thus the point (x, y, z) is restricted to lie on the level surface S with equation $g(x, y, z) = k$. Instead of the level curves in Figure 1, we consider the level surfaces $f(x, y, z) = c$ and argue that if the maximum value of f is $f(x_0, y_0, z_0) = c$, then the level surface $f(x, y, z) = c$ is tangent to the level surface $g(x, y, z) = k$ and so the corresponding gradient vectors are parallel.

This intuitive argument can be made precise as follows. Suppose that a function f has an extreme value at a point $P(x_0, y_0, z_0)$ on the surface S and let C be a curve with vector equation $\mathbf{r}(t) = \langle x(t), y(t), z(t) \rangle$ that lies on S and passes through P. If t_0 is the parameter value corresponding to the point P, then $\mathbf{r}(t_0) = \langle x_0, y_0, z_0 \rangle$. The composite function $h(t) = f(x(t), y(t), z(t))$ represents the values that f takes on the curve C. Since f has an extreme value at (x_0, y_0, z_0), it follows that h has an extreme value at t_0, so $h'(t_0) = 0$. But if f is differentiable, we can use the Chain Rule to write

$$\begin{aligned} 0 &= h'(t_0) \\ &= f_x(x_0, y_0, z_0)x'(t_0) + f_y(x_0, y_0, z_0)y'(t_0) + f_z(x_0, y_0, z_0)z'(t_0) \\ &= \nabla f(x_0, y_0, z_0) \cdot \mathbf{r}'(t_0) \end{aligned}$$

This shows that the gradient vector $\nabla f(x_0, y_0, z_0)$ is orthogonal to the tangent vector $\mathbf{r}'(t_0)$ to every such curve C. But we already know from Section 14.6 that the gradient vector of g, $\nabla g(x_0, y_0, z_0)$, is also orthogonal to $\mathbf{r}'(t_0)$ for every such curve. (See Equation 14.6.18.) This means that the gradient vectors $\nabla f(x_0, y_0, z_0)$ and $\nabla g(x_0, y_0, z_0)$ must be parallel. Therefore, if $\nabla g(x_0, y_0, z_0) \neq \mathbf{0}$, there is a number λ such that

1
$$\nabla f(x_0, y_0, z_0) = \lambda \nabla g(x_0, y_0, z_0)$$

The number λ in Equation 1 is called a **Lagrange multiplier**. The procedure based on Equation 1 is as follows.

Lagrange multipliers are named after the French-Italian mathematician Joseph-Louis Lagrange (1736–1813). See page 289 for a biographical sketch of Lagrange.

In deriving Lagrange's method we assumed that $\nabla g \neq \mathbf{0}$. In each of our examples you can check that $\nabla g \neq \mathbf{0}$ at all points where $g(x, y, z) = k$. See Exercise 25 for what can go wrong if $\nabla g = \mathbf{0}$.

Method of Lagrange Multipliers To find the maximum and minimum values of $f(x, y, z)$ subject to the constraint $g(x, y, z) = k$ [assuming that these extreme values exist and $\nabla g \neq \mathbf{0}$ on the surface $g(x, y, z) = k$]:

(a) Find all values of x, y, z, and λ such that

$$\nabla f(x, y, z) = \lambda \nabla g(x, y, z)$$

and

$$g(x, y, z) = k$$

(b) Evaluate f at all the points (x, y, z) that result from step (a). The largest of these values is the maximum value of f; the smallest is the minimum value of f.

If we write the vector equation $\nabla f = \lambda \nabla g$ in terms of components, then the equations in step (a) become

$$f_x = \lambda g_x \qquad f_y = \lambda g_y \qquad f_z = \lambda g_z \qquad g(x, y, z) = k$$

This is a system of four equations in the four unknowns x, y, z, and λ, but it is not necessary to find explicit values for λ.

For functions of two variables the method of Lagrange multipliers is similar to the method just described. To find the extreme values of $f(x, y)$ subject to the constraint $g(x, y) = k$, we look for values of x, y, and λ such that

$$\nabla f(x, y) = \lambda\, \nabla g(x, y) \qquad \text{and} \qquad g(x, y) = k$$

This amounts to solving three equations in three unknowns:

$$f_x = \lambda g_x \qquad f_y = \lambda g_y \qquad g(x, y) = k$$

Our first illustration of Lagrange's method is to reconsider the problem given in Example 14.7.6.

EXAMPLE 1 A rectangular box without a lid is to be made from 12 m^2 of cardboard. Find the maximum volume of such a box.

SOLUTION As in Example 14.7.6, we let x, y, and z be the length, width, and height, respectively, of the box in meters. Then we wish to maximize

$$V = xyz$$

subject to the constraint

$$g(x, y, z) = 2xz + 2yz + xy = 12$$

Using the method of Lagrange multipliers, we look for values of x, y, z, and λ such that $\nabla V = \lambda\, \nabla g$ and $g(x, y, z) = 12$. This gives the equations

$$V_x = \lambda g_x$$
$$V_y = \lambda g_y$$
$$V_z = \lambda g_z$$
$$2xz + 2yz + xy = 12$$

which become

$$\boxed{2} \qquad\qquad yz = \lambda(2z + y)$$

$$\boxed{3} \qquad\qquad xz = \lambda(2z + x)$$

$$\boxed{4} \qquad\qquad xy = \lambda(2x + 2y)$$

$$\boxed{5} \qquad\qquad 2xz + 2yz + xy = 12$$

There are no general rules for solving systems of equations. Sometimes some ingenuity is required. In the present example you might notice that if we multiply (2) by x, (3) by y, and (4) by z, then the left sides of these equations will be identical. Doing this, we have

$$\boxed{6} \qquad\qquad xyz = \lambda(2xz + xy)$$

$$\boxed{7} \qquad\qquad xyz = \lambda(2yz + xy)$$

$$\boxed{8} \qquad\qquad xyz = \lambda(2xz + 2yz)$$

Another method for solving the system of equations (2–5) is to solve each of Equations 2, 3, and 4 for λ and then to equate the resulting expressions.

We observe that $\lambda \neq 0$ because $\lambda = 0$ would imply $yz = xz = xy = 0$ from (2), (3), and (4) and this would contradict (5). Therefore, from (6) and (7), we have

$$2xz + xy = 2yz + xy$$

which gives $xz = yz$. But $z \neq 0$ (since $z = 0$ would give $V = 0$), so $x = y$. From (7) and (8) we have

$$2yz + xy = 2xz + 2yz$$

which gives $2xz = xy$ and so (since $x \neq 0$) $y = 2z$. If we now put $x = y = 2z$ in (5), we get

$$4z^2 + 4z^2 + 4z^2 = 12$$

Since x, y, and z are all positive, we therefore have $z = 1$ and so $x = 2$ and $y = 2$. This agrees with our answer in Section 14.7.   ∎

EXAMPLE 2 Find the extreme values of the function $f(x, y) = x^2 + 2y^2$ on the circle $x^2 + y^2 = 1$.

SOLUTION We are asked for the extreme values of f subject to the constraint $g(x, y) = x^2 + y^2 = 1$. Using Lagrange multipliers, we solve the equations $\nabla f = \lambda \nabla g$ and $g(x, y) = 1$, which can be written as

$$f_x = \lambda g_x \qquad f_y = \lambda g_y \qquad g(x, y) = 1$$

or as

9 $\qquad 2x = 2x\lambda$

10 $\qquad 4y = 2y\lambda$

11 $\qquad x^2 + y^2 = 1$

From (9) we have $x = 0$ or $\lambda = 1$. If $x = 0$, then (11) gives $y = \pm 1$. If $\lambda = 1$, then $y = 0$ from (10), so then (11) gives $x = \pm 1$. Therefore f has possible extreme values at the points $(0, 1)$, $(0, -1)$, $(1, 0)$, and $(-1, 0)$. Evaluating f at these four points, we find that

$$f(0, 1) = 2 \qquad f(0, -1) = 2 \qquad f(1, 0) = 1 \qquad f(-1, 0) = 1$$

Therefore the maximum value of f on the circle $x^2 + y^2 = 1$ is $f(0, \pm 1) = 2$ and the minimum value is $f(\pm 1, 0) = 1$. In geometric terms, these correspond to the highest and lowest points on the curve C in Figure 2, where C consists of those points on the paraboloid $z = x^2 + 2y^2$ that are directly above the constraint circle $x^2 + y^2 = 1$.   ∎

EXAMPLE 3 Find the extreme values of $f(x, y) = x^2 + 2y^2$ on the disk $x^2 + y^2 \leq 1$.

SOLUTION According to the procedure in (14.7.9), we compare the values of f at the critical points with values at the points on the boundary. Since $f_x = 2x$ and $f_y = 4y$, the only critical point is $(0, 0)$. We compare the value of f at that point with the extreme values on the boundary from Example 2:

$$f(0, 0) = 0 \qquad f(\pm 1, 0) = 1 \qquad f(0, \pm 1) = 2$$

Therefore the maximum value of f on the disk $x^2 + y^2 \leq 1$ is $f(0, \pm 1) = 2$ and the minimum value is $f(0, 0) = 0$.   ∎

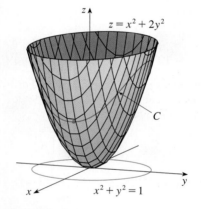

FIGURE 2

The geometry behind the use of Lagrange multipliers in Example 2 is shown in Figure 3. The extreme values of $f(x, y) = x^2 + 2y^2$ correspond to the level curves that touch the circle $x^2 + y^2 = 1$.

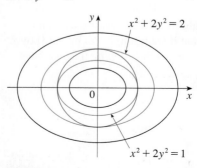

FIGURE 3

EXAMPLE 4 Find the points on the sphere $x^2 + y^2 + z^2 = 4$ that are closest to and farthest from the point $(3, 1, -1)$.

SOLUTION The distance from a point (x, y, z) to the point $(3, 1, -1)$ is

$$d = \sqrt{(x-3)^2 + (y-1)^2 + (z+1)^2}$$

but the algebra is simpler if we instead maximize and minimize the square of the distance:

$$d^2 = f(x, y, z) = (x-3)^2 + (y-1)^2 + (z+1)^2$$

The constraint is that the point (x, y, z) lies on the sphere, that is,

$$g(x, y, z) = x^2 + y^2 + z^2 = 4$$

According to the method of Lagrange multipliers, we solve $\nabla f = \lambda \, \nabla g$, $g = 4$. This gives

12          $$2(x - 3) = 2x\lambda$$

13          $$2(y - 1) = 2y\lambda$$

14          $$2(z + 1) = 2z\lambda$$

15          $$x^2 + y^2 + z^2 = 4$$

The simplest way to solve these equations is to solve for x, y, and z in terms of λ from (12), (13), and (14), and then substitute these values into (15). From (12) we have

$$x - 3 = x\lambda \quad \text{or} \quad x(1 - \lambda) = 3 \quad \text{or} \quad x = \frac{3}{1 - \lambda}$$

[Note that $1 - \lambda \neq 0$ because $\lambda = 1$ is impossible from (12).] Similarly, (13) and (14) give

$$y = \frac{1}{1 - \lambda} \qquad z = -\frac{1}{1 - \lambda}$$

Therefore, from (15), we have

$$\frac{3^2}{(1 - \lambda)^2} + \frac{1^2}{(1 - \lambda)^2} + \frac{(-1)^2}{(1 - \lambda)^2} = 4$$

which gives $(1 - \lambda)^2 = \frac{11}{4}$, $1 - \lambda = \pm\sqrt{11}/2$, so

$$\lambda = 1 \pm \frac{\sqrt{11}}{2}$$

These values of λ then give the corresponding points (x, y, z):

$$\left(\frac{6}{\sqrt{11}}, \frac{2}{\sqrt{11}}, -\frac{2}{\sqrt{11}} \right) \quad \text{and} \quad \left(-\frac{6}{\sqrt{11}}, -\frac{2}{\sqrt{11}}, \frac{2}{\sqrt{11}} \right)$$

It's easy to see that f has a smaller value at the first of these points, so the closest point is $\left(6/\sqrt{11}, 2/\sqrt{11}, -2/\sqrt{11} \right)$ and the farthest is $\left(-6/\sqrt{11}, -2/\sqrt{11}, 2/\sqrt{11} \right)$. ∎

Figure 4 shows the sphere and the nearest point P in Example 4. Can you see how to find the coordinates of P without using calculus?

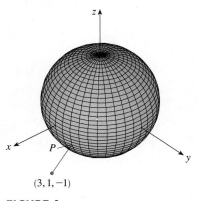

$(3, 1, -1)$

FIGURE 4

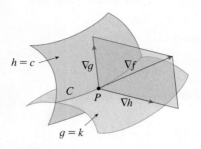

$h = c$

∇g ∇f

C P ∇h

$g = k$

FIGURE 5

Two Constraints

Suppose now that we want to find the maximum and minimum values of a function $f(x, y, z)$ subject to two constraints (side conditions) of the form $g(x, y, z) = k$ and $h(x, y, z) = c$. Geometrically, this means that we are looking for the extreme values of f when (x, y, z) is restricted to lie on the curve of intersection C of the level surfaces $g(x, y, z) = k$ and $h(x, y, z) = c$. (See Figure 5.) Suppose f has such an extreme value at a point $P(x_0, y_0, z_0)$. We know from the beginning of this section that ∇f is orthogonal to C at P. But we also know that ∇g is orthogonal to $g(x, y, z) = k$ and ∇h is orthogonal to $h(x, y, z) = c$, so ∇g and ∇h are both orthogonal to C. This means that the gradient vector $\nabla f(x_0, y_0, z_0)$ is in the plane determined by $\nabla g(x_0, y_0, z_0)$ and $\nabla h(x_0, y_0, z_0)$. (We assume that these gradient vectors are not zero and not parallel.) So there are numbers λ and μ (called Lagrange multipliers) such that

$$\boxed{16} \qquad \nabla f(x_0, y_0, z_0) = \lambda \, \nabla g(x_0, y_0, z_0) + \mu \, \nabla h(x_0, y_0, z_0)$$

In this case Lagrange's method is to look for extreme values by solving five equations in the five unknowns x, y, z, λ, and μ. These equations are obtained by writing Equation 16 in terms of its components and using the constraint equations:

$$f_x = \lambda \, g_x + \mu h_x$$

$$f_y = \lambda \, g_y + \mu h_y$$

$$f_z = \lambda \, g_z + \mu h_z$$

$$g(x, y, z) = k$$

$$h(x, y, z) = c$$

The cylinder $x^2 + y^2 = 1$ intersects the plane $x - y + z = 1$ in an ellipse (Figure 6). Example 5 asks for the maximum value of f when (x, y, z) is restricted to lie on the ellipse.

EXAMPLE 5 Find the maximum value of the function $f(x, y, z) = x + 2y + 3z$ on the curve of intersection of the plane $x - y + z = 1$ and the cylinder $x^2 + y^2 = 1$.

SOLUTION We maximize the function $f(x, y, z) = x + 2y + 3z$ subject to the constraints $g(x, y, z) = x - y + z = 1$ and $h(x, y, z) = x^2 + y^2 = 1$. The Lagrange condition is $\nabla f = \lambda \, \nabla g + \mu \, \nabla h$, so we solve the equations

$$\boxed{17} \qquad 1 = \lambda + 2x\mu$$

$$\boxed{18} \qquad 2 = -\lambda + 2y\mu$$

$$\boxed{19} \qquad 3 = \lambda$$

$$\boxed{20} \qquad x - y + z = 1$$

$$\boxed{21} \qquad x^2 + y^2 = 1$$

Putting $\lambda = 3$ [from (19)] in (17), we get $2x\mu = -2$, so $x = -1/\mu$. Similarly, (18) gives $y = 5/(2\mu)$. Substitution in (21) then gives

$$\frac{1}{\mu^2} + \frac{25}{4\mu^2} = 1$$

FIGURE 6

and so $\mu^2 = \frac{29}{4}$, $\mu = \pm\sqrt{29}/2$. Then $x = \mp 2/\sqrt{29}$, $y = \pm 5/\sqrt{29}$, and, from (20), $z = 1 - x + y = 1 \pm 7/\sqrt{29}$. The corresponding values of f are

$$\mp \frac{2}{\sqrt{29}} + 2\left(\pm\frac{5}{\sqrt{29}}\right) + 3\left(1 \pm \frac{7}{\sqrt{29}}\right) = 3 \pm \sqrt{29}$$

Therefore the maximum value of f on the given curve is $3 + \sqrt{29}$.   ■

14.8   EXERCISES

1. Pictured are a contour map of f and a curve with equation $g(x, y) = 8$. Estimate the maximum and minimum values of f subject to the constraint that $g(x, y) = 8$. Explain your reasoning.

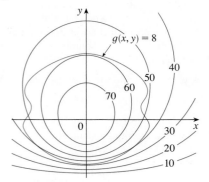

2. (a) Use a graphing calculator or computer to graph the circle $x^2 + y^2 = 1$. On the same screen, graph several curves of the form $x^2 + y = c$ until you find two that just touch the circle. What is the significance of the values of c for these two curves?
   (b) Use Lagrange multipliers to find the extreme values of $f(x, y) = x^2 + y$ subject to the constraint $x^2 + y^2 = 1$. Compare your answers with those in part (a).

3–14 Each of these extreme value problems has a solution with both a maximum value and a minimum value. Use Lagrange multipliers to find the extreme values of the function subject to the given constraint.

3. $f(x, y) = x^2 - y^2$; $x^2 + y^2 = 1$

4. $f(x, y) = 3x + y$; $x^2 + y^2 = 10$

5. $f(x, y) = xy$; $4x^2 + y^2 = 8$

6. $f(x, y) = xe^y$; $x^2 + y^2 = 2$

7. $f(x, y, z) = 2x + 2y + z$; $x^2 + y^2 + z^2 = 9$

8. $f(x, y, z) = e^{xyz}$; $2x^2 + y^2 + z^2 = 24$

9. $f(x, y, z) = xy^2z$; $x^2 + y^2 + z^2 = 4$

10. $f(x, y, z) = \ln(x^2 + 1) + \ln(y^2 + 1) + \ln(z^2 + 1)$; $x^2 + y^2 + z^2 = 12$

11. $f(x, y, z) = x^2 + y^2 + z^2$; $x^4 + y^4 + z^4 = 1$

12. $f(x, y, z) = x^4 + y^4 + z^4$; $x^2 + y^2 + z^2 = 1$

13. $f(x, y, z, t) = x + y + z + t$; $x^2 + y^2 + z^2 + t^2 = 1$

14. $f(x_1, x_2, \ldots, x_n) = x_1 + x_2 + \cdots + x_n$; $x_1^2 + x_2^2 + \cdots + x_n^2 = 1$

15. The method of Lagrange multipliers assumes that the extreme values exist, but that is not always the case. Show that the problem of finding the minimum value of $f(x, y) = x^2 + y^2$ subject to the constraint $xy = 1$ can be solved using Lagrange multipliers, but f does not have a maximum value with that constraint.

16. Find the minimum value of $f(x, y, z) = x^2 + 2y^2 + 3z^2$ subject to the constraint $x + 2y + 3z = 10$. Show that f has no maximum value with this constraint.

17–20 Find the extreme values of f subject to both constraints.

17. $f(x, y, z) = x + y + z$; $x^2 + z^2 = 2$, $x + y = 1$

18. $f(x, y, z) = z$; $x^2 + y^2 = z^2$, $x + y + z = 24$

19. $f(x, y, z) = yz + xy$; $xy = 1$, $y^2 + z^2 = 1$

20. $f(x, y, z) = x^2 + y^2 + z^2$; $x - y = 1$, $y^2 - z^2 = 1$

21–23 Find the extreme values of f on the region described by the inequality.

21. $f(x, y) = x^2 + y^2 + 4x - 4y$, $x^2 + y^2 \leq 9$

22. $f(x, y) = 2x^2 + 3y^2 - 4x - 5$, $x^2 + y^2 \leq 16$

23. $f(x, y) = e^{-xy}$, $x^2 + 4y^2 \leq 1$

24. Consider the problem of maximizing the function $f(x, y) = 2x + 3y$ subject to the constraint $\sqrt{x} + \sqrt{y} = 5$.
   (a) Try using Lagrange multipliers to solve the problem.
   (b) Does $f(25, 0)$ give a larger value than the one in part (a)?
   (c) Solve the problem by graphing the constraint equation and several level curves of f.
   (d) Explain why the method of Lagrange multipliers fails to solve the problem.
   (e) What is the significance of $f(9, 4)$?

25. Consider the problem of minimizing the function
$f(x, y) = x$ on the curve $y^2 + x^4 - x^3 = 0$ (a piriform).
  (a) Try using Lagrange multipliers to solve the problem.
  (b) Show that the minimum value is $f(0, 0) = 0$ but the Lagrange condition $\nabla f(0, 0) = \lambda \nabla g(0, 0)$ is not satisfied for any value of λ.
  (c) Explain why Lagrange multipliers fail to find the minimum value in this case.

CAS 26. (a) If your computer algebra system plots implicitly defined curves, use it to estimate the minimum and maximum values of $f(x, y) = x^3 + y^3 + 3xy$ subject to the constraint $(x - 3)^2 + (y - 3)^2 = 9$ by graphical methods.
  (b) Solve the problem in part (a) with the aid of Lagrange multipliers. Use your CAS to solve the equations numerically. Compare your answers with those in part (a).

27. The total production P of a certain product depends on the amount L of labor used and the amount K of capital investment. In Sections 14.1 and 14.3 we discussed how the Cobb-Douglas model $P = bL^\alpha K^{1-\alpha}$ follows from certain economic assumptions, where b and α are positive constants and $\alpha < 1$. If the cost of a unit of labor is m and the cost of a unit of capital is n, and the company can spend only p dollars as its total budget, then maximizing the production P is subject to the constraint $mL + nK = p$. Show that the maximum production occurs when

$$L = \frac{\alpha p}{m} \quad \text{and} \quad K = \frac{(1 - \alpha)p}{n}$$

28. Referring to Exercise 27, we now suppose that the production is fixed at $bL^\alpha K^{1-\alpha} = Q$, where Q is a constant. What values of L and K minimize the cost function $C(L, K) = mL + nK$?

29. Use Lagrange multipliers to prove that the rectangle with maximum area that has a given perimeter p is a square.

30. Use Lagrange multipliers to prove that the triangle with maximum area that has a given perimeter p is equilateral.
  Hint: Use Heron's formula for the area:

$$A = \sqrt{s(s - x)(s - y)(s - z)}$$

where $s = p/2$ and x, y, z are the lengths of the sides.

31–43 Use Lagrange multipliers to give an alternate solution to the indicated exercise in Section 14.7.

31. Exercise 41
32. Exercise 42
33. Exercise 43
34. Exercise 44
35. Exercise 45
36. Exercise 46
37. Exercise 47
38. Exercise 48
39. Exercise 49
40. Exercise 50
41. Exercise 51
42. Exercise 52
43. Exercise 55

44. Find the maximum and minimum volumes of a rectangular box whose surface area is 1500 cm^2 and whose total edge length is 200 cm.

45. The plane $x + y + 2z = 2$ intersects the paraboloid $z = x^2 + y^2$ in an ellipse. Find the points on this ellipse that are nearest to and farthest from the origin.

46. The plane $4x - 3y + 8z = 5$ intersects the cone $z^2 = x^2 + y^2$ in an ellipse.
  (a) Graph the cone and the plane, and observe the elliptical intersection.
  (b) Use Lagrange multipliers to find the highest and lowest points on the ellipse.

CAS 47–48 Find the maximum and minimum values of f subject to the given constraints. Use a computer algebra system to solve the system of equations that arises in using Lagrange multipliers. (If your CAS finds only one solution, you may need to use additional commands.)

47. $f(x, y, z) = ye^{x-z}$; $9x^2 + 4y^2 + 36z^2 = 36$, $xy + yz = 1$
48. $f(x, y, z) = x + y + z$; $x^2 - y^2 = z$, $x^2 + z^2 = 4$

49. (a) Find the maximum value of

$$f(x_1, x_2, \ldots, x_n) = \sqrt[n]{x_1 x_2 \cdots x_n}$$

given that $x_1, x_2, \ldots, x_n$ are positive numbers and $x_1 + x_2 + \cdots + x_n = c$, where c is a constant.
  (b) Deduce from part (a) that if $x_1, x_2, \ldots, x_n$ are positive numbers, then

$$\sqrt[n]{x_1 x_2 \cdots x_n} \leq \frac{x_1 + x_2 + \cdots + x_n}{n}$$

This inequality says that the geometric mean of n numbers is no larger than the arithmetic mean of the numbers. Under what circumstances are these two means equal?

50. (a) Maximize $\sum_{i=1}^{n} x_i y_i$ subject to the constraints $\sum_{i=1}^{n} x_i^2 = 1$ and $\sum_{i=1}^{n} y_i^2 = 1$.
  (b) Put

$$x_i = \frac{a_i}{\sqrt{\sum a_j^2}} \quad \text{and} \quad y_i = \frac{b_i}{\sqrt{\sum b_j^2}}$$

to show that

$$\sum a_i b_i \leq \sqrt{\sum a_j^2} \sqrt{\sum b_j^2}$$

for any numbers $a_1, \ldots, a_n, b_1, \ldots, b_n$. This inequality is known as the Cauchy-Schwarz Inequality.

APPLIED PROJECT

ROCKET SCIENCE

Courtesy of Orbital Sciences Corporation

Many rockets, such as the *Pegasus XL* currently used to launch satellites and the *Saturn V* that first put men on the moon, are designed to use three stages in their ascent into space. A large first stage initially propels the rocket until its fuel is consumed, at which point the stage is jettisoned to reduce the mass of the rocket. The smaller second and third stages function similarly in order to place the rocket's payload into orbit about the earth. (With this design, at least two stages are required in order to reach the necessary velocities, and using three stages has proven to be a good compromise between cost and performance.) Our goal here is to determine the individual masses of the three stages, which are to be designed to minimize the total mass of the rocket while enabling it to reach a desired velocity.

For a single-stage rocket consuming fuel at a constant rate, the change in velocity resulting from the acceleration of the rocket vehicle has been modeled by

$$\Delta V = -c \ln\left(1 - \frac{(1 - S)M_r}{P + M_r}\right)$$

where M_r is the mass of the rocket engine including initial fuel, P is the mass of the payload, S is a *structural factor* determined by the design of the rocket (specifically, it is the ratio of the mass of the rocket vehicle without fuel to the total mass of the rocket with payload), and c is the (constant) speed of exhaust relative to the rocket.

Now consider a rocket with three stages and a payload of mass A. Assume that outside forces are negligible and that c and S remain constant for each stage. If M_i is the mass of the ith stage, we can initially consider the rocket engine to have mass M_1 and its payload to have mass $M_2 + M_3 + A$; the second and third stages can be handled similarly.

1. Show that the velocity attained after all three stages have been jettisoned is given by

$$v_f = c\left[\ln\left(\frac{M_1 + M_2 + M_3 + A}{SM_1 + M_2 + M_3 + A}\right) + \ln\left(\frac{M_2 + M_3 + A}{SM_2 + M_3 + A}\right) + \ln\left(\frac{M_3 + A}{SM_3 + A}\right)\right]$$

2. We wish to minimize the total mass $M = M_1 + M_2 + M_3$ of the rocket engine subject to the constraint that the desired velocity v_f from Problem 1 is attained. The method of Lagrange multipliers is appropriate here, but difficult to implement using the current expressions. To simplify, we define variables N_i so that the constraint equation may be expressed as $v_f = c(\ln N_1 + \ln N_2 + \ln N_3)$. Since M is now difficult to express in terms of the N_i's, we wish to use a simpler function that will be minimized at the same place as M. Show that

$$\frac{M_1 + M_2 + M_3 + A}{M_2 + M_3 + A} = \frac{(1 - S)N_1}{1 - SN_1}$$

$$\frac{M_2 + M_3 + A}{M_3 + A} = \frac{(1 - S)N_2}{1 - SN_2}$$

$$\frac{M_3 + A}{A} = \frac{(1 - S)N_3}{1 - SN_3}$$

and conclude that

$$\frac{M + A}{A} = \frac{(1 - S)^3 N_1 N_2 N_3}{(1 - SN_1)(1 - SN_2)(1 - SN_3)}$$

3. Verify that $\ln((M + A)/A)$ is minimized at the same location as M; use Lagrange multipliers and the results of Problem 2 to find expressions for the values of N_i where the minimum occurs subject to the constraint $v_f = c(\ln N_1 + \ln N_2 + \ln N_3)$. [*Hint:* Use properties of logarithms to help simplify the expressions.]

4. Find an expression for the minimum value of M as a function of v_f.

5. If we want to put a three-stage rocket into orbit 100 miles above the earth's surface, a final velocity of approximately 17,500 mi/h is required. Suppose that each stage is built with a structural factor $S = 0.2$ and an exhaust speed of $c = 6000$ mi/h.
   (a) Find the minimum total mass M of the rocket engines as a function of A.
   (b) Find the mass of each individual stage as a function of A. (They are not equally sized!)

6. The same rocket would require a final velocity of approximately 24,700 mi/h in order to escape earth's gravity. Find the mass of each individual stage that would minimize the total mass of the rocket engines and allow the rocket to propel a 500-pound probe into deep space.

APPLIED PROJECT   HYDRO-TURBINE OPTIMIZATION

At a hydroelectric generating station (once operated by the Katahdin Paper Company) in Millinocket, Maine, water is piped from a dam to the power station. The rate at which the water flows through the pipe varies, depending on external conditions.

The power station has three different hydroelectric turbines, each with a known (and unique) power function that gives the amount of electric power generated as a function of the water flow arriving at the turbine. The incoming water can be apportioned in different volumes to each turbine, so the goal is to determine how to distribute water among the turbines to give the maximum total energy production for any rate of flow.

Using experimental evidence and *Bernoulli's equation*, the following quadratic models were determined for the power output of each turbine, along with the allowable flows of operation:

$$KW_1 = (-18.89 + 0.1277Q_1 - 4.08 \cdot 10^{-5}Q_1^2)(170 - 1.6 \cdot 10^{-6}Q_T^2)$$

$$KW_2 = (-24.51 + 0.1358Q_2 - 4.69 \cdot 10^{-5}Q_2^2)(170 - 1.6 \cdot 10^{-6}Q_T^2)$$

$$KW_3 = (-27.02 + 0.1380Q_3 - 3.84 \cdot 10^{-5}Q_3^2)(170 - 1.6 \cdot 10^{-6}Q_T^2)$$

$$250 \leqslant Q_1 \leqslant 1110, \quad 250 \leqslant Q_2 \leqslant 1110, \quad 250 \leqslant Q_3 \leqslant 1225$$

where

Q_i = flow through turbine i in cubic feet per second

KW_i = power generated by turbine i in kilowatts

Q_T = total flow through the station in cubic feet per second

1. If all three turbines are being used, we wish to determine the flow Q_i to each turbine that will give the maximum total energy production. Our limitations are that the flows must sum to the total incoming flow and the given domain restrictions must be observed. Consequently, use Lagrange multipliers to find the values for the individual flows (as functions of Q_T) that maximize the total energy production $KW_1 + KW_2 + KW_3$ subject to the constraints $Q_1 + Q_2 + Q_3 = Q_T$ and the domain restrictions on each Q_i.

2. For which values of Q_T is your result valid?

3. For an incoming flow of 2500 ft³/s, determine the distribution to the turbines and verify (by trying some nearby distributions) that your result is indeed a maximum.

4. Until now we have assumed that all three turbines are operating; is it possible in some situations that more power could be produced by using only one turbine? Make a graph of the three power functions and use it to help decide if an incoming flow of 1000 ft³/s should be

distributed to all three turbines or routed to just one. (If you determine that only one turbine should be used, which one would it be?) What if the flow is only 600 ft³/s?

5. Perhaps for some flow levels it would be advantageous to use two turbines. If the incoming flow is 1500 ft³/s, which two turbines would you recommend using? Use Lagrange multipliers to determine how the flow should be distributed between the two turbines to maximize the energy produced. For this flow, is using two turbines more efficient than using all three?

6. If the incoming flow is 3400 ft³/s, what would you recommend to the station management?

14 REVIEW

CONCEPT CHECK

Answers to the Concept Check can be found on the back endpapers.

1. (a) What is a function of two variables?
   (b) Describe three methods for visualizing a function of two variables.

2. What is a function of three variables? How can you visualize such a function?

3. What does

$$\lim_{(x,\,y)\to(a,\,b)} f(x, y) = L$$

mean? How can you show that such a limit does not exist?

4. (a) What does it mean to say that f is continuous at (a, b)?
   (b) If f is continuous on $\mathbb{R}^2$, what can you say about its graph?

5. (a) Write expressions for the partial derivatives $f_x(a, b)$ and $f_y(a, b)$ as limits.
   (b) How do you interpret $f_x(a, b)$ and $f_y(a, b)$ geometrically? How do you interpret them as rates of change?
   (c) If $f(x, y)$ is given by a formula, how do you calculate f_x and f_y?

6. What does Clairaut's Theorem say?

7. How do you find a tangent plane to each of the following types of surfaces?
   (a) A graph of a function of two variables, $z = f(x, y)$
   (b) A level surface of a function of three variables, $F(x, y, z) = k$

8. Define the linearization of f at (a, b). What is the corresponding linear approximation? What is the geometric interpretation of the linear approximation?

9. (a) What does it mean to say that f is differentiable at (a, b)?
   (b) How do you usually verify that f is differentiable?

10. If $z = f(x, y)$, what are the differentials dx, dy, and dz?

11. State the Chain Rule for the case where $z = f(x, y)$ and x and y are functions of one variable. What if x and y are functions of two variables?

12. If z is defined implicitly as a function of x and y by an equation of the form $F(x, y, z) = 0$, how do you find $\partial z/\partial x$ and $\partial z/\partial y$?

13. (a) Write an expression as a limit for the directional derivative of f at (x_0, y_0) in the direction of a unit vector $\mathbf{u} = \langle a, b \rangle$. How do you interpret it as a rate? How do you interpret it geometrically?
   (b) If f is differentiable, write an expression for $D_{\mathbf{u}} f(x_0, y_0)$ in terms of f_x and f_y.

14. (a) Define the gradient vector ∇f for a function f of two or three variables.
   (b) Express $D_{\mathbf{u}} f$ in terms of ∇f.
   (c) Explain the geometric significance of the gradient.

15. What do the following statements mean?
   (a) f has a local maximum at (a, b).
   (b) f has an absolute maximum at (a, b).
   (c) f has a local minimum at (a, b).
   (d) f has an absolute minimum at (a, b).
   (e) f has a saddle point at (a, b).

16. (a) If f has a local maximum at (a, b), what can you say about its partial derivatives at (a, b)?
   (b) What is a critical point of f?

17. State the Second Derivatives Test.

18. (a) What is a closed set in $\mathbb{R}^2$? What is a bounded set?
   (b) State the Extreme Value Theorem for functions of two variables.
   (c) How do you find the values that the Extreme Value Theorem guarantees?

19. Explain how the method of Lagrange multipliers works in finding the extreme values of $f(x, y, z)$ subject to the constraint $g(x, y, z) = k$. What if there is a second constraint $h(x, y, z) = c$?

TRUE-FALSE QUIZ

Determine whether the statement is true or false. If it is true, explain why. If it is false, explain why or give an example that disproves the statement.

1. $f_y(a, b) = \lim\limits_{y \to b} \dfrac{f(a, y) - f(a, b)}{y - b}$

2. There exists a function f with continuous second-order partial derivatives such that $f_x(x, y) = x + y^2$ and $f_y(x, y) = x - y^2$.

3. $f_{xy} = \dfrac{\partial^2 f}{\partial x\, \partial y}$

4. $D_{\mathbf{k}} f(x, y, z) = f_z(x, y, z)$

5. If $f(x, y) \to L$ as $(x, y) \to (a, b)$ along every straight line through (a, b), then $\lim_{(x, y) \to (a, b)} f(x, y) = L$.

6. If $f_x(a, b)$ and $f_y(a, b)$ both exist, then f is differentiable at (a, b).

7. If f has a local minimum at (a, b) and f is differentiable at (a, b), then $\nabla f(a, b) = \mathbf{0}$.

8. If f is a function, then

$$\lim_{(x, y) \to (2, 5)} f(x, y) = f(2, 5)$$

9. If $f(x, y) = \ln y$, then $\nabla f(x, y) = 1/y$.

10. If $(2, 1)$ is a critical point of f and

$$f_{xx}(2, 1)\, f_{yy}(2, 1) < [f_{xy}(2, 1)]^2$$

then f has a saddle point at $(2, 1)$.

11. If $f(x, y) = \sin x + \sin y$, then $-\sqrt{2} \leqslant D_{\mathbf{u}} f(x, y) \leqslant \sqrt{2}$.

12. If $f(x, y)$ has two local maxima, then f must have a local minimum.

EXERCISES

1–2 Find and sketch the domain of the function.

1. $f(x, y) = \ln(x + y + 1)$

2. $f(x, y) = \sqrt{4 - x^2 - y^2} + \sqrt{1 - x^2}$

3–4 Sketch the graph of the function.

3. $f(x, y) = 1 - y^2$

4. $f(x, y) = x^2 + (y - 2)^2$

5–6 Sketch several level curves of the function.

5. $f(x, y) = \sqrt{4x^2 + y^2}$

6. $f(x, y) = e^x + y$

7. Make a rough sketch of a contour map for the function whose graph is shown.

8. The contour map of a function f is shown.
(a) Estimate the value of $f(3, 2)$.

(b) Is $f_x(3, 2)$ positive or negative? Explain.
(c) Which is greater, $f_y(2, 1)$ or $f_y(2, 2)$? Explain.

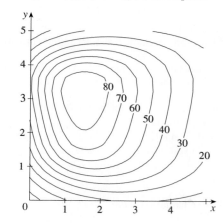

9–10 Evaluate the limit or show that it does not exist.

9. $\lim\limits_{(x, y) \to (1, 1)} \dfrac{2xy}{x^2 + 2y^2}$

10. $\lim\limits_{(x, y) \to (0, 0)} \dfrac{2xy}{x^2 + 2y^2}$

11. A metal plate is situated in the xy-plane and occupies the rectangle $0 \leqslant x \leqslant 10$, $0 \leqslant y \leqslant 8$, where x and y are measured in meters. The temperature at the point (x, y) in the plate is $T(x, y)$, where T is measured in degrees Celsius. Temperatures at equally spaced points were measured and recorded in the table.
(a) Estimate the values of the partial derivatives $T_x(6, 4)$ and $T_y(6, 4)$. What are the units?

(b) Estimate the value of $D_{\mathbf{u}} T(6, 4)$, where $\mathbf{u} = (\mathbf{i} + \mathbf{j})/\sqrt{2}$. Interpret your result.

(c) Estimate the value of $T_{xy}(6, 4)$.

y x	0	2	4	6	8
0	30	38	45	51	55
2	52	56	60	62	61
4	78	74	72	68	66
6	98	87	80	75	71
8	96	90	86	80	75
10	92	92	91	87	78

12. Find a linear approximation to the temperature function $T(x, y)$ in Exercise 11 near the point $(6, 4)$. Then use it to estimate the temperature at the point $(5, 3.8)$.

13–17 Find the first partial derivatives.

13. $f(x, y) = (5y^3 + 2x^2 y)^8$

14. $g(u, v) = \dfrac{u + 2v}{u^2 + v^2}$

15. $F(\alpha, \beta) = \alpha^2 \ln(\alpha^2 + \beta^2)$

16. $G(x, y, z) = e^{xz} \sin(y/z)$

17. $S(u, v, w) = u \arctan(v\sqrt{w})$

18. The speed of sound traveling through ocean water is a function of temperature, salinity, and pressure. It has been modeled by the function

$$C = 1449.2 + 4.6T - 0.055T^2 + 0.00029T^3$$
$$+ (1.34 - 0.01T)(S - 35) + 0.016D$$

where C is the speed of sound (in meters per second), T is the temperature (in degrees Celsius), S is the salinity (the concentration of salts in parts per thousand, which means the number of grams of dissolved solids per 1000 g of water), and D is the depth below the ocean surface (in meters). Compute $\partial C/\partial T$, $\partial C/\partial S$, and $\partial C/\partial D$ when $T = 10°C$, $S = 35$ parts per thousand, and $D = 100$ m. Explain the physical significance of these partial derivatives.

19–22 Find all second partial derivatives of f.

19. $f(x, y) = 4x^3 - xy^2$

20. $z = xe^{-2y}$

21. $f(x, y, z) = x^k y^l z^m$

22. $v = r \cos(s + 2t)$

23. If $z = xy + xe^{y/x}$, show that $x \dfrac{\partial z}{\partial x} + y \dfrac{\partial z}{\partial y} = xy + z$.

24. If $z = \sin(x + \sin t)$, show that

$$\frac{\partial z}{\partial x} \frac{\partial^2 z}{\partial x\, \partial t} = \frac{\partial z}{\partial t} \frac{\partial^2 z}{\partial x^2}$$

25–29 Find equations of (a) the tangent plane and (b) the normal line to the given surface at the specified point.

25. $z = 3x^2 - y^2 + 2x$, $(1, -2, 1)$

26. $z = e^x \cos y$, $(0, 0, 1)$

27. $x^2 + 2y^2 - 3z^2 = 3$, $(2, -1, 1)$

28. $xy + yz + zx = 3$, $(1, 1, 1)$

29. $\sin(xyz) = x + 2y + 3z$, $(2, -1, 0)$

30. Use a computer to graph the surface $z = x^2 + y^4$ and its tangent plane and normal line at $(1, 1, 2)$ on the same screen. Choose the domain and viewpoint so that you get a good view of all three objects.

31. Find the points on the hyperboloid $x^2 + 4y^2 - z^2 = 4$ where the tangent plane is parallel to the plane $2x + 2y + z = 5$.

32. Find du if $u = \ln(1 + se^{2t})$.

33. Find the linear approximation of the function $f(x, y, z) = x^3\sqrt{y^2 + z^2}$ at the point $(2, 3, 4)$ and use it to estimate the number $(1.98)^3\sqrt{(3.01)^2 + (3.97)^2}$.

34. The two legs of a right triangle are measured as 5 m and 12 m with a possible error in measurement of at most 0.2 cm in each. Use differentials to estimate the maximum error in the calculated value of (a) the area of the triangle and (b) the length of the hypotenuse.

35. If $u = x^2 y^3 + z^4$, where $x = p + 3p^2$, $y = pe^p$, and $z = p \sin p$, use the Chain Rule to find du/dp.

36. If $v = x^2 \sin y + ye^{xy}$, where $x = s + 2t$ and $y = st$, use the Chain Rule to find $\partial v/\partial s$ and $\partial v/\partial t$ when $s = 0$ and $t = 1$.

37. Suppose $z = f(x, y)$, where $x = g(s, t)$, $y = h(s, t)$, $g(1, 2) = 3$, $g_s(1, 2) = -1$, $g_t(1, 2) = 4$, $h(1, 2) = 6$, $h_s(1, 2) = -5$, $h_t(1, 2) = 10$, $f_x(3, 6) = 7$, and $f_y(3, 6) = 8$. Find $\partial z/\partial s$ and $\partial z/\partial t$ when $s = 1$ and $t = 2$.

38. Use a tree diagram to write out the Chain Rule for the case where $w = f(t, u, v)$, $t = t(p, q, r, s)$, $u = u(p, q, r, s)$, and $v = v(p, q, r, s)$ are all differentiable functions.

39. If $z = y + f(x^2 - y^2)$, where f is differentiable, show that

$$y \frac{\partial z}{\partial x} + x \frac{\partial z}{\partial y} = x$$

40. The length x of a side of a triangle is increasing at a rate of 3 in/s, the length y of another side is decreasing at a rate of 2 in/s, and the contained angle θ is increasing at a rate of 0.05 radian/s. How fast is the area of the triangle changing when $x = 40$ in, $y = 50$ in, and $\theta = \pi/6$?

41. If $z = f(u, v)$, where $u = xy$, $v = y/x$, and f has continuous second partial derivatives, show that

$$x^2 \frac{\partial^2 z}{\partial x^2} - y^2 \frac{\partial^2 z}{\partial y^2} = -4uv \frac{\partial^2 z}{\partial u\, \partial v} + 2v \frac{\partial z}{\partial v}$$

42. If $\cos(xyz) = 1 + x^2y^2 + z^2$, find $\dfrac{\partial z}{\partial x}$ and $\dfrac{\partial z}{\partial y}$.

43. Find the gradient of the function $f(x, y, z) = x^2 e^{yz^2}$.

44. (a) When is the directional derivative of f a maximum?
(b) When is it a minimum?
(c) When is it 0?
(d) When is it half of its maximum value?

45–46 Find the directional derivative of f at the given point in the indicated direction.

45. $f(x, y) = x^2 e^{-y}$, $(-2, 0)$,
in the direction toward the point $(2, -3)$

46. $f(x, y, z) = x^2 y + x\sqrt{1 + z}$, $(1, 2, 3)$,
in the direction of $\mathbf{v} = 2\mathbf{i} + \mathbf{j} - 2\mathbf{k}$

47. Find the maximum rate of change of $f(x, y) = x^2 y + \sqrt{y}$ at the point $(2, 1)$. In which direction does it occur?

48. Find the direction in which $f(x, y, z) = ze^{xy}$ increases most rapidly at the point $(0, 1, 2)$. What is the maximum rate of increase?

49. The contour map shows wind speed in knots during Hurricane Andrew on August 24, 1992. Use it to estimate the value of the directional derivative of the wind speed at Homestead, Florida, in the direction of the eye of the hurricane.

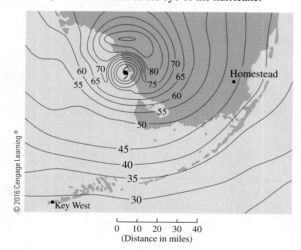

50. Find parametric equations of the tangent line at the point $(-2, 2, 4)$ to the curve of intersection of the surface $z = 2x^2 - y^2$ and the plane $z = 4$.

51–54 Find the local maximum and minimum values and saddle points of the function. If you have three-dimensional graphing software, graph the function with a domain and viewpoint that reveal all the important aspects of the function.

51. $f(x, y) = x^2 - xy + y^2 + 9x - 6y + 10$

52. $f(x, y) = x^3 - 6xy + 8y^3$

53. $f(x, y) = 3xy - x^2y - xy^2$

54. $f(x, y) = (x^2 + y)e^{y/2}$

55–56 Find the absolute maximum and minimum values of f on the set D.

55. $f(x, y) = 4xy^2 - x^2y^2 - xy^3$; D is the closed triangular region in the xy-plane with vertices $(0, 0)$, $(0, 6)$, and $(6, 0)$

56. $f(x, y) = e^{-x^2-y^2}(x^2 + 2y^2)$; D is the disk $x^2 + y^2 \leq 4$

57. Use a graph or level curves or both to estimate the local maximum and minimum values and saddle points of $f(x, y) = x^3 - 3x + y^4 - 2y^2$. Then use calculus to find these values precisely.

58. Use a graphing calculator or computer (or Newton's method or a computer algebra system) to find the critical points of $f(x, y) = 12 + 10y - 2x^2 - 8xy - y^4$ correct to three decimal places. Then classify the critical points and find the highest point on the graph.

59–62 Use Lagrange multipliers to find the maximum and minimum values of f subject to the given constraint(s).

59. $f(x, y) = x^2 y$; $x^2 + y^2 = 1$

60. $f(x, y) = \dfrac{1}{x} + \dfrac{1}{y}$; $\dfrac{1}{x^2} + \dfrac{1}{y^2} = 1$

61. $f(x, y, z) = xyz$; $x^2 + y^2 + z^2 = 3$

62. $f(x, y, z) = x^2 + 2y^2 + 3z^2$;
$x + y + z = 1$, $x - y + 2z = 2$

63. Find the points on the surface $xy^2z^3 = 2$ that are closest to the origin.

64. A package in the shape of a rectangular box can be mailed by the US Postal Service if the sum of its length and girth (the perimeter of a cross-section perpendicular to the length) is at most 108 in. Find the dimensions of the package with largest volume that can be mailed.

65. A pentagon is formed by placing an isosceles triangle on a rectangle, as shown in the figure. If the pentagon has fixed perimeter P, find the lengths of the sides of the pentagon that maximize the area of the pentagon.

Problems Plus

1. A rectangle with length L and width W is cut into four smaller rectangles by two lines parallel to the sides. Find the maximum and minimum values of the sum of the squares of the areas of the smaller rectangles.

2. Marine biologists have determined that when a shark detects the presence of blood in the water, it will swim in the direction in which the concentration of the blood increases most rapidly. Based on certain tests, the concentration of blood (in parts per million) at a point $P(x, y)$ on the surface of seawater is approximated by

$$C(x, y) = e^{-(x^2+2y^2)/10^4}$$

where x and y are measured in meters in a rectangular coordinate system with the blood source at the origin.
   (a) Identify the level curves of the concentration function and sketch several members of this family together with a path that a shark will follow to the source.
   (b) Suppose a shark is at the point (x_0, y_0) when it first detects the presence of blood in the water. Find an equation of the shark's path by setting up and solving a differential equation.

3. A long piece of galvanized sheet metal with width w is to be bent into a symmetric form with three straight sides to make a rain gutter. A cross-section is shown in the figure.
   (a) Determine the dimensions that allow the maximum possible flow; that is, find the dimensions that give the maximum possible cross-sectional area.
   (b) Would it be better to bend the metal into a gutter with a semicircular cross-section?

4. For what values of the number r is the function

$$f(x, y, z) = \begin{cases} \dfrac{(x + y + z)^r}{x^2 + y^2 + z^2} & \text{if } (x, y, z) \neq (0, 0, 0) \\ 0 & \text{if } (x, y, z) = (0, 0, 0) \end{cases}$$

continuous on $\mathbb{R}^3$?

5. Suppose f is a differentiable function of one variable. Show that all tangent planes to the surface $z = xf(y/x)$ intersect in a common point.

6. (a) Newton's method for approximating a root of an equation $f(x) = 0$ (see Section 4.8) can be adapted to approximating a solution of a system of equations $f(x, y) = 0$ and $g(x, y) = 0$. The surfaces $z = f(x, y)$ and $z = g(x, y)$ intersect in a curve that intersects the xy-plane at the point (r, s), which is the solution of the system. If an initial approximation (x_1, y_1) is close to this point, then the tangent planes to the surfaces at (x_1, y_1) intersect in a straight line that intersects the xy-plane in a point (x_2, y_2), which should be closer to (r, s). (Compare with Figure 4.8.2.) Show that

$$x_2 = x_1 - \frac{fg_y - f_y g}{f_x g_y - f_y g_x} \qquad \text{and} \qquad y_2 = y_1 - \frac{f_x g - f g_x}{f_x g_y - f_y g_x}$$

where f, g, and their partial derivatives are evaluated at (x_1, y_1). If we continue this procedure, we obtain successive approximations (x_n, y_n).
   (b) It was Thomas Simpson (1710–1761) who formulated Newton's method as we know it today and who extended it to functions of two variables as in part (a). (See the biogra-

phy of Simpson on page 520.) The example that he gave to illustrate the method was to solve the system of equations

$$x^x + y^y = 1000 \qquad x^y + y^x = 100$$

In other words, he found the points of intersection of the curves in the figure. Use the method of part (a) to find the coordinates of the points of intersection correct to six decimal places.

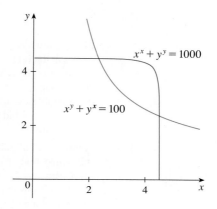

7. If the ellipse $x^2/a^2 + y^2/b^2 = 1$ is to enclose the circle $x^2 + y^2 = 2y$, what values of a and b minimize the area of the ellipse?

8. Show that the maximum value of the function

$$f(x, y) = \frac{(ax + by + c)^2}{x^2 + y^2 + 1}$$

is $a^2 + b^2 + c^2$.

Hint: One method for attacking this problem is to use the Cauchy-Schwarz Inequality:

$$|\mathbf{a} \cdot \mathbf{b}| \leq |\mathbf{a}||\mathbf{b}|$$

(See Exercise 12.3.61.)

15 Multiple Integrals

© Juan Gaertner / Shutterstock.com

Tumors, like the one shown, have been modeled as "bumpy spheres." In Exercise 47 in Section 15.8 you are asked to compute the volume enclosed by such a surface.

IN THIS CHAPTER WE EXTEND the idea of a definite integral to double and triple integrals of functions of two or three variables. These ideas are then used to compute volumes, masses, and centroids of more general regions than we were able to consider in Chapters 6 and 8. We also use double integrals to calculate probabilities when two random variables are involved.

We will see that polar coordinates are useful in computing double integrals over some types of regions. In a similar way, we will introduce two new coordinate systems in three-dimensional space—cylindrical coordinates and spherical coordinates—that greatly simplify the computation of triple integrals over certain commonly occurring solid regions.

15.1 Double Integrals over Rectangles

In much the same way that our attempt to solve the area problem led to the definition of a definite integral, we now seek to find the volume of a solid and in the process we arrive at the definition of a double integral.

■ Review of the Definite Integral

First let's recall the basic facts concerning definite integrals of functions of a single variable. If $f(x)$ is defined for $a \leq x \leq b$, we start by dividing the interval $[a, b]$ into n subintervals $[x_{i-1}, x_i]$ of equal width $\Delta x = (b - a)/n$ and we choose sample points x_i^* in these subintervals. Then we form the Riemann sum

$$\boxed{1} \qquad \sum_{i=1}^{n} f(x_i^*) \, \Delta x$$

and take the limit of such sums as $n \to \infty$ to obtain the definite integral of f from a to b:

$$\boxed{2} \qquad \int_a^b f(x) \, dx = \lim_{n \to \infty} \sum_{i=1}^{n} f(x_i^*) \, \Delta x$$

In the special case where $f(x) \geq 0$, the Riemann sum can be interpreted as the sum of the areas of the approximating rectangles in Figure 1, and $\int_a^b f(x) \, dx$ represents the area under the curve $y = f(x)$ from a to b.

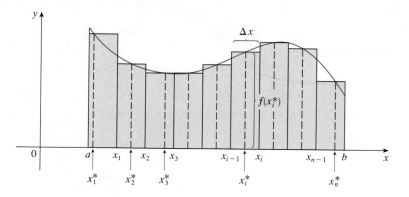

FIGURE 1

■ Volumes and Double Integrals

In a similar manner we consider a function f of two variables defined on a closed rectangle

$$R = [a, b] \times [c, d] = \{(x, y) \in \mathbb{R}^2 \mid a \leq x \leq b, c \leq y \leq d\}$$

and we first suppose that $f(x, y) \geq 0$. The graph of f is a surface with equation $z = f(x, y)$. Let S be the solid that lies above R and under the graph of f, that is,

$$S = \{(x, y, z) \in \mathbb{R}^3 \mid 0 \leq z \leq f(x, y), \ (x, y) \in R\}$$

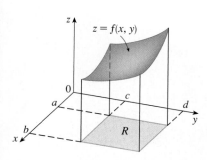

FIGURE 2

(See Figure 2.) Our goal is to find the volume of S.

The first step is to divide the rectangle R into subrectangles. We accomplish this by dividing the interval $[a, b]$ into m subintervals $[x_{i-1}, x_i]$ of equal width $\Delta x = (b - a)/m$ and dividing $[c, d]$ into n subintervals $[y_{j-1}, y_j]$ of equal width $\Delta y = (d - c)/n$. By drawing lines parallel to the coordinate axes through the endpoints of these subintervals, as in Figure 3, we form the subrectangles

$$R_{ij} = [x_{i-1}, x_i] \times [y_{j-1}, y_j] = \{(x, y) \mid x_{i-1} \leq x \leq x_i, \ y_{j-1} \leq y \leq y_j\}$$

each with area $\Delta A = \Delta x \, \Delta y$.

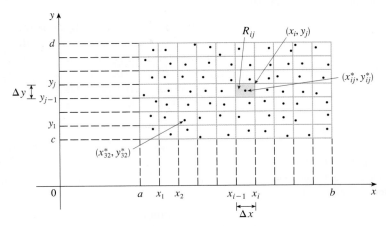

FIGURE 3

Dividing R into subrectangles

If we choose a **sample point** (x_{ij}^*, y_{ij}^*) in each R_{ij}, then we can approximate the part of S that lies above each R_{ij} by a thin rectangular box (or "column") with base R_{ij} and height $f(x_{ij}^*, y_{ij}^*)$ as shown in Figure 4. (Compare with Figure 1.) The volume of this box is the height of the box times the area of the base rectangle:

$$f(x_{ij}^*, y_{ij}^*) \, \Delta A$$

If we follow this procedure for all the rectangles and add the volumes of the corresponding boxes, we get an approximation to the total volume of S:

$$\boxed{3} \qquad V \approx \sum_{i=1}^{m} \sum_{j=1}^{n} f(x_{ij}^*, y_{ij}^*) \, \Delta A$$

(See Figure 5.) This double sum means that for each subrectangle we evaluate f at the chosen point and multiply by the area of the subrectangle, and then we add the results.

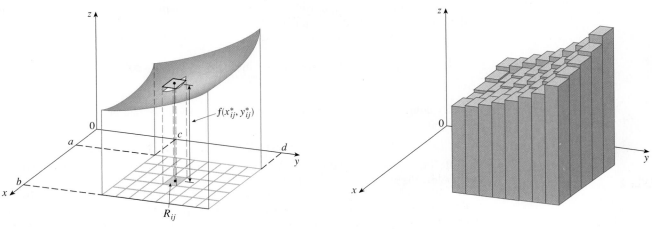

FIGURE 4

FIGURE 5

The meaning of the double limit in Equation 4 is that we can make the double sum as close as we like to the number V [for any choice of (x_{ij}^*, y_{ij}^*) in R_{ij}] by taking m and n sufficiently large.

Our intuition tells us that the approximation given in (3) becomes better as m and n become larger and so we would expect that

$$\boxed{4} \qquad V = \lim_{m, n \to \infty} \sum_{i=1}^{m} \sum_{j=1}^{n} f(x_{ij}^*, y_{ij}^*) \, \Delta A$$

We use the expression in Equation 4 to define the **volume** of the solid S that lies under the graph of f and above the rectangle R. (It can be shown that this definition is consistent with our formula for volume in Section 6.2.)

Limits of the type that appear in Equation 4 occur frequently, not just in finding volumes but in a variety of other situations as well—as we will see in Section 15.4—even when f is not a positive function. So we make the following definition.

Notice the similarity between Definition 5 and the definition of a single integral in Equation 2.

> $\boxed{5}$ **Definition** The **double integral** of f over the rectangle R is
>
> $$\iint_R f(x, y) \, dA = \lim_{m, n \to \infty} \sum_{i=1}^{m} \sum_{j=1}^{n} f(x_{ij}^*, y_{ij}^*) \, \Delta A$$
>
> if this limit exists.

Although we have defined the double integral by dividing R into equal-sized subrectangles, we could have used subrectangles R_{ij} of unequal size. But then we would have to ensure that all of their dimensions approach 0 in the limiting process.

The precise meaning of the limit in Definition 5 is that for every number $\varepsilon > 0$ there is an integer N such that

$$\left| \iint_R f(x, y) \, dA - \sum_{i=1}^{m} \sum_{j=1}^{n} f(x_{ij}^*, y_{ij}^*) \, \Delta A \right| < \varepsilon$$

for all integers m and n greater than N and for any choice of sample points (x_{ij}^*, y_{ij}^*) in R_{ij}.

A function f is called **integrable** if the limit in Definition 5 exists. It is shown in courses on advanced calculus that all continuous functions are integrable. In fact, the double integral of f exists provided that f is "not too discontinuous." In particular, if f is bounded on R, [that is, there is a constant M such that $|f(x, y)| \leq M$ for all (x, y) in R], and f is continuous there, except on a finite number of smooth curves, then f is integrable over R.

The sample point (x_{ij}^*, y_{ij}^*) can be chosen to be any point in the subrectangle R_{ij}, but if we choose it to be the upper right-hand corner of R_{ij} [namely (x_i, y_j), see Figure 3], then the expression for the double integral looks simpler:

$$\boxed{6} \qquad \iint_R f(x, y) \, dA = \lim_{m, n \to \infty} \sum_{i=1}^{m} \sum_{j=1}^{n} f(x_i, y_j) \, \Delta A$$

By comparing Definitions 4 and 5, we see that a volume can be written as a double integral:

> If $f(x, y) \geq 0$, then the volume V of the solid that lies above the rectangle R and below the surface $z = f(x, y)$ is
>
> $$V = \iint_R f(x, y) \, dA$$

The sum in Definition 5,

$$\sum_{i=1}^{m} \sum_{j=1}^{n} f(x_{ij}^*, y_{ij}^*) \, \Delta A$$

is called a **double Riemann sum** and is used as an approximation to the value of the double integral. [Notice how similar it is to the Riemann sum in (1) for a function of a single variable.] If f happens to be a *positive* function, then the double Riemann sum represents the sum of volumes of columns, as in Figure 5, and is an approximation to the volume under the graph of f.

EXAMPLE 1 Estimate the volume of the solid that lies above the square $R = [0, 2] \times [0, 2]$ and below the elliptic paraboloid $z = 16 - x^2 - 2y^2$. Divide R into four equal squares and choose the sample point to be the upper right corner of each square R_{ij}. Sketch the solid and the approximating rectangular boxes.

SOLUTION The squares are shown in Figure 6. The paraboloid is the graph of $f(x, y) = 16 - x^2 - 2y^2$ and the area of each square is $\Delta A = 1$. Approximating the volume by the Riemann sum with $m = n = 2$, we have

$$V \approx \sum_{i=1}^{2} \sum_{j=1}^{2} f(x_i, y_j) \, \Delta A$$

$$= f(1, 1) \, \Delta A + f(1, 2) \, \Delta A + f(2, 1) \, \Delta A + f(2, 2) \, \Delta A$$

$$= 13(1) + 7(1) + 10(1) + 4(1) = 34$$

This is the volume of the approximating rectangular boxes shown in Figure 7. ∎

We get better approximations to the volume in Example 1 if we increase the number of squares. Figure 8 shows how the columns start to look more like the actual solid and the corresponding approximations become more accurate when we use 16, 64, and 256 squares. In Example 7 we will be able to show that the exact volume is 48.

FIGURE 6

FIGURE 7

FIGURE 8
The Riemann sum approximations to the volume under $z = 16 - x^2 - 2y^2$ become more accurate as m and n increase.

(a) $m = n = 4$, $V \approx 41.5$ (b) $m = n = 8$, $V \approx 44.875$ (c) $m = n = 16$, $V \approx 46.46875$

EXAMPLE 2 If $R = \{(x, y) \mid -1 \le x \le 1, -2 \le y \le 2\}$, evaluate the integral

$$\iint_R \sqrt{1 - x^2} \, dA$$

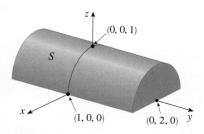

FIGURE 9

SOLUTION It would be very difficult to evaluate this integral directly from Definition 5 but, because $\sqrt{1 - x^2} \geqslant 0$, we can compute the integral by interpreting it as a volume. If $z = \sqrt{1 - x^2}$, then $x^2 + z^2 = 1$ and $z \geqslant 0$, so the given double integral represents the volume of the solid S that lies below the circular cylinder $x^2 + z^2 = 1$ and above the rectangle R. (See Figure 9.) The volume of S is the area of a semicircle with radius 1 times the length of the cylinder. Thus

$$\iint_R \sqrt{1 - x^2} \, dA = \tfrac{1}{2}\pi(1)^2 \times 4 = 2\pi \qquad \blacksquare$$

◼ The Midpoint Rule

The methods that we used for approximating single integrals (the Midpoint Rule, the Trapezoidal Rule, Simpson's Rule) all have counterparts for double integrals. Here we consider only the Midpoint Rule for double integrals. This means that we use a double Riemann sum to approximate the double integral, where the sample point (x_{ij}^*, y_{ij}^*) in R_{ij} is chosen to be the center $(\bar{x}_i, \bar{y}_j)$ of R_{ij}. In other words, $\bar{x}_i$ is the midpoint of $[x_{i-1}, x_i]$ and $\bar{y}_j$ is the midpoint of $[y_{j-1}, y_j]$.

Midpoint Rule for Double Integrals

$$\iint_R f(x, y) \, dA \approx \sum_{i=1}^{m} \sum_{j=1}^{n} f(\bar{x}_i, \bar{y}_j) \, \Delta A$$

where $\bar{x}_i$ is the midpoint of $[x_{i-1}, x_i]$ and $\bar{y}_j$ is the midpoint of $[y_{j-1}, y_j]$.

EXAMPLE 3 Use the Midpoint Rule with $m = n = 2$ to estimate the value of the integral $\iint_R (x - 3y^2) \, dA$, where $R = \{(x, y) \mid 0 \leqslant x \leqslant 2, 1 \leqslant y \leqslant 2\}$.

SOLUTION In using the Midpoint Rule with $m = n = 2$, we evaluate $f(x, y) = x - 3y^2$ at the centers of the four subrectangles shown in Figure 10. So $\bar{x}_1 = \tfrac{1}{2}$, $\bar{x}_2 = \tfrac{3}{2}$, $\bar{y}_1 = \tfrac{5}{4}$, and $\bar{y}_2 = \tfrac{7}{4}$. The area of each subrectangle is $\Delta A = \tfrac{1}{2}$. Thus

$$\iint_R (x - 3y^2) \, dA \approx \sum_{i=1}^{2} \sum_{j=1}^{2} f(\bar{x}_i, \bar{y}_j) \, \Delta A$$

$$= f(\bar{x}_1, \bar{y}_1) \, \Delta A + f(\bar{x}_1, \bar{y}_2) \, \Delta A + f(\bar{x}_2, \bar{y}_1) \, \Delta A + f(\bar{x}_2, \bar{y}_2) \, \Delta A$$

$$= f\left(\tfrac{1}{2}, \tfrac{5}{4}\right) \Delta A + f\left(\tfrac{1}{2}, \tfrac{7}{4}\right) \Delta A + f\left(\tfrac{3}{2}, \tfrac{5}{4}\right) \Delta A + f\left(\tfrac{3}{2}, \tfrac{7}{4}\right) \Delta A$$

$$= \left(-\tfrac{67}{16}\right)\tfrac{1}{2} + \left(-\tfrac{139}{16}\right)\tfrac{1}{2} + \left(-\tfrac{51}{16}\right)\tfrac{1}{2} + \left(-\tfrac{123}{16}\right)\tfrac{1}{2}$$

$$= -\tfrac{95}{8} = -11.875$$

Thus we have

$$\iint_R (x - 3y^2) \, dA \approx -11.875 \qquad \blacksquare$$

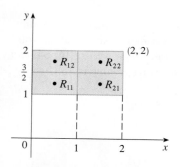

FIGURE 10

NOTE In Example 5 we will see that the exact value of the double integral in Example 3 is -12. (Remember that the interpretation of a double integral as a volume is valid only when the integrand f is a *positive* function. The integrand in Example 3 is not a positive function, so its integral is not a volume. In Examples 5 and 6 we will discuss how to interpret integrals of functions that are not always positive in terms of volumes.) If we keep dividing each subrectangle in Figure 10 into four smaller ones with similar shape, we get the Midpoint Rule approximations displayed in the chart in the margin. Notice how these approximations approach the exact value of the double integral, -12.

Number of subrectangles	Midpoint Rule approximation
1	-11.5000
4	-11.8750
16	-11.9687
64	-11.9922
256	-11.9980
1024	-11.9995

■ Iterated Integrals

Recall that it is usually difficult to evaluate single integrals directly from the definition of an integral, but the Fundamental Theorem of Calculus provides a much easier method. The evaluation of double integrals from first principles is even more difficult, but here we see how to express a double integral as an iterated integral, which can then be evaluated by calculating two single integrals.

Suppose that f is a function of two variables that is integrable on the rectangle $R = [a, b] \times [c, d]$. We use the notation $\int_c^d f(x, y)\, dy$ to mean that x is held fixed and $f(x, y)$ is integrated with respect to y from $y = c$ to $y = d$. This procedure is called *partial integration with respect to* y. (Notice its similarity to partial differentiation.) Now $\int_c^d f(x, y)\, dy$ is a number that depends on the value of x, so it defines a function of x:

$$A(x) = \int_c^d f(x, y)\, dy$$

If we now integrate the function A with respect to x from $x = a$ to $x = b$, we get

$$\boxed{7} \qquad \int_a^b A(x)\, dx = \int_a^b \left[\int_c^d f(x, y)\, dy \right] dx$$

The integral on the right side of Equation 7 is called an **iterated integral**. Usually the brackets are omitted. Thus

$$\boxed{8} \qquad \int_a^b \int_c^d f(x, y)\, dy\, dx = \int_a^b \left[\int_c^d f(x, y)\, dy \right] dx$$

means that we first integrate with respect to y from c to d and then with respect to x from a to b.

Similarly, the iterated integral

$$\boxed{9} \qquad \int_c^d \int_a^b f(x, y)\, dx\, dy = \int_c^d \left[\int_a^b f(x, y)\, dx \right] dy$$

means that we first integrate with respect to x (holding y fixed) from $x = a$ to $x = b$ and then we integrate the resulting function of y with respect to y from $y = c$ to $y = d$. Notice that in both Equations 8 and 9 we work *from the inside out*.

EXAMPLE 4 Evaluate the iterated integrals.

(a) $\displaystyle\int_0^3 \int_1^2 x^2 y\, dy\, dx$ (b) $\displaystyle\int_1^2 \int_0^3 x^2 y\, dx\, dy$

SOLUTION

(a) Regarding x as a constant, we obtain

$$\int_1^2 x^2 y\, dy = \left[x^2 \frac{y^2}{2} \right]_{y=1}^{y=2} = x^2 \left(\frac{2^2}{2} \right) - x^2 \left(\frac{1^2}{2} \right) = \tfrac{3}{2} x^2$$

Thus the function A in the preceding discussion is given by $A(x) = \tfrac{3}{2} x^2$ in this example. We now integrate this function of x from 0 to 3:

$$\int_0^3 \int_1^2 x^2 y\, dy\, dx = \int_0^3 \left[\int_1^2 x^2 y\, dy \right] dx$$

$$= \int_0^3 \tfrac{3}{2} x^2\, dx = \frac{x^3}{2} \Big]_0^3 = \frac{27}{2}$$

(b) Here we first integrate with respect to x:

$$\int_1^2 \int_0^3 x^2 y \, dx \, dy = \int_1^2 \left[\int_0^3 x^2 y \, dx \right] dy = \int_1^2 \left[\frac{x^3}{3} y \right]_{x=0}^{x=3} dy$$

$$= \int_1^2 9y \, dy = 9 \frac{y^2}{2} \bigg]_1^2 = \frac{27}{2}$$ ∎

Notice that in Example 4 we obtained the same answer whether we integrated with respect to y or x first. In general, it turns out (see Theorem 10) that the two iterated integrals in Equations 8 and 9 are always equal; that is, the order of integration does not matter. (This is similar to Clairaut's Theorem on the equality of the mixed partial derivatives.)

The following theorem gives a practical method for evaluating a double integral by expressing it as an iterated integral (in either order).

Theorem 10 is named after the Italian mathematician Guido Fubini (1879–1943), who proved a very general version of this theorem in 1907. But the version for continuous functions was known to the French mathematician Augustin-Louis Cauchy almost a century earlier.

> **10 Fubini's Theorem** If f is continuous on the rectangle $R = \{(x, y) \mid a \leqslant x \leqslant b, c \leqslant y \leqslant d\}$, then
>
> $$\iint_R f(x, y) \, dA = \int_a^b \int_c^d f(x, y) \, dy \, dx = \int_c^d \int_a^b f(x, y) \, dx \, dy$$
>
> More generally, this is true if we assume that f is bounded on R, f is discontinuous only on a finite number of smooth curves, and the iterated integrals exist.

The proof of Fubini's Theorem is too difficult to include in this book, but we can at least give an intuitive indication of why it is true for the case where $f(x, y) \geqslant 0$. Recall that if f is positive, then we can interpret the double integral $\iint_R f(x, y) \, dA$ as the volume V of the solid S that lies above R and under the surface $z = f(x, y)$. But we have another formula that we used for volume in Chapter 6, namely,

$$V = \int_a^b A(x) \, dx$$

FIGURE 11

Visual 15.1 illustrates Fubini's Theorem by showing an animation of Figures 11 and 12.

where $A(x)$ is the area of a cross-section of S in the plane through x perpendicular to the x-axis. From Figure 11 you can see that $A(x)$ is the area under the curve C whose equation is $z = f(x, y)$, where x is held constant and $c \leqslant y \leqslant d$. Therefore

$$A(x) = \int_c^d f(x, y) \, dy$$

and we have

$$\iint_R f(x, y) \, dA = V = \int_a^b A(x) \, dx = \int_a^b \int_c^d f(x, y) \, dy \, dx$$

A similar argument, using cross-sections perpendicular to the y-axis as in Figure 12, shows that

$$\iint_R f(x, y) \, dA = \int_c^d \int_a^b f(x, y) \, dx \, dy$$

FIGURE 12

EXAMPLE 5 Evaluate the double integral $\iint_R (x - 3y^2)\, dA$, where $R = \{(x, y) \mid 0 \leqslant x \leqslant 2,\, 1 \leqslant y \leqslant 2\}$. (Compare with Example 3.)

SOLUTION 1 Fubini's Theorem gives

$$\iint\limits_R (x - 3y^2)\, dA = \int_0^2 \int_1^2 (x - 3y^2)\, dy\, dx = \int_0^2 \Big[xy - y^3 \Big]_{y=1}^{y=2} dx$$

$$= \int_0^2 (x - 7)\, dx = \frac{x^2}{2} - 7x \bigg]_0^2 = -12$$

SOLUTION 2 Again applying Fubini's Theorem, but this time integrating with respect to x first, we have

$$\iint\limits_R (x - 3y^2)\, dA = \int_1^2 \int_0^2 (x - 3y^2)\, dx\, dy$$

$$= \int_1^2 \left[\frac{x^2}{2} - 3xy^2 \right]_{x=0}^{x=2} dy$$

$$= \int_1^2 (2 - 6y^2)\, dy = 2y - 2y^3 \Big]_1^2 = -12$$

Notice the negative answer in Example 5; nothing is wrong with that. The function f is not a positive function, so its integral doesn't represent a volume. From Figure 13 we see that f is always negative on R, so the value of the integral is the *negative* of the volume that lies *above* the graph of f and *below* R.

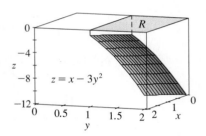

$z = x - 3y^2$

FIGURE 13

For a function f that takes on both positive and negative values, $\iint_R f(x, y)\, dA$ is a difference of volumes: $V_1 - V_2$, where V_1 is the volume above R and below the graph of f, and V_2 is the volume below R and above the graph. The fact that the integral in Example 6 is 0 means that these two volumes V_1 and V_2 are equal. (See Figure 14.)

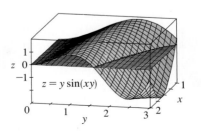

$z = y \sin(xy)$

FIGURE 14

EXAMPLE 6 Evaluate $\iint_R y \sin(xy)\, dA$, where $R = [1, 2] \times [0, \pi]$.

SOLUTION If we first integrate with respect to x, we get

$$\iint\limits_R y \sin(xy)\, dA = \int_0^\pi \int_1^2 y \sin(xy)\, dx\, dy$$

$$= \int_0^\pi \Big[-\cos(xy) \Big]_{x=1}^{x=2} dy$$

$$= \int_0^\pi (-\cos 2y + \cos y)\, dy$$

$$= -\tfrac{1}{2} \sin 2y + \sin y \Big]_0^\pi = 0$$

NOTE If we reverse the order of integration and first integrate with respect to y in Example 6, we get

$$\iint\limits_R y \sin(xy)\, dA = \int_1^2 \int_0^\pi y \sin(xy)\, dy\, dx$$

but this order of integration is much more difficult than the method given in the example because it involves integration by parts twice. Therefore, when we evaluate double integrals it is wise to choose the order of integration that gives simpler integrals.

EXAMPLE 7 Find the volume of the solid S that is bounded by the elliptic paraboloid $x^2 + 2y^2 + z = 16$, the planes $x = 2$ and $y = 2$, and the three coordinate planes.

SOLUTION We first observe that S is the solid that lies under the surface $z = 16 - x^2 - 2y^2$ and above the square $R = [0, 2] \times [0, 2]$. (See Figure 15.) This solid was considered in Example 1, but we are now in a position to evaluate the double integral using Fubini's Theorem. Therefore

$$V = \iint_R (16 - x^2 - 2y^2) \, dA = \int_0^2 \int_0^2 (16 - x^2 - 2y^2) \, dx \, dy$$

$$= \int_0^2 \left[16x - \tfrac{1}{3}x^3 - 2y^2 x \right]_{x=0}^{x=2} dy$$

$$= \int_0^2 \left(\tfrac{88}{3} - 4y^2 \right) dy = \left[\tfrac{88}{3}y - \tfrac{4}{3}y^3 \right]_0^2 = 48 \qquad \blacksquare$$

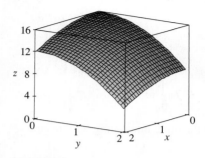

FIGURE 15

In the special case where $f(x, y)$ can be factored as the product of a function of x only and a function of y only, the double integral of f can be written in a particularly simple form. To be specific, suppose that $f(x, y) = g(x)h(y)$ and $R = [a, b] \times [c, d]$. Then Fubini's Theorem gives

$$\iint_R f(x, y) \, dA = \int_c^d \int_a^b g(x)h(y) \, dx \, dy = \int_c^d \left[\int_a^b g(x)h(y) \, dx \right] dy$$

In the inner integral, y is a constant, so $h(y)$ is a constant and we can write

$$\int_c^d \left[\int_a^b g(x)h(y) \, dx \right] dy = \int_c^d \left[h(y) \left(\int_a^b g(x) \, dx \right) \right] dy = \int_a^b g(x) \, dx \int_c^d h(y) \, dy$$

since $\int_a^b g(x) \, dx$ is a constant. Therefore, in this case the double integral of f can be written as the product of two single integrals:

$$\boxed{11} \qquad \iint_R g(x) h(y) \, dA = \int_a^b g(x) \, dx \int_c^d h(y) \, dy \qquad \text{where } R = [a, b] \times [c, d]$$

EXAMPLE 8 If $R = [0, \pi/2] \times [0, \pi/2]$, then, by Equation 11,

$$\iint_R \sin x \cos y \, dA = \int_0^{\pi/2} \sin x \, dx \int_0^{\pi/2} \cos y \, dy$$

$$= \left[-\cos x \right]_0^{\pi/2} \left[\sin y \right]_0^{\pi/2} = 1 \cdot 1 = 1 \qquad \blacksquare$$

The function $f(x, y) = \sin x \cos y$ in Example 8 is positive on R, so the integral represents the volume of the solid that lies above R and below the graph of f shown in Figure 16.

FIGURE 16

Average Value

Recall from Section 6.5 that the average value of a function f of one variable defined on an interval $[a, b]$ is

$$f_{ave} = \frac{1}{b - a} \int_a^b f(x) \, dx$$

In a similar fashion we define the **average value** of a function f of two variables defined on a rectangle R to be

$$f_{ave} = \frac{1}{A(R)} \iint_R f(x, y) \, dA$$

where $A(R)$ is the area of R.

If $f(x, y) \geqslant 0$, the equation

$$A(R) \times f_{ave} = \iint_R f(x, y) \, dA$$

says that the box with base R and height f_{ave} has the same volume as the solid that lies under the graph of f. [If $z = f(x, y)$ describes a mountainous region and you chop off the tops of the mountains at height f_{ave}, then you can use them to fill in the valleys so that the region becomes completely flat. See Figure 17.]

FIGURE 17

EXAMPLE 9 The contour map in Figure 18 shows the snowfall, in inches, that fell on the state of Colorado on December 20 and 21, 2006. (The state is in the shape of a rectangle that measures 388 mi west to east and 276 mi south to north.) Use the contour map to estimate the average snowfall for the entire state of Colorado on those days.

FIGURE 18

SOLUTION Let's place the origin at the southwest corner of the state. Then $0 \leqslant x \leqslant 388$, $0 \leqslant y \leqslant 276$, and $f(x, y)$ is the snowfall, in inches, at a location x miles to the east and y miles to the north of the origin. If R is the rectangle that represents Colorado, then the average snowfall for the state on December 20–21 was

$$f_{ave} = \frac{1}{A(R)} \iint_R f(x, y) \, dA$$

where $A(R) = 388 \cdot 276$. To estimate the value of this double integral, let's use the Midpoint Rule with $m = n = 4$. In other words, we divide R into 16 subrectangles of equal size, as in Figure 19. The area of each subrectangle is

$$\Delta A = \tfrac{1}{16}(388)(276) = 6693 \text{ mi}^2$$

FIGURE 19

Using the contour map to estimate the value of f at the center of each subrectangle, we get

$$\iint\limits_{R} f(x, y)\, dA \approx \sum_{i=1}^{4} \sum_{j=1}^{4} f(\bar{x}_i, \bar{y}_j)\, \Delta A$$

$$\approx \Delta A[0 + 15 + 8 + 7 + 2 + 25 + 18.5 + 11$$

$$+ 4.5 + 28 + 17 + 13.5 + 12 + 15 + 17.5 + 13]$$

$$= (6693)(207)$$

Therefore
$$f_{\text{ave}} \approx \frac{(6693)(207)}{(388)(276)} \approx 12.9$$

On December 20–21, 2006, Colorado received an average of approximately 13 inches of snow.

15.1 EXERCISES

1. (a) Estimate the volume of the solid that lies below the surface $z = xy$ and above the rectangle

   $$R = \{(x, y) \mid 0 \leqslant x \leqslant 6, 0 \leqslant y \leqslant 4\}$$

   Use a Riemann sum with $m = 3$, $n = 2$, and take the sample point to be the upper right corner of each square.
   (b) Use the Midpoint Rule to estimate the volume of the solid in part (a).

2. If $R = [0, 4] \times [-1, 2]$, use a Riemann sum with $m = 2$, $n = 3$ to estimate the value of $\iint_R (1 - xy^2) \, dA$. Take the sample points to be (a) the lower right corners and (b) the upper left corners of the rectangles.

3. (a) Use a Riemann sum with $m = n = 2$ to estimate the value of $\iint_R xe^{-xy} \, dA$, where $R = [0, 2] \times [0, 1]$. Take the sample points to be upper right corners.
   (b) Use the Midpoint Rule to estimate the integral in part (a).

4. (a) Estimate the volume of the solid that lies below the surface $z = 1 + x^2 + 3y$ and above the rectangle $R = [1, 2] \times [0, 3]$. Use a Riemann sum with $m = n = 2$ and choose the sample points to be lower left corners.
   (b) Use the Midpoint Rule to estimate the volume in part (a).

5. Let V be the volume of the solid that lies under the graph of $f(x, y) = \sqrt{52 - x^2 - y^2}$ and above the rectangle given by $2 \leqslant x \leqslant 4$, $2 \leqslant y \leqslant 6$. Use the lines $x = 3$ and $y = 4$ to divide R into subrectangles. Let L and U be the Riemann sums computed using lower left corners and upper right corners, respectively. Without calculating the numbers V, L, and U, arrange them in increasing order and explain your reasoning.

6. A 20-ft-by-30-ft swimming pool is filled with water. The depth is measured at 5-ft intervals, starting at one corner of the pool, and the values are recorded in the table. Estimate the volume of water in the pool.

	0	5	10	15	20	25	30
0	2	3	4	6	7	8	8
5	2	3	4	7	8	10	8
10	2	4	6	8	10	12	10
15	2	3	4	5	6	8	7
20	2	2	2	2	3	4	4

7. A contour map is shown for a function f on the square $R = [0, 4] \times [0, 4]$.
   (a) Use the Midpoint Rule with $m = n = 2$ to estimate the value of $\iint_R f(x, y) \, dA$.
   (b) Estimate the average value of f.

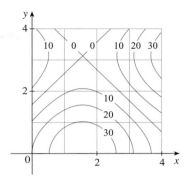

8. The contour map shows the temperature, in degrees Fahrenheit, at 4:00 PM on February 26, 2007, in Colorado. (The state measures 388 mi west to east and 276 mi south to north.) Use the Midpoint Rule with $m = n = 4$ to estimate the average temperature in Colorado at that time.

9–11 Evaluate the double integral by first identifying it as the volume of a solid.

9. $\iint_R \sqrt{2} \, dA$, $R = \{(x, y) \mid 2 \leqslant x \leqslant 6, -1 \leqslant y \leqslant 5\}$

10. $\iint_R (2x + 1) \, dA$, $R = \{(x, y) \mid 0 \leqslant x \leqslant 2, 0 \leqslant y \leqslant 4\}$

11. $\iint_R (4 - 2y) \, dA$, $R = [0, 1] \times [0, 1]$

12. The integral $\iint_R \sqrt{9 - y^2} \, dA$, where $R = [0, 4] \times [0, 2]$, represents the volume of a solid. Sketch the solid.

13–14 Find $\int_0^2 f(x, y) \, dx$ and $\int_0^3 f(x, y) \, dy$

13. $f(x, y) = x + 3x^2y^2$ 14. $f(x, y) = y\sqrt{x + 2}$

15–26 Calculate the iterated integral.

15. $\displaystyle\int_1^4 \int_0^2 (6x^2y - 2x) \, dy \, dx$ 16. $\displaystyle\int_0^1 \int_0^1 (x + y)^2 \, dx \, dy$

17. $\int_0^1 \int_1^2 (x + e^{-y}) \, dx \, dy$

18. $\int_0^{\pi/6} \int_0^{\pi/2} (\sin x + \sin y) \, dy \, dx$

19. $\int_{-3}^3 \int_0^{\pi/2} (y + y^2 \cos x) \, dx \, dy$

20. $\int_1^3 \int_1^5 \frac{\ln y}{xy} \, dy \, dx$

21. $\int_1^4 \int_1^2 \left(\frac{x}{y} + \frac{y}{x} \right) dy \, dx$

22. $\int_0^1 \int_0^2 ye^{x-y} \, dx \, dy$

23. $\int_0^3 \int_0^{\pi/2} t^2 \sin^3 \phi \, d\phi \, dt$

24. $\int_0^1 \int_0^1 xy\sqrt{x^2 + y^2} \, dy \, dx$

25. $\int_0^1 \int_0^1 v(u + v^2)^4 \, du \, dv$

26. $\int_0^1 \int_0^1 \sqrt{s + t} \, ds \, dt$

27–34 Calculate the double integral.

27. $\iint_R x \sec^2 y \, dA, \quad R = \{(x, y) \mid 0 \le x \le 2, 0 \le y \le \pi/4\}$

28. $\iint_R (y + xy^{-2}) \, dA, \quad R = \{(x, y) \mid 0 \le x \le 2, 1 \le y \le 2\}$

29. $\iint_R \frac{xy^2}{x^2 + 1} \, dA, \quad R = \{(x, y) \mid 0 \le x \le 1, -3 \le y \le 3\}$

30. $\iint_R \frac{\tan \theta}{\sqrt{1 - t^2}} \, dA, \quad R = \{(\theta, t) \mid 0 \le \theta \le \pi/3, 0 \le t \le \frac{1}{2}\}$

31. $\iint_R x \sin(x + y) \, dA, \quad R = [0, \pi/6] \times [0, \pi/3]$

32. $\iint_R \frac{x}{1 + xy} \, dA, \quad R = [0, 1] \times [0, 1]$

33. $\iint_R ye^{-xy} \, dA, \quad R = [0, 2] \times [0, 3]$

34. $\iint_R \frac{1}{1 + x + y} \, dA, \quad R = [1, 3] \times [1, 2]$

35–36 Sketch the solid whose volume is given by the iterated integral.

35. $\int_0^1 \int_0^1 (4 - x - 2y) \, dx \, dy$

36. $\int_0^1 \int_0^1 (2 - x^2 - y^2) \, dy \, dx$

---

37. Find the volume of the solid that lies under the plane $4x + 6y - 2z + 15 = 0$ and above the rectangle $R = \{(x, y) \mid -1 \le x \le 2, -1 \le y \le 1\}$.

38. Find the volume of the solid that lies under the hyperbolic paraboloid $z = 3y^2 - x^2 + 2$ and above the rectangle $R = [-1, 1] \times [1, 2]$.

39. Find the volume of the solid lying under the elliptic paraboloid $x^2/4 + y^2/9 + z = 1$ and above the rectangle $R = [-1, 1] \times [-2, 2]$.

40. Find the volume of the solid enclosed by the surface $z = x^2 + xy^2$ and the planes $z = 0, x = 0, x = 5$, and $y = \pm2$.

41. Find the volume of the solid enclosed by the surface $z = 1 + x^2ye^y$ and the planes $z = 0, x = \pm1, y = 0$, and $y = 1$.

42. Find the volume of the solid in the first octant bounded by the cylinder $z = 16 - x^2$ and the plane $y = 5$.

43. Find the volume of the solid enclosed by the paraboloid $z = 2 + x^2 + (y - 2)^2$ and the planes $z = 1, x = 1$, $x = -1, y = 0$, and $y = 4$.

44. Graph the solid that lies between the surface $z = 2xy/(x^2 + 1)$ and the plane $z = x + 2y$ and is bounded by the planes $x = 0, x = 2, y = 0$, and $y = 4$. Then find its volume.

CAS **45.** Use a computer algebra system to find the exact value of the integral $\iint_R x^5y^3e^{xy} \, dA$, where $R = [0, 1] \times [0, 1]$. Then use the CAS to draw the solid whose volume is given by the integral.

CAS **46.** Graph the solid that lies between the surfaces $z = e^{-x^2} \cos(x^2 + y^2)$ and $z = 2 - x^2 - y^2$ for $|x| \le 1$, $|y| \le 1$. Use a computer algebra system to approximate the volume of this solid correct to four decimal places.

47–48 Find the average value of f over the given rectangle.

47. $f(x, y) = x^2y$,
R has vertices $(-1, 0), (-1, 5), (1, 5), (1, 0)$

48. $f(x, y) = e^y\sqrt{x + e^y}, \quad R = [0, 4] \times [0, 1]$

49–50 Use symmetry to evaluate the double integral.

49. $\iint_R \frac{xy}{1 + x^4} \, dA, \quad R = \{(x, y) \mid -1 \le x \le 1, 0 \le y \le 1\}$

50. $\iint_R (1 + x^2 \sin y + y^2 \sin x) \, dA, \quad R = [-\pi, \pi] \times [-\pi, \pi]$

CAS **51.** Use a CAS to compute the iterated integrals

$$\int_0^1 \int_0^1 \frac{x - y}{(x + y)^3} \, dy \, dx \quad \text{and} \quad \int_0^1 \int_0^1 \frac{x - y}{(x + y)^3} \, dx \, dy$$

Do the answers contradict Fubini's Theorem? Explain what is happening.

52. (a) In what way are the theorems of Fubini and Clairaut similar?
(b) If $f(x, y)$ is continuous on $[a, b] \times [c, d]$ and

$$g(x, y) = \int_a^x \int_c^y f(s, t) \, dt \, ds$$

for $a < x < b, c < y < d$, show that $g_{xy} = g_{yx} = f(x, y)$.

Eleanor Cook

15.2 Double Integrals over General Regions

For single integrals, the region over which we integrate is always an interval. But for double integrals, we want to be able to integrate a function f not just over rectangles but also over regions D of more general shape, such as the one illustrated in Figure 1. We suppose that D is a bounded region, which means that D can be enclosed in a rectangular region R as in Figure 2. Then we define a new function F with domain R by

$$\boxed{1} \qquad F(x, y) = \begin{cases} f(x, y) & \text{if } (x, y) \text{ is in } D \\ 0 & \text{if } (x, y) \text{ is in } R \text{ but not in } D \end{cases}$$

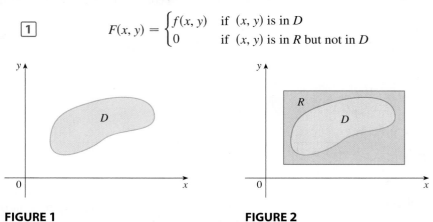

FIGURE 1   **FIGURE 2**

If F is integrable over R, then we define the **double integral of f over D** by

$$\boxed{2} \qquad \iint\limits_{D} f(x, y) \, dA = \iint\limits_{R} F(x, y) \, dA \qquad \text{where } F \text{ is given by Equation 1}$$

Definition 2 makes sense because R is a rectangle and so $\iint_R F(x, y) \, dA$ has been previously defined in Section 15.1. The procedure that we have used is reasonable because the values of $F(x, y)$ are 0 when (x, y) lies outside D and so they contribute nothing to the integral. This means that it doesn't matter what rectangle R we use as long as it contains D.

In the case where $f(x, y) \geqslant 0$, we can still interpret $\iint_D f(x, y) \, dA$ as the volume of the solid that lies above D and under the surface $z = f(x, y)$ (the graph of f). You can see that this is reasonable by comparing the graphs of f and F in Figures 3 and 4 and remembering that $\iint_R F(x, y) \, dA$ is the volume under the graph of F.

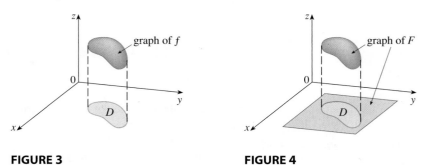

FIGURE 3   **FIGURE 4**

Figure 4 also shows that F is likely to have discontinuities at the boundary points of D. Nonetheless, if f is continuous on D and the boundary curve of D is "well behaved" (in a sense outside the scope of this book), then it can be shown that $\iint_R F(x, y) \, dA$ exists and therefore $\iint_D f(x, y)\, dA$ exists. In particular, this is the case for the following two types of regions.

A plane region D is said to be of **type I** if it lies between the graphs of two continuous functions of x, that is,

$$D = \{(x, y) \mid a \leqslant x \leqslant b,\ g_1(x) \leqslant y \leqslant g_2(x)\}$$

where g_1 and g_2 are continuous on $[a, b]$. Some examples of type I regions are shown in Figure 5.

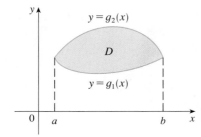

FIGURE 5
Some type I regions

In order to evaluate $\iint_D f(x, y)\, dA$ when D is a region of type I, we choose a rectangle $R = [a, b] \times [c, d]$ that contains D, as in Figure 6, and we let F be the function given by Equation 1; that is, F agrees with f on D and F is 0 outside D. Then, by Fubini's Theorem,

$$\iint_D f(x, y)\, dA = \iint_R F(x, y)\, dA = \int_a^b \int_c^d F(x, y)\, dy\, dx$$

Observe that $F(x, y) = 0$ if $y < g_1(x)$ or $y > g_2(x)$ because (x, y) then lies outside D. Therefore

$$\int_c^d F(x, y)\, dy = \int_{g_1(x)}^{g_2(x)} F(x, y)\, dy = \int_{g_1(x)}^{g_2(x)} f(x, y)\, dy$$

because $F(x, y) = f(x, y)$ when $g_1(x) \leqslant y \leqslant g_2(x)$. Thus we have the following formula that enables us to evaluate the double integral as an iterated integral.

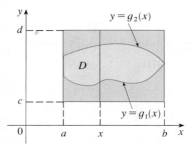

FIGURE 6

---

3 If f is continuous on a type I region D such that

$$D = \{(x, y) \mid a \leqslant x \leqslant b,\ g_1(x) \leqslant y \leqslant g_2(x)\}$$

then

$$\iint_D f(x, y)\, dA = \int_a^b \int_{g_1(x)}^{g_2(x)} f(x, y)\, dy\, dx$$

The integral on the right side of (3) is an iterated integral that is similar to the ones we considered in the preceding section, except that in the inner integral we regard x as being constant not only in $f(x, y)$ but also in the limits of integration, $g_1(x)$ and $g_2(x)$.

We also consider plane regions of **type II**, which can be expressed as

4 $$D = \{(x, y) \mid c \leqslant y \leqslant d,\ h_1(y) \leqslant x \leqslant h_2(y)\}$$

where h_1 and h_2 are continuous. Two such regions are illustrated in Figure 7.

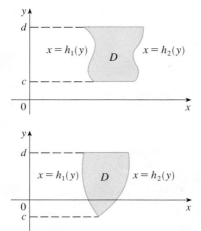

FIGURE 7
Some type II regions

Using the same methods that were used in establishing (3), we can show that

$$\boxed{5} \qquad \iint_D f(x, y)\, dA = \int_c^d \int_{h_1(y)}^{h_2(y)} f(x, y)\, dx\, dy$$

where D is a type II region given by Equation 4.

EXAMPLE 1 Evaluate $\iint_D (x + 2y)\, dA$, where D is the region bounded by the parabolas $y = 2x^2$ and $y = 1 + x^2$.

SOLUTION The parabolas intersect when $2x^2 = 1 + x^2$, that is, $x^2 = 1$, so $x = \pm 1$. We note that the region D, sketched in Figure 8, is a type I region but not a type II region and we can write

$$D = \left\{(x, y) \mid -1 \leqslant x \leqslant 1,\ 2x^2 \leqslant y \leqslant 1 + x^2\right\}$$

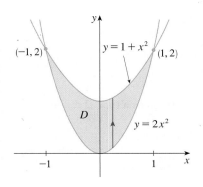

FIGURE 8

Since the lower boundary is $y = 2x^2$ and the upper boundary is $y = 1 + x^2$, Equation 3 gives

$$\iint_D (x + 2y)\, dA = \int_{-1}^1 \int_{2x^2}^{1+x^2} (x + 2y)\, dy\, dx$$

$$= \int_{-1}^1 \left[xy + y^2\right]_{y=2x^2}^{y=1+x^2} dx$$

$$= \int_{-1}^1 \left[x(1 + x^2) + (1 + x^2)^2 - x(2x^2) - (2x^2)^2\right] dx$$

$$= \int_{-1}^1 \left(-3x^4 - x^3 + 2x^2 + x + 1\right) dx$$

$$= -3\frac{x^5}{5} - \frac{x^4}{4} + 2\frac{x^3}{3} + \frac{x^2}{2} + x\Bigg]_{-1}^1 = \frac{32}{15} \qquad \blacksquare$$

NOTE When we set up a double integral as in Example 1, it is essential to draw a diagram. Often it is helpful to draw a vertical arrow as in Figure 8. Then the limits of integration for the *inner* integral can be read from the diagram as follows: The arrow starts at the lower boundary $y = g_1(x)$, which gives the lower limit in the integral, and the arrow ends at the upper boundary $y = g_2(x)$, which gives the upper limit of integration. For a type II region the arrow is drawn horizontally from the left boundary to the right boundary.

EXAMPLE 2 Find the volume of the solid that lies under the paraboloid $z = x^2 + y^2$ and above the region D in the xy-plane bounded by the line $y = 2x$ and the parabola $y = x^2$.

SOLUTION 1 From Figure 9 we see that D is a type I region and

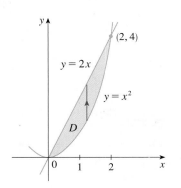

FIGURE 9
D as a type I region

$$D = \left\{(x, y) \mid 0 \leqslant x \leqslant 2,\ x^2 \leqslant y \leqslant 2x\right\}$$

Figure 10 shows the solid whose volume is calculated in Example 2. It lies above the xy-plane, below the paraboloid $z = x^2 + y^2$, and between the plane $y = 2x$ and the parabolic cylinder $y = x^2$.

FIGURE 10

FIGURE 11
D as a type II region

Therefore the volume under $z = x^2 + y^2$ and above D is

$$V = \iint_D (x^2 + y^2)\, dA = \int_0^2 \int_{x^2}^{2x} (x^2 + y^2)\, dy\, dx$$

$$= \int_0^2 \left[x^2 y + \frac{y^3}{3} \right]_{y=x^2}^{y=2x} dx$$

$$= \int_0^2 \left[x^2(2x) + \frac{(2x)^3}{3} - x^2 x^2 - \frac{(x^2)^3}{3} \right] dx$$

$$= \int_0^2 \left(-\frac{x^6}{3} - x^4 + \frac{14x^3}{3} \right) dx$$

$$= -\frac{x^7}{21} - \frac{x^5}{5} + \frac{7x^4}{6} \Big]_0^2 = \frac{216}{35}$$

SOLUTION 2 From Figure 11 we see that D can also be written as a type II region:

$$D = \left\{ (x, y) \mid 0 \leqslant y \leqslant 4, \tfrac{1}{2}y \leqslant x \leqslant \sqrt{y} \right\}$$

Therefore another expression for V is

$$V = \iint_D (x^2 + y^2)\, dA = \int_0^4 \int_{\frac{1}{2}y}^{\sqrt{y}} (x^2 + y^2)\, dx\, dy$$

$$= \int_0^4 \left[\frac{x^3}{3} + y^2 x \right]_{x=\frac{1}{2}y}^{x=\sqrt{y}} dy = \int_0^4 \left(\frac{y^{3/2}}{3} + y^{5/2} - \frac{y^3}{24} - \frac{y^3}{2} \right) dy$$

$$= \tfrac{2}{15} y^{5/2} + \tfrac{2}{7} y^{7/2} - \tfrac{13}{96} y^4 \Big]_0^4 = \tfrac{216}{35} \qquad \blacksquare$$

EXAMPLE 3 Evaluate $\iint_D xy\, dA$, where D is the region bounded by the line $y = x - 1$ and the parabola $y^2 = 2x + 6$.

SOLUTION The region D is shown in Figure 12. Again D is both type I and type II, but the description of D as a type I region is more complicated because the lower boundary consists of two parts. Therefore we prefer to express D as a type II region:

$$D = \left\{ (x, y) \mid -2 \leqslant y \leqslant 4, \tfrac{1}{2}y^2 - 3 \leqslant x \leqslant y + 1 \right\}$$

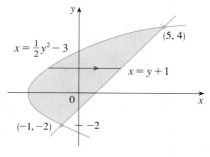

FIGURE 12   (a) D as a type I region   (b) D as a type II region

Then (5) gives

$$\iint_D xy\, dA = \int_{-2}^{4} \int_{\frac{1}{2}y^2-3}^{y+1} xy\, dx\, dy = \int_{-2}^{4} \left[\frac{x^2}{2}\, y \right]_{x=\frac{1}{2}y^2-3}^{x=y+1} dy$$

$$= \tfrac{1}{2} \int_{-2}^{4} y\Big[(y+1)^2 - (\tfrac{1}{2}y^2-3)^2\Big]\, dy$$

$$= \tfrac{1}{2} \int_{-2}^{4} \left(-\frac{y^5}{4} + 4y^3 + 2y^2 - 8y \right) dy$$

$$= \frac{1}{2}\left[-\frac{y^6}{24} + y^4 + 2\,\frac{y^3}{3} - 4y^2 \right]_{-2}^{4} = 36$$

If we had expressed D as a type I region using Figure 12(a), then we would have obtained

$$\iint_D xy\, dA = \int_{-3}^{-1} \int_{-\sqrt{2x+6}}^{\sqrt{2x+6}} xy\, dy\, dx + \int_{-1}^{5} \int_{x-1}^{\sqrt{2x+6}} xy\, dy\, dx$$

but this would have involved more work than the other method.   ∎

EXAMPLE 4 Find the volume of the tetrahedron bounded by the planes $x + 2y + z = 2$, $x = 2y$, $x = 0$, and $z = 0$.

SOLUTION In a question such as this, it's wise to draw two diagrams: one of the three-dimensional solid and another of the plane region D over which it lies. Figure 13 shows the tetrahedron T bounded by the coordinate planes $x = 0$, $z = 0$, the vertical plane $x = 2y$, and the plane $x + 2y + z = 2$. Since the plane $x + 2y + z = 2$ intersects the xy-plane (whose equation is $z = 0$) in the line $x + 2y = 2$, we see that T lies above the triangular region D in the xy-plane bounded by the lines $x = 2y$, $x + 2y = 2$, and $x = 0$. (See Figure 14.)

The plane $x + 2y + z = 2$ can be written as $z = 2 - x - 2y$, so the required volume lies under the graph of the function $z = 2 - x - 2y$ and above

$$D = \{(x, y) \mid 0 \le x \le 1,\ x/2 \le y \le 1 - x/2\}$$

Therefore

$$V = \iint_D (2 - x - 2y)\, dA$$

$$= \int_0^1 \int_{x/2}^{1-x/2} (2 - x - 2y)\, dy\, dx$$

$$= \int_0^1 \Big[2y - xy - y^2 \Big]_{y=x/2}^{y=1-x/2} dx$$

$$= \int_0^1 \left[2 - x - x\left(1 - \frac{x}{2}\right) - \left(1 - \frac{x}{2}\right)^2 - x + \frac{x^2}{2} + \frac{x^2}{4} \right] dx$$

$$= \int_0^1 (x^2 - 2x + 1)\, dx = \frac{x^3}{3} - x^2 + x\, \bigg]_0^1 = \frac{1}{3}$$ ∎

FIGURE 13

FIGURE 14

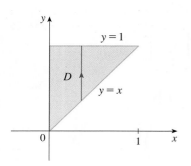

FIGURE 15
D as a type I region

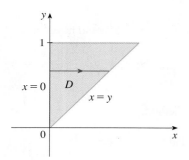

FIGURE 16
D as a type II region

EXAMPLE 5 Evaluate the iterated integral $\int_0^1 \int_x^1 \sin(y^2)\, dy\, dx$.

SOLUTION If we try to evaluate the integral as it stands, we are faced with the task of first evaluating $\int \sin(y^2)\, dy$. But it's impossible to do so in finite terms since $\int \sin(y^2)\, dy$ is not an elementary function. (See the end of Section 7.5.) So we must change the order of integration. This is accomplished by first expressing the given iterated integral as a double integral. Using (3) backward, we have

$$\int_0^1 \int_x^1 \sin(y^2)\, dy\, dx = \iint_D \sin(y^2)\, dA$$

where
$$D = \left\{ (x, y) \mid 0 \leqslant x \leqslant 1,\ x \leqslant y \leqslant 1 \right\}$$

We sketch this region D in Figure 15. Then from Figure 16 we see that an alternative description of D is

$$D = \left\{ (x, y) \mid 0 \leqslant y \leqslant 1,\ 0 \leqslant x \leqslant y \right\}$$

This enables us to use (5) to express the double integral as an iterated integral in the reverse order:

$$\int_0^1 \int_x^1 \sin(y^2)\, dy\, dx = \iint_D \sin(y^2)\, dA$$

$$= \int_0^1 \int_0^y \sin(y^2)\, dx\, dy = \int_0^1 \left[x \sin(y^2) \right]_{x=0}^{x=y} dy$$

$$= \int_0^1 y \sin(y^2)\, dy = -\tfrac{1}{2} \cos(y^2) \Big]_0^1 = \tfrac{1}{2}(1 - \cos 1) \qquad ■$$

■ Properties of Double Integrals

We assume that all of the following integrals exist. For rectangular regions D the first three properties can be proved in the same manner as in Section 5.2. And then for general regions the properties follow from Definition 2.

$$\boxed{6} \qquad \iint_D [f(x, y) + g(x, y)]\, dA = \iint_D f(x, y)\, dA + \iint_D g(x, y)\, dA$$

$$\boxed{7} \qquad \iint_D cf(x, y)\, dA = c \iint_D f(x, y)\, dA \qquad \text{where } c \text{ is a constant}$$

If $f(x, y) \geqslant g(x, y)$ for all (x, y) in D, then

$$\boxed{8} \qquad \iint_D f(x, y)\, dA \geqslant \iint_D g(x, y)\, dA$$

The next property of double integrals is similar to the property of single integrals given by the equation $\int_a^b f(x)\, dx = \int_a^c f(x)\, dx + \int_c^b f(x)\, dx$.

If $D = D_1 \cup D_2$, where D_1 and D_2 don't overlap except perhaps on their boundaries (see Figure 17), then

$$\boxed{9} \qquad \iint_D f(x, y)\, dA = \iint_{D_1} f(x, y)\, dA + \iint_{D_2} f(x, y)\, dA$$

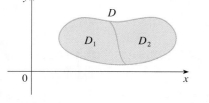

FIGURE 17

Property 9 can be used to evaluate double integrals over regions D that are neither type I nor type II but can be expressed as a union of regions of type I or type II. Figure 18 illustrates this procedure. (See Exercises 57 and 58.)

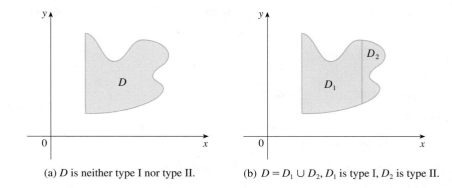

FIGURE 18

(a) D is neither type I nor type II.

(b) $D = D_1 \cup D_2$, D_1 is type I, D_2 is type II.

The next property of integrals says that if we integrate the constant function $f(x, y) = 1$ over a region D, we get the area of D:

$$\boxed{10} \qquad \iint\limits_{D} 1 \, dA = A(D)$$

FIGURE 19
Cylinder with base D and height 1

Figure 19 illustrates why Equation 10 is true: A solid cylinder whose base is D and whose height is 1 has volume $A(D) \cdot 1 = A(D)$, but we know that we can also write its volume as $\iint_D 1 \, dA$.

Finally, we can combine Properties 7, 8, and 10 to prove the following property. (See Exercise 63.)

$\boxed{11}$ If $m \leq f(x, y) \leq M$ for all (x, y) in D, then

$$mA(D) \leq \iint\limits_{D} f(x, y) \, dA \leq MA(D)$$

EXAMPLE 6 Use Property 11 to estimate the integral $\iint_D e^{\sin x \cos y} \, dA$, where D is the disk with center the origin and radius 2.

SOLUTION Since $-1 \leq \sin x \leq 1$ and $-1 \leq \cos y \leq 1$, we have $-1 \leq \sin x \cos y \leq 1$ and therefore

$$e^{-1} \leq e^{\sin x \cos y} \leq e^{1} = e$$

Thus, using $m = e^{-1} = 1/e$, $M = e$, and $A(D) = \pi(2)^2$ in Property 11, we obtain

$$\frac{4\pi}{e} \leq \iint\limits_{D} e^{\sin x \cos y} \, dA \leq 4\pi e \qquad \blacksquare$$

15.2 EXERCISES

1–6 Evaluate the iterated integral.

1. $\int_1^5 \int_0^x (8x - 2y)\, dy\, dx$

2. $\int_0^2 \int_0^{y^2} x^2 y\, dx\, dy$

3. $\int_0^1 \int_0^y x e^{y^3}\, dx\, dy$

4. $\int_0^{\pi/2} \int_0^x x \sin y\, dy\, dx$

5. $\int_0^1 \int_0^{s^2} \cos(s^3)\, dt\, ds$

6. $\int_0^1 \int_0^{e^v} \sqrt{1 + e^v}\, dw\, dv$

7–10 Evaluate the double integral.

7. $\iint_D \dfrac{y}{x^2 + 1}\, dA, \quad D = \{(x, y) \mid 0 \leqslant x \leqslant 4, 0 \leqslant y \leqslant \sqrt{x}\}$

8. $\iint_D (2x + y)\, dA, \quad D = \{(x, y) \mid 1 \leqslant y \leqslant 2, y - 1 \leqslant x \leqslant 1\}$

9. $\iint_D e^{-y^2}\, dA, \quad D = \{(x, y) \mid 0 \leqslant y \leqslant 3, 0 \leqslant x \leqslant y\}$

10. $\iint_D y\sqrt{x^2 - y^2}\, dA, \quad D = \{(x, y) \mid 0 \leqslant x \leqslant 2, 0 \leqslant y \leqslant x\}$

11. Draw an example of a region that is
(a) type I but not type II
(b) type II but not type I

12. Draw an example of a region that is
(a) both type I and type II
(b) neither type I nor type II

13–14 Express D as a region of type I and also as a region of type II. Then evaluate the double integral in two ways.

13. $\iint_D x\, dA, \quad D$ is enclosed by the lines $y = x, y = 0, x = 1$

14. $\iint_D xy\, dA, \quad D$ is enclosed by the curves $y = x^2, y = 3x$

15–16 Set up iterated integrals for both orders of integration. Then evaluate the double integral using the easier order and explain why it's easier.

15. $\iint_D y\, dA, \quad D$ is bounded by $y = x - 2, x = y^2$

16. $\iint_D y^2 e^{xy}\, dA, \quad D$ is bounded by $y = x, y = 4, x = 0$

17–22 Evaluate the double integral.

17. $\iint_D x \cos y\, dA, \quad D$ is bounded by $y = 0, \ y = x^2, \ x = 1$

18. $\iint_D (x^2 + 2y)\, dA, \quad D$ is bounded by $y = x, y = x^3, x \geqslant 0$

19. $\iint_D y^2\, dA,$
D is the triangular region with vertices $(0, 1), (1, 2), (4, 1)$

20. $\iint_D xy\, dA, \quad D$ is enclosed by the quarter-circle
$y = \sqrt{1 - x^2}, x \geqslant 0$, and the axes

21. $\iint_D (2x - y)\, dA,$
D is bounded by the circle with center the origin and radius 2

22. $\iint_D y\, dA, \quad D$ is the triangular region with vertices $(0, 0),$
$(1, 1)$, and $(4, 0)$

23–32 Find the volume of the given solid.

23. Under the plane $3x + 2y - z = 0$ and above the region enclosed by the parabolas $y = x^2$ and $x = y^2$

24. Under the surface $z = 1 + x^2 y^2$ and above the region enclosed by $x = y^2$ and $x = 4$

25. Under the surface $z = xy$ and above the triangle with vertices $(1, 1), (4, 1),$ and $(1, 2)$

26. Enclosed by the paraboloid $z = x^2 + y^2 + 1$ and the planes $x = 0, y = 0, z = 0$, and $x + y = 2$

27. The tetrahedron enclosed by the coordinate planes and the plane $2x + y + z = 4$

28. Bounded by the planes $z = x, y = x, x + y = 2$, and $z = 0$

29. Enclosed by the cylinders $z = x^2, y = x^2$ and the planes $z = 0, y = 4$

30. Bounded by the cylinder $y^2 + z^2 = 4$ and the planes $x = 2y,$ $x = 0, z = 0$ in the first octant

31. Bounded by the cylinder $x^2 + y^2 = 1$ and the planes $y = z,$ $x = 0, z = 0$ in the first octant

32. Bounded by the cylinders $x^2 + y^2 = r^2$ and $y^2 + z^2 = r^2$

33. Use a graphing calculator or computer to estimate the x-coordinates of the points of intersection of the curves $y = x^4$ and $y = 3x - x^2$. If D is the region bounded by these curves, estimate $\iint_D x\, dA$.

34. Find the approximate volume of the solid in the first octant that is bounded by the planes $y = x$, $z = 0$, and $z = x$ and the cylinder $y = \cos x$. (Use a graphing device to estimate the points of intersection.)

35–38 Find the volume of the solid by subtracting two volumes.

35. The solid enclosed by the parabolic cylinders $y = 1 - x^2$, $y = x^2 - 1$ and the planes $x + y + z = 2$, $2x + 2y - z + 10 = 0$

36. The solid enclosed by the parabolic cylinder $y = x^2$ and the planes $z = 3y$, $z = 2 + y$

37. The solid under the plane $z = 3$, above the plane $z = y$, and between the parabolic cylinders $y = x^2$ and $y = 1 - x^2$

38. The solid in the first octant under the plane $z = x + y$, above the surface $z = xy$, and enclosed by the surfaces $x = 0$, $y = 0$, and $x^2 + y^2 = 4$

39–40 Sketch the solid whose volume is given by the iterated integral.

39. $\displaystyle\int_0^1 \int_0^{1-x} (1 - x - y) \, dy \, dx$ **40.** $\displaystyle\int_0^1 \int_0^{1-x^2} (1 - x) \, dy \, dx$

CAS 41–44 Use a computer algebra system to find the exact volume of the solid.

41. Under the surface $z = x^3y^4 + xy^2$ and above the region bounded by the curves $y = x^3 - x$ and $y = x^2 + x$ for $x \geq 0$

42. Between the paraboloids $z = 2x^2 + y^2$ and $z = 8 - x^2 - 2y^2$ and inside the cylinder $x^2 + y^2 = 1$

43. Enclosed by $z = 1 - x^2 - y^2$ and $z = 0$

44. Enclosed by $z = x^2 + y^2$ and $z = 2y$

45–50 Sketch the region of integration and change the order of integration.

45. $\displaystyle\int_0^1 \int_0^y f(x, y) \, dx \, dy$ **46.** $\displaystyle\int_0^2 \int_{x^2}^4 f(x, y) \, dy \, dx$

47. $\displaystyle\int_0^{\pi/2} \int_0^{\cos x} f(x, y) \, dy \, dx$ **48.** $\displaystyle\int_{-2}^2 \int_0^{\sqrt{4-y^2}} f(x, y) \, dx \, dy$

49. $\displaystyle\int_1^2 \int_0^{\ln x} f(x, y) \, dy \, dx$ **50.** $\displaystyle\int_0^1 \int_{\arctan x}^{\pi/4} f(x, y) \, dy \, dx$

51–56 Evaluate the integral by reversing the order of integration.

51. $\displaystyle\int_0^1 \int_{3y}^3 e^{x^2} \, dx \, dy$ **52.** $\displaystyle\int_0^1 \int_{x^2}^1 \sqrt{y} \, \sin y \, dy \, dx$

53. $\displaystyle\int_0^1 \int_{\sqrt{x}}^1 \sqrt{y^3 + 1} \, dy \, dx$

54. $\displaystyle\int_0^2 \int_{y/2}^1 y \cos(x^3 - 1) \, dx \, dy$

55. $\displaystyle\int_0^1 \int_{\arcsin y}^{\pi/2} \cos x \sqrt{1 + \cos^2 x} \, dx \, dy$

56. $\displaystyle\int_0^8 \int_{\sqrt[3]{y}}^2 e^{x^4} \, dx \, dy$

57–58 Express D as a union of regions of type I or type II and evaluate the integral.

57. $\displaystyle\iint_D x^2 \, dA$ **58.** $\displaystyle\iint_D y \, dA$

59–60 Use Property 11 to estimate the value of the integral.

59. $\displaystyle\iint_S \sqrt{4 - x^2y^2} \, dA$, $S = \{(x, y) \mid x^2 + y^2 \leq 1, x \geq 0\}$

60. $\displaystyle\iint_T \sin^4(x + y) \, dA$, T is the triangle enclosed by the lines $y = 0$, $y = 2x$, and $x = 1$

61–62 Find the averge value of f over the region D.

61. $f(x, y) = xy$, D is the triangle with vertices $(0, 0)$, $(1, 0)$, and $(1, 3)$

62. $f(x, y) = x \sin y$, D is enclosed by the curves $y = 0$, $y = x^2$, and $x = 1$

63. Prove Property 11.

64. In evaluating a double integral over a region D, a sum of iterated integrals was obtained as follows:

$$\iint_D f(x, y) \, dA = \int_0^1 \int_0^{2y} f(x, y) \, dx \, dy + \int_1^3 \int_0^{3-y} f(x, y) \, dx \, dy$$

Sketch the region D and express the double integral as an iterated integral with reversed order of integration.

65–69 Use geometry or symmetry, or both, to evaluate the double integral.

65. $\iint\limits_{D} (x + 2)\, dA,$

$D = \{(x, y) \mid 0 \le y \le \sqrt{9 - x^2}\}$

66. $\iint\limits_{D} \sqrt{R^2 - x^2 - y^2}\, dA,$

D is the disk with center the origin and radius R

67. $\iint\limits_{D} (2x + 3y)\, dA,$

D is the rectangle $0 \le x \le a, 0 \le y \le b$

68. $\iint\limits_{D} (2 + x^2 y^3 - y^2 \sin x)\, dA,$

$D = \{(x, y) \mid |x| + |y| \le 1\}$

69. $\iint\limits_{D} \left(ax^3 + by^3 + \sqrt{a^2 - x^2}\right) dA,$

$D = [-a, a] \times [-b, b]$

CAS 70. Graph the solid bounded by the plane $x + y + z = 1$ and the paraboloid $z = 4 - x^2 - y^2$ and find its exact volume. (Use your CAS to do the graphing, to find the equations of the boundary curves of the region of integration, and to evaluate the double integral.)

15.3   Double Integrals in Polar Coordinates

Suppose that we want to evaluate a double integral $\iint_R f(x, y)\, dA$, where R is one of the regions shown in Figure 1. In either case the description of R in terms of rectangular coordinates is rather complicated, but R is easily described using polar coordinates.

FIGURE 1

(a) $R = \{(r, \theta) \mid 0 \le r \le 1, 0 \le \theta \le 2\pi\}$ (b) $R = \{(r, \theta) \mid 1 \le r \le 2, 0 \le \theta \le \pi\}$

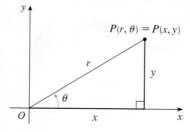

FIGURE 2

Recall from Figure 2 that the polar coordinates (r, θ) of a point are related to the rectangular coordinates (x, y) by the equations

$$r^2 = x^2 + y^2 \qquad x = r \cos \theta \qquad y = r \sin \theta$$

(See Section 10.3.)

The regions in Figure 1 are special cases of a **polar rectangle**

$$R = \{(r, \theta) \mid a \le r \le b, \alpha \le \theta \le \beta\}$$

which is shown in Figure 3. In order to compute the double integral $\iint_R f(x, y)\, dA$, where R is a polar rectangle, we divide the interval $[a, b]$ into m subintervals $[r_{i-1}, r_i]$ of equal width $\Delta r = (b - a)/m$ and we divide the interval $[\alpha, \beta]$ into n subintervals $[\theta_{j-1}, \theta_j]$ of equal width $\Delta \theta = (\beta - \alpha)/n$. Then the circles $r = r_i$ and the rays $\theta = \theta_j$ divide the polar rectangle R into the small polar rectangles R_{ij} shown in Figure 4.

FIGURE 3 Polar rectangle

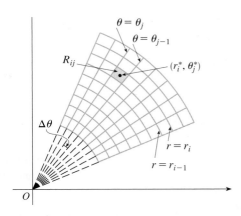

FIGURE 4 Dividing R into polar subrectangles

The "center" of the polar subrectangle

$$R_{ij} = \left\{ (r, \theta) \mid r_{i-1} \leq r \leq r_i, \; \theta_{j-1} \leq \theta \leq \theta_j \right\}$$

has polar coordinates

$$r_i^* = \tfrac{1}{2}(r_{i-1} + r_i) \qquad \theta_j^* = \tfrac{1}{2}(\theta_{j-1} + \theta_j)$$

We compute the area of R_{ij} using the fact that the area of a sector of a circle with radius r and central angle θ is $\tfrac{1}{2}r^2\theta$. Subtracting the areas of two such sectors, each of which has central angle $\Delta\theta = \theta_j - \theta_{j-1}$, we find that the area of R_{ij} is

$$\Delta A_i = \tfrac{1}{2}r_i^2 \Delta\theta - \tfrac{1}{2}r_{i-1}^2 \Delta\theta = \tfrac{1}{2}\left(r_i^2 - r_{i-1}^2\right)\Delta\theta$$

$$= \tfrac{1}{2}(r_i + r_{i-1})(r_i - r_{i-1})\,\Delta\theta = r_i^*\,\Delta r\,\Delta\theta$$

Although we have defined the double integral $\iint_R f(x, y)\, dA$ in terms of ordinary rectangles, it can be shown that, for continuous functions f, we always obtain the same answer using polar rectangles. The rectangular coordinates of the center of R_{ij} are $(r_i^* \cos\theta_j^*, \, r_i^* \sin\theta_j^*)$, so a typical Riemann sum is

$$\boxed{1} \quad \sum_{i=1}^{m}\sum_{j=1}^{n} f(r_i^* \cos\theta_j^*, \, r_i^* \sin\theta_j^*)\,\Delta A_i = \sum_{i=1}^{m}\sum_{j=1}^{n} f(r_i^* \cos\theta_j^*, \, r_i^* \sin\theta_j^*)\, r_i^*\,\Delta r\,\Delta\theta$$

If we write $g(r, \theta) = r f(r\cos\theta, r\sin\theta)$, then the Riemann sum in Equation 1 can be written as

$$\sum_{i=1}^{m}\sum_{j=1}^{n} g(r_i^*, \theta_j^*)\,\Delta r\,\Delta\theta$$

which is a Riemann sum for the double integral

$$\int_{\alpha}^{\beta} \int_a^b g(r, \theta)\, dr\, d\theta$$

Therefore we have

$$\iint\limits_R f(x, y)\, dA = \lim_{m,n\to\infty} \sum_{i=1}^{m}\sum_{j=1}^{n} f(r_i^* \cos\theta_j^*, \, r_i^* \sin\theta_j^*)\,\Delta A_i$$

$$= \lim_{m,n\to\infty} \sum_{i=1}^{m}\sum_{j=1}^{n} g(r_i^*, \theta_j^*)\,\Delta r\,\Delta\theta = \int_{\alpha}^{\beta} \int_a^b g(r, \theta)\, dr\, d\theta$$

$$= \int_{\alpha}^{\beta} \int_a^b f(r\cos\theta, \, r\sin\theta)\, r\, dr\, d\theta$$

> **2** **Change to Polar Coordinates in a Double Integral** If f is continuous on a polar rectangle R given by $0 \leqslant a \leqslant r \leqslant b$, $\alpha \leqslant \theta \leqslant \beta$, where $0 \leqslant \beta - \alpha \leqslant 2\pi$, then
> $$\iint_R f(x, y) \, dA = \int_\alpha^\beta \int_a^b f(r \cos \theta, r \sin \theta) \, r \, dr \, d\theta$$

The formula in (2) says that we convert from rectangular to polar coordinates in a double integral by writing $x = r \cos \theta$ and $y = r \sin \theta$, using the appropriate limits of integration for r and θ, and replacing dA by $r \, dr \, d\theta$. Be careful not to forget the additional factor r on the right side of Formula 2. A classical method for remembering this is shown in Figure 5, where the "infinitesimal" polar rectangle can be thought of as an ordinary rectangle with dimensions $r \, d\theta$ and dr and therefore has "area" $dA = r \, dr \, d\theta$.

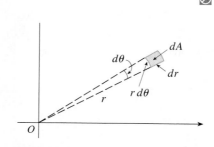

FIGURE 5

EXAMPLE 1 Evaluate $\iint_R (3x + 4y^2) \, dA$, where R is the region in the upper half-plane bounded by the circles $x^2 + y^2 = 1$ and $x^2 + y^2 = 4$.

SOLUTION The region R can be described as

$$R = \{(x, y) \mid y \geqslant 0, \ 1 \leqslant x^2 + y^2 \leqslant 4\}$$

It is the half-ring shown in Figure 1(b), and in polar coordinates it is given by $1 \leqslant r \leqslant 2$, $0 \leqslant \theta \leqslant \pi$. Therefore, by Formula 2,

$$\iint_R (3x + 4y^2) \, dA = \int_0^\pi \int_1^2 (3r \cos \theta + 4r^2 \sin^2\theta) \, r \, dr \, d\theta$$

$$= \int_0^\pi \int_1^2 (3r^2 \cos \theta + 4r^3 \sin^2\theta) \, dr \, d\theta$$

$$= \int_0^\pi \left[r^3 \cos \theta + r^4 \sin^2\theta \right]_{r=1}^{r=2} d\theta = \int_0^\pi (7 \cos \theta + 15 \sin^2\theta) \, d\theta$$

$$= \int_0^\pi \left[7 \cos \theta + \tfrac{15}{2}(1 - \cos 2\theta) \right] d\theta$$

$$= 7 \sin \theta + \frac{15\theta}{2} - \frac{15}{4} \sin 2\theta \Big]_0^\pi = \frac{15\pi}{2} \qquad \blacksquare$$

Here we use the trigonometric identity
$$\sin^2\theta = \tfrac{1}{2}(1 - \cos 2\theta)$$

See Section 7.2 for advice on integrating trigonometric functions.

EXAMPLE 2 Find the volume of the solid bounded by the plane $z = 0$ and the paraboloid $z = 1 - x^2 - y^2$.

SOLUTION If we put $z = 0$ in the equation of the paraboloid, we get $x^2 + y^2 = 1$. This means that the plane intersects the paraboloid in the circle $x^2 + y^2 = 1$, so the solid lies under the paraboloid and above the circular disk D given by $x^2 + y^2 \leqslant 1$ [see Figures 6 and 1(a)]. In polar coordinates D is given by $0 \leqslant r \leqslant 1$, $0 \leqslant \theta \leqslant 2\pi$. Since $1 - x^2 - y^2 = 1 - r^2$, the volume is

$$V = \iint_D (1 - x^2 - y^2) \, dA = \int_0^{2\pi} \int_0^1 (1 - r^2) \, r \, dr \, d\theta$$

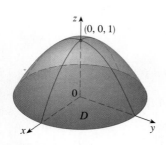

FIGURE 6

$$= \int_0^{2\pi} d\theta \int_0^1 (r - r^3) \, dr = 2\pi \left[\frac{r^2}{2} - \frac{r^4}{4} \right]_0^1 = \frac{\pi}{2}$$

If we had used rectangular coordinates instead of polar coordinates, then we would have obtained

$$V = \iint_D (1 - x^2 - y^2)\, dA = \int_{-1}^{1} \int_{-\sqrt{1-x^2}}^{\sqrt{1-x^2}} (1 - x^2 - y^2)\, dy\, dx$$

which is not easy to evaluate because it involves finding $\int (1 - x^2)^{3/2}\, dx$. ∎

What we have done so far can be extended to the more complicated type of region shown in Figure 7. It's similar to the type II rectangular regions considered in Section 15.2. In fact, by combining Formula 2 in this section with Formula 15.2.5, we obtain the following formula.

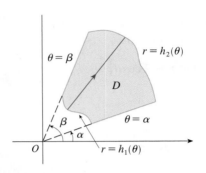

$\theta = \beta$ $r = h_2(\theta)$

D

β $\theta = \alpha$

α

O $r = h_1(\theta)$

FIGURE 7
$D = \{(r, \theta) \mid \alpha \le \theta \le \beta,\, h_1(\theta) \le r \le h_2(\theta)\}$

> **3** If f is continuous on a polar region of the form
>
> $$D = \{(r, \theta) \mid \alpha \le \theta \le \beta,\ h_1(\theta) \le r \le h_2(\theta)\}$$
>
> then
>
> $$\iint_D f(x, y)\, dA = \int_{\alpha}^{\beta} \int_{h_1(\theta)}^{h_2(\theta)} f(r\cos\theta,\, r\sin\theta)\, r\, dr\, d\theta$$

In particular, taking $f(x, y) = 1$, $h_1(\theta) = 0$, and $h_2(\theta) = h(\theta)$ in this formula, we see that the area of the region D bounded by $\theta = \alpha$, $\theta = \beta$, and $r = h(\theta)$ is

$$A(D) = \iint_D 1\, dA = \int_{\alpha}^{\beta} \int_{0}^{h(\theta)} r\, dr\, d\theta$$

$$= \int_{\alpha}^{\beta} \left[\frac{r^2}{2} \right]_{0}^{h(\theta)} d\theta = \int_{\alpha}^{\beta} \tfrac{1}{2} [h(\theta)]^2\, d\theta$$

and this agrees with Formula 10.4.3.

EXAMPLE 3 Use a double integral to find the area enclosed by one loop of the four-leaved rose $r = \cos 2\theta$.

SOLUTION From the sketch of the curve in Figure 8, we see that a loop is given by the region

$$D = \{(r, \theta) \mid -\pi/4 \le \theta \le \pi/4,\, 0 \le r \le \cos 2\theta\}$$

So the area is

$$A(D) = \iint_D dA = \int_{-\pi/4}^{\pi/4} \int_{0}^{\cos 2\theta} r\, dr\, d\theta$$

$$= \int_{-\pi/4}^{\pi/4} \left[\tfrac{1}{2} r^2 \right]_{0}^{\cos 2\theta} d\theta = \tfrac{1}{2} \int_{-\pi/4}^{\pi/4} \cos^2 2\theta\, d\theta$$

$$= \tfrac{1}{4} \int_{-\pi/4}^{\pi/4} (1 + \cos 4\theta)\, d\theta = \tfrac{1}{4} \left[\theta + \tfrac{1}{4} \sin 4\theta \right]_{-\pi/4}^{\pi/4} = \frac{\pi}{8}$$ ∎

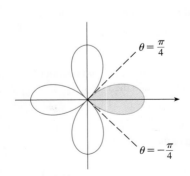

$\theta = \dfrac{\pi}{4}$

$\theta = -\dfrac{\pi}{4}$

FIGURE 8

$(x-1)^2 + y^2 = 1$
(or $r = 2\cos\theta$)

D

FIGURE 9

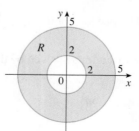

FIGURE 10

EXAMPLE 4 Find the volume of the solid that lies under the paraboloid $z = x^2 + y^2$, above the xy-plane, and inside the cylinder $x^2 + y^2 = 2x$.

SOLUTION The solid lies above the disk D whose boundary circle has equation $x^2 + y^2 = 2x$ or, after completing the square,

$$(x - 1)^2 + y^2 = 1$$

(See Figures 9 and 10.)

In polar coordinates we have $x^2 + y^2 = r^2$ and $x = r\cos\theta$, so the boundary circle becomes $r^2 = 2r\cos\theta$, or $r = 2\cos\theta$. Thus the disk D is given by

$$D = \left\{(r, \theta) \mid -\pi/2 \leq \theta \leq \pi/2, 0 \leq r \leq 2\cos\theta\right\}$$

and, by Formula 3, we have

$$V = \iint_D (x^2 + y^2)\, dA = \int_{-\pi/2}^{\pi/2} \int_0^{2\cos\theta} r^2\, r\, dr\, d\theta = \int_{-\pi/2}^{\pi/2} \left[\frac{r^4}{4}\right]_0^{2\cos\theta} d\theta$$

$$= 4 \int_{-\pi/2}^{\pi/2} \cos^4\theta\, d\theta = 8 \int_0^{\pi/2} \cos^4\theta\, d\theta = 8 \int_0^{\pi/2} \left(\frac{1 + \cos 2\theta}{2}\right)^2 d\theta$$

$$= 2 \int_0^{\pi/2} \left[1 + 2\cos 2\theta + \tfrac{1}{2}(1 + \cos 4\theta)\right] d\theta$$

$$= 2\left[\tfrac{3}{2}\theta + \sin 2\theta + \tfrac{1}{8}\sin 4\theta\right]_0^{\pi/2} = 2\left(\frac{3}{2}\right)\left(\frac{\pi}{2}\right) = \frac{3\pi}{2} \qquad\blacksquare$$

15.3 EXERCISES

1–4 A region R is shown. Decide whether to use polar coordinates or rectangular coordinates and write $\iint_R f(x, y)\, dA$ as an iterated integral, where f is an arbitrary continuous function on R.

1.

2.

3.

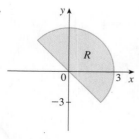

4.

5–6 Sketch the region whose area is given by the integral and evaluate the integral.

5. $\int_{\pi/4}^{3\pi/4} \int_1^2 r\, dr\, d\theta$

6. $\int_{\pi/2}^{\pi} \int_0^{2\sin\theta} r\, dr\, d\theta$

7–14 Evaluate the given integral by changing to polar coordinates.

7. $\iint_D x^2 y\, dA$, where D is the top half of the disk with center the origin and radius 5

8. $\iint_R (2x - y)\, dA$, where R is the region in the first quadrant enclosed by the circle $x^2 + y^2 = 4$ and the lines $x = 0$ and $y = x$

9. $\iint_R \sin(x^2 + y^2)\, dA$, where R is the region in the first quadrant between the circles with center the origin and radii 1 and 3

10. $\iint_R \dfrac{y^2}{x^2 + y^2}\, dA$, where R is the region that lies between the circles $x^2 + y^2 = a^2$ and $x^2 + y^2 = b^2$ with $0 < a < b$

11. $\iint_D e^{-x^2 - y^2}\, dA$, where D is the region bounded by the semicircle $x = \sqrt{4 - y^2}$ and the y-axis

12. $\iint_D \cos\sqrt{x^2 + y^2}\, dA$, where D is the disk with center the origin and radius 2

13. $\iint_R \arctan(y/x)\, dA$,
where $R = \{(x, y) \mid 1 \leqslant x^2 + y^2 \leqslant 4,\ 0 \leqslant y \leqslant x\}$

14. $\iint_D x\, dA$, where D is the region in the first quadrant that lies between the circles $x^2 + y^2 = 4$ and $x^2 + y^2 = 2x$

15–18 Use a double integral to find the area of the region.

15. One loop of the rose $r = \cos 3\theta$

16. The region enclosed by both of the cardioids $r = 1 + \cos\theta$ and $r = 1 - \cos\theta$

17. The region inside the circle $(x - 1)^2 + y^2 = 1$ and outside the circle $x^2 + y^2 = 1$

18. The region inside the cardioid $r = 1 + \cos\theta$ and outside the circle $r = 3\cos\theta$

19–27 Use polar coordinates to find the volume of the given solid.

19. Under the paraboloid $z = x^2 + y^2$ and above the disk $x^2 + y^2 \leqslant 25$

20. Below the cone $z = \sqrt{x^2 + y^2}$ and above the ring $1 \leqslant x^2 + y^2 \leqslant 4$

21. Below the plane $2x + y + z = 4$ and above the disk $x^2 + y^2 \leqslant 1$

22. Inside the sphere $x^2 + y^2 + z^2 = 16$ and outside the cylinder $x^2 + y^2 = 4$

23. A sphere of radius a

24. Bounded by the paraboloid $z = 1 + 2x^2 + 2y^2$ and the plane $z = 7$ in the first octant

25. Above the cone $z = \sqrt{x^2 + y^2}$ and below the sphere $x^2 + y^2 + z^2 = 1$

26. Bounded by the paraboloids $z = 6 - x^2 - y^2$ and $z = 2x^2 + 2y^2$

27. Inside both the cylinder $x^2 + y^2 = 4$ and the ellipsoid $4x^2 + 4y^2 + z^2 = 64$

---

28. (a) A cylindrical drill with radius r_1 is used to bore a hole through the center of a sphere of radius r_2. Find the volume of the ring-shaped solid that remains.
(b) Express the volume in part (a) in terms of the height h of the ring. Notice that the volume depends only on h, not on r_1 or r_2.

29–32 Evaluate the iterated integral by converting to polar coordinates.

29. $\int_0^2 \int_0^{\sqrt{4-x^2}} e^{-x^2-y^2}\, dy\, dx$

30. $\int_0^a \int_{-\sqrt{a^2-y^2}}^{\sqrt{a^2-y^2}} (2x + y)\, dx\, dy$

31. $\int_0^{1/2} \int_{\sqrt{3}\,y}^{\sqrt{1-y^2}} xy^2\, dx\, dy$

32. $\int_0^2 \int_0^{\sqrt{2x-x^2}} \sqrt{x^2 + y^2}\, dy\, dx$

33–34 Express the double integral in terms of a single integral with respect to r. Then use your calculator to evaluate the integral correct to four decimal places.

33. $\iint_D e^{(x^2+y^2)^2}\, dA$, where D is the disk with center the origin and radius 1

34. $\iint_D xy\sqrt{1 + x^2 + y^2}\, dA$, where D is the portion of the disk $x^2 + y^2 \leqslant 1$ that lies in the first quadrant

---

35. A swimming pool is circular with a 40-ft diameter. The depth is constant along east-west lines and increases linearly from 2 ft at the south end to 7 ft at the north end. Find the volume of water in the pool.

36. An agricultural sprinkler distributes water in a circular pattern of radius 100 ft. It supplies water to a depth of e^{-r} feet per hour at a distance of r feet from the sprinkler.
(a) If $0 < R \leqslant 100$, what is the total amount of water supplied per hour to the region inside the circle of radius R centered at the sprinkler?
(b) Determine an expression for the average amount of water per hour per square foot supplied to the region inside the circle of radius R.

37. Find the average value of the function $f(x, y) = 1/\sqrt{x^2 + y^2}$ on the annular region $a^2 \leqslant x^2 + y^2 \leqslant b^2$, where $0 < a < b$.

38. Let D be the disk with center the origin and radius a. What is the average distance from points in D to the origin?

39. Use polar coordinates to combine the sum

$$\int_{1/\sqrt{2}}^1 \int_{\sqrt{1-x^2}}^x xy\, dy\, dx + \int_1^{\sqrt{2}} \int_0^x xy\, dy\, dx + \int_{\sqrt{2}}^2 \int_0^{\sqrt{4-x^2}} xy\, dy\, dx$$

into one double integral. Then evaluate the double integral.

40. (a) We define the improper integral (over the entire plane $\mathbb{R}^2$)

$$I = \iint_{\mathbb{R}^2} e^{-(x^2+y^2)}\, dA$$

$$= \int_{-\infty}^{\infty} \int_{-\infty}^{\infty} e^{-(x^2+y^2)}\, dy\, dx$$

$$= \lim_{a\to\infty} \iint_{D_a} e^{-(x^2+y^2)}\, dA$$

where D_a is the disk with radius a and center the origin. Show that

$$\int_{-\infty}^{\infty} \int_{-\infty}^{\infty} e^{-(x^2+y^2)}\, dA = \pi$$

(b) An equivalent definition of the improper integral in part (a) is

$$\iint\limits_{\mathbb{R}^2} e^{-(x^2+y^2)} \, dA = \lim_{a \to \infty} \iint\limits_{S_a} e^{-(x^2+y^2)} \, dA$$

where S_a is the square with vertices $(\pm a, \pm a)$. Use this to show that

$$\int_{-\infty}^{\infty} e^{-x^2} \, dx \int_{-\infty}^{\infty} e^{-y^2} \, dy = \pi$$

(c) Deduce that

$$\int_{-\infty}^{\infty} e^{-x^2} \, dx = \sqrt{\pi}$$

(d) By making the change of variable $t = \sqrt{2}\, x$, show that

$$\int_{-\infty}^{\infty} e^{-x^2/2} \, dx = \sqrt{2\pi}$$

(This is a fundamental result for probability and statistics.)

41. Use the result of Exercise 40 part (c) to evaluate the following integrals.

(a) $\int_0^{\infty} x^2 e^{-x^2} \, dx$ (b) $\int_0^{\infty} \sqrt{x}\, e^{-x} \, dx$

15.4 Applications of Double Integrals

We have already seen one application of double integrals: computing volumes. Another geometric application is finding areas of surfaces and this will be done in the next section. In this section we explore physical applications such as computing mass, electric charge, center of mass, and moment of inertia. We will see that these physical ideas are also important when applied to probability density functions of two random variables.

◾ Density and Mass

In Section 8.3 we were able to use single integrals to compute moments and the center of mass of a thin plate or lamina with constant density. But now, equipped with the double integral, we can consider a lamina with variable density. Suppose the lamina occupies a region D of the xy-plane and its **density** (in units of mass per unit area) at a point (x, y) in D is given by $\rho(x, y)$, where ρ is a continuous function on D. This means that

$$\rho(x, y) = \lim \frac{\Delta m}{\Delta A}$$

where Δm and ΔA are the mass and area of a small rectangle that contains (x, y) and the limit is taken as the dimensions of the rectangle approach 0. (See Figure 1.)

To find the total mass m of the lamina we divide a rectangle R containing D into subrectangles R_{ij} of the same size (as in Figure 2) and consider $\rho(x, y)$ to be 0 outside D. If we choose a point (x_{ij}^*, y_{ij}^*) in R_{ij}, then the mass of the part of the lamina that occupies R_{ij} is approximately $\rho(x_{ij}^*, y_{ij}^*)\, \Delta A$, where ΔA is the area of R_{ij}. If we add all such masses, we get an approximation to the total mass:

$$m \approx \sum_{i=1}^{k} \sum_{j=1}^{l} \rho(x_{ij}^*, y_{ij}^*)\, \Delta A$$

If we now increase the number of subrectangles, we obtain the total mass m of the lamina as the limiting value of the approximations:

FIGURE 1

FIGURE 2

$$\boxed{1} \qquad m = \lim_{k, l \to \infty} \sum_{i=1}^{k} \sum_{j=1}^{l} \rho(x_{ij}^*, y_{ij}^*)\, \Delta A = \iint\limits_{D} \rho(x, y) \, dA$$

Physicists also consider other types of density that can be treated in the same manner. For example, if an electric charge is distributed over a region D and the charge density (in units of charge per unit area) is given by $\sigma(x, y)$ at a point (x, y) in D, then the total charge Q is given by

$$\boxed{2} \qquad Q = \iint_D \sigma(x, y)\, dA$$

EXAMPLE 1 Charge is distributed over the triangular region D in Figure 3 so that the charge density at (x, y) is $\sigma(x, y) = xy$, measured in coulombs per square meter (C/m^2). Find the total charge.

SOLUTION From Equation 2 and Figure 3 we have

$$Q = \iint_D \sigma(x, y)\, dA = \int_0^1 \int_{1-x}^1 xy\, dy\, dx$$

$$= \int_0^1 \left[x \frac{y^2}{2} \right]_{y=1-x}^{y=1} dx = \int_0^1 \frac{x}{2} [1^2 - (1-x)^2]\, dx$$

$$= \tfrac{1}{2} \int_0^1 (2x^2 - x^3)\, dx = \frac{1}{2} \left[\frac{2x^3}{3} - \frac{x^4}{4} \right]_0^1 = \frac{5}{24}$$

Thus the total charge is $\frac{5}{24}$ C. ∎

Moments and Centers of Mass

In Section 8.3 we found the center of mass of a lamina with constant density; here we consider a lamina with variable density. Suppose the lamina occupies a region D and has density function $\rho(x, y)$. Recall from Chapter 8 that we defined the moment of a particle about an axis as the product of its mass and its directed distance from the axis. We divide D into small rectangles as in Figure 2. Then the mass of R_{ij} is approximately $\rho(x_{ij}^*, y_{ij}^*) \Delta A$, so we can approximate the moment of R_{ij} with respect to the x-axis by

$$[\rho(x_{ij}^*, y_{ij}^*) \Delta A] y_{ij}^*$$

If we now add these quantities and take the limit as the number of subrectangles becomes large, we obtain the **moment** of the entire lamina **about the x-axis**:

$$\boxed{3} \qquad M_x = \lim_{m, n \to \infty} \sum_{i=1}^m \sum_{j=1}^n y_{ij}^* \rho(x_{ij}^*, y_{ij}^*) \Delta A = \iint_D y \rho(x, y)\, dA$$

Similarly, the **moment about the y-axis** is

$$\boxed{4} \qquad M_y = \lim_{m, n \to \infty} \sum_{i=1}^m \sum_{j=1}^n x_{ij}^* \rho(x_{ij}^*, y_{ij}^*) \Delta A = \iint_D x \rho(x, y)\, dA$$

As before, we define the center of mass $(\bar{x}, \bar{y})$ so that $m\bar{x} = M_y$ and $m\bar{y} = M_x$. The physical significance is that the lamina behaves as if its entire mass is concentrated at its center of mass. Thus the lamina balances horizontally when supported at its center of mass (see Figure 4).

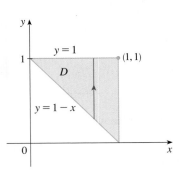

$y = 1$
$(1, 1)$
D
$y = 1 - x$

FIGURE 3

$(\bar{x}, \bar{y})$
D

FIGURE 4

5 The coordinates $(\bar{x}, \bar{y})$ of the center of mass of a lamina occupying the region D and having density function $\rho(x, y)$ are

$$\bar{x} = \frac{M_y}{m} = \frac{1}{m} \iint_D x\rho(x, y) \, dA \qquad \bar{y} = \frac{M_x}{m} = \frac{1}{m} \iint_D y\rho(x, y) \, dA$$

where the mass m is given by

$$m = \iint_D \rho(x, y) \, dA$$

EXAMPLE 2 Find the mass and center of mass of a triangular lamina with vertices $(0, 0)$, $(1, 0)$, and $(0, 2)$ if the density function is $\rho(x, y) = 1 + 3x + y$.

SOLUTION The triangle is shown in Figure 5. (Note that the equation of the upper boundary is $y = 2 - 2x$.) The mass of the lamina is

$$m = \iint_D \rho(x, y) \, dA = \int_0^1 \int_0^{2-2x} (1 + 3x + y) \, dy \, dx$$

$$= \int_0^1 \left[y + 3xy + \frac{y^2}{2} \right]_{y=0}^{y=2-2x} dx$$

$$= 4 \int_0^1 (1 - x^2) \, dx = 4 \left[x - \frac{x^3}{3} \right]_0^1 = \frac{8}{3}$$

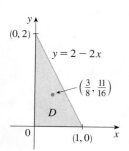

(0, 2)

$y = 2 - 2x$

$\left(\frac{3}{8}, \frac{11}{16} \right)$

D

0

(1, 0) x

FIGURE 5

Then the formulas in (5) give

$$\bar{x} = \frac{1}{m} \iint_D x\rho(x, y) \, dA = \frac{3}{8} \int_0^1 \int_0^{2-2x} (x + 3x^2 + xy) \, dy \, dx$$

$$= \frac{3}{8} \int_0^1 \left[xy + 3x^2 y + x \frac{y^2}{2} \right]_{y=0}^{y=2-2x} dx$$

$$= \frac{3}{2} \int_0^1 (x - x^3) \, dx = \frac{3}{2} \left[\frac{x^2}{2} - \frac{x^4}{4} \right]_0^1 = \frac{3}{8}$$

$$\bar{y} = \frac{1}{m} \iint_D y\rho(x, y) \, dA = \frac{3}{8} \int_0^1 \int_0^{2-2x} (y + 3xy + y^2) \, dy \, dx$$

$$= \frac{3}{8} \int_0^1 \left[\frac{y^2}{2} + 3x \frac{y^2}{2} + \frac{y^3}{3} \right]_{y=0}^{y=2-2x} dx = \frac{1}{4} \int_0^1 (7 - 9x - 3x^2 + 5x^3) \, dx$$

$$= \frac{1}{4} \left[7x - 9\frac{x^2}{2} - x^3 + 5\frac{x^4}{4} \right]_0^1 = \frac{11}{16}$$

The center of mass is at the point $\left(\frac{3}{8}, \frac{11}{16} \right)$. ∎

EXAMPLE 3 The density at any point on a semicircular lamina is proportional to the distance from the center of the circle. Find the center of mass of the lamina.

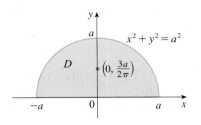

FIGURE 6

SOLUTION Let's place the lamina as the upper half of the circle $x^2 + y^2 = a^2$. (See Figure 6.) Then the distance from a point (x, y) to the center of the circle (the origin) is $\sqrt{x^2 + y^2}$. Therefore the density function is

$$\rho(x, y) = K\sqrt{x^2 + y^2}$$

where K is some constant. Both the density function and the shape of the lamina suggest that we convert to polar coordinates. Then $\sqrt{x^2 + y^2} = r$ and the region D is given by $0 \le r \le a$, $0 \le \theta \le \pi$. Thus the mass of the lamina is

$$m = \iint_D \rho(x, y)\, dA = \iint_D K\sqrt{x^2 + y^2}\, dA$$

$$= \int_0^\pi \int_0^a (Kr)\, r\, dr\, d\theta = K \int_0^\pi d\theta \int_0^a r^2\, dr$$

$$= K\pi \left. \frac{r^3}{3} \right]_0^a = \frac{K\pi a^3}{3}$$

Both the lamina and the density function are symmetric with respect to the y-axis, so the center of mass must lie on the y-axis, that is, $\bar{x} = 0$. The y-coordinate is given by

Compare the location of the center of mass in Example 3 with Example 8.3.4, where we found that the center of mass of a lamina with the same shape but uniform density is located at the point $(0, 4a/(3\pi))$.

$$\bar{y} = \frac{1}{m} \iint_D y\rho(x, y)\, dA = \frac{3}{K\pi a^3} \int_0^\pi \int_0^a r\sin\theta\,(Kr)\, r\, dr\, d\theta$$

$$= \frac{3}{\pi a^3} \int_0^\pi \sin\theta\, d\theta \int_0^a r^3\, dr = \frac{3}{\pi a^3} \left[-\cos\theta \right]_0^\pi \left[\frac{r^4}{4} \right]_0^a$$

$$= \frac{3}{\pi a^3}\, \frac{2a^4}{4} = \frac{3a}{2\pi}$$

Therefore the center of mass is located at the point $(0, 3a/(2\pi))$. ∎

■ Moment of Inertia

The **moment of inertia** (also called the **second moment**) of a particle of mass m about an axis is defined to be mr^2, where r is the distance from the particle to the axis. We extend this concept to a lamina with density function $\rho(x, y)$ and occupying a region D by proceeding as we did for ordinary moments. We divide D into small rectangles, approximate the moment of inertia of each subrectangle about the x-axis, and take the limit of the sum as the number of subrectangles becomes large. The result is the **moment of inertia** of the lamina **about the x-axis**:

6
$$I_x = \lim_{m,\, n \to \infty} \sum_{i=1}^m \sum_{j=1}^n (y_{ij}^*)^2\, \rho(x_{ij}^*, y_{ij}^*)\, \Delta A = \iint_D y^2\, \rho(x, y)\, dA$$

Similarly, the **moment of inertia about the y-axis** is

7
$$I_y = \lim_{m,\, n \to \infty} \sum_{i=1}^m \sum_{j=1}^n (x_{ij}^*)^2\, \rho(x_{ij}^*, y_{ij}^*)\, \Delta A = \iint_D x^2\, \rho(x, y)\, dA$$

It is also of interest to consider the **moment of inertia about the origin**, also called the **polar moment of inertia**:

$$\boxed{8} \quad I_0 = \lim_{m, n \to \infty} \sum_{i=1}^{m} \sum_{j=1}^{n} \left[(x_{ij}^*)^2 + (y_{ij}^*)^2\right] \rho(x_{ij}^*, y_{ij}^*) \, \Delta A = \iint_D (x^2 + y^2) \rho(x, y) \, dA$$

Note that $I_0 = I_x + I_y$.

EXAMPLE 4 Find the moments of inertia I_x, I_y, and I_0 of a homogeneous disk D with density $\rho(x, y) = \rho$, center the origin, and radius a.

SOLUTION The boundary of D is the circle $x^2 + y^2 = a^2$ and in polar coordinates D is described by $0 \le \theta \le 2\pi$, $0 \le r \le a$. Let's compute I_0 first:

$$I_0 = \iint_D (x^2 + y^2)\rho \, dA = \rho \int_0^{2\pi} \int_0^a r^2 r \, dr \, d\theta$$

$$= \rho \int_0^{2\pi} d\theta \int_0^a r^3 \, dr = 2\pi\rho \left[\frac{r^4}{4}\right]_0^a = \frac{\pi\rho a^4}{2}$$

Instead of computing I_x and I_y directly, we use the facts that $I_x + I_y = I_0$ and $I_x = I_y$ (from the symmetry of the problem). Thus

$$I_x = I_y = \frac{I_0}{2} = \frac{\pi\rho a^4}{4}$$
∎

In Example 4 notice that the mass of the disk is

$$m = \text{density} \times \text{area} = \rho(\pi a^2)$$

so the moment of inertia of the disk about the origin (like a wheel about its axle) can be written as

$$I_0 = \frac{\pi\rho a^4}{2} = \tfrac{1}{2}(\rho\pi a^2)a^2 = \tfrac{1}{2}ma^2$$

Thus if we increase the mass or the radius of the disk, we thereby increase the moment of inertia. In general, the moment of inertia plays much the same role in rotational motion that mass plays in linear motion. The moment of inertia of a wheel is what makes it difficult to start or stop the rotation of the wheel, just as the mass of a car is what makes it difficult to start or stop the motion of the car.

The **radius of gyration of a lamina about an axis** is the number R such that

$$\boxed{9} \qquad\qquad mR^2 = I$$

where m is the mass of the lamina and I is the moment of inertia about the given axis. Equation 9 says that if the mass of the lamina were concentrated at a distance R from the axis, then the moment of inertia of this "point mass" would be the same as the moment of inertia of the lamina.

In particular, the radius of gyration $\bar{\bar{y}}$ with respect to the x-axis and the radius of gyration $\bar{\bar{x}}$ with respect to the y-axis are given by the equations

$$\boxed{10} \qquad\qquad m\bar{\bar{y}}^2 = I_x \qquad\qquad m\bar{\bar{x}}^2 = I_y$$

Thus $(\bar{\bar{x}}, \bar{\bar{y}})$ is the point at which the mass of the lamina can be concentrated without changing the moments of inertia with respect to the coordinate axes. (Note the analogy with the center of mass.)

EXAMPLE 5 Find the radius of gyration about the x-axis of the disk in Example 4.

SOLUTION As noted, the mass of the disk is $m = \rho\pi a^2$, so from Equations 10 we have

$$\bar{\bar{y}}^2 = \frac{I_x}{m} = \frac{\frac{1}{4}\pi\rho a^4}{\rho\pi a^2} = \frac{a^2}{4}$$

Therefore the radius of gyration about the x-axis is $\bar{\bar{y}} = \frac{1}{2}a$, which is half the radius of the disk. ∎

■ Probability

In Section 8.5 we considered the *probability density function* f of a continuous random variable X. This means that $f(x) \geqslant 0$ for all x, $\int_{-\infty}^{\infty} f(x)\,dx = 1$, and the probability that X lies between a and b is found by integrating f from a to b:

$$P(a \leqslant X \leqslant b) = \int_a^b f(x)\,dx$$

Now we consider a pair of continuous random variables X and Y, such as the lifetimes of two components of a machine or the height and weight of an adult female chosen at random. The **joint density function** of X and Y is a function f of two variables such that the probability that (X, Y) lies in a region D is

$$P((X, Y) \in D) = \iint\limits_{D} f(x, y)\,dA$$

In particular, if the region is a rectangle, the probability that X lies between a and b and Y lies between c and d is

$$P(a \leqslant X \leqslant b,\ c \leqslant Y \leqslant d) = \int_a^b \int_c^d f(x, y)\,dy\,dx$$

(See Figure 7.)

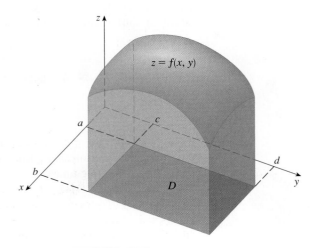

FIGURE 7

The probability that X lies between a and b and Y lies between c and d is the volume that lies above the rectangle $D = [a, b] \times [c, d]$ and below the graph of the joint density function.

Because probabilities aren't negative and are measured on a scale from 0 to 1, the joint density function has the following properties:

$$f(x, y) \geqslant 0 \qquad \iint_{\mathbb{R}^2} f(x, y) \, dA = 1$$

As in Exercise 15.3.40, the double integral over $\mathbb{R}^2$ is an improper integral defined as the limit of double integrals over expanding circles or squares, and we can write

$$\iint_{\mathbb{R}^2} f(x, y) \, dA = \int_{-\infty}^{\infty} \int_{-\infty}^{\infty} f(x, y) \, dx \, dy = 1$$

EXAMPLE 6 If the joint density function for X and Y is given by

$$f(x, y) = \begin{cases} C(x + 2y) & \text{if } 0 \leqslant x \leqslant 10, \ 0 \leqslant y \leqslant 10 \\ 0 & \text{otherwise} \end{cases}$$

find the value of the constant C. Then find $P(X \leqslant 7, Y \geqslant 2)$.

SOLUTION We find the value of C by ensuring that the double integral of f is equal to 1. Because $f(x, y) = 0$ outside the rectangle $[0, 10] \times [0, 10]$, we have

$$\int_{-\infty}^{\infty} \int_{-\infty}^{\infty} f(x, y) \, dy \, dx = \int_0^{10} \int_0^{10} C(x + 2y) \, dy \, dx = C \int_0^{10} \left[xy + y^2 \right]_{y=0}^{y=10} dx$$

$$= C \int_0^{10} (10x + 100) \, dx = 1500C$$

Therefore $1500C = 1$ and so $C = \frac{1}{1500}$.

Now we can compute the probability that X is at most 7 and Y is at least 2:

$$P(X \leqslant 7, Y \geqslant 2) = \int_{-\infty}^{7} \int_2^{\infty} f(x, y) \, dy \, dx = \int_0^7 \int_2^{10} \tfrac{1}{1500}(x + 2y) \, dy \, dx$$

$$= \tfrac{1}{1500} \int_0^7 \left[xy + y^2 \right]_{y=2}^{y=10} dx = \tfrac{1}{1500} \int_0^7 (8x + 96) \, dx$$

$$= \tfrac{868}{1500} \approx 0.5787 \qquad \blacksquare$$

Suppose X is a random variable with probability density function $f_1(x)$ and Y is a random variable with density function $f_2(y)$. Then X and Y are called **independent random variables** if their joint density function is the product of their individual density functions:

$$f(x, y) = f_1(x) f_2(y)$$

In Section 8.5 we modeled waiting times by using exponential density functions

$$f(t) = \begin{cases} 0 & \text{if } t < 0 \\ \mu^{-1} e^{-t/\mu} & \text{if } t \geqslant 0 \end{cases}$$

where μ is the mean waiting time. In the next example we consider a situation with two independent waiting times.

EXAMPLE 7 The manager of a movie theater determines that the average time movie-goers wait in line to buy a ticket for this week's film is 10 minutes and the average time they wait to buy popcorn is 5 minutes. Assuming that the waiting times are independent, find the probability that a moviegoer waits a total of less than 20 minutes before taking his or her seat.

SOLUTION Assuming that both the waiting time X for the ticket purchase and the waiting time Y in the refreshment line are modeled by exponential probability density functions, we can write the individual density functions as

$$f_1(x) = \begin{cases} 0 & \text{if } x < 0 \\ \frac{1}{10}e^{-x/10} & \text{if } x \geqslant 0 \end{cases} \qquad f_2(y) = \begin{cases} 0 & \text{if } y < 0 \\ \frac{1}{5}e^{-y/5} & \text{if } y \geqslant 0 \end{cases}$$

Since X and Y are independent, the joint density function is the product:

$$f(x, y) = f_1(x)f_2(y) = \begin{cases} \frac{1}{50}e^{-x/10}e^{-y/5} & \text{if } x \geqslant 0, y \geqslant 0 \\ 0 & \text{otherwise} \end{cases}$$

We are asked for the probability that $X + Y < 20$:

$$P(X + Y < 20) = P((X, Y) \in D)$$

where D is the triangular region shown in Figure 8. Thus

$$P(X + Y < 20) = \iint_D f(x, y)\, dA = \int_0^{20} \int_0^{20-x} \frac{1}{50}e^{-x/10}e^{-y/5}\, dy\, dx$$

$$= \frac{1}{50}\int_0^{20} \left[e^{-x/10}(-5)e^{-y/5} \right]_{y=0}^{y=20-x} dx$$

$$= \frac{1}{10}\int_0^{20} e^{-x/10}\left(1 - e^{(x-20)/5}\right) dx$$

$$= \frac{1}{10}\int_0^{20} \left(e^{-x/10} - e^{-4}e^{x/10}\right) dx$$

$$= 1 + e^{-4} - 2e^{-2} \approx 0.7476$$

This means that about 75% of the moviegoers wait less than 20 minutes before taking their seats.

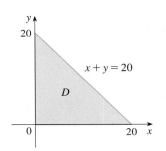

FIGURE 8

Expected Values

Recall from Section 8.5 that if X is a random variable with probability density function f, then its *mean* is

$$\mu = \int_{-\infty}^{\infty} x f(x)\, dx$$

Now if X and Y are random variables with joint density function f, we define the **X-mean** and **Y-mean**, also called the **expected values** of X and Y, to be

$$\boxed{11} \qquad \mu_1 = \iint_{\mathbb{R}^2} x f(x, y)\, dA \qquad \mu_2 = \iint_{\mathbb{R}^2} y f(x, y)\, dA$$

Notice how closely the expressions for μ_1 and μ_2 in (11) resemble the moments M_x and M_y of a lamina with density function ρ in Equations 3 and 4. In fact, we can think of probability as being like continuously distributed mass. We calculate probability the way we calculate mass—by integrating a density function. And because the total "probability mass" is 1, the expressions for $\bar{x}$ and $\bar{y}$ in (5) show that we can think of the expected values of X and Y, μ_1 and μ_2, as the coordinates of the "center of mass" of the probability distribution.

In the next example we deal with normal distributions. As in Section 8.5, a single random variable is *normally distributed* if its probability density function is of the form

$$f(x) = \frac{1}{\sigma\sqrt{2\pi}}\, e^{-(x-\mu)^2/(2\sigma^2)}$$

where μ is the mean and σ is the standard deviation.

EXAMPLE 8 A factory produces (cylindrically shaped) roller bearings that are sold as having diameter 4.0 cm and length 6.0 cm. In fact, the diameters X are normally distributed with mean 4.0 cm and standard deviation 0.01 cm while the lengths Y are normally distributed with mean 6.0 cm and standard deviation 0.01 cm. Assuming that X and Y are independent, write the joint density function and graph it. Find the probability that a bearing randomly chosen from the production line has either length or diameter that differs from the mean by more than 0.02 cm.

SOLUTION We are given that X and Y are normally distributed with $\mu_1 = 4.0$, $\mu_2 = 6.0$, and $\sigma_1 = \sigma_2 = 0.01$. So the individual density functions for X and Y are

$$f_1(x) = \frac{1}{0.01\sqrt{2\pi}}\, e^{-(x-4)^2/0.0002} \qquad f_2(y) = \frac{1}{0.01\sqrt{2\pi}}\, e^{-(y-6)^2/0.0002}$$

Since X and Y are independent, the joint density function is the product:

$$f(x, y) = f_1(x)f_2(y)$$

$$= \frac{1}{0.0002\pi}\, e^{-(x-4)^2/0.0002}e^{-(y-6)^2/0.0002}$$

$$= \frac{5000}{\pi}\, e^{-5000[(x-4)^2+(y-6)^2]}$$

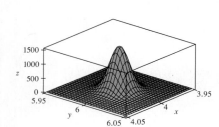

FIGURE 9
Graph of the bivariate normal joint density function in Example 8

A graph of this function is shown in Figure 9.
    Let's first calculate the probability that both X and Y differ from their means by less than 0.02 cm. Using a calculator or computer to estimate the integral, we have

$$P(3.98 < X < 4.02,\ 5.98 < Y < 6.02) = \int_{3.98}^{4.02}\int_{5.98}^{6.02} f(x, y)\, dy\, dx$$

$$= \frac{5000}{\pi} \int_{3.98}^{4.02}\int_{5.98}^{6.02} e^{-5000[(x-4)^2+(y-6)^2]}\, dy\, dx$$

$$\approx 0.91$$

Then the probability that either X or Y differs from its mean by more than 0.02 cm is approximately

$$1 - 0.91 = 0.09 \qquad \blacksquare$$

15.4 EXERCISES

1. Electric charge is distributed over the rectangle $0 \le x \le 5$, $2 \le y \le 5$ so that the charge density at (x, y) is $\sigma(x, y) = 2x + 4y$ (measured in coulombs per square meter). Find the total charge on the rectangle.

2. Electric charge is distributed over the disk $x^2 + y^2 \le 1$ so that the charge density at (x, y) is $\sigma(x, y) = \sqrt{x^2 + y^2}$

(measured in coulombs per square meter). Find the total charge on the disk.

3–10 Find the mass and center of mass of the lamina that occupies the region D and has the given density function ρ.

3. $D = \{(x, y) \mid 1 \le x \le 3,\ 1 \le y \le 4\};\ \rho(x, y) = ky^2$

4. $D = \{(x, y) \mid 0 \leqslant x \leqslant a, 0 \leqslant y \leqslant b\};$
$\rho(x, y) = 1 + x^2 + y^2$

5. D is the triangular region with vertices $(0, 0), (2, 1), (0, 3)$;
$\rho(x, y) = x + y$

6. D is the triangular region enclosed by the lines $y = 0$,
$y = 2x$, and $x + 2y = 1$; $\rho(x, y) = x$

7. D is bounded by $y = 1 - x^2$ and $y = 0$; $\rho(x, y) = ky$

8. D is bounded by $y = x + 2$ and $y = x^2$; $\rho(x, y) = kx^2$

9. D is bounded by the curves $y = e^{-x}, y = 0, x = 0, x = 1$;
$\rho(x, y) = xy$

10. D is enclosed by the curves $y = 0$ and $y = \cos x$,
$-\pi/2 \leqslant x \leqslant \pi/2$; $\rho(x, y) = y$

---

11. A lamina occupies the part of the disk $x^2 + y^2 \leqslant 1$ in the first quadrant. Find its center of mass if the density at any point is proportional to its distance from the x-axis.

12. Find the center of mass of the lamina in Exercise 11 if the density at any point is proportional to the square of its distance from the origin.

13. The boundary of a lamina consists of the semicircles $y = \sqrt{1 - x^2}$ and $y = \sqrt{4 - x^2}$ together with the portions of the x-axis that join them. Find the center of mass of the lamina if the density at any point is proportional to its distance from the origin.

14. Find the center of mass of the lamina in Exercise 13 if the density at any point is inversely proportional to its distance from the origin.

15. Find the center of mass of a lamina in the shape of an isosceles right triangle with equal sides of length a if the density at any point is proportional to the square of the distance from the vertex opposite the hypotenuse.

16. A lamina occupies the region inside the circle $x^2 + y^2 = 2y$ but outside the circle $x^2 + y^2 = 1$. Find the center of mass if the density at any point is inversely proportional to its distance from the origin.

17. Find the moments of inertia I_x, I_y, I_0 for the lamina of Exercise 3.

18. Find the moments of inertia I_x, I_y, I_0 for the lamina of Exercise 6.

19. Find the moments of inertia I_x, I_y, I_0 for the lamina of Exercise 15.

20. Consider a square fan blade with sides of length 2 and the lower left corner placed at the origin. If the density of the blade is $\rho(x, y) = 1 + 0.1x$, is it more difficult to rotate the blade about the x-axis or the y-axis?

21–24 A lamina with constant density $\rho(x, y) = \rho$ occupies the given region. Find the moments of inertia I_x and I_y and the radii of gyration $\bar{\bar{x}}$ and $\bar{\bar{y}}$.

21. The rectangle $0 \leqslant x \leqslant b, 0 \leqslant y \leqslant h$

22. The triangle with vertices $(0, 0), (b, 0)$, and $(0, h)$

23. The part of the disk $x^2 + y^2 \leqslant a^2$ in the first quadrant

24. The region under the curve $y = \sin x$ from $x = 0$ to $x = \pi$

CAS **25–26** Use a computer algebra system to find the mass, center of mass, and moments of inertia of the lamina that occupies the region D and has the given density function.

25. D is enclosed by the right loop of the four-leaved rose $r = \cos 2\theta$; $\rho(x, y) = x^2 + y^2$

26. $D = \{(x, y) \mid 0 \leqslant y \leqslant xe^{-x}, 0 \leqslant x \leqslant 2\};$ $\rho(x, y) = x^2y^2$

---

27. The joint density function for a pair of random variables X and Y is

$$f(x, y) = \begin{cases} Cx(1 + y) & \text{if } 0 \leqslant x \leqslant 1, 0 \leqslant y \leqslant 2 \\ 0 & \text{otherwise} \end{cases}$$

(a) Find the value of the constant C.
(b) Find $P(X \leqslant 1, Y \leqslant 1)$.
(c) Find $P(X + Y \leqslant 1)$.

28. (a) Verify that

$$f(x, y) = \begin{cases} 4xy & \text{if } 0 \leqslant x \leqslant 1, 0 \leqslant y \leqslant 1 \\ 0 & \text{otherwise} \end{cases}$$

is a joint density function.
(b) If X and Y are random variables whose joint density function is the function f in part (a), find
(i) $P\left(X \geqslant \frac{1}{2}\right)$ (ii) $P\left(X \geqslant \frac{1}{2}, Y \leqslant \frac{1}{2}\right)$
(c) Find the expected values of X and Y.

29. Suppose X and Y are random variables with joint density function

$$f(x, y) = \begin{cases} 0.1e^{-(0.5x+0.2y)} & \text{if } x \geqslant 0, y \geqslant 0 \\ 0 & \text{otherwise} \end{cases}$$

(a) Verify that f is indeed a joint density function.
(b) Find the following probabilities.
(i) $P(Y \geqslant 1)$ (ii) $P(X \leqslant 2, Y \leqslant 4)$
(c) Find the expected values of X and Y.

30. (a) A lamp has two bulbs, each of a type with average lifetime 1000 hours. Assuming that we can model the probability of failure of a bulb by an exponential density function with mean $\mu = 1000$, find the probability that both of the lamp's bulbs fail within 1000 hours.
(b) Another lamp has just one bulb of the same type as in part (a). If one bulb burns out and is replaced by a bulb of the same type, find the probability that the two bulbs fail within a total of 1000 hours.

CAS 31. Suppose that X and Y are independent random variables, where X is normally distributed with mean 45 and standard deviation 0.5 and Y is normally distributed with mean 20 and standard deviation 0.1.
(a) Find $P(40 \leqslant X \leqslant 50, 20 \leqslant Y \leqslant 25)$.
(b) Find $P(4(X - 45)^2 + 100(Y - 20)^2 \leqslant 2)$.

32. Xavier and Yolanda both have classes that end at noon and they agree to meet every day after class. They arrive at the coffee shop independently. Xavier's arrival time is X and Yolanda's arrival time is Y, where X and Y are measured in minutes after noon. The individual density functions are

$$f_1(x) = \begin{cases} e^{-x} & \text{if } x \geqslant 0 \\ 0 & \text{if } x < 0 \end{cases} \qquad f_2(y) = \begin{cases} \frac{1}{50}y & \text{if } 0 \leqslant y \leqslant 10 \\ 0 & \text{otherwise} \end{cases}$$

(Xavier arrives sometime after noon and is more likely to arrive promptly than late. Yolanda always arrives by 12:10 PM and is more likely to arrive late than promptly.) After Yolanda arrives, she'll wait for up to half an hour for Xavier, but he won't wait for her. Find the probability that they meet.

33. When studying the spread of an epidemic, we assume that the probability that an infected individual will spread the disease to an uninfected individual is a function of the distance between them. Consider a circular city of radius 10 miles in which the population is uniformly distributed. For an uninfected individual at a fixed point $A(x_0, y_0)$, assume that the probability function is given by

$$f(P) = \tfrac{1}{20}[20 - d(P, A)]$$

where $d(P, A)$ denotes the distance between points P and A.
(a) Suppose the exposure of a person to the disease is the sum of the probabilities of catching the disease from all members of the population. Assume that the infected people are uniformly distributed throughout the city, with k infected individuals per square mile. Find a double integral that represents the exposure of a person residing at A.
(b) Evaluate the integral for the case in which A is the center of the city and for the case in which A is located on the edge of the city. Where would you prefer to live?

15.5 Surface Area

In Section 16.6 we will deal with areas of more general surfaces, called parametric surfaces, and so this section need not be covered if that later section will be covered.

In this section we apply double integrals to the problem of computing the area of a surface. In Section 8.2 we found the area of a very special type of surface—a surface of revolution—by the methods of single-variable calculus. Here we compute the area of a surface with equation $z = f(x, y)$, the graph of a function of two variables.

Let S be a surface with equation $z = f(x, y)$, where f has continuous partial derivatives. For simplicity in deriving the surface area formula, we assume that $f(x, y) \geqslant 0$ and the domain D of f is a rectangle. We divide D into small rectangles R_{ij} with area $\Delta A = \Delta x \, \Delta y$. If (x_i, y_j) is the corner of R_{ij} closest to the origin, let $P_{ij}(x_i, y_j, f(x_i, y_j))$ be the point on S directly above it (see Figure 1). The tangent plane to S at P_{ij} is an approximation to S near P_{ij}. So the area ΔT_{ij} of the part of this tangent plane (a parallelogram) that lies directly above R_{ij} is an approximation to the area ΔS_{ij} of the part of S that lies directly above R_{ij}. Thus the sum $\sum \sum \Delta T_{ij}$ is an approximation to the total area of S, and this approximation appears to improve as the number of rectangles increases. Therefore we define the **surface area** of S to be

FIGURE 1

$$\boxed{1} \qquad A(S) = \lim_{m, n \to \infty} \sum_{i=1}^{m} \sum_{j=1}^{n} \Delta T_{ij}$$

To find a formula that is more convenient than Equation 1 for computational purposes, we let $\mathbf{a}$ and $\mathbf{b}$ be the vectors that start at P_{ij} and lie along the sides of the parallelogram with area ΔT_{ij}. (See Figure 2.) Then $\Delta T_{ij} = |\mathbf{a} \times \mathbf{b}|$. Recall from Section 14.3 that $f_x(x_i, y_j)$ and $f_y(x_i, y_j)$ are the slopes of the tangent lines through P_{ij} in the directions of $\mathbf{a}$ and $\mathbf{b}$. Therefore

$$\mathbf{a} = \Delta x \, \mathbf{i} + f_x(x_i, y_j) \, \Delta x \, \mathbf{k}$$

$$\mathbf{b} = \Delta y \, \mathbf{j} + f_y(x_i, y_j) \, \Delta y \, \mathbf{k}$$

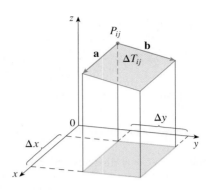

FIGURE 2

and

$$\mathbf{a} \times \mathbf{b} = \begin{vmatrix} \mathbf{i} & \mathbf{j} & \mathbf{k} \\ \Delta x & 0 & f_x(x_i, y_j)\,\Delta x \\ 0 & \Delta y & f_y(x_i, y_j)\,\Delta y \end{vmatrix}$$

$$= -f_x(x_i, y_j)\,\Delta x\,\Delta y\,\mathbf{i} - f_y(x_i, y_j)\,\Delta x\,\Delta y\,\mathbf{j} + \Delta x\,\Delta y\,\mathbf{k}$$

$$= [-f_x(x_i, y_j)\mathbf{i} - f_y(x_i, y_j)\mathbf{j} + \mathbf{k}]\,\Delta A$$

Thus
$$\Delta T_{ij} = |\mathbf{a} \times \mathbf{b}| = \sqrt{[f_x(x_i, y_j)]^2 + [f_y(x_i, y_j)]^2 + 1}\;\Delta A$$

From Definition 1 we then have

$$A(S) = \lim_{m,n \to \infty} \sum_{i=1}^{m} \sum_{j=1}^{n} \Delta T_{ij}$$

$$= \lim_{m,n \to \infty} \sum_{i=1}^{m} \sum_{j=1}^{n} \sqrt{[f_x(x_i, y_j)]^2 + [f_y(x_i, y_j)]^2 + 1}\;\Delta A$$

and by the definition of a double integral we get the following formula.

> **2** The area of the surface with equation $z = f(x, y)$, $(x, y) \in D$, where f_x and f_y are continuous, is
> $$A(S) = \iint_D \sqrt{[f_x(x, y)]^2 + [f_y(x, y)]^2 + 1}\;dA$$

We will verify in Section 16.6 that this formula is consistent with our previous formula for the area of a surface of revolution. If we use the alternative notation for partial derivatives, we can rewrite Formula 2 as follows:

3
$$A(S) = \iint_D \sqrt{1 + \left(\frac{\partial z}{\partial x}\right)^2 + \left(\frac{\partial z}{\partial y}\right)^2}\;dA$$

Notice the similarity between the surface area formula in Equation 3 and the arc length formula from Section 8.1:

$$L = \int_a^b \sqrt{1 + \left(\frac{dy}{dx}\right)^2}\;dx$$

EXAMPLE 1 Find the surface area of the part of the surface $z = x^2 + 2y$ that lies above the triangular region T in the xy-plane with vertices $(0, 0)$, $(1, 0)$, and $(1, 1)$.

SOLUTION The region T is shown in Figure 3 and is described by

$$T = \{(x, y) \mid 0 \le x \le 1,\ 0 \le y \le x\}$$

FIGURE 3

FIGURE 4

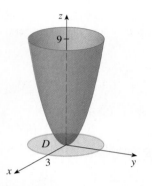

FIGURE 5

Using Formula 2 with $f(x, y) = x^2 + 2y$, we get

$$A = \iint_T \sqrt{(2x)^2 + (2)^2 + 1} \, dA = \int_0^1 \int_0^x \sqrt{4x^2 + 5} \, dy \, dx$$

$$= \int_0^1 x\sqrt{4x^2 + 5} \, dx = \tfrac{1}{8} \cdot \tfrac{2}{3}(4x^2 + 5)^{3/2} \Big]_0^1 = \tfrac{1}{12}\left(27 - 5\sqrt{5}\right)$$

Figure 4 shows the portion of the surface whose area we have just computed. ∎

EXAMPLE 2 Find the area of the part of the paraboloid $z = x^2 + y^2$ that lies under the plane $z = 9$.

SOLUTION The plane intersects the paraboloid in the circle $x^2 + y^2 = 9$, $z = 9$. Therefore the given surface lies above the disk D with center the origin and radius 3. (See Figure 5.) Using Formula 3, we have

$$A = \iint_D \sqrt{1 + \left(\frac{\partial z}{\partial x}\right)^2 + \left(\frac{\partial z}{\partial y}\right)^2} \, dA = \iint_D \sqrt{1 + (2x)^2 + (2y)^2} \, dA$$

$$= \iint_D \sqrt{1 + 4(x^2 + y^2)} \, dA$$

Converting to polar coordinates, we obtain

$$A = \int_0^{2\pi} \int_0^3 \sqrt{1 + 4r^2} \, r \, dr \, d\theta = \int_0^{2\pi} d\theta \int_0^3 \tfrac{1}{8}\sqrt{1 + 4r^2}\,(8r) \, dr$$

$$= 2\pi\left(\tfrac{1}{8}\right)\tfrac{2}{3}(1 + 4r^2)^{3/2} \Big]_0^3 = \frac{\pi}{6}\left(37\sqrt{37} - 1\right)$$ ∎

15.5 EXERCISES

1–12 Find the area of the surface.

1. The part of the plane $5x + 3y - z + 6 = 0$ that lies above the rectangle $[1, 4] \times [2, 6]$

2. The part of the plane $6x + 4y + 2z = 1$ that lies inside the cylinder $x^2 + y^2 = 25$

3. The part of the plane $3x + 2y + z = 6$ that lies in the first octant

4. The part of the surface $2y + 4z - x^2 = 5$ that lies above the triangle with vertices $(0, 0)$, $(2, 0)$, and $(2, 4)$

5. The part of the paraboloid $z = 1 - x^2 - y^2$ that lies above the plane $z = -2$

6. The part of the cylinder $x^2 + z^2 = 4$ that lies above the square with vertices $(0, 0)$, $(1, 0)$, $(0, 1)$, and $(1, 1)$

7. The part of the hyperbolic paraboloid $z = y^2 - x^2$ that lies between the cylinders $x^2 + y^2 = 1$ and $x^2 + y^2 = 4$

8. The surface $z = \tfrac{2}{3}(x^{3/2} + y^{3/2})$, $0 \leqslant x \leqslant 1$, $0 \leqslant y \leqslant 1$

9. The part of the surface $z = xy$ that lies within the cylinder $x^2 + y^2 = 1$

10. The part of the sphere $x^2 + y^2 + z^2 = 4$ that lies above the plane $z = 1$

11. The part of the sphere $x^2 + y^2 + z^2 = a^2$ that lies within the cylinder $x^2 + y^2 = ax$ and above the xy-plane

12. The part of the sphere $x^2 + y^2 + z^2 = 4z$ that lies inside the paraboloid $z = x^2 + y^2$

13–14 Find the area of the surface correct to four decimal places by expressing the area in terms of a single integral and using your calculator to estimate the integral.

13. The part of the surface $z = 1/(1 + x^2 + y^2)$ that lies above the disk $x^2 + y^2 \leqslant 1$

14. The part of the surface $z = \cos(x^2 + y^2)$ that lies inside the cylinder $x^2 + y^2 = 1$

15. (a) Use the Midpoint Rule for double integrals (see Section 15.1) with four squares to estimate the surface area of the portion of the paraboloid $z = x^2 + y^2$ that lies above the square $[0, 1] \times [0, 1]$.

    CAS (b) Use a computer algebra system to approximate the surface area in part (a) to four decimal places. Compare with the answer to part (a).

16. (a) Use the Midpoint Rule for double integrals with $m = n = 2$ to estimate the area of the surface $z = xy + x^2 + y^2$, $0 \le x \le 2$, $0 \le y \le 2$.

    CAS (b) Use a computer algebra system to approximate the surface area in part (a) to four decimal places. Compare with the answer to part (a).

CAS **17.** Find the exact area of the surface $z = 1 + 2x + 3y + 4y^2$, $1 \le x \le 4$, $0 \le y \le 1$.

CAS **18.** Find the exact area of the surface

$$z = 1 + x + y + x^2 \qquad -2 \le x \le 1 \quad -1 \le y \le 1$$

Illustrate by graphing the surface.

CAS **19.** Find, to four decimal places, the area of the part of the surface $z = 1 + x^2 y^2$ that lies above the disk $x^2 + y^2 \le 1$.

CAS **20.** Find, to four decimal places, the area of the part of the surface $z = (1 + x^2)/(1 + y^2)$ that lies above the square $|x| + |y| \le 1$. Illustrate by graphing this part of the surface.

21. Show that the area of the part of the plane $z = ax + by + c$ that projects onto a region D in the xy-plane with area $A(D)$ is $\sqrt{a^2 + b^2 + 1}\, A(D)$.

22. If you attempt to use Formula 2 to find the area of the top half of the sphere $x^2 + y^2 + z^2 = a^2$, you have a slight problem because the double integral is improper. In fact, the integrand has an infinite discontinuity at every point of the boundary circle $x^2 + y^2 = a^2$. However, the integral can be computed as the limit of the integral over the disk $x^2 + y^2 \le t^2$ as $t \to a^-$. Use this method to show that the area of a sphere of radius a is $4\pi a^2$.

23. Find the area of the finite part of the paraboloid $y = x^2 + z^2$ cut off by the plane $y = 25$. [*Hint:* Project the surface onto the xz-plane.]

24. The figure shows the surface created when the cylinder $y^2 + z^2 = 1$ intersects the cylinder $x^2 + z^2 = 1$. Find the area of this surface.

15.6 Triple Integrals

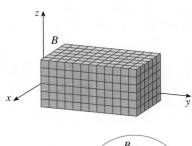

Just as we defined single integrals for functions of one variable and double integrals for functions of two variables, so we can define triple integrals for functions of three variables. Let's first deal with the simplest case where f is defined on a rectangular box:

$$\boxed{1} \qquad B = \{(x, y, z) \mid a \le x \le b,\ c \le y \le d,\ r \le z \le s\}$$

The first step is to divide B into sub-boxes. We do this by dividing the interval $[a, b]$ into l subintervals $[x_{i-1}, x_i]$ of equal width Δx, dividing $[c, d]$ into m subintervals of width Δy, and dividing $[r, s]$ into n subintervals of width Δz. The planes through the endpoints of these subintervals parallel to the coordinate planes divide the box B into lmn sub-boxes

$$B_{ijk} = [x_{i-1}, x_i] \times [y_{j-1}, y_j] \times [z_{k-1}, z_k]$$

which are shown in Figure 1. Each sub-box has volume $\Delta V = \Delta x\, \Delta y\, \Delta z$.

    Then we form the **triple Riemann sum**

$$\boxed{2} \qquad \sum_{i=1}^{l} \sum_{j=1}^{m} \sum_{k=1}^{n} f(x_{ijk}^*, y_{ijk}^*, z_{ijk}^*)\, \Delta V$$

where the sample point $(x_{ijk}^*, y_{ijk}^*, z_{ijk}^*)$ is in B_{ijk}. By analogy with the definition of a double integral (15.1.5), we define the triple integral as the limit of the triple Riemann sums in (2).

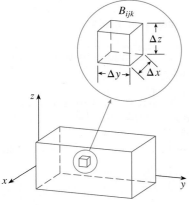

FIGURE 1

3 **Definition** The **triple integral** of f over the box B is

$$\iiint\limits_B f(x, y, z)\, dV = \lim_{l, m, n \to \infty} \sum_{i=1}^{l} \sum_{j=1}^{m} \sum_{k=1}^{n} f(x_{ijk}^*, y_{ijk}^*, z_{ijk}^*)\, \Delta V$$

if this limit exists.

Again, the triple integral always exists if f is continuous. We can choose the sample point to be any point in the sub-box, but if we choose it to be the point (x_i, y_j, z_k) we get a simpler-looking expression for the triple integral:

$$\iiint\limits_B f(x, y, z)\, dV = \lim_{l, m, n \to \infty} \sum_{i=1}^{l} \sum_{j=1}^{m} \sum_{k=1}^{n} f(x_i, y_j, z_k)\, \Delta V$$

Just as for double integrals, the practical method for evaluating triple integrals is to express them as iterated integrals as follows.

4 **Fubini's Theorem for Triple Integrals** If f is continuous on the rectangular box $B = [a, b] \times [c, d] \times [r, s]$, then

$$\iiint\limits_B f(x, y, z)\, dV = \int_r^s \int_c^d \int_a^b f(x, y, z)\, dx\, dy\, dz$$

The iterated integral on the right side of Fubini's Theorem means that we integrate first with respect to x (keeping y and z fixed), then we integrate with respect to y (keeping z fixed), and finally we integrate with respect to z. There are five other possible orders in which we can integrate, all of which give the same value. For instance, if we integrate with respect to y, then z, and then x, we have

$$\iiint\limits_B f(x, y, z)\, dV = \int_a^b \int_r^s \int_c^d f(x, y, z)\, dy\, dz\, dx$$

EXAMPLE 1 Evaluate the triple integral $\iiint_B xyz^2\, dV$, where B is the rectangular box given by

$$B = \{(x, y, z) \mid 0 \leqslant x \leqslant 1,\ -1 \leqslant y \leqslant 2,\ 0 \leqslant z \leqslant 3\}$$

SOLUTION We could use any of the six possible orders of integration. If we choose to integrate with respect to x, then y, and then z, we obtain

$$\iiint\limits_B xyz^2\, dV = \int_0^3 \int_{-1}^2 \int_0^1 xyz^2\, dx\, dy\, dz = \int_0^3 \int_{-1}^2 \left[\frac{x^2 yz^2}{2} \right]_{x=0}^{x=1} dy\, dz$$

$$= \int_0^3 \int_{-1}^2 \frac{yz^2}{2}\, dy\, dz = \int_0^3 \left[\frac{y^2 z^2}{4} \right]_{y=-1}^{y=2} dz$$

$$= \int_0^3 \frac{3z^2}{4}\, dz = \frac{z^3}{4} \Big]_0^3 = \frac{27}{4}$$

Now we define the **triple integral over a general bounded region** E in three-dimensional space (a solid) by much the same procedure that we used for double integrals (15.2.2). We enclose E in a box B of the type given by Equation 1. Then we define F so that it agrees with f on E but is 0 for points in B that are outside E. By definition,

$$\iiint_E f(x, y, z)\, dV = \iiint_B F(x, y, z)\, dV$$

This integral exists if f is continuous and the boundary of E is "reasonably smooth." The triple integral has essentially the same properties as the double integral (Properties 6–9 in Section 15.2).

We restrict our attention to continuous functions f and to certain simple types of regions. A solid region E is said to be of **type 1** if it lies between the graphs of two continuous functions of x and y, that is,

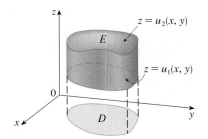

FIGURE 2
A type 1 solid region

$$\boxed{5} \qquad E = \left\{ (x, y, z) \mid (x, y) \in D,\ u_1(x, y) \le z \le u_2(x, y) \right\}$$

where D is the projection of E onto the xy-plane as shown in Figure 2. Notice that the upper boundary of the solid E is the surface with equation $z = u_2(x, y)$, while the lower boundary is the surface $z = u_1(x, y)$.

By the same sort of argument that led to (15.2.3), it can be shown that if E is a type 1 region given by Equation 5, then

$$\boxed{6} \qquad \iiint_E f(x, y, z)\, dV = \iint_D \left[\int_{u_1(x,y)}^{u_2(x,y)} f(x, y, z)\, dz \right] dA$$

The meaning of the inner integral on the right side of Equation 6 is that x and y are held fixed, and therefore $u_1(x, y)$ and $u_2(x, y)$ are regarded as constants, while $f(x, y, z)$ is integrated with respect to z.

In particular, if the projection D of E onto the xy-plane is a type I plane region (as in Figure 3), then

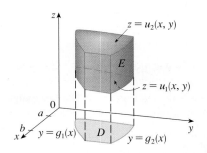

FIGURE 3
A type 1 solid region where the projection D is a type I plane region

$$E = \left\{ (x, y, z) \mid a \le x \le b,\ g_1(x) \le y \le g_2(x),\ u_1(x, y) \le z \le u_2(x, y) \right\}$$

and Equation 6 becomes

$$\boxed{7} \qquad \iiint_E f(x, y, z)\, dV = \int_a^b \int_{g_1(x)}^{g_2(x)} \int_{u_1(x,y)}^{u_2(x,y)} f(x, y, z)\, dz\, dy\, dx$$

If, on the other hand, D is a type II plane region (as in Figure 4), then

$$E = \left\{ (x, y, z) \mid c \le y \le d,\ h_1(y) \le x \le h_2(y),\ u_1(x, y) \le z \le u_2(x, y) \right\}$$

and Equation 6 becomes

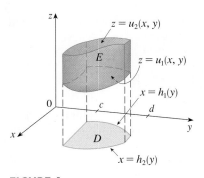

FIGURE 4
A type 1 solid region with a type II projection

$$\boxed{8} \qquad \iiint_E f(x, y, z)\, dV = \int_c^d \int_{h_1(y)}^{h_2(y)} \int_{u_1(x,y)}^{u_2(x,y)} f(x, y, z)\, dz\, dx\, dy$$

FIGURE 5

FIGURE 6

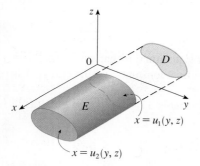

FIGURE 7

A type 2 region

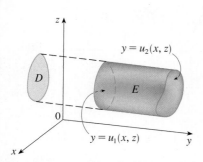

FIGURE 8

A type 3 region

EXAMPLE 2 Evaluate $\iiint_E z \, dV$, where E is the solid tetrahedron bounded by the four planes $x = 0$, $y = 0$, $z = 0$, and $x + y + z = 1$.

SOLUTION When we set up a triple integral it's wise to draw *two* diagrams: one of the solid region E (see Figure 5) and one of its projection D onto the xy-plane (see Figure 6). The lower boundary of the tetrahedron is the plane $z = 0$ and the upper boundary is the plane $x + y + z = 1$ (or $z = 1 - x - y$), so we use $u_1(x, y) = 0$ and $u_2(x, y) = 1 - x - y$ in Formula 7. Notice that the planes $x + y + z = 1$ and $z = 0$ intersect in the line $x + y = 1$ (or $y = 1 - x$) in the xy-plane. So the projection of E is the triangular region shown in Figure 6, and we have

$$\boxed{9} \quad E = \{(x, y, z) \mid 0 \leqslant x \leqslant 1, \ 0 \leqslant y \leqslant 1 - x, \ 0 \leqslant z \leqslant 1 - x - y\}$$

This description of E as a type 1 region enables us to evaluate the integral as follows:

$$\iiint_E z \, dV = \int_0^1 \int_0^{1-x} \int_0^{1-x-y} z \, dz \, dy \, dx = \int_0^1 \int_0^{1-x} \left[\frac{z^2}{2} \right]_{z=0}^{z=1-x-y} dy \, dx$$

$$= \frac{1}{2} \int_0^1 \int_0^{1-x} (1 - x - y)^2 \, dy \, dx = \frac{1}{2} \int_0^1 \left[-\frac{(1 - x - y)^3}{3} \right]_{y=0}^{y=1-x} dx$$

$$= \frac{1}{6} \int_0^1 (1 - x)^3 \, dx = \frac{1}{6} \left[-\frac{(1 - x)^4}{4} \right]_0^1 = \frac{1}{24}$$

A solid region E is of **type 2** if it is of the form

$$E = \{(x, y, z) \mid (y, z) \in D, \ u_1(y, z) \leqslant x \leqslant u_2(y, z)\}$$

where, this time, D is the projection of E onto the yz-plane (see Figure 7). The back surface is $x = u_1(y, z)$, the front surface is $x = u_2(y, z)$, and we have

$$\boxed{10} \quad \iiint_E f(x, y, z) \, dV = \iint_D \left[\int_{u_1(y, z)}^{u_2(y, z)} f(x, y, z) \, dx \right] dA$$

Finally, a **type 3** region is of the form

$$E = \{(x, y, z) \mid (x, z) \in D, \ u_1(x, z) \leqslant y \leqslant u_2(x, z)\}$$

where D is the projection of E onto the xz-plane, $y = u_1(x, z)$ is the left surface, and $y = u_2(x, z)$ is the right surface (see Figure 8). For this type of region we have

$$\boxed{11} \quad \iiint_E f(x, y, z) \, dV = \iint_D \left[\int_{u_1(x, z)}^{u_2(x, z)} f(x, y, z) \, dy \right] dA$$

In each of Equations 10 and 11 there may be two possible expressions for the integral depending on whether D is a type I or type II plane region (and corresponding to Equations 7 and 8).

EXAMPLE 3 Evaluate $\iiint_E \sqrt{x^2 + z^2} \, dV$, where E is the region bounded by the paraboloid $y = x^2 + z^2$ and the plane $y = 4$.

SOLUTION The solid E is shown in Figure 9. If we regard it as a type 1 region, then we need to consider its projection D_1 onto the xy-plane, which is the parabolic region in Figure 10. (The trace of $y = x^2 + z^2$ in the plane $z = 0$ is the parabola $y = x^2$.)

TEC Visual 15.6 illustrates how solid regions (including the one in Figure 9) project onto coordinate planes.

FIGURE 9
Region of integration

FIGURE 10
Projection onto xy-plane

From $y = x^2 + z^2$ we obtain $z = \pm\sqrt{y - x^2}$, so the lower boundary surface of E is $z = -\sqrt{y - x^2}$ and the upper surface is $z = \sqrt{y - x^2}$. Therefore the description of E as a type 1 region is

$$E = \left\{(x, y, z) \mid -2 \leqslant x \leqslant 2,\ x^2 \leqslant y \leqslant 4,\ -\sqrt{y - x^2} \leqslant z \leqslant \sqrt{y - x^2}\right\}$$

and so we obtain

$$\iiint\limits_{E} \sqrt{x^2 + z^2}\, dV = \int_{-2}^{2} \int_{x^2}^{4} \int_{-\sqrt{y-x^2}}^{\sqrt{y-x^2}} \sqrt{x^2 + z^2}\; dz\, dy\, dx$$

Although this expression is correct, it is extremely difficult to evaluate. So let's instead consider E as a type 3 region. As such, its projection D_3 onto the xz-plane is the disk $x^2 + z^2 \leqslant 4$ shown in Figure 11.

Then the left boundary of E is the paraboloid $y = x^2 + z^2$ and the right boundary is the plane $y = 4$, so taking $u_1(x, z) = x^2 + z^2$ and $u_2(x, z) = 4$ in Equation 11, we have

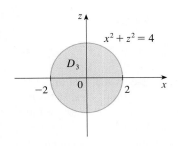

FIGURE 11
Projection onto xz-plane

$$\iiint\limits_{E} \sqrt{x^2 + z^2}\, dV = \iint\limits_{D_3} \left[\int_{x^2+z^2}^{4} \sqrt{x^2 + z^2}\; dy \right] dA = \iint\limits_{D_3} (4 - x^2 - z^2)\sqrt{x^2 + z^2}\; dA$$

Although this integral could be written as

$$\int_{-2}^{2} \int_{-\sqrt{4-x^2}}^{\sqrt{4-x^2}} (4 - x^2 - z^2)\sqrt{x^2 + z^2}\; dz\, dx$$

⊘ The most difficult step in evaluating a triple integral is setting up an expression for the region of integration (such as Equation 9 in Example 2). Remember that the limits of integration in the inner integral contain at most two variables, the limits of integration in the middle integral contain at most one variable, and the limits of integration in the outer integral must be constants.

it's easier to convert to polar coordinates in the xz-plane: $x = r\cos\theta$, $z = r\sin\theta$. This gives

$$\iiint\limits_{E} \sqrt{x^2 + z^2}\, dV = \iint\limits_{D_3} (4 - x^2 - z^2)\sqrt{x^2 + z^2}\; dA$$

$$= \int_0^{2\pi} \int_0^2 (4 - r^2)r\, r\, dr\, d\theta = \int_0^{2\pi} d\theta \int_0^2 (4r^2 - r^4)\, dr$$

$$= 2\pi \left[\frac{4r^3}{3} - \frac{r^5}{5} \right]_0^2 = \frac{128\pi}{15} \qquad \blacksquare$$

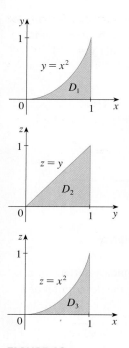

FIGURE 12
Projections of E

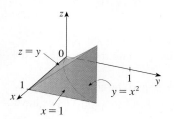

FIGURE 13
The solid E

EXAMPLE 4 Express the iterated integral $\int_0^1 \int_0^{x^2} \int_0^y f(x, y, z)\, dz\, dy\, dx$ as a triple integral and then rewrite it as an iterated integral in a different order, integrating first with respect to x, then z, and then y.

SOLUTION We can write

$$\int_0^1 \int_0^{x^2} \int_0^y f(x, y, z)\, dz\, dy\, dx = \iiint_E f(x, y, z)\, dV$$

where $E = \{(x, y, z) \mid 0 \leqslant x \leqslant 1,\, 0 \leqslant y \leqslant x^2,\, 0 \leqslant z \leqslant y\}$. This description of E enables us to write projections onto the three coordinate planes as follows:

on the xy-plane: $D_1 = \{(x, y) \mid 0 \leqslant x \leqslant 1,\, 0 \leqslant y \leqslant x^2\}$

$\qquad\qquad\qquad\qquad = \{(x, y) \mid 0 \leqslant y \leqslant 1,\, \sqrt{y} \leqslant x \leqslant 1\}$

on the yz-plane: $D_2 = \{(y, z) \mid 0 \leqslant y \leqslant 1,\, 0 \leqslant z \leqslant y\}$

on the xz-plane: $D_3 = \{(x, z) \mid 0 \leqslant x \leqslant 1,\, 0 \leqslant z \leqslant x^2\}$

From the resulting sketches of the projections in Figure 12 we sketch the solid E in Figure 13. We see that it is the solid enclosed by the planes $z = 0$, $x = 1$, $y = z$ and the parabolic cylinder $y = x^2$ $\left(\text{or } x = \sqrt{y}\,\right)$.

If we integrate first with respect to x, then z, and then y, we use an alternate description of E:

$$E = \left\{(x, y, z) \mid 0 \leqslant y \leqslant 1,\, 0 \leqslant z \leqslant y,\, \sqrt{y} \leqslant x \leqslant 1\right\}$$

Thus

$$\iiint_E f(x, y, z)\, dV = \int_0^1 \int_0^y \int_{\sqrt{y}}^1 f(x, y, z)\, dx\, dz\, dy$$ ∎

▪ Applications of Triple Integrals

Recall that if $f(x) \geqslant 0$, then the single integral $\int_a^b f(x)\, dx$ represents the area under the curve $y = f(x)$ from a to b, and if $f(x, y) \geqslant 0$, then the double integral $\iint_D f(x, y)\, dA$ represents the volume under the surface $z = f(x, y)$ and above D. The corresponding interpretation of a triple integral $\iiint_E f(x, y, z)\, dV$, where $f(x, y, z) \geqslant 0$, is not very useful because it would be the "hypervolume" of a four-dimensional object and, of course, that is very difficult to visualize. (Remember that E is just the *domain* of the function f; the graph of f lies in four-dimensional space.) Nonetheless, the triple integral $\iiint_E f(x, y, z)\, dV$ can be interpreted in different ways in different physical situations, depending on the physical interpretations of x, y, z, and $f(x, y, z)$.

Let's begin with the special case where $f(x, y, z) = 1$ for all points in E. Then the triple integral does represent the volume of E:

12
$$V(E) = \iiint_E dV$$

For example, you can see this in the case of a type 1 region by putting $f(x, y, z) = 1$ in Formula 6:

$$\iiint_E 1\, dV = \iint_D \left[\int_{u_1(x, y)}^{u_2(x, y)} dz\right] dA = \iint_D [u_2(x, y) - u_1(x, y)]\, dA$$

and from Section 15.2 we know this represents the volume that lies between the surfaces $z = u_1(x, y)$ and $z = u_2(x, y)$.

EXAMPLE 5 Use a triple integral to find the volume of the tetrahedron T bounded by the planes $x + 2y + z = 2$, $x = 2y$, $x = 0$, and $z = 0$.

SOLUTION The tetrahedron T and its projection D onto the xy-plane are shown in Figures 14 and 15. The lower boundary of T is the plane $z = 0$ and the upper boundary is the plane $x + 2y + z = 2$, that is, $z = 2 - x - 2y$.

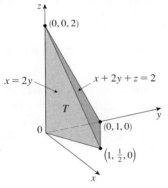

FIGURE 14　　　　　　　　　　　　**FIGURE 15**

Therefore we have

$$V(T) = \iiint_T dV = \int_0^1 \int_{x/2}^{1-x/2} \int_0^{2-x-2y} dz\, dy\, dx$$

$$= \int_0^1 \int_{x/2}^{1-x/2} (2 - x - 2y)\, dy\, dx = \tfrac{1}{3}$$

by the same calculation as in Example 15.2.4.

(Notice that it is not necessary to use triple integrals to compute volumes. They simply give an alternative method for setting up the calculation.) ■

All the applications of double integrals in Section 15.4 can be immediately extended to triple integrals. For example, if the density function of a solid object that occupies the region E is $\rho(x, y, z)$, in units of mass per unit volume, at any given point (x, y, z), then its **mass** is

$$\boxed{13} \qquad\qquad m = \iiint_E \rho(x, y, z)\, dV$$

and its **moments** about the three coordinate planes are

$$\boxed{14} \qquad M_{yz} = \iiint_E x\rho(x, y, z)\, dV \qquad M_{xz} = \iiint_E y\rho(x, y, z)\, dV$$

$$M_{xy} = \iiint_E z\rho(x, y, z)\, dV$$

The **center of mass** is located at the point $(\bar{x}, \bar{y}, \bar{z})$, where

$$\boxed{15} \qquad\qquad \bar{x} = \frac{M_{yz}}{m} \qquad \bar{y} = \frac{M_{xz}}{m} \qquad \bar{z} = \frac{M_{xy}}{m}$$

If the density is constant, the center of mass of the solid is called the **centroid** of E. The **moments of inertia** about the three coordinate axes are

$$\boxed{16} \quad I_x = \iiint\limits_E (y^2 + z^2)\rho(x, y, z)\, dV \qquad I_y = \iiint\limits_E (x^2 + z^2)\rho(x, y, z)\, dV$$

$$I_z = \iiint\limits_E (x^2 + y^2)\rho(x, y, z)\, dV$$

As in Section 15.4, the total **electric charge** on a solid object occupying a region E and having charge density $\sigma(x, y, z)$ is

$$Q = \iiint\limits_E \sigma(x, y, z)\, dV$$

If we have three continuous random variables X, Y, and Z, their **joint density function** is a function of three variables such that the probability that (X, Y, Z) lies in E is

$$P((X, Y, Z) \in E) = \iiint\limits_E f(x, y, z)\, dV$$

In particular,

$$P(a \leqslant X \leqslant b,\ c \leqslant Y \leqslant d,\ r \leqslant Z \leqslant s) = \int_a^b \int_c^d \int_r^s f(x, y, z)\, dz\, dy\, dx$$

The joint density function satisfies

$$f(x, y, z) \geqslant 0 \qquad \int_{-\infty}^{\infty} \int_{-\infty}^{\infty} \int_{-\infty}^{\infty} f(x, y, z)\, dz\, dy\, dx = 1$$

EXAMPLE 6 Find the center of mass of a solid of constant density that is bounded by the parabolic cylinder $x = y^2$ and the planes $x = z$, $z = 0$, and $x = 1$.

SOLUTION The solid E and its projection onto the xy-plane are shown in Figure 16. The lower and upper surfaces of E are the planes $z = 0$ and $z = x$, so we describe E as a type 1 region:

$$E = \left\{ (x, y, z) \mid -1 \leqslant y \leqslant 1,\ y^2 \leqslant x \leqslant 1,\ 0 \leqslant z \leqslant x \right\}$$

Then, if the density is $\rho(x, y, z) = \rho$, the mass is

$$m = \iiint\limits_E \rho\, dV = \int_{-1}^{1} \int_{y^2}^{1} \int_0^x \rho\, dz\, dx\, dy$$

$$= \rho \int_{-1}^{1} \int_{y^2}^{1} x\, dx\, dy = \rho \int_{-1}^{1} \left[\frac{x^2}{2} \right]_{x=y^2}^{x=1} dy$$

$$= \frac{\rho}{2} \int_{-1}^{1} (1 - y^4)\, dy = \rho \int_0^1 (1 - y^4)\, dy$$

$$= \rho \left[y - \frac{y^5}{5} \right]_0^1 = \frac{4\rho}{5}$$

$z = x$

E

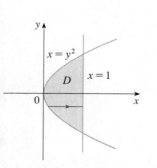

$x = y^2$

D $x = 1$

FIGURE 16

Because of the symmetry of E and ρ about the xz-plane, we can immediately say that $M_{xz} = 0$ and therefore $\bar{y} = 0$. The other moments are

$$M_{yz} = \iiint_E x\rho \, dV = \int_{-1}^{1} \int_{y^2}^{1} \int_{0}^{x} x\rho \, dz \, dx \, dy$$

$$= \rho \int_{-1}^{1} \int_{y^2}^{1} x^2 \, dx \, dy = \rho \int_{-1}^{1} \left[\frac{x^3}{3} \right]_{x=y^2}^{x=1} dy$$

$$= \frac{2\rho}{3} \int_{0}^{1} (1 - y^6) \, dy = \frac{2\rho}{3} \left[y - \frac{y^7}{7} \right]_{0}^{1} = \frac{4\rho}{7}$$

$$M_{xy} = \iiint_E z\rho \, dV = \int_{-1}^{1} \int_{y^2}^{1} \int_{0}^{x} z\rho \, dz \, dx \, dy$$

$$= \rho \int_{-1}^{1} \int_{y^2}^{1} \left[\frac{z^2}{2} \right]_{z=0}^{z=x} dx \, dy = \frac{\rho}{2} \int_{-1}^{1} \int_{y^2}^{1} x^2 \, dx \, dy$$

$$= \frac{\rho}{3} \int_{0}^{1} (1 - y^6) \, dy = \frac{2\rho}{7}$$

Therefore the center of mass is

$$(\bar{x}, \bar{y}, \bar{z}) = \left(\frac{M_{yz}}{m}, \frac{M_{xz}}{m}, \frac{M_{xy}}{m} \right) = \left(\tfrac{5}{7}, 0, \tfrac{5}{14} \right)$$

■

15.6 EXERCISES

1. Evaluate the integral in Example 1, integrating first with respect to y, then z, and then x.

2. Evaluate the integral $\iiint_E (xy + z^2) \, dV$, where

$$E = \{(x, y, z) \mid 0 \leqslant x \leqslant 2, 0 \leqslant y \leqslant 1, 0 \leqslant z \leqslant 3\}$$

using three different orders of integration.

3–8 Evaluate the iterated integral.

3. $\int_{0}^{2} \int_{0}^{z^2} \int_{0}^{y-z} (2x - y) \, dx \, dy \, dz$

4. $\int_{0}^{1} \int_{y}^{2y} \int_{0}^{x+y} 6xy \, dz \, dx \, dy$

5. $\int_{1}^{2} \int_{0}^{2z} \int_{0}^{\ln x} xe^{-y} \, dy \, dx \, dz$

6. $\int_{0}^{1} \int_{0}^{1} \int_{0}^{\sqrt{1-z^2}} \frac{z}{y+1} \, dx \, dz \, dy$

7. $\int_{0}^{\pi} \int_{0}^{1} \int_{0}^{\sqrt{1-z^2}} z \sin x \, dy \, dz \, dx$

8. $\int_{0}^{1} \int_{0}^{1} \int_{0}^{2-x^2-y^2} xye^z \, dz \, dy \, dx$

9–18 Evaluate the triple integral.

9. $\iiint_E y \, dV$, where

$$E = \{(x, y, z) \mid 0 \leqslant x \leqslant 3, 0 \leqslant y \leqslant x, x - y \leqslant z \leqslant x + y\}$$

10. $\iiint_E e^{z/y} \, dV$, where

$$E = \{(x, y, z) \mid 0 \leqslant y \leqslant 1, y \leqslant x \leqslant 1, 0 \leqslant z \leqslant xy\}$$

11. $\iiint_E \dfrac{z}{x^2 + z^2} \, dV$, where

$$E = \{(x, y, z) \mid 1 \leqslant y \leqslant 4, y \leqslant z \leqslant 4, 0 \leqslant x \leqslant z\}$$

12. $\iiint_E \sin y \, dV$, where E lies below the plane $z = x$ and above the triangular region with vertices $(0, 0, 0)$, $(\pi, 0, 0)$, and $(0, \pi, 0)$

13. $\iiint_E 6xy \, dV$, where E lies under the plane $z = 1 + x + y$ and above the region in the xy-plane bounded by the curves $y = \sqrt{x}$, $y = 0$, and $x = 1$

14. $\iiint_E (x - y) \, dV$, where E is enclosed by the surfaces $z = x^2 - 1$, $z = 1 - x^2$, $y = 0$, and $y = 2$

15. $\iiint_T y^2 \, dV$, where T is the solid tetrahedron with vertices $(0, 0, 0)$, $(2, 0, 0)$, $(0, 2, 0)$, and $(0, 0, 2)$

16. $\iiint_T xz \, dV$, where T is the solid tetrahedron with vertices $(0, 0, 0)$, $(1, 0, 1)$, $(0, 1, 1)$, and $(0, 0, 1)$

17. $\iiint_E x \, dV$, where E is bounded by the paraboloid $x = 4y^2 + 4z^2$ and the plane $x = 4$

18. $\iiint_E z \, dV$, where E is bounded by the cylinder $y^2 + z^2 = 9$ and the planes $x = 0$, $y = 3x$, and $z = 0$ in the first octant

---

19–22 Use a triple integral to find the volume of the given solid.

19. The tetrahedron enclosed by the coordinate planes and the plane $2x + y + z = 4$

20. The solid enclosed by the paraboloids $y = x^2 + z^2$ and $y = 8 - x^2 - z^2$

21. The solid enclosed by the cylinder $y = x^2$ and the planes $z = 0$ and $y + z = 1$

22. The solid enclosed by the cylinder $x^2 + z^2 = 4$ and the planes $y = -1$ and $y + z = 4$

---

23. (a) Express the volume of the wedge in the first octant that is cut from the cylinder $y^2 + z^2 = 1$ by the planes $y = x$ and $x = 1$ as a triple integral.

CAS (b) Use either the Table of Integrals (on Reference Pages 6–10) or a computer algebra system to find the exact value of the triple integral in part (a).

24. (a) In the **Midpoint Rule for triple integrals** we use a triple Riemann sum to approximate a triple integral over a box B, where $f(x, y, z)$ is evaluated at the center $(\bar{x}_i, \bar{y}_j, \bar{z}_k)$ of the box B_{ijk}. Use the Midpoint Rule to estimate $\iiint_B \sqrt{x^2 + y^2 + z^2} \, dV$, where B is the cube defined by $0 \le x \le 4$, $0 \le y \le 4$, $0 \le z \le 4$. Divide B into eight cubes of equal size.

CAS (b) Use a computer algebra system to approximate the integral in part (a) correct to the nearest integer. Compare with the answer to part (a).

25–26 Use the Midpoint Rule for triple integrals (Exercise 24) to estimate the value of the integral. Divide B into eight subboxes of equal size.

25. $\iiint_B \cos(xyz) \, dV$, where $B = \{(x, y, z) \mid 0 \le x \le 1, \ 0 \le y \le 1, 0 \le z \le 1\}$

26. $\iiint_B \sqrt{x} e^{xyz} \, dV$, where $B = \{(x, y, z) \mid 0 \le x \le 4, \ 0 \le y \le 1, 0 \le z \le 2\}$

27–28 Sketch the solid whose volume is given by the iterated integral.

27. $\int_0^1 \int_0^{1-x} \int_0^{2-2z} dy \, dz \, dx$ **28.** $\int_0^2 \int_0^{2-y} \int_0^{4-y^2} dx \, dz \, dy$

29–32 Express the integral $\iiint_E f(x, y, z) \, dV$ as an iterated integral in six different ways, where E is the solid bounded by the given surfaces.

29. $y = 4 - x^2 - 4z^2$, $y = 0$

30. $y^2 + z^2 = 9$, $x = -2$, $x = 2$

31. $y = x^2$, $z = 0$, $y + 2z = 4$

32. $x = 2$, $y = 2$, $z = 0$, $x + y - 2z = 2$

---

33. The figure shows the region of integration for the integral

$$\int_0^1 \int_{\sqrt{x}}^1 \int_0^{1-y} f(x, y, z) \, dz \, dy \, dx$$

Rewrite this integral as an equivalent iterated integral in the five other orders.

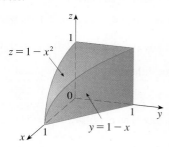

34. The figure shows the region of integration for the integral

$$\int_0^1 \int_0^{1-x^2} \int_0^{1-x} f(x, y, z) \, dy \, dz \, dx$$

Rewrite this integral as an equivalent iterated integral in the five other orders.

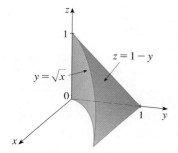

35–36 Write five other iterated integrals that are equal to the given iterated integral.

35. $\int_0^1 \int_y^1 \int_0^y f(x, y, z) \, dz \, dx \, dy$

36. $\int_0^1 \int_y^1 \int_0^z f(x, y, z) \, dx \, dz \, dy$

37–38 Evaluate the triple integral using only geometric interpretation and symmetry.

37. $\iiint_C (4 + 5x^2yz^2) \, dV$, where C is the cylindrical region
$x^2 + y^2 \leqslant 4, \, -2 \leqslant z \leqslant 2$

38. $\iiint_B (z^3 + \sin y + 3) \, dV$, where B is the unit ball
$x^2 + y^2 + z^2 \leqslant 1$

39–42 Find the mass and center of mass of the solid E with the given density function ρ.

39. E lies above the xy-plane and below the paraboloid
$z = 1 - x^2 - y^2$; $\quad \rho(x, y, z) = 3$

40. E is bounded by the parabolic cylinder $z = 1 - y^2$ and the planes $x + z = 1$, $x = 0$, and $z = 0$; $\quad \rho(x, y, z) = 4$

41. E is the cube given by $0 \leqslant x \leqslant a, \, 0 \leqslant y \leqslant a, \, 0 \leqslant z \leqslant a$;
$\rho(x, y, z) = x^2 + y^2 + z^2$

42. E is the tetrahedron bounded by the planes $x = 0, \, y = 0$,
$z = 0, \, x + y + z = 1$; $\quad \rho(x, y, z) = y$

43–46 Assume that the solid has constant density k.

43. Find the moments of inertia for a cube with side length L if one vertex is located at the origin and three edges lie along the coordinate axes.

44. Find the moments of inertia for a rectangular brick with dimensions a, b, and c and mass M if the center of the brick is situated at the origin and the edges are parallel to the coordinate axes.

45. Find the moment of inertia about the z-axis of the solid cylinder $x^2 + y^2 \leqslant a^2$, $0 \leqslant z \leqslant h$.

46. Find the moment of inertia about the z-axis of the solid cone $\sqrt{x^2 + y^2} \leqslant z \leqslant h$.

47–48 Set up, but do not evaluate, integral expressions for (a) the mass, (b) the center of mass, and (c) the moment of inertia about the z-axis.

47. The solid of Exercise 21; $\quad \rho(x, y, z) = \sqrt{x^2 + y^2}$

48. The hemisphere $x^2 + y^2 + z^2 \leqslant 1$, $z \geqslant 0$;
$\rho(x, y, z) = \sqrt{x^2 + y^2 + z^2}$

CAS **49.** Let E be the solid in the first octant bounded by the cylinder $x^2 + y^2 = 1$ and the planes $y = z$, $x = 0$, and $z = 0$ with the density function $\rho(x, y, z) = 1 + x + y + z$. Use a computer algebra system to find the exact values of the following quantities for E.
(a) The mass
(b) The center of mass
(c) The moment of inertia about the z-axis

CAS **50.** If E is the solid of Exercise 18 with density function $\rho(x, y, z) = x^2 + y^2$, find the following quantities, correct to three decimal places.
(a) The mass
(b) The center of mass
(c) The moment of inertia about the z-axis

51. The joint density function for random variables X, Y, and Z is $f(x, y, z) = Cxyz$ if $0 \leqslant x \leqslant 2$, $0 \leqslant y \leqslant 2$, $0 \leqslant z \leqslant 2$, and $f(x, y, z) = 0$ otherwise.
(a) Find the value of the constant C.
(b) Find $P(X \leqslant 1, Y \leqslant 1, Z \leqslant 1)$.
(c) Find $P(X + Y + Z \leqslant 1)$.

52. Suppose X, Y, and Z are random variables with joint density function $f(x, y, z) = Ce^{-(0.5x+0.2y+0.1z)}$ if $x \geqslant 0$, $y \geqslant 0$, $z \geqslant 0$, and $f(x, y, z) = 0$ otherwise.
(a) Find the value of the constant C.
(b) Find $P(X \leqslant 1, Y \leqslant 1)$.
(c) Find $P(X \leqslant 1, Y \leqslant 1, Z \leqslant 1)$.

53–54 The **average value** of a function $f(x, y, z)$ over a solid region E is defined to be

$$f_{\text{ave}} = \frac{1}{V(E)} \iiint_E f(x, y, z) \, dV$$

where $V(E)$ is the volume of E. For instance, if ρ is a density function, then ρ_{ave} is the average density of E.

53. Find the average value of the function $f(x, y, z) = xyz$ over the cube with side length L that lies in the first octant with one vertex at the origin and edges parallel to the coordinate axes.

54. Find the average height of the points in the solid hemisphere $x^2 + y^2 + z^2 \leqslant 1$, $z \geqslant 0$.

---

55. (a) Find the region E for which the triple integral

$$\iiint_E (1 - x^2 - 2y^2 - 3z^2) \, dV$$

is a maximum.

CAS (b) Use a computer algebra system to calculate the exact maximum value of the triple integral in part (a).

15.7 Triple Integrals in Cylindrical Coordinates

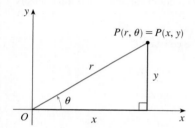

FIGURE 1

In plane geometry the polar coordinate system is used to give a convenient description of certain curves and regions. (See Section 10.3.) Figure 1 enables us to recall the connection between polar and Cartesian coordinates. If the point P has Cartesian coordinates (x, y) and polar coordinates (r, θ), then, from the figure,

$$x = r \cos \theta \qquad y = r \sin \theta$$

$$r^2 = x^2 + y^2 \qquad \tan \theta = \frac{y}{x}$$

In three dimensions there is a coordinate system, called *cylindrical coordinates*, that is similar to polar coordinates and gives convenient descriptions of some commonly occurring surfaces and solids. As we will see, some triple integrals are much easier to evaluate in cylindrical coordinates.

■ Cylindrical Coordinates

In the **cylindrical coordinate system**, a point P in three-dimensional space is represented by the ordered triple (r, θ, z), where r and θ are polar coordinates of the projection of P onto the xy-plane and z is the directed distance from the xy-plane to P. (See Figure 2.)

To convert from cylindrical to rectangular coordinates, we use the equations

FIGURE 2
The cylindrical coordinates of a point

$$\boxed{1} \qquad x = r \cos \theta \qquad y = r \sin \theta \qquad z = z$$

whereas to convert from rectangular to cylindrical coordinates, we use

$$\boxed{2} \qquad r^2 = x^2 + y^2 \qquad \tan \theta = \frac{y}{x} \qquad z = z$$

EXAMPLE 1

(a) Plot the point with cylindrical coordinates $(2, 2\pi/3, 1)$ and find its rectangular coordinates.

(b) Find cylindrical coordinates of the point with rectangular coordinates $(3, -3, -7)$.

SOLUTION

(a) The point with cylindrical coordinates $(2, 2\pi/3, 1)$ is plotted in Figure 3. From Equations 1, its rectangular coordinates are

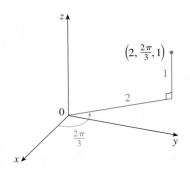

FIGURE 3

$$x = 2 \cos \frac{2\pi}{3} = 2\left(-\frac{1}{2}\right) = -1$$

$$y = 2 \sin \frac{2\pi}{3} = 2\left(\frac{\sqrt{3}}{2}\right) = \sqrt{3}$$

$$z = 1$$

So the point is $\left(-1, \sqrt{3}, 1\right)$ in rectangular coordinates.

(b) From Equations 2 we have

$$r = \sqrt{3^2 + (-3)^2} = 3\sqrt{2}$$

$$\tan \theta = \frac{-3}{3} = -1 \qquad \text{so} \qquad \theta = \frac{7\pi}{4} + 2n\pi$$

$$z = -7$$

Therefore one set of cylindrical coordinates is $\left(3\sqrt{2}, 7\pi/4, -7\right)$. Another is $\left(3\sqrt{2}, -\pi/4, -7\right)$. As with polar coordinates, there are infinitely many choices. ∎

Cylindrical coordinates are useful in problems that involve symmetry about an axis, and the z-axis is chosen to coincide with this axis of symmetry. For instance, the axis of the circular cylinder with Cartesian equation $x^2 + y^2 = c^2$ is the z-axis. In cylindrical coordinates this cylinder has the very simple equation $r = c$. (See Figure 4.) This is the reason for the name "cylindrical" coordinates.

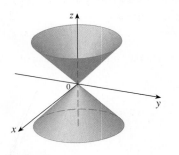

FIGURE 4

$r = c$, a cylinder

EXAMPLE 2 Describe the surface whose equation in cylindrical coordinates is $z = r$.

SOLUTION The equation says that the z-value, or height, of each point on the surface is the same as r, the distance from the point to the z-axis. Because θ doesn't appear, it can vary. So any horizontal trace in the plane $z = k$ ($k > 0$) is a circle of radius k. These traces suggest that the surface is a cone. This prediction can be confirmed by converting the equation into rectangular coordinates. From the first equation in (2) we have

$$z^2 = r^2 = x^2 + y^2$$

We recognize the equation $z^2 = x^2 + y^2$ (by comparison with Table 1 in Section 12.6) as being a circular cone whose axis is the z-axis (see Figure 5). ∎

FIGURE 5

$z = r$, a cone

FIGURE 6

FIGURE 7
Volume element in cylindrical
coordinates: $dV = r \, dz \, dr \, d\theta$

Evaluating Triple Integrals with Cylindrical Coordinates

Suppose that E is a type 1 region whose projection D onto the xy-plane is conveniently described in polar coordinates (see Figure 6). In particular, suppose that f is continuous and

$$E = \{(x, y, z) \mid (x, y) \in D, \; u_1(x, y) \le z \le u_2(x, y)\}$$

where D is given in polar coordinates by

$$D = \{(r, \theta) \mid \alpha \le \theta \le \beta, \; h_1(\theta) \le r \le h_2(\theta)\}$$

We know from Equation 15.6.6 that

$$\boxed{3} \qquad \iiint_E f(x, y, z) \, dV = \iint_D \left[\int_{u_1(x, y)}^{u_2(x, y)} f(x, y, z) \, dz \right] dA$$

But we also know how to evaluate double integrals in polar coordinates. In fact, combining Equation 3 with Equation 15.3.3, we obtain

$$\boxed{4} \qquad \iiint_E f(x, y, z) \, dV = \int_\alpha^\beta \int_{h_1(\theta)}^{h_2(\theta)} \int_{u_1(r\cos\theta,\, r\sin\theta)}^{u_2(r\cos\theta,\, r\sin\theta)} f(r\cos\theta, r\sin\theta, z) \, r \, dz \, dr \, d\theta$$

Formula 4 is the **formula for triple integration in cylindrical coordinates**. It says that we convert a triple integral from rectangular to cylindrical coordinates by writing $x = r\cos\theta$, $y = r\sin\theta$, leaving z as it is, using the appropriate limits of integration for z, r, and θ, and replacing dV by $r \, dz \, dr \, d\theta$. (Figure 7 shows how to remember this.) It is worthwhile to use this formula when E is a solid region easily described in cylindrical coordinates, and especially when the function $f(x, y, z)$ involves the expression $x^2 + y^2$.

EXAMPLE 3 A solid E lies within the cylinder $x^2 + y^2 = 1$, below the plane $z = 4$, and above the paraboloid $z = 1 - x^2 - y^2$. (See Figure 8.) The density at any point is proportional to its distance from the axis of the cylinder. Find the mass of E.

SOLUTION In cylindrical coordinates the cylinder is $r = 1$ and the paraboloid is $z = 1 - r^2$, so we can write

$$E = \{(r, \theta, z) \mid 0 \le \theta \le 2\pi, \; 0 \le r \le 1, \; 1 - r^2 \le z \le 4\}$$

Since the density at (x, y, z) is proportional to the distance from the z-axis, the density function is

$$f(x, y, z) = K\sqrt{x^2 + y^2} = Kr$$

where K is the proportionality constant. Therefore, from Formula 15.6.13, the mass of E is

$$m = \iiint_E K\sqrt{x^2 + y^2} \, dV = \int_0^{2\pi} \int_0^1 \int_{1-r^2}^4 (Kr) \, r \, dz \, dr \, d\theta$$

$$= \int_0^{2\pi} \int_0^1 Kr^2[4 - (1 - r^2)] \, dr \, d\theta = K \int_0^{2\pi} d\theta \int_0^1 (3r^2 + r^4) \, dr$$

$$= 2\pi K \left[r^3 + \frac{r^5}{5} \right]_0^1 = \frac{12\pi K}{5}$$

FIGURE 8

EXAMPLE 4 Evaluate $\int_{-2}^{2} \int_{-\sqrt{4-x^2}}^{\sqrt{4-x^2}} \int_{\sqrt{x^2+y^2}}^{2} (x^2 + y^2) \, dz \, dy \, dx$.

SOLUTION This iterated integral is a triple integral over the solid region

$$E = \left\{ (x, y, z) \mid -2 \leqslant x \leqslant 2, \ -\sqrt{4 - x^2} \leqslant y \leqslant \sqrt{4 - x^2}, \ \sqrt{x^2 + y^2} \leqslant z \leqslant 2 \right\}$$

and the projection of E onto the xy-plane is the disk $x^2 + y^2 \leqslant 4$. The lower surface of E is the cone $z = \sqrt{x^2 + y^2}$ and its upper surface is the plane $z = 2$. (See Figure 9.) This region has a much simpler description in cylindrical coordinates:

$$E = \left\{ (r, \theta, z) \mid 0 \leqslant \theta \leqslant 2\pi, \ 0 \leqslant r \leqslant 2, \ r \leqslant z \leqslant 2 \right\}$$

Therefore we have

$$\int_{-2}^{2} \int_{-\sqrt{4-x^2}}^{\sqrt{4-x^2}} \int_{\sqrt{x^2+y^2}}^{2} (x^2 + y^2) \, dz \, dy \, dx = \iiint_E (x^2 + y^2) \, dV$$

$$= \int_0^{2\pi} \int_0^2 \int_r^2 r^2 \, r \, dz \, dr \, d\theta$$

$$= \int_0^{2\pi} d\theta \int_0^2 r^3 (2 - r) \, dr$$

$$= 2\pi \left[\tfrac{1}{2} r^4 - \tfrac{1}{5} r^5 \right]_0^2 = \tfrac{16}{5}\pi \quad \blacksquare$$

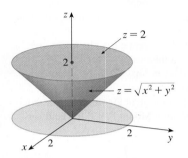

FIGURE 9

15.7 EXERCISES

1–2 Plot the point whose cylindrical coordinates are given. Then find the rectangular coordinates of the point.

1. (a) $(4, \pi/3, -2)$ (b) $(2, -\pi/2, 1)$

2. (a) $\left(\sqrt{2}, 3\pi/4, 2\right)$ (b) $(1, 1, 1)$

3–4 Change from rectangular to cylindrical coordinates.

3. (a) $(-1, 1, 1)$ (b) $\left(-2, 2\sqrt{3}, 3\right)$

4. (a) $\left(-\sqrt{2}, \sqrt{2}, 1\right)$ (b) $(2, 2, 2)$

5–6 Describe in words the surface whose equation is given.

5. $r = 2$ **6.** $\theta = \pi/6$

7–8 Identify the surface whose equation is given.

7. $r^2 + z^2 = 4$ **8.** $r = 2 \sin \theta$

9–10 Write the equations in cylindrical coordinates.

9. (a) $x^2 - x + y^2 + z^2 = 1$ (b) $z = x^2 - y^2$

10. (a) $2x^2 + 2y^2 - z^2 = 4$ (b) $2x - y + z = 1$

11–12 Sketch the solid described by the given inequalities.

11. $r^2 \leqslant z \leqslant 8 - r^2$

12. $0 \leqslant \theta \leqslant \pi/2, \quad r \leqslant z \leqslant 2$

13. A cylindrical shell is 20 cm long, with inner radius 6 cm and outer radius 7 cm. Write inequalities that describe the shell in an appropriate coordinate system. Explain how you have positioned the coordinate system with respect to the shell.

14. Use a graphing device to draw the solid enclosed by the paraboloids $z = x^2 + y^2$ and $z = 5 - x^2 - y^2$.

15–16 Sketch the solid whose volume is given by the integral and evaluate the integral.

15. $\int_{-\pi/2}^{\pi/2} \int_0^2 \int_0^{r^2} r \, dz \, dr \, d\theta$ **16.** $\int_0^2 \int_0^{2\pi} \int_0^r r \, dz \, d\theta \, dr$

17–28 Use cylindrical coordinates.

17. Evaluate $\iiint_E \sqrt{x^2 + y^2} \, dV$, where E is the region that lies inside the cylinder $x^2 + y^2 = 16$ and between the planes $z = -5$ and $z = 4$.

18. Evaluate $\iiint_E z \, dV$, where E is enclosed by the paraboloid $z = x^2 + y^2$ and the plane $z = 4$.

19. Evaluate $\iiint_E (x + y + z)\, dV$, where E is the solid in the first octant that lies under the paraboloid $z = 4 - x^2 - y^2$.

20. Evaluate $\iiint_E (x - y)\, dV$, where E is the solid that lies between the cylinders $x^2 + y^2 = 1$ and $x^2 + y^2 = 16$, above the xy-plane, and below the plane $z = y + 4$.

21. Evaluate $\iiint_E x^2\, dV$, where E is the solid that lies within the cylinder $x^2 + y^2 = 1$, above the plane $z = 0$, and below the cone $z^2 = 4x^2 + 4y^2$.

22. Find the volume of the solid that lies within both the cylinder $x^2 + y^2 = 1$ and the sphere $x^2 + y^2 + z^2 = 4$.

23. Find the volume of the solid that is enclosed by the cone $z = \sqrt{x^2 + y^2}$ and the sphere $x^2 + y^2 + z^2 = 2$.

24. Find the volume of the solid that lies between the paraboloid $z = x^2 + y^2$ and the sphere $x^2 + y^2 + z^2 = 2$.

25. (a) Find the volume of the region E that lies between the paraboloid $z = 24 - x^2 - y^2$ and the cone $z = 2\sqrt{x^2 + y^2}$.
   (b) Find the centroid of E (the center of mass in the case where the density is constant).

26. (a) Find the volume of the solid that the cylinder $r = a \cos \theta$ cuts out of the sphere of radius a centered at the origin.
   (b) Illustrate the solid of part (a) by graphing the sphere and the cylinder on the same screen.

27. Find the mass and center of mass of the solid S bounded by the paraboloid $z = 4x^2 + 4y^2$ and the plane $z = a$ $(a > 0)$ if S has constant density K.

28. Find the mass of a ball B given by $x^2 + y^2 + z^2 \leqslant a^2$ if the density at any point is proportional to its distance from the z-axis.

29–30 Evaluate the integral by changing to cylindrical coordinates.

29. $\displaystyle\int_{-2}^{2} \int_{-\sqrt{4-y^2}}^{\sqrt{4-y^2}} \int_{\sqrt{x^2+y^2}}^{2} xz\, dz\, dx\, dy$

30. $\displaystyle\int_{-3}^{3} \int_{0}^{\sqrt{9-x^2}} \int_{0}^{9-x^2-y^2} \sqrt{x^2 + y^2}\, dz\, dy\, dx$

31. When studying the formation of mountain ranges, geologists estimate the amount of work required to lift a mountain from sea level. Consider a mountain that is essentially in the shape of a right circular cone. Suppose that the weight density of the material in the vicinity of a point P is $g(P)$ and the height is $h(P)$.
   (a) Find a definite integral that represents the total work done in forming the mountain.
   (b) Assume that Mount Fuji in Japan is in the shape of a right circular cone with radius 62,000 ft, height 12,400 ft, and density a constant 200 lb/ft³. How much work was done in forming Mount Fuji if the land was initially at sea level?

© S.R. Lee Photo Traveller / Shutterstock.com

DISCOVERY PROJECT   THE INTERSECTION OF THREE CYLINDERS

The figure shows the solid enclosed by three circular cylinders with the same diameter that intersect at right angles. In this project we compute its volume and determine how its shape changes if the cylinders have different diameters.

1. Sketch carefully the solid enclosed by the three cylinders $x^2 + y^2 = 1$, $x^2 + z^2 = 1$, and $y^2 + z^2 = 1$. Indicate the positions of the coordinate axes and label the faces with the equations of the corresponding cylinders.

2. Find the volume of the solid in Problem 1.

CAS 3. Use a computer algebra system to draw the edges of the solid.

4. What happens to the solid in Problem 1 if the radius of the first cylinder is different from 1? Illustrate with a hand-drawn sketch or a computer graph.

5. If the first cylinder is $x^2 + y^2 = a^2$, where $a < 1$, set up, but do not evaluate, a double integral for the volume of the solid. What if $a > 1$?

15.8 Triple Integrals in Spherical Coordinates

Another useful coordinate system in three dimensions is the *spherical coordinate system*. It simplifies the evaluation of triple integrals over regions bounded by spheres or cones.

■ Spherical Coordinates

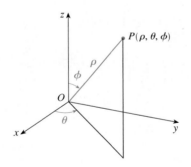

FIGURE 1
The spherical coordinates of a point

The **spherical coordinates** (ρ, θ, ϕ) of a point P in space are shown in Figure 1, where $\rho = |OP|$ is the distance from the origin to P, θ is the same angle as in cylindrical coordinates, and ϕ is the angle between the positive z-axis and the line segment OP. Note that

$$\rho \geqslant 0 \qquad 0 \leqslant \phi \leqslant \pi$$

The spherical coordinate system is especially useful in problems where there is symmetry about a point, and the origin is placed at this point. For example, the sphere with center the origin and radius c has the simple equation $\rho = c$ (see Figure 2); this is the reason for the name "spherical" coordinates. The graph of the equation $\theta = c$ is a vertical half-plane (see Figure 3), and the equation $\phi = c$ represents a half-cone with the z-axis as its axis (see Figure 4).

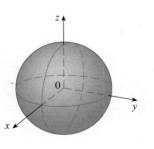

FIGURE 2 $\rho = c$, a sphere

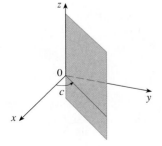

FIGURE 3 $\theta = c$, a half-plane

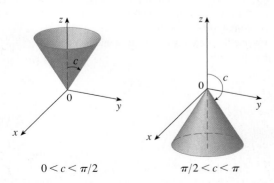

$$0 < c < \pi/2 \qquad\qquad \pi/2 < c < \pi$$

FIGURE 4 $\phi = c$, a half-cone

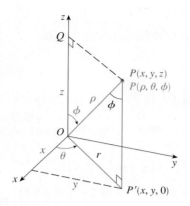

FIGURE 5

The relationship between rectangular and spherical coordinates can be seen from Figure 5. From triangles OPQ and OPP' we have

$$z = \rho \cos \phi \qquad r = \rho \sin \phi$$

But $x = r \cos \theta$ and $y = r \sin \theta$, so to convert from spherical to rectangular coordinates, we use the equations

1    $$x = \rho \sin \phi \, \cos \theta \qquad y = \rho \sin \phi \, \sin \theta \qquad z = \rho \cos \phi$$

Also, the distance formula shows that

2    $$\rho^2 = x^2 + y^2 + z^2$$

We use this equation in converting from rectangular to spherical coordinates.

EXAMPLE 1 The point $(2, \pi/4, \pi/3)$ is given in spherical coordinates. Plot the point and find its rectangular coordinates.

SOLUTION We plot the point in Figure 6. From Equations 1 we have

$$x = \rho \sin \phi \, \cos \theta = 2 \sin \frac{\pi}{3} \, \cos \frac{\pi}{4} = 2\left(\frac{\sqrt{3}}{2}\right)\left(\frac{1}{\sqrt{2}}\right) = \sqrt{\frac{3}{2}}$$

$$y = \rho \sin \phi \, \sin \theta = 2 \sin \frac{\pi}{3} \, \sin \frac{\pi}{4} = 2\left(\frac{\sqrt{3}}{2}\right)\left(\frac{1}{\sqrt{2}}\right) = \sqrt{\frac{3}{2}}$$

$$z = \rho \cos \phi = 2 \cos \frac{\pi}{3} = 2\left(\tfrac{1}{2}\right) = 1$$

FIGURE 6

Thus the point $(2, \pi/4, \pi/3)$ is $\left(\sqrt{3/2}, \sqrt{3/2}, 1\right)$ in rectangular coordinates.

EXAMPLE 2 The point $\left(0, 2\sqrt{3}, -2\right)$ is given in rectangular coordinates. Find spherical coordinates for this point.

SOLUTION From Equation 2 we have

$$\rho = \sqrt{x^2 + y^2 + z^2} = \sqrt{0 + 12 + 4} = 4$$

and so Equations 1 give

$$\cos \phi = \frac{z}{\rho} = \frac{-2}{4} = -\frac{1}{2} \qquad \phi = \frac{2\pi}{3}$$

$$\cos \theta = \frac{x}{\rho \sin \phi} = 0 \qquad \theta = \frac{\pi}{2}$$

⊘ **WARNING** There is not universal agreement on the notation for spherical coordinates. Most books on physics reverse the meanings of θ and ϕ and use r in place of ρ.

TEC In Module 15.8 you can investigate families of surfaces in cylindrical and spherical coordinates.

(Note that $\theta \neq 3\pi/2$ because $y = 2\sqrt{3} > 0$.) Therefore spherical coordinates of the given point are $(4, \pi/2, 2\pi/3)$.

■ Evaluating Triple Integrals with Spherical Coordinates

In the spherical coordinate system the counterpart of a rectangular box is a **spherical wedge**

$$E = \{(\rho, \theta, \phi) \mid a \leqslant \rho \leqslant b,\ \alpha \leqslant \theta \leqslant \beta,\ c \leqslant \phi \leqslant d\}$$

where $a \geqslant 0$ and $\beta - \alpha \leqslant 2\pi$, and $d - c \leqslant \pi$. Although we defined triple integrals by dividing solids into small boxes, it can be shown that dividing a solid into small spherical wedges always gives the same result. So we divide E into smaller spherical wedges E_{ijk} by means of equally spaced spheres $\rho = \rho_i$, half-planes $\theta = \theta_j$, and half-cones $\phi = \phi_k$. Figure 7 shows that E_{ijk} is approximately a rectangular box with dimensions $\Delta\rho$, $\rho_i\,\Delta\phi$ (arc of a circle with radius ρ_i, angle $\Delta\phi$), and $\rho_i \sin\phi_k\,\Delta\theta$ (arc of a circle with radius $\rho_i \sin\phi_k$, angle $\Delta\theta$). So an approximation to the volume of E_{ijk} is given by

$$\Delta V_{ijk} \approx (\Delta\rho)(\rho_i\,\Delta\phi)(\rho_i \sin\phi_k\,\Delta\theta) = \rho_i^2 \sin\phi_k\,\Delta\rho\,\Delta\theta\,\Delta\phi$$

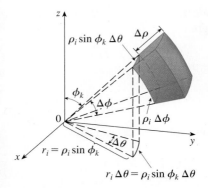

FIGURE 7

In fact, it can be shown, with the aid of the Mean Value Theorem (Exercise 49), that the volume of E_{ijk} is given exactly by

$$\Delta V_{ijk} = \tilde{\rho}_i^2 \sin\tilde{\phi}_k\,\Delta\rho\,\Delta\theta\,\Delta\phi$$

where $(\tilde{\rho}_i, \tilde{\theta}_j, \tilde{\phi}_k)$ is some point in E_{ijk}. Let $(x_{ijk}^*, y_{ijk}^*, z_{ijk}^*)$ be the rectangular coordinates of this point. Then

$$\iiint_E f(x, y, z)\,dV = \lim_{l, m, n \to \infty} \sum_{i=1}^{l} \sum_{j=1}^{m} \sum_{k=1}^{n} f(x_{ijk}^*, y_{ijk}^*, z_{ijk}^*)\,\Delta V_{ijk}$$

$$= \lim_{l, m, n \to \infty} \sum_{i=1}^{l} \sum_{j=1}^{m} \sum_{k=1}^{n} f(\tilde{\rho}_i \sin\tilde{\phi}_k \cos\tilde{\theta}_j,\, \tilde{\rho}_i \sin\tilde{\phi}_k \sin\tilde{\theta}_j,\, \tilde{\rho}_i \cos\tilde{\phi}_k)\,\tilde{\rho}_i^2 \sin\tilde{\phi}_k\,\Delta\rho\Delta\theta\Delta\phi$$

But this sum is a Riemann sum for the function

$$F(\rho, \theta, \phi) = f(\rho \sin\phi \cos\theta,\, \rho \sin\phi \sin\theta,\, \rho \cos\phi)\,\rho^2 \sin\phi$$

Consequently, we have arrived at the following **formula for triple integration in spherical coordinates**.

---

3 $\displaystyle\iiint_E f(x, y, z)\,dV$

$$= \int_c^d \int_\alpha^\beta \int_a^b f(\rho \sin\phi \cos\theta,\, \rho \sin\phi \sin\theta,\, \rho \cos\phi)\,\rho^2 \sin\phi\,d\rho\,d\theta\,d\phi$$

where E is a spherical wedge given by

$$E = \{(\rho, \theta, \phi) \mid a \leqslant \rho \leqslant b,\ \alpha \leqslant \theta \leqslant \beta,\ c \leqslant \phi \leqslant d\}$$

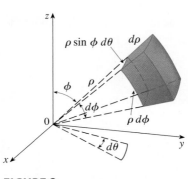

FIGURE 8

Volume element in spherical coordinates: $dV = \rho^2 \sin\phi\,d\rho\,d\theta\,d\phi$

Formula 3 says that we convert a triple integral from rectangular coordinates to spherical coordinates by writing

$$x = \rho \sin\phi \cos\theta \qquad y = \rho \sin\phi \sin\theta \qquad z = \rho \cos\phi$$

using the appropriate limits of integration, and replacing dV by $\rho^2 \sin\phi\,d\rho\,d\theta\,d\phi$. This is illustrated in Figure 8.

This formula can be extended to include more general spherical regions such as

$$E = \{(\rho, \theta, \phi) \mid \alpha \leqslant \theta \leqslant \beta, \ c \leqslant \phi \leqslant d, \ g_1(\theta, \phi) \leqslant \rho \leqslant g_2(\theta, \phi)\}$$

In this case the formula is the same as in (3) except that the limits of integration for ρ are $g_1(\theta, \phi)$ and $g_2(\theta, \phi)$.

Usually, spherical coordinates are used in triple integrals when surfaces such as cones and spheres form the boundary of the region of integration.

EXAMPLE 3 Evaluate $\iiint_B e^{(x^2+y^2+z^2)^{3/2}} \, dV$, where B is the unit ball:

$$B = \{(x, y, z) \mid x^2 + y^2 + z^2 \leqslant 1\}$$

SOLUTION Since the boundary of B is a sphere, we use spherical coordinates:

$$B = \{(\rho, \theta, \phi) \mid 0 \leqslant \rho \leqslant 1, \ 0 \leqslant \theta \leqslant 2\pi, \ 0 \leqslant \phi \leqslant \pi\}$$

In addition, spherical coordinates are appropriate because

$$x^2 + y^2 + z^2 = \rho^2$$

Thus (3) gives

$$\iiint_B e^{(x^2+y^2+z^2)^{3/2}} \, dV = \int_0^\pi \int_0^{2\pi} \int_0^1 e^{(\rho^2)^{3/2}} \rho^2 \sin\phi \, d\rho \, d\theta \, d\phi$$

$$= \int_0^\pi \sin\phi \, d\phi \int_0^{2\pi} d\theta \int_0^1 \rho^2 e^{\rho^3} \, d\rho$$

$$= \left[-\cos\phi\right]_0^\pi (2\pi) \left[\tfrac{1}{3}e^{\rho^3}\right]_0^1 = \tfrac{4}{3}\pi(e - 1) \qquad \blacksquare$$

NOTE It would have been extremely awkward to evaluate the integral in Example 3 without spherical coordinates. In rectangular coordinates the iterated integral would have been

$$\int_{-1}^1 \int_{-\sqrt{1-x^2}}^{\sqrt{1-x^2}} \int_{-\sqrt{1-x^2-y^2}}^{\sqrt{1-x^2-y^2}} e^{(x^2+y^2+z^2)^{3/2}} \, dz \, dy \, dx$$

EXAMPLE 4 Use spherical coordinates to find the volume of the solid that lies above the cone $z = \sqrt{x^2 + y^2}$ and below the sphere $x^2 + y^2 + z^2 = z$. (See Figure 9.)

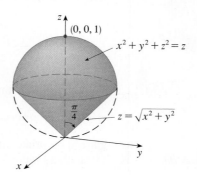

FIGURE 9

Figure 10 gives another look (this time drawn by Maple) at the solid of Example 4.

FIGURE 10

SOLUTION Notice that the sphere passes through the origin and has center $(0, 0, \frac{1}{2})$. We write the equation of the sphere in spherical coordinates as

$$\rho^2 = \rho \cos \phi \qquad \text{or} \qquad \rho = \cos \phi$$

The equation of the cone can be written as

$$\rho \cos \phi = \sqrt{\rho^2 \sin^2\phi \, \cos^2\theta + \rho^2 \sin^2\phi \, \sin^2\theta} = \rho \sin \phi$$

This gives $\sin \phi = \cos \phi$, or $\phi = \pi/4$. Therefore the description of the solid E in spherical coordinates is

$$E = \{(\rho, \theta, \phi) \mid 0 \le \theta \le 2\pi, \ 0 \le \phi \le \pi/4, \ 0 \le \rho \le \cos \phi\}$$

Figure 11 shows how E is swept out if we integrate first with respect to ρ, then ϕ, and then θ. The volume of E is

$$V(E) = \iiint\limits_E dV = \int_0^{2\pi} \int_0^{\pi/4} \int_0^{\cos \phi} \rho^2 \sin \phi \, d\rho \, d\phi \, d\theta$$

$$= \int_0^{2\pi} d\theta \int_0^{\pi/4} \sin \phi \left[\frac{\rho^3}{3} \right]_{\rho=0}^{\rho=\cos \phi} d\phi$$

$$= \frac{2\pi}{3} \int_0^{\pi/4} \sin \phi \cos^3\phi \, d\phi = \frac{2\pi}{3} \left[-\frac{\cos^4\phi}{4} \right]_0^{\pi/4} = \frac{\pi}{8}$$

TEC Visual 15.8 shows an animation of Figure 11.

ρ varies from 0 to $\cos \phi$ while ϕ and θ are constant.

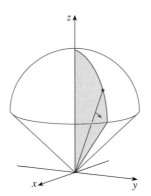

ϕ varies from 0 to $\pi/4$ while θ is constant.

θ varies from 0 to 2π.

FIGURE 11

15.8 EXERCISES

1–2 Plot the point whose spherical coordinates are given. Then find the rectangular coordinates of the point.

1. (a) $(6, \pi/3, \pi/6)$ (b) $(3, \pi/2, 3\pi/4)$

2. (a) $(2, \pi/2, \pi/2)$ (b) $(4, -\pi/4, \pi/3)$

3–4 Change from rectangular to spherical coordinates.

3. (a) $(0, -2, 0)$ (b) $(-1, 1, -\sqrt{2})$

4. (a) $(1, 0, \sqrt{3})$ (b) $(\sqrt{3}, -1, 2\sqrt{3})$

5–6 Describe in words the surface whose equation is given.

5. $\phi = \pi/3$

6. $\rho^2 - 3\rho + 2 = 0$

7–8 Identify the surface whose equation is given.

7. $\rho \cos \phi = 1$

8. $\rho = \cos \phi$

9–10 Write the equation in spherical coordinates.

9. (a) $x^2 + y^2 + z^2 = 9$ (b) $x^2 - y^2 - z^2 = 1$

10. (a) $z = x^2 + y^2$ (b) $z = x^2 - y^2$

11–14 Sketch the solid described by the given inequalities.

11. $\rho \leq 1$, $0 \leq \phi \leq \pi/6$, $0 \leq \theta \leq \pi$

12. $1 \leq \rho \leq 2$, $\pi/2 \leq \phi \leq \pi$

13. $2 \leq \rho \leq 4$, $0 \leq \phi \leq \pi/3$, $0 \leq \theta \leq \pi$

14. $\rho \leq 2$, $\rho \leq \csc \phi$

---

15. A solid lies above the cone $z = \sqrt{x^2 + y^2}$ and below the sphere $x^2 + y^2 + z^2 = z$. Write a description of the solid in terms of inequalities involving spherical coordinates.

16. (a) Find inequalities that describe a hollow ball with diameter 30 cm and thickness 0.5 cm. Explain how you have positioned the coordinate system that you have chosen.
(b) Suppose the ball is cut in half. Write inequalities that describe one of the halves.

17–18 Sketch the solid whose volume is given by the integral and evaluate the integral.

17. $\int_0^{\pi/6} \int_0^{\pi/2} \int_0^3 \rho^2 \sin \phi \, d\rho \, d\theta \, d\phi$

18. $\int_0^{\pi/4} \int_0^{2\pi} \int_0^{\sec \phi} \rho^2 \sin \phi \, d\rho \, d\theta \, d\phi$

19–20 Set up the triple integral of an arbitrary continuous function $f(x, y, z)$ in cylindrical or spherical coordinates over the solid shown.

19.

20.

21–34 Use spherical coordinates.

21. Evaluate $\iiint_B (x^2 + y^2 + z^2)^2 \, dV$, where B is the ball with center the origin and radius 5.

22. Evaluate $\iiint_E y^2 z^2 \, dV$, where E lies above the cone $\phi = \pi/3$ and below the sphere $\rho = 1$.

23. Evaluate $\iiint_E (x^2 + y^2) \, dV$, where E lies between the spheres $x^2 + y^2 + z^2 = 4$ and $x^2 + y^2 + z^2 = 9$.

24. Evaluate $\iiint_E y^2 \, dV$, where E is the solid hemisphere $x^2 + y^2 + z^2 \leq 9$, $y \geq 0$.

25. Evaluate $\iiint_E xe^{x^2+y^2+z^2} \, dV$, where E is the portion of the unit ball $x^2 + y^2 + z^2 \leq 1$ that lies in the first octant.

26. Evaluate $\iiint_E \sqrt{x^2 + y^2 + z^2} \, dV$, where E lies above the cone $z = \sqrt{x^2 + y^2}$ and between the spheres $x^2 + y^2 + z^2 = 1$ and $x^2 + y^2 + z^2 = 4$.

27. Find the volume of the part of the ball $\rho \leq a$ that lies between the cones $\phi = \pi/6$ and $\phi = \pi/3$.

28. Find the average distance from a point in a ball of radius a to its center.

29. (a) Find the volume of the solid that lies above the cone $\phi = \pi/3$ and below the sphere $\rho = 4 \cos \phi$.
(b) Find the centroid of the solid in part (a).

30. Find the volume of the solid that lies within the sphere $x^2 + y^2 + z^2 = 4$, above the xy-plane, and below the cone $z = \sqrt{x^2 + y^2}$.

31. (a) Find the centroid of the solid in Example 4. (Assume constant density K.)
(b) Find the moment of inertia about the z-axis for this solid.

32. Let H be a solid hemisphere of radius a whose density at any point is proportional to its distance from the center of the base.
(a) Find the mass of H.
(b) Find the center of mass of H.
(c) Find the moment of inertia of H about its axis.

33. (a) Find the centroid of a solid homogeneous hemisphere of radius a.
(b) Find the moment of inertia of the solid in part (a) about a diameter of its base.

34. Find the mass and center of mass of a solid hemisphere of radius a if the density at any point is proportional to its distance from the base.

35–40 Use cylindrical or spherical coordinates, whichever seems more appropriate.

35. Find the volume and centroid of the solid E that lies above the cone $z = \sqrt{x^2 + y^2}$ and below the sphere $x^2 + y^2 + z^2 = 1$.

36. Find the volume of the smaller wedge cut from a sphere of radius a by two planes that intersect along a diameter at an angle of $\pi/6$.

37. A solid cylinder with constant density has base radius a and height h.
(a) Find the moment of inertia of the cylinder about its axis.
(b) Find the moment of inertia of the cylinder about a diameter of its base.

38. A solid right circular cone with constant density has base radius a and height h.
  (a) Find the moment of inertia of the cone about its axis.
  (b) Find the moment of inertia of the cone about a diameter of its base.

CAS 39. Evaluate $\iiint_E z\, dV$, where E lies above the paraboloid $z = x^2 + y^2$ and below the plane $z = 2y$. Use either the Table of Integrals (on Reference Pages 6–10) or a computer algebra system to evaluate the integral.

CAS 40. (a) Find the volume enclosed by the torus $\rho = \sin\phi$.
  (b) Use a computer to draw the torus.

41–43 Evaluate the integral by changing to spherical coordinates.

41. $\displaystyle\int_0^1 \int_0^{\sqrt{1-x^2}} \int_{\sqrt{x^2+y^2}}^{\sqrt{2-x^2-y^2}} xy\, dz\, dy\, dx$

42. $\displaystyle\int_{-a}^{a} \int_{-\sqrt{a^2-y^2}}^{\sqrt{a^2-y^2}} \int_{-\sqrt{a^2-x^2-y^2}}^{\sqrt{a^2-x^2-y^2}} (x^2 z + y^2 z + z^3)\, dz\, dx\, dy$

43. $\displaystyle\int_{-2}^{2} \int_{-\sqrt{4-x^2}}^{\sqrt{4-x^2}} \int_{2-\sqrt{4-x^2-y^2}}^{2+\sqrt{4-x^2-y^2}} (x^2 + y^2 + z^2)^{3/2}\, dz\, dy\, dx$

44. A model for the density δ of the earth's atmosphere near its surface is

$$\delta = 619.09 - 0.000097\rho$$

where ρ (the distance from the center of the earth) is measured in meters and δ is measured in kilograms per cubic meter. If we take the surface of the earth to be a sphere with radius 6370 km, then this model is a reasonable one for $6.370 \times 10^6 \le \rho \le 6.375 \times 10^6$. Use this model to estimate the mass of the atmosphere between the ground and an altitude of 5 km.

45. Use a graphing device to draw a silo consisting of a cylinder with radius 3 and height 10 surmounted by a hemisphere.

46. The latitude and longitude of a point P in the Northern Hemisphere are related to spherical coordinates ρ, θ, ϕ as follows. We take the origin to be the center of the earth and the positive z-axis to pass through the North Pole. The positive x-axis passes through the point where the prime meridian (the meridian through Greenwich, England) intersects the equator. Then the latitude of P is $\alpha = 90° - \phi°$ and the longitude is $\beta = 360° - \theta°$. Find the great-circle

distance from Los Angeles (lat. 34.06° N, long. 118.25° W) to Montréal (lat. 45.50° N, long. 73.60° W). Take the radius of the earth to be 3960 mi. (A *great circle* is the circle of intersection of a sphere and a plane through the center of the sphere.)

CAS 47. The surfaces $\rho = 1 + \frac{1}{5}\sin m\theta \sin n\phi$ have been used as models for tumors. The "bumpy sphere" with $m = 6$ and $n = 5$ is shown. Use a computer algebra system to find the volume it encloses.

48. Show that

$$\int_{-\infty}^{\infty} \int_{-\infty}^{\infty} \int_{-\infty}^{\infty} \sqrt{x^2 + y^2 + z^2}\; e^{-(x^2+y^2+z^2)}\, dx\, dy\, dz = 2\pi$$

(The improper triple integral is defined as the limit of a triple integral over a solid sphere as the radius of the sphere increases indefinitely.)

49. (a) Use cylindrical coordinates to show that the volume of the solid bounded above by the sphere $r^2 + z^2 = a^2$ and below by the cone $z = r\cot\phi_0$ (or $\phi = \phi_0$), where $0 < \phi_0 < \pi/2$, is

$$V = \frac{2\pi a^3}{3}(1 - \cos\phi_0)$$

  (b) Deduce that the volume of the spherical wedge given by $\rho_1 \le \rho \le \rho_2$, $\theta_1 \le \theta \le \theta_2$, $\phi_1 \le \phi \le \phi_2$ is

$$\Delta V = \frac{\rho_2^3 - \rho_1^3}{3}(\cos\phi_1 - \cos\phi_2)(\theta_2 - \theta_1)$$

  (c) Use the Mean Value Theorem to show that the volume in part (b) can be written as

$$\Delta V = \tilde{\rho}^2 \sin\bar{\phi}\, \Delta\rho\, \Delta\theta\, \Delta\phi$$

where $\tilde{\rho}$ lies between ρ_1 and ρ_2, $\bar{\phi}$ lies between ϕ_1 and ϕ_2, $\Delta\rho = \rho_2 - \rho_1$, $\Delta\theta = \theta_2 - \theta_1$, and $\Delta\phi = \phi_2 - \phi_1$.

APPLIED PROJECT   ROLLER DERBY

Suppose that a solid ball (a marble), a hollow ball (a squash ball), a solid cylinder (a steel bar), and a hollow cylinder (a lead pipe) roll down a slope. Which of these objects reaches the bottom first? (Make a guess before proceeding.)

To answer this question, we consider a ball or cylinder with mass m, radius r, and moment of inertia I (about the axis of rotation). If the vertical drop is h, then the potential energy at the top is mgh. Suppose the object reaches the bottom with velocity v and angular velocity ω, so $v = \omega r$. The kinetic energy at the bottom consists of two parts: $\frac{1}{2}mv^2$ from translation (moving down the slope) and $\frac{1}{2}I\omega^2$ from rotation. If we assume that energy loss from rolling friction is negligible, then conservation of energy gives

$$mgh = \tfrac{1}{2}mv^2 + \tfrac{1}{2}I\omega^2$$

1. Show that

$$v^2 = \frac{2gh}{1 + I^*} \qquad \text{where } I^* = \frac{I}{mr^2}$$

2. If $y(t)$ is the vertical distance traveled at time t, then the same reasoning as used in Problem 1 shows that $v^2 = 2gy/(1 + I^*)$ at any time t. Use this result to show that y satisfies the differential equation

$$\frac{dy}{dt} = \sqrt{\frac{2g}{1 + I^*}}\,(\sin \alpha)\sqrt{y}$$

   where α is the angle of inclination of the plane.

3. By solving the differential equation in Problem 2, show that the total travel time is

$$T = \sqrt{\frac{2h(1 + I^*)}{g \sin^2\alpha}}$$

   This shows that the object with the smallest value of I^* wins the race.

4. Show that $I^* = \frac{1}{2}$ for a solid cylinder and $I^* = 1$ for a hollow cylinder.

5. Calculate I^* for a partly hollow ball with inner radius a and outer radius r. Express your answer in terms of $b = a/r$. What happens as $a \to 0$ and as $a \to r$?

6. Show that $I^* = \frac{2}{5}$ for a solid ball and $I^* = \frac{2}{3}$ for a hollow ball. Thus the objects finish in the following order: solid ball, solid cylinder, hollow ball, hollow cylinder.

15.9   Change of Variables in Multiple Integrals

In one-dimensional calculus we often use a change of variable (a substitution) to simplify an integral. By reversing the roles of x and u, we can write the Substitution Rule (5.5.6) as

$$\boxed{1} \qquad \int_a^b f(x)\, dx = \int_c^d f(g(u))\, g'(u)\, du$$

where $x = g(u)$ and $a = g(c)$, $b = g(d)$. Another way of writing Formula 1 is as follows:

$$\boxed{2} \qquad \int_a^b f(x)\, dx = \int_c^d f(x(u))\, \frac{dx}{du}\, du$$

A change of variables can also be useful in double integrals. We have already seen one example of this: conversion to polar coordinates. The new variables r and θ are related to the old variables x and y by the equations

$$x = r\cos\theta \qquad y = r\sin\theta$$

and the change of variables formula (15.3.2) can be written as

$$\iint_R f(x, y)\, dA = \iint_S f(r\cos\theta,\, r\sin\theta)\, r\, dr\, d\theta$$

where S is the region in the $r\theta$-plane that corresponds to the region R in the xy-plane.

More generally, we consider a change of variables that is given by a **transformation** T from the uv-plane to the xy-plane:

$$T(u, v) = (x, y)$$

where x and y are related to u and v by the equations

$$\boxed{3} \qquad\qquad x = g(u, v) \qquad y = h(u, v)$$

or, as we sometimes write,

$$x = x(u, v) \qquad y = y(u, v)$$

We usually assume that T is a C^1 **transformation**, which means that g and h have continuous first-order partial derivatives.

A transformation T is really just a function whose domain and range are both subsets of $\mathbb{R}^2$. If $T(u_1, v_1) = (x_1, y_1)$, then the point (x_1, y_1) is called the **image** of the point (u_1, v_1). If no two points have the same image, T is called **one-to-one**. Figure 1 shows the effect of a transformation T on a region S in the uv-plane. T transforms S into a region R in the xy-plane called the **image of S**, consisting of the images of all points in S.

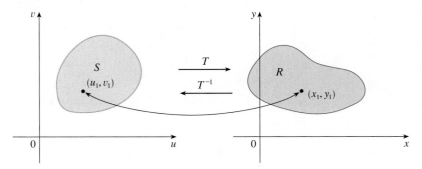

FIGURE 1

If T is a one-to-one transformation, then it has an **inverse transformation** T^{-1} from the xy-plane to the uv-plane and it may be possible to solve Equations 3 for u and v in terms of x and y:

$$u = G(x, y) \qquad v = H(x, y)$$

EXAMPLE 1 A transformation is defined by the equations

$$x = u^2 - v^2 \qquad y = 2uv$$

Find the image of the square $S = \{(u, v) \mid 0 \le u \le 1,\ 0 \le v \le 1\}$.

SOLUTION The transformation maps the boundary of S into the boundary of the image. So we begin by finding the images of the sides of S. The first side, S_1, is given by $v = 0$

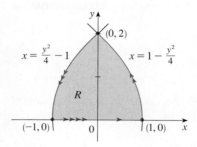

FIGURE 2

$(0 \leqslant u \leqslant 1)$. (See Figure 2.) From the given equations we have $x = u^2$, $y = 0$, and so $0 \leqslant x \leqslant 1$. Thus S_1 is mapped into the line segment from $(0, 0)$ to $(1, 0)$ in the xy-plane. The second side, S_2, is $u = 1$ $(0 \leqslant v \leqslant 1)$ and, putting $u = 1$ in the given equations, we get

$$x = 1 - v^2 \qquad y = 2v$$

Eliminating v, we obtain

$$\boxed{4} \qquad\qquad x = 1 - \frac{y^2}{4} \qquad 0 \leqslant x \leqslant 1$$

which is part of a parabola. Similarly, S_3 is given by $v = 1$ $(0 \leqslant u \leqslant 1)$, whose image is the parabolic arc

$$\boxed{5} \qquad\qquad x = \frac{y^2}{4} - 1 \qquad -1 \leqslant x \leqslant 0$$

Finally, S_4 is given by $u = 0$ $(0 \leqslant v \leqslant 1)$ whose image is $x = -v^2$, $y = 0$, that is, $-1 \leqslant x \leqslant 0$. (Notice that as we move around the square in the counterclockwise direction, we also move around the parabolic region in the counterclockwise direction.) The image of S is the region R (shown in Figure 2) bounded by the x-axis and the parabolas given by Equations 4 and 5. ∎

Now let's see how a change of variables affects a double integral. We start with a small rectangle S in the uv-plane whose lower left corner is the point (u_0, v_0) and whose dimensions are Δu and Δv. (See Figure 3.)

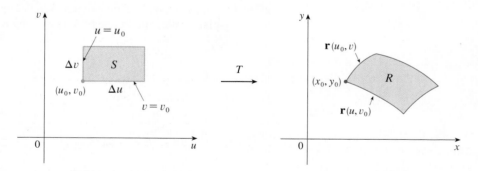

FIGURE 3

The image of S is a region R in the xy-plane, one of whose boundary points is $(x_0, y_0) = T(u_0, v_0)$. The vector

$$\mathbf{r}(u, v) = g(u, v)\,\mathbf{i} + h(u, v)\,\mathbf{j}$$

is the position vector of the image of the point (u, v). The equation of the lower side of S is $v = v_0$, whose image curve is given by the vector function $\mathbf{r}(u, v_0)$. The tangent vector at (x_0, y_0) to this image curve is

$$\mathbf{r}_u = g_u(u_0, v_0)\,\mathbf{i} + h_u(u_0, v_0)\,\mathbf{j} = \frac{\partial x}{\partial u}\,\mathbf{i} + \frac{\partial y}{\partial u}\,\mathbf{j}$$

Similarly, the tangent vector at (x_0, y_0) to the image curve of the left side of S (namely,

$u = u_0)$ is

$$\mathbf{r}_v = g_v(u_0, v_0)\,\mathbf{i} + h_v(u_0, v_0)\,\mathbf{j} = \frac{\partial x}{\partial v}\,\mathbf{i} + \frac{\partial y}{\partial v}\,\mathbf{j}$$

We can approximate the image region $R = T(S)$ by a parallelogram determined by the secant vectors

$$\mathbf{a} = \mathbf{r}(u_0 + \Delta u, v_0) - \mathbf{r}(u_0, v_0) \qquad \mathbf{b} = \mathbf{r}(u_0, v_0 + \Delta v) - \mathbf{r}(u_0, v_0)$$

shown in Figure 4. But

$$\mathbf{r}_u = \lim_{\Delta u \to 0} \frac{\mathbf{r}(u_0 + \Delta u, v_0) - \mathbf{r}(u_0, v_0)}{\Delta u}$$

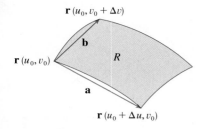

$\mathbf{r}\,(u_0, v_0 + \Delta v)$

$\mathbf{r}\,(u_0, v_0)$ R $\mathbf{b}$ $\mathbf{a}$

$\mathbf{r}\,(u_0 + \Delta u, v_0)$

FIGURE 4

and so

$$\mathbf{r}(u_0 + \Delta u, v_0) - \mathbf{r}(u_0, v_0) \approx \Delta u\,\mathbf{r}_u$$

Similarly

$$\mathbf{r}(u_0, v_0 + \Delta v) - \mathbf{r}(u_0, v_0) \approx \Delta v\,\mathbf{r}_v$$

This means that we can approximate R by a parallelogram determined by the vectors $\Delta u\,\mathbf{r}_u$ and $\Delta v\,\mathbf{r}_v$. (See Figure 5.) Therefore we can approximate the area of R by the area of this parallelogram, which, from Section 12.4, is

$$\boxed{6} \qquad \left|(\Delta u\,\mathbf{r}_u) \times (\Delta v\,\mathbf{r}_v)\right| = \left|\mathbf{r}_u \times \mathbf{r}_v\right| \Delta u\, \Delta v$$

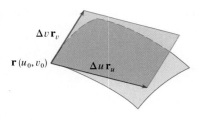

$\Delta v\,\mathbf{r}_v$

$\mathbf{r}\,(u_0, v_0)$ $\Delta u\,\mathbf{r}_u$

FIGURE 5

Computing the cross product, we obtain

$$\mathbf{r}_u \times \mathbf{r}_v = \begin{vmatrix} \mathbf{i} & \mathbf{j} & \mathbf{k} \\ \dfrac{\partial x}{\partial u} & \dfrac{\partial y}{\partial u} & 0 \\ \dfrac{\partial x}{\partial v} & \dfrac{\partial y}{\partial v} & 0 \end{vmatrix} = \begin{vmatrix} \dfrac{\partial x}{\partial u} & \dfrac{\partial y}{\partial u} \\ \dfrac{\partial x}{\partial v} & \dfrac{\partial y}{\partial v} \end{vmatrix} \mathbf{k} = \begin{vmatrix} \dfrac{\partial x}{\partial u} & \dfrac{\partial x}{\partial v} \\ \dfrac{\partial y}{\partial u} & \dfrac{\partial y}{\partial v} \end{vmatrix} \mathbf{k}$$

The determinant that arises in this calculation is called the *Jacobian* of the transformation and is given a special notation.

The Jacobian is named after the German mathematician Carl Gustav Jacob Jacobi (1804–1851). Although the French mathematician Cauchy first used these special determinants involving partial derivatives, Jacobi developed them into a method for evaluating multiple integrals.

> **7 Definition** The **Jacobian** of the transformation T given by $x = g(u, v)$ and $y = h(u, v)$ is
>
> $$\frac{\partial(x, y)}{\partial(u, v)} = \begin{vmatrix} \dfrac{\partial x}{\partial u} & \dfrac{\partial x}{\partial v} \\ \dfrac{\partial y}{\partial u} & \dfrac{\partial y}{\partial v} \end{vmatrix} = \frac{\partial x}{\partial u}\frac{\partial y}{\partial v} - \frac{\partial x}{\partial v}\frac{\partial y}{\partial u}$$

With this notation we can use Equation 6 to give an approximation to the area ΔA of R:

$$\boxed{8} \qquad \Delta A \approx \left|\frac{\partial(x, y)}{\partial(u, v)}\right| \Delta u\, \Delta v$$

where the Jacobian is evaluated at (u_0, v_0).

Next we divide a region S in the uv-plane into rectangles S_{ij} and call their images in the xy-plane R_{ij}. (See Figure 6.)

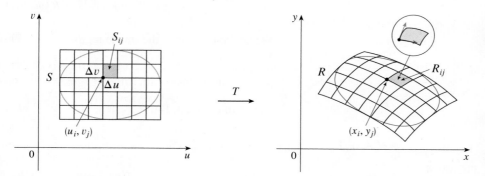

FIGURE 6

Applying the approximation (8) to each R_{ij}, we approximate the double integral of f over R as follows:

$$\iint\limits_{R} f(x, y) \, dA \approx \sum_{i=1}^{m} \sum_{j=1}^{n} f(x_i, y_j) \, \Delta A$$

$$\approx \sum_{i=1}^{m} \sum_{j=1}^{n} f(g(u_i, v_j), h(u_i, v_j)) \left| \frac{\partial(x, y)}{\partial(u, v)} \right| \Delta u \, \Delta v$$

where the Jacobian is evaluated at (u_i, v_j). Notice that this double sum is a Riemann sum for the integral

$$\iint\limits_{S} f(g(u, v), h(u, v)) \left| \frac{\partial(x, y)}{\partial(u, v)} \right| du \, dv$$

The foregoing argument suggests that the following theorem is true. (A full proof is given in books on advanced calculus.)

---

9   **Change of Variables in a Double Integral**   Suppose that T is a C^1 transformation whose Jacobian is nonzero and that T maps a region S in the uv-plane onto a region R in the xy-plane. Suppose that f is continuous on R and that R and S are type I or type II plane regions. Suppose also that T is one-to-one, except perhaps on the boundary of S. Then

$$\iint\limits_{R} f(x, y) \, dA = \iint\limits_{S} f(x(u, v), y(u, v)) \left| \frac{\partial(x, y)}{\partial(u, v)} \right| du \, dv$$

Theorem 9 says that we change from an integral in x and y to an integral in u and v by expressing x and y in terms of u and v and writing

$$dA = \left| \frac{\partial(x, y)}{\partial(u, v)} \right| du \, dv$$

Notice the similarity between Theorem 9 and the one-dimensional formula in Equation 2. Instead of the derivative dx/du, we have the absolute value of the Jacobian, that is, $\left| \partial(x, y)/\partial(u, v) \right|$.

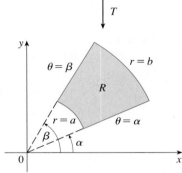

FIGURE 7
The polar coordinate transformation

As a first illustration of Theorem 9, we show that the formula for integration in polar coordinates is just a special case. Here the transformation T from the $r\theta$-plane to the xy-plane is given by

$$x = g(r, \theta) = r \cos \theta \qquad y = h(r, \theta) = r \sin \theta$$

and the geometry of the transformation is shown in Figure 7. T maps an ordinary rectangle in the $r\theta$-plane to a polar rectangle in the xy-plane. The Jacobian of T is

$$\frac{\partial(x, y)}{\partial(r, \theta)} = \begin{vmatrix} \dfrac{\partial x}{\partial r} & \dfrac{\partial x}{\partial \theta} \\ \dfrac{\partial y}{\partial r} & \dfrac{\partial y}{\partial \theta} \end{vmatrix} = \begin{vmatrix} \cos \theta & -r \sin \theta \\ \sin \theta & r \cos \theta \end{vmatrix} = r \cos^2\theta + r \sin^2\theta = r > 0$$

Thus Theorem 9 gives

$$\iint_R f(x, y) \, dx \, dy = \iint_S f(r \cos \theta, r \sin \theta) \left| \frac{\partial(x, y)}{\partial(r, \theta)} \right| dr \, d\theta$$

$$= \int_\alpha^\beta \int_a^b f(r \cos \theta, r \sin \theta) \, r \, dr \, d\theta$$

which is the same as Formula 15.3.2.

EXAMPLE 2 Use the change of variables $x = u^2 - v^2$, $y = 2uv$ to evaluate the integral $\iint_R y \, dA$, where R is the region bounded by the x-axis and the parabolas $y^2 = 4 - 4x$ and $y^2 = 4 + 4x$, $y \geq 0$.

SOLUTION The region R is pictured in Figure 2 (on page 1054). In Example 1 we discovered that $T(S) = R$, where S is the square $[0, 1] \times [0, 1]$. Indeed, the reason for making the change of variables to evaluate the integral is that S is a much simpler region than R. First we need to compute the Jacobian:

$$\frac{\partial(x, y)}{\partial(u, v)} = \begin{vmatrix} \dfrac{\partial x}{\partial u} & \dfrac{\partial x}{\partial v} \\ \dfrac{\partial y}{\partial u} & \dfrac{\partial y}{\partial v} \end{vmatrix} = \begin{vmatrix} 2u & -2v \\ 2v & 2u \end{vmatrix} = 4u^2 + 4v^2 > 0$$

Therefore, by Theorem 9,

$$\iint_R y \, dA = \iint_S 2uv \left| \frac{\partial(x, y)}{\partial(u, v)} \right| dA = \int_0^1 \int_0^1 (2uv) 4(u^2 + v^2) \, du \, dv$$

$$= 8 \int_0^1 \int_0^1 (u^3 v + uv^3) \, du \, dv = 8 \int_0^1 \left[\tfrac{1}{4} u^4 v + \tfrac{1}{2} u^2 v^3 \right]_{u=0}^{u=1} dv$$

$$= \int_0^1 (2v + 4v^3) \, dv = \left[v^2 + v^4 \right]_0^1 = 2 \qquad \blacksquare$$

NOTE Example 2 was not a very difficult problem to solve because we were given a suitable change of variables. If we are not supplied with a transformation, then the first step is to think of an appropriate change of variables. If $f(x, y)$ is difficult to integrate, then the form of $f(x, y)$ may suggest a transformation. If the region of integration R is awkward, then the transformation should be chosen so that the corresponding region S in the uv-plane has a convenient description.

EXAMPLE 3 Evaluate the integral $\iint_R e^{(x+y)/(x-y)}\, dA$, where R is the trapezoidal region with vertices $(1, 0)$, $(2, 0)$, $(0, -2)$, and $(0, -1)$.

SOLUTION Since it isn't easy to integrate $e^{(x+y)/(x-y)}$, we make a change of variables suggested by the form of this function:

$$\boxed{10} \qquad\qquad u = x + y \qquad v = x - y$$

These equations define a transformation T^{-1} from the xy-plane to the uv-plane. Theorem 9 talks about a transformation T from the uv-plane to the xy-plane. It is obtained by solving Equations 10 for x and y:

$$\boxed{11} \qquad\qquad x = \tfrac{1}{2}(u + v) \qquad y = \tfrac{1}{2}(u - v)$$

The Jacobian of T is

$$\frac{\partial(x, y)}{\partial(u, v)} = \begin{vmatrix} \dfrac{\partial x}{\partial u} & \dfrac{\partial x}{\partial v} \\[2mm] \dfrac{\partial y}{\partial u} & \dfrac{\partial y}{\partial v} \end{vmatrix} = \begin{vmatrix} \tfrac{1}{2} & \tfrac{1}{2} \\[2mm] \tfrac{1}{2} & -\tfrac{1}{2} \end{vmatrix} = -\tfrac{1}{2}$$

To find the region S in the uv-plane corresponding to R, we note that the sides of R lie on the lines

$$y = 0 \qquad x - y = 2 \qquad x = 0 \qquad x - y = 1$$

and, from either Equations 10 or Equations 11, the image lines in the uv-plane are

$$u = v \qquad v = 2 \qquad u = -v \qquad v = 1$$

Thus the region S is the trapezoidal region with vertices $(1, 1)$, $(2, 2)$, $(-2, 2)$, and $(-1, 1)$ shown in Figure 8. Since

$$S = \left\{ (u, v) \mid 1 \leqslant v \leqslant 2,\ -v \leqslant u \leqslant v \right\}$$

Theorem 9 gives

$$\iint_R e^{(x+y)/(x-y)}\, dA = \iint_S e^{u/v} \left| \frac{\partial(x, y)}{\partial(u, v)} \right| du\, dv$$

$$= \int_1^2 \int_{-v}^{v} e^{u/v} \left(\tfrac{1}{2}\right) du\, dv = \tfrac{1}{2} \int_1^2 \left[v e^{u/v} \right]_{u=-v}^{u=v} dv$$

$$= \tfrac{1}{2} \int_1^2 (e - e^{-1}) v\, dv = \tfrac{3}{4}(e - e^{-1}) \qquad\blacksquare$$

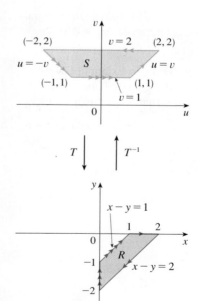

FIGURE 8

Triple Integrals

There is a similar change of variables formula for triple integrals. Let T be a transformation that maps a region S in uvw-space onto a region R in xyz-space by means of the equations

$$x = g(u, v, w) \qquad y = h(u, v, w) \qquad z = k(u, v, w)$$

The **Jacobian** of T is the following 3×3 determinant:

$$\boxed{12} \qquad \frac{\partial(x, y, z)}{\partial(u, v, w)} = \begin{vmatrix} \dfrac{\partial x}{\partial u} & \dfrac{\partial x}{\partial v} & \dfrac{\partial x}{\partial w} \\[2mm] \dfrac{\partial y}{\partial u} & \dfrac{\partial y}{\partial v} & \dfrac{\partial y}{\partial w} \\[2mm] \dfrac{\partial z}{\partial u} & \dfrac{\partial z}{\partial v} & \dfrac{\partial z}{\partial w} \end{vmatrix}$$

Under hypotheses similar to those in Theorem 9, we have the following formula for triple integrals:

$$\boxed{13} \qquad \iiint\limits_{R} f(x, y, z)\, dV = \iiint\limits_{S} f(x(u, v, w),\, y(u, v, w),\, z(u, v, w)) \left| \frac{\partial(x, y, z)}{\partial(u, v, w)} \right| du\, dv\, dw$$

EXAMPLE 4 Use Formula 13 to derive the formula for triple integration in spherical coordinates.

SOLUTION Here the change of variables is given by

$$x = \rho \sin \phi \, \cos \theta \qquad y = \rho \sin \phi \, \sin \theta \qquad z = \rho \cos \phi$$

We compute the Jacobian as follows:

$$\frac{\partial(x, y, z)}{\partial(\rho, \theta, \phi)} = \begin{vmatrix} \sin \phi \, \cos \theta & -\rho \sin \phi \, \sin \theta & \rho \cos \phi \, \cos \theta \\ \sin \phi \, \sin \theta & \rho \sin \phi \, \cos \theta & \rho \cos \phi \, \sin \theta \\ \cos \phi & 0 & -\rho \sin \phi \end{vmatrix}$$

$$= \cos \phi \begin{vmatrix} -\rho \sin \phi \, \sin \theta & \rho \cos \phi \, \cos \theta \\ \rho \sin \phi \, \cos \theta & \rho \cos \phi \, \sin \theta \end{vmatrix} - \rho \sin \phi \begin{vmatrix} \sin \phi \, \cos \theta & -\rho \sin \phi \, \sin \theta \\ \sin \phi \, \sin \theta & \rho \sin \phi \, \cos \theta \end{vmatrix}$$

$$= \cos \phi \, (-\rho^2 \sin \phi \, \cos \phi \, \sin^2\theta - \rho^2 \sin \phi \, \cos \phi \, \cos^2\theta)$$

$$\qquad - \rho \sin \phi \, (\rho \sin^2\phi \, \cos^2\theta + \rho \sin^2\phi \, \sin^2\theta)$$

$$= -\rho^2 \sin \phi \, \cos^2\phi - \rho^2 \sin \phi \, \sin^2\phi = -\rho^2 \sin \phi$$

Since $0 \leq \phi \leq \pi$, we have $\sin \phi \geq 0$. Therefore

$$\left| \frac{\partial(x, y, z)}{\partial(\rho, \theta, \phi)} \right| = \left| -\rho^2 \sin \phi \right| = \rho^2 \sin \phi$$

and Formula 13 gives

$$\iiint\limits_{R} f(x, y, z)\, dV = \iiint\limits_{S} f(\rho \sin \phi \, \cos \theta,\, \rho \sin \phi \, \sin \theta,\, \rho \cos \phi)\, \rho^2 \sin \phi \, d\rho \, d\theta \, d\phi$$

which is equivalent to Formula 15.8.3. ∎

15.9 EXERCISES

1–6 Find the Jacobian of the transformation.

1. $x = 2u + v, \quad y = 4u - v$

2. $x = u^2 + uv, \quad y = uv^2$

3. $x = s \cos t, \quad y = s \sin t$

4. $x = pe^q, \quad y = qe^p$

5. $x = uv, \quad y = vw, \quad z = wu$

6. $x = u + vw, \quad y = v + wu, \quad z = w + uv$

7–10 Find the image of the set S under the given transformation.

7. $S = \{(u, v) \mid 0 \le u \le 3, \; 0 \le v \le 2\};$
$x = 2u + 3v, \; y = u - v$

8. S is the square bounded by the lines $u = 0, u = 1, v = 0,$
$v = 1; \quad x = v, \; y = u(1 + v^2)$

9. S is the triangular region with vertices $(0, 0), (1, 1), (0, 1);$
$x = u^2, \; y = v$

10. S is the disk given by $u^2 + v^2 \le 1; \quad x = au, \; y = bv$

11–14 A region R in the xy-plane is given. Find equations for a transformation T that maps a rectangular region S in the uv-plane onto R, where the sides of S are parallel to the u- and v-axes.

11. R is bounded by $y = 2x - 1, y = 2x + 1, y = 1 - x,$
$y = 3 - x$

12. R is the parallelogram with vertices $(0, 0), (4, 3), (2, 4),$
$(-2, 1)$

13. R lies between the circles $x^2 + y^2 = 1$ and $x^2 + y^2 = 2$ in the first quadrant

14. R is bounded by the hyperbolas $y = 1/x, y = 4/x$ and the lines $y = x, y = 4x$ in the first quadrant

---

15–20 Use the given transformation to evaluate the integral.

15. $\iint_R (x - 3y) \, dA,$ where R is the triangular region with vertices $(0, 0), (2, 1),$ and $(1, 2); \quad x = 2u + v, \; y = u + 2v$

16. $\iint_R (4x + 8y) \, dA,$ where R is the parallelogram with vertices $(-1, 3), (1, -3), (3, -1),$ and $(1, 5);$
$x = \frac{1}{4}(u + v), \; y = \frac{1}{4}(v - 3u)$

17. $\iint_R x^2 \, dA,$ where R is the region bounded by the ellipse $9x^2 + 4y^2 = 36; \quad x = 2u, \; y = 3v$

18. $\iint_R (x^2 - xy + y^2) \, dA,$ where R is the region bounded by the ellipse $x^2 - xy + y^2 = 2;$
$x = \sqrt{2} \, u - \sqrt{2/3} \, v, \; y = \sqrt{2} \, u + \sqrt{2/3} \, v$

19. $\iint_R xy \, dA,$ where R is the region in the first quadrant bounded by the lines $y = x$ and $y = 3x$ and the hyperbolas $xy = 1, xy = 3; \quad x = u/v, \; y = v$

20. $\iint_R y^2 \, dA,$ where R is the region bounded by the curves $xy = 1, xy = 2, xy^2 = 1, xy^2 = 2; \quad u = xy, \; v = xy^2.$ Illustrate by using a graphing calculator or computer to draw R.

---

21. (a) Evaluate $\iiint_E dV,$ where E is the solid enclosed by the ellipsoid $x^2/a^2 + y^2/b^2 + z^2/c^2 = 1.$ Use the transformation $x = au, \; y = bv, \; z = cw.$

(b) The earth is not a perfect sphere; rotation has resulted in flattening at the poles. So the shape can be approximated by an ellipsoid with $a = b = 6378$ km and $c = 6356$ km. Use part (a) to estimate the volume of the earth.

(c) If the solid of part (a) has constant density k, find its moment of inertia about the z-axis.

22. An important problem in thermodynamics is to find the work done by an ideal Carnot engine. A cycle consists of alternating expansion and compression of gas in a piston. The work done by the engine is equal to the area of the region R enclosed by two isothermal curves $xy = a, xy = b$ and two adiabatic curves $xy^{1.4} = c, xy^{1.4} = d,$ where $0 < a < b$ and $0 < c < d.$ Compute the work done by determining the area of $R.$

23–27 Evaluate the integral by making an appropriate change of variables.

23. $\iint_R \dfrac{x - 2y}{3x - y} \, dA,$ where R is the parallelogram enclosed by the lines $x - 2y = 0, x - 2y = 4, 3x - y = 1,$ and $3x - y = 8$

24. $\iint_R (x + y)e^{x^2 - y^2} \, dA,$ where R is the rectangle enclosed by the lines $x - y = 0, x - y = 2, x + y = 0,$ and $x + y = 3$

25. $\iint_R \cos\left(\dfrac{y - x}{y + x}\right) dA,$ where R is the trapezoidal region with vertices $(1, 0), (2, 0), (0, 2),$ and $(0, 1)$

26. $\iint_R \sin(9x^2 + 4y^2) \, dA,$ where R is the region in the first quadrant bounded by the ellipse $9x^2 + 4y^2 = 1$

27. $\iint_R e^{x+y} \, dA,$ where R is given by the inequality $|x| + |y| \le 1$

---

28. Let f be continuous on $[0, 1]$ and let R be the triangular region with vertices $(0, 0), (1, 0),$ and $(0, 1).$ Show that

$$\iint_R f(x + y) \, dA = \int_0^1 uf(u) \, du$$

15 REVIEW

CONCEPT CHECK

1. Suppose f is a continuous function defined on a rectangle $R = [a, b] \times [c, d]$.
   (a) Write an expression for a double Riemann sum of f. If $f(x, y) \geq 0$, what does the sum represent?
   (b) Write the definition of $\iint_R f(x, y)\, dA$ as a limit.
   (c) What is the geometric interpretation of $\iint_R f(x, y)\, dA$ if $f(x, y) \geq 0$? What if f takes on both positive and negative values?
   (d) How do you evaluate $\iint_R f(x, y)\, dA$?
   (e) What does the Midpoint Rule for double integrals say?
   (f) Write an expression for the average value of f.

2. (a) How do you define $\iint_D f(x, y)\, dA$ if D is a bounded region that is not a rectangle?
   (b) What is a type I region? How do you evaluate $\iint_D f(x, y)\, dA$ if D is a type I region?
   (c) What is a type II region? How do you evaluate $\iint_D f(x, y)\, dA$ if D is a type II region?
   (d) What properties do double integrals have?

3. How do you change from rectangular coordinates to polar coordinates in a double integral? Why would you want to make the change?

4. If a lamina occupies a plane region D and has density function $\rho(x, y)$, write expressions for each of the following in terms of double integrals.
   (a) The mass
   (b) The moments about the axes
   (c) The center of mass
   (d) The moments of inertia about the axes and the origin

5. Let f be a joint density function of a pair of continuous random variables X and Y.
   (a) Write a double integral for the probability that X lies between a and b and Y lies between c and d.

   (b) What properties does f possess?
   (c) What are the expected values of X and Y?

6. Write an expression for the area of a surface with equation $z = f(x, y)$, $(x, y) \in D$.

7. (a) Write the definition of the triple integral of f over a rectangular box B.
   (b) How do you evaluate $\iiint_B f(x, y, z)\, dV$?
   (c) How do you define $\iiint_E f(x, y, z)\, dV$ if E is a bounded solid region that is not a box?
   (d) What is a type 1 solid region? How do you evaluate $\iiint_E f(x, y, z)\, dV$ if E is such a region?
   (e) What is a type 2 solid region? How do you evaluate $\iiint_E f(x, y, z)\, dV$ if E is such a region?
   (f) What is a type 3 solid region? How do you evaluate $\iiint_E f(x, y, z)\, dV$ if E is such a region?

8. Suppose a solid object occupies the region E and has density function $\rho(x, y, z)$. Write expressions for each of the following.
   (a) The mass
   (b) The moments about the coordinate planes
   (c) The coordinates of the center of mass
   (d) The moments of inertia about the axes

9. (a) How do you change from rectangular coordinates to cylindrical coordinates in a triple integral?
   (b) How do you change from rectangular coordinates to spherical coordinates in a triple integral?
   (c) In what situations would you change to cylindrical or spherical coordinates?

10. (a) If a transformation T is given by $x = g(u, v)$, $y = h(u, v)$, what is the Jacobian of T?
   (b) How do you change variables in a double integral?
   (c) How do you change variables in a triple integral?

TRUE-FALSE QUIZ

Determine whether the statement is true or false. If it is true, explain why. If it is false, explain why or give an example that disproves the statement.

1. $\displaystyle\int_{-1}^{2}\int_{0}^{6} x^2 \sin(x - y)\, dx\, dy = \int_{0}^{6}\int_{-1}^{2} x^2 \sin(x - y)\, dy\, dx$

2. $\displaystyle\int_{0}^{1}\int_{0}^{x} \sqrt{x + y^2}\, dy\, dx = \int_{0}^{x}\int_{0}^{1} \sqrt{x + y^2}\, dx\, dy$

3. $\displaystyle\int_{1}^{2}\int_{3}^{4} x^2 e^y\, dy\, dx = \int_{1}^{2} x^2\, dx \int_{3}^{4} e^y\, dy$

4. $\displaystyle\int_{-1}^{1}\int_{0}^{1} e^{x^2+y^2} \sin y\, dx\, dy = 0$

5. If f is continuous on $[0, 1]$, then
$$\int_{0}^{1}\int_{0}^{1} f(x) f(y)\, dy\, dx = \left[\int_{0}^{1} f(x)\, dx\right]^2$$

6. $\displaystyle\int_{1}^{4}\int_{0}^{1} \left(x^2 + \sqrt{y}\right) \sin(x^2 y^2)\, dx\, dy \leq 9$

7. If D is the disk given by $x^2 + y^2 \leq 4$, then
$$\iint_D \sqrt{4 - x^2 - y^2}\, dA = \tfrac{16}{3}\pi$$

8. The integral $\iiint_E k r^3\, dz\, dr\, d\theta$ represents the moment of inertia about the z-axis of a solid E with constant density k.

9. The integral
$$\int_{0}^{2\pi}\int_{0}^{2}\int_{r}^{2} dz\, dr\, d\theta$$
represents the volume enclosed by the cone $z = \sqrt{x^2 + y^2}$ and the plane $z = 2$.

EXERCISES

1. A contour map is shown for a function f on the square $R = [0, 3] \times [0, 3]$. Use a Riemann sum with nine terms to estimate the value of $\iint_R f(x, y) \, dA$. Take the sample points to be the upper right corners of the squares.

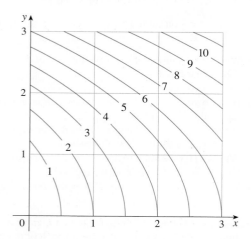

2. Use the Midpoint Rule to estimate the integral in Exercise 1.

3–8 Calculate the iterated integral.

3. $\int_1^2 \int_0^2 (y + 2xe^y) \, dx \, dy$

4. $\int_0^1 \int_0^1 ye^{xy} \, dx \, dy$

5. $\int_0^1 \int_0^x \cos(x^2) \, dy \, dx$

6. $\int_0^1 \int_x^{e^x} 3xy^2 \, dy \, dx$

7. $\int_0^\pi \int_0^1 \int_0^{\sqrt{1-y^2}} y \sin x \, dz \, dy \, dx$

8. $\int_0^1 \int_0^y \int_x^1 6xyz \, dz \, dx \, dy$

9–10 Write $\iint_R f(x, y) \, dA$ as an iterated integral, where R is the region shown and f is an arbitrary continuous function on R.

9.

10.
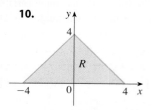

11. The cylindrical coordinates of a point are $(2\sqrt{3}, \pi/3, 2)$. Find the rectangular and spherical coordinates of the point.

12. The rectangular coordinates of a point are $(2, 2, -1)$. Find the cylindrical and spherical coordinates of the point.

13. The spherical coordinates of a point are $(8, \pi/4, \pi/6)$. Find the rectangular and cylindrical coordinates of the point.

14. Identify the surfaces whose equations are given.
(a) $\theta = \pi/4$ (b) $\phi = \pi/4$

15. Write the equation in cylindrical coordinates and in spherical coordinates.
(a) $x^2 + y^2 + z^2 = 4$ (b) $x^2 + y^2 = 4$

16. Sketch the solid consisting of all points with spherical coordinates (ρ, θ, ϕ) such that $0 \le \theta \le \pi/2$, $0 \le \phi \le \pi/6$, and $0 \le \rho \le 2 \cos \phi$.

17. Describe the region whose area is given by the integral
$$\int_0^{\pi/2} \int_0^{\sin 2\theta} r \, dr \, d\theta$$

18. Describe the solid whose volume is given by the integral
$$\int_0^{\pi/2} \int_0^{\pi/2} \int_1^2 \rho^2 \sin \phi \, d\rho \, d\phi \, d\theta$$
and evaluate the integral.

19–20 Calculate the iterated integral by first reversing the order of integration.

19. $\int_0^1 \int_x^1 \cos(y^2) \, dy \, dx$

20. $\int_0^1 \int_{\sqrt{y}}^1 \frac{ye^{x^2}}{x^3} \, dx \, dy$

21–34 Calculate the value of the multiple integral.

21. $\iint_R ye^{xy} \, dA$, where $R = \{(x, y) \mid 0 \le x \le 2, \ 0 \le y \le 3\}$

22. $\iint_D xy \, dA$, where $D = \{(x, y) \mid 0 \le y \le 1, \ y^2 \le x \le y + 2\}$

23. $\iint_D \frac{y}{1 + x^2} \, dA$, where D is bounded by $y = \sqrt{x}$, $y = 0$, $x = 1$

24. $\iint_D \frac{1}{1 + x^2} \, dA$, where D is the triangular region with vertices $(0, 0)$, $(1, 1)$, and $(0, 1)$

25. $\iint_D y \, dA$, where D is the region in the first quadrant bounded by the parabolas $x = y^2$ and $x = 8 - y^2$

26. $\iint_D y \, dA$, where D is the region in the first quadrant that lies above the hyperbola $xy = 1$ and the line $y = x$ and below the line $y = 2$

27. $\iint_D (x^2 + y^2)^{3/2} \, dA$, where D is the region in the first quadrant bounded by the lines $y = 0$ and $y = \sqrt{3}\, x$ and the circle $x^2 + y^2 = 9$

28. $\iint_D x \, dA$, where D is the region in the first quadrant that lies between the circles $x^2 + y^2 = 1$ and $x^2 + y^2 = 2$

29. $\iiint_E xy \, dV$, where $E = \{(x, y, z) \mid 0 \le x \le 3, \ 0 \le y \le x, \ 0 \le z \le x + y\}$

30. $\iiint_T xy \, dV$, where T is the solid tetrahedron with vertices $(0, 0, 0)$, $(\frac{1}{3}, 0, 0)$, $(0, 1, 0)$, and $(0, 0, 1)$

31. $\iiint_E y^2z^2 \, dV$, where E is bounded by the paraboloid $x = 1 - y^2 - z^2$ and the plane $x = 0$

32. $\iiint_E z \, dV$, where E is bounded by the planes $y = 0$, $z = 0$, $x + y = 2$ and the cylinder $y^2 + z^2 = 1$ in the first octant

33. $\iiint_E yz \, dV$, where E lies above the plane $z = 0$, below the plane $z = y$, and inside the cylinder $x^2 + y^2 = 4$

34. $\iiint_H z^3 \sqrt{x^2 + y^2 + z^2} \, dV$, where H is the solid hemisphere that lies above the xy-plane and has center the origin and radius 1

35–40 Find the volume of the given solid.

35. Under the paraboloid $z = x^2 + 4y^2$ and above the rectangle $R = [0, 2] \times [1, 4]$

36. Under the surface $z = x^2 y$ and above the triangle in the xy-plane with vertices $(1, 0)$, $(2, 1)$, and $(4, 0)$

37. The solid tetrahedron with vertices $(0, 0, 0)$, $(0, 0, 1)$, $(0, 2, 0)$, and $(2, 2, 0)$

38. Bounded by the cylinder $x^2 + y^2 = 4$ and the planes $z = 0$ and $y + z = 3$

39. One of the wedges cut from the cylinder $x^2 + 9y^2 = a^2$ by the planes $z = 0$ and $z = mx$

40. Above the paraboloid $z = x^2 + y^2$ and below the half-cone $z = \sqrt{x^2 + y^2}$

---

41. Consider a lamina that occupies the region D bounded by the parabola $x = 1 - y^2$ and the coordinate axes in the first quadrant with density function $\rho(x, y) = y$.
(a) Find the mass of the lamina.
(b) Find the center of mass.
(c) Find the moments of inertia and radii of gyration about the x- and y-axes.

42. A lamina occupies the part of the disk $x^2 + y^2 \leq a^2$ that lies in the first quadrant.
(a) Find the centroid of the lamina.
(b) Find the center of mass of the lamina if the density function is $\rho(x, y) = xy^2$.

43. (a) Find the centroid of a solid right circular cone with height h and base radius a. (Place the cone so that its base is in the xy-plane with center the origin and its axis along the positive z-axis.)
(b) If the cone has density function $\rho(x, y, z) = \sqrt{x^2 + y^2}$, find the moment of inertia of the cone about its axis (the z-axis).

44. Find the area of the part of the cone $z^2 = a^2(x^2 + y^2)$ between the planes $z = 1$ and $z = 2$.

45. Find the area of the part of the surface $z = x^2 + y$ that lies above the triangle with vertices $(0, 0)$, $(1, 0)$, and $(0, 2)$.

CAS 46. Graph the surface $z = x \sin y$, $-3 \leq x \leq 3$, $-\pi \leq y \leq \pi$, and find its surface area correct to four decimal places.

47. Use polar coordinates to evaluate
$$\int_0^3 \int_{-\sqrt{9-x^2}}^{\sqrt{9-x^2}} (x^3 + xy^2) \, dy \, dx$$

48. Use spherical coordinates to evaluate
$$\int_{-2}^{2} \int_0^{\sqrt{4-y^2}} \int_{-\sqrt{4-x^2-y^2}}^{\sqrt{4-x^2-y^2}} y^2 \sqrt{x^2 + y^2 + z^2} \, dz \, dx \, dy$$

49. If D is the region bounded by the curves $y = 1 - x^2$ and $y = e^x$, find the approximate value of the integral $\iint_D y^2 \, dA$. (Use a graphing device to estimate the points of intersection of the curves.)

CAS 50. Find the center of mass of the solid tetrahedron with vertices $(0, 0, 0)$, $(1, 0, 0)$, $(0, 2, 0)$, $(0, 0, 3)$ and density function $\rho(x, y, z) = x^2 + y^2 + z^2$.

51. The joint density function for random variables X and Y is
$$f(x, y) = \begin{cases} C(x + y) & \text{if } 0 \leq x \leq 3, \ 0 \leq y \leq 2 \\ 0 & \text{otherwise} \end{cases}$$
(a) Find the value of the constant C.
(b) Find $P(X \leq 2, Y \geq 1)$.
(c) Find $P(X + Y \leq 1)$.

52. A lamp has three bulbs, each of a type with average lifetime 800 hours. If we model the probability of failure of a bulb by an exponential density function with mean 800, find the probability that all three bulbs fail within a total of 1000 hours.

53. Rewrite the integral
$$\int_{-1}^{1} \int_{x^2}^{1} \int_0^{1-y} f(x, y, z) \, dz \, dy \, dx$$
as an iterated integral in the order $dx \, dy \, dz$.

54. Give five other iterated integrals that are equal to
$$\int_0^2 \int_0^{y^3} \int_0^{y^2} f(x, y, z) \, dz \, dx \, dy$$

55. Use the transformation $u = x - y$, $v = x + y$ to evaluate
$$\iint_R \frac{x - y}{x + y} \, dA$$
where R is the square with vertices $(0, 2)$, $(1, 1)$, $(2, 2)$, and $(1, 3)$.

56. Use the transformation $x = u^2$, $y = v^2$, $z = w^2$ to find the volume of the region bounded by the surface $\sqrt{x} + \sqrt{y} + \sqrt{z} = 1$ and the coordinate planes.

57. Use the change of variables formula and an appropriate transformation to evaluate $\iint_R xy \, dA$, where R is the square with vertices $(0, 0)$, $(1, 1)$, $(2, 0)$, and $(1, -1)$.

58. The **Mean Value Theorem for double integrals** says that if f is a continuous function on a plane region D that is of type I or II, then there exists a point (x_0, y_0) in D such that
$$\iint_D f(x, y) \, dA = f(x_0, y_0) A(D)$$
Use the Extreme Value Theorem (14.7.8) and Property 15.2.11 of integrals to prove this theorem. (Use the proof of the single-variable version in Section 6.5 as a guide.)

59. Suppose that f is continuous on a disk that contains the point (a, b). Let D_r be the closed disk with center (a, b) and radius r. Use the Mean Value Theorem for double integrals (see Exercise 58) to show that

$$\lim_{r \to 0} \frac{1}{\pi r^2} \iint_{D_r} f(x, y) \, dA = f(a, b)$$

60. (a) Evaluate $\displaystyle\iint_{D} \frac{1}{(x^2 + y^2)^{n/2}} \, dA$, where n is an integer and D is the region bounded by the circles with center the origin and radii r and R, $0 < r < R$.

(b) For what values of n does the integral in part (a) have a limit as $r \to 0^+$?

(c) Find $\displaystyle\iiint_{E} \frac{1}{(x^2 + y^2 + z^2)^{n/2}} \, dV$, where E is the region bounded by the spheres with center the origin and radii r and R, $0 < r < R$.

(d) For what values of n does the integral in part (c) have a limit as $r \to 0^+$?

Problems Plus

1. If $[\![x]\!]$ denotes the greatest integer in x, evaluate the integral

$$\iint_R [\![x + y]\!]\, dA$$

where $R = \{(x, y) \mid 1 \leqslant x \leqslant 3,\ 2 \leqslant y \leqslant 5\}$.

2. Evaluate the integral

$$\int_0^1 \int_0^1 e^{\max\{x^2,\, y^2\}}\, dy\, dx$$

where $\max\{x^2, y^2\}$ means the larger of the numbers x^2 and y^2.

3. Find the average value of the function $f(x) = \int_x^1 \cos(t^2)\, dt$ on the interval $[0, 1]$.

4. If $\mathbf{a}$, $\mathbf{b}$, and $\mathbf{c}$ are constant vectors, $\mathbf{r}$ is the position vector $x\mathbf{i} + y\mathbf{j} + z\mathbf{k}$, and E is given by the inequalities $0 \leqslant \mathbf{a} \cdot \mathbf{r} \leqslant \alpha$, $0 \leqslant \mathbf{b} \cdot \mathbf{r} \leqslant \beta$, $0 \leqslant \mathbf{c} \cdot \mathbf{r} \leqslant \gamma$, show that

$$\iiint_E (\mathbf{a} \cdot \mathbf{r})(\mathbf{b} \cdot \mathbf{r})(\mathbf{c} \cdot \mathbf{r})\, dV = \frac{(\alpha\beta\gamma)^2}{8\,|\mathbf{a} \cdot (\mathbf{b} \times \mathbf{c})|}$$

5. The double integral $\displaystyle\int_0^1 \int_0^1 \frac{1}{1 - xy}\, dx\, dy$ is an improper integral and could be defined as the limit of double integrals over the rectangle $[0, t] \times [0, t]$ as $t \to 1^-$. But if we expand the integrand as a geometric series, we can express the integral as the sum of an infinite series. Show that

$$\int_0^1 \int_0^1 \frac{1}{1 - xy}\, dx\, dy = \sum_{n=1}^{\infty} \frac{1}{n^2}$$

6. Leonhard Euler was able to find the exact sum of the series in Problem 5. In 1736 he proved that

$$\sum_{n=1}^{\infty} \frac{1}{n^2} = \frac{\pi^2}{6}$$

In this problem we ask you to prove this fact by evaluating the double integral in Problem 5. Start by making the change of variables

$$x = \frac{u - v}{\sqrt{2}} \qquad y = \frac{u + v}{\sqrt{2}}$$

This gives a rotation about the origin through the angle $\pi/4$. You will need to sketch the corresponding region in the uv-plane.

[*Hint:* If, in evaluating the integral, you encounter either of the expressions $(1 - \sin\theta)/\cos\theta$ or $(\cos\theta)/(1 + \sin\theta)$, you might like to use the identity $\cos\theta = \sin((\pi/2) - \theta)$ and the corresponding identity for $\sin\theta$.]

7. (a) Show that

$$\int_0^1 \int_0^1 \int_0^1 \frac{1}{1 - xyz}\, dx\, dy\, dz = \sum_{n=1}^{\infty} \frac{1}{n^3}$$

(Nobody has ever been able to find the exact value of the sum of this series.)

(b) Show that

$$\int_0^1 \int_0^1 \int_0^1 \frac{1}{1 + xyz}\, dx\, dy\, dz = \sum_{n=1}^{\infty} \frac{(-1)^{n-1}}{n^3}$$

Use this equation to evaluate the triple integral correct to two decimal places.

8. Show that

$$\int_0^\infty \frac{\arctan \pi x - \arctan x}{x}\, dx = \frac{\pi}{2} \ln \pi$$

by first expressing the integral as an iterated integral.

9. (a) Show that when Laplace's equation

$$\frac{\partial^2 u}{\partial x^2} + \frac{\partial^2 u}{\partial y^2} + \frac{\partial^2 u}{\partial z^2} = 0$$

is written in cylindrical coordinates, it becomes

$$\frac{\partial^2 u}{\partial r^2} + \frac{1}{r}\frac{\partial u}{\partial r} + \frac{1}{r^2}\frac{\partial^2 u}{\partial \theta^2} + \frac{\partial^2 u}{\partial z^2} = 0$$

(b) Show that when Laplace's equation is written in spherical coordinates, it becomes

$$\frac{\partial^2 u}{\partial \rho^2} + \frac{2}{\rho}\frac{\partial u}{\partial \rho} + \frac{\cot \phi}{\rho^2}\frac{\partial u}{\partial \phi} + \frac{1}{\rho^2}\frac{\partial^2 u}{\partial \phi^2} + \frac{1}{\rho^2 \sin^2\phi}\frac{\partial^2 u}{\partial \theta^2} = 0$$

10. (a) A lamina has constant density ρ and takes the shape of a disk with center the origin and radius R. Use Newton's Law of Gravitation (see Section 13.4) to show that the magnitude of the force of attraction that the lamina exerts on a body with mass m located at the point $(0, 0, d)$ on the positive z-axis is

$$F = 2\pi Gm\rho d\left(\frac{1}{d} - \frac{1}{\sqrt{R^2 + d^2}}\right)$$

[*Hint:* Divide the disk as in Figure 15.3.4 and first compute the vertical component of the force exerted by the polar subrectangle R_{ij}.]

(b) Show that the magnitude of the force of attraction of a lamina with density ρ that occupies an entire plane on an object with mass m located at a distance d from the plane is

$$F = 2\pi Gm\rho$$

Notice that this expression does not depend on d.

11. If f is continuous, show that

$$\int_0^x \int_0^y \int_0^z f(t)\, dt\, dz\, dy = \frac{1}{2}\int_0^x (x - t)^2 f(t)\, dt$$

12. Evaluate $\displaystyle\lim_{n\to\infty} n^{-2} \sum_{i=1}^{n} \sum_{j=1}^{n^2} \frac{1}{\sqrt{n^2 + ni + j}}$.

13. The plane

$$\frac{x}{a} + \frac{y}{b} + \frac{z}{c} = 1 \qquad a > 0, \quad b > 0, \quad c > 0$$

cuts the solid ellipsoid

$$\frac{x^2}{a^2} + \frac{y^2}{b^2} + \frac{z^2}{c^2} \leqslant 1$$

into two pieces. Find the volume of the smaller piece.

16 Vector Calculus

Parametric surfaces, which are studied in Section 16.6, are frequently used by programmers in creating the sophisticated software used in the development of computer animated films like the *Shrek* series. The software employs parametric and other types of surfaces to create 3D models of the characters and objects in a scene. Color, texture, and lighting is then rendered to bring the scene to life.

IN THIS CHAPTER WE STUDY the calculus of vector fields. (These are functions that assign vectors to points in space.) In particular we define line integrals (which can be used to find the work done by a force field in moving an object along a curve). Then we define surface integrals (which can be used to find the rate of fluid flow across a surface). The connections between these new types of integrals and the single, double, and triple integrals that we have already met are given by the higher-dimensional versions of the Fundamental Theorem of Calculus: Green's Theorem, Stokes' Theorem, and the Divergence Theorem.

16.1   Vector Fields

The vectors in Figure 1 are air velocity vectors that indicate the wind speed and direction at points 10 m above the surface elevation in the San Francisco Bay area. We see at a glance from the largest arrows in part (a) that the greatest wind speeds at that time occurred as the winds entered the bay across the Golden Gate Bridge. Part (b) shows the very different wind pattern 12 hours earlier. Associated with every point in the air we can imagine a wind velocity vector. This is an example of a *velocity vector field*.

© 2016 Cengage Learning ®

(a) 6:00 PM, March 1, 2010

(b) 6:00 AM, March 1, 2010

FIGURE 1 Velocity vector fields showing San Francisco Bay wind patterns

Other examples of velocity vector fields are illustrated in Figure 2: ocean currents and flow past an airfoil.

© 2016 Cengage Learning ®

ONERA photograph, Werle, 1974

(a) Ocean currents off the coast of Nova Scotia

(b) Airflow past an inclined airfoil

FIGURE 2
Velocity vector fields

Another type of vector field, called a *force field,* associates a force vector with each point in a region. An example is the gravitational force field that we will look at in Example 4.

In general, a vector field is a function whose domain is a set of points in $\mathbb{R}^2$ (or $\mathbb{R}^3$) and whose range is a set of vectors in V_2 (or V_3).

> **1 Definition** Let D be a set in $\mathbb{R}^2$ (a plane region). A **vector field on** $\mathbb{R}^2$ is a function **F** that assigns to each point (x, y) in D a two-dimensional vector $\mathbf{F}(x, y)$.

The best way to picture a vector field is to draw the arrow representing the vector $\mathbf{F}(x, y)$ starting at the point (x, y). Of course, it's impossible to do this for all points (x, y), but we can gain a reasonable impression of **F** by doing it for a few representative points in D as in Figure 3. Since $\mathbf{F}(x, y)$ is a two-dimensional vector, we can write it in terms of its **component functions** P and Q as follows:

$$\mathbf{F}(x, y) = P(x, y)\,\mathbf{i} + Q(x, y)\,\mathbf{j} = \langle P(x, y), Q(x, y)\rangle$$

or, for short,

$$\mathbf{F} = P\,\mathbf{i} + Q\,\mathbf{j}$$

Notice that P and Q are scalar functions of two variables and are sometimes called **scalar fields** to distinguish them from vector fields.

> **2 Definition** Let E be a subset of $\mathbb{R}^3$. A **vector field on** $\mathbb{R}^3$ is a function **F** that assigns to each point (x, y, z) in E a three-dimensional vector $\mathbf{F}(x, y, z)$.

A vector field **F** on $\mathbb{R}^3$ is pictured in Figure 4. We can express it in terms of its component functions P, Q, and R as

$$\mathbf{F}(x, y, z) = P(x, y, z)\,\mathbf{i} + Q(x, y, z)\,\mathbf{j} + R(x, y, z)\,\mathbf{k}$$

As with the vector functions in Section 13.1, we can define continuity of vector fields and show that **F** is continuous if and only if its component functions P, Q, and R are continuous.

We sometimes identify a point (x, y, z) with its position vector $\mathbf{x} = \langle x, y, z\rangle$ and write $\mathbf{F}(\mathbf{x})$ instead of $\mathbf{F}(x, y, z)$. Then **F** becomes a function that assigns a vector $\mathbf{F}(\mathbf{x})$ to a vector **x**.

EXAMPLE 1 A vector field on $\mathbb{R}^2$ is defined by $\mathbf{F}(x, y) = -y\,\mathbf{i} + x\,\mathbf{j}$. Describe **F** by sketching some of the vectors $\mathbf{F}(x, y)$ as in Figure 3.

SOLUTION Since $\mathbf{F}(1, 0) = \mathbf{j}$, we draw the vector $\mathbf{j} = \langle 0, 1\rangle$ starting at the point $(1, 0)$ in Figure 5. Since $\mathbf{F}(0, 1) = -\mathbf{i}$, we draw the vector $\langle -1, 0\rangle$ with starting point $(0, 1)$. Continuing in this way, we calculate several other representative values of $\mathbf{F}(x, y)$ in the table and draw the corresponding vectors to represent the vector field in Figure 5.

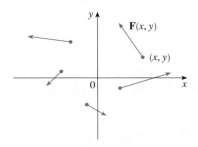

FIGURE 3
Vector field on $\mathbb{R}^2$

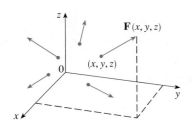

FIGURE 4
Vector field on $\mathbb{R}^3$

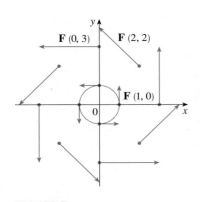

FIGURE 5
$\mathbf{F}(x, y) = -y\,\mathbf{i} + x\,\mathbf{j}$

(x, y)	$\mathbf{F}(x, y)$	(x, y)	$\mathbf{F}(x, y)$
$(1, 0)$	$\langle 0, 1\rangle$	$(-1, 0)$	$\langle 0, -1\rangle$
$(2, 2)$	$\langle -2, 2\rangle$	$(-2, -2)$	$\langle 2, -2\rangle$
$(3, 0)$	$\langle 0, 3\rangle$	$(-3, 0)$	$\langle 0, -3\rangle$
$(0, 1)$	$\langle -1, 0\rangle$	$(0, -1)$	$\langle 1, 0\rangle$
$(-2, 2)$	$\langle -2, -2\rangle$	$(2, -2)$	$\langle 2, 2\rangle$
$(0, 3)$	$\langle -3, 0\rangle$	$(0, -3)$	$\langle 3, 0\rangle$

It appears from Figure 5 that each arrow is tangent to a circle with center the origin. To confirm this, we take the dot product of the position vector $\mathbf{x} = x\,\mathbf{i} + y\,\mathbf{j}$ with the vector $\mathbf{F}(\mathbf{x}) = \mathbf{F}(x, y)$:

$$\mathbf{x} \cdot \mathbf{F}(\mathbf{x}) = (x\,\mathbf{i} + y\,\mathbf{j}) \cdot (-y\,\mathbf{i} + x\,\mathbf{j}) = -xy + yx = 0$$

This shows that $\mathbf{F}(x, y)$ is perpendicular to the position vector $\langle x, y \rangle$ and is therefore tangent to a circle with center the origin and radius $|\mathbf{x}| = \sqrt{x^2 + y^2}$. Notice also that

$$|\mathbf{F}(x, y)| = \sqrt{(-y)^2 + x^2} = \sqrt{x^2 + y^2} = |\mathbf{x}|$$

so the magnitude of the vector $\mathbf{F}(x, y)$ is equal to the radius of the circle. ∎

Some computer algebra systems are capable of plotting vector fields in two or three dimensions. They give a better impression of the vector field than is possible by hand because the computer can plot a large number of representative vectors. Figure 6 shows a computer plot of the vector field in Example 1; Figures 7 and 8 show two other vector fields. Notice that the computer scales the lengths of the vectors so they are not too long and yet are proportional to their true lengths.

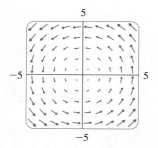

FIGURE 6
$\mathbf{F}(x, y) = \langle -y, x \rangle$

FIGURE 7
$\mathbf{F}(x, y) = \langle y, \sin x \rangle$

FIGURE 8
$\mathbf{F}(x, y) = \langle \ln(1 + y^2), \ln(1 + x^2) \rangle$

EXAMPLE 2 Sketch the vector field on $\mathbb{R}^3$ given by $\mathbf{F}(x, y, z) = z\,\mathbf{k}$.

SOLUTION The sketch is shown in Figure 9. Notice that all vectors are vertical and point upward above the xy-plane or downward below it. The magnitude increases with the distance from the xy-plane.

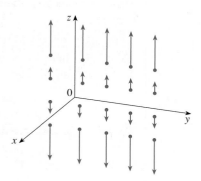

FIGURE 9
$\mathbf{F}(x, y, z) = z\,\mathbf{k}$

We were able to draw the vector field in Example 2 by hand because of its particularly simple formula. Most three-dimensional vector fields, however, are virtually impossible to sketch by hand and so we need to resort to computer software. Examples are shown in Figures 10, 11, and 12. Notice that the vector fields in Figures 10 and 11 have similar formulas, but all the vectors in Figure 11 point in the general direction of the negative y-axis because their y-components are all -2. If the vector field in Figure 12 represents a velocity field, then a particle would be swept upward and would spiral around the z-axis in the clockwise direction as viewed from above.

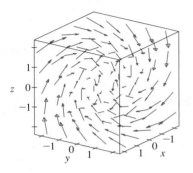

FIGURE 10
$\mathbf{F}(x, y, z) = y\,\mathbf{i} + z\,\mathbf{j} + x\,\mathbf{k}$

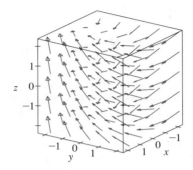

FIGURE 11
$\mathbf{F}(x, y, z) = y\,\mathbf{i} - 2\,\mathbf{j} + x\,\mathbf{k}$

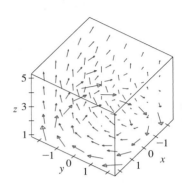

FIGURE 12
$\mathbf{F}(x, y, z) = \dfrac{y}{z}\,\mathbf{i} - \dfrac{x}{z}\,\mathbf{j} + \dfrac{z}{4}\,\mathbf{k}$

TEC In Visual 16.1 you can rotate the vector fields in Figures 10–12 as well as additional fields.

FIGURE 13
Velocity field in fluid flow

EXAMPLE 3 Imagine a fluid flowing steadily along a pipe and let $\mathbf{V}(x, y, z)$ be the velocity vector at a point (x, y, z). Then $\mathbf{V}$ assigns a vector to each point (x, y, z) in a certain domain E (the interior of the pipe) and so $\mathbf{V}$ is a vector field on $\mathbb{R}^3$ called a **velocity field**. A possible velocity field is illustrated in Figure 13. The speed at any given point is indicated by the length of the arrow.

Velocity fields also occur in other areas of physics. For instance, the vector field in Example 1 could be used as the velocity field describing the counterclockwise rotation of a wheel. We have seen other examples of velocity fields in Figures 1 and 2. ∎

EXAMPLE 4 Newton's Law of Gravitation states that the magnitude of the gravitational force between two objects with masses m and M is

$$|\mathbf{F}| = \frac{mMG}{r^2}$$

where r is the distance between the objects and G is the gravitational constant. (This is an example of an inverse square law.) Let's assume that the object with mass M is located at the origin in $\mathbb{R}^3$. (For instance, M could be the mass of the earth and the origin would be at its center.) Let the position vector of the object with mass m be $\mathbf{x} = \langle x, y, z \rangle$. Then $r = |\mathbf{x}|$, so $r^2 = |\mathbf{x}|^2$. The gravitational force exerted on this second object acts toward the origin, and the unit vector in this direction is

$$-\frac{\mathbf{x}}{|\mathbf{x}|}$$

Therefore the gravitational force acting on the object at $\mathbf{x} = \langle x, y, z \rangle$ is

$$\boxed{3} \qquad\qquad \mathbf{F}(\mathbf{x}) = -\frac{mMG}{|\mathbf{x}|^3}\,\mathbf{x}$$

[Physicists often use the notation $\mathbf{r}$ instead of $\mathbf{x}$ for the position vector, so you may see Formula 3 written in the form $\mathbf{F} = -(mMG/r^3)\mathbf{r}$.] The function given by Equation 3 is

FIGURE 14
Gravitational force field

an example of a vector field, called the **gravitational field**, because it associates a vector [the force $\mathbf{F(x)}$] with every point $\mathbf{x}$ in space.

Formula 3 is a compact way of writing the gravitational field, but we can also write it in terms of its component functions by using the facts that $\mathbf{x} = x\,\mathbf{i} + y\,\mathbf{j} + z\,\mathbf{k}$ and $|\mathbf{x}| = \sqrt{x^2 + y^2 + z^2}$:

$$\mathbf{F}(x, y, z) = \frac{-mMGx}{(x^2 + y^2 + z^2)^{3/2}}\,\mathbf{i} + \frac{-mMGy}{(x^2 + y^2 + z^2)^{3/2}}\,\mathbf{j} + \frac{-mMGz}{(x^2 + y^2 + z^2)^{3/2}}\,\mathbf{k}$$

The gravitational field $\mathbf{F}$ is pictured in Figure 14. ∎

EXAMPLE 5 Suppose an electric charge Q is located at the origin. According to Coulomb's Law, the electric force $\mathbf{F(x)}$ exerted by this charge on a charge q located at a point (x, y, z) with position vector $\mathbf{x} = \langle x, y, z \rangle$ is

$$\boxed{4} \qquad\qquad \mathbf{F(x)} = \frac{\varepsilon qQ}{|\mathbf{x}|^3}\,\mathbf{x}$$

where ε is a constant (that depends on the units used). For like charges, we have $qQ > 0$ and the force is repulsive; for unlike charges, we have $qQ < 0$ and the force is attractive. Notice the similarity between Formulas 3 and 4. Both vector fields are examples of **force fields**.

Instead of considering the electric force $\mathbf{F}$, physicists often consider the force per unit charge:

$$\mathbf{E(x)} = \frac{1}{q}\,\mathbf{F(x)} = \frac{\varepsilon Q}{|\mathbf{x}|^3}\,\mathbf{x}$$

Then $\mathbf{E}$ is a vector field on $\mathbb{R}^3$ called the **electric field** of Q. ∎

■ **Gradient Fields**

If f is a scalar function of two variables, recall from Section 14.6 that its gradient ∇f (or grad f) is defined by

$$\nabla f(x, y) = f_x(x, y)\,\mathbf{i} + f_y(x, y)\,\mathbf{j}$$

Therefore ∇f is really a vector field on $\mathbb{R}^2$ and is called a **gradient vector field**. Likewise, if f is a scalar function of three variables, its gradient is a vector field on $\mathbb{R}^3$ given by

$$\nabla f(x, y, z) = f_x(x, y, z)\,\mathbf{i} + f_y(x, y, z)\,\mathbf{j} + f_z(x, y, z)\,\mathbf{k}$$

EXAMPLE 6 Find the gradient vector field of $f(x, y) = x^2 y - y^3$. Plot the gradient vector field together with a contour map of f. How are they related?

SOLUTION The gradient vector field is given by

$$\nabla f(x, y) = \frac{\partial f}{\partial x}\,\mathbf{i} + \frac{\partial f}{\partial y}\,\mathbf{j} = 2xy\,\mathbf{i} + (x^2 - 3y^2)\,\mathbf{j}$$

FIGURE 15

Figure 15 shows a contour map of f with the gradient vector field. Notice that the gradient vectors are perpendicular to the level curves, as we would expect from

Section 14.6. Notice also that the gradient vectors are long where the level curves are close to each other and short where the curves are farther apart. That's because the length of the gradient vector is the value of the directional derivative of f and closely spaced level curves indicate a steep graph. ∎

A vector field **F** is called a **conservative vector field** if it is the gradient of some scalar function, that is, if there exists a function f such that $\mathbf{F} = \nabla f$. In this situation f is called a **potential function** for **F**.

Not all vector fields are conservative, but such fields do arise frequently in physics. For example, the gravitational field **F** in Example 4 is conservative because if we define

$$f(x, y, z) = \frac{mMG}{\sqrt{x^2 + y^2 + z^2}}$$

then

$$\nabla f(x, y, z) = \frac{\partial f}{\partial x}\,\mathbf{i} + \frac{\partial f}{\partial y}\,\mathbf{j} + \frac{\partial f}{\partial z}\,\mathbf{k}$$

$$= \frac{-mMGx}{(x^2 + y^2 + z^2)^{3/2}}\,\mathbf{i} + \frac{-mMGy}{(x^2 + y^2 + z^2)^{3/2}}\,\mathbf{j} + \frac{-mMGz}{(x^2 + y^2 + z^2)^{3/2}}\,\mathbf{k}$$

$$= \mathbf{F}(x, y, z)$$

In Sections 16.3 and 16.5 we will learn how to tell whether or not a given vector field is conservative.

16.1 EXERCISES

1–10 Sketch the vector field **F** by drawing a diagram like Figure 5 or Figure 9.

1. $\mathbf{F}(x, y) = 0.3\,\mathbf{i} - 0.4\,\mathbf{j}$ **2.** $\mathbf{F}(x, y) = \frac{1}{2}x\,\mathbf{i} + y\,\mathbf{j}$

3. $\mathbf{F}(x, y) = -\frac{1}{2}\,\mathbf{i} + (y - x)\,\mathbf{j}$

4. $\mathbf{F}(x, y) = y\,\mathbf{i} + (x + y)\,\mathbf{j}$

5. $\mathbf{F}(x, y) = \dfrac{y\,\mathbf{i} + x\,\mathbf{j}}{\sqrt{x^2 + y^2}}$

6. $\mathbf{F}(x, y) = \dfrac{y\,\mathbf{i} - x\,\mathbf{j}}{\sqrt{x^2 + y^2}}$

7. $\mathbf{F}(x, y, z) = \mathbf{i}$

8. $\mathbf{F}(x, y, z) = z\,\mathbf{i}$

9. $\mathbf{F}(x, y, z) = -y\,\mathbf{i}$

10. $\mathbf{F}(x, y, z) = \mathbf{i} + \mathbf{k}$

11–14 Match the vector fields **F** with the plots labeled I–IV. Give reasons for your choices.

11. $\mathbf{F}(x, y) = \langle x, -y \rangle$

12. $\mathbf{F}(x, y) = \langle y, x - y \rangle$

13. $\mathbf{F}(x, y) = \langle y, y + 2 \rangle$

14. $\mathbf{F}(x, y) = \langle \cos(x + y), x \rangle$

I

II

III

IV
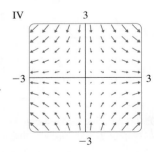

15–18 Match the vector fields **F** on $\mathbb{R}^3$ with the plots labeled I–IV. Give reasons for your choices.

15. $\mathbf{F}(x, y, z) = \mathbf{i} + 2\mathbf{j} + 3\mathbf{k}$

16. $\mathbf{F}(x, y, z) = \mathbf{i} + 2\mathbf{j} + z\mathbf{k}$

17. $\mathbf{F}(x, y, z) = x\mathbf{i} + y\mathbf{j} + 3\mathbf{k}$

18. $\mathbf{F}(x, y, z) = x\mathbf{i} + y\mathbf{j} + z\mathbf{k}$

I

II

III

IV
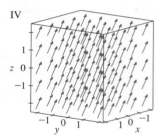

CAS 19. If you have a CAS that plots vector fields (the command is `fieldplot` in Maple and `PlotVectorField` or `VectorPlot` in Mathematica), use it to plot

$$\mathbf{F}(x, y) = (y^2 - 2xy)\mathbf{i} + (3xy - 6x^2)\mathbf{j}$$

Explain the appearance by finding the set of points (x, y) such that $\mathbf{F}(x, y) = \mathbf{0}$.

CAS 20. Let $\mathbf{F}(\mathbf{x}) = (r^2 - 2r)\mathbf{x}$, where $\mathbf{x} = \langle x, y \rangle$ and $r = |\mathbf{x}|$. Use a CAS to plot this vector field in various domains until you can see what is happening. Describe the appearance of the plot and explain it by finding the points where $\mathbf{F}(\mathbf{x}) = \mathbf{0}$.

21–24 Find the gradient vector field of f.

21. $f(x, y) = y \sin(xy)$ **22.** $f(s, t) = \sqrt{2s + 3t}$

23. $f(x, y, z) = \sqrt{x^2 + y^2 + z^2}$ **24.** $f(x, y, z) = x^2 y e^{y/z}$

25–26 Find the gradient vector field ∇f of f and sketch it.

25. $f(x, y) = \frac{1}{2}(x - y)^2$ **26.** $f(x, y) = \frac{1}{2}(x^2 - y^2)$

CAS 27–28 Plot the gradient vector field of f together with a contour map of f. Explain how they are related to each other.

27. $f(x, y) = \ln(1 + x^2 + 2y^2)$ **28.** $f(x, y) = \cos x - 2 \sin y$

29–32 Match the functions f with the plots of their gradient vector fields labeled I–IV. Give reasons for your choices.

29. $f(x, y) = x^2 + y^2$ **30.** $f(x, y) = x(x + y)$

31. $f(x, y) = (x + y)^2$ **32.** $f(x, y) = \sin\sqrt{x^2 + y^2}$

I

II

III

IV
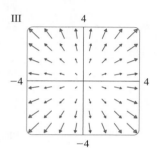

33. A particle moves in a velocity field $\mathbf{V}(x, y) = \langle x^2, x + y^2 \rangle$. If it is at position $(2, 1)$ at time $t = 3$, estimate its location at time $t = 3.01$.

34. At time $t = 1$, a particle is located at position $(1, 3)$. If it moves in a velocity field

$$\mathbf{F}(x, y) = \langle xy - 2, y^2 - 10 \rangle$$

find its approximate location at time $t = 1.05$.

35. The **flow lines** (or **streamlines**) of a vector field are the paths followed by a particle whose velocity field is the given vector field. Thus the vectors in a vector field are tangent to the flow lines.
(a) Use a sketch of the vector field $\mathbf{F}(x, y) = x\mathbf{i} - y\mathbf{j}$ to draw some flow lines. From your sketches, can you guess the equations of the flow lines?
(b) If parametric equations of a flow line are $x = x(t)$, $y = y(t)$, explain why these functions satisfy the differential equations $dx/dt = x$ and $dy/dt = -y$. Then solve the differential equations to find an equation of the flow line that passes through the point $(1, 1)$.

36. (a) Sketch the vector field $\mathbf{F}(x, y) = \mathbf{i} + x\mathbf{j}$ and then sketch some flow lines. What shape do these flow lines appear to have?
(b) If parametric equations of the flow lines are $x = x(t)$, $y = y(t)$, what differential equations do these functions satisfy? Deduce that $dy/dx = x$.
(c) If a particle starts at the origin in the velocity field given by $\mathbf{F}$, find an equation of the path it follows.

16.2 Line Integrals

In this section we define an integral that is similar to a single integral except that instead of integrating over an interval $[a, b]$, we integrate over a curve C. Such integrals are called *line integrals,* although "curve integrals" would be better terminology. They were invented in the early 19th century to solve problems involving fluid flow, forces, electricity, and magnetism.

We start with a plane curve C given by the parametric equations

$$\boxed{1} \qquad x = x(t) \qquad y = y(t) \qquad a \leqslant t \leqslant b$$

or, equivalently, by the vector equation $\mathbf{r}(t) = x(t)\,\mathbf{i} + y(t)\,\mathbf{j}$, and we assume that C is a smooth curve. [This means that $\mathbf{r}'$ is continuous and $\mathbf{r}'(t) \neq \mathbf{0}$. See Section 13.3.] If we divide the parameter interval $[a, b]$ into n subintervals $[t_{i-1}, t_i]$ of equal width and we let $x_i = x(t_i)$ and $y_i = y(t_i)$, then the corresponding points $P_i(x_i, y_i)$ divide C into n subarcs with lengths $\Delta s_1, \Delta s_2, \ldots, \Delta s_n$. (See Figure 1.) We choose any point $P_i^*(x_i^*, y_i^*)$ in the ith subarc. (This corresponds to a point t_i^* in $[t_{i-1}, t_i]$.) Now if f is any function of two variables whose domain includes the curve C, we evaluate f at the point (x_i^*, y_i^*), multiply by the length Δs_i of the subarc, and form the sum

$$\sum_{i=1}^{n} f(x_i^*, y_i^*)\, \Delta s_i$$

which is similar to a Riemann sum. Then we take the limit of these sums and make the following definition by analogy with a single integral.

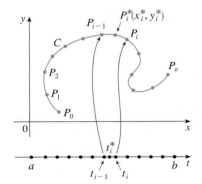

FIGURE 1

> $\boxed{2}$ **Definition** If f is defined on a smooth curve C given by Equations 1, then the **line integral of f along C** is
>
> $$\int_C f(x, y)\, ds = \lim_{n \to \infty} \sum_{i=1}^{n} f(x_i^*, y_i^*)\, \Delta s_i$$
>
> if this limit exists.

In Section 10.2 we found that the length of C is

$$L = \int_a^b \sqrt{\left(\frac{dx}{dt}\right)^2 + \left(\frac{dy}{dt}\right)^2}\; dt$$

A similar type of argument can be used to show that if f is a continuous function, then the limit in Definition 2 always exists and the following formula can be used to evaluate the line integral:

$$\boxed{3} \qquad \int_C f(x, y)\, ds = \int_a^b f(x(t), y(t)) \sqrt{\left(\frac{dx}{dt}\right)^2 + \left(\frac{dy}{dt}\right)^2}\; dt$$

The value of the line integral does not depend on the parametrization of the curve, provided that the curve is traversed exactly once as t increases from a to b.

The arc length function s is discussed in Section 13.3.

If $s(t)$ is the length of C between $\mathbf{r}(a)$ and $\mathbf{r}(t)$, then

$$\frac{ds}{dt} = \sqrt{\left(\frac{dx}{dt}\right)^2 + \left(\frac{dy}{dt}\right)^2}$$

So the way to remember Formula 3 is to express everything in terms of the parameter t: Use the parametric equations to express x and y in terms of t and write ds as

$$ds = \sqrt{\left(\frac{dx}{dt}\right)^2 + \left(\frac{dy}{dt}\right)^2} \, dt$$

In the special case where C is the line segment that joins $(a, 0)$ to $(b, 0)$, using x as the parameter, we can write the parametric equations of C as follows: $x = x$, $y = 0$, $a \leqslant x \leqslant b$. Formula 3 then becomes

$$\int_C f(x, y) \, ds = \int_a^b f(x, 0) \, dx$$

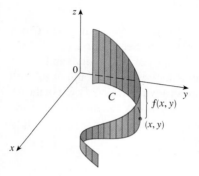

FIGURE 2

and so the line integral reduces to an ordinary single integral in this case.

Just as for an ordinary single integral, we can interpret the line integral of a *positive* function as an area. In fact, if $f(x, y) \geqslant 0$, $\int_C f(x, y) \, ds$ represents the area of one side of the "fence" or "curtain" in Figure 2, whose base is C and whose height above the point (x, y) is $f(x, y)$.

EXAMPLE 1 Evaluate $\int_C (2 + x^2 y) \, ds$, where C is the upper half of the unit circle $x^2 + y^2 = 1$.

SOLUTION In order to use Formula 3, we first need parametric equations to represent C. Recall that the unit circle can be parametrized by means of the equations

$$x = \cos t \qquad y = \sin t$$

and the upper half of the circle is described by the parameter interval $0 \leqslant t \leqslant \pi$. (See Figure 3.) Therefore Formula 3 gives

FIGURE 3

$$\int_C (2 + x^2 y) \, ds = \int_0^\pi (2 + \cos^2 t \sin t) \sqrt{\left(\frac{dx}{dt}\right)^2 + \left(\frac{dy}{dt}\right)^2} \, dt$$

$$= \int_0^\pi (2 + \cos^2 t \sin t) \sqrt{\sin^2 t + \cos^2 t} \, dt$$

$$= \int_0^\pi (2 + \cos^2 t \sin t) \, dt = \left[2t - \frac{\cos^3 t}{3} \right]_0^\pi$$

$$= 2\pi + \tfrac{2}{3}$$ ∎

Suppose now that C is a **piecewise-smooth curve**; that is, C is a union of a finite number of smooth curves $C_1, C_2, \ldots, C_n$, where, as illustrated in Figure 4, the initial point of C_{i+1} is the terminal point of C_i. Then we define the integral of f along C as the sum of the integrals of f along each of the smooth pieces of C:

FIGURE 4
A piecewise-smooth curve

$$\int_C f(x, y) \, ds = \int_{C_1} f(x, y) \, ds + \int_{C_2} f(x, y) \, ds + \cdots + \int_{C_n} f(x, y) \, ds$$

EXAMPLE 2 Evaluate $\int_C 2x \, ds$, where C consists of the arc C_1 of the parabola $y = x^2$ from $(0, 0)$ to $(1, 1)$ followed by the vertical line segment C_2 from $(1, 1)$ to $(1, 2)$.

SOLUTION The curve C is shown in Figure 5. C_1 is the graph of a function of x, so we can choose x as the parameter and the equations for C_1 become

$$x = x \qquad y = x^2 \qquad 0 \le x \le 1$$

Therefore

$$\int_{C_1} 2x \, ds = \int_0^1 2x \sqrt{\left(\frac{dx}{dx}\right)^2 + \left(\frac{dy}{dx}\right)^2} \, dx = \int_0^1 2x\sqrt{1 + 4x^2} \, dx$$

$$= \tfrac{1}{4} \cdot \tfrac{2}{3}(1 + 4x^2)^{3/2}\Big]_0^1 = \frac{5\sqrt{5} - 1}{6}$$

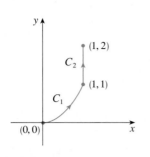

FIGURE 5
$C = C_1 \cup C_2$

On C_2 we choose y as the parameter, so the equations of C_2 are

$$x = 1 \qquad y = y \qquad 1 \le y \le 2$$

and

$$\int_{C_2} 2x \, ds = \int_1^2 2(1) \sqrt{\left(\frac{dx}{dy}\right)^2 + \left(\frac{dy}{dy}\right)^2} \, dy = \int_1^2 2 \, dy = 2$$

Thus

$$\int_C 2x \, ds = \int_{C_1} 2x \, ds + \int_{C_2} 2x \, ds = \frac{5\sqrt{5} - 1}{6} + 2 \qquad \blacksquare$$

Any physical interpretation of a line integral $\int_C f(x, y) \, ds$ depends on the physical interpretation of the function f. Suppose that $\rho(x, y)$ represents the linear density at a point (x, y) of a thin wire shaped like a curve C. Then the mass of the part of the wire from P_{i-1} to P_i in Figure 1 is approximately $\rho(x_i^*, y_i^*)\,\Delta s_i$ and so the total mass of the wire is approximately $\Sigma \, \rho(x_i^*, y_i^*)\,\Delta s_i$. By taking more and more points on the curve, we obtain the **mass** m of the wire as the limiting value of these approximations:

$$m = \lim_{n \to \infty} \sum_{i=1}^{n} \rho(x_i^*, y_i^*)\,\Delta s_i = \int_C \rho(x, y) \, ds$$

[For example, if $f(x, y) = 2 + x^2 y$ represents the density of a semicircular wire, then the integral in Example 1 would represent the mass of the wire.] The **center of mass** of the wire with density function ρ is located at the point $(\bar{x}, \bar{y})$, where

$$\boxed{4} \qquad \bar{x} = \frac{1}{m} \int_C x \rho(x, y) \, ds \qquad \bar{y} = \frac{1}{m} \int_C y \rho(x, y) \, ds$$

Other physical interpretations of line integrals will be discussed later in this chapter.

EXAMPLE 3 A wire takes the shape of the semicircle $x^2 + y^2 = 1$, $y \ge 0$, and is thicker near its base than near the top. Find the center of mass of the wire if the linear density at any point is proportional to its distance from the line $y = 1$.

SOLUTION As in Example 1 we use the parametrization $x = \cos t$, $y = \sin t$, $0 \le t \le \pi$, and find that $ds = dt$. The linear density is

$$\rho(x, y) = k(1 - y)$$

where k is a constant, and so the mass of the wire is

$$m = \int_C k(1 - y)\, ds = \int_0^\pi k(1 - \sin t)\, dt = k\big[t + \cos t\big]_0^\pi = k(\pi - 2)$$

From Equations 4 we have

$$\bar{y} = \frac{1}{m} \int_C y\rho(x, y)\, ds = \frac{1}{k(\pi - 2)} \int_C y k(1 - y)\, ds$$

$$= \frac{1}{\pi - 2} \int_0^\pi (\sin t - \sin^2 t)\, dt = \frac{1}{\pi - 2}\big[-\cos t - \tfrac{1}{2}t + \tfrac{1}{4}\sin 2t\big]_0^\pi$$

$$= \frac{4 - \pi}{2(\pi - 2)}$$

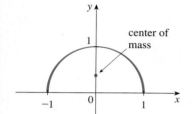

By symmetry we see that $\bar{x} = 0$, so the center of mass is

$$\left(0, \frac{4 - \pi}{2(\pi - 2)}\right) \approx (0, 0.38)$$

FIGURE 6

See Figure 6. ∎

Two other line integrals are obtained by replacing Δs_i by either $\Delta x_i = x_i - x_{i-1}$ or $\Delta y_i = y_i - y_{i-1}$ in Definition 2. They are called the **line integrals of f along C with respect to x and y:**

$$\boxed{5} \qquad \int_C f(x, y)\, dx = \lim_{n \to \infty} \sum_{i=1}^{n} f(x_i^*, y_i^*)\, \Delta x_i$$

$$\boxed{6} \qquad \int_C f(x, y)\, dy = \lim_{n \to \infty} \sum_{i=1}^{n} f(x_i^*, y_i^*)\, \Delta y_i$$

When we want to distinguish the original line integral $\int_C f(x, y)\, ds$ from those in Equations 5 and 6, we call it the **line integral with respect to arc length**.

The following formulas say that line integrals with respect to x and y can also be evaluated by expressing everything in terms of t: $x = x(t)$, $y = y(t)$, $dx = x'(t)\, dt$, $dy = y'(t)\, dt$.

$$\boxed{7} \qquad \int_C f(x, y)\, dx = \int_a^b f(x(t), y(t))\, x'(t)\, dt$$

$$\int_C f(x, y)\, dy = \int_a^b f(x(t), y(t))\, y'(t)\, dt$$

It frequently happens that line integrals with respect to x and y occur together. When this happens, it's customary to abbreviate by writing

$$\int_C P(x, y)\, dx + \int_C Q(x, y)\, dy = \int_C P(x, y)\, dx + Q(x, y)\, dy$$

When we are setting up a line integral, sometimes the most difficult thing is to think of a parametric representation for a curve whose geometric description is given. In particular, we often need to parametrize a line segment, so it's useful to remember that a vector representation of the line segment that starts at $\mathbf{r}_0$ and ends at $\mathbf{r}_1$ is given by

8
$$\mathbf{r}(t) = (1 - t)\mathbf{r}_0 + t\,\mathbf{r}_1 \qquad 0 \leqslant t \leqslant 1$$

(See Equation 12.5.4.)

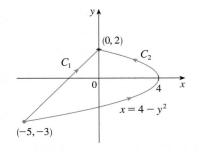

C_1

$(0, 2)$

C_2

0 4 x

$x = 4 - y^2$

$(-5, -3)$

FIGURE 7

EXAMPLE 4 Evaluate $\int_C y^2\,dx + x\,dy$, where (a) $C = C_1$ is the line segment from $(-5, -3)$ to $(0, 2)$ and (b) $C = C_2$ is the arc of the parabola $x = 4 - y^2$ from $(-5, -3)$ to $(0, 2)$. (See Figure 7.)

SOLUTION

(a) A parametric representation for the line segment is

$$x = 5t - 5 \qquad y = 5t - 3 \qquad 0 \leqslant t \leqslant 1$$

(Use Equation 8 with $\mathbf{r}_0 = \langle -5, -3 \rangle$ and $\mathbf{r}_1 = \langle 0, 2 \rangle$.) Then $dx = 5\,dt$, $dy = 5\,dt$, and Formulas 7 give

$$\int_{C_1} y^2\,dx + x\,dy = \int_0^1 (5t - 3)^2(5\,dt) + (5t - 5)(5\,dt)$$

$$= 5\int_0^1 (25t^2 - 25t + 4)\,dt$$

$$= 5\left[\frac{25t^3}{3} - \frac{25t^2}{2} + 4t\right]_0^1 = -\frac{5}{6}$$

(b) Since the parabola is given as a function of y, let's take y as the parameter and write C_2 as

$$x = 4 - y^2 \qquad y = y \qquad -3 \leqslant y \leqslant 2$$

Then $dx = -2y\,dy$ and by Formulas 7 we have

$$\int_{C_2} y^2\,dx + x\,dy = \int_{-3}^2 y^2(-2y)\,dy + (4 - y^2)\,dy$$

$$= \int_{-3}^2 (-2y^3 - y^2 + 4)\,dy$$

$$= \left[-\frac{y^4}{2} - \frac{y^3}{3} + 4y\right]_{-3}^2 = 40\tfrac{5}{6}$$ ■

    Notice that we got different answers in parts (a) and (b) of Example 4 even though the two curves had the same endpoints. Thus, in general, the value of a line integral depends not just on the endpoints of the curve but also on the path. (But see Section 16.3 for conditions under which the integral is independent of the path.)

    Notice also that the answers in Example 4 depend on the direction, or orientation, of the curve. If $-C_1$ denotes the line segment from $(0, 2)$ to $(-5, -3)$, you can verify, using the parametrization

$$x = -5t \qquad y = 2 - 5t \qquad 0 \leqslant t \leqslant 1$$

that

$$\int_{-C_1} y^2\,dx + x\,dy = \tfrac{5}{6}$$

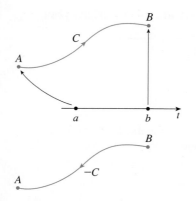

FIGURE 8

In general, a given parametrization $x = x(t)$, $y = y(t)$, $a \leqslant t \leqslant b$, determines an **orientation** of a curve C, with the positive direction corresponding to increasing values of the parameter t. (See Figure 8, where the initial point A corresponds to the parameter value a and the terminal point B corresponds to $t = b$.)

If $-C$ denotes the curve consisting of the same points as C but with the opposite orientation (from initial point B to terminal point A in Figure 8), then we have

$$\int_{-C} f(x, y)\, dx = -\int_C f(x, y)\, dx \qquad \int_{-C} f(x, y)\, dy = -\int_C f(x, y)\, dy$$

But if we integrate with respect to arc length, the value of the line integral does *not* change when we reverse the orientation of the curve:

$$\int_{-C} f(x, y)\, ds = \int_C f(x, y)\, ds$$

This is because Δs_i is always positive, whereas Δx_i and Δy_i change sign when we reverse the orientation of C.

Line Integrals in Space

We now suppose that C is a smooth space curve given by the parametric equations

$$x = x(t) \qquad y = y(t) \qquad z = z(t) \qquad a \leqslant t \leqslant b$$

or by a vector equation $\mathbf{r}(t) = x(t)\,\mathbf{i} + y(t)\,\mathbf{j} + z(t)\,\mathbf{k}$. If f is a function of three variables that is continuous on some region containing C, then we define the **line integral of f along C** (with respect to arc length) in a manner similar to that for plane curves:

$$\int_C f(x, y, z)\, ds = \lim_{n \to \infty} \sum_{i=1}^{n} f(x_i^*, y_i^*, z_i^*)\, \Delta s_i$$

We evaluate it using a formula similar to Formula 3:

$$\boxed{9} \quad \int_C f(x, y, z)\, ds = \int_a^b f(x(t), y(t), z(t)) \sqrt{\left(\frac{dx}{dt}\right)^2 + \left(\frac{dy}{dt}\right)^2 + \left(\frac{dz}{dt}\right)^2}\, dt$$

Observe that the integrals in both Formulas 3 and 9 can be written in the more compact vector notation

$$\int_a^b f(\mathbf{r}(t)) \, |\mathbf{r}'(t)|\, dt$$

For the special case $f(x, y, z) = 1$, we get

$$\int_C ds = \int_a^b |\mathbf{r}'(t)|\, dt = L$$

where L is the length of the curve C (see Formula 13.3.3).

Line integrals along C with respect to x, y, and z can also be defined. For example,

$$\int_C f(x, y, z)\, dz = \lim_{n \to \infty} \sum_{i=1}^{n} f(x_i^*, y_i^*, z_i^*)\, \Delta z_i$$

$$= \int_a^b f(x(t), y(t), z(t))\, z'(t)\, dt$$

Therefore, as with line integrals in the plane, we evaluate integrals of the form

$$\boxed{10} \qquad \int_C P(x, y, z)\, dx + Q(x, y, z)\, dy + R(x, y, z)\, dz$$

by expressing everything (x, y, z, dx, dy, dz) in terms of the parameter t.

EXAMPLE 5 Evaluate $\int_C y \sin z\, ds$, where C is the circular helix given by the equations $x = \cos t$, $y = \sin t$, $z = t$, $0 \leqslant t \leqslant 2\pi$. (See Figure 9.)

SOLUTION Formula 9 gives

$$\int_C y \sin z\, ds = \int_0^{2\pi} (\sin t) \sin t \sqrt{\left(\frac{dx}{dt}\right)^2 + \left(\frac{dy}{dt}\right)^2 + \left(\frac{dz}{dt}\right)^2}\, dt$$

$$= \int_0^{2\pi} \sin^2 t \sqrt{\sin^2 t + \cos^2 t + 1}\, dt = \sqrt{2} \int_0^{2\pi} \tfrac{1}{2}(1 - \cos 2t)\, dt$$

$$= \frac{\sqrt{2}}{2} \left[t - \tfrac{1}{2} \sin 2t \right]_0^{2\pi} = \sqrt{2}\, \pi \qquad \blacksquare$$

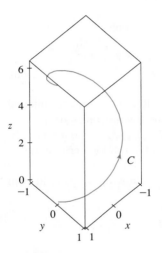

FIGURE 9

EXAMPLE 6 Evaluate $\int_C y\, dx + z\, dy + x\, dz$, where C consists of the line segment C_1 from $(2, 0, 0)$ to $(3, 4, 5)$, followed by the vertical line segment C_2 from $(3, 4, 5)$ to $(3, 4, 0)$.

SOLUTION The curve C is shown in Figure 10. Using Equation 8, we write C_1 as

$$\mathbf{r}(t) = (1 - t)\langle 2, 0, 0 \rangle + t\langle 3, 4, 5 \rangle = \langle 2 + t, 4t, 5t \rangle$$

or, in parametric form, as

$$x = 2 + t \qquad y = 4t \qquad z = 5t \qquad 0 \leqslant t \leqslant 1$$

Thus

$$\int_{C_1} y\, dx + z\, dy + x\, dz = \int_0^1 (4t)\, dt + (5t)4\, dt + (2 + t)5\, dt$$

$$= \int_0^1 (10 + 29t)\, dt = 10t + 29 \frac{t^2}{2} \bigg]_0^1 = 24.5$$

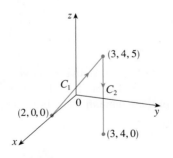

FIGURE 10

Likewise, C_2 can be written in the form

$$\mathbf{r}(t) = (1 - t)\langle 3, 4, 5 \rangle + t\langle 3, 4, 0 \rangle = \langle 3, 4, 5 - 5t \rangle$$

or

$$x = 3 \qquad y = 4 \qquad z = 5 - 5t \qquad 0 \leqslant t \leqslant 1$$

Then $dx = 0 = dy$, so

$$\int_{C_2} y\,dx + z\,dy + x\,dz = \int_0^1 3(-5)\,dt = -15$$

Adding the values of these integrals, we obtain

$$\int_C y\,dx + z\,dy + x\,dz = 24.5 - 15 = 9.5$$

■

■ Line Integrals of Vector Fields

Recall from Section 6.4 that the work done by a variable force $f(x)$ in moving a particle from a to b along the x-axis is $W = \int_a^b f(x)\,dx$. Then in Section 12.3 we found that the work done by a constant force $\mathbf{F}$ in moving an object from a point P to another point Q in space is $W = \mathbf{F} \cdot \mathbf{D}$, where $\mathbf{D} = \overrightarrow{PQ}$ is the displacement vector.

Now suppose that $\mathbf{F} = P\,\mathbf{i} + Q\,\mathbf{j} + R\,\mathbf{k}$ is a continuous force field on $\mathbb{R}^3$, such as the gravitational field of Example 16.1.4 or the electric force field of Example 16.1.5. (A force field on $\mathbb{R}^2$ could be regarded as a special case where $R = 0$ and P and Q depend only on x and y.) We wish to compute the work done by this force in moving a particle along a smooth curve C.

We divide C into subarcs $P_{i-1}P_i$ with lengths Δs_i by dividing the parameter interval $[a, b]$ into subintervals of equal width. (See Figure 1 for the two-dimensional case or Figure 11 for the three-dimensional case.) Choose a point $P_i^*(x_i^*, y_i^*, z_i^*)$ on the ith subarc corresponding to the parameter value t_i^*. If Δs_i is small, then as the particle moves from P_{i-1} to P_i along the curve, it proceeds approximately in the direction of $\mathbf{T}(t_i^*)$, the unit tangent vector at P_i^*. Thus the work done by the force $\mathbf{F}$ in moving the particle from P_{i-1} to P_i is approximately

$$\mathbf{F}(x_i^*,\, y_i^*,\, z_i^*) \cdot [\Delta s_i\,\mathbf{T}(t_i^*)] = [\mathbf{F}(x_i^*,\, y_i^*,\, z_i^*) \cdot \mathbf{T}(t_i^*)]\,\Delta s_i$$

and the total work done in moving the particle along C is approximately

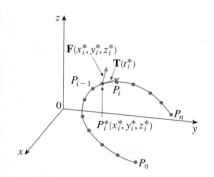

FIGURE 11

$$\boxed{11} \qquad \sum_{i=1}^{n} [\mathbf{F}(x_i^*,\, y_i^*,\, z_i^*) \cdot \mathbf{T}(x_i^*,\, y_i^*,\, z_i^*)]\,\Delta s_i$$

where $\mathbf{T}(x, y, z)$ is the unit tangent vector at the point (x, y, z) on C. Intuitively, we see that these approximations ought to become better as n becomes larger. Therefore we define the **work** W done by the force field $\mathbf{F}$ as the limit of the Riemann sums in (11), namely,

$$\boxed{12} \qquad W = \int_C \mathbf{F}(x, y, z) \cdot \mathbf{T}(x, y, z)\,ds = \int_C \mathbf{F} \cdot \mathbf{T}\,ds$$

Equation 12 says that *work is the line integral with respect to arc length of the tangential component of the force.*

If the curve C is given by the vector equation $\mathbf{r}(t) = x(t)\,\mathbf{i} + y(t)\,\mathbf{j} + z(t)\,\mathbf{k}$, then $\mathbf{T}(t) = \mathbf{r}'(t)/|\mathbf{r}'(t)|$, so using Equation 9 we can rewrite Equation 12 in the form

$$W = \int_a^b \left[\mathbf{F}(\mathbf{r}(t)) \cdot \frac{\mathbf{r}'(t)}{|\mathbf{r}'(t)|} \right] |\mathbf{r}'(t)|\,dt = \int_a^b \mathbf{F}(\mathbf{r}(t)) \cdot \mathbf{r}'(t)\,dt$$

This integral is often abbreviated as $\int_C \mathbf{F} \cdot d\mathbf{r}$ and occurs in other areas of physics as well. Therefore we make the following definition for the line integral of *any* continuous vector field.

> **13** **Definition** Let $\mathbf{F}$ be a continuous vector field defined on a smooth curve C given by a vector function $\mathbf{r}(t)$, $a \leqslant t \leqslant b$. Then the **line integral of F along C** is
>
> $$\int_C \mathbf{F} \cdot d\mathbf{r} = \int_a^b \mathbf{F}(\mathbf{r}(t)) \cdot \mathbf{r}'(t)\, dt = \int_C \mathbf{F} \cdot \mathbf{T}\, ds$$

When using Definition 13, bear in mind that $\mathbf{F}(\mathbf{r}(t))$ is just an abbreviation for the vector field $\mathbf{F}(x(t), y(t), z(t))$, so we evaluate $\mathbf{F}(\mathbf{r}(t))$ simply by putting $x = x(t)$, $y = y(t)$, and $z = z(t)$ in the expression for $\mathbf{F}(x, y, z)$. Notice also that we can formally write $d\mathbf{r} = \mathbf{r}'(t)\, dt$.

Figure 12 shows the force field and the curve in Example 7. The work done is negative because the field impedes movement along the curve.

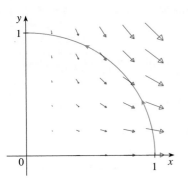

FIGURE 12

EXAMPLE 7 Find the work done by the force field $\mathbf{F}(x, y) = x^2\, \mathbf{i} - xy\, \mathbf{j}$ in moving a particle along the quarter-circle $\mathbf{r}(t) = \cos t\, \mathbf{i} + \sin t\, \mathbf{j}$, $0 \leqslant t \leqslant \pi/2$.

SOLUTION Since $x = \cos t$ and $y = \sin t$, we have

$$\mathbf{F}(\mathbf{r}(t)) = \cos^2 t\, \mathbf{i} - \cos t \sin t\, \mathbf{j}$$

and

$$\mathbf{r}'(t) = -\sin t\, \mathbf{i} + \cos t\, \mathbf{j}$$

Therefore the work done is

$$\int_C \mathbf{F} \cdot d\mathbf{r} = \int_0^{\pi/2} \mathbf{F}(\mathbf{r}(t)) \cdot \mathbf{r}'(t)\, dt = \int_0^{\pi/2} (-2\cos^2 t \sin t)\, dt$$

$$= 2\, \frac{\cos^3 t}{3} \bigg]_0^{\pi/2} = -\frac{2}{3}$$

NOTE Even though $\int_C \mathbf{F} \cdot d\mathbf{r} = \int_C \mathbf{F} \cdot \mathbf{T}\, ds$ and integrals with respect to arc length are unchanged when orientation is reversed, it is still true that

$$\int_{-C} \mathbf{F} \cdot d\mathbf{r} = -\int_C \mathbf{F} \cdot d\mathbf{r}$$

because the unit tangent vector $\mathbf{T}$ is replaced by its negative when C is replaced by $-C$.

Figure 13 shows the twisted cubic C in Example 8 and some typical vectors acting at three points on C.

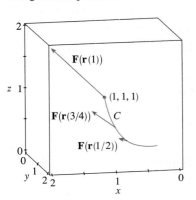

FIGURE 13

EXAMPLE 8 Evaluate $\int_C \mathbf{F} \cdot d\mathbf{r}$, where $\mathbf{F}(x, y, z) = xy\, \mathbf{i} + yz\, \mathbf{j} + zx\, \mathbf{k}$ and C is the twisted cubic given by

$$x = t \qquad y = t^2 \qquad z = t^3 \qquad 0 \leqslant t \leqslant 1$$

SOLUTION We have

$$\mathbf{r}(t) = t\, \mathbf{i} + t^2\, \mathbf{j} + t^3\, \mathbf{k}$$

$$\mathbf{r}'(t) = \mathbf{i} + 2t\, \mathbf{j} + 3t^2\, \mathbf{k}$$

$$\mathbf{F}(\mathbf{r}(t)) = t^3\, \mathbf{i} + t^5\, \mathbf{j} + t^4\, \mathbf{k}$$

Thus
$$\int_C \mathbf{F} \cdot d\mathbf{r} = \int_0^1 \mathbf{F}(\mathbf{r}(t)) \cdot \mathbf{r}'(t)\, dt$$

$$= \int_0^1 (t^3 + 5t^6)\, dt = \frac{t^4}{4} + \frac{5t^7}{7}\bigg]_0^1 = \frac{27}{28} \qquad \blacksquare$$

Finally, we note the connection between line integrals of vector fields and line integrals of scalar fields. Suppose the vector field $\mathbf{F}$ on $\mathbb{R}^3$ is given in component form by the equation $\mathbf{F} = P\,\mathbf{i} + Q\,\mathbf{j} + R\,\mathbf{k}$. We use Definition 13 to compute its line integral along C:

$$\int_C \mathbf{F} \cdot d\mathbf{r} = \int_a^b \mathbf{F}(\mathbf{r}(t)) \cdot \mathbf{r}'(t)\, dt$$

$$= \int_a^b (P\,\mathbf{i} + Q\,\mathbf{j} + R\,\mathbf{k}) \cdot (x'(t)\,\mathbf{i} + y'(t)\,\mathbf{j} + z'(t)\,\mathbf{k})\, dt$$

$$= \int_a^b \left[P(x(t), y(t), z(t))\,x'(t) + Q(x(t), y(t), z(t))\,y'(t) + R(x(t), y(t), z(t))\,z'(t) \right] dt$$

But this last integral is precisely the line integral in (10). Therefore we have

$$\int_C \mathbf{F} \cdot d\mathbf{r} = \int_C P\,dx + Q\,dy + R\,dz \qquad \text{where } \mathbf{F} = P\,\mathbf{i} + Q\,\mathbf{j} + R\,\mathbf{k}$$

For example, the integral $\int_C y\,dx + z\,dy + x\,dz$ in Example 6 could be expressed as $\int_C \mathbf{F} \cdot d\mathbf{r}$ where

$$\mathbf{F}(x, y, z) = y\,\mathbf{i} + z\,\mathbf{j} + x\,\mathbf{k}$$

16.2 EXERCISES

1–16 Evaluate the line integral, where C is the given curve.

1. $\int_C y\,ds$, $C: x = t^2,\ y = 2t,\ 0 \leqslant t \leqslant 3$

2. $\int_C (x/y)\,ds$, $C: x = t^3,\ y = t^4,\ 1 \leqslant t \leqslant 2$

3. $\int_C xy^4\,ds$, C is the right half of the circle $x^2 + y^2 = 16$

4. $\int_C xe^y\,ds$, C is the line segment from $(2, 0)$ to $(5, 4)$

5. $\int_C (x^2 y + \sin x)\,dy$,
C is the arc of the parabola $y = x^2$ from $(0, 0)$ to (π, π^2)

6. $\int_C e^x\,dx$,
C is the arc of the curve $x = y^3$ from $(-1, -1)$ to $(1, 1)$

7. $\int_C (x + 2y)\,dx + x^2\,dy$, C consists of line segments from $(0, 0)$ to $(2, 1)$ and from $(2, 1)$ to $(3, 0)$

8. $\int_C x^2\,dx + y^2\,dy$, C consists of the arc of the circle $x^2 + y^2 = 4$ from $(2, 0)$ to $(0, 2)$ followed by the line segment from $(0, 2)$ to $(4, 3)$

9. $\int_C x^2 y\,ds$,
$C: x = \cos t,\ y = \sin t,\ z = t,\ 0 \leqslant t \leqslant \pi/2$

10. $\int_C y^2 z\,ds$,
C is the line segment from $(3, 1, 2)$ to $(1, 2, 5)$

11. $\int_C xe^{yz}\,ds$,
C is the line segment from $(0, 0, 0)$ to $(1, 2, 3)$

12. $\int_C (x^2 + y^2 + z^2)\,ds$,
$C: x = t,\ y = \cos 2t,\ z = \sin 2t,\ 0 \leqslant t \leqslant 2\pi$

13. $\int_C xye^{yz}\,dy$, $C: x = t,\ y = t^2,\ z = t^3,\ 0 \leqslant t \leqslant 1$

14. $\int_C y\,dx + z\,dy + x\,dz$,
$C: x = \sqrt{t},\ y = t,\ z = t^2,\ 1 \leqslant t \leqslant 4$

15. $\int_C z^2\,dx + x^2\,dy + y^2\,dz$,
C is the line segment from $(1, 0, 0)$ to $(4, 1, 2)$

16. $\int_C (y + z)\,dx + (x + z)\,dy + (x + y)\,dz$,
C consists of line segments from $(0, 0, 0)$ to $(1, 0, 1)$ and from $(1, 0, 1)$ to $(0, 1, 2)$

17. Let **F** be the vector field shown in the figure.
   (a) If C_1 is the vertical line segment from $(-3, -3)$ to $(-3, 3)$, determine whether $\int_{C_1} \mathbf{F} \cdot d\mathbf{r}$ is positive, negative, or zero.
   (b) If C_2 is the counterclockwise-oriented circle with radius 3 and center the origin, determine whether $\int_{C_2} \mathbf{F} \cdot d\mathbf{r}$ is positive, negative, or zero.

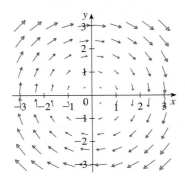

18. The figure shows a vector field **F** and two curves C_1 and C_2. Are the line integrals of **F** over C_1 and C_2 positive, negative, or zero? Explain.

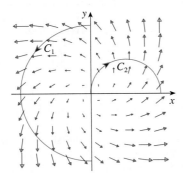

19–22 Evaluate the line integral $\int_C \mathbf{F} \cdot d\mathbf{r}$, where C is given by the vector function $\mathbf{r}(t)$.

19. $\mathbf{F}(x, y) = xy^2\,\mathbf{i} - x^2\,\mathbf{j}$,
$\mathbf{r}(t) = t^3\,\mathbf{i} + t^2\,\mathbf{j}, \quad 0 \leq t \leq 1$

20. $\mathbf{F}(x, y, z) = (x + y^2)\,\mathbf{i} + xz\,\mathbf{j} + (y + z)\,\mathbf{k}$,
$\mathbf{r}(t) = t^2\,\mathbf{i} + t^3\,\mathbf{j} - 2t\,\mathbf{k}, \quad 0 \leq t \leq 2$

21. $\mathbf{F}(x, y, z) = \sin x\,\mathbf{i} + \cos y\,\mathbf{j} + xz\,\mathbf{k}$,
$\mathbf{r}(t) = t^3\,\mathbf{i} - t^2\,\mathbf{j} + t\,\mathbf{k}, \quad 0 \leq t \leq 1$

22. $\mathbf{F}(x, y, z) = x\,\mathbf{i} + y\,\mathbf{j} + xy\,\mathbf{k}$,
$\mathbf{r}(t) = \cos t\,\mathbf{i} + \sin t\,\mathbf{j} + t\,\mathbf{k}, \quad 0 \leq t \leq \pi$

23–26 Use a calculator to evaluate the line integral correct to four decimal places.

23. $\int_C \mathbf{F} \cdot d\mathbf{r}$, where $\mathbf{F}(x, y) = \sqrt{x + y}\,\mathbf{i} + (y/x)\,\mathbf{j}$ and $\mathbf{r}(t) = \sin^2 t\,\mathbf{i} + \sin t \cos t\,\mathbf{j}, \quad \pi/6 \leq t \leq \pi/3$

24. $\int_C \mathbf{F} \cdot d\mathbf{r}$, where $\mathbf{F}(x, y, z) = yze^x\,\mathbf{i} + zxe^y\,\mathbf{j} + xye^z\,\mathbf{k}$ and $\mathbf{r}(t) = \sin t\,\mathbf{i} + \cos t\,\mathbf{j} + \tan t\,\mathbf{k}, \quad 0 \leq t \leq \pi/4$

25. $\int_C xy \arctan z\, ds$, where C has parametric equations $x = t^2, \ y = t^3, \ z = \sqrt{t}, \ 1 \leq t \leq 2$

26. $\int_C z \ln(x + y)\, ds$, where C has parametric equations $x = 1 + 3t, \ y = 2 + t^2, \ z = t^4, \ -1 \leq t \leq 1$

CAS 27–28 Use a graph of the vector field **F** and the curve C to guess whether the line integral of **F** over C is positive, negative, or zero. Then evaluate the line integral.

27. $\mathbf{F}(x, y) = (x - y)\,\mathbf{i} + xy\,\mathbf{j}$,
C is the arc of the circle $x^2 + y^2 = 4$ traversed counterclockwise from $(2, 0)$ to $(0, -2)$

28. $\mathbf{F}(x, y) = \dfrac{x}{\sqrt{x^2 + y^2}}\,\mathbf{i} + \dfrac{y}{\sqrt{x^2 + y^2}}\,\mathbf{j}$,
C is the parabola $y = 1 + x^2$ from $(-1, 2)$ to $(1, 2)$

29. (a) Evaluate the line integral $\int_C \mathbf{F} \cdot d\mathbf{r}$, where $\mathbf{F}(x, y) = e^{x-1}\,\mathbf{i} + xy\,\mathbf{j}$ and C is given by $\mathbf{r}(t) = t^2\,\mathbf{i} + t^3\,\mathbf{j}, \ 0 \leq t \leq 1$.
   (b) Illustrate part (a) by using a graphing calculator or computer to graph C and the vectors from the vector field corresponding to $t = 0, 1/\sqrt{2}$, and 1 (as in Figure 13).

30. (a) Evaluate the line integral $\int_C \mathbf{F} \cdot d\mathbf{r}$, where $\mathbf{F}(x, y, z) = x\,\mathbf{i} - z\,\mathbf{j} + y\,\mathbf{k}$ and C is given by $\mathbf{r}(t) = 2t\,\mathbf{i} + 3t\,\mathbf{j} - t^2\,\mathbf{k}, \ -1 \leq t \leq 1$.
   (b) Illustrate part (a) by using a computer to graph C and the vectors from the vector field corresponding to $t = \pm 1$ and $\pm\frac{1}{2}$ (as in Figure 13).

CAS 31. Find the exact value of $\int_C x^3 y^2 z\, ds$, where C is the curve with parametric equations $x = e^{-t} \cos 4t, \ y = e^{-t} \sin 4t, \ z = e^{-t}, \ 0 \leq t \leq 2\pi$.

32. (a) Find the work done by the force field $\mathbf{F}(x, y) = x^2\,\mathbf{i} + xy\,\mathbf{j}$ on a particle that moves once around the circle $x^2 + y^2 = 4$ oriented in the counterclockwise direction.
   CAS (b) Use a computer algebra system to graph the force field and circle on the same screen. Use the graph to explain your answer to part (a).

33. A thin wire is bent into the shape of a semicircle $x^2 + y^2 = 4, \ x \geq 0$. If the linear density is a constant k, find the mass and center of mass of the wire.

34. A thin wire has the shape of the first-quadrant part of the circle with center the origin and radius a. If the density function is $\rho(x, y) = kxy$, find the mass and center of mass of the wire.

35. (a) Write the formulas similar to Equations 4 for the center of mass $(\bar{x}, \bar{y}, \bar{z})$ of a thin wire in the shape of a space curve C if the wire has density function $\rho(x, y, z)$.

(b) Find the center of mass of a wire in the shape of the helix $x = 2 \sin t$, $y = 2 \cos t$, $z = 3t$, $0 \leqslant t \leqslant 2\pi$, if the density is a constant k.

36. Find the mass and center of mass of a wire in the shape of the helix $x = t$, $y = \cos t$, $z = \sin t$, $0 \leqslant t \leqslant 2\pi$, if the density at any point is equal to the square of the distance from the origin.

37. If a wire with linear density $\rho(x, y)$ lies along a plane curve C, its **moments of inertia** about the x- and y-axes are defined as

$$I_x = \int_C y^2 \rho(x, y)\, ds \qquad I_y = \int_C x^2 \rho(x, y)\, ds$$

Find the moments of inertia for the wire in Example 3.

38. If a wire with linear density $\rho(x, y, z)$ lies along a space curve C, its **moments of inertia** about the x-, y-, and z-axes are defined as

$$I_x = \int_C (y^2 + z^2)\rho(x, y, z)\, ds$$

$$I_y = \int_C (x^2 + z^2)\rho(x, y, z)\, ds$$

$$I_z = \int_C (x^2 + y^2)\rho(x, y, z)\, ds$$

Find the moments of inertia for the wire in Exercise 35.

39. Find the work done by the force field

$$\mathbf{F}(x, y) = x\,\mathbf{i} + (y + 2)\,\mathbf{j}$$

in moving an object along an arch of the cycloid

$$\mathbf{r}(t) = (t - \sin t)\,\mathbf{i} + (1 - \cos t)\,\mathbf{j} \qquad 0 \leqslant t \leqslant 2\pi$$

40. Find the work done by the force field $\mathbf{F}(x, y) = x^2\,\mathbf{i} + ye^x\,\mathbf{j}$ on a particle that moves along the parabola $x = y^2 + 1$ from $(1, 0)$ to $(2, 1)$.

41. Find the work done by the force field

$$\mathbf{F}(x, y, z) = \langle x - y^2, y - z^2, z - x^2 \rangle$$

on a particle that moves along the line segment from $(0, 0, 1)$ to $(2, 1, 0)$.

42. The force exerted by an electric charge at the origin on a charged particle at a point (x, y, z) with position vector $\mathbf{r} = \langle x, y, z \rangle$ is $\mathbf{F}(\mathbf{r}) = K\mathbf{r}/|\mathbf{r}|^3$ where K is a constant. (See Example 16.1.5.) Find the work done as the particle moves along a straight line from $(2, 0, 0)$ to $(2, 1, 5)$.

43. The position of an object with mass m at time t is $\mathbf{r}(t) = at^2\,\mathbf{i} + bt^3\,\mathbf{j}$, $0 \leqslant t \leqslant 1$.
(a) What is the force acting on the object at time t?
(b) What is the work done by the force during the time interval $0 \leqslant t \leqslant 1$?

44. An object with mass m moves with position function $\mathbf{r}(t) = a \sin t\,\mathbf{i} + b \cos t\,\mathbf{j} + ct\,\mathbf{k}$, $0 \leqslant t \leqslant \pi/2$. Find the work done on the object during this time period.

45. A 160-lb man carries a 25-lb can of paint up a helical staircase that encircles a silo with a radius of 20 ft. If the silo is 90 ft high and the man makes exactly three complete revolutions climbing to the top, how much work is done by the man against gravity?

46. Suppose there is a hole in the can of paint in Exercise 45 and 9 lb of paint leaks steadily out of the can during the man's ascent. How much work is done?

47. (a) Show that a constant force field does zero work on a particle that moves once uniformly around the circle $x^2 + y^2 = 1$.
(b) Is this also true for a force field $\mathbf{F}(\mathbf{x}) = k\mathbf{x}$, where k is a constant and $\mathbf{x} = \langle x, y \rangle$?

48. The base of a circular fence with radius 10 m is given by $x = 10 \cos t$, $y = 10 \sin t$. The height of the fence at position (x, y) is given by the function $h(x, y) = 4 + 0.01(x^2 - y^2)$, so the height varies from 3 m to 5 m. Suppose that 1 L of paint covers 100 m². Sketch the fence and determine how much paint you will need if you paint both sides of the fence.

49. If C is a smooth curve given by a vector function $\mathbf{r}(t)$, $a \leqslant t \leqslant b$, and $\mathbf{v}$ is a constant vector, show that

$$\int_C \mathbf{v} \cdot d\mathbf{r} = \mathbf{v} \cdot [\mathbf{r}(b) - \mathbf{r}(a)]$$

50. If C is a smooth curve given by a vector function $\mathbf{r}(t)$, $a \leqslant t \leqslant b$, show that

$$\int_C \mathbf{r} \cdot d\mathbf{r} = \tfrac{1}{2}\Big[|\mathbf{r}(b)|^2 - |\mathbf{r}(a)|^2\Big]$$

51. An object moves along the curve C shown in the figure from $(1, 2)$ to $(9, 8)$. The lengths of the vectors in the force field $\mathbf{F}$ are measured in newtons by the scales on the axes. Estimate the work done by $\mathbf{F}$ on the object.

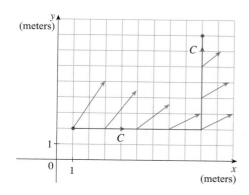

52. Experiments show that a steady current I in a long wire produces a magnetic field $\mathbf{B}$ that is tangent to any circle that lies in the plane perpendicular to the wire and whose center is the axis of the wire (as in the figure). *Ampère's Law* relates the electric current to its magnetic effects and states that

$$\int_C \mathbf{B} \cdot d\mathbf{r} = \mu_0 I$$

where I is the net current that passes through any surface bounded by a closed curve C, and μ_0 is a constant called the permeability of free space. By taking C to be a circle with radius r, show that the magnitude $B = |\mathbf{B}|$ of the magnetic field at a distance r from the center of the wire is

$$B = \frac{\mu_0 I}{2\pi r}$$

16.3   The Fundamental Theorem for Line Integrals

Recall from Section 5.3 that Part 2 of the Fundamental Theorem of Calculus can be written as

$$\boxed{1}\qquad \int_a^b F'(x)\, dx = F(b) - F(a)$$

where F' is continuous on $[a, b]$. We also called Equation 1 the Net Change Theorem: The integral of a rate of change is the net change.

If we think of the gradient vector ∇f of a function f of two or three variables as a sort of derivative of f, then the following theorem can be regarded as a version of the Fundamental Theorem for line integrals.

> $\boxed{2}$ **Theorem** Let C be a smooth curve given by the vector function $\mathbf{r}(t)$, $a \le t \le b$. Let f be a differentiable function of two or three variables whose gradient vector ∇f is continuous on C. Then
> $$\int_C \nabla f \cdot d\mathbf{r} = f(\mathbf{r}(b)) - f(\mathbf{r}(a))$$

NOTE Theorem 2 says that we can evaluate the line integral of a conservative vector field (the gradient vector field of the potential function f) simply by knowing the value of f at the endpoints of C. In fact, Theorem 2 says that the line integral of ∇f is the net change in f. If f is a function of two variables and C is a plane curve with initial point $A(x_1, y_1)$ and terminal point $B(x_2, y_2)$, as in Figure 1(a), then Theorem 2 becomes

$$\int_C \nabla f \cdot d\mathbf{r} = f(x_2, y_2) - f(x_1, y_1)$$

If f is a function of three variables and C is a space curve joining the point $A(x_1, y_1, z_1)$ to the point $B(x_2, y_2, z_2)$, as in Figure 1(b), then we have

$$\int_C \nabla f \cdot d\mathbf{r} = f(x_2, y_2, z_2) - f(x_1, y_1, z_1)$$

Let's prove Theorem 2 for this case.

(a)

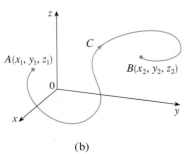

(b)

FIGURE 1

PROOF OF THEOREM 2 Using Definition 16.2.13, we have

$$\int_C \nabla f \cdot d\mathbf{r} = \int_a^b \nabla f(\mathbf{r}(t)) \cdot \mathbf{r}'(t) \, dt$$

$$= \int_a^b \left(\frac{\partial f}{\partial x} \frac{dx}{dt} + \frac{\partial f}{\partial y} \frac{dy}{dt} + \frac{\partial f}{\partial z} \frac{dz}{dt} \right) dt$$

$$= \int_a^b \frac{d}{dt} f(\mathbf{r}(t)) \, dt \qquad \text{(by the Chain Rule)}$$

$$= f(\mathbf{r}(b)) - f(\mathbf{r}(a))$$

The last step follows from the Fundamental Theorem of Calculus (Equation 1). ∎

Although we have proved Theorem 2 for smooth curves, it is also true for piecewise-smooth curves. This can be seen by subdividing C into a finite number of smooth curves and adding the resulting integrals.

EXAMPLE 1 Find the work done by the gravitational field

$$\mathbf{F}(\mathbf{x}) = -\frac{mMG}{|\mathbf{x}|^3} \mathbf{x}$$

in moving a particle with mass m from the point $(3, 4, 12)$ to the point $(2, 2, 0)$ along a piecewise-smooth curve C. (See Example 16.1.4.)

SOLUTION From Section 16.1 we know that $\mathbf{F}$ is a conservative vector field and, in fact, $\mathbf{F} = \nabla f$, where

$$f(x, y, z) = \frac{mMG}{\sqrt{x^2 + y^2 + z^2}}$$

Therefore, by Theorem 2, the work done is

$$W = \int_C \mathbf{F} \cdot d\mathbf{r} = \int_C \nabla f \cdot d\mathbf{r}$$

$$= f(2, 2, 0) - f(3, 4, 12)$$

$$= \frac{mMG}{\sqrt{2^2 + 2^2}} - \frac{mMG}{\sqrt{3^2 + 4^2 + 12^2}} = mMG \left(\frac{1}{2\sqrt{2}} - \frac{1}{13} \right)$$ ∎

■ Independence of Path

Suppose C_1 and C_2 are two piecewise-smooth curves (which are called **paths**) that have the same initial point A and terminal point B. We know from Example 16.2.4 that, in general, $\int_{C_1} \mathbf{F} \cdot d\mathbf{r} \neq \int_{C_2} \mathbf{F} \cdot d\mathbf{r}$. But one implication of Theorem 2 is that

$$\int_{C_1} \nabla f \cdot d\mathbf{r} = \int_{C_2} \nabla f \cdot d\mathbf{r}$$

whenever ∇f is continuous. In other words, the line integral of a *conservative* vector field depends only on the initial point and terminal point of a curve.

In general, if $\mathbf{F}$ is a continuous vector field with domain D, we say that the line integral $\int_C \mathbf{F} \cdot d\mathbf{r}$ is **independent of path** if $\int_{C_1} \mathbf{F} \cdot d\mathbf{r} = \int_{C_2} \mathbf{F} \cdot d\mathbf{r}$ for any two paths C_1 and C_2 in D that have the same initial points and the same terminal points. With this terminology we can say that *line integrals of conservative vector fields are independent of path.*

FIGURE 2
A closed curve

FIGURE 3

A curve is called **closed** if its terminal point coincides with its initial point, that is, $\mathbf{r}(b) = \mathbf{r}(a)$. (See Figure 2.) If $\int_C \mathbf{F} \cdot d\mathbf{r}$ is independent of path in D and C is any closed path in D, we can choose any two points A and B on C and regard C as being composed of the path C_1 from A to B followed by the path C_2 from B to A. (See Figure 3.) Then

$$\int_C \mathbf{F} \cdot d\mathbf{r} = \int_{C_1} \mathbf{F} \cdot d\mathbf{r} + \int_{C_2} \mathbf{F} \cdot d\mathbf{r} = \int_{C_1} \mathbf{F} \cdot d\mathbf{r} - \int_{-C_2} \mathbf{F} \cdot d\mathbf{r} = 0$$

since C_1 and $-C_2$ have the same initial and terminal points.

Conversely, if it is true that $\int_C \mathbf{F} \cdot d\mathbf{r} = 0$ whenever C is a closed path in D, then we demonstrate independence of path as follows. Take any two paths C_1 and C_2 from A to B in D and define C to be the curve consisting of C_1 followed by $-C_2$. Then

$$0 = \int_C \mathbf{F} \cdot d\mathbf{r} = \int_{C_1} \mathbf{F} \cdot d\mathbf{r} + \int_{-C_2} \mathbf{F} \cdot d\mathbf{r} = \int_{C_1} \mathbf{F} \cdot d\mathbf{r} - \int_{C_2} \mathbf{F} \cdot d\mathbf{r}$$

and so $\int_{C_1} \mathbf{F} \cdot d\mathbf{r} = \int_{C_2} \mathbf{F} \cdot d\mathbf{r}$. Thus we have proved the following theorem.

> **3 Theorem** $\int_C \mathbf{F} \cdot d\mathbf{r}$ is independent of path in D if and only if $\int_C \mathbf{F} \cdot d\mathbf{r} = 0$ for every closed path C in D.

Since we know that the line integral of any conservative vector field $\mathbf{F}$ is independent of path, it follows that $\int_C \mathbf{F} \cdot d\mathbf{r} = 0$ for any closed path. The physical interpretation is that the work done by a conservative force field (such as the gravitational or electric field in Section 16.1) as it moves an object around a closed path is 0.

The following theorem says that the *only* vector fields that are independent of path are conservative. It is stated and proved for plane curves, but there is a similar version for space curves. We assume that D is **open**, which means that for every point P in D there is a disk with center P that lies entirely in D. (So D doesn't contain any of its boundary points.) In addition, we assume that D is **connected**: this means that any two points in D can be joined by a path that lies in D.

> **4 Theorem** Suppose $\mathbf{F}$ is a vector field that is continuous on an open connected region D. If $\int_C \mathbf{F} \cdot d\mathbf{r}$ is independent of path in D, then $\mathbf{F}$ is a conservative vector field on D; that is, there exists a function f such that $\nabla f = \mathbf{F}$.

PROOF Let $A(a, b)$ be a fixed point in D. We construct the desired potential function f by defining

$$f(x, y) = \int_{(a, b)}^{(x, y)} \mathbf{F} \cdot d\mathbf{r}$$

for any point (x, y) in D. Since $\int_C \mathbf{F} \cdot d\mathbf{r}$ is independent of path, it does not matter which path C from (a, b) to (x, y) is used to evaluate $f(x, y)$. Since D is open, there exists a disk contained in D with center (x, y). Choose any point (x_1, y) in the disk with $x_1 < x$ and let C consist of any path C_1 from (a, b) to (x_1, y) followed by the horizontal line segment C_2 from (x_1, y) to (x, y). (See Figure 4.) Then

$$f(x, y) = \int_{C_1} \mathbf{F} \cdot d\mathbf{r} + \int_{C_2} \mathbf{F} \cdot d\mathbf{r} = \int_{(a, b)}^{(x_1, y)} \mathbf{F} \cdot d\mathbf{r} + \int_{C_2} \mathbf{F} \cdot d\mathbf{r}$$

Notice that the first of these integrals does not depend on x, so

$$\frac{\partial}{\partial x} f(x, y) = 0 + \frac{\partial}{\partial x} \int_{C_2} \mathbf{F} \cdot d\mathbf{r}$$

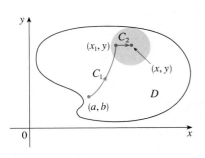

FIGURE 4

If we write $\mathbf{F} = P\,\mathbf{i} + Q\,\mathbf{j}$, then

$$\int_{C_2} \mathbf{F} \cdot d\mathbf{r} = \int_{C_2} P\,dx + Q\,dy$$

On C_2, y is constant, so $dy = 0$. Using t as the parameter, where $x_1 \leqslant t \leqslant x$, we have

$$\frac{\partial}{\partial x}\,f(x, y) = \frac{\partial}{\partial x}\int_{C_2} P\,dx + Q\,dy = \frac{\partial}{\partial x}\int_{x_1}^{x} P(t, y)\,dt = P(x, y)$$

by Part 1 of the Fundamental Theorem of Calculus (see Section 5.3). A similar argument, using a vertical line segment (see Figure 5), shows that

$$\frac{\partial}{\partial y}\,f(x, y) = \frac{\partial}{\partial y}\int_{C_2} P\,dx + Q\,dy = \frac{\partial}{\partial y}\int_{y_1}^{y} Q(x, t)\,dt = Q(x, y)$$

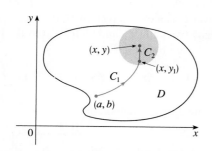

FIGURE 5

Thus
$$\mathbf{F} = P\,\mathbf{i} + Q\,\mathbf{j} = \frac{\partial f}{\partial x}\,\mathbf{i} + \frac{\partial f}{\partial y}\,\mathbf{j} = \nabla f$$

which says that $\mathbf{F}$ is conservative. ∎

The question remains: how is it possible to determine whether or not a vector field $\mathbf{F}$ is conservative? Suppose it is known that $\mathbf{F} = P\,\mathbf{i} + Q\,\mathbf{j}$ is conservative, where P and Q have continuous first-order partial derivatives. Then there is a function f such that $\mathbf{F} = \nabla f$, that is,

$$P = \frac{\partial f}{\partial x} \qquad \text{and} \qquad Q = \frac{\partial f}{\partial y}$$

Therefore, by Clairaut's Theorem,

$$\frac{\partial P}{\partial y} = \frac{\partial^2 f}{\partial y\,\partial x} = \frac{\partial^2 f}{\partial x\,\partial y} = \frac{\partial Q}{\partial x}$$

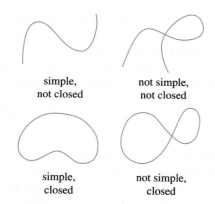

simple,
not closed

not simple,
not closed

simple,
closed

not simple,
closed

FIGURE 6
Types of curves

> **5** **Theorem** If $\mathbf{F}(x, y) = P(x, y)\,\mathbf{i} + Q(x, y)\,\mathbf{j}$ is a conservative vector field, where P and Q have continuous first-order partial derivatives on a domain D, then throughout D we have
>
> $$\frac{\partial P}{\partial y} = \frac{\partial Q}{\partial x}$$

simply-connected region

regions that are not simply-connected

FIGURE 7

The converse of Theorem 5 is true only for a special type of region. To explain this, we first need the concept of a **simple curve**, which is a curve that doesn't intersect itself anywhere between its endpoints. [See Figure 6; $\mathbf{r}(a) = \mathbf{r}(b)$ for a simple closed curve, but $\mathbf{r}(t_1) \neq \mathbf{r}(t_2)$ when $a < t_1 < t_2 < b$.]

In Theorem 4 we needed an open connected region. For the next theorem we need a stronger condition. A **simply-connected region** in the plane is a connected region D such that every simple closed curve in D encloses only points that are in D. Notice from Figure 7 that, intuitively speaking, a simply-connected region contains no hole and can't consist of two separate pieces.

In terms of simply-connected regions, we can now state a partial converse to Theorem 5 that gives a convenient method for verifying that a vector field on $\mathbb{R}^2$ is conservative. The proof will be sketched in the next section as a consequence of Green's Theorem.

> **6 Theorem** Let $\mathbf{F} = P\,\mathbf{i} + Q\,\mathbf{j}$ be a vector field on an open simply-connected region D. Suppose that P and Q have continuous first-order partial derivatives and
>
> $$\frac{\partial P}{\partial y} = \frac{\partial Q}{\partial x} \qquad \text{throughout } D$$
>
> Then $\mathbf{F}$ is conservative.

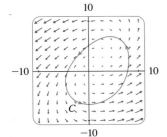

10

−10   10

C

−10

FIGURE 8

Figures 8 and 9 show the vector fields in Examples 2 and 3, respectively. The vectors in Figure 8 that start on the closed curve C all appear to point in roughly the same direction as C. So it looks as if $\int_C \mathbf{F} \cdot d\mathbf{r} > 0$ and therefore $\mathbf{F}$ is not conservative. The calculation in Example 2 confirms this impression. Some of the vectors near the curves C_1 and C_2 in Figure 9 point in approximately the same direction as the curves, whereas others point in the opposite direction. So it appears plausible that line integrals around all closed paths are 0. Example 3 shows that $\mathbf{F}$ is indeed conservative.

2

−2   2

C_1 C_2

−2

FIGURE 9

EXAMPLE 2 Determine whether or not the vector field

$$\mathbf{F}(x, y) = (x - y)\,\mathbf{i} + (x - 2)\,\mathbf{j}$$

is conservative.

SOLUTION Let $P(x, y) = x - y$ and $Q(x, y) = x - 2$. Then

$$\frac{\partial P}{\partial y} = -1 \qquad \frac{\partial Q}{\partial x} = 1$$

Since $\partial P/\partial y \neq \partial Q/\partial x$, $\mathbf{F}$ is not conservative by Theorem 5.   ∎

EXAMPLE 3 Determine whether or not the vector field

$$\mathbf{F}(x, y) = (3 + 2xy)\,\mathbf{i} + (x^2 - 3y^2)\,\mathbf{j}$$

is conservative.

SOLUTION Let $P(x, y) = 3 + 2xy$ and $Q(x, y) = x^2 - 3y^2$. Then

$$\frac{\partial P}{\partial y} = 2x = \frac{\partial Q}{\partial x}$$

Also, the domain of $\mathbf{F}$ is the entire plane ($D = \mathbb{R}^2$), which is open and simply-connected. Therefore we can apply Theorem 6 and conclude that $\mathbf{F}$ is conservative.   ∎

In Example 3, Theorem 6 told us that $\mathbf{F}$ is conservative, but it did not tell us how to find the (potential) function f such that $\mathbf{F} = \nabla f$. The proof of Theorem 4 gives us a clue as to how to find f. We use "partial integration" as in the following example.

EXAMPLE 4
(a) If $\mathbf{F}(x, y) = (3 + 2xy)\,\mathbf{i} + (x^2 - 3y^2)\,\mathbf{j}$, find a function f such that $\mathbf{F} = \nabla f$.
(b) Evaluate the line integral $\int_C \mathbf{F} \cdot d\mathbf{r}$, where C is the curve given by

$$\mathbf{r}(t) = e^t \sin t\,\mathbf{i} + e^t \cos t\,\mathbf{j} \qquad 0 \le t \le \pi$$

SOLUTION
(a) From Example 3 we know that $\mathbf{F}$ is conservative and so there exists a function f with $\nabla f = \mathbf{F}$, that is,

$$\boxed{7} \qquad\qquad f_x(x, y) = 3 + 2xy$$

$$\boxed{8} \qquad\qquad f_y(x, y) = x^2 - 3y^2$$

Integrating (7) with respect to x, we obtain

$$\boxed{9} \qquad f(x, y) = 3x + x^2 y + g(y)$$

Notice that the constant of integration is a constant with respect to x, that is, a function of y, which we have called $g(y)$. Next we differentiate both sides of (9) with respect to y:

$$\boxed{10} \qquad f_y(x, y) = x^2 + g'(y)$$

Comparing (8) and (10), we see that

$$g'(y) = -3y^2$$

Integrating with respect to y, we have

$$g(y) = -y^3 + K$$

where K is a constant. Putting this in (9), we have

$$f(x, y) = 3x + x^2 y - y^3 + K$$

as the desired potential function.

(b) To use Theorem 2 all we have to know are the initial and terminal points of C, namely, $\mathbf{r}(0) = (0, 1)$ and $\mathbf{r}(\pi) = (0, -e^\pi)$. In the expression for $f(x, y)$ in part (a), any value of the constant K will do, so let's choose $K = 0$. Then we have

$$\int_C \mathbf{F} \cdot d\mathbf{r} = \int_C \nabla f \cdot d\mathbf{r} = f(0, -e^\pi) - f(0, 1) = e^{3\pi} - (-1) = e^{3\pi} + 1$$

This method is much shorter than the straightforward method for evaluating line integrals that we learned in Section 16.2.   ∎

A criterion for determining whether or not a vector field $\mathbf{F}$ on $\mathbb{R}^3$ is conservative is given in Section 16.5. Meanwhile, the next example shows that the technique for finding the potential function is much the same as for vector fields on $\mathbb{R}^2$.

EXAMPLE 5   If $\mathbf{F}(x, y, z) = y^2\,\mathbf{i} + (2xy + e^{3z})\,\mathbf{j} + 3ye^{3z}\,\mathbf{k}$, find a function f such that $\nabla f = \mathbf{F}$.

SOLUTION   If there is such a function f, then

$$\boxed{11} \qquad f_x(x, y, z) = y^2$$

$$\boxed{12} \qquad f_y(x, y, z) = 2xy + e^{3z}$$

$$\boxed{13} \qquad f_z(x, y, z) = 3ye^{3z}$$

Integrating (11) with respect to x, we get

$$\boxed{14} \qquad f(x, y, z) = xy^2 + g(y, z)$$

where $g(y, z)$ is a constant with respect to x. Then differentiating (14) with respect to y, we have

$$f_y(x, y, z) = 2xy + g_y(y, z)$$

and comparison with (12) gives

$$g_y(y, z) = e^{3z}$$

Thus $g(y, z) = ye^{3z} + h(z)$ and we rewrite (14) as

$$f(x, y, z) = xy^2 + ye^{3z} + h(z)$$

Finally, differentiating with respect to z and comparing with (13), we obtain $h'(z) = 0$ and therefore $h(z) = K$, a constant. The desired function is

$$f(x, y, z) = xy^2 + ye^{3z} + K$$

It is easily verified that $\nabla f = \mathbf{F}$. ∎

■ Conservation of Energy

Let's apply the ideas of this chapter to a continuous force field $\mathbf{F}$ that moves an object along a path C given by $\mathbf{r}(t)$, $a \leqslant t \leqslant b$, where $\mathbf{r}(a) = A$ is the initial point and $\mathbf{r}(b) = B$ is the terminal point of C. According to Newton's Second Law of Motion (see Section 13.4), the force $\mathbf{F}(\mathbf{r}(t))$ at a point on C is related to the acceleration $\mathbf{a}(t) = \mathbf{r}''(t)$ by the equation

$$\mathbf{F}(\mathbf{r}(t)) = m\mathbf{r}''(t)$$

So the work done by the force on the object is

$$W = \int_C \mathbf{F} \cdot d\mathbf{r} = \int_a^b \mathbf{F}(\mathbf{r}(t)) \cdot \mathbf{r}'(t) \, dt = \int_a^b m\mathbf{r}''(t) \cdot \mathbf{r}'(t) \, dt$$

$$= \frac{m}{2} \int_a^b \frac{d}{dt} [\mathbf{r}'(t) \cdot \mathbf{r}'(t)] \, dt \qquad \text{(Theorem 13.2.3, Formula 4)}$$

$$= \frac{m}{2} \int_a^b \frac{d}{dt} |\mathbf{r}'(t)|^2 \, dt = \frac{m}{2} \left[|\mathbf{r}'(t)|^2 \right]_a^b \qquad \text{(Fundamental Theorem of Calculus)}$$

$$= \frac{m}{2} \left(|\mathbf{r}'(b)|^2 - |\mathbf{r}'(a)|^2 \right)$$

Therefore

$$\boxed{15} \qquad \qquad W = \tfrac{1}{2}m |\mathbf{v}(b)|^2 - \tfrac{1}{2}m |\mathbf{v}(a)|^2$$

where $\mathbf{v} = \mathbf{r}'$ is the velocity.

The quantity $\tfrac{1}{2}m |\mathbf{v}(t)|^2$, that is, half the mass times the square of the speed, is called the **kinetic energy** of the object. Therefore we can rewrite Equation 15 as

$$\boxed{16} \qquad \qquad W = K(B) - K(A)$$

which says that the work done by the force field along C is equal to the change in kinetic energy at the endpoints of C.

Now let's further assume that $\mathbf{F}$ is a conservative force field; that is, we can write $\mathbf{F} = \nabla f$. In physics, the **potential energy** of an object at the point (x, y, z) is defined as $P(x, y, z) = -f(x, y, z)$, so we have $\mathbf{F} = -\nabla P$. Then by Theorem 2 we have

$$W = \int_C \mathbf{F} \cdot d\mathbf{r} = -\int_C \nabla P \cdot d\mathbf{r} = -[P(\mathbf{r}(b)) - P(\mathbf{r}(a))] = P(A) - P(B)$$

Comparing this equation with Equation 16, we see that

$$P(A) + K(A) = P(B) + K(B)$$

which says that if an object moves from one point A to another point B under the influence of a conservative force field, then the sum of its potential energy and its kinetic energy remains constant. This is called the **Law of Conservation of Energy** and it is the reason the vector field is called *conservative*.

16.3 EXERCISES

1. The figure shows a curve C and a contour map of a function f whose gradient is continuous. Find $\int_C \nabla f \cdot d\mathbf{r}$.

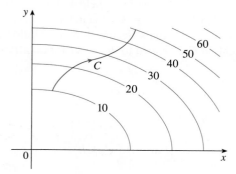

2. A table of values of a function f with continuous gradient is given. Find $\int_C \nabla f \cdot d\mathbf{r}$, where C has parametric equations

$$x = t^2 + 1 \qquad y = t^3 + t \qquad 0 \leqslant t \leqslant 1$$

x \ y	0	1	2
0	1	6	4
1	3	5	7
2	8	2	9

3–10 Determine whether or not $\mathbf{F}$ is a conservative vector field. If it is, find a function f such that $\mathbf{F} = \nabla f$.

3. $\mathbf{F}(x, y) = (xy + y^2)\,\mathbf{i} + (x^2 + 2xy)\,\mathbf{j}$

4. $\mathbf{F}(x, y) = (y^2 - 2x)\,\mathbf{i} + 2xy\,\mathbf{j}$

5. $\mathbf{F}(x, y) = y^2 e^{xy}\,\mathbf{i} + (1 + xy)e^{xy}\,\mathbf{j}$

6. $\mathbf{F}(x, y) = ye^x\,\mathbf{i} + (e^x + e^y)\,\mathbf{j}$

7. $\mathbf{F}(x, y) = (ye^x + \sin y)\,\mathbf{i} + (e^x + x \cos y)\,\mathbf{j}$

8. $\mathbf{F}(x, y) = (2xy + y^{-2})\,\mathbf{i} + (x^2 - 2xy^{-3})\,\mathbf{j}, \quad y > 0$

9. $\mathbf{F}(x, y) = (y^2 \cos x + \cos y)\,\mathbf{i} + (2y \sin x - x \sin y)\,\mathbf{j}$

10. $\mathbf{F}(x, y) = (\ln y + y/x)\,\mathbf{i} + (\ln x + x/y)\,\mathbf{j}$

11. The figure shows the vector field $\mathbf{F}(x, y) = \langle 2xy, x^2 \rangle$ and three curves that start at $(1, 2)$ and end at $(3, 2)$.
   (a) Explain why $\int_C \mathbf{F} \cdot d\mathbf{r}$ has the same value for all three curves.
   (b) What is this common value?

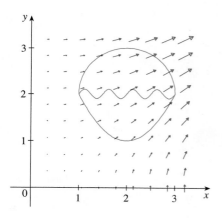

12–18 (a) Find a function f such that $\mathbf{F} = \nabla f$ and (b) use part (a) to evaluate $\int_C \mathbf{F} \cdot d\mathbf{r}$ along the given curve C.

12. $\mathbf{F}(x, y) = (3 + 2xy^2)\,\mathbf{i} + 2x^2 y\,\mathbf{j}$,
   C is the arc of the hyperbola $y = 1/x$ from $(1, 1)$ to $\left(4, \frac{1}{4}\right)$

13. $\mathbf{F}(x, y) = x^2 y^3\,\mathbf{i} + x^3 y^2\,\mathbf{j}$,
   C: $\mathbf{r}(t) = \langle t^3 - 2t, t^3 + 2t \rangle$, $\quad 0 \leqslant t \leqslant 1$

14. $\mathbf{F}(x, y) = (1 + xy)e^{xy}\,\mathbf{i} + x^2 e^{xy}\,\mathbf{j}$,
   C: $\mathbf{r}(t) = \cos t\,\mathbf{i} + 2 \sin t\,\mathbf{j}$, $\quad 0 \leqslant t \leqslant \pi/2$

15. $\mathbf{F}(x, y, z) = yz\,\mathbf{i} + xz\,\mathbf{j} + (xy + 2z)\,\mathbf{k}$,
   C is the line segment from $(1, 0, -2)$ to $(4, 6, 3)$

16. $\mathbf{F}(x, y, z) = (y^2z + 2xz^2)\,\mathbf{i} + 2xyz\,\mathbf{j} + (xy^2 + 2x^2z)\,\mathbf{k}$,
C: $x = \sqrt{t}$, $y = t + 1$, $z = t^2$, $0 \le t \le 1$

17. $\mathbf{F}(x, y, z) = yze^{xz}\,\mathbf{i} + e^{xz}\,\mathbf{j} + xye^{xz}\,\mathbf{k}$,
C: $\mathbf{r}(t) = (t^2 + 1)\,\mathbf{i} + (t^2 - 1)\,\mathbf{j} + (t^2 - 2t)\,\mathbf{k}$,
$0 \le t \le 2$

18. $\mathbf{F}(x, y, z) = \sin y\,\mathbf{i} + (x \cos y + \cos z)\,\mathbf{j} - y \sin z\,\mathbf{k}$,
C: $\mathbf{r}(t) = \sin t\,\mathbf{i} + t\,\mathbf{j} + 2t\,\mathbf{k}$, $0 \le t \le \pi/2$

19–20 Show that the line integral is independent of path and evaluate the integral.

19. $\int_C 2xe^{-y}\,dx + (2y - x^2e^{-y})\,dy$,
C is any path from $(1, 0)$ to $(2, 1)$

20. $\int_C \sin y\,dx + (x \cos y - \sin y)\,dy$,
C is any path from $(2, 0)$ to $(1, \pi)$

---

21. Suppose you're asked to determine the curve that requires the least work for a force field $\mathbf{F}$ to move a particle from one point to another point. You decide to check first whether $\mathbf{F}$ is conservative, and indeed it turns out that it is. How would you reply to the request?

22. Suppose an experiment determines that the amount of work required for a force field $\mathbf{F}$ to move a particle from the point $(1, 2)$ to the point $(5, -3)$ along a curve C_1 is 1.2 J and the work done by $\mathbf{F}$ in moving the particle along another curve C_2 between the same two points is 1.4 J. What can you say about $\mathbf{F}$? Why?

23–24 Find the work done by the force field $\mathbf{F}$ in moving an object from P to Q.

23. $\mathbf{F}(x, y) = x^3\,\mathbf{i} + y^3\,\mathbf{j}$; $P(1, 0)$, $Q(2, 2)$

24. $\mathbf{F}(x, y) = (2x + y)\,\mathbf{i} + x\,\mathbf{j}$; $P(1, 1)$, $Q(4, 3)$

25–26 Is the vector field shown in the figure conservative? Explain.

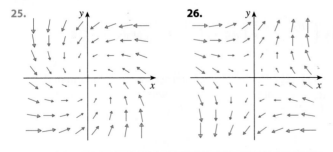

25.

26.

CAS **27.** If $\mathbf{F}(x, y) = \sin y\,\mathbf{i} + (1 + x \cos y)\,\mathbf{j}$, use a plot to guess whether $\mathbf{F}$ is conservative. Then determine whether your guess is correct.

28. Let $\mathbf{F} = \nabla f$, where $f(x, y) = \sin(x - 2y)$. Find curves C_1 and C_2 that are not closed and satisfy the equation.

(a) $\int_{C_1} \mathbf{F} \cdot d\mathbf{r} = 0$ (b) $\int_{C_2} \mathbf{F} \cdot d\mathbf{r} = 1$

29. Show that if the vector field $\mathbf{F} = P\,\mathbf{i} + Q\,\mathbf{j} + R\,\mathbf{k}$ is conservative and P, Q, R have continuous first-order partial derivatives, then

$$\frac{\partial P}{\partial y} = \frac{\partial Q}{\partial x} \qquad \frac{\partial P}{\partial z} = \frac{\partial R}{\partial x} \qquad \frac{\partial Q}{\partial z} = \frac{\partial R}{\partial y}$$

30. Use Exercise 29 to show that the line integral $\int_C y\,dx + x\,dy + xyz\,dz$ is not independent of path.

31–34 Determine whether or not the given set is (a) open, (b) connected, and (c) simply-connected.

31. $\{(x, y) \mid 0 < y < 3\}$ **32.** $\{(x, y) \mid 1 < |x| < 2\}$

33. $\{(x, y) \mid 1 \le x^2 + y^2 \le 4,\ y \ge 0\}$

34. $\{(x, y) \mid (x, y) \ne (2, 3)\}$

---

35. Let $\mathbf{F}(x, y) = \dfrac{-y\,\mathbf{i} + x\,\mathbf{j}}{x^2 + y^2}$.

(a) Show that $\partial P/\partial y = \partial Q/\partial x$.

(b) Show that $\int_C \mathbf{F} \cdot d\mathbf{r}$ is not independent of path. [*Hint:* Compute $\int_{C_1} \mathbf{F} \cdot d\mathbf{r}$ and $\int_{C_2} \mathbf{F} \cdot d\mathbf{r}$, where C_1 and C_2 are the upper and lower halves of the circle $x^2 + y^2 = 1$ from $(1, 0)$ to $(-1, 0)$.] Does this contradict Theorem 6?

36. (a) Suppose that $\mathbf{F}$ is an inverse square force field, that is,

$$\mathbf{F}(\mathbf{r}) = \frac{c\mathbf{r}}{|\mathbf{r}|^3}$$

for some constant c, where $\mathbf{r} = x\,\mathbf{i} + y\,\mathbf{j} + z\,\mathbf{k}$. Find the work done by $\mathbf{F}$ in moving an object from a point P_1 along a path to a point P_2 in terms of the distances d_1 and d_2 from these points to the origin.

(b) An example of an inverse square field is the gravitational field $\mathbf{F} = -(mMG)\mathbf{r}/|\mathbf{r}|^3$ discussed in Example 16.1.4. Use part (a) to find the work done by the gravitational field when the earth moves from aphelion (at a maximum distance of 1.52×10^8 km from the sun) to perihelion (at a minimum distance of 1.47×10^8 km). (Use the values $m = 5.97 \times 10^{24}$ kg, $M = 1.99 \times 10^{30}$ kg, and $G = 6.67 \times 10^{-11}$ N·m^2/kg^2.)

(c) Another example of an inverse square field is the electric force field $\mathbf{F} = \varepsilon qQ\mathbf{r}/|\mathbf{r}|^3$ discussed in Example 16.1.5. Suppose that an electron with a charge of -1.6×10^{-19} C is located at the origin. A positive unit charge is positioned a distance 10^{-12} m from the electron and moves to a position half that distance from the electron. Use part (a) to find the work done by the electric force field. (Use the value $\varepsilon = 8.985 \times 10^9$.)

16.4 Green's Theorem

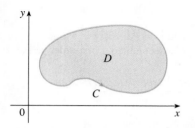

FIGURE 1

Green's Theorem gives the relationship between a line integral around a simple closed curve C and a double integral over the plane region D bounded by C. (See Figure 1. We assume that D consists of all points inside C as well as all points on C.) In stating Green's Theorem we use the convention that the **positive orientation** of a simple closed curve C refers to a single *counterclockwise* traversal of C. Thus if C is given by the vector function $\mathbf{r}(t)$, $a \leqslant t \leqslant b$, then the region D is always on the left as the point $\mathbf{r}(t)$ traverses C. (See Figure 2.)

(a) Positive orientation   (b) Negative orientation

FIGURE 2

> **Green's Theorem** Let C be a positively oriented, piecewise-smooth, simple closed curve in the plane and let D be the region bounded by C. If P and Q have continuous partial derivatives on an open region that contains D, then
>
> $$\int_C P\,dx + Q\,dy = \iint_D \left(\frac{\partial Q}{\partial x} - \frac{\partial P}{\partial y} \right) dA$$

Recall that the left side of this equation is another way of writing $\int_C \mathbf{F} \cdot d\mathbf{r}$, where $\mathbf{F} = P\,\mathbf{i} + Q\,\mathbf{j}$.

NOTE The notation

$$\oint_C P\,dx + Q\,dy \qquad \text{or} \qquad \oint_C P\,dx + Q\,dy$$

is sometimes used to indicate that the line integral is calculated using the positive orientation of the closed curve C. Another notation for the positively oriented boundary curve of D is ∂D, so the equation in Green's Theorem can be written as

$$\boxed{1} \qquad \iint_D \left(\frac{\partial Q}{\partial x} - \frac{\partial P}{\partial y} \right) dA = \int_{\partial D} P\,dx + Q\,dy$$

Green's Theorem should be regarded as the counterpart of the Fundamental Theorem of Calculus for double integrals. Compare Equation 1 with the statement of the Fundamental Theorem of Calculus, Part 2, in the following equation:

$$\int_a^b F'(x)\,dx = F(b) - F(a)$$

In both cases there is an integral involving derivatives (F', $\partial Q/\partial x$, and $\partial P/\partial y$) on the left side of the equation. And in both cases the right side involves the values of the original functions (F, Q, and P) only on the *boundary* of the domain. (In the one-dimensional case, the domain is an interval $[a, b]$ whose boundary consists of just two points, a and b.)

Green's Theorem is not easy to prove in general, but we can give a proof for the special case where the region is both type I and type II (see Section 15.2). Let's call such regions **simple regions**.

PROOF OF GREEN'S THEOREM FOR THE CASE IN WHICH D IS A SIMPLE REGION Notice that Green's Theorem will be proved if we can show that

$$\boxed{2} \qquad \int_C P \, dx = -\iint_D \frac{\partial P}{\partial y} \, dA$$

and

$$\boxed{3} \qquad \int_C Q \, dy = \iint_D \frac{\partial Q}{\partial x} \, dA$$

We prove Equation 2 by expressing D as a type I region:

$$D = \{(x, y) \mid a \leqslant x \leqslant b, \ g_1(x) \leqslant y \leqslant g_2(x)\}$$

where g_1 and g_2 are continuous functions. This enables us to compute the double integral on the right side of Equation 2 as follows:

$$\boxed{4} \qquad \iint_D \frac{\partial P}{\partial y} \, dA = \int_a^b \int_{g_1(x)}^{g_2(x)} \frac{\partial P}{\partial y}(x, y) \, dy \, dx = \int_a^b [P(x, g_2(x)) - P(x, g_1(x))] \, dx$$

where the last step follows from the Fundamental Theorem of Calculus.

Now we compute the left side of Equation 2 by breaking up C as the union of the four curves C_1, C_2, C_3, and C_4 shown in Figure 3. On C_1 we take x as the parameter and write the parametric equations as $x = x$, $y = g_1(x)$, $a \leqslant x \leqslant b$. Thus

$$\int_{C_1} P(x, y) \, dx = \int_a^b P(x, g_1(x)) \, dx$$

Observe that C_3 goes from right to left but $-C_3$ goes from left to right, so we can write the parametric equations of $-C_3$ as $x = x$, $y = g_2(x)$, $a \leqslant x \leqslant b$. Therefore

$$\int_{C_3} P(x, y) \, dx = -\int_{-C_3} P(x, y) \, dx = -\int_a^b P(x, g_2(x)) \, dx$$

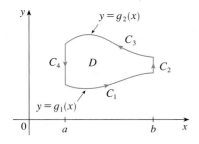

FIGURE 3

On C_2 or C_4 (either of which might reduce to just a single point), x is constant, so $dx = 0$ and

$$\int_{C_2} P(x, y) \, dx = 0 = \int_{C_4} P(x, y) \, dx$$

Hence

$$\int_C P(x, y) \, dx = \int_{C_1} P(x, y) \, dx + \int_{C_2} P(x, y) \, dx + \int_{C_3} P(x, y) \, dx + \int_{C_4} P(x, y) \, dx$$

$$= \int_a^b P(x, g_1(x)) \, dx - \int_a^b P(x, g_2(x)) \, dx$$

Comparing this expression with the one in Equation 4, we see that

$$\int_C P(x, y)\, dx = -\iint_D \frac{\partial P}{\partial y}\, dA$$

Equation 3 can be proved in much the same way by expressing D as a type II region (see Exercise 30). Then, by adding Equations 2 and 3, we obtain Green's Theorem. ∎

EXAMPLE 1 Evaluate $\int_C x^4\, dx + xy\, dy$, where C is the triangular curve consisting of the line segments from $(0, 0)$ to $(1, 0)$, from $(1, 0)$ to $(0, 1)$, and from $(0, 1)$ to $(0, 0)$.

SOLUTION Although the given line integral could be evaluated as usual by the methods of Section 16.2, that would involve setting up three separate integrals along the three sides of the triangle, so let's use Green's Theorem instead. Notice that the region D enclosed by C is simple and C has positive orientation (see Figure 4). If we let $P(x, y) = x^4$ and $Q(x, y) = xy$, then we have

$$\int_C x^4\, dx + xy\, dy = \iint_D \left(\frac{\partial Q}{\partial x} - \frac{\partial P}{\partial y} \right) dA = \int_0^1 \int_0^{1-x} (y - 0)\, dy\, dx$$

$$= \int_0^1 \left[\tfrac{1}{2} y^2 \right]_{y=0}^{y=1-x} dx = \tfrac{1}{2} \int_0^1 (1 - x)^2\, dx$$

$$= -\tfrac{1}{6}(1 - x)^3 \Big]_0^1 = \tfrac{1}{6}$$

∎

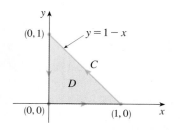

FIGURE 4

EXAMPLE 2 Evaluate $\oint_C (3y - e^{\sin x})\, dx + \left(7x + \sqrt{y^4 + 1}\right) dy$, where C is the circle $x^2 + y^2 = 9$.

SOLUTION The region D bounded by C is the disk $x^2 + y^2 \le 9$, so let's change to polar coordinates after applying Green's Theorem:

$$\oint_C (3y - e^{\sin x})\, dx + \left(7x + \sqrt{y^4 + 1}\right) dy$$

$$= \iint_D \left[\frac{\partial}{\partial x}\left(7x + \sqrt{y^4 + 1}\right) - \frac{\partial}{\partial y}(3y - e^{\sin x}) \right] dA$$

$$= \int_0^{2\pi} \int_0^3 (7 - 3)\, r\, dr\, d\theta = 4 \int_0^{2\pi} d\theta \int_0^3 r\, dr = 36\pi$$

∎

Instead of using polar coordinates, we could simply use the fact that D is a disk of radius 3 and write

$$\iint_D 4\, dA = 4 \cdot \pi (3)^2 = 36\pi$$

In Examples 1 and 2 we found that the double integral was easier to evaluate than the line integral. (Try setting up the line integral in Example 2 and you'll soon be convinced!) But sometimes it's easier to evaluate the line integral, and Green's Theorem is used in the reverse direction. For instance, if it is known that $P(x, y) = Q(x, y) = 0$ on the curve C, then Green's Theorem gives

$$\iint_D \left(\frac{\partial Q}{\partial x} - \frac{\partial P}{\partial y} \right) dA = \int_C P\, dx + Q\, dy = 0$$

no matter what values P and Q assume in the region D.

Another application of the reverse direction of Green's Theorem is in computing areas. Since the area of D is $\iint_D 1\, dA$, we wish to choose P and Q so that

$$\frac{\partial Q}{\partial x} - \frac{\partial P}{\partial y} = 1$$

There are several possibilities:

$$P(x, y) = 0 \qquad\qquad P(x, y) = -y \qquad\qquad P(x, y) = -\tfrac{1}{2}y$$

$$Q(x, y) = x \qquad\qquad Q(x, y) = 0 \qquad\qquad Q(x, y) = \tfrac{1}{2}x$$

Then Green's Theorem gives the following formulas for the area of D:

$$\boxed{5} \qquad \boxed{A = \oint_C x \, dy = -\oint_C y \, dx = \tfrac{1}{2}\oint_C x \, dy - y \, dx}$$

EXAMPLE 3 Find the area enclosed by the ellipse $\dfrac{x^2}{a^2} + \dfrac{y^2}{b^2} = 1$.

SOLUTION The ellipse has parametric equations $x = a \cos t$ and $y = b \sin t$, where $0 \le t \le 2\pi$. Using the third formula in Equation 5, we have

$$A = \tfrac{1}{2}\int_C x \, dy - y \, dx$$

$$= \tfrac{1}{2}\int_0^{2\pi} (a \cos t)(b \cos t) \, dt - (b \sin t)(-a \sin t) \, dt$$

$$= \frac{ab}{2}\int_0^{2\pi} dt = \pi ab \qquad\blacksquare$$

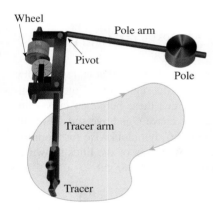

Wheel

Pole arm

Pivot

Pole

Tracer arm

Tracer

FIGURE 5
A Keuffel and Esser polar planimeter

Formula 5 can be used to explain how planimeters work. A **planimeter** is a mechanical instrument used for measuring the area of a region by tracing its boundary curve. These devices are useful in all the sciences: in biology for measuring the area of leaves or wings, in medicine for measuring the size of cross-sections of organs or tumors, in forestry for estimating the size of forested regions from photographs.

Figure 5 shows the operation of a polar planimeter: the pole is fixed and, as the tracer is moved along the boundary curve of the region, the wheel partly slides and partly rolls perpendicular to the tracer arm. The planimeter measures the distance that the wheel rolls and this is proportional to the area of the enclosed region. The explanation as a consequence of Formula 5 can be found in the following articles:

- R. W. Gatterman, "The planimeter as an example of Green's Theorem" *Amer. Math. Monthly,* Vol. 88 (1981), pp. 701–4.
- Tanya Leise, "As the planimeter wheel turns" *College Math. Journal,* Vol. 38 (2007), pp. 24–31.

■ Extended Versions of Green's Theorem

Although we have proved Green's Theorem only for the case where D is simple, we can now extend it to the case where D is a finite union of simple regions. For example, if D is the region shown in Figure 6, then we can write $D = D_1 \cup D_2$, where D_1 and D_2 are both simple. The boundary of D_1 is $C_1 \cup C_3$ and the boundary of D_2 is $C_2 \cup (-C_3)$ so, applying Green's Theorem to D_1 and D_2 separately, we get

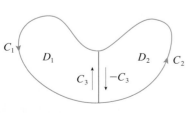

C_1 D_1 D_2 C_2

C_3 $-C_3$

FIGURE 6

$$\int_{C_1 \cup C_3} P \, dx + Q \, dy = \iint_{D_1} \left(\frac{\partial Q}{\partial x} - \frac{\partial P}{\partial y} \right) dA$$

$$\int_{C_2 \cup (-C_3)} P \, dx + Q \, dy = \iint_{D_2} \left(\frac{\partial Q}{\partial x} - \frac{\partial P}{\partial y} \right) dA$$

FIGURE 7

FIGURE 8

FIGURE 9

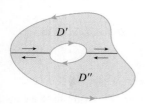

FIGURE 10

If we add these two equations, the line integrals along C_3 and $-C_3$ cancel, so we get

$$\int_{C_1 \cup C_2} P\,dx + Q\,dy = \iint_D \left(\frac{\partial Q}{\partial x} - \frac{\partial P}{\partial y} \right) dA$$

which is Green's Theorem for $D = D_1 \cup D_2$, since its boundary is $C = C_1 \cup C_2$.

The same sort of argument allows us to establish Green's Theorem for any finite union of nonoverlapping simple regions (see Figure 7).

EXAMPLE 4 Evaluate $\oint_C y^2\,dx + 3xy\,dy$, where C is the boundary of the semiannular region D in the upper half-plane between the circles $x^2 + y^2 = 1$ and $x^2 + y^2 = 4$.

SOLUTION Notice that although D is not simple, the y-axis divides it into two simple regions (see Figure 8). In polar coordinates we can write

$$D = \left\{ (r, \theta) \mid 1 \leqslant r \leqslant 2,\ 0 \leqslant \theta \leqslant \pi \right\}$$

Therefore Green's Theorem gives

$$\oint_C y^2\,dx + 3xy\,dy = \iint_D \left[\frac{\partial}{\partial x}(3xy) - \frac{\partial}{\partial y}(y^2) \right] dA$$

$$= \iint_D y\,dA = \int_0^\pi \int_1^2 (r\sin\theta)\,r\,dr\,d\theta$$

$$= \int_0^\pi \sin\theta\,d\theta \int_1^2 r^2\,dr = \left[-\cos\theta \right]_0^\pi \left[\tfrac{1}{3} r^3 \right]_1^2 = \frac{14}{3} \quad\blacksquare$$

Green's Theorem can be extended to apply to regions with holes, that is, regions that are not simply-connected. Observe that the boundary C of the region D in Figure 9 consists of two simple closed curves C_1 and C_2. We assume that these boundary curves are oriented so that the region D is always on the left as the curve C is traversed. Thus the positive direction is counterclockwise for the outer curve C_1 but clockwise for the inner curve C_2. If we divide D into two regions D' and D'' by means of the lines shown in Figure 10 and then apply Green's Theorem to each of D' and D'', we get

$$\iint_D \left(\frac{\partial Q}{\partial x} - \frac{\partial P}{\partial y} \right) dA = \iint_{D'} \left(\frac{\partial Q}{\partial x} - \frac{\partial P}{\partial y} \right) dA + \iint_{D''} \left(\frac{\partial Q}{\partial x} - \frac{\partial P}{\partial y} \right) dA$$

$$= \int_{\partial D'} P\,dx + Q\,dy + \int_{\partial D''} P\,dx + Q\,dy$$

Since the line integrals along the common boundary lines are in opposite directions, they cancel and we get

$$\iint_D \left(\frac{\partial Q}{\partial x} - \frac{\partial P}{\partial y} \right) dA = \int_{C_1} P\,dx + Q\,dy + \int_{C_2} P\,dx + Q\,dy = \int_C P\,dx + Q\,dy$$

which is Green's Theorem for the region D.

EXAMPLE 5 If $\mathbf{F}(x, y) = (-y\,\mathbf{i} + x\,\mathbf{j})/(x^2 + y^2)$, show that $\int_C \mathbf{F} \cdot d\mathbf{r} = 2\pi$ for every positively oriented simple closed path that encloses the origin.

SOLUTION Since C is an *arbitrary* closed path that encloses the origin, it's difficult to compute the given integral directly. So let's consider a counterclockwise-oriented circle C'

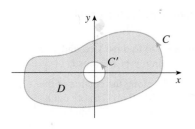

FIGURE 11

with center the origin and radius a, where a is chosen to be small enough that C' lies inside C. (See Figure 11.) Let D be the region bounded by C and C'. Then its positively oriented boundary is $C \cup (-C')$ and so the general version of Green's Theorem gives

$$\int_C P\,dx + Q\,dy + \int_{-C'} P\,dx + Q\,dy = \iint_D \left(\frac{\partial Q}{\partial x} - \frac{\partial P}{\partial y} \right) dA$$

$$= \iint_D \left[\frac{y^2 - x^2}{(x^2 + y^2)^2} - \frac{y^2 - x^2}{(x^2 + y^2)^2} \right] dA = 0$$

Therefore

$$\int_C P\,dx + Q\,dy = \int_{C'} P\,dx + Q\,dy$$

that is,

$$\int_C \mathbf{F} \cdot d\mathbf{r} = \int_{C'} \mathbf{F} \cdot d\mathbf{r}$$

We now easily compute this last integral using the parametrization given by $\mathbf{r}(t) = a \cos t\,\mathbf{i} + a \sin t\,\mathbf{j}$, $0 \le t \le 2\pi$. Thus

$$\int_C \mathbf{F} \cdot d\mathbf{r} = \int_{C'} \mathbf{F} \cdot d\mathbf{r} = \int_0^{2\pi} \mathbf{F}(\mathbf{r}(t)) \cdot \mathbf{r}'(t)\,dt$$

$$= \int_0^{2\pi} \frac{(-a \sin t)(-a \sin t) + (a \cos t)(a \cos t)}{a^2 \cos^2 t + a^2 \sin^2 t}\,dt = \int_0^{2\pi} dt = 2\pi \quad \blacksquare$$

We end this section by using Green's Theorem to discuss a result that was stated in the preceding section.

SKETCH OF PROOF OF THEOREM 16.3.6 We're assuming that $\mathbf{F} = P\,\mathbf{i} + Q\,\mathbf{j}$ is a vector field on an open simply-connected region D, that P and Q have continuous first-order partial derivatives, and that

$$\frac{\partial P}{\partial y} = \frac{\partial Q}{\partial x} \qquad \text{throughout } D$$

If C is any simple closed path in D and R is the region that C encloses, then Green's Theorem gives

$$\oint_C \mathbf{F} \cdot d\mathbf{r} = \oint_C P\,dx + Q\,dy = \iint_R \left(\frac{\partial Q}{\partial x} - \frac{\partial P}{\partial y} \right) dA = \iint_R 0\,dA = 0$$

A curve that is not simple crosses itself at one or more points and can be broken up into a number of simple curves. We have shown that the line integrals of $\mathbf{F}$ around these simple curves are all 0 and, adding these integrals, we see that $\int_C \mathbf{F} \cdot d\mathbf{r} = 0$ for any closed curve C. Therefore $\int_C \mathbf{F} \cdot d\mathbf{r}$ is independent of path in D by Theorem 16.3.3. It follows that $\mathbf{F}$ is a conservative vector field. $\blacksquare$

16.4 EXERCISES

1–4 Evaluate the line integral by two methods: (a) directly and (b) using Green's Theorem.

1. $\oint_C y^2\,dx + x^2 y\,dy$,
   C is the rectangle with vertices $(0, 0)$, $(5, 0)$, $(5, 4)$, and $(0, 4)$

2. $\oint_C y\,dx - x\,dy$,
   C is the circle with center the origin and radius 4

3. $\oint_C xy\,dx + x^2 y^3\,dy$,
   C is the triangle with vertices $(0, 0)$, $(1, 0)$, and $(1, 2)$

4. $\oint_C x^2y^2\,dx + xy\,dy$, C consists of the arc of the parabola $y = x^2$ from $(0, 0)$ to $(1, 1)$ and the line segments from $(1, 1)$ to $(0, 1)$ and from $(0, 1)$ to $(0, 0)$

5–10 Use Green's Theorem to evaluate the line integral along the given positively oriented curve.

5. $\int_C ye^x\,dx + 2e^x\,dy$,
C is the rectangle with vertices $(0, 0)$, $(3, 0)$, $(3, 4)$, and $(0, 4)$

6. $\int_C (x^2 + y^2)\,dx + (x^2 - y^2)\,dy$,
C is the triangle with vertices $(0, 0)$, $(2, 1)$, and $(0, 1)$

7. $\int_C (y + e^{\sqrt{x}})\,dx + (2x + \cos y^2)\,dy$,
C is the boundary of the region enclosed by the parabolas $y = x^2$ and $x = y^2$

8. $\int_C y^4\,dx + 2xy^3\,dy$, C is the ellipse $x^2 + 2y^2 = 2$

9. $\int_C y^3\,dx - x^3\,dy$, C is the circle $x^2 + y^2 = 4$

10. $\int_C (1 - y^3)\,dx + (x^3 + e^{y^2})\,dy$, C is the boundary of the region between the circles $x^2 + y^2 = 4$ and $x^2 + y^2 = 9$

11–14 Use Green's Theorem to evaluate $\int_C \mathbf{F} \cdot d\mathbf{r}$. (Check the orientation of the curve before applying the theorem.)

11. $\mathbf{F}(x, y) = \langle y \cos x - xy \sin x, xy + x \cos x \rangle$,
C is the triangle from $(0, 0)$ to $(0, 4)$ to $(2, 0)$ to $(0, 0)$

12. $\mathbf{F}(x, y) = \langle e^{-x} + y^2, e^{-y} + x^2 \rangle$,
C consists of the arc of the curve $y = \cos x$ from $(-\pi/2, 0)$ to $(\pi/2, 0)$ and the line segment from $(\pi/2, 0)$ to $(-\pi/2, 0)$

13. $\mathbf{F}(x, y) = \langle y - \cos y, x \sin y \rangle$,
C is the circle $(x - 3)^2 + (y + 4)^2 = 4$ oriented clockwise

14. $\mathbf{F}(x, y) = \langle \sqrt{x^2 + 1}, \tan^{-1}x \rangle$, C is the triangle from $(0, 0)$ to $(1, 1)$ to $(0, 1)$ to $(0, 0)$

CAS **15–16** Verify Green's Theorem by using a computer algebra system to evaluate both the line integral and the double integral.

15. $P(x, y) = x^3y^4$, $Q(x, y) = x^5y^4$,
C consists of the line segment from $(-\pi/2, 0)$ to $(\pi/2, 0)$ followed by the arc of the curve $y = \cos x$ from $(\pi/2, 0)$ to $(-\pi/2, 0)$

16. $P(x, y) = 2x - x^3y^5$, $Q(x, y) = x^3y^8$,
C is the ellipse $4x^2 + y^2 = 4$

17. Use Green's Theorem to find the work done by the force $\mathbf{F}(x, y) = x(x + y)\,\mathbf{i} + xy^2\,\mathbf{j}$ in moving a particle from the origin along the x-axis to $(1, 0)$, then along the line segment to $(0, 1)$, and then back to the origin along the y-axis.

18. A particle starts at the origin, moves along the x-axis to $(5, 0)$, then along the quarter-circle $x^2 + y^2 = 25$, $x \geq 0$, $y \geq 0$ to the point $(0, 5)$, and then down the y-axis back to the origin. Use Green's Theorem to find

the work done on this particle by the force field $\mathbf{F}(x, y) = \langle \sin x, \sin y + xy^2 + \frac{1}{3}x^3 \rangle$.

19. Use one of the formulas in (5) to find the area under one arch of the cycloid $x = t - \sin t$, $y = 1 - \cos t$.

20. If a circle C with radius 1 rolls along the outside of the circle $x^2 + y^2 = 16$, a fixed point P on C traces out a curve called an *epicycloid*, with parametric equations $x = 5 \cos t - \cos 5t$, $y = 5 \sin t - \sin 5t$. Graph the epicycloid and use (5) to find the area it encloses.

21. (a) If C is the line segment connecting the point (x_1, y_1) to the point (x_2, y_2), show that
$$\int_C x\,dy - y\,dx = x_1y_2 - x_2y_1$$
(b) If the vertices of a polygon, in counterclockwise order, are (x_1, y_1), (x_2, y_2), $\ldots$, (x_n, y_n), show that the area of the polygon is
$$A = \tfrac{1}{2}[(x_1y_2 - x_2y_1) + (x_2y_3 - x_3y_2) + \cdots$$
$$+ (x_{n-1}y_n - x_ny_{n-1}) + (x_ny_1 - x_1y_n)]$$
(c) Find the area of the pentagon with vertices $(0, 0)$, $(2, 1)$, $(1, 3)$, $(0, 2)$, and $(-1, 1)$.

22. Let D be a region bounded by a simple closed path C in the xy-plane. Use Green's Theorem to prove that the coordinates of the centroid $(\bar{x}, \bar{y})$ of D are
$$\bar{x} = \frac{1}{2A}\oint_C x^2\,dy \qquad \bar{y} = -\frac{1}{2A}\oint_C y^2\,dx$$
where A is the area of D.

23. Use Exercise 22 to find the centroid of a quarter-circular region of radius a.

24. Use Exercise 22 to find the centroid of the triangle with vertices $(0, 0)$, $(a, 0)$, and (a, b), where $a > 0$ and $b > 0$.

25. A plane lamina with constant density $\rho(x, y) = \rho$ occupies a region in the xy-plane bounded by a simple closed path C. Show that its moments of inertia about the axes are
$$I_x = -\frac{\rho}{3}\oint_C y^3\,dx \qquad I_y = \frac{\rho}{3}\oint_C x^3\,dy$$

26. Use Exercise 25 to find the moment of inertia of a circular disk of radius a with constant density ρ about a diameter. (Compare with Example 15.4.4.)

27. Use the method of Example 5 to calculate $\int_C \mathbf{F} \cdot d\mathbf{r}$, where
$$\mathbf{F}(x, y) = \frac{2xy\,\mathbf{i} + (y^2 - x^2)\,\mathbf{j}}{(x^2 + y^2)^2}$$
and C is any positively oriented simple closed curve that encloses the origin.

28. Calculate $\int_C \mathbf{F} \cdot d\mathbf{r}$, where $\mathbf{F}(x, y) = \langle x^2 + y, 3x - y^2 \rangle$ and C is the positively oriented boundary curve of a region D that has area 6.

29. If $\mathbf{F}$ is the vector field of Example 5, show that $\int_C \mathbf{F} \cdot d\mathbf{r} = 0$ for every simple closed path that does not pass through or enclose the origin.

30. Complete the proof of the special case of Green's Theorem by proving Equation 3.

31. Use Green's Theorem to prove the change of variables formula for a double integral (Formula 15.9.9) for the case where $f(x, y) = 1$:

$$\iint_R dx\, dy = \iint_S \left| \frac{\partial(x, y)}{\partial(u, v)} \right| du\, dv$$

Here R is the region in the xy-plane that corresponds to the region S in the uv-plane under the transformation given by $x = g(u, v)$, $y = h(u, v)$.

[*Hint:* Note that the left side is $A(R)$ and apply the first part of Equation 5. Convert the line integral over ∂R to a line integral over ∂S and apply Green's Theorem in the uv-plane.]

16.5 Curl and Divergence

In this section we define two operations that can be performed on vector fields and that play a basic role in the applications of vector calculus to fluid flow and electricity and magnetism. Each operation resembles differentiation, but one produces a vector field whereas the other produces a scalar field.

■ Curl

If $\mathbf{F} = P\,\mathbf{i} + Q\,\mathbf{j} + R\,\mathbf{k}$ is a vector field on $\mathbb{R}^3$ and the partial derivatives of P, Q, and R all exist, then the **curl** of $\mathbf{F}$ is the vector field on $\mathbb{R}^3$ defined by

$$\boxed{1} \qquad \text{curl } \mathbf{F} = \left(\frac{\partial R}{\partial y} - \frac{\partial Q}{\partial z} \right) \mathbf{i} + \left(\frac{\partial P}{\partial z} - \frac{\partial R}{\partial x} \right) \mathbf{j} + \left(\frac{\partial Q}{\partial x} - \frac{\partial P}{\partial y} \right) \mathbf{k}$$

As an aid to our memory, let's rewrite Equation 1 using operator notation. We introduce the vector differential operator ∇ ("del") as

$$\nabla = \mathbf{i}\, \frac{\partial}{\partial x} + \mathbf{j}\, \frac{\partial}{\partial y} + \mathbf{k}\, \frac{\partial}{\partial z}$$

It has meaning when it operates on a scalar function to produce the gradient of f:

$$\nabla f = \mathbf{i}\, \frac{\partial f}{\partial x} + \mathbf{j}\, \frac{\partial f}{\partial y} + \mathbf{k}\, \frac{\partial f}{\partial z} = \frac{\partial f}{\partial x}\,\mathbf{i} + \frac{\partial f}{\partial y}\,\mathbf{j} + \frac{\partial f}{\partial z}\,\mathbf{k}$$

If we think of ∇ as a vector with components $\partial/\partial x$, $\partial/\partial y$, and $\partial/\partial z$, we can also consider the formal cross product of ∇ with the vector field $\mathbf{F}$ as follows:

$$\nabla \times \mathbf{F} = \begin{vmatrix} \mathbf{i} & \mathbf{j} & \mathbf{k} \\ \dfrac{\partial}{\partial x} & \dfrac{\partial}{\partial y} & \dfrac{\partial}{\partial z} \\ P & Q & R \end{vmatrix}$$

$$= \left(\frac{\partial R}{\partial y} - \frac{\partial Q}{\partial z} \right) \mathbf{i} + \left(\frac{\partial P}{\partial z} - \frac{\partial R}{\partial x} \right) \mathbf{j} + \left(\frac{\partial Q}{\partial x} - \frac{\partial P}{\partial y} \right) \mathbf{k}$$

$$= \text{curl } \mathbf{F}$$

So the easiest way to remember Definition 1 is by means of the symbolic expression

$$\boxed{2} \qquad \text{curl } \mathbf{F} = \nabla \times \mathbf{F}$$

EXAMPLE 1   If $\mathbf{F}(x, y, z) = xz\,\mathbf{i} + xyz\,\mathbf{j} - y^2\,\mathbf{k}$, find curl $\mathbf{F}$.

SOLUTION   Using Equation 2, we have

$$\text{curl } \mathbf{F} = \nabla \times \mathbf{F} = \begin{vmatrix} \mathbf{i} & \mathbf{j} & \mathbf{k} \\ \dfrac{\partial}{\partial x} & \dfrac{\partial}{\partial y} & \dfrac{\partial}{\partial z} \\ xz & xyz & -y^2 \end{vmatrix}$$

$$= \left[\frac{\partial}{\partial y}(-y^2) - \frac{\partial}{\partial z}(xyz) \right]\mathbf{i} - \left[\frac{\partial}{\partial x}(-y^2) - \frac{\partial}{\partial z}(xz) \right]\mathbf{j}$$

$$+ \left[\frac{\partial}{\partial x}(xyz) - \frac{\partial}{\partial y}(xz) \right]\mathbf{k}$$

$$= (-2y - xy)\,\mathbf{i} - (0 - x)\,\mathbf{j} + (yz - 0)\,\mathbf{k}$$

$$= -y(2 + x)\,\mathbf{i} + x\,\mathbf{j} + yz\,\mathbf{k} \qquad\blacksquare$$

CAS Most computer algebra systems have commands that compute the curl and divergence of vector fields. If you have access to a CAS, use these commands to check the answers to the examples and exercises in this section.

Recall that the gradient of a function f of three variables is a vector field on $\mathbb{R}^3$ and so we can compute its curl. The following theorem says that the curl of a gradient vector field is $\mathbf{0}$.

> **3**   **Theorem**   If f is a function of three variables that has continuous second-order partial derivatives, then
> $$\text{curl}(\nabla f) = \mathbf{0}$$

PROOF   We have

Notice the similarity to what we know from Section 12.4: $\mathbf{a} \times \mathbf{a} = \mathbf{0}$ for every three-dimensional vector $\mathbf{a}$.

$$\text{curl}(\nabla f) = \nabla \times (\nabla f) = \begin{vmatrix} \mathbf{i} & \mathbf{j} & \mathbf{k} \\ \dfrac{\partial}{\partial x} & \dfrac{\partial}{\partial y} & \dfrac{\partial}{\partial z} \\ \dfrac{\partial f}{\partial x} & \dfrac{\partial f}{\partial y} & \dfrac{\partial f}{\partial z} \end{vmatrix}$$

$$= \left(\frac{\partial^2 f}{\partial y\,\partial z} - \frac{\partial^2 f}{\partial z\,\partial y} \right)\mathbf{i} + \left(\frac{\partial^2 f}{\partial z\,\partial x} - \frac{\partial^2 f}{\partial x\,\partial z} \right)\mathbf{j} + \left(\frac{\partial^2 f}{\partial x\,\partial y} - \frac{\partial^2 f}{\partial y\,\partial x} \right)\mathbf{k}$$

$$= 0\,\mathbf{i} + 0\,\mathbf{j} + 0\,\mathbf{k} = \mathbf{0}$$

by Clairaut's Theorem. $\qquad\blacksquare$

Since a conservative vector field is one for which $\mathbf{F} = \nabla f$, Theorem 3 can be rephrased as follows:

Compare this with Exercise 16.3.29.

$$\text{If } \mathbf{F} \text{ is conservative, then curl } \mathbf{F} = \mathbf{0}.$$

This gives us a way of verifying that a vector field is not conservative.

EXAMPLE 2 Show that the vector field $\mathbf{F}(x, y, z) = xz\,\mathbf{i} + xyz\,\mathbf{j} - y^2\,\mathbf{k}$ is not conservative.

SOLUTION In Example 1 we showed that

$$\operatorname{curl}\mathbf{F} = -y(2 + x)\,\mathbf{i} + x\,\mathbf{j} + yz\,\mathbf{k}$$

This shows that curl $\mathbf{F} \neq \mathbf{0}$ and so, by the remarks preceding this example, $\mathbf{F}$ is not conservative. ∎

The converse of Theorem 3 is not true in general, but the following theorem says the converse is true if $\mathbf{F}$ is defined everywhere. (More generally it is true if the domain is simply-connected, that is, "has no hole.") Theorem 4 is the three-dimensional version of Theorem 16.3.6. Its proof requires Stokes' Theorem and is sketched at the end of Section 16.8.

---

4 Theorem If $\mathbf{F}$ is a vector field defined on all of $\mathbb{R}^3$ whose component functions have continuous partial derivatives and curl $\mathbf{F} = \mathbf{0}$, then $\mathbf{F}$ is a conservative vector field.

EXAMPLE 3
(a) Show that

$$\mathbf{F}(x, y, z) = y^2z^3\,\mathbf{i} + 2xyz^3\,\mathbf{j} + 3xy^2z^2\,\mathbf{k}$$

is a conservative vector field.
(b) Find a function f such that $\mathbf{F} = \nabla f$.

SOLUTION
(a) We compute the curl of $\mathbf{F}$:

$$\operatorname{curl}\mathbf{F} = \nabla \times \mathbf{F} = \begin{vmatrix} \mathbf{i} & \mathbf{j} & \mathbf{k} \\ \dfrac{\partial}{\partial x} & \dfrac{\partial}{\partial y} & \dfrac{\partial}{\partial z} \\ y^2z^3 & 2xyz^3 & 3xy^2z^2 \end{vmatrix}$$

$$= (6xyz^2 - 6xyz^2)\,\mathbf{i} - (3y^2z^2 - 3y^2z^2)\,\mathbf{j} + (2yz^3 - 2yz^3)\,\mathbf{k}$$

$$= \mathbf{0}$$

Since curl $\mathbf{F} = \mathbf{0}$ and the domain of $\mathbf{F}$ is $\mathbb{R}^3$, $\mathbf{F}$ is a conservative vector field by Theorem 4.

(b) The technique for finding f was given in Section 16.3. We have

$$\boxed{5} \qquad\qquad f_x(x, y, z) = y^2z^3$$

$$\boxed{6} \qquad\qquad f_y(x, y, z) = 2xyz^3$$

$$\boxed{7} \qquad\qquad f_z(x, y, z) = 3xy^2z^2$$

Integrating (5) with respect to x, we obtain

$$\boxed{8} \qquad\qquad f(x, y, z) = xy^2z^3 + g(y, z)$$

Differentiating (8) with respect to y, we get $f_y(x, y, z) = 2xyz^3 + g_y(y, z)$, so comparison with (6) gives $g_y(y, z) = 0$. Thus $g(y, z) = h(z)$ and

$$f_z(x, y, z) = 3xy^2z^2 + h'(z)$$

Then (7) gives $h'(z) = 0$. Therefore

$$f(x, y, z) = xy^2z^3 + K$$ ∎

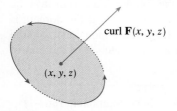

curl $\mathbf{F}(x, y, z)$

(x, y, z)

FIGURE 1

The reason for the name *curl* is that the curl vector is associated with rotations. One connection is explained in Exercise 37. Another occurs when $\mathbf{F}$ represents the velocity field in fluid flow (see Example 16.1.3). Particles near (x, y, z) in the fluid tend to rotate about the axis that points in the direction of curl $\mathbf{F}(x, y, z)$, and the length of this curl vector is a measure of how quickly the particles move around the axis (see Figure 1). If curl $\mathbf{F} = \mathbf{0}$ at a point P, then the fluid is free from rotations at P and $\mathbf{F}$ is called **irrotational** at P. In other words, there is no whirlpool or eddy at P. If curl $\mathbf{F} = \mathbf{0}$, then a tiny paddle wheel moves with the fluid but doesn't rotate about its axis. If curl $\mathbf{F} \neq \mathbf{0}$, the paddle wheel rotates about its axis. We give a more detailed explanation in Section 16.8 as a consequence of Stokes' Theorem.

■ Divergence

If $\mathbf{F} = P\,\mathbf{i} + Q\,\mathbf{j} + R\,\mathbf{k}$ is a vector field on $\mathbb{R}^3$ and $\partial P/\partial x$, $\partial Q/\partial y$, and $\partial R/\partial z$ exist, then the **divergence of F** is the function of three variables defined by

9
$$\text{div } \mathbf{F} = \frac{\partial P}{\partial x} + \frac{\partial Q}{\partial y} + \frac{\partial R}{\partial z}$$

Observe that curl $\mathbf{F}$ is a vector field but div $\mathbf{F}$ is a scalar field. In terms of the gradient operator $\nabla = (\partial/\partial x)\,\mathbf{i} + (\partial/\partial y)\,\mathbf{j} + (\partial/\partial z)\,\mathbf{k}$, the divergence of $\mathbf{F}$ can be written symbolically as the dot product of ∇ and $\mathbf{F}$:

10
$$\text{div } \mathbf{F} = \nabla \cdot \mathbf{F}$$

EXAMPLE 4 If $\mathbf{F}(x, y, z) = xz\,\mathbf{i} + xyz\,\mathbf{j} - y^2\,\mathbf{k}$, find div $\mathbf{F}$.

SOLUTION By the definition of divergence (Equation 9 or 10) we have

$$\text{div } \mathbf{F} = \nabla \cdot \mathbf{F} = \frac{\partial}{\partial x}(xz) + \frac{\partial}{\partial y}(xyz) + \frac{\partial}{\partial z}(-y^2) = z + xz$$ ∎

If $\mathbf{F}$ is a vector field on $\mathbb{R}^3$, then curl $\mathbf{F}$ is also a vector field on $\mathbb{R}^3$. As such, we can compute its divergence. The next theorem shows that the result is 0.

11 **Theorem** If $\mathbf{F} = P\,\mathbf{i} + Q\,\mathbf{j} + R\,\mathbf{k}$ is a vector field on $\mathbb{R}^3$ and P, Q, and R have continuous second-order partial derivatives, then

$$\text{div curl } \mathbf{F} = 0$$

Note the analogy with the scalar triple product: $\mathbf{a} \cdot (\mathbf{a} \times \mathbf{b}) = 0$.

PROOF Using the definitions of divergence and curl, we have

$$\text{div curl } \mathbf{F} = \nabla \cdot (\nabla \times \mathbf{F})$$

$$= \frac{\partial}{\partial x}\left(\frac{\partial R}{\partial y} - \frac{\partial Q}{\partial z}\right) + \frac{\partial}{\partial y}\left(\frac{\partial P}{\partial z} - \frac{\partial R}{\partial x}\right) + \frac{\partial}{\partial z}\left(\frac{\partial Q}{\partial x} - \frac{\partial P}{\partial y}\right)$$

$$= \frac{\partial^2 R}{\partial x\,\partial y} - \frac{\partial^2 Q}{\partial x\,\partial z} + \frac{\partial^2 P}{\partial y\,\partial z} - \frac{\partial^2 R}{\partial y\,\partial x} + \frac{\partial^2 Q}{\partial z\,\partial x} - \frac{\partial^2 P}{\partial z\,\partial y}$$

$$= 0$$

because the terms cancel in pairs by Clairaut's Theorem. ∎

EXAMPLE 5 Show that the vector field $\mathbf{F}(x, y, z) = xz\,\mathbf{i} + xyz\,\mathbf{j} - y^2\,\mathbf{k}$ can't be written as the curl of another vector field, that is, $\mathbf{F} \neq \text{curl } \mathbf{G}$.

SOLUTION In Example 4 we showed that

$$\text{div } \mathbf{F} = z + xz$$

and therefore div $\mathbf{F} \neq 0$. If it were true that $\mathbf{F} = \text{curl } \mathbf{G}$, then Theorem 11 would give

$$\text{div } \mathbf{F} = \text{div curl } \mathbf{G} = 0$$

which contradicts div $\mathbf{F} \neq 0$. Therefore $\mathbf{F}$ is not the curl of another vector field. ∎

The reason for this interpretation of div $\mathbf{F}$ will be explained at the end of Section 16.9 as a consequence of the Divergence Theorem.

Again, the reason for the name *divergence* can be understood in the context of fluid flow. If $\mathbf{F}(x, y, z)$ is the velocity of a fluid (or gas), then div $\mathbf{F}(x, y, z)$ represents the net rate of change (with respect to time) of the mass of fluid (or gas) flowing from the point (x, y, z) per unit volume. In other words, div $\mathbf{F}(x, y, z)$ measures the tendency of the fluid to diverge from the point (x, y, z). If div $\mathbf{F} = 0$, then $\mathbf{F}$ is said to be **incompressible**.

Another differential operator occurs when we compute the divergence of a gradient vector field ∇f. If f is a function of three variables, we have

$$\text{div}(\nabla f) = \nabla \cdot (\nabla f) = \frac{\partial^2 f}{\partial x^2} + \frac{\partial^2 f}{\partial y^2} + \frac{\partial^2 f}{\partial z^2}$$

and this expression occurs so often that we abbreviate it as $\nabla^2 f$. The operator

$$\nabla^2 = \nabla \cdot \nabla$$

is called the **Laplace operator** because of its relation to **Laplace's equation**

$$\nabla^2 f = \frac{\partial^2 f}{\partial x^2} + \frac{\partial^2 f}{\partial y^2} + \frac{\partial^2 f}{\partial z^2} = 0$$

We can also apply the Laplace operator ∇^2 to a vector field

$$\mathbf{F} = P\,\mathbf{i} + Q\,\mathbf{j} + R\,\mathbf{k}$$

in terms of its components:

$$\nabla^2 \mathbf{F} = \nabla^2 P\,\mathbf{i} + \nabla^2 Q\,\mathbf{j} + \nabla^2 R\,\mathbf{k}$$

■ Vector Forms of Green's Theorem

The curl and divergence operators allow us to rewrite Green's Theorem in versions that will be useful in our later work. We suppose that the plane region D, its boundary curve C, and the functions P and Q satisfy the hypotheses of Green's Theorem. Then we consider the vector field $\mathbf{F} = P\,\mathbf{i} + Q\,\mathbf{j}$. Its line integral is

$$\oint_C \mathbf{F} \cdot d\mathbf{r} = \oint_C P\,dx + Q\,dy$$

and, regarding $\mathbf{F}$ as a vector field on $\mathbb{R}^3$ with third component 0, we have

$$\operatorname{curl} \mathbf{F} = \begin{vmatrix} \mathbf{i} & \mathbf{j} & \mathbf{k} \\ \dfrac{\partial}{\partial x} & \dfrac{\partial}{\partial y} & \dfrac{\partial}{\partial z} \\ P(x,y) & Q(x,y) & 0 \end{vmatrix} = \left(\frac{\partial Q}{\partial x} - \frac{\partial P}{\partial y} \right) \mathbf{k}$$

Therefore

$$(\operatorname{curl} \mathbf{F}) \cdot \mathbf{k} = \left(\frac{\partial Q}{\partial x} - \frac{\partial P}{\partial y} \right) \mathbf{k} \cdot \mathbf{k} = \frac{\partial Q}{\partial x} - \frac{\partial P}{\partial y}$$

and we can now rewrite the equation in Green's Theorem in the vector form

12
$$\oint_C \mathbf{F} \cdot d\mathbf{r} = \iint_D (\operatorname{curl} \mathbf{F}) \cdot \mathbf{k}\, dA$$

Equation 12 expresses the line integral of the tangential component of $\mathbf{F}$ along C as the double integral of the vertical component of curl $\mathbf{F}$ over the region D enclosed by C. We now derive a similar formula involving the *normal* component of $\mathbf{F}$.

If C is given by the vector equation

$$\mathbf{r}(t) = x(t)\,\mathbf{i} + y(t)\,\mathbf{j} \qquad a \leqslant t \leqslant b$$

then the unit tangent vector (see Section 13.2) is

$$\mathbf{T}(t) = \frac{x'(t)}{|\mathbf{r}'(t)|}\,\mathbf{i} + \frac{y'(t)}{|\mathbf{r}'(t)|}\,\mathbf{j}$$

You can verify that the outward unit normal vector to C is given by

$$\mathbf{n}(t) = \frac{y'(t)}{|\mathbf{r}'(t)|}\,\mathbf{i} - \frac{x'(t)}{|\mathbf{r}'(t)|}\,\mathbf{j}$$

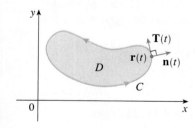

FIGURE 2

(See Figure 2.) Then, from Equation 16.2.3, we have

$$\oint_C \mathbf{F} \cdot \mathbf{n}\, ds = \int_a^b (\mathbf{F} \cdot \mathbf{n})(t)\,|\mathbf{r}'(t)|\,dt$$

$$= \int_a^b \left[\frac{P(x(t), y(t))\,y'(t)}{|\mathbf{r}'(t)|} - \frac{Q(x(t), y(t))\,x'(t)}{|\mathbf{r}'(t)|} \right] |\mathbf{r}'(t)|\,dt$$

$$= \int_a^b P(x(t), y(t))\,y'(t)\,dt - Q(x(t), y(t))\,x'(t)\,dt$$

$$= \int_C P\,dy - Q\,dx = \iint_D \left(\frac{\partial P}{\partial x} + \frac{\partial Q}{\partial y} \right) dA$$

by Green's Theorem. But the integrand in this double integral is just the divergence of **F**. So we have a second vector form of Green's Theorem.

$$\boxed{13} \qquad \oint_C \mathbf{F} \cdot \mathbf{n}\, ds = \iint_D \text{div } \mathbf{F}(x, y)\, dA$$

This version says that the line integral of the normal component of **F** along C is equal to the double integral of the divergence of **F** over the region D enclosed by C.

16.5 EXERCISES

1–8 Find (a) the curl and (b) the divergence of the vector field.

1. $\mathbf{F}(x, y, z) = xy^2z^2\,\mathbf{i} + x^2yz^2\,\mathbf{j} + x^2y^2z\,\mathbf{k}$

2. $\mathbf{F}(x, y, z) = x^3yz^2\,\mathbf{j} + y^4z^3\,\mathbf{k}$

3. $\mathbf{F}(x, y, z) = xye^z\,\mathbf{i} + yze^x\,\mathbf{k}$

4. $\mathbf{F}(x, y, z) = \sin yz\,\mathbf{i} + \sin zx\,\mathbf{j} + \sin xy\,\mathbf{k}$

5. $\mathbf{F}(x, y, z) = \dfrac{\sqrt{x}}{1 + z}\,\mathbf{i} + \dfrac{\sqrt{y}}{1 + x}\,\mathbf{j} + \dfrac{\sqrt{z}}{1 + y}\,\mathbf{k}$

6. $\mathbf{F}(x, y, z) = \ln(2y + 3z)\,\mathbf{i} + \ln(x + 3z)\,\mathbf{j} + \ln(x + 2y)\,\mathbf{k}$

7. $\mathbf{F}(x, y, z) = \langle e^x \sin y, e^y \sin z, e^z \sin x \rangle$

8. $\mathbf{F}(x, y, z) = \langle \arctan(xy), \arctan(yz), \arctan(zx) \rangle$

9–11 The vector field **F** is shown in the xy-plane and looks the same in all other horizontal planes. (In other words, **F** is independent of z and its z-component is 0.)
(a) Is div **F** positive, negative, or zero? Explain.
(b) Determine whether curl **F** $= \mathbf{0}$. If not, in which direction does curl **F** point?

9.

10.

11.
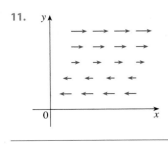

12. Let f be a scalar field and **F** a vector field. State whether each expression is meaningful. If not, explain why. If so, state whether it is a scalar field or a vector field.
(a) curl f (b) grad f
(c) div **F**   (d) curl(grad f)
(e) grad **F**   (f) grad(div **F**)
(g) div(grad f)   (h) grad(div f)
(i) curl(curl **F**)   (j) div(div **F**)
(k) (grad f) $\times$ (div **F**)   (l) div(curl(grad f))

13–18 Determine whether or not the vector field is conservative. If it is conservative, find a function f such that $\mathbf{F} = \nabla f$.

13. $\mathbf{F}(x, y, z) = y^2z^3\,\mathbf{i} + 2xyz^3\,\mathbf{j} + 3xy^2z^2\,\mathbf{k}$

14. $\mathbf{F}(x, y, z) = xyz^4\,\mathbf{i} + x^2z^4\,\mathbf{j} + 4x^2yz^3\,\mathbf{k}$

15. $\mathbf{F}(x, y, z) = z \cos y\,\mathbf{i} + xz \sin y\,\mathbf{j} + x \cos y\,\mathbf{k}$

16. $\mathbf{F}(x, y, z) = \mathbf{i} + \sin z\,\mathbf{j} + y \cos z\,\mathbf{k}$

17. $\mathbf{F}(x, y, z) = e^{yz}\,\mathbf{i} + xze^{yz}\,\mathbf{j} + xye^{yz}\,\mathbf{k}$

18. $\mathbf{F}(x, y, z) = e^x \sin yz\,\mathbf{i} + ze^x \cos yz\,\mathbf{j} + ye^x \cos yz\,\mathbf{k}$

---

19. Is there a vector field **G** on $\mathbb{R}^3$ such that curl **G** $= \langle x \sin y, \cos y, z - xy \rangle$? Explain.

20. Is there a vector field **G** on $\mathbb{R}^3$ such that curl **G** $= \langle x, y, z \rangle$? Explain.

21. Show that any vector field of the form

$$\mathbf{F}(x, y, z) = f(x)\,\mathbf{i} + g(y)\,\mathbf{j} + h(z)\,\mathbf{k}$$

where f, g, h are differentiable functions, is irrotational.

22. Show that any vector field of the form

$$\mathbf{F}(x, y, z) = f(y, z)\,\mathbf{i} + g(x, z)\,\mathbf{j} + h(x, y)\,\mathbf{k}$$

is incompressible.

23–29 Prove the identity, assuming that the appropriate partial derivatives exist and are continuous. If f is a scalar field and $\mathbf{F}$, $\mathbf{G}$ are vector fields, then $f\mathbf{F}$, $\mathbf{F} \cdot \mathbf{G}$, and $\mathbf{F} \times \mathbf{G}$ are defined by

$$(f\mathbf{F})(x, y, z) = f(x, y, z)\,\mathbf{F}(x, y, z)$$

$$(\mathbf{F} \cdot \mathbf{G})(x, y, z) = \mathbf{F}(x, y, z) \cdot \mathbf{G}(x, y, z)$$

$$(\mathbf{F} \times \mathbf{G})(x, y, z) = \mathbf{F}(x, y, z) \times \mathbf{G}(x, y, z)$$

23. $\operatorname{div}(\mathbf{F} + \mathbf{G}) = \operatorname{div} \mathbf{F} + \operatorname{div} \mathbf{G}$

24. $\operatorname{curl}(\mathbf{F} + \mathbf{G}) = \operatorname{curl} \mathbf{F} + \operatorname{curl} \mathbf{G}$

25. $\operatorname{div}(f\mathbf{F}) = f \operatorname{div} \mathbf{F} + \mathbf{F} \cdot \nabla f$

26. $\operatorname{curl}(f\mathbf{F}) = f \operatorname{curl} \mathbf{F} + (\nabla f) \times \mathbf{F}$

27. $\operatorname{div}(\mathbf{F} \times \mathbf{G}) = \mathbf{G} \cdot \operatorname{curl} \mathbf{F} - \mathbf{F} \cdot \operatorname{curl} \mathbf{G}$

28. $\operatorname{div}(\nabla f \times \nabla g) = 0$

29. $\operatorname{curl}(\operatorname{curl} \mathbf{F}) = \operatorname{grad}(\operatorname{div} \mathbf{F}) - \nabla^2 \mathbf{F}$

30–32 Let $\mathbf{r} = x\,\mathbf{i} + y\,\mathbf{j} + z\,\mathbf{k}$ and $r = |\mathbf{r}|$.

30. Verify each identity.
(a) $\nabla \cdot \mathbf{r} = 3$ (b) $\nabla \cdot (r\mathbf{r}) = 4r$
(c) $\nabla^2 r^3 = 12r$

31. Verify each identity.
(a) $\nabla r = \mathbf{r}/r$ (b) $\nabla \times \mathbf{r} = \mathbf{0}$
(c) $\nabla(1/r) = -\mathbf{r}/r^3$ (d) $\nabla \ln r = \mathbf{r}/r^2$

32. If $\mathbf{F} = \mathbf{r}/r^p$, find div $\mathbf{F}$. Is there a value of p for which div $\mathbf{F} = 0$?

---

33. Use Green's Theorem in the form of Equation 13 to prove **Green's first identity**:

$$\iint_D f \nabla^2 g\, dA = \oint_C f(\nabla g) \cdot \mathbf{n}\, ds - \iint_D \nabla f \cdot \nabla g\, dA$$

where D and C satisfy the hypotheses of Green's Theorem and the appropriate partial derivatives of f and g exist and are continuous. (The quantity $\nabla g \cdot \mathbf{n} = D_{\mathbf{n}} g$ occurs in the line integral. This is the directional derivative in the direction of the normal vector $\mathbf{n}$ and is called the **normal derivative** of g.)

34. Use Green's first identity (Exercise 33) to prove **Green's second identity**:

$$\iint_D (f \nabla^2 g - g \nabla^2 f)\, dA = \oint_C (f\nabla g - g\nabla f) \cdot \mathbf{n}\, ds$$

where D and C satisfy the hypotheses of Green's Theorem and the appropriate partial derivatives of f and g exist and are continuous.

35. Recall from Section 14.3 that a function g is called *harmonic* on D if it satisfies Laplace's equation, that is, $\nabla^2 g = 0$ on D. Use Green's first identity (with the same hypotheses as in

Exercise 33) to show that if g is harmonic on D, then $\oint_C D_{\mathbf{n}} g\, ds = 0$. Here $D_{\mathbf{n}} g$ is the normal derivative of g defined in Exercise 33.

36. Use Green's first identity to show that if f is harmonic on D, and if $f(x, y) = 0$ on the boundary curve C, then $\iint_D |\nabla f|^2\, dA = 0$. (Assume the same hypotheses as in Exercise 33.)

37. This exercise demonstrates a connection between the curl vector and rotations. Let B be a rigid body rotating about the z-axis. The rotation can be described by the vector $\mathbf{w} = \omega\mathbf{k}$, where ω is the angular speed of B, that is, the tangential speed of any point P in B divided by the distance d from the axis of rotation. Let $\mathbf{r} = \langle x, y, z \rangle$ be the position vector of P.
(a) By considering the angle θ in the figure, show that the velocity field of B is given by $\mathbf{v} = \mathbf{w} \times \mathbf{r}$.
(b) Show that $\mathbf{v} = -\omega y\,\mathbf{i} + \omega x\,\mathbf{j}$.
(c) Show that curl $\mathbf{v} = 2\mathbf{w}$.

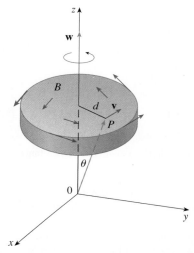

38. Maxwell's equations relating the electric field $\mathbf{E}$ and magnetic field $\mathbf{H}$ as they vary with time in a region containing no charge and no current can be stated as follows:

$$\operatorname{div} \mathbf{E} = 0 \qquad\qquad \operatorname{div} \mathbf{H} = 0$$

$$\operatorname{curl} \mathbf{E} = -\frac{1}{c}\frac{\partial \mathbf{H}}{\partial t} \qquad \operatorname{curl} \mathbf{H} = \frac{1}{c}\frac{\partial \mathbf{E}}{\partial t}$$

where c is the speed of light. Use these equations to prove the following:

(a) $\nabla \times (\nabla \times \mathbf{E}) = -\dfrac{1}{c^2}\dfrac{\partial^2 \mathbf{E}}{\partial t^2}$

(b) $\nabla \times (\nabla \times \mathbf{H}) = -\dfrac{1}{c^2}\dfrac{\partial^2 \mathbf{H}}{\partial t^2}$

(c) $\nabla^2 \mathbf{E} = \dfrac{1}{c^2}\dfrac{\partial^2 \mathbf{E}}{\partial t^2}$ [*Hint:* Use Exercise 29.]

(d) $\nabla^2 \mathbf{H} = \dfrac{1}{c^2}\dfrac{\partial^2 \mathbf{H}}{\partial t^2}$

39. We have seen that all vector fields of the form $\mathbf{F} = \nabla g$ satisfy the equation curl $\mathbf{F} = \mathbf{0}$ and that all vector fields of the form $\mathbf{F} = \text{curl } \mathbf{G}$ satisfy the equation div $\mathbf{F} = 0$ (assuming continuity of the appropriate partial derivatives). This suggests the question: are there any equations that all functions of the form $f = \text{div } \mathbf{G}$ must satisfy? Show that the answer to this question is "No" by proving that *every* continuous function f on $\mathbb{R}^3$ is the divergence of some vector field.

[*Hint:* Let $\mathbf{G}(x, y, z) = \langle g(x, y, z), 0, 0 \rangle$, where $g(x, y, z) = \int_0^x f(t, y, z) \, dt$.]

16.6 Parametric Surfaces and Their Areas

So far we have considered special types of surfaces: cylinders, quadric surfaces, graphs of functions of two variables, and level surfaces of functions of three variables. Here we use vector functions to describe more general surfaces, called *parametric surfaces*, and compute their areas. Then we take the general surface area formula and see how it applies to special surfaces.

■ Parametric Surfaces

In much the same way that we describe a space curve by a vector function $\mathbf{r}(t)$ of a single parameter t, we can describe a surface by a vector function $\mathbf{r}(u, v)$ of two parameters u and v. We suppose that

$$\boxed{1} \qquad \mathbf{r}(u, v) = x(u, v)\,\mathbf{i} + y(u, v)\,\mathbf{j} + z(u, v)\,\mathbf{k}$$

is a vector-valued function defined on a region D in the uv-plane. So x, y, and z, the component functions of $\mathbf{r}$, are functions of the two variables u and v with domain D. The set of all points (x, y, z) in $\mathbb{R}^3$ such that

$$\boxed{2} \qquad x = x(u, v) \qquad y = y(u, v) \qquad z = z(u, v)$$

and (u, v) varies throughout D, is called a **parametric surface** S and Equations 2 are called **parametric equations** of S. Each choice of u and v gives a point on S; by making all choices, we get all of S. In other words, the surface S is traced out by the tip of the position vector $\mathbf{r}(u, v)$ as (u, v) moves throughout the region D. (See Figure 1.)

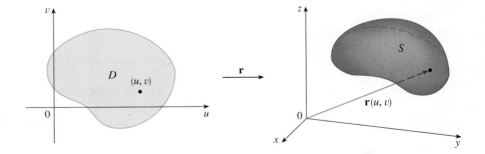

FIGURE 1
A parametric surface

EXAMPLE 1 Identify and sketch the surface with vector equation

$$\mathbf{r}(u, v) = 2 \cos u\,\mathbf{i} + v\,\mathbf{j} + 2 \sin u\,\mathbf{k}$$

SOLUTION The parametric equations for this surface are

$$x = 2 \cos u \qquad y = v \qquad z = 2 \sin u$$

FIGURE 2

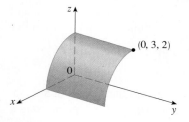

FIGURE 3

TEC Visual 16.6 shows animated versions of Figures 4 and 5, with moving grid curves, for several parametric surfaces.

So for any point (x, y, z) on the surface, we have

$$x^2 + z^2 = 4\cos^2 u + 4\sin^2 u = 4$$

This means that vertical cross-sections parallel to the xz-plane (that is, with y constant) are all circles with radius 2. Since $y = v$ and no restriction is placed on v, the surface is a circular cylinder with radius 2 whose axis is the y-axis (see Figure 2). ∎

In Example 1 we placed no restrictions on the parameters u and v and so we obtained the entire cylinder. If, for instance, we restrict u and v by writing the parameter domain as

$$0 \leqslant u \leqslant \pi/2 \qquad 0 \leqslant v \leqslant 3$$

then $x \geqslant 0$, $z \geqslant 0$, $0 \leqslant y \leqslant 3$, and we get the quarter-cylinder with length 3 illustrated in Figure 3.

If a parametric surface S is given by a vector function $\mathbf{r}(u, v)$, then there are two useful families of curves that lie on S, one family with u constant and the other with v constant. These families correspond to vertical and horizontal lines in the uv-plane. If we keep u constant by putting $u = u_0$, then $\mathbf{r}(u_0, v)$ becomes a vector function of the single parameter v and defines a curve C_1 lying on S. (See Figure 4.)

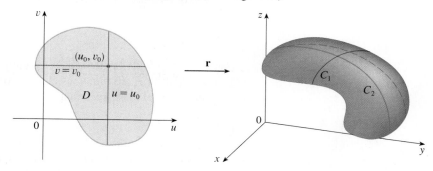

FIGURE 4

Similarly, if we keep v constant by putting $v = v_0$, we get a curve C_2 given by $\mathbf{r}(u, v_0)$ that lies on S. We call these curves **grid curves**. (In Example 1, for instance, the grid curves obtained by letting u be constant are horizontal lines whereas the grid curves with v constant are circles.) In fact, when a computer graphs a parametric surface, it usually depicts the surface by plotting these grid curves, as we see in the following example.

EXAMPLE 2 Use a computer algebra system to graph the surface

$$\mathbf{r}(u, v) = \langle (2 + \sin v)\cos u, (2 + \sin v)\sin u, u + \cos v \rangle$$

Which grid curves have u constant? Which have v constant?

SOLUTION We graph the portion of the surface with parameter domain $0 \leqslant u \leqslant 4\pi$, $0 \leqslant v \leqslant 2\pi$ in Figure 5. It has the appearance of a spiral tube. To identify the grid curves, we write the corresponding parametric equations:

$$x = (2 + \sin v)\cos u \qquad y = (2 + \sin v)\sin u \qquad z = u + \cos v$$

If v is constant, then $\sin v$ and $\cos v$ are constant, so the parametric equations resemble those of the helix in Example 13.1.4. Thus the grid curves with v constant are the spiral curves in Figure 5. We deduce that the grid curves with u constant must be the curves

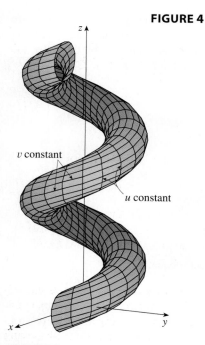

v constant

u constant

FIGURE 5

that look like circles in the figure. Further evidence for this assertion is that if u is kept constant, $u = u_0$, then the equation $z = u_0 + \cos v$ shows that the z-values vary from $u_0 - 1$ to $u_0 + 1$. ∎

In Examples 1 and 2 we were given a vector equation and asked to graph the corresponding parametric surface. In the following examples, however, we are given the more challenging problem of finding a vector function to represent a given surface. In the rest of this chapter we will often need to do exactly that.

EXAMPLE 3 Find a vector function that represents the plane that passes through the point P_0 with position vector $\mathbf{r}_0$ and that contains two nonparallel vectors $\mathbf{a}$ and $\mathbf{b}$.

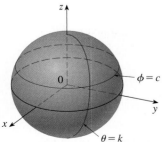

FIGURE 6

SOLUTION If P is any point in the plane, we can get from P_0 to P by moving a certain distance in the direction of $\mathbf{a}$ and another distance in the direction of $\mathbf{b}$. So there are scalars u and v such that $\overrightarrow{P_0P} = u\mathbf{a} + v\mathbf{b}$. (Figure 6 illustrates how this works, by means of the Parallelogram Law, for the case where u and v are positive. See also Exercise 12.2.46.) If $\mathbf{r}$ is the position vector of P, then

$$\mathbf{r} = \overrightarrow{OP_0} + \overrightarrow{P_0P} = \mathbf{r}_0 + u\mathbf{a} + v\mathbf{b}$$

So the vector equation of the plane can be written as

$$\mathbf{r}(u, v) = \mathbf{r}_0 + u\mathbf{a} + v\mathbf{b}$$

where u and v are real numbers.

If we write $\mathbf{r} = \langle x, y, z \rangle$, $\mathbf{r}_0 = \langle x_0, y_0, z_0 \rangle$, $\mathbf{a} = \langle a_1, a_2, a_3 \rangle$, and $\mathbf{b} = \langle b_1, b_2, b_3 \rangle$, then we can write the parametric equations of the plane through the point (x_0, y_0, z_0) as follows:

$$x = x_0 + ua_1 + vb_1 \qquad y = y_0 + ua_2 + vb_2 \qquad z = z_0 + ua_3 + vb_3 \qquad ∎$$

EXAMPLE 4 Find a parametric representation of the sphere

$$x^2 + y^2 + z^2 = a^2$$

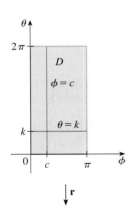

FIGURE 7

SOLUTION The sphere has a simple representation $\rho = a$ in spherical coordinates, so let's choose the angles ϕ and θ in spherical coordinates as the parameters (see Section 15.8). Then, putting $\rho = a$ in the equations for conversion from spherical to rectangular coordinates (Equations 15.8.1), we obtain

$$x = a \sin \phi \cos \theta \qquad y = a \sin \phi \sin \theta \qquad z = a \cos \phi$$

as the parametric equations of the sphere. The corresponding vector equation is

$$\mathbf{r}(\phi, \theta) = a \sin \phi \cos \theta \, \mathbf{i} + a \sin \phi \sin \theta \, \mathbf{j} + a \cos \phi \, \mathbf{k}$$

We have $0 \le \phi \le \pi$ and $0 \le \theta \le 2\pi$, so the parameter domain is the rectangle $D = [0, \pi] \times [0, 2\pi]$. The grid curves with ϕ constant are the circles of constant latitude (including the equator). The grid curves with θ constant are the meridians (semicircles), which connect the north and south poles (see Figure 7). ∎

NOTE We saw in Example 4 that the grid curves for a sphere are curves of constant latitude or constant longitude. For a general parametric surface we are really making a map and the grid curves are similar to lines of latitude and longitude. Describing a point on a parametric surface (like the one in Figure 5) by giving specific values of u and v is like giving the latitude and longitude of a point.

One of the uses of parametric surfaces is in computer graphics. Figure 8 shows the result of trying to graph the sphere $x^2 + y^2 + z^2 = 1$ by solving the equation for z and graphing the top and bottom hemispheres separately. Part of the sphere appears to be missing because of the rectangular grid system used by the computer. The much better picture in Figure 9 was produced by a computer using the parametric equations found in Example 4.

FIGURE 8

FIGURE 9

EXAMPLE 5 Find a parametric representation for the cylinder

$$x^2 + y^2 = 4 \qquad 0 \le z \le 1$$

SOLUTION The cylinder has a simple representation $r = 2$ in cylindrical coordinates, so we choose as parameters θ and z in cylindrical coordinates. Then the parametric equations of the cylinder are

$$x = 2\cos\theta \qquad y = 2\sin\theta \qquad z = z$$

where $0 \le \theta \le 2\pi$ and $0 \le z \le 1$. ∎

EXAMPLE 6 Find a vector function that represents the elliptic paraboloid $z = x^2 + 2y^2$.

SOLUTION If we regard x and y as parameters, then the parametric equations are simply

$$x = x \qquad y = y \qquad z = x^2 + 2y^2$$

and the vector equation is

$$\mathbf{r}(x, y) = x\,\mathbf{i} + y\,\mathbf{j} + (x^2 + 2y^2)\,\mathbf{k}$$ ∎

TEC In Module 16.6 you can investigate several families of parametric surfaces.

In general, a surface given as the graph of a function of x and y, that is, with an equation of the form $z = f(x, y)$, can always be regarded as a parametric surface by taking x and y as parameters and writing the parametric equations as

$$x = x \qquad y = y \qquad z = f(x, y)$$

Parametric representations (also called parametrizations) of surfaces are not unique. The next example shows two ways to parametrize a cone.

EXAMPLE 7 Find a parametric representation for the surface $z = 2\sqrt{x^2 + y^2}$, that is, the top half of the cone $z^2 = 4x^2 + 4y^2$.

SOLUTION 1 One possible representation is obtained by choosing x and y as parameters:

$$x = x \qquad y = y \qquad z = 2\sqrt{x^2 + y^2}$$

So the vector equation is

$$\mathbf{r}(x, y) = x\,\mathbf{i} + y\,\mathbf{j} + 2\sqrt{x^2 + y^2}\,\mathbf{k}$$

SOLUTION 2 Another representation results from choosing as parameters the polar coordinates r and θ. A point (x, y, z) on the cone satisfies $x = r\cos\theta$, $y = r\sin\theta$, and

For some purposes the parametric representations in Solutions 1 and 2 are equally good, but Solution 2 might be preferable in certain situations. If we are interested only in the part of the cone that lies below the plane $z = 1$, for instance, all we have to do in Solution 2 is change the parameter domain to

$$0 \leqslant r \leqslant \tfrac{1}{2} \qquad 0 \leqslant \theta \leqslant 2\pi$$

FIGURE 10

FIGURE 11

$z = 2\sqrt{x^2 + y^2} = 2r$. So a vector equation for the cone is

$$\mathbf{r}(r, \theta) = r \cos \theta \, \mathbf{i} + r \sin \theta \, \mathbf{j} + 2r \, \mathbf{k}$$

where $r \geqslant 0$ and $0 \leqslant \theta \leqslant 2\pi$. ∎

■ Surfaces of Revolution

Surfaces of revolution can be represented parametrically and thus graphed using a computer. For instance, let's consider the surface S obtained by rotating the curve $y = f(x)$, $a \leqslant x \leqslant b$, about the x-axis, where $f(x) \geqslant 0$. Let θ be the angle of rotation as shown in Figure 10. If (x, y, z) is a point on S, then

$$\boxed{3} \qquad x = x \qquad y = f(x) \cos \theta \qquad z = f(x) \sin \theta$$

Therefore we take x and θ as parameters and regard Equations 3 as parametric equations of S. The parameter domain is given by $a \leqslant x \leqslant b$, $0 \leqslant \theta \leqslant 2\pi$.

EXAMPLE 8 Find parametric equations for the surface generated by rotating the curve $y = \sin x$, $0 \leqslant x \leqslant 2\pi$, about the x-axis. Use these equations to graph the surface of revolution.

SOLUTION From Equations 3, the parametric equations are

$$x = x \qquad y = \sin x \cos \theta \qquad z = \sin x \sin \theta$$

and the parameter domain is $0 \leqslant x \leqslant 2\pi$, $0 \leqslant \theta \leqslant 2\pi$. Using a computer to plot these equations and and then rotating the image, we obtain the graph in Figure 11. ∎

We can adapt Equations 3 to represent a surface obtained through revolution about the y- or z-axis (see Exercise 30).

■ Tangent Planes

We now find the tangent plane to a parametric surface S traced out by a vector function

$$\mathbf{r}(u, v) = x(u, v) \, \mathbf{i} + y(u, v) \, \mathbf{j} + z(u, v) \, \mathbf{k}$$

at a point P_0 with position vector $\mathbf{r}(u_0, v_0)$. If we keep u constant by putting $u = u_0$, then $\mathbf{r}(u_0, v)$ becomes a vector function of the single parameter v and defines a grid curve C_1 lying on S. (See Figure 12.) The tangent vector to C_1 at P_0 is obtained by taking the partial derivative of $\mathbf{r}$ with respect to v:

$$\boxed{4} \qquad \mathbf{r}_v = \frac{\partial x}{\partial v}(u_0, v_0) \, \mathbf{i} + \frac{\partial y}{\partial v}(u_0, v_0) \, \mathbf{j} + \frac{\partial z}{\partial v}(u_0, v_0) \, \mathbf{k}$$

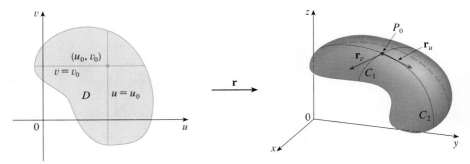

FIGURE 12

Similarly, if we keep v constant by putting $v = v_0$, we get a grid curve C_2 given by $\mathbf{r}(u, v_0)$ that lies on S, and its tangent vector at P_0 is

$$\boxed{5} \qquad \mathbf{r}_u = \frac{\partial x}{\partial u}(u_0, v_0)\mathbf{i} + \frac{\partial y}{\partial u}(u_0, v_0)\mathbf{j} + \frac{\partial z}{\partial u}(u_0, v_0)\mathbf{k}$$

If $\mathbf{r}_u \times \mathbf{r}_v$ is not $\mathbf{0}$, then the surface S is called **smooth** (it has no "corners"). For a smooth surface, the **tangent plane** is the plane that contains the tangent vectors $\mathbf{r}_u$ and $\mathbf{r}_v$, and the vector $\mathbf{r}_u \times \mathbf{r}_v$ is a normal vector to the tangent plane.

Figure 13 shows the self-intersecting surface in Example 9 and its tangent plane at $(1, 1, 3)$.

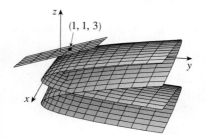

FIGURE 13

EXAMPLE 9 Find the tangent plane to the surface with parametric equations $x = u^2$, $y = v^2$, $z = u + 2v$ at the point $(1, 1, 3)$.

SOLUTION We first compute the tangent vectors:

$$\mathbf{r}_u = \frac{\partial x}{\partial u}\mathbf{i} + \frac{\partial y}{\partial u}\mathbf{j} + \frac{\partial z}{\partial u}\mathbf{k} = 2u\,\mathbf{i} + \mathbf{k}$$

$$\mathbf{r}_v = \frac{\partial x}{\partial v}\mathbf{i} + \frac{\partial y}{\partial v}\mathbf{j} + \frac{\partial z}{\partial v}\mathbf{k} = 2v\,\mathbf{j} + 2\,\mathbf{k}$$

Thus a normal vector to the tangent plane is

$$\mathbf{r}_u \times \mathbf{r}_v = \begin{vmatrix} \mathbf{i} & \mathbf{j} & \mathbf{k} \\ 2u & 0 & 1 \\ 0 & 2v & 2 \end{vmatrix} = -2v\,\mathbf{i} - 4u\,\mathbf{j} + 4uv\,\mathbf{k}$$

Notice that the point $(1, 1, 3)$ corresponds to the parameter values $u = 1$ and $v = 1$, so the normal vector there is

$$-2\,\mathbf{i} - 4\,\mathbf{j} + 4\,\mathbf{k}$$

Therefore an equation of the tangent plane at $(1, 1, 3)$ is

$$-2(x - 1) - 4(y - 1) + 4(z - 3) = 0$$

or

$$x + 2y - 2z + 3 = 0 \qquad \blacksquare$$

■ **Surface Area**

Now we define the surface area of a general parametric surface given by Equation 1. For simplicity we start by considering a surface whose parameter domain D is a rectangle, and we divide it into subrectangles R_{ij}. Let's choose (u_i^*, v_j^*) to be the lower left corner of R_{ij}. (See Figure 14.)

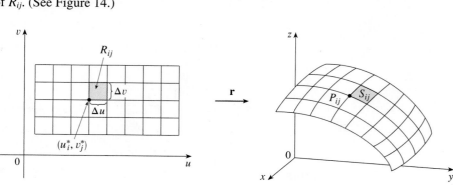

FIGURE 14
The image of the subrectangle R_{ij} is the patch S_{ij}.

The part S_{ij} of the surface S that corresponds to R_{ij} is called a *patch* and has the point P_{ij} with position vector $\mathbf{r}(u_i^*, v_j^*)$ as one of its corners. Let

$$\mathbf{r}_u^* = \mathbf{r}_u(u_i^*, v_j^*) \qquad \text{and} \qquad \mathbf{r}_v^* = \mathbf{r}_v(u_i^*, v_j^*)$$

be the tangent vectors at P_{ij} as given by Equations 5 and 4.

Figure 15(a) shows how the two edges of the patch that meet at P_{ij} can be approximated by vectors. These vectors, in turn, can be approximated by the vectors $\Delta u\, \mathbf{r}_u^*$ and $\Delta v\, \mathbf{r}_v^*$ because partial derivatives can be approximated by difference quotients. So we approximate S_{ij} by the parallelogram determined by the vectors $\Delta u\, \mathbf{r}_u^*$ and $\Delta v\, \mathbf{r}_v^*$. This parallelogram is shown in Figure 15(b) and lies in the tangent plane to S at P_{ij}. The area of this parallelogram is

$$\left| (\Delta u\, \mathbf{r}_u^*) \times (\Delta v\, \mathbf{r}_v^*) \right| = \left| \mathbf{r}_u^* \times \mathbf{r}_v^* \right| \Delta u\, \Delta v$$

and so an approximation to the area of S is

$$\sum_{i=1}^{m} \sum_{j=1}^{n} \left| \mathbf{r}_u^* \times \mathbf{r}_v^* \right| \Delta u\, \Delta v$$

Our intuition tells us that this approximation gets better as we increase the number of subrectangles, and we recognize the double sum as a Riemann sum for the double integral $\iint_D |\mathbf{r}_u \times \mathbf{r}_v|\, du\, dv$. This motivates the following definition.

S_{ij}

P_{ij}

(a)

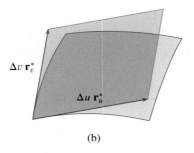

$\Delta v\, \mathbf{r}_v^*$

$\Delta u\, \mathbf{r}_u^*$

(b)

FIGURE 15
Approximating a patch by a parallelogram

---

6 **Definition** If a smooth parametric surface S is given by the equation

$$\mathbf{r}(u, v) = x(u, v)\, \mathbf{i} + y(u, v)\, \mathbf{j} + z(u, v)\, \mathbf{k} \qquad (u, v) \in D$$

and S is covered just once as (u, v) ranges throughout the parameter domain D, then the **surface area** of S is

$$A(S) = \iint_D |\mathbf{r}_u \times \mathbf{r}_v|\, dA$$

where $\quad \mathbf{r}_u = \dfrac{\partial x}{\partial u}\, \mathbf{i} + \dfrac{\partial y}{\partial u}\, \mathbf{j} + \dfrac{\partial z}{\partial u}\, \mathbf{k} \qquad \mathbf{r}_v = \dfrac{\partial x}{\partial v}\, \mathbf{i} + \dfrac{\partial y}{\partial v}\, \mathbf{j} + \dfrac{\partial z}{\partial v}\, \mathbf{k}$

EXAMPLE 10 Find the surface area of a sphere of radius a.

SOLUTION In Example 4 we found the parametric representation

$$x = a \sin \phi\, \cos \theta \qquad y = a \sin \phi\, \sin \theta \qquad z = a \cos \phi$$

where the parameter domain is

$$D = \{(\phi, \theta) \mid 0 \leqslant \phi \leqslant \pi,\ 0 \leqslant \theta \leqslant 2\pi\}$$

We first compute the cross product of the tangent vectors:

$$\mathbf{r}_\phi \times \mathbf{r}_\theta = \begin{vmatrix} \mathbf{i} & \mathbf{j} & \mathbf{k} \\ \dfrac{\partial x}{\partial \phi} & \dfrac{\partial y}{\partial \phi} & \dfrac{\partial z}{\partial \phi} \\ \dfrac{\partial x}{\partial \theta} & \dfrac{\partial y}{\partial \theta} & \dfrac{\partial z}{\partial \theta} \end{vmatrix} = \begin{vmatrix} \mathbf{i} & \mathbf{j} & \mathbf{k} \\ a \cos \phi\, \cos \theta & a \cos \phi\, \sin \theta & -a \sin \phi \\ -a \sin \phi\, \sin \theta & a \sin \phi\, \cos \theta & 0 \end{vmatrix}$$

$$= a^2 \sin^2\phi\, \cos \theta\, \mathbf{i} + a^2 \sin^2\phi\, \sin \theta\, \mathbf{j} + a^2 \sin \phi\, \cos \phi\, \mathbf{k}$$

Thus

$$|\mathbf{r}_\phi \times \mathbf{r}_\theta| = \sqrt{a^4 \sin^4\phi \, \cos^2\theta + a^4 \sin^4\phi \, \sin^2\theta + a^4 \sin^2\phi \, \cos^2\phi}$$

$$= \sqrt{a^4 \sin^4\phi + a^4 \sin^2\phi \, \cos^2\phi} = a^2 \sqrt{\sin^2\phi} = a^2 \sin \phi$$

since $\sin \phi \geq 0$ for $0 \leq \phi \leq \pi$. Therefore, by Definition 6, the area of the sphere is

$$A = \iint_D |\mathbf{r}_\phi \times \mathbf{r}_\theta| \, dA = \int_0^{2\pi} \int_0^\pi a^2 \sin \phi \, d\phi \, d\theta$$

$$= a^2 \int_0^{2\pi} d\theta \int_0^\pi \sin \phi \, d\phi = a^2(2\pi)2 = 4\pi a^2 \qquad \blacksquare$$

◼ Surface Area of the Graph of a Function

For the special case of a surface S with equation $z = f(x, y)$, where (x, y) lies in D and f has continuous partial derivatives, we take x and y as parameters. The parametric equations are

$$x = x \qquad y = y \qquad z = f(x, y)$$

so

$$\mathbf{r}_x = \mathbf{i} + \left(\frac{\partial f}{\partial x}\right)\mathbf{k} \qquad \mathbf{r}_y = \mathbf{j} + \left(\frac{\partial f}{\partial y}\right)\mathbf{k}$$

and

$$\boxed{7} \qquad \mathbf{r}_x \times \mathbf{r}_y = \begin{vmatrix} \mathbf{i} & \mathbf{j} & \mathbf{k} \\ 1 & 0 & \dfrac{\partial f}{\partial x} \\ 0 & 1 & \dfrac{\partial f}{\partial y} \end{vmatrix} = -\frac{\partial f}{\partial x}\mathbf{i} - \frac{\partial f}{\partial y}\mathbf{j} + \mathbf{k}$$

Thus we have

$$\boxed{8} \qquad |\mathbf{r}_x \times \mathbf{r}_y| = \sqrt{\left(\frac{\partial f}{\partial x}\right)^2 + \left(\frac{\partial f}{\partial y}\right)^2 + 1} = \sqrt{1 + \left(\frac{\partial z}{\partial x}\right)^2 + \left(\frac{\partial z}{\partial y}\right)^2}$$

and the surface area formula in Definition 6 becomes

Notice the similarity between the surface area formula in Equation 9 and the arc length formula

$$L = \int_a^b \sqrt{1 + \left(\frac{dy}{dx}\right)^2} \, dx$$

from Section 8.1.

$$\boxed{9} \qquad A(S) = \iint_D \sqrt{1 + \left(\frac{\partial z}{\partial x}\right)^2 + \left(\frac{\partial z}{\partial y}\right)^2} \, dA$$

EXAMPLE 11 Find the area of the part of the paraboloid $z = x^2 + y^2$ that lies under the plane $z = 9$.

SOLUTION The plane intersects the paraboloid in the circle $x^2 + y^2 = 9$, $z = 9$. Therefore the given surface lies above the disk D with center the origin and radius 3. (See

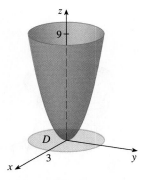

FIGURE 16

Figure 16.) Using Formula 9, we have

$$A = \iint_D \sqrt{1 + \left(\frac{\partial z}{\partial x}\right)^2 + \left(\frac{\partial z}{\partial y}\right)^2} \, dA$$

$$= \iint_D \sqrt{1 + (2x)^2 + (2y)^2} \, dA$$

$$= \iint_D \sqrt{1 + 4(x^2 + y^2)} \, dA$$

Converting to polar coordinates, we obtain

$$A = \int_0^{2\pi} \int_0^3 \sqrt{1 + 4r^2} \, r \, dr \, d\theta = \int_0^{2\pi} d\theta \int_0^3 r\sqrt{1 + 4r^2} \, dr$$

$$= 2\pi \left(\tfrac{1}{8}\right) \tfrac{2}{3} (1 + 4r^2)^{3/2} \Big]_0^3 = \frac{\pi}{6} \left(37\sqrt{37} - 1\right) \qquad \blacksquare$$

The question remains whether our definition of surface area (6) is consistent with the surface area formula from single-variable calculus (8.2.4).

We consider the surface S obtained by rotating the curve $y = f(x)$, $a \leqslant x \leqslant b$, about the x-axis, where $f(x) \geqslant 0$ and f' is continuous. From Equations 3 we know that parametric equations of S are

$$x = x \qquad y = f(x) \cos\theta \qquad z = f(x) \sin\theta \qquad a \leqslant x \leqslant b \qquad 0 \leqslant \theta \leqslant 2\pi$$

To compute the surface area of S we need the tangent vectors

$$\mathbf{r}_x = \mathbf{i} + f'(x) \cos\theta \, \mathbf{j} + f'(x) \sin\theta \, \mathbf{k}$$

$$\mathbf{r}_\theta = -f(x) \sin\theta \, \mathbf{j} + f(x) \cos\theta \, \mathbf{k}$$

Thus

$$\mathbf{r}_x \times \mathbf{r}_\theta = \begin{vmatrix} \mathbf{i} & \mathbf{j} & \mathbf{k} \\ 1 & f'(x) \cos\theta & f'(x) \sin\theta \\ 0 & -f(x) \sin\theta & f(x) \cos\theta \end{vmatrix}$$

$$= f(x)f'(x) \, \mathbf{i} - f(x) \cos\theta \, \mathbf{j} - f(x) \sin\theta \, \mathbf{k}$$

and so

$$|\mathbf{r}_x \times \mathbf{r}_\theta| = \sqrt{[f(x)]^2[f'(x)]^2 + [f(x)]^2 \cos^2\theta + [f(x)]^2 \sin^2\theta}$$

$$= \sqrt{[f(x)]^2[1 + [f'(x)]^2]} = f(x)\sqrt{1 + [f'(x)]^2}$$

because $f(x) \geqslant 0$. Therefore the area of S is

$$A = \iint_D |\mathbf{r}_x \times \mathbf{r}_\theta| \, dA$$

$$= \int_0^{2\pi} \int_a^b f(x)\sqrt{1 + [f'(x)]^2} \, dx \, d\theta$$

$$= 2\pi \int_a^b f(x)\sqrt{1 + [f'(x)]^2} \, dx$$

This is precisely the formula that was used to define the area of a surface of revolution in single-variable calculus (8.2.4).

16.6 EXERCISES

1–2 Determine whether the points P and Q lie on the given surface.

1. $\mathbf{r}(u, v) = \langle u + v, u - 2v, 3 + u - v \rangle$
   $P(4, -5, 1)$, $Q(0, 4, 6)$

2. $\mathbf{r}(u, v) = \langle 1 + u - v, u + v^2, u^2 - v^2 \rangle$
   $P(1, 2, 1)$, $Q(2, 3, 3)$

3–6 Identify the surface with the given vector equation.

3. $\mathbf{r}(u, v) = (u + v)\,\mathbf{i} + (3 - v)\,\mathbf{j} + (1 + 4u + 5v)\,\mathbf{k}$

4. $\mathbf{r}(u, v) = u^2\,\mathbf{i} + u \cos v\,\mathbf{j} + u \sin v\,\mathbf{k}$

5. $\mathbf{r}(s, t) = \langle s \cos t, s \sin t, s \rangle$

6. $\mathbf{r}(s, t) = \langle 3 \cos t, s, \sin t \rangle$, $-1 \leqslant s \leqslant 1$

7–12 Use a computer to graph the parametric surface. Get a printout and indicate on it which grid curves have u constant and which have v constant.

7. $\mathbf{r}(u, v) = \langle u^2, v^2, u + v \rangle$,
   $-1 \leqslant u \leqslant 1,\ -1 \leqslant v \leqslant 1$

8. $\mathbf{r}(u, v) = \langle u, v^3, -v \rangle$,
   $-2 \leqslant u \leqslant 2,\ -2 \leqslant v \leqslant 2$

9. $\mathbf{r}(u, v) = \langle u^3, u \sin v, u \cos v \rangle$,
   $-1 \leqslant u \leqslant 1,\ 0 \leqslant v \leqslant 2\pi$

10. $\mathbf{r}(u, v) = \langle u, \sin(u + v), \sin v \rangle$,
   $-\pi \leqslant u \leqslant \pi,\ -\pi \leqslant v \leqslant \pi$

11. $x = \sin v$, $y = \cos u \sin 4v$, $z = \sin 2u \sin 4v$,
   $0 \leqslant u \leqslant 2\pi,\ -\pi/2 \leqslant v \leqslant \pi/2$

12. $x = \cos u$, $y = \sin u \sin v$, $z = \cos v$,
   $0 \leqslant u \leqslant 2\pi,\ 0 \leqslant v \leqslant 2\pi$

13–18 Match the equations with the graphs labeled I–VI and give reasons for your answers. Determine which families of grid curves have u constant and which have v constant.

13. $\mathbf{r}(u, v) = u \cos v\,\mathbf{i} + u \sin v\,\mathbf{j} + v\,\mathbf{k}$

14. $\mathbf{r}(u, v) = uv^2\,\mathbf{i} + u^2 v\,\mathbf{j} + (u^2 - v^2)\,\mathbf{k}$

15. $\mathbf{r}(u, v) = (u^3 - u)\,\mathbf{i} + v^2\,\mathbf{j} + u^2\,\mathbf{k}$

16. $x = (1 - u)(3 + \cos v) \cos 4\pi u$,

   $y = (1 - u)(3 + \cos v) \sin 4\pi u$,

   $z = 3u + (1 - u) \sin v$

17. $x = \cos^3 u \cos^3 v$, $y = \sin^3 u \cos^3 v$, $z = \sin^3 v$

18. $x = \sin u$, $y = \cos u \sin v$, $z = \sin v$

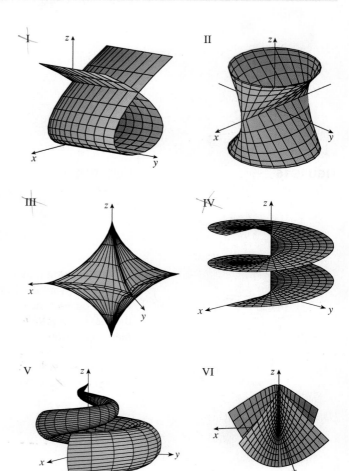

19–26 Find a parametric representation for the surface.

19. The plane through the origin that contains the vectors $\mathbf{i} - \mathbf{j}$ and $\mathbf{j} - \mathbf{k}$

20. The plane that passes through the point $(0, -1, 5)$ and contains the vectors $\langle 2, 1, 4 \rangle$ and $\langle -3, 2, 5 \rangle$

21. The part of the hyperboloid $4x^2 - 4y^2 - z^2 = 4$ that lies in front of the yz-plane

22. The part of the ellipsoid $x^2 + 2y^2 + 3z^2 = 1$ that lies to the left of the xz-plane

23. The part of the sphere $x^2 + y^2 + z^2 = 4$ that lies above the cone $z = \sqrt{x^2 + y^2}$

24. The part of the cylinder $x^2 + z^2 = 9$ that lies above the xy-plane and between the planes $y = -4$ and $y = 4$

25. The part of the sphere $x^2 + y^2 + z^2 = 36$ that lies between the planes $z = 0$ and $z = 3\sqrt{3}$

26. The part of the plane $z = x + 3$ that lies inside the cylinder $x^2 + y^2 = 1$

27–28 Use a graphing device to produce a graph that looks like the given one.

27.

28.

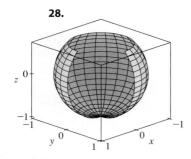

---

29. Find parametric equations for the surface obtained by rotating the curve $y = 1/(1 + x^2)$, $-2 \le x \le 2$, about the x-axis and use them to graph the surface.

30. Find parametric equations for the surface obtained by rotating the curve $x = 1/y$, $y \ge 1$, about the y-axis and use them to graph the surface.

31. (a) What happens to the spiral tube in Example 2 (see Figure 5) if we replace $\cos u$ by $\sin u$ and $\sin u$ by $\cos u$?
   (b) What happens if we replace $\cos u$ by $\cos 2u$ and $\sin u$ by $\sin 2u$?

32. The surface with parametric equations

$$x = 2 \cos \theta + r \cos(\theta/2)$$

$$y = 2 \sin \theta + r \cos(\theta/2)$$

$$z = r \sin(\theta/2)$$

where $-\frac{1}{2} \le r \le \frac{1}{2}$ and $0 \le \theta \le 2\pi$, is called a **Möbius strip**. Graph this surface with several viewpoints. What is unusual about it?

33–36 Find an equation of the tangent plane to the given parametric surface at the specified point.

33. $x = u + v$, $\quad y = 3u^2$, $\quad z = u - v$; $\quad (2, 3, 0)$

34. $x = u^2 + 1$, $\quad y = v^3 + 1$, $\quad z = u + v$; $\quad (5, 2, 3)$

35. $\mathbf{r}(u, v) = u \cos v\, \mathbf{i} + u \sin v\, \mathbf{j} + v\, \mathbf{k}$; $\quad u = 1$, $v = \pi/3$

36. $\mathbf{r}(u, v) = \sin u\, \mathbf{i} + \cos u \sin v\, \mathbf{j} + \sin v\, \mathbf{k}$; $u = \pi/6$, $v = \pi/6$

37–38 Find an equation of the tangent plane to the given parametric surface at the specified point. Graph the surface and the tangent plane.

37. $\mathbf{r}(u, v) = u^2 \mathbf{i} + 2u \sin v\, \mathbf{j} + u \cos v\, \mathbf{k}$; $\quad u = 1$, $v = 0$

38. $\mathbf{r}(u, v) = (1 - u^2 - v^2)\, \mathbf{i} - v\, \mathbf{j} - u\, \mathbf{k}$; $\quad (-1, -1, -1)$

39–50 Find the area of the surface.

39. The part of the plane $3x + 2y + z = 6$ that lies in the first octant

40. The part of the plane with vector equation $\mathbf{r}(u, v) = \langle u + v, 2 - 3u, 1 + u - v \rangle$ that is given by $0 \le u \le 2$, $-1 \le v \le 1$

41. The part of the plane $x + 2y + 3z = 1$ that lies inside the cylinder $x^2 + y^2 = 3$

42. The part of the cone $z = \sqrt{x^2 + y^2}$ that lies between the plane $y = x$ and the cylinder $y = x^2$

43. The surface $z = \frac{2}{3}(x^{3/2} + y^{3/2})$, $0 \le x \le 1$, $0 \le y \le 1$

44. The part of the surface $z = 4 - 2x^2 + y$ that lies above the triangle with vertices $(0, 0)$, $(1, 0)$, and $(1, 1)$

45. The part of the surface $z = xy$ that lies within the cylinder $x^2 + y^2 = 1$

46. The part of the surface $x = z^2 + y$ that lies between the planes $y = 0$, $y = 2$, $z = 0$, and $z = 2$

47. The part of the paraboloid $y = x^2 + z^2$ that lies within the cylinder $x^2 + z^2 = 16$

48. The helicoid (or spiral ramp) with vector equation $\mathbf{r}(u, v) = u \cos v\, \mathbf{i} + u \sin v\, \mathbf{j} + v\, \mathbf{k}$, $0 \le u \le 1$, $0 \le v \le \pi$

49. The surface with parametric equations $x = u^2$, $y = uv$, $z = \frac{1}{2}v^2$, $0 \le u \le 1$, $0 \le v \le 2$

50. The part of the sphere $x^2 + y^2 + z^2 = b^2$ that lies inside the cylinder $x^2 + y^2 = a^2$, where $0 < a < b$

---

51. If the equation of a surface S is $z = f(x, y)$, where $x^2 + y^2 \le R^2$, and you know that $|f_x| \le 1$ and $|f_y| \le 1$, what can you say about $A(S)$?

52–53 Find the area of the surface correct to four decimal places by expressing the area in terms of a single integral and using your calculator to estimate the integral.

52. The part of the surface $z = \cos(x^2 + y^2)$ that lies inside the cylinder $x^2 + y^2 = 1$

53. The part of the surface $z = \ln(x^2 + y^2 + 2)$ that lies above the disk $x^2 + y^2 \le 1$

---

54. Find, to four decimal places, the area of the part of the surface $z = (1 + x^2)/(1 + y^2)$ that lies above the square $|x| + |y| \le 1$. Illustrate by graphing this part of the surface.

55. (a) Use the Midpoint Rule for double integrals (see Section 15.1) with six squares to estimate the area of the surface $z = 1/(1 + x^2 + y^2)$, $0 \le x \le 6$, $0 \le y \le 4$.
   (b) Use a computer algebra system to approximate the surface area in part (a) to four decimal places. Compare with the answer to part (a).

CAS **56.** Find the area of the surface with vector equation
$\mathbf{r}(u, v) = \langle \cos^3 u \, \cos^3 v, \, \sin^3 u \, \cos^3 v, \, \sin^3 v \rangle$, $0 \le u \le \pi$,
$0 \le v \le 2\pi$. State your answer correct to four decimal
places.

CAS **57.** Find the exact area of the surface $z = 1 + 2x + 3y + 4y^2$,
$1 \le x \le 4, 0 \le y \le 1$.

58. (a) Set up, but do not evaluate, a double integral for
the area of the surface with parametric equations
$x = au \cos v$, $y = bu \sin v$, $z = u^2$, $0 \le u \le 2$,
$0 \le v \le 2\pi$.

(b) Eliminate the parameters to show that the surface is an
elliptic paraboloid and set up another double integral
for the surface area.

(c) Use the parametric equations in part (a) with $a = 2$ and
$b = 3$ to graph the surface.

CAS (d) For the case $a = 2$, $b = 3$, use a computer algebra
system to find the surface area correct to four decimal
places.

59. (a) Show that the parametric equations $x = a \sin u \cos v$,
$y = b \sin u \sin v$, $z = c \cos u$, $0 \le u \le \pi$,
$0 \le v \le 2\pi$, represent an ellipsoid.

(b) Use the parametric equations in part (a) to graph the
ellipsoid for the case $a = 1$, $b = 2$, $c = 3$.

(c) Set up, but do not evaluate, a double integral for the surface area of the ellipsoid in part (b).

60. (a) Show that the parametric equations $x = a \cosh u \cos v$,
$y = b \cosh u \sin v$, $z = c \sinh u$, represent a hyperboloid of one sheet.

(b) Use the parametric equations in part (a) to graph the
hyperboloid for the case $a = 1$, $b = 2$, $c = 3$.

(c) Set up, but do not evaluate, a double integral for the surface area of the part of the hyperboloid in part (b) that
lies between the planes $z = -3$ and $z = 3$.

61. Find the area of the part of the sphere $x^2 + y^2 + z^2 = 4z$
that lies inside the paraboloid $z = x^2 + y^2$.

62. The figure shows the surface created when the cylinder
$y^2 + z^2 = 1$ intersects the cylinder $x^2 + z^2 = 1$. Find the
area of this surface.

63. Find the area of the part of the sphere $x^2 + y^2 + z^2 = a^2$
that lies inside the cylinder $x^2 + y^2 = ax$.

64. (a) Find a parametric representation for the torus obtained
by rotating about the z-axis the circle in the xz-plane
with center $(b, 0, 0)$ and radius $a < b$. [*Hint:* Take as
parameters the angles θ and α shown in the figure.]

(b) Use the parametric equations found in part (a) to graph
the torus for several values of a and b.

(c) Use the parametric representation from part (a) to find
the surface area of the torus.

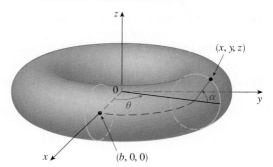

16.7 Surface Integrals

The relationship between surface integrals and surface area is much the same as the relationship between line integrals and arc length. Suppose f is a function of three variables
whose domain includes a surface S. We will define the surface integral of f over S in
such a way that, in the case where $f(x, y, z) = 1$, the value of the surface integral is equal
to the surface area of S. We start with parametric surfaces and then deal with the special
case where S is the graph of a function of two variables.

■ Parametric Surfaces

Suppose that a surface S has a vector equation

$$\mathbf{r}(u, v) = x(u, v) \, \mathbf{i} + y(u, v) \, \mathbf{j} + z(u, v) \, \mathbf{k} \qquad (u, v) \in D$$

We first assume that the parameter domain D is a rectangle and we divide it into subrect-

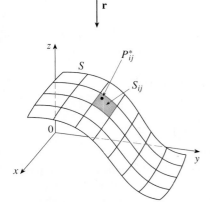

FIGURE 1

We assume that the surface is covered only once as (u, v) ranges throughout D. The value of the surface integral does not depend on the parametrization that is used.

angles R_{ij} with dimensions Δu and Δv. Then the surface S is divided into corresponding patches S_{ij} as in Figure 1. We evaluate f at a point P_{ij}^* in each patch, multiply by the area ΔS_{ij} of the patch, and form the Riemann sum

$$\sum_{i=1}^{m} \sum_{j=1}^{n} f(P_{ij}^*) \, \Delta S_{ij}$$

Then we take the limit as the number of patches increases and define the **surface integral of f over the surface S** as

$$\boxed{1} \qquad \iint_S f(x, y, z) \, dS = \lim_{m, n \to \infty} \sum_{i=1}^{m} \sum_{j=1}^{n} f(P_{ij}^*) \, \Delta S_{ij}$$

Notice the analogy with the definition of a line integral (16.2.2) and also the analogy with the definition of a double integral (15.1.5).

To evaluate the surface integral in Equation 1 we approximate the patch area ΔS_{ij} by the area of an approximating parallelogram in the tangent plane. In our discussion of surface area in Section 16.6 we made the approximation

$$\Delta S_{ij} \approx |\mathbf{r}_u \times \mathbf{r}_v| \, \Delta u \, \Delta v$$

where

$$\mathbf{r}_u = \frac{\partial x}{\partial u} \mathbf{i} + \frac{\partial y}{\partial u} \mathbf{j} + \frac{\partial z}{\partial u} \mathbf{k} \qquad \mathbf{r}_v = \frac{\partial x}{\partial v} \mathbf{i} + \frac{\partial y}{\partial v} \mathbf{j} + \frac{\partial z}{\partial v} \mathbf{k}$$

are the tangent vectors at a corner of S_{ij}. If the components are continuous and $\mathbf{r}_u$ and $\mathbf{r}_v$ are nonzero and nonparallel in the interior of D, it can be shown from Definition 1, even when D is not a rectangle, that

$$\boxed{2} \qquad \iint_S f(x, y, z) \, dS = \iint_D f(\mathbf{r}(u, v)) \, |\mathbf{r}_u \times \mathbf{r}_v| \, dA$$

This should be compared with the formula for a line integral:

$$\int_C f(x, y, z) \, ds = \int_a^b f(\mathbf{r}(t)) \, |\mathbf{r}'(t)| \, dt$$

Observe also that

$$\iint_S 1 \, dS = \iint_D |\mathbf{r}_u \times \mathbf{r}_v| \, dA = A(S)$$

Formula 2 allows us to compute a surface integral by converting it into a double integral over the parameter domain D. When using this formula, remember that $f(\mathbf{r}(u, v))$ is evaluated by writing $x = x(u, v)$, $y = y(u, v)$, and $z = z(u, v)$ in the formula for $f(x, y, z)$.

EXAMPLE 1 Compute the surface integral $\iint_S x^2 \, dS$, where S is the unit sphere $x^2 + y^2 + z^2 = 1$.

SOLUTION As in Example 16.6.4, we use the parametric representation

$$x = \sin\phi \, \cos\theta \qquad y = \sin\phi \, \sin\theta \qquad z = \cos\phi \qquad 0 \le \phi \le \pi \qquad 0 \le \theta \le 2\pi$$

that is,
$$\mathbf{r}(\phi, \theta) = \sin\phi \, \cos\theta \, \mathbf{i} + \sin\phi \, \sin\theta \, \mathbf{j} + \cos\phi \, \mathbf{k}$$

As in Example 16.6.10 , we can compute that
$$|\mathbf{r}_\phi \times \mathbf{r}_\theta| = \sin\phi$$

Therefore, by Formula 2,

$$\iint_S x^2 \, dS = \iint_D (\sin\phi \, \cos\theta)^2 \, |\mathbf{r}_\phi \times \mathbf{r}_\theta| \, dA$$

Here we use the identities
$$\cos^2\theta = \tfrac{1}{2}(1 + \cos 2\theta)$$
$$\sin^2\phi = 1 - \cos^2\phi$$

Instead, we could use Formulas 64 and 67 in the Table of Integrals.

$$= \int_0^{2\pi} \int_0^\pi \sin^2\phi \, \cos^2\theta \, \sin\phi \, d\phi \, d\theta = \int_0^{2\pi} \cos^2\theta \, d\theta \, \int_0^\pi \sin^3\phi \, d\phi$$

$$= \int_0^{2\pi} \tfrac{1}{2}(1 + \cos 2\theta) \, d\theta \, \int_0^\pi (\sin\phi - \sin\phi \, \cos^2\phi) \, d\phi$$

$$= \tfrac{1}{2}\left[\theta + \tfrac{1}{2}\sin 2\theta\right]_0^{2\pi} \left[-\cos\phi + \tfrac{1}{3}\cos^3\phi\right]_0^\pi = \frac{4\pi}{3}$$ ∎

Surface integrals have applications similar to those for the integrals we have previously considered. For example, if a thin sheet (say, of aluminum foil) has the shape of a surface S and the density (mass per unit area) at the point (x, y, z) is $\rho(x, y, z)$, then the total **mass** of the sheet is

$$m = \iint_S \rho(x, y, z) \, dS$$

and the **center of mass** is $(\bar{x}, \bar{y}, \bar{z})$, where

$$\bar{x} = \frac{1}{m}\iint_S x\rho(x, y, z) \, dS \qquad \bar{y} = \frac{1}{m}\iint_S y\rho(x, y, z) \, dS \qquad \bar{z} = \frac{1}{m}\iint_S z\rho(x, y, z) \, dS$$

Moments of inertia can also be defined as before (see Exercise 41).

■ Graphs of Functions

Any surface S with equation $z = g(x, y)$ can be regarded as a parametric surface with parametric equations

$$x = x \qquad y = y \qquad z = g(x, y)$$

and so we have
$$\mathbf{r}_x = \mathbf{i} + \left(\frac{\partial g}{\partial x}\right)\mathbf{k} \qquad \mathbf{r}_y = \mathbf{j} + \left(\frac{\partial g}{\partial y}\right)\mathbf{k}$$

Thus

$$\boxed{3} \qquad \mathbf{r}_x \times \mathbf{r}_y = -\frac{\partial g}{\partial x}\mathbf{i} - \frac{\partial g}{\partial y}\mathbf{j} + \mathbf{k}$$

and
$$|\mathbf{r}_x \times \mathbf{r}_y| = \sqrt{\left(\frac{\partial z}{\partial x}\right)^2 + \left(\frac{\partial z}{\partial y}\right)^2 + 1}$$

Therefore, in this case, Formula 2 becomes

$$
\boxed{4} \qquad \iint_S f(x, y, z)\, dS = \iint_D f(x, y, g(x, y)) \sqrt{\left(\frac{\partial z}{\partial x}\right)^2 + \left(\frac{\partial z}{\partial y}\right)^2 + 1}\; dA
$$

Similar formulas apply when it is more convenient to project S onto the yz-plane or xz-plane. For instance, if S is a surface with equation $y = h(x, z)$ and D is its projection onto the xz-plane, then

$$
\iint_S f(x, y, z)\, dS = \iint_D f(x, h(x, z), z) \sqrt{\left(\frac{\partial y}{\partial x}\right)^2 + \left(\frac{\partial y}{\partial z}\right)^2 + 1}\; dA
$$

EXAMPLE 2 Evaluate $\iint_S y\, dS$, where S is the surface $z = x + y^2, 0 \leqslant x \leqslant 1$, $0 \leqslant y \leqslant 2$. (See Figure 2.)

SOLUTION Since

$$
\frac{\partial z}{\partial x} = 1 \qquad \text{and} \qquad \frac{\partial z}{\partial y} = 2y
$$

Formula 4 gives

$$
\iint_S y\, dS = \iint_D y \sqrt{1 + \left(\frac{\partial z}{\partial x}\right)^2 + \left(\frac{\partial z}{\partial y}\right)^2}\; dA
$$

$$
= \int_0^1 \int_0^2 y \sqrt{1 + 1 + 4y^2}\; dy\, dx
$$

$$
= \int_0^1 dx \; \sqrt{2} \int_0^2 y \sqrt{1 + 2y^2}\; dy
$$

$$
= \sqrt{2} \left(\tfrac{1}{4}\right)\tfrac{2}{3}(1 + 2y^2)^{3/2}\Big]_0^2 = \frac{13\sqrt{2}}{3} \qquad \blacksquare
$$

FIGURE 2

If S is a piecewise-smooth surface, that is, a finite union of smooth surfaces $S_1, S_2, \ldots,$ S_n that intersect only along their boundaries, then the surface integral of f over S is defined by

$$
\iint_S f(x, y, z)\, dS = \iint_{S_1} f(x, y, z)\, dS + \cdots + \iint_{S_n} f(x, y, z)\, dS
$$

EXAMPLE 3 Evaluate $\iint_S z\, dS$, where S is the surface whose sides S_1 are given by the cylinder $x^2 + y^2 = 1$, whose bottom S_2 is the disk $x^2 + y^2 \leqslant 1$ in the plane $z = 0$, and whose top S_3 is the part of the plane $z = 1 + x$ that lies above S_2.

SOLUTION The surface S is shown in Figure 3. (We have changed the usual position of the axes to get a better look at S.) For S_1 we use θ and z as parameters (see Example 16.6.5) and write its parametric equations as

$$
x = \cos\theta \qquad y = \sin\theta \qquad z = z
$$

where

$$
0 \leqslant \theta \leqslant 2\pi \qquad \text{and} \qquad 0 \leqslant z \leqslant 1 + x = 1 + \cos\theta
$$

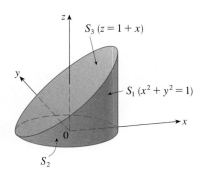

FIGURE 3

Therefore

$$\mathbf{r}_\theta \times \mathbf{r}_z = \begin{vmatrix} \mathbf{i} & \mathbf{j} & \mathbf{k} \\ -\sin\theta & \cos\theta & 0 \\ 0 & 0 & 1 \end{vmatrix} = \cos\theta\,\mathbf{i} + \sin\theta\,\mathbf{j}$$

and

$$|\mathbf{r}_\theta \times \mathbf{r}_z| = \sqrt{\cos^2\theta + \sin^2\theta} = 1$$

Thus the surface integral over S_1 is

$$\iint_{S_1} z\,dS = \iint_D z\,|\mathbf{r}_\theta \times \mathbf{r}_z|\,dA$$

$$= \int_0^{2\pi}\int_0^{1+\cos\theta} z\,dz\,d\theta = \int_0^{2\pi} \tfrac{1}{2}(1+\cos\theta)^2\,d\theta$$

$$= \tfrac{1}{2}\int_0^{2\pi}\left[1 + 2\cos\theta + \tfrac{1}{2}(1+\cos 2\theta)\right]d\theta$$

$$= \tfrac{1}{2}\left[\tfrac{3}{2}\theta + 2\sin\theta + \tfrac{1}{4}\sin 2\theta\right]_0^{2\pi} = \frac{3\pi}{2}$$

Since S_2 lies in the plane $z = 0$, we have

$$\iint_{S_2} z\,dS = \iint_{S_2} 0\,dS = 0$$

The top surface S_3 lies above the unit disk D and is part of the plane $z = 1 + x$. So, taking $g(x, y) = 1 + x$ in Formula 4 and converting to polar coordinates, we have

$$\iint_{S_3} z\,dS = \iint_D (1+x)\sqrt{1 + \left(\frac{\partial z}{\partial x}\right)^2 + \left(\frac{\partial z}{\partial y}\right)^2}\,dA$$

$$= \int_0^{2\pi}\int_0^1 (1 + r\cos\theta)\sqrt{1 + 1 + 0}\,r\,dr\,d\theta$$

$$= \sqrt{2}\int_0^{2\pi}\int_0^1 (r + r^2\cos\theta)\,dr\,d\theta$$

$$= \sqrt{2}\int_0^{2\pi}\left(\tfrac{1}{2} + \tfrac{1}{3}\cos\theta\right)d\theta$$

$$= \sqrt{2}\left[\frac{\theta}{2} + \frac{\sin\theta}{3}\right]_0^{2\pi} = \sqrt{2}\,\pi$$

Therefore

$$\iint_S z\,dS = \iint_{S_1} z\,dS + \iint_{S_2} z\,dS + \iint_{S_3} z\,dS$$

$$= \frac{3\pi}{2} + 0 + \sqrt{2}\,\pi = \left(\tfrac{3}{2} + \sqrt{2}\right)\pi$$

FIGURE 4
A Möbius strip

TEC Visual 16.7 shows a Möbius strip with a normal vector that can be moved along the surface.

Oriented Surfaces

To define surface integrals of vector fields, we need to rule out nonorientable surfaces such as the Möbius strip shown in Figure 4. [It is named after the German geometer August Möbius (1790–1868).] You can construct one for yourself by taking a long rectangular strip of paper, giving it a half-twist, and taping the short edges together as in Figure 5. If an ant were to crawl along the Möbius strip starting at a point P, it would end up on the "other side" of the strip (that is, with its upper side pointing in the opposite direction). Then, if the ant continued to crawl in the same direction, it would end up back at the same point P without ever having crossed an edge. (If you have constructed a Möbius strip, try drawing a pencil line down the middle.) Therefore a Möbius strip really has only one side. You can graph the Möbius strip using the parametric equations in Exercise 16.6.32.

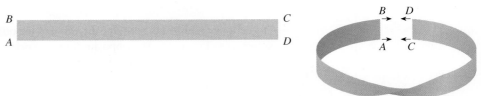

FIGURE 5
Constructing a Möbius strip

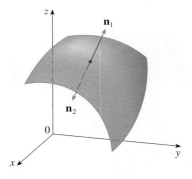

FIGURE 6

From now on we consider only orientable (two-sided) surfaces. We start with a surface S that has a tangent plane at every point (x, y, z) on S (except at any boundary point). There are two unit normal vectors $\mathbf{n}_1$ and $\mathbf{n}_2 = -\mathbf{n}_1$ at (x, y, z). (See Figure 6.)

If it is possible to choose a unit normal vector $\mathbf{n}$ at every such point (x, y, z) so that $\mathbf{n}$ varies continuously over S, then S is called an **oriented surface** and the given choice of $\mathbf{n}$ provides S with an **orientation**. There are two possible orientations for any orientable surface (see Figure 7).

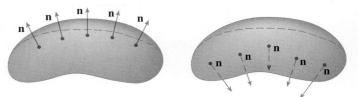

FIGURE 7
The two orientations
of an orientable surface

For a surface $z = g(x, y)$ given as the graph of g, we use Equation 3 to associate with the surface a natural orientation given by the unit normal vector

$$\boxed{5} \qquad \mathbf{n} = \frac{-\dfrac{\partial g}{\partial x}\mathbf{i} - \dfrac{\partial g}{\partial y}\mathbf{j} + \mathbf{k}}{\sqrt{1 + \left(\dfrac{\partial g}{\partial x}\right)^2 + \left(\dfrac{\partial g}{\partial y}\right)^2}}$$

Since the $\mathbf{k}$-component is positive, this gives the *upward* orientation of the surface.

If S is a smooth orientable surface given in parametric form by a vector function $\mathbf{r}(u, v)$, then it is automatically supplied with the orientation of the unit normal vector

$$\boxed{6} \qquad \mathbf{n} = \frac{\mathbf{r}_u \times \mathbf{r}_v}{|\mathbf{r}_u \times \mathbf{r}_v|}$$

and the opposite orientation is given by $-\mathbf{n}$. For instance, in Example 16.6.4 we found

the parametric representation

$$\mathbf{r}(\phi, \theta) = a \sin \phi \, \cos \theta \, \mathbf{i} + a \sin \phi \, \sin \theta \, \mathbf{j} + a \cos \phi \, \mathbf{k}$$

for the sphere $x^2 + y^2 + z^2 = a^2$. Then in Example 16.6.10 we found that

$$\mathbf{r}_\phi \times \mathbf{r}_\theta = a^2 \sin^2\phi \, \cos \theta \, \mathbf{i} + a^2 \sin^2\phi \, \sin \theta \, \mathbf{j} + a^2 \sin \phi \, \cos \phi \, \mathbf{k}$$

and

$$|\mathbf{r}_\phi \times \mathbf{r}_\theta| = a^2 \sin \phi$$

So the orientation induced by $\mathbf{r}(\phi, \theta)$ is defined by the unit normal vector

$$\mathbf{n} = \frac{\mathbf{r}_\phi \times \mathbf{r}_\theta}{|\mathbf{r}_\phi \times \mathbf{r}_\theta|} = \sin \phi \, \cos \theta \, \mathbf{i} + \sin \phi \, \sin \theta \, \mathbf{j} + \cos \phi \, \mathbf{k} = \frac{1}{a} \mathbf{r}(\phi, \theta)$$

Observe that $\mathbf{n}$ points in the same direction as the position vector, that is, outward from the sphere (see Figure 8). The opposite (inward) orientation would have been obtained (see Figure 9) if we had reversed the order of the parameters because $\mathbf{r}_\theta \times \mathbf{r}_\phi = -\mathbf{r}_\phi \times \mathbf{r}_\theta$.

FIGURE 8
Positive orientation

FIGURE 9
Negative orientation

For a **closed surface**, that is, a surface that is the boundary of a solid region E, the convention is that the **positive orientation** is the one for which the normal vectors point *outward* from E, and inward-pointing normals give the negative orientation (see Figures 8 and 9).

■ Surface Integrals of Vector Fields

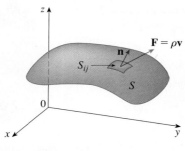

FIGURE 10

Suppose that S is an oriented surface with unit normal vector $\mathbf{n}$, and imagine a fluid with density $\rho(x, y, z)$ and velocity field $\mathbf{v}(x, y, z)$ flowing through S. (Think of S as an imaginary surface that doesn't impede the fluid flow, like a fishing net across a stream.) Then the rate of flow (mass per unit time) per unit area is $\rho \mathbf{v}$. If we divide S into small patches S_{ij}, as in Figure 10 (compare with Figure 1), then S_{ij} is nearly planar and so we can approximate the mass of fluid per unit time crossing S_{ij} in the direction of the normal $\mathbf{n}$ by the quantity

$$(\rho \mathbf{v} \cdot \mathbf{n}) A(S_{ij})$$

where ρ, $\mathbf{v}$, and $\mathbf{n}$ are evaluated at some point on S_{ij}. (Recall that the component of the vector $\rho \mathbf{v}$ in the direction of the unit vector $\mathbf{n}$ is $\rho \mathbf{v} \cdot \mathbf{n}$.) By summing these quantities and taking the limit we get, according to Definition 1, the surface integral of the function $\rho \mathbf{v} \cdot \mathbf{n}$ over S:

$$\boxed{7} \qquad \iint_S \rho \mathbf{v} \cdot \mathbf{n} \, dS = \iint_S \rho(x, y, z) \mathbf{v}(x, y, z) \cdot \mathbf{n}(x, y, z) \, dS$$

and this is interpreted physically as the rate of flow through S.

If we write $\mathbf{F} = \rho\mathbf{v}$, then $\mathbf{F}$ is also a vector field on $\mathbb{R}^3$ and the integral in Equation 7 becomes

$$\iint_S \mathbf{F} \cdot \mathbf{n} \, dS$$

A surface integral of this form occurs frequently in physics, even when $\mathbf{F}$ is not $\rho\mathbf{v}$, and is called the *surface integral* (or *flux integral*) of $\mathbf{F}$ over S.

> **8   Definition** If $\mathbf{F}$ is a continuous vector field defined on an oriented surface S with unit normal vector $\mathbf{n}$, then the **surface integral of F over S** is
>
> $$\iint_S \mathbf{F} \cdot d\mathbf{S} = \iint_S \mathbf{F} \cdot \mathbf{n} \, dS$$
>
> This integral is also called the **flux** of $\mathbf{F}$ across S.

In words, Definition 8 says that the surface integral of a vector field over S is equal to the surface integral of its normal component over S (as previously defined).

If S is given by a vector function $\mathbf{r}(u, v)$, then $\mathbf{n}$ is given by Equation 6, and from Definition 8 and Equation 2 we have

$$\iint_S \mathbf{F} \cdot d\mathbf{S} = \iint_S \mathbf{F} \cdot \frac{\mathbf{r}_u \times \mathbf{r}_v}{|\mathbf{r}_u \times \mathbf{r}_v|} \, dS$$

$$= \iint_D \left[\mathbf{F}(\mathbf{r}(u, v)) \cdot \frac{\mathbf{r}_u \times \mathbf{r}_v}{|\mathbf{r}_u \times \mathbf{r}_v|} \right] |\mathbf{r}_u \times \mathbf{r}_v| \, dA$$

where D is the parameter domain. Thus we have

Compare Equation 9 to the similar expression for evaluating line integrals of vector fields in Definition 16.2.13:

$$\int_C \mathbf{F} \cdot d\mathbf{r} = \int_a^b \mathbf{F}(\mathbf{r}(t)) \cdot \mathbf{r}'(t) \, dt$$

> **9**
>
> $$\iint_S \mathbf{F} \cdot d\mathbf{S} = \iint_D \mathbf{F} \cdot (\mathbf{r}_u \times \mathbf{r}_v) \, dA$$

EXAMPLE 4 Find the flux of the vector field $\mathbf{F}(x, y, z) = z\,\mathbf{i} + y\,\mathbf{j} + x\,\mathbf{k}$ across the unit sphere $x^2 + y^2 + z^2 = 1$.

Figure 11 shows the vector field $\mathbf{F}$ in Example 4 at points on the unit sphere.

SOLUTION As in Example 1, we use the parametric representation

$$\mathbf{r}(\phi, \theta) = \sin\phi\,\cos\theta\,\mathbf{i} + \sin\phi\,\sin\theta\,\mathbf{j} + \cos\phi\,\mathbf{k} \qquad 0 \le \phi \le \pi \qquad 0 \le \theta \le 2\pi$$

Then

$$\mathbf{F}(\mathbf{r}(\phi, \theta)) = \cos\phi\,\mathbf{i} + \sin\phi\,\sin\theta\,\mathbf{j} + \sin\phi\,\cos\theta\,\mathbf{k}$$

and, from Example 16.6.10,

$$\mathbf{r}_\phi \times \mathbf{r}_\theta = \sin^2\phi\,\cos\theta\,\mathbf{i} + \sin^2\phi\,\sin\theta\,\mathbf{j} + \sin\phi\,\cos\phi\,\mathbf{k}$$

Therefore

$$\mathbf{F}(\mathbf{r}(\phi, \theta)) \cdot (\mathbf{r}_\phi \times \mathbf{r}_\theta) = \cos\phi\,\sin^2\phi\,\cos\theta + \sin^3\phi\,\sin^2\theta + \sin^2\phi\,\cos\phi\,\cos\theta$$

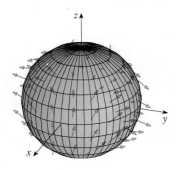

FIGURE 11

and, by Formula 9, the flux is

$$\iint_S \mathbf{F} \cdot d\mathbf{S} = \iint_D \mathbf{F} \cdot (\mathbf{r}_\phi \times \mathbf{r}_\theta)\, dA$$

$$= \int_0^{2\pi} \int_0^{\pi} (2 \sin^2\phi \, \cos\phi \, \cos\theta + \sin^3\phi \, \sin^2\theta)\, d\phi \, d\theta$$

$$= 2 \int_0^{\pi} \sin^2\phi \, \cos\phi \, d\phi \int_0^{2\pi} \cos\theta \, d\theta + \int_0^{\pi} \sin^3\phi \, d\phi \int_0^{2\pi} \sin^2\theta \, d\theta$$

$$= 0 + \int_0^{\pi} \sin^3\phi \, d\phi \int_0^{2\pi} \sin^2\theta \, d\theta \qquad \left(\text{since } \int_0^{2\pi} \cos\theta \, d\theta = 0\right)$$

$$= \frac{4\pi}{3}$$

by the same calculation as in Example 1.   ∎

If, for instance, the vector field in Example 4 is a velocity field describing the flow of a fluid with density 1, then the answer, $4\pi/3$, represents the rate of flow through the unit sphere in units of mass per unit time.

In the case of a surface S given by a graph $z = g(x, y)$, we can think of x and y as parameters and use Equation 3 to write

$$\mathbf{F} \cdot (\mathbf{r}_x \times \mathbf{r}_y) = (P\mathbf{i} + Q\mathbf{j} + R\mathbf{k}) \cdot \left(-\frac{\partial g}{\partial x}\mathbf{i} - \frac{\partial g}{\partial y}\mathbf{j} + \mathbf{k}\right)$$

Thus Formula 9 becomes

[10]
$$\iint_S \mathbf{F} \cdot d\mathbf{S} = \iint_D \left(-P\frac{\partial g}{\partial x} - Q\frac{\partial g}{\partial y} + R\right) dA$$

This formula assumes the upward orientation of S; for a downward orientation we multiply by -1. Similar formulas can be worked out if S is given by $y = h(x, z)$ or $x = k(y, z)$. (See Exercises 37 and 38.)

EXAMPLE 5 Evaluate $\iint_S \mathbf{F} \cdot d\mathbf{S}$, where $\mathbf{F}(x, y, z) = y\,\mathbf{i} + x\,\mathbf{j} + z\,\mathbf{k}$ and S is the boundary of the solid region E enclosed by the paraboloid $z = 1 - x^2 - y^2$ and the plane $z = 0$.

SOLUTION S consists of a parabolic top surface S_1 and a circular bottom surface S_2. (See Figure 12.) Since S is a closed surface, we use the convention of positive (outward) orientation. This means that S_1 is oriented upward and we can use Equation 10 with D being the projection of S_1 onto the xy-plane, namely, the disk $x^2 + y^2 \leqslant 1$. Since

$$P(x, y, z) = y \qquad Q(x, y, z) = x \qquad R(x, y, z) = z = 1 - x^2 - y^2$$

on S_1 and

$$\frac{\partial g}{\partial x} = -2x \qquad \frac{\partial g}{\partial y} = -2y$$

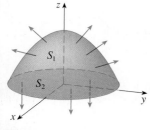

FIGURE 12

we have

$$\iint\limits_{S_1} \mathbf{F} \cdot d\mathbf{S} = \iint\limits_{D} \left(-P \frac{\partial g}{\partial x} - Q \frac{\partial g}{\partial y} + R \right) dA$$

$$= \iint\limits_{D} \left[-y(-2x) - x(-2y) + 1 - x^2 - y^2 \right] dA$$

$$= \iint\limits_{D} (1 + 4xy - x^2 - y^2)\, dA$$

$$= \int_0^{2\pi} \int_0^1 (1 + 4r^2 \cos\theta \sin\theta - r^2)\, r\, dr\, d\theta$$

$$= \int_0^{2\pi} \int_0^1 (r - r^3 + 4r^3 \cos\theta \sin\theta)\, dr\, d\theta$$

$$= \int_0^{2\pi} \left(\tfrac{1}{4} + \cos\theta \sin\theta \right) d\theta = \tfrac{1}{4}(2\pi) + 0 = \frac{\pi}{2}$$

The disk S_2 is oriented downward, so its unit normal vector is $\mathbf{n} = -\mathbf{k}$ and we have

$$\iint\limits_{S_2} \mathbf{F} \cdot d\mathbf{S} = \iint\limits_{S_2} \mathbf{F} \cdot (-\mathbf{k})\, dS = \iint\limits_{D} (-z)\, dA = \iint\limits_{D} 0\, dA = 0$$

since $z = 0$ on S_2. Finally, we compute, by definition, $\iint_S \mathbf{F} \cdot d\mathbf{S}$ as the sum of the surface integrals of $\mathbf{F}$ over the pieces S_1 and S_2:

$$\iint\limits_{S} \mathbf{F} \cdot d\mathbf{S} = \iint\limits_{S_1} \mathbf{F} \cdot d\mathbf{S} + \iint\limits_{S_2} \mathbf{F} \cdot d\mathbf{S} = \frac{\pi}{2} + 0 = \frac{\pi}{2} \qquad \blacksquare$$

Although we motivated the surface integral of a vector field using the example of fluid flow, this concept also arises in other physical situations. For instance, if $\mathbf{E}$ is an electric field (see Example 16.1.5), then the surface integral

$$\iint\limits_{S} \mathbf{E} \cdot d\mathbf{S}$$

is called the **electric flux** of $\mathbf{E}$ through the surface S. One of the important laws of electrostatics is **Gauss's Law**, which says that the net charge enclosed by a closed surface S is

$$\boxed{11} \qquad\qquad Q = \varepsilon_0 \iint\limits_{S} \mathbf{E} \cdot d\mathbf{S}$$

where ε_0 is a constant (called the permittivity of free space) that depends on the units used. (In the SI system, $\varepsilon_0 \approx 8.8542 \times 10^{-12} \text{ C}^2/\text{N} \cdot \text{m}^2$.) Therefore, if the vector field $\mathbf{F}$ in Example 4 represents an electric field, we can conclude that the charge enclosed by S is $Q = \tfrac{4}{3}\pi\varepsilon_0$.

Another application of surface integrals occurs in the study of heat flow. Suppose the temperature at a point (x, y, z) in a body is $u(x, y, z)$. Then the **heat flow** is defined as the vector field

$$\mathbf{F} = -K\, \nabla u$$

where K is an experimentally determined constant called the **conductivity** of the substance. The rate of heat flow across the surface S in the body is then given by the surface integral

$$\iint_S \mathbf{F} \cdot d\mathbf{S} = -K \iint_S \nabla u \cdot d\mathbf{S}$$

EXAMPLE 6 The temperature u in a metal ball is proportional to the square of the distance from the center of the ball. Find the rate of heat flow across a sphere S of radius a with center at the center of the ball.

SOLUTION Taking the center of the ball to be at the origin, we have

$$u(x, y, z) = C(x^2 + y^2 + z^2)$$

where C is the proportionality constant. Then the heat flow is

$$\mathbf{F}(x, y, z) = -K \nabla u = -KC(2x\,\mathbf{i} + 2y\,\mathbf{j} + 2z\,\mathbf{k})$$

where K is the conductivity of the metal. Instead of using the usual parametrization of the sphere as in Example 4, we observe that the outward unit normal to the sphere $x^2 + y^2 + z^2 = a^2$ at the point (x, y, z) is

$$\mathbf{n} = \frac{1}{a}(x\,\mathbf{i} + y\,\mathbf{j} + z\,\mathbf{k})$$

and so

$$\mathbf{F} \cdot \mathbf{n} = -\frac{2KC}{a}(x^2 + y^2 + z^2)$$

But on S we have $x^2 + y^2 + z^2 = a^2$, so $\mathbf{F} \cdot \mathbf{n} = -2aKC$. Therefore the rate of heat flow across S is

$$\iint_S \mathbf{F} \cdot d\mathbf{S} = \iint_S \mathbf{F} \cdot \mathbf{n}\, dS = -2aKC \iint_S dS$$

$$= -2aKCA(S) = -2aKC(4\pi a^2) = -8KC\pi a^3 \qquad \blacksquare$$

16.7 EXERCISES

1. Let S be the surface of the box enclosed by the planes $x = \pm 1$, $y = \pm 1$, $z = \pm 1$. Approximate $\iint_S \cos(x + 2y + 3z)\, dS$ by using a Riemann sum as in Definition 1, taking the patches S_{ij} to be the squares that are the faces of the box S and the points P_{ij}^* to be the centers of the squares.

2. A surface S consists of the cylinder $x^2 + y^2 = 1$, $-1 \leqslant z \leqslant 1$, together with its top and bottom disks. Suppose you know that f is a continuous function with

   $$f(\pm 1, 0, 0) = 2 \qquad f(0, \pm 1, 0) = 3 \qquad f(0, 0, \pm 1) = 4$$

   Estimate the value of $\iint_S f(x, y, z)\, dS$ by using a Riemann sum, taking the patches S_{ij} to be four quarter-cylinders and the top and bottom disks.

3. Let H be the hemisphere $x^2 + y^2 + z^2 = 50$, $z \geqslant 0$, and suppose f is a continuous function with $f(3, 4, 5) = 7$, $f(3, -4, 5) = 8$, $f(-3, 4, 5) = 9$, and $f(-3, -4, 5) = 12$. By dividing H into four patches, estimate the value of $\iint_H f(x, y, z)\, dS$.

4. Suppose that $f(x, y, z) = g(\sqrt{x^2 + y^2 + z^2})$, where g is a function of one variable such that $g(2) = -5$. Evaluate $\iint_S f(x, y, z)\, dS$, where S is the sphere $x^2 + y^2 + z^2 = 4$.

5–20 Evaluate the surface integral.

5. $\iint_S (x + y + z)\, dS$,
   S is the parallelogram with parametric equations $x = u + v$, $y = u - v$, $z = 1 + 2u + v$, $0 \leqslant u \leqslant 2$, $0 \leqslant v \leqslant 1$

6. $\iint_S xyz \, dS$,
S is the cone with parametric equations $x = u \cos v$,
$y = u \sin v$, $z = u$, $0 \leqslant u \leqslant 1$, $0 \leqslant v \leqslant \pi/2$

7. $\iint_S y \, dS$, S is the helicoid with vector equation
$\mathbf{r}(u, v) = \langle u \cos v, u \sin v, v \rangle$, $0 \leqslant u \leqslant 1$, $0 \leqslant v \leqslant \pi$

8. $\iint_S (x^2 + y^2) \, dS$,
S is the surface with vector equation
$\mathbf{r}(u, v) = \langle 2uv, u^2 - v^2, u^2 + v^2 \rangle$, $u^2 + v^2 \leqslant 1$

9. $\iint_S x^2 yz \, dS$,
S is the part of the plane $z = 1 + 2x + 3y$ that lies above
the rectangle $[0, 3] \times [0, 2]$

10. $\iint_S xz \, dS$,
S is the part of the plane $2x + 2y + z = 4$ that lies in the
first octant

11. $\iint_S x \, dS$,
S is the triangular region with vertices $(1, 0, 0)$, $(0, -2, 0)$,
and $(0, 0, 4)$

12. $\iint_S y \, dS$,
S is the surface $z = \frac{2}{3}(x^{3/2} + y^{3/2})$, $0 \leqslant x \leqslant 1$, $0 \leqslant y \leqslant 1$

13. $\iint_S z^2 \, dS$,
S is the part of the paraboloid $x = y^2 + z^2$ given by
$0 \leqslant x \leqslant 1$

14. $\iint_S y^2 z^2 \, dS$,
S is the part of the cone $y = \sqrt{x^2 + z^2}$ given by $0 \leqslant y \leqslant 5$

15. $\iint_S x \, dS$,
S is the surface $y = x^2 + 4z$, $0 \leqslant x \leqslant 1$, $0 \leqslant z \leqslant 1$

16. $\iint_S y^2 \, dS$,
S is the part of the sphere $x^2 + y^2 + z^2 = 1$ that lies above
the cone $z = \sqrt{x^2 + y^2}$

17. $\iint_S (x^2 z + y^2 z) \, dS$,
S is the hemisphere $x^2 + y^2 + z^2 = 4$, $z \geqslant 0$

18. $\iint_S (x + y + z) \, dS$,
S is the part of the half-cylinder $x^2 + z^2 = 1$, $z \geqslant 0$, that
lies between the planes $y = 0$ and $y = 2$

19. $\iint_S xz \, dS$,
S is the boundary of the region enclosed by the cylinder
$y^2 + z^2 = 9$ and the planes $x = 0$ and $x + y = 5$

20. $\iint_S (x^2 + y^2 + z^2) \, dS$,
S is the part of the cylinder $x^2 + y^2 = 9$ between the planes
$z = 0$ and $z = 2$, together with its top and bottom disks

21–32 Evaluate the surface integral $\iint_S \mathbf{F} \cdot d\mathbf{S}$ for the given
vector field $\mathbf{F}$ and the oriented surface S. In other words, find
the flux of $\mathbf{F}$ across S. For closed surfaces, use the positive
(outward) orientation.

21. $\mathbf{F}(x, y, z) = ze^{xy} \mathbf{i} - 3ze^{xy} \mathbf{j} + xy \mathbf{k}$,
S is the parallelogram of Exercise 5 with upward orientation

22. $\mathbf{F}(x, y, z) = z \mathbf{i} + y \mathbf{j} + x \mathbf{k}$,
S is the helicoid of Exercise 7 with upward orientation

23. $\mathbf{F}(x, y, z) = xy \mathbf{i} + yz \mathbf{j} + zx \mathbf{k}$, S is the part of the
paraboloid $z = 4 - x^2 - y^2$ that lies above the square
$0 \leqslant x \leqslant 1$, $0 \leqslant y \leqslant 1$, and has upward orientation

24. $\mathbf{F}(x, y, z) = -x \mathbf{i} - y \mathbf{j} + z^3 \mathbf{k}$, S is the part of the cone
$z = \sqrt{x^2 + y^2}$ between the planes $z = 1$ and $z = 3$ with
downward orientation

25. $\mathbf{F}(x, y, z) = x \mathbf{i} + y \mathbf{j} + z^2 \mathbf{k}$, S is the sphere with radius 1
and center the origin

26. $\mathbf{F}(x, y, z) = y \mathbf{i} - x \mathbf{j} + 2z \mathbf{k}$, S is the hemisphere
$x^2 + y^2 + z^2 = 4$, $z \geqslant 0$, oriented downward

27. $\mathbf{F}(x, y, z) = y \mathbf{j} - z \mathbf{k}$,
S consists of the paraboloid $y = x^2 + z^2$, $0 \leqslant y \leqslant 1$,
and the disk $x^2 + z^2 \leqslant 1$, $y = 1$

28. $\mathbf{F}(x, y, z) = yz \mathbf{i} + zx \mathbf{j} + xy \mathbf{k}$,
S is the surface $z = x \sin y$, $0 \leqslant x \leqslant 2$, $0 \leqslant y \leqslant \pi$, with
upward orientation

29. $\mathbf{F}(x, y, z) = x \mathbf{i} + 2y \mathbf{j} + 3z \mathbf{k}$,
S is the cube with vertices $(\pm 1, \pm 1, \pm 1)$

30. $\mathbf{F}(x, y, z) = x \mathbf{i} + y \mathbf{j} + 5 \mathbf{k}$, S is the boundary of the
region enclosed by the cylinder $x^2 + z^2 = 1$ and the planes
$y = 0$ and $x + y = 2$

31. $\mathbf{F}(x, y, z) = x^2 \mathbf{i} + y^2 \mathbf{j} + z^2 \mathbf{k}$, S is the boundary of the
solid half-cylinder $0 \leqslant z \leqslant \sqrt{1 - y^2}$, $0 \leqslant x \leqslant 2$

32. $\mathbf{F}(x, y, z) = y \mathbf{i} + (z - y) \mathbf{j} + x \mathbf{k}$,
S is the surface of the tetrahedron with vertices $(0, 0, 0)$,
$(1, 0, 0)$, $(0, 1, 0)$, and $(0, 0, 1)$

CAS 33. Evaluate $\iint_S (x^2 + y^2 + z^2) \, dS$ correct to four decimal places, where S is the surface $z = xe^y$, $0 \leqslant x \leqslant 1$,
$0 \leqslant y \leqslant 1$.

CAS 34. Find the exact value of $\iint_S xyz \, dS$, where S is the surface
$z = x^2 y^2$, $0 \leqslant x \leqslant 1$, $0 \leqslant y \leqslant 2$.

CAS 35. Find the value of $\iint_S x^2 y^2 z^2 \, dS$ correct to four decimal places, where S is the part of the paraboloid
$z = 3 - 2x^2 - y^2$ that lies above the xy-plane.

CAS 36. Find the flux of

$$\mathbf{F}(x, y, z) = \sin(xyz) \, \mathbf{i} + x^2 y \mathbf{j} + z^2 e^{x/5} \mathbf{k}$$

across the part of the cylinder $4y^2 + z^2 = 4$ that lies above
the xy-plane and between the planes $x = -2$ and $x = 2$
with upward orientation. Illustrate by using a computer
algebra system to draw the cylinder and the vector field on
the same screen.

37. Find a formula for $\iint_S \mathbf{F} \cdot d\mathbf{S}$ similar to Formula 10 for
the case where S is given by $y = h(x, z)$ and $\mathbf{n}$ is the unit
normal that points toward the left.

38. Find a formula for $\iint_S \mathbf{F} \cdot d\mathbf{S}$ similar to Formula 10 for the case where S is given by $x = k(y, z)$ and $\mathbf{n}$ is the unit normal that points forward (that is, toward the viewer when the axes are drawn in the usual way).

39. Find the center of mass of the hemisphere $x^2 + y^2 + z^2 = a^2$, $z \geqslant 0$, if it has constant density.

40. Find the mass of a thin funnel in the shape of a cone $z = \sqrt{x^2 + y^2}$, $1 \leqslant z \leqslant 4$, if its density function is $\rho(x, y, z) = 10 - z$.

41. (a) Give an integral expression for the moment of inertia I_z about the z-axis of a thin sheet in the shape of a surface S if the density function is ρ.
(b) Find the moment of inertia about the z-axis of the funnel in Exercise 40.

42. Let S be the part of the sphere $x^2 + y^2 + z^2 = 25$ that lies above the plane $z = 4$. If S has constant density k, find (a) the center of mass and (b) the moment of inertia about the z-axis.

43. A fluid has density 870 kg/m^3 and flows with velocity $\mathbf{v} = z\,\mathbf{i} + y^2\,\mathbf{j} + x^2\,\mathbf{k}$, where x, y, and z are measured in meters and the components of $\mathbf{v}$ in meters per second. Find the rate of flow outward through the cylinder $x^2 + y^2 = 4$, $0 \leqslant z \leqslant 1$.

44. Seawater has density 1025 kg/m^3 and flows in a velocity field $\mathbf{v} = y\,\mathbf{i} + x\,\mathbf{j}$, where x, y, and z are measured in meters and the components of $\mathbf{v}$ in meters per second. Find the rate of flow outward through the hemisphere $x^2 + y^2 + z^2 = 9$, $z \geqslant 0$.

45. Use Gauss's Law to find the charge contained in the solid hemisphere $x^2 + y^2 + z^2 \leqslant a^2$, $z \geqslant 0$, if the electric field is
$$\mathbf{E}(x, y, z) = x\,\mathbf{i} + y\,\mathbf{j} + 2z\,\mathbf{k}$$

46. Use Gauss's Law to find the charge enclosed by the cube with vertices $(\pm 1, \pm 1, \pm 1)$ if the electric field is
$$\mathbf{E}(x, y, z) = x\,\mathbf{i} + y\,\mathbf{j} + z\,\mathbf{k}$$

47. The temperature at the point (x, y, z) in a substance with conductivity $K = 6.5$ is $u(x, y, z) = 2y^2 + 2z^2$. Find the rate of heat flow inward across the cylindrical surface $y^2 + z^2 = 6$, $0 \leqslant x \leqslant 4$.

48. The temperature at a point in a ball with conductivity K is inversely proportional to the distance from the center of the ball. Find the rate of heat flow across a sphere S of radius a with center at the center of the ball.

49. Let $\mathbf{F}$ be an inverse square field, that is, $\mathbf{F}(\mathbf{r}) = c\mathbf{r}/|\mathbf{r}|^3$ for some constant c, where $\mathbf{r} = x\,\mathbf{i} + y\,\mathbf{j} + z\,\mathbf{k}$. Show that the flux of $\mathbf{F}$ across a sphere S with center the origin is independent of the radius of S.

16.8 Stokes' Theorem

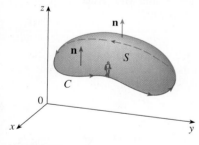

FIGURE 1

Stokes' Theorem can be regarded as a higher-dimensional version of Green's Theorem. Whereas Green's Theorem relates a double integral over a plane region D to a line integral around its plane boundary curve, Stokes' Theorem relates a surface integral over a surface S to a line integral around the boundary curve of S (which is a space curve). Figure 1 shows an oriented surface with unit normal vector $\mathbf{n}$. The orientation of S induces the **positive orientation of the boundary curve** C shown in the figure. This means that if you walk in the positive direction around C with your head pointing in the direction of $\mathbf{n}$, then the surface will always be on your left.

> **Stokes' Theorem** Let S be an oriented piecewise-smooth surface that is bounded by a simple, closed, piecewise-smooth boundary curve C with positive orientation. Let $\mathbf{F}$ be a vector field whose components have continuous partial derivatives on an open region in $\mathbb{R}^3$ that contains S. Then
> $$\int_C \mathbf{F} \cdot d\mathbf{r} = \iint_S \text{curl } \mathbf{F} \cdot d\mathbf{S}$$

Since

$$\int_C \mathbf{F} \cdot d\mathbf{r} = \int_C \mathbf{F} \cdot \mathbf{T}\, ds \qquad \text{and} \qquad \iint_S \text{curl } \mathbf{F} \cdot d\mathbf{S} = \iint_S \text{curl } \mathbf{F} \cdot \mathbf{n}\, dS$$

Stokes' Theorem says that the line integral around the boundary curve of S of the tangential component of $\mathbf{F}$ is equal to the surface integral over S of the normal component of the curl of $\mathbf{F}$.

The positively oriented boundary curve of the oriented surface S is often written as ∂S, so Stokes' Theorem can be expressed as

$$\boxed{1} \qquad \iint_S \text{curl } \mathbf{F} \cdot d\mathbf{S} = \int_{\partial S} \mathbf{F} \cdot d\mathbf{r}$$

There is an analogy among Stokes' Theorem, Green's Theorem, and the Fundamental Theorem of Calculus. As before, there is an integral involving derivatives on the left side of Equation 1 (recall that curl $\mathbf{F}$ is a sort of derivative of $\mathbf{F}$) and the right side involves the values of $\mathbf{F}$ only on the *boundary* of S.

In fact, in the special case where the surface S is flat and lies in the xy-plane with upward orientation, the unit normal is $\mathbf{k}$, the surface integral becomes a double integral, and Stokes' Theorem becomes

$$\int_C \mathbf{F} \cdot d\mathbf{r} = \iint_S \text{curl } \mathbf{F} \cdot d\mathbf{S} = \iint_S (\text{curl } \mathbf{F}) \cdot \mathbf{k} \, dA$$

This is precisely the vector form of Green's Theorem given in Equation 16.5.12. Thus we see that Green's Theorem is really a special case of Stokes' Theorem.

Although Stokes' Theorem is too difficult for us to prove in its full generality, we can give a proof when S is a graph and $\mathbf{F}$, S, and C are well behaved.

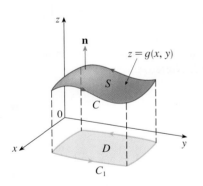

FIGURE 2

PROOF OF A SPECIAL CASE OF STOKES' THEOREM We assume that the equation of S is $z = g(x, y)$, $(x, y) \in D$, where g has continuous second-order partial derivatives and D is a simple plane region whose boundary curve C_1 corresponds to C. If the orientation of S is upward, then the positive orientation of C corresponds to the positive orientation of C_1. (See Figure 2.) We are also given that $\mathbf{F} = P\mathbf{i} + Q\mathbf{j} + R\mathbf{k}$, where the partial derivatives of P, Q, and R are continuous.

Since S is a graph of a function, we can apply Formula 16.7.10 with $\mathbf{F}$ replaced by curl $\mathbf{F}$. The result is

$$\boxed{2} \quad \iint_S \text{curl } \mathbf{F} \cdot d\mathbf{S}$$

$$= \iint_D \left[-\left(\frac{\partial R}{\partial y} - \frac{\partial Q}{\partial z} \right) \frac{\partial z}{\partial x} - \left(\frac{\partial P}{\partial z} - \frac{\partial R}{\partial x} \right) \frac{\partial z}{\partial y} + \left(\frac{\partial Q}{\partial x} - \frac{\partial P}{\partial y} \right) \right] dA$$

where the partial derivatives of P, Q, and R are evaluated at $(x, y, g(x, y))$. If

$$x = x(t) \qquad y = y(t) \qquad a \leqslant t \leqslant b$$

is a parametric representation of C_1, then a parametric representation of C is

$$x = x(t) \qquad y = y(t) \qquad z = g(x(t), y(t)) \qquad a \leqslant t \leqslant b$$

This allows us, with the aid of the Chain Rule, to evaluate the line integral as follows:

$$\int_C \mathbf{F} \cdot d\mathbf{r} = \int_a^b \left(P\frac{dx}{dt} + Q\frac{dy}{dt} + R\frac{dz}{dt} \right) dt$$

$$= \int_a^b \left[P\frac{dx}{dt} + Q\frac{dy}{dt} + R\left(\frac{\partial z}{\partial x}\frac{dx}{dt} + \frac{\partial z}{\partial y}\frac{dy}{dt} \right) \right] dt$$

$$= \int_a^b \left[\left(P + R\frac{\partial z}{\partial x} \right)\frac{dx}{dt} + \left(Q + R\frac{\partial z}{\partial y} \right)\frac{dy}{dt} \right] dt$$

$$= \int_{C_1} \left(P + R\frac{\partial z}{\partial x} \right) dx + \left(Q + R\frac{\partial z}{\partial y} \right) dy$$

$$= \iint_D \left[\frac{\partial}{\partial x}\left(Q + R\frac{\partial z}{\partial y} \right) - \frac{\partial}{\partial y}\left(P + R\frac{\partial z}{\partial x} \right) \right] dA$$

where we have used Green's Theorem in the last step. Then, using the Chain Rule again and remembering that P, Q, and R are functions of x, y, and z and that z is itself a function of x and y, we get

$$\int_C \mathbf{F} \cdot d\mathbf{r} = \iint_D \left[\left(\frac{\partial Q}{\partial x} + \frac{\partial Q}{\partial z}\frac{\partial z}{\partial x} + \frac{\partial R}{\partial x}\frac{\partial z}{\partial y} + \frac{\partial R}{\partial z}\frac{\partial z}{\partial x}\frac{\partial z}{\partial y} + R\frac{\partial^2 z}{\partial x\,\partial y} \right) \right.$$

$$\left. - \left(\frac{\partial P}{\partial y} + \frac{\partial P}{\partial z}\frac{\partial z}{\partial y} + \frac{\partial R}{\partial y}\frac{\partial z}{\partial x} + \frac{\partial R}{\partial z}\frac{\partial z}{\partial y}\frac{\partial z}{\partial x} + R\frac{\partial^2 z}{\partial y\,\partial x} \right) \right] dA$$

Four of the terms in this double integral cancel and the remaining six terms can be arranged to coincide with the right side of Equation 2. Therefore

$$\int_C \mathbf{F} \cdot d\mathbf{r} = \iint_S \operatorname{curl} \mathbf{F} \cdot d\mathbf{S}$$ ∎

EXAMPLE 1 Evaluate $\int_C \mathbf{F} \cdot d\mathbf{r}$, where $\mathbf{F}(x, y, z) = -y^2\,\mathbf{i} + x\,\mathbf{j} + z^2\,\mathbf{k}$ and C is the curve of intersection of the plane $y + z = 2$ and the cylinder $x^2 + y^2 = 1$. (Orient C to be counterclockwise when viewed from above.)

SOLUTION The curve C (an ellipse) is shown in Figure 3. Although $\int_C \mathbf{F} \cdot d\mathbf{r}$ could be evaluated directly, it's easier to use Stokes' Theorem. We first compute

$$\operatorname{curl} \mathbf{F} = \begin{vmatrix} \mathbf{i} & \mathbf{j} & \mathbf{k} \\ \dfrac{\partial}{\partial x} & \dfrac{\partial}{\partial y} & \dfrac{\partial}{\partial z} \\ -y^2 & x & z^2 \end{vmatrix} = (1 + 2y)\,\mathbf{k}$$

Although there are many surfaces with boundary C, the most convenient choice is the elliptical region S in the plane $y + z = 2$ that is bounded by C. If we orient S upward, then C has the induced positive orientation. The projection D of S onto the xy-plane is

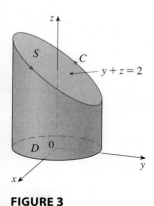

FIGURE 3

the disk $x^2 + y^2 \le 1$ and so using Equation 16.7.10 with $z = g(x, y) = 2 - y$, we have

$$\int_C \mathbf{F} \cdot d\mathbf{r} = \iint_S \text{curl } \mathbf{F} \cdot d\mathbf{S} = \iint_D (1 + 2y) \, dA$$

$$= \int_0^{2\pi} \int_0^1 (1 + 2r \sin \theta) \, r \, dr \, d\theta$$

$$= \int_0^{2\pi} \left[\frac{r^2}{2} + 2 \frac{r^3}{3} \sin \theta \right]_0^1 d\theta = \int_0^{2\pi} \left(\tfrac{1}{2} + \tfrac{2}{3} \sin \theta \right) d\theta$$

$$= \tfrac{1}{2}(2\pi) + 0 = \pi$$

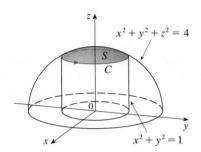

FIGURE 4

EXAMPLE 2 Use Stokes' Theorem to compute the integral $\iint_S \text{curl } \mathbf{F} \cdot d\mathbf{S}$, where $\mathbf{F}(x, y, z) = xz\,\mathbf{i} + yz\,\mathbf{j} + xy\,\mathbf{k}$ and S is the part of the sphere $x^2 + y^2 + z^2 = 4$ that lies inside the cylinder $x^2 + y^2 = 1$ and above the xy-plane. (See Figure 4.)

SOLUTION To find the boundary curve C we solve the equations $x^2 + y^2 + z^2 = 4$ and $x^2 + y^2 = 1$. Subtracting, we get $z^2 = 3$ and so $z = \sqrt{3}$ (since $z > 0$). Thus C is the circle given by the equations $x^2 + y^2 = 1, z = \sqrt{3}$. A vector equation of C is

$$\mathbf{r}(t) = \cos t\,\mathbf{i} + \sin t\,\mathbf{j} + \sqrt{3}\,\mathbf{k} \qquad 0 \le t \le 2\pi$$

so

$$\mathbf{r}'(t) = -\sin t\,\mathbf{i} + \cos t\,\mathbf{j}$$

Also, we have

$$\mathbf{F}(\mathbf{r}(t)) = \sqrt{3}\,\cos t\,\mathbf{i} + \sqrt{3}\,\sin t\,\mathbf{j} + \cos t \sin t\,\mathbf{k}$$

Therefore, by Stokes' Theorem,

$$\iint_S \text{curl } \mathbf{F} \cdot d\mathbf{S} = \int_C \mathbf{F} \cdot d\mathbf{r} = \int_0^{2\pi} \mathbf{F}(\mathbf{r}(t)) \cdot \mathbf{r}'(t) \, dt$$

$$= \int_0^{2\pi} \left(-\sqrt{3}\,\cos t \sin t + \sqrt{3}\,\sin t \cos t \right) dt$$

$$= \sqrt{3} \int_0^{2\pi} 0 \, dt = 0$$

Note that in Example 2 we computed a surface integral simply by knowing the values of $\mathbf{F}$ on the boundary curve C. This means that if we have another oriented surface with the same boundary curve C, then we get exactly the same value for the surface integral!

In general, if S_1 and S_2 are oriented surfaces with the same oriented boundary curve C and both satisfy the hypotheses of Stokes' Theorem, then

$$\boxed{3} \qquad \iint_{S_1} \text{curl } \mathbf{F} \cdot d\mathbf{S} = \int_C \mathbf{F} \cdot d\mathbf{r} = \iint_{S_2} \text{curl } \mathbf{F} \cdot d\mathbf{S}$$

This fact is useful when it is difficult to integrate over one surface but easy to integrate over the other.

We now use Stokes' Theorem to throw some light on the meaning of the curl vector. Suppose that C is an oriented closed curve and $\mathbf{v}$ represents the velocity field in fluid flow. Consider the line integral

$$\int_C \mathbf{v} \cdot d\mathbf{r} = \int_C \mathbf{v} \cdot \mathbf{T} \, ds$$

and recall that $\mathbf{v} \cdot \mathbf{T}$ is the component of $\mathbf{v}$ in the direction of the unit tangent vector $\mathbf{T}$. This means that the closer the direction of $\mathbf{v}$ is to the direction of $\mathbf{T}$, the larger the value of $\mathbf{v} \cdot \mathbf{T}$. Thus $\int_C \mathbf{v} \cdot d\mathbf{r}$ is a measure of the tendency of the fluid to move around C and is called the **circulation** of $\mathbf{v}$ around C. (See Figure 5.)

FIGURE 5   (a) $\int_C \mathbf{v} \cdot d\mathbf{r} > 0$, positive circulation   (b) $\int_C \mathbf{v} \cdot d\mathbf{r} < 0$, negative circulation

Now let $P_0(x_0, y_0, z_0)$ be a point in the fluid and let S_a be a small disk with radius a and center P_0. Then $(\text{curl } \mathbf{F})(P) \approx (\text{curl } \mathbf{F})(P_0)$ for all points P on S_a because curl $\mathbf{F}$ is continuous. Thus, by Stokes' Theorem, we get the following approximation to the circulation around the boundary circle C_a:

$$\int_{C_a} \mathbf{v} \cdot d\mathbf{r} = \iint_{S_a} \text{curl } \mathbf{v} \cdot d\mathbf{S} = \iint_{S_a} \text{curl } \mathbf{v} \cdot \mathbf{n} \, dS$$

$$\approx \iint_{S_a} \text{curl } \mathbf{v}(P_0) \cdot \mathbf{n}(P_0) \, dS = \text{curl } \mathbf{v}(P_0) \cdot \mathbf{n}(P_0) \pi a^2$$

Imagine a tiny paddle wheel placed in the fluid at a point P, as in Figure 6; the paddle wheel rotates fastest when its axis is parallel to curl $\mathbf{v}$.

This approximation becomes better as $a \to 0$ and we have

$$\boxed{4} \qquad \text{curl } \mathbf{v}(P_0) \cdot \mathbf{n}(P_0) = \lim_{a \to 0} \frac{1}{\pi a^2} \int_{C_a} \mathbf{v} \cdot d\mathbf{r}$$

Equation 4 gives the relationship between the curl and the circulation. It shows that curl $\mathbf{v} \cdot \mathbf{n}$ is a measure of the rotating effect of the fluid about the axis $\mathbf{n}$. The curling effect is greatest about the axis parallel to curl $\mathbf{v}$.

Finally, we mention that Stokes' Theorem can be used to prove Theorem 16.5.4 (which states that if curl $\mathbf{F} = \mathbf{0}$ on all of $\mathbb{R}^3$, then $\mathbf{F}$ is conservative). From our previous work (Theorems 16.3.3 and 16.3.4), we know that $\mathbf{F}$ is conservative if $\int_C \mathbf{F} \cdot d\mathbf{r} = 0$ for every closed path C. Given C, suppose we can find an orientable surface S whose boundary is C. (This can be done, but the proof requires advanced techniques.) Then Stokes' Theorem gives

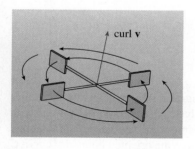

FIGURE 6

$$\int_C \mathbf{F} \cdot d\mathbf{r} = \iint_S \text{curl } \mathbf{F} \cdot d\mathbf{S} = \iint_S \mathbf{0} \cdot d\mathbf{S} = 0$$

A curve that is not simple can be broken into a number of simple curves, and the integrals around these simple curves are all 0. Adding these integrals, we obtain $\int_C \mathbf{F} \cdot d\mathbf{r} = 0$ for any closed curve C.

16.8 EXERCISES

1. A hemisphere H and a portion P of a paraboloid are shown. Suppose $\mathbf{F}$ is a vector field on $\mathbb{R}^3$ whose components have continuous partial derivatives. Explain why

$$\iint_H \operatorname{curl} \mathbf{F} \cdot d\mathbf{S} = \iint_P \operatorname{curl} \mathbf{F} \cdot d\mathbf{S}$$

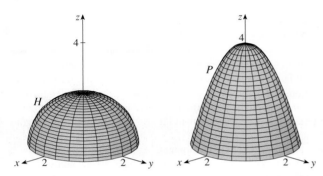

2–6 Use Stokes' Theorem to evaluate $\iint_S \operatorname{curl} \mathbf{F} \cdot d\mathbf{S}$.

2. $\mathbf{F}(x, y, z) = x^2 \sin z\, \mathbf{i} + y^2\, \mathbf{j} + xy\, \mathbf{k}$,
   S is the part of the paraboloid $z = 1 - x^2 - y^2$ that lies above the xy-plane, oriented upward

3. $\mathbf{F}(x, y, z) = ze^y\, \mathbf{i} + x \cos y\, \mathbf{j} + xz \sin y\, \mathbf{k}$,
   S is the hemisphere $x^2 + y^2 + z^2 = 16$, $y \geq 0$, oriented in the direction of the positive y-axis

4. $\mathbf{F}(x, y, z) = \tan^{-1}(x^2 yz^2)\, \mathbf{i} + x^2 y\, \mathbf{j} + x^2 z^2\, \mathbf{k}$,
   S is the cone $x = \sqrt{y^2 + z^2}$, $0 \leq x \leq 2$, oriented in the direction of the positive x-axis

5. $\mathbf{F}(x, y, z) = xyz\, \mathbf{i} + xy\, \mathbf{j} + x^2 yz\, \mathbf{k}$,
   S consists of the top and the four sides (but not the bottom) of the cube with vertices $(\pm 1, \pm 1, \pm 1)$, oriented outward

6. $\mathbf{F}(x, y, z) = e^{xy}\, \mathbf{i} + e^{xz}\, \mathbf{j} + x^2 z\, \mathbf{k}$,
   S is the half of the ellipsoid $4x^2 + y^2 + 4z^2 = 4$ that lies to the right of the xz-plane, oriented in the direction of the positive y-axis

---

7–10 Use Stokes' Theorem to evaluate $\int_C \mathbf{F} \cdot d\mathbf{r}$. In each case C is oriented counterclockwise as viewed from above.

7. $\mathbf{F}(x, y, z) = (x + y^2)\, \mathbf{i} + (y + z^2)\, \mathbf{j} + (z + x^2)\, \mathbf{k}$,
   C is the triangle with vertices $(1, 0, 0)$, $(0, 1, 0)$, and $(0, 0, 1)$

8. $\mathbf{F}(x, y, z) = \mathbf{i} + (x + yz)\, \mathbf{j} + \left(xy - \sqrt{z}\,\right) \mathbf{k}$,
   C is the boundary of the part of the plane $3x + 2y + z = 1$ in the first octant

9. $\mathbf{F}(x, y, z) = xy\, \mathbf{i} + yz\, \mathbf{j} + zx\, \mathbf{k}$, C is the boundary of the part of the paraboloid $z = 1 - x^2 - y^2$ in the first octant

10. $\mathbf{F}(x, y, z) = 2y\, \mathbf{i} + xz\, \mathbf{j} + (x + y)\, \mathbf{k}$, C is the curve of intersection of the plane $z = y + 2$ and the cylinder $x^2 + y^2 = 1$

---

11. (a) Use Stokes' Theorem to evaluate $\int_C \mathbf{F} \cdot d\mathbf{r}$, where

    $$\mathbf{F}(x, y, z) = x^2 z\, \mathbf{i} + xy^2\, \mathbf{j} + z^2\, \mathbf{k}$$

    and C is the curve of intersection of the plane $x + y + z = 1$ and the cylinder $x^2 + y^2 = 9$, oriented counterclockwise as viewed from above.

    (b) Graph both the plane and the cylinder with domains chosen so that you can see the curve C and the surface that you used in part (a).

    (c) Find parametric equations for C and use them to graph C.

12. (a) Use Stokes' Theorem to evaluate $\int_C \mathbf{F} \cdot d\mathbf{r}$, where $\mathbf{F}(x, y, z) = x^2 y\, \mathbf{i} + \frac{1}{3} x^3\, \mathbf{j} + xy\, \mathbf{k}$ and C is the curve of intersection of the hyperbolic paraboloid $z = y^2 - x^2$ and the cylinder $x^2 + y^2 = 1$, oriented counterclockwise as viewed from above.

    (b) Graph both the hyperbolic paraboloid and the cylinder with domains chosen so that you can see the curve C and the surface that you used in part (a).

    (c) Find parametric equations for C and use them to graph C.

13–15 Verify that Stokes' Theorem is true for the given vector field $\mathbf{F}$ and surface S.

13. $\mathbf{F}(x, y, z) = -y\, \mathbf{i} + x\, \mathbf{j} - 2\, \mathbf{k}$,
    S is the cone $z^2 = x^2 + y^2$, $0 \leq z \leq 4$, oriented downward

14. $\mathbf{F}(x, y, z) = -2yz\, \mathbf{i} + y\, \mathbf{j} + 3x\, \mathbf{k}$,
    S is the part of the paraboloid $z = 5 - x^2 - y^2$ that lies above the plane $z = 1$, oriented upward

15. $\mathbf{F}(x, y, z) = y\, \mathbf{i} + z\, \mathbf{j} + x\, \mathbf{k}$,
    S is the hemisphere $x^2 + y^2 + z^2 = 1$, $y \geq 0$, oriented in the direction of the positive y-axis

---

16. Let C be a simple closed smooth curve that lies in the plane $x + y + z = 1$. Show that the line integral

    $$\int_C z\, dx - 2x\, dy + 3y\, dz$$

    depends only on the area of the region enclosed by C and not on the shape of C or its location in the plane.

17. A particle moves along line segments from the origin to the points $(1, 0, 0)$, $(1, 2, 1)$, $(0, 2, 1)$, and back to the origin under the influence of the force field

    $$\mathbf{F}(x, y, z) = z^2\, \mathbf{i} + 2xy\, \mathbf{j} + 4y^2\, \mathbf{k}$$

    Find the work done.

18. Evaluate

$$\int_C (y + \sin x)\, dx + (z^2 + \cos y)\, dy + x^3\, dz$$

where C is the curve $\mathbf{r}(t) = \langle \sin t, \cos t, \sin 2t \rangle$, $0 \le t \le 2\pi$. [*Hint:* Observe that C lies on the surface $z = 2xy$.]

19. If S is a sphere and $\mathbf{F}$ satisfies the hypotheses of Stokes' Theorem, show that $\iint_S \operatorname{curl} \mathbf{F} \cdot d\mathbf{S} = 0$.

20. Suppose S and C satisfy the hypotheses of Stokes' Theorem and f, g have continuous second-order partial derivatives. Use Exercises 24 and 26 in Section 16.5 to show the following.

(a) $\int_C (f\, \nabla g) \cdot d\mathbf{r} = \iint_S (\nabla f \times \nabla g) \cdot d\mathbf{S}$

(b) $\int_C (f\, \nabla f) \cdot d\mathbf{r} = 0$

(c) $\int_C (f\, \nabla g + g\, \nabla f) \cdot d\mathbf{r} = 0$

WRITING PROJECT   THREE MEN AND TWO THEOREMS

The photograph shows a stained-glass window at Cambridge University in honor of George Green.

Courtesy of the Masters and Fellows of Gonville and Caius College, Cambridge University, England

Although two of the most important theorems in vector calculus are named after George Green and George Stokes, a third man, William Thomson (also known as Lord Kelvin), played a large role in the formulation, dissemination, and application of both of these results. All three men were interested in how the two theorems could help to explain and predict physical phenomena in electricity and magnetism and fluid flow. The basic facts of the story are given in the margin notes on pages 1097 and 1135.

Write a report on the historical origins of Green's Theorem and Stokes' Theorem. Explain the similarities and relationship between the theorems. Discuss the roles that Green, Thomson, and Stokes played in discovering these theorems and making them widely known. Show how both theorems arose from the investigation of electricity and magnetism and were later used to study a variety of physical problems.

The dictionary edited by Gillispie [2] is a good source for both biographical and scientific information. The book by Hutchinson [5] gives an account of Stokes' life and the book by Thompson [8] is a biography of Lord Kelvin. The articles by Grattan-Guinness [3] and Gray [4] and the book by Cannell [1] give background on the extraordinary life and works of Green. Additional historical and mathematical information is found in the books by Katz [6] and Kline [7].

1. D. M. Cannell, *George Green, Mathematician and Physicist 1793–1841: The Background to His Life and Work* (Philadelphia: Society for Industrial and Applied Mathematics, 2001).

2. C. C. Gillispie, ed., *Dictionary of Scientific Biography* (New York: Scribner's, 1974). See the article on Green by P. J. Wallis in Volume XV and the articles on Thomson by Jed Buchwald and on Stokes by E. M. Parkinson in Volume XIII.

3. I. Grattan-Guinness, "Why Did George Green Write his Essay of 1828 on Electricity and Magnetism?" *Amer. Math. Monthly*, Vol. 102 (1995), pp. 387–96.

4. J. Gray, "There Was a Jolly Miller." *The New Scientist*, Vol. 139 (1993), pp. 24–27.

5. G. E. Hutchinson, *The Enchanted Voyage and Other Studies* (Westport, CT: Greenwood Press, 1978).

6. Victor Katz, *A History of Mathematics: An Introduction* (New York: HarperCollins, 1993), pp. 678–80.

7. Morris Kline, *Mathematical Thought from Ancient to Modern Times* (New York: Oxford University Press, 1972), pp. 683–85.

8. Sylvanus P. Thompson, *The Life of Lord Kelvin* (New York: Chelsea, 1976).

16.9 The Divergence Theorem

In Section 16.5 we rewrote Green's Theorem in a vector version as

$$\int_C \mathbf{F} \cdot \mathbf{n} \, ds = \iint_D \operatorname{div} \mathbf{F}(x, y) \, dA$$

where C is the positively oriented boundary curve of the plane region D. If we were seeking to extend this theorem to vector fields on $\mathbb{R}^3$, we might make the guess that

$$\boxed{1} \qquad \iint_S \mathbf{F} \cdot \mathbf{n} \, dS = \iiint_E \operatorname{div} \mathbf{F}(x, y, z) \, dV$$

where S is the boundary surface of the solid region E. It turns out that Equation 1 is true, under appropriate hypotheses, and is called the Divergence Theorem. Notice its similarity to Green's Theorem and Stokes' Theorem in that it relates the integral of a derivative of a function (div $\mathbf{F}$ in this case) over a region to the integral of the original function $\mathbf{F}$ over the boundary of the region.

At this stage you may wish to review the various types of regions over which we were able to evaluate triple integrals in Section 15.6. We state and prove the Divergence Theorem for regions E that are simultaneously of types 1, 2, and 3 and we call such regions **simple solid regions**. (For instance, regions bounded by ellipsoids or rectangular boxes are simple solid regions.) The boundary of E is a closed surface, and we use the convention, introduced in Section 16.7, that the positive orientation is outward; that is, the unit normal vector $\mathbf{n}$ is directed outward from E.

The Divergence Theorem is sometimes called Gauss's Theorem after the great German mathematician Karl Friedrich Gauss (1777–1855), who discovered this theorem during his investigation of electrostatics. In Eastern Europe the Divergence Theorem is known as Ostrogradsky's Theorem after the Russian mathematician Mikhail Ostrogradsky (1801–1862), who published this result in 1826.

> **The Divergence Theorem** Let E be a simple solid region and let S be the boundary surface of E, given with positive (outward) orientation. Let $\mathbf{F}$ be a vector field whose component functions have continuous partial derivatives on an open region that contains E. Then
>
> $$\iint_S \mathbf{F} \cdot d\mathbf{S} = \iiint_E \operatorname{div} \mathbf{F} \, dV$$

Thus the Divergence Theorem states that, under the given conditions, the flux of $\mathbf{F}$ across the boundary surface of E is equal to the triple integral of the divergence of $\mathbf{F}$ over E.

PROOF Let $\mathbf{F} = P\mathbf{i} + Q\mathbf{j} + R\mathbf{k}$. Then

$$\operatorname{div} \mathbf{F} = \frac{\partial P}{\partial x} + \frac{\partial Q}{\partial y} + \frac{\partial R}{\partial z}$$

so

$$\iiint_E \operatorname{div} \mathbf{F} \, dV = \iiint_E \frac{\partial P}{\partial x} \, dV + \iiint_E \frac{\partial Q}{\partial y} \, dV + \iiint_E \frac{\partial R}{\partial z} \, dV$$

If $\mathbf{n}$ is the unit outward normal of S, then the surface integral on the left side of the Divergence Theorem is

$$\iint_S \mathbf{F} \cdot d\mathbf{S} = \iint_S \mathbf{F} \cdot \mathbf{n} \, dS = \iint_S (P\mathbf{i} + Q\mathbf{j} + R\mathbf{k}) \cdot \mathbf{n} \, dS$$

$$= \iint_S P\mathbf{i} \cdot \mathbf{n} \, dS + \iint_S Q\mathbf{j} \cdot \mathbf{n} \, dS + \iint_S R\mathbf{k} \cdot \mathbf{n} \, dS$$

Therefore, to prove the Divergence Theorem, it suffices to prove the following three equations:

$$\boxed{2} \qquad \iint_S P\,\mathbf{i} \cdot \mathbf{n}\,dS = \iiint_E \frac{\partial P}{\partial x}\,dV$$

$$\boxed{3} \qquad \iint_S Q\,\mathbf{j} \cdot \mathbf{n}\,dS = \iiint_E \frac{\partial Q}{\partial y}\,dV$$

$$\boxed{4} \qquad \iint_S R\,\mathbf{k} \cdot \mathbf{n}\,dS = \iiint_E \frac{\partial R}{\partial z}\,dV$$

To prove Equation 4 we use the fact that E is a type 1 region:

$$E = \big\{(x, y, z) \mid (x, y) \in D,\ u_1(x, y) \leqslant z \leqslant u_2(x, y)\big\}$$

where D is the projection of E onto the xy-plane. By Equation 15.6.6, we have

$$\iiint_E \frac{\partial R}{\partial z}\,dV = \iint_D \left[\int_{u_1(x,\,y)}^{u_2(x,\,y)} \frac{\partial R}{\partial z}(x, y, z)\,dz \right] dA$$

and therefore, by the Fundamental Theorem of Calculus,

$$\boxed{5} \qquad \iiint_E \frac{\partial R}{\partial z}\,dV = \iint_D \big[R(x, y, u_2(x, y)) - R(x, y, u_1(x, y)) \big]\,dA$$

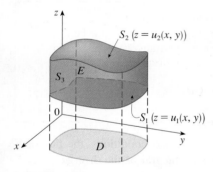

FIGURE 1

The boundary surface S consists of three pieces: the bottom surface S_1, the top surface S_2, and possibly a vertical surface S_3, which lies above the boundary curve of D. (See Figure 1. It might happen that S_3 doesn't appear, as in the case of a sphere.) Notice that on S_3 we have $\mathbf{k} \cdot \mathbf{n} = 0$, because $\mathbf{k}$ is vertical and $\mathbf{n}$ is horizontal, and so

$$\iint_{S_3} R\,\mathbf{k} \cdot \mathbf{n}\,dS = \iint_{S_3} 0\,dS = 0$$

Thus, regardless of whether there is a vertical surface, we can write

$$\boxed{6} \qquad \iint_S R\,\mathbf{k} \cdot \mathbf{n}\,dS = \iint_{S_1} R\,\mathbf{k} \cdot \mathbf{n}\,dS + \iint_{S_2} R\,\mathbf{k} \cdot \mathbf{n}\,dS$$

The equation of S_2 is $z = u_2(x, y)$, $(x, y) \in D$, and the outward normal $\mathbf{n}$ points upward, so from Equation 16.7.10 (with $\mathbf{F}$ replaced by $R\,\mathbf{k}$) we have

$$\iint_{S_2} R\,\mathbf{k} \cdot \mathbf{n}\,dS = \iint_D R(x, y, u_2(x, y))\,dA$$

On S_1 we have $z = u_1(x, y)$, but here the outward normal $\mathbf{n}$ points downward, so we multiply by -1:

$$\iint_{S_1} R\,\mathbf{k} \cdot \mathbf{n}\,dS = -\iint_D R(x, y, u_1(x, y))\,dA$$

Therefore Equation 6 gives

$$\iint_S R\,\mathbf{k} \cdot \mathbf{n}\,dS = \iint_D \big[R(x, y, u_2(x, y)) - R(x, y, u_1(x, y)) \big]\,dA$$

Comparison with Equation 5 shows that

$$\iint_S R\,\mathbf{k}\cdot\mathbf{n}\,dS = \iiint_E \frac{\partial R}{\partial z}\,dV$$

Notice that the method of proof of the Divergence Theorem is very similar to that of Green's Theorem.

Equations 2 and 3 are proved in a similar manner using the expressions for E as a type 2 or type 3 region, respectively. ∎

EXAMPLE 1 Find the flux of the vector field $\mathbf{F}(x, y, z) = z\,\mathbf{i} + y\,\mathbf{j} + x\,\mathbf{k}$ over the unit sphere $x^2 + y^2 + z^2 = 1$.

SOLUTION First we compute the divergence of $\mathbf{F}$:

$$\operatorname{div}\mathbf{F} = \frac{\partial}{\partial x}(z) + \frac{\partial}{\partial y}(y) + \frac{\partial}{\partial z}(x) = 1$$

The unit sphere S is the boundary of the unit ball B given by $x^2 + y^2 + z^2 \leqslant 1$. Thus the Divergence Theorem gives the flux as

The solution in Example 1 should be compared with the solution in Example 16.7.4.

$$\iint_S \mathbf{F}\cdot d\mathbf{S} = \iiint_B \operatorname{div}\mathbf{F}\,dV = \iiint_B 1\,dV = V(B) = \tfrac{4}{3}\pi(1)^3 = \frac{4\pi}{3}$$ ∎

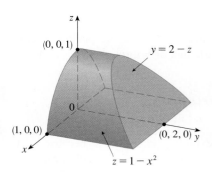

FIGURE 2

EXAMPLE 2 Evaluate $\iint_S \mathbf{F}\cdot d\mathbf{S}$, where

$$\mathbf{F}(x, y, z) = xy\,\mathbf{i} + \left(y^2 + e^{xz^2}\right)\mathbf{j} + \sin(xy)\,\mathbf{k}$$

and S is the surface of the region E bounded by the parabolic cylinder $z = 1 - x^2$ and the planes $z = 0$, $y = 0$, and $y + z = 2$. (See Figure 2.)

SOLUTION It would be extremely difficult to evaluate the given surface integral directly. (We would have to evaluate four surface integrals corresponding to the four pieces of S.) Furthermore, the divergence of $\mathbf{F}$ is much less complicated than $\mathbf{F}$ itself:

$$\operatorname{div}\mathbf{F} = \frac{\partial}{\partial x}(xy) + \frac{\partial}{\partial y}\left(y^2 + e^{xz^2}\right) + \frac{\partial}{\partial z}(\sin xy) = y + 2y = 3y$$

Therefore we use the Divergence Theorem to transform the given surface integral into a triple integral. The easiest way to evaluate the triple integral is to express E as a type 3 region:

$$E = \left\{(x, y, z) \mid -1 \leqslant x \leqslant 1,\ 0 \leqslant z \leqslant 1 - x^2,\ 0 \leqslant y \leqslant 2 - z\right\}$$

Then we have

$$\iint_S \mathbf{F}\cdot d\mathbf{S} = \iiint_E \operatorname{div}\mathbf{F}\,dV = \iiint_E 3y\,dV$$

$$= 3\int_{-1}^{1}\int_{0}^{1-x^2}\int_{0}^{2-z} y\,dy\,dz\,dx = 3\int_{-1}^{1}\int_{0}^{1-x^2}\frac{(2-z)^2}{2}\,dz\,dx$$

$$= \frac{3}{2}\int_{-1}^{1}\left[-\frac{(2-z)^3}{3}\right]_0^{1-x^2} dx = -\tfrac{1}{2}\int_{-1}^{1}\left[(x^2+1)^3 - 8\right]dx$$

$$= -\int_{0}^{1}(x^6 + 3x^4 + 3x^2 - 7)\,dx = \frac{184}{35}$$ ∎

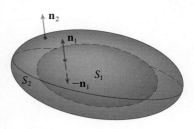

FIGURE 3

Although we have proved the Divergence Theorem only for simple solid regions, it can be proved for regions that are finite unions of simple solid regions. (The procedure is similar to the one we used in Section 16.4 to extend Green's Theorem.)

For example, let's consider the region E that lies between the closed surfaces S_1 and S_2, where S_1 lies inside S_2. Let $\mathbf{n}_1$ and $\mathbf{n}_2$ be outward normals of S_1 and S_2. Then the boundary surface of E is $S = S_1 \cup S_2$ and its normal $\mathbf{n}$ is given by $\mathbf{n} = -\mathbf{n}_1$ on S_1 and $\mathbf{n} = \mathbf{n}_2$ on S_2. (See Figure 3.) Applying the Divergence Theorem to S, we get

$$\boxed{7} \qquad \iiint_E \operatorname{div} \mathbf{F} \, dV = \iint_S \mathbf{F} \cdot d\mathbf{S} = \iint_S \mathbf{F} \cdot \mathbf{n} \, dS$$

$$= \iint_{S_1} \mathbf{F} \cdot (-\mathbf{n}_1) \, dS + \iint_{S_2} \mathbf{F} \cdot \mathbf{n}_2 \, dS$$

$$= -\iint_{S_1} \mathbf{F} \cdot d\mathbf{S} + \iint_{S_2} \mathbf{F} \cdot d\mathbf{S}$$

EXAMPLE 3 In Example 16.1.5 we considered the electric field

$$\mathbf{E}(\mathbf{x}) = \frac{\varepsilon Q}{|\mathbf{x}|^3} \mathbf{x}$$

where the electric charge Q is located at the origin and $\mathbf{x} = \langle x, y, z \rangle$ is a position vector. Use the Divergence Theorem to show that the electric flux of $\mathbf{E}$ through any closed surface S_2 that encloses the origin is

$$\iint_{S_2} \mathbf{E} \cdot d\mathbf{S} = 4\pi\varepsilon Q$$

SOLUTION The difficulty is that we don't have an explicit equation for S_2 because it is *any* closed surface enclosing the origin. The simplest such surface would be a sphere, so we let S_1 be a small sphere with radius a and center the origin. You can verify that $\operatorname{div} \mathbf{E} = 0$. (See Exercise 23.) Therefore Equation 7 gives

$$\iint_{S_2} \mathbf{E} \cdot d\mathbf{S} = \iint_{S_1} \mathbf{E} \cdot d\mathbf{S} + \iiint_E \operatorname{div} \mathbf{E} \, dV = \iint_{S_1} \mathbf{E} \cdot d\mathbf{S} = \iint_{S_1} \mathbf{E} \cdot \mathbf{n} \, dS$$

The point of this calculation is that we can compute the surface integral over S_1 because S_1 is a sphere. The normal vector at $\mathbf{x}$ is $\mathbf{x}/|\mathbf{x}|$. Therefore

$$\mathbf{E} \cdot \mathbf{n} = \frac{\varepsilon Q}{|\mathbf{x}|^3} \mathbf{x} \cdot \left(\frac{\mathbf{x}}{|\mathbf{x}|} \right) = \frac{\varepsilon Q}{|\mathbf{x}|^4} \mathbf{x} \cdot \mathbf{x} = \frac{\varepsilon Q}{|\mathbf{x}|^2} = \frac{\varepsilon Q}{a^2}$$

since the equation of S_1 is $|\mathbf{x}| = a$. Thus we have

$$\iint_{S_2} \mathbf{E} \cdot d\mathbf{S} = \iint_{S_1} \mathbf{E} \cdot \mathbf{n} \, dS = \frac{\varepsilon Q}{a^2} \iint_{S_1} dS = \frac{\varepsilon Q}{a^2} A(S_1) = \frac{\varepsilon Q}{a^2} 4\pi a^2 = 4\pi\varepsilon Q$$

This shows that the electric flux of $\mathbf{E}$ is $4\pi\varepsilon Q$ through *any* closed surface S_2 that contains the origin. [This is a special case of Gauss's Law (Equation 16.7.11) for a single charge. The relationship between ε and ε_0 is $\varepsilon = 1/(4\pi\varepsilon_0)$.] ∎

Another application of the Divergence Theorem occurs in fluid flow. Let $\mathbf{v}(x, y, z)$ be the velocity field of a fluid with constant density ρ. Then $\mathbf{F} = \rho\mathbf{v}$ is the rate of flow per unit area. If $P_0(x_0, y_0, z_0)$ is a point in the fluid and B_a is a ball with center P_0 and very small radius a, then div $\mathbf{F}(P) \approx$ div $\mathbf{F}(P_0)$ for all points P in B_a since div $\mathbf{F}$ is continuous. We approximate the flux over the boundary sphere S_a as follows:

$$\iint_{S_a} \mathbf{F} \cdot d\mathbf{S} = \iiint_{B_a} \text{div } \mathbf{F} \, dV \approx \iiint_{B_a} \text{div } \mathbf{F}(P_0) \, dV = \text{div } \mathbf{F}(P_0)V(B_a)$$

This approximation becomes better as $a \to 0$ and suggests that

$$\boxed{8} \qquad \text{div } \mathbf{F}(P_0) = \lim_{a \to 0} \frac{1}{V(B_a)} \iint_{S_a} \mathbf{F} \cdot d\mathbf{S}$$

Equation 8 says that div $\mathbf{F}(P_0)$ is the net rate of outward flux per unit volume at P_0. (This is the reason for the name *divergence*.) If div $\mathbf{F}(P) > 0$, the net flow is outward near P and P is called a **source**. If div $\mathbf{F}(P) < 0$, the net flow is inward near P and P is called a **sink**.

For the vector field in Figure 4, it appears that the vectors that end near P_1 are shorter than the vectors that start near P_1. Thus the net flow is outward near P_1, so div $\mathbf{F}(P_1) > 0$ and P_1 is a source. Near P_2, on the other hand, the incoming arrows are longer than the outgoing arrows. Here the net flow is inward, so div $\mathbf{F}(P_2) < 0$ and P_2 is a sink. We can use the formula for $\mathbf{F}$ to confirm this impression. Since $\mathbf{F} = x^2\,\mathbf{i} + y^2\,\mathbf{j}$, we have div $\mathbf{F} = 2x + 2y$, which is positive when $y > -x$. So the points above the line $y = -x$ are sources and those below are sinks.

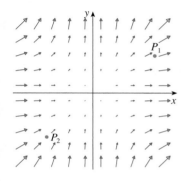

FIGURE 4
The vector field $\mathbf{F} = x^2\,\mathbf{i} + y^2\,\mathbf{j}$

16.9 EXERCISES

1–4 Verify that the Divergence Theorem is true for the vector field $\mathbf{F}$ on the region E.

1. $\mathbf{F}(x, y, z) = 3x\,\mathbf{i} + xy\,\mathbf{j} + 2xz\,\mathbf{k}$,
E is the cube bounded by the planes $x = 0$, $x = 1$, $y = 0$, $y = 1$, $z = 0$, and $z = 1$

2. $\mathbf{F}(x, y, z) = y^2z^3\,\mathbf{i} + 2yz\,\mathbf{j} + 4z^2\,\mathbf{k}$,
E is the solid enclosed by the paraboloid $z = x^2 + y^2$ and the plane $z = 9$

3. $\mathbf{F}(x, y, z) = \langle z, y, x \rangle$,
E is the solid ball $x^2 + y^2 + z^2 \le 16$

4. $\mathbf{F}(x, y, z) = \langle x^2, -y, z \rangle$,
E is the solid cylinder $y^2 + z^2 \le 9, 0 \le x \le 2$

5–15 Use the Divergence Theorem to calculate the surface integral $\iint_S \mathbf{F} \cdot d\mathbf{S}$; that is, calculate the flux of $\mathbf{F}$ across S.

5. $\mathbf{F}(x, y, z) = xye^z\,\mathbf{i} + xy^2z^3\,\mathbf{j} - ye^z\,\mathbf{k}$,
S is the surface of the box bounded by the coordinate planes and the planes $x = 3$, $y = 2$, and $z = 1$

6. $\mathbf{F}(x, y, z) = x^2yz\,\mathbf{i} + xy^2z\,\mathbf{j} + xyz^2\,\mathbf{k}$,
S is the surface of the box enclosed by the planes $x = 0$, $x = a$, $y = 0$, $y = b$, $z = 0$, and $z = c$, where a, b, and c are positive numbers

7. $\mathbf{F}(x, y, z) = 3xy^2\,\mathbf{i} + xe^z\,\mathbf{j} + z^3\,\mathbf{k}$,
S is the surface of the solid bounded by the cylinder $y^2 + z^2 = 1$ and the planes $x = -1$ and $x = 2$

8. $\mathbf{F}(x, y, z) = (x^3 + y^3)\,\mathbf{i} + (y^3 + z^3)\,\mathbf{j} + (z^3 + x^3)\,\mathbf{k}$,
S is the sphere with center the origin and radius 2

9. $\mathbf{F}(x, y, z) = xe^y\,\mathbf{i} + (z - e^y)\,\mathbf{j} - xy\,\mathbf{k}$,
S is the ellipsoid $x^2 + 2y^2 + 3z^2 = 4$

10. $\mathbf{F}(x, y, z) = z\,\mathbf{i} + y\,\mathbf{j} + zx\,\mathbf{k}$,
S is the surface of the tetrahedron enclosed by the coordinate planes and the plane

$$\frac{x}{a} + \frac{y}{b} + \frac{z}{c} = 1$$

where a, b, and c are positive numbers

11. $\mathbf{F}(x, y, z) = (2x^3 + y^3)\,\mathbf{i} + (y^3 + z^3)\,\mathbf{j} + 3y^2z\,\mathbf{k}$,
S is the surface of the solid bounded by the paraboloid $z = 1 - x^2 - y^2$ and the xy-plane

12. $\mathbf{F}(x, y, z) = (xy + 2xz)\,\mathbf{i} + (x^2 + y^2)\,\mathbf{j} + (xy - z^2)\,\mathbf{k}$,
S is the surface of the solid bounded by the cylinder $x^2 + y^2 = 4$ and the planes $z = y - 2$ and $z = 0$

13. $\mathbf{F} = |\mathbf{r}|\,\mathbf{r}$, where $\mathbf{r} = x\,\mathbf{i} + y\,\mathbf{j} + z\,\mathbf{k}$,
S consists of the hemisphere $z = \sqrt{1 - x^2 - y^2}$ and the disk $x^2 + y^2 \le 1$ in the xy-plane

14. $\mathbf{F} = |\mathbf{r}|^2 \mathbf{r}$, where $\mathbf{r} = x\,\mathbf{i} + y\,\mathbf{j} + z\,\mathbf{k}$,
S is the sphere with radius R and center the origin

CAS 15. $\mathbf{F}(x, y, z) = e^y \tan z\,\mathbf{i} + y\sqrt{3 - x^2}\,\mathbf{j} + x \sin y\,\mathbf{k}$,
S is the surface of the solid that lies above the xy-plane
and below the surface $z = 2 - x^4 - y^4$, $-1 \leqslant x \leqslant 1$,
$-1 \leqslant y \leqslant 1$

CAS 16. Use a computer algebra system to plot the vector field
$\mathbf{F}(x, y, z) = \sin x \cos^2 y\,\mathbf{i} + \sin^3 y \cos^4 z\,\mathbf{j} + \sin^5 z \cos^6 x\,\mathbf{k}$
in the cube cut from the first octant by the planes $x = \pi/2$,
$y = \pi/2$, and $z = \pi/2$. Then compute the flux across the
surface of the cube.

17. Use the Divergence Theorem to evaluate $\iint_S \mathbf{F} \cdot d\mathbf{S}$, where
$\mathbf{F}(x, y, z) = z^2 x\,\mathbf{i} + \left(\frac{1}{3}y^3 + \tan z\right)\mathbf{j} + (x^2 z + y^2)\,\mathbf{k}$
and S is the top half of the sphere $x^2 + y^2 + z^2 = 1$.
[*Hint:* Note that S is not a closed surface. First compute
integrals over S_1 and S_2, where S_1 is the disk $x^2 + y^2 \leqslant 1$,
oriented downward, and $S_2 = S \cup S_1$.]

18. Let $\mathbf{F}(x, y, z) = z \tan^{-1}(y^2)\,\mathbf{i} + z^3 \ln(x^2 + 1)\,\mathbf{j} + z\,\mathbf{k}$.
Find the flux of $\mathbf{F}$ across the part of the paraboloid
$x^2 + y^2 + z = 2$ that lies above the plane $z = 1$ and is
oriented upward.

19. A vector field $\mathbf{F}$ is shown. Use the interpretation of divergence derived in this section to determine whether div $\mathbf{F}$
is positive or negative at P_1 and at P_2.

20. (a) Are the points P_1 and P_2 sources or sinks for the vector
field $\mathbf{F}$ shown in the figure? Give an explanation based
solely on the picture.
(b) Given that $\mathbf{F}(x, y) = \langle x, y^2 \rangle$, use the definition of divergence to verify your answer to part (a).

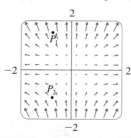

CAS 21–22 Plot the vector field and guess where div $\mathbf{F} > 0$ and
where div $\mathbf{F} < 0$. Then calculate div $\mathbf{F}$ to check your guess.

21. $\mathbf{F}(x, y) = \langle xy, x + y^2 \rangle$ **22.** $\mathbf{F}(x, y) = \langle x^2, y^2 \rangle$

23. Verify that div $\mathbf{E} = 0$ for the electric field $\mathbf{E}(\mathbf{x}) = \dfrac{\varepsilon Q}{|\mathbf{x}|^3}\,\mathbf{x}$.

24. Use the Divergence Theorem to evaluate
$$\iint_S (2x + 2y + z^2)\,dS$$
where S is the sphere $x^2 + y^2 + z^2 = 1$.

25–30 Prove each identity, assuming that S and E satisfy the
conditions of the Divergence Theorem and the scalar functions
and components of the vector fields have continuous second-order
partial derivatives.

25. $\displaystyle\iint_S \mathbf{a} \cdot \mathbf{n}\,dS = 0$, where $\mathbf{a}$ is a constant vector

26. $V(E) = \dfrac{1}{3}\displaystyle\iint_S \mathbf{F} \cdot d\mathbf{S}$, where $\mathbf{F}(x, y, z) = x\,\mathbf{i} + y\,\mathbf{j} + z\,\mathbf{k}$

27. $\displaystyle\iint_S \text{curl } \mathbf{F} \cdot d\mathbf{S} = 0$ **28.** $\displaystyle\iint_S D_{\mathbf{n}} f\,dS = \iiint_E \nabla^2 f\,dV$

29. $\displaystyle\iint_S (f \nabla g) \cdot \mathbf{n}\,dS = \iiint_E (f \nabla^2 g + \nabla f \cdot \nabla g)\,dV$

30. $\displaystyle\iint_S (f \nabla g - g \nabla f) \cdot \mathbf{n}\,dS = \iiint_E (f \nabla^2 g - g \nabla^2 f)\,dV$

31. Suppose S and E satisfy the conditions of the Divergence Theorem and f is a scalar function with continuous partial derivatives. Prove that
$$\iint_S f\mathbf{n}\,dS = \iiint_E \nabla f\,dV$$
These surface and triple integrals of vector functions are
vectors defined by integrating each component function.
[*Hint:* Start by applying the Divergence Theorem to $\mathbf{F} = f\mathbf{c}$,
where $\mathbf{c}$ is an arbitrary constant vector.]

32. A solid occupies a region E with surface S and is immersed
in a liquid with constant density ρ. We set up a coordinate
system so that the xy-plane coincides with the surface of the
liquid, and positive values of z are measured downward into the
liquid. Then the pressure at depth z is $p = \rho gz$, where g is the
acceleration due to gravity (see Section 8.3). The total buoyant
force on the solid due to the pressure distribution is given by
the surface integral
$$\mathbf{F} = -\iint_S p\mathbf{n}\,dS$$
where $\mathbf{n}$ is the outer unit normal. Use the result of Exercise 31
to show that $\mathbf{F} = -W\mathbf{k}$, where W is the weight of the liquid
displaced by the solid. (Note that $\mathbf{F}$ is directed upward because
z is directed downward.) The result is *Archimedes' Principle:*
The buoyant force on an object equals the weight of the displaced liquid.

16.10 Summary

The main results of this chapter are all higher-dimensional versions of the Fundamental Theorem of Calculus. To help you remember them, we collect them together here (without hypotheses) so that you can see more easily their essential similarity. Notice that in each case we have an integral of a "derivative" over a region on the left side, and the right side involves the values of the original function only on the *boundary* of the region.

Fundamental Theorem of Calculus

$$\int_a^b F'(x)\, dx = F(b) - F(a)$$

Fundamental Theorem for Line Integrals

$$\int_C \nabla f \cdot d\mathbf{r} = f(\mathbf{r}(b)) - f(\mathbf{r}(a))$$

Green's Theorem

$$\iint_D \left(\frac{\partial Q}{\partial x} - \frac{\partial P}{\partial y} \right) dA = \int_C P\, dx + Q\, dy$$

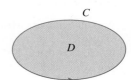

Stokes' Theorem

$$\iint_S \operatorname{curl} \mathbf{F} \cdot d\mathbf{S} = \int_C \mathbf{F} \cdot d\mathbf{r}$$

Divergence Theorem

$$\iiint_E \operatorname{div} \mathbf{F}\, dV = \iint_S \mathbf{F} \cdot d\mathbf{S}$$

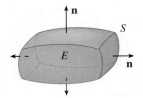

16   REVIEW

CONCEPT CHECK

Answers to the Concept Check can be found on the back endpapers.

1. What is a vector field? Give three examples that have physical meaning.

2. (a) What is a conservative vector field?
   (b) What is a potential function?

3. (a) Write the definition of the line integral of a scalar function f along a smooth curve C with respect to arc length.
   (b) How do you evaluate such a line integral?
   (c) Write expressions for the mass and center of mass of a thin wire shaped like a curve C if the wire has linear density function $\rho(x, y)$.
   (d) Write the definitions of the line integrals along C of a scalar function f with respect to x, y, and z.
   (e) How do you evaluate these line integrals?

4. (a) Define the line integral of a vector field $\mathbf{F}$ along a smooth curve C given by a vector function $\mathbf{r}(t)$.
   (b) If $\mathbf{F}$ is a force field, what does this line integral represent?
   (c) If $\mathbf{F} = \langle P, Q, R \rangle$, what is the connection between the line integral of $\mathbf{F}$ and the line integrals of the component functions P, Q, and R?

5. State the Fundamental Theorem for Line Integrals.

6. (a) What does it mean to say that $\int_C \mathbf{F} \cdot d\mathbf{r}$ is independent of path?
   (b) If you know that $\int_C \mathbf{F} \cdot d\mathbf{r}$ is independent of path, what can you say about $\mathbf{F}$?

7. State Green's Theorem.

8. Write expressions for the area enclosed by a curve C in terms of line integrals around C.

9. Suppose $\mathbf{F}$ is a vector field on $\mathbb{R}^3$.
   (a) Define curl $\mathbf{F}$.   (b) Define div $\mathbf{F}$.

   (c) If $\mathbf{F}$ is a velocity field in fluid flow, what are the physical interpretations of curl $\mathbf{F}$ and div $\mathbf{F}$?

10. If $\mathbf{F} = P\,\mathbf{i} + Q\,\mathbf{j}$, how do you determine whether $\mathbf{F}$ is conservative? What if $\mathbf{F}$ is a vector field on $\mathbb{R}^3$?

11. (a) What is a parametric surface? What are its grid curves?
    (b) Write an expression for the area of a parametric surface.
    (c) What is the area of a surface given by an equation $z = g(x, y)$?

12. (a) Write the definition of the surface integral of a scalar function f over a surface S.
    (b) How do you evaluate such an integral if S is a parametric surface given by a vector function $\mathbf{r}(u, v)$?
    (c) What if S is given by an equation $z = g(x, y)$?
    (d) If a thin sheet has the shape of a surface S, and the density at (x, y, z) is $\rho(x, y, z)$, write expressions for the mass and center of mass of the sheet.

13. (a) What is an oriented surface? Give an example of a non-orientable surface.
    (b) Define the surface integral (or flux) of a vector field $\mathbf{F}$ over an oriented surface S with unit normal vector $\mathbf{n}$.
    (c) How do you evaluate such an integral if S is a parametric surface given by a vector function $\mathbf{r}(u, v)$?
    (d) What if S is given by an equation $z = g(x, y)$?

14. State Stokes' Theorem.

15. State the Divergence Theorem.

16. In what ways are the Fundamental Theorem for Line Integrals, Green's Theorem, Stokes' Theorem, and the Divergence Theorem similar?

TRUE-FALSE QUIZ

Determine whether the statement is true or false. If it is true, explain why. If it is false, explain why or give an example that disproves the statement.

1. If $\mathbf{F}$ is a vector field, then div $\mathbf{F}$ is a vector field.

2. If $\mathbf{F}$ is a vector field, then curl $\mathbf{F}$ is a vector field.

3. If f has continuous partial derivatives of all orders on $\mathbb{R}^3$, then $\operatorname{div}(\operatorname{curl} \nabla f) = 0$.

4. If f has continuous partial derivatives on $\mathbb{R}^3$ and C is any circle, then $\int_C \nabla f \cdot d\mathbf{r} = 0$.

5. If $\mathbf{F} = P\,\mathbf{i} + Q\,\mathbf{j}$ and $P_y = Q_x$ in an open region D, then $\mathbf{F}$ is conservative.

6. $\int_{-C} f(x, y)\, ds = -\int_C f(x, y)\, ds$

7. If $\mathbf{F}$ and $\mathbf{G}$ are vector fields and div $\mathbf{F} = $ div $\mathbf{G}$, then $\mathbf{F} = \mathbf{G}$.

8. The work done by a conservative force field in moving a particle around a closed path is zero.

9. If $\mathbf{F}$ and $\mathbf{G}$ are vector fields, then
$$\operatorname{curl}(\mathbf{F} + \mathbf{G}) = \operatorname{curl} \mathbf{F} + \operatorname{curl} \mathbf{G}$$

10. If $\mathbf{F}$ and $\mathbf{G}$ are vector fields, then
$$\operatorname{curl}(\mathbf{F} \cdot \mathbf{G}) = \operatorname{curl} \mathbf{F} \cdot \operatorname{curl} \mathbf{G}$$

11. If S is a sphere and $\mathbf{F}$ is a constant vector field, then $\iint_S \mathbf{F} \cdot d\mathbf{S} = 0$.

12. There is a vector field $\mathbf{F}$ such that
$$\operatorname{curl} \mathbf{F} = x\,\mathbf{i} + y\,\mathbf{j} + z\,\mathbf{k}$$

13. The area of the region bounded by the positively oriented, piecewise smooth, simple closed curve C is $A = \oint_C y\, dx$.

EXERCISES

1. A vector field $\mathbf{F}$, a curve C, and a point P are shown.
  (a) Is $\int_C \mathbf{F} \cdot d\mathbf{r}$ positive, negative, or zero? Explain.
  (b) Is div $\mathbf{F}(P)$ positive, negative, or zero? Explain.

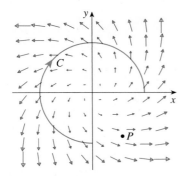

2–9 Evaluate the line integral.

2. $\int_C x \, ds$,
C is the arc of the parabola $y = x^2$ from $(0, 0)$ to $(1, 1)$

3. $\int_C yz \cos x \, ds$,
$C: x = t, \ y = 3 \cos t, \ z = 3 \sin t, \ 0 \le t \le \pi$

4. $\int_C y \, dx + (x + y^2) \, dy$, C is the ellipse $4x^2 + 9y^2 = 36$
with counterclockwise orientation

5. $\int_C y^3 \, dx + x^2 \, dy$, C is the arc of the parabola $x = 1 - y^2$
from $(0, -1)$ to $(0, 1)$

6. $\int_C \sqrt{xy} \, dx + e^y \, dy + xz \, dz$,
C is given by $\mathbf{r}(t) = t^4 \mathbf{i} + t^2 \mathbf{j} + t^3 \mathbf{k}, 0 \le t \le 1$

7. $\int_C xy \, dx + y^2 \, dy + yz \, dz$,
C is the line segment from $(1, 0, -1)$, to $(3, 4, 2)$

8. $\int_C \mathbf{F} \cdot d\mathbf{r}$, where $\mathbf{F}(x, y) = xy \, \mathbf{i} + x^2 \mathbf{j}$ and C is given by
$\mathbf{r}(t) = \sin t \, \mathbf{i} + (1 + t) \mathbf{j}, 0 \le t \le \pi$

9. $\int_C \mathbf{F} \cdot d\mathbf{r}$, where $\mathbf{F}(x, y, z) = e^z \mathbf{i} + xz \mathbf{j} + (x + y) \mathbf{k}$ and
C is given by $\mathbf{r}(t) = t^2 \mathbf{i} + t^3 \mathbf{j} - t \mathbf{k}, 0 \le t \le 1$

10. Find the work done by the force field

$$\mathbf{F}(x, y, z) = z \, \mathbf{i} + x \, \mathbf{j} + y \, \mathbf{k}$$

in moving a particle from the point $(3, 0, 0)$ to the point
$(0, \pi/2, 3)$ along
  (a) a straight line
  (b) the helix $x = 3 \cos t, \ y = t, \ z = 3 \sin t$

11–12 Show that $\mathbf{F}$ is a conservative vector field. Then find a
function f such that $\mathbf{F} = \nabla f$.

11. $\mathbf{F}(x, y) = (1 + xy)e^{xy} \, \mathbf{i} + (e^y + x^2 e^{xy}) \, \mathbf{j}$

12. $\mathbf{F}(x, y, z) = \sin y \, \mathbf{i} + x \cos y \, \mathbf{j} - \sin z \, \mathbf{k}$

13–14 Show that $\mathbf{F}$ is conservative and use this fact to evaluate
$\int_C \mathbf{F} \cdot d\mathbf{r}$ along the given curve.

13. $\mathbf{F}(x, y) = (4x^3 y^2 - 2xy^3) \, \mathbf{i} + (2x^4 y - 3x^2 y^2 + 4y^3) \, \mathbf{j}$,
$C: \mathbf{r}(t) = (t + \sin \pi t) \, \mathbf{i} + (2t + \cos \pi t) \, \mathbf{j}, \ 0 \le t \le 1$

14. $\mathbf{F}(x, y, z) = e^y \, \mathbf{i} + (xe^y + e^z) \, \mathbf{j} + ye^z \, \mathbf{k}$,
C is the line segment from $(0, 2, 0)$ to $(4, 0, 3)$

15. Verify that Green's Theorem is true for the line integral
$\int_C xy^2 \, dx - x^2 y \, dy$, where C consists of the parabola $y = x^2$
from $(-1, 1)$ to $(1, 1)$ and the line segment from $(1, 1)$
to $(-1, 1)$.

16. Use Green's Theorem to evaluate

$$\int_C \sqrt{1 + x^3} \, dx + 2xy \, dy$$

where C is the triangle with vertices $(0, 0)$, $(1, 0)$, and $(1, 3)$.

17. Use Green's Theorem to evaluate $\int_C x^2 y \, dx - xy^2 \, dy$,
where C is the circle $x^2 + y^2 = 4$ with counterclockwise
orientation.

18. Find curl $\mathbf{F}$ and div $\mathbf{F}$ if

$$\mathbf{F}(x, y, z) = e^{-x} \sin y \, \mathbf{i} + e^{-y} \sin z \, \mathbf{j} + e^{-z} \sin x \, \mathbf{k}$$

19. Show that there is no vector field $\mathbf{G}$ such that

$$\text{curl } \mathbf{G} = 2x \, \mathbf{i} + 3yz \, \mathbf{j} - xz^2 \, \mathbf{k}$$

20. If $\mathbf{F}$ and $\mathbf{G}$ are vector fields whose component functions have
continuous first partial derivatives, show that

$$\text{curl}(\mathbf{F} \times \mathbf{G}) = \mathbf{F} \, \text{div } \mathbf{G} - \mathbf{G} \, \text{div } \mathbf{F} + (\mathbf{G} \cdot \nabla)\mathbf{F} - (\mathbf{F} \cdot \nabla)\mathbf{G}$$

21. If C is any piecewise-smooth simple closed plane curve
and f and g are differentiable functions, show that
$\int_C f(x) \, dx + g(y) \, dy = 0$.

22. If f and g are twice differentiable functions, show that

$$\nabla^2(fg) = f\nabla^2 g + g\nabla^2 f + 2\nabla f \cdot \nabla g$$

23. If f is a harmonic function, that is, $\nabla^2 f = 0$, show that the line
integral $\int f_y \, dx - f_x \, dy$ is independent of path in any simple
region D.

24. (a) Sketch the curve C with parametric equations

$$x = \cos t \qquad y = \sin t \qquad z = \sin t \qquad 0 \le t \le 2\pi$$

  (b) Find $\int_C 2xe^{2y} \, dx + (2x^2 e^{2y} + 2y \cot z) \, dy - y^2 \csc^2 z \, dz$.

25. Find the area of the part of the surface $z = x^2 + 2y$ that lies above the triangle with vertices $(0, 0)$, $(1, 0)$, and $(1, 2)$.

26. (a) Find an equation of the tangent plane at the point $(4, -2, 1)$ to the parametric surface S given by

$$\mathbf{r}(u, v) = v^2\,\mathbf{i} - uv\,\mathbf{j} + u^2\,\mathbf{k} \qquad 0 \le u \le 3,\ -3 \le v \le 3$$

(b) Use a computer to graph the surface S and the tangent plane found in part (a).

(c) Set up, but do not evaluate, an integral for the surface area of S.

(d) If

$$\mathbf{F}(x, y, z) = \frac{z^2}{1 + x^2}\,\mathbf{i} + \frac{x^2}{1 + y^2}\,\mathbf{j} + \frac{y^2}{1 + z^2}\,\mathbf{k}$$

find $\iint_S \mathbf{F} \cdot d\mathbf{S}$ correct to four decimal places.

27–30 Evaluate the surface integral.

27. $\iint_S z\,dS$, where S is the part of the paraboloid $z = x^2 + y^2$ that lies under the plane $z = 4$

28. $\iint_S (x^2z + y^2z)\,dS$, where S is the part of the plane $z = 4 + x + y$ that lies inside the cylinder $x^2 + y^2 = 4$

29. $\iint_S \mathbf{F} \cdot d\mathbf{S}$, where $\mathbf{F}(x, y, z) = xz\,\mathbf{i} - 2y\,\mathbf{j} + 3x\,\mathbf{k}$ and S is the sphere $x^2 + y^2 + z^2 = 4$ with outward orientation

30. $\iint_S \mathbf{F} \cdot d\mathbf{S}$, where $\mathbf{F}(x, y, z) = x^2\,\mathbf{i} + xy\,\mathbf{j} + z\,\mathbf{k}$ and S is the part of the paraboloid $z = x^2 + y^2$ below the plane $z = 1$ with upward orientation

---

31. Verify that Stokes' Theorem is true for the vector field $\mathbf{F}(x, y, z) = x^2\,\mathbf{i} + y^2\,\mathbf{j} + z^2\,\mathbf{k}$, where S is the part of the paraboloid $z = 1 - x^2 - y^2$ that lies above the xy-plane and S has upward orientation.

32. Use Stokes' Theorem to evaluate $\iint_S \text{curl } \mathbf{F} \cdot d\mathbf{S}$, where $\mathbf{F}(x, y, z) = x^2yz\,\mathbf{i} + yz^2\,\mathbf{j} + z^3e^{xy}\,\mathbf{k}$, S is the part of the sphere $x^2 + y^2 + z^2 = 5$ that lies above the plane $z = 1$, and S is oriented upward.

33. Use Stokes' Theorem to evaluate $\int_C \mathbf{F} \cdot d\mathbf{r}$, where $\mathbf{F}(x, y, z) = xy\,\mathbf{i} + yz\,\mathbf{j} + zx\,\mathbf{k}$, and C is the triangle with vertices $(1, 0, 0)$, $(0, 1, 0)$, and $(0, 0, 1)$, oriented counterclockwise as viewed from above.

34. Use the Divergence Theorem to calculate the surface integral $\iint_S \mathbf{F} \cdot d\mathbf{S}$, where $\mathbf{F}(x, y, z) = x^3\,\mathbf{i} + y^3\,\mathbf{j} + z^3\,\mathbf{k}$ and S is the surface of the solid bounded by the cylinder $x^2 + y^2 = 1$ and the planes $z = 0$ and $z = 2$.

35. Verify that the Divergence Theorem is true for the vector field $\mathbf{F}(x, y, z) = x\,\mathbf{i} + y\,\mathbf{j} + z\,\mathbf{k}$, where E is the unit ball $x^2 + y^2 + z^2 \le 1$.

36. Compute the outward flux of

$$\mathbf{F}(x, y, z) = \frac{x\,\mathbf{i} + y\,\mathbf{j} + z\,\mathbf{k}}{(x^2 + y^2 + z^2)^{3/2}}$$

through the ellipsoid $4x^2 + 9y^2 + 6z^2 = 36$.

37. Let

$$\mathbf{F}(x, y, z) = (3x^2yz - 3y)\,\mathbf{i} + (x^3z - 3x)\,\mathbf{j} + (x^3y + 2z)\,\mathbf{k}$$

Evaluate $\int_C \mathbf{F} \cdot d\mathbf{r}$, where C is the curve with initial point $(0, 0, 2)$ and terminal point $(0, 3, 0)$ shown in the figure.

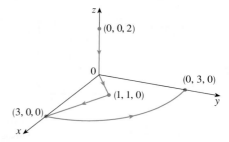

38. Let

$$\mathbf{F}(x, y) = \frac{(2x^3 + 2xy^2 - 2y)\,\mathbf{i} + (2y^3 + 2x^2y + 2x)\,\mathbf{j}}{x^2 + y^2}$$

Evaluate $\oint_C \mathbf{F} \cdot d\mathbf{r}$, where C is shown in the figure.

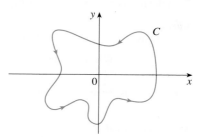

39. Find $\iint_S \mathbf{F} \cdot \mathbf{n}\,dS$, where $\mathbf{F}(x, y, z) = x\,\mathbf{i} + y\,\mathbf{j} + z\,\mathbf{k}$ and S is the outwardly oriented surface shown in the figure (the boundary surface of a cube with a unit corner cube removed).

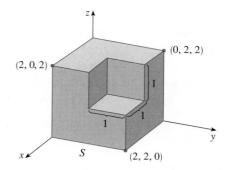

40. If the components of $\mathbf{F}$ have continuous second partial derivatives and S is the boundary surface of a simple solid region, show that $\iint_S \text{curl } \mathbf{F} \cdot d\mathbf{S} = 0$.

41. If $\mathbf{a}$ is a constant vector, $\mathbf{r} = x\,\mathbf{i} + y\,\mathbf{j} + z\,\mathbf{k}$, and S is an oriented, smooth surface with a simple, closed, smooth, positively oriented boundary curve C, show that

$$\iint_S 2\mathbf{a} \cdot d\mathbf{S} = \int_C (\mathbf{a} \times \mathbf{r}) \cdot d\mathbf{r}$$

Problems Plus

1. Let S be a smooth parametric surface and let P be a point such that each line that starts at P intersects S at most once. The **solid angle** $\Omega(S)$ subtended by S at P is the set of lines starting at P and passing through S. Let $S(a)$ be the intersection of $\Omega(S)$ with the surface of the sphere with center P and radius a. Then the measure of the solid angle (in *steradians*) is defined to be

$$|\Omega(S)| = \frac{\text{area of } S(a)}{a^2}$$

Apply the Divergence Theorem to the part of $\Omega(S)$ between $S(a)$ and S to show that

$$|\Omega(S)| = \iint_S \frac{\mathbf{r} \cdot \mathbf{n}}{r^3}\, dS$$

where $\mathbf{r}$ is the radius vector from P to any point on S, $r = |\mathbf{r}|$, and the unit normal vector $\mathbf{n}$ is directed away from P.

This shows that the definition of the measure of a solid angle is independent of the radius a of the sphere. Thus the measure of the solid angle is equal to the area subtended on a *unit* sphere. (Note the analogy with the definition of radian measure.) The total solid angle subtended by a sphere at its center is thus 4π steradians.

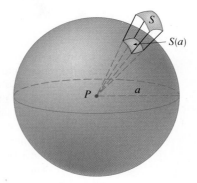

2. Find the positively oriented simple closed curve C for which the value of the line integral

$$\int_C (y^3 - y)\, dx - 2x^3\, dy$$

is a maximum.

3. Let C be a simple closed piecewise-smooth space curve that lies in a plane with unit normal vector $\mathbf{n} = \langle a, b, c \rangle$ and has positive orientation with respect to $\mathbf{n}$. Show that the plane area enclosed by C is

$$\tfrac{1}{2} \int_C (bz - cy)\, dx + (cx - az)\, dy + (ay - bx)\, dz$$

4. Investigate the shape of the surface with parametric equations $x = \sin u$, $y = \sin v$, $z = \sin(u + v)$. Start by graphing the surface from several points of view. Explain the appearance of the graphs by determining the traces in the horizontal planes $z = 0$, $z = \pm 1$, and $z = \pm\frac{1}{2}$.

5. Prove the following identity:

$$\nabla(\mathbf{F} \cdot \mathbf{G}) = (\mathbf{F} \cdot \nabla)\mathbf{G} + (\mathbf{G} \cdot \nabla)\mathbf{F} + \mathbf{F} \times \operatorname{curl} \mathbf{G} + \mathbf{G} \times \operatorname{curl} \mathbf{F}$$

6. The figure depicts the sequence of events in each cylinder of a four-cylinder internal combustion engine. Each piston moves up and down and is connected by a pivoted arm to a rotating crankshaft. Let $P(t)$ and $V(t)$ be the pressure and volume within a cylinder at time t, where $a \leqslant t \leqslant b$ gives the time required for a complete cycle. The graph shows how P and V vary through one cycle of a four-stroke engine.

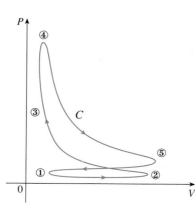

During the intake stroke (from ① to ②) a mixture of air and gasoline at atmospheric pressure is drawn into a cylinder through the intake valve as the piston moves downward. Then the piston rapidly compresses the mix with the valves closed in the compression stroke (from ② to ③) during which the pressure rises and the volume decreases. At ③ the sparkplug ignites the fuel, raising the temperature and pressure at almost constant volume to ④. Then, with valves closed, the rapid expansion forces the piston downward during the power stroke (from ④ to ⑤). The exhaust valve opens, temperature and pressure drop, and mechanical energy stored in a rotating flywheel pushes the piston upward, forcing the waste products out of the exhaust valve in the exhaust stroke. The exhaust valve closes and the intake valve opens. We're now back at ① and the cycle starts again.

(a) Show that the work done on the piston during one cycle of a four-stroke engine is $W = \int_C P \, dV$, where C is the curve in the PV-plane shown in the figure.

[*Hint:* Let $x(t)$ be the distance from the piston to the top of the cylinder and note that the force on the piston is $\mathbf{F} = AP(t) \, \mathbf{i}$, where A is the area of the top of the piston. Then $W = \int_{C_1} \mathbf{F} \cdot d\mathbf{r}$, where C_1 is given by $\mathbf{r}(t) = x(t) \, \mathbf{i}$, $a \leqslant t \leqslant b$. An alternative approach is to work directly with Riemann sums.]

(b) Use Formula 16.4.5 to show that the work is the difference of the areas enclosed by the two loops of C.

Appendixes

A   Numbers, Inequalities, and Absolute Values

Calculus is based on the real number system. We start with the **integers**:

$$\ldots, \quad -3, \quad -2, \quad -1, \quad 0, \quad 1, \quad 2, \quad 3, \quad 4, \quad \ldots$$

Then we construct the **rational numbers**, which are ratios of integers. Thus any rational number r can be expressed as

$$r = \frac{m}{n} \qquad \text{where } m \text{ and } n \text{ are integers and } n \neq 0$$

Examples are

$$\tfrac{1}{2} \qquad -\tfrac{3}{7} \qquad 46 = \tfrac{46}{1} \qquad 0.17 = \tfrac{17}{100}$$

(Recall that division by 0 is always ruled out, so expressions like $\frac{3}{0}$ and $\frac{0}{0}$ are undefined.) Some real numbers, such as $\sqrt{2}$, can't be expressed as a ratio of integers and are therefore called **irrational numbers**. It can be shown, with varying degrees of difficulty, that the following are also irrational numbers:

$$\sqrt{3} \qquad \sqrt{5} \qquad \sqrt[3]{2} \qquad \pi \qquad \sin 1° \qquad \log_{10} 2$$

The set of all real numbers is usually denoted by the symbol $\mathbb{R}$. When we use the word *number* without qualification, we mean "real number."

Every number has a decimal representation. If the number is rational, then the corresponding decimal is repeating. For example,

$$\tfrac{1}{2} = 0.5000\ldots = 0.5\overline{0} \qquad\qquad \tfrac{2}{3} = 0.66666\ldots = 0.\overline{6}$$

$$\tfrac{157}{495} = 0.317171717\ldots = 0.3\overline{17} \qquad\qquad \tfrac{9}{7} = 1.285714285714\ldots = 1.\overline{285714}$$

(The bar indicates that the sequence of digits repeats forever.) On the other hand, if the number is irrational, the decimal is nonrepeating:

$$\sqrt{2} = 1.414213562373095\ldots \qquad\qquad \pi = 3.141592653589793\ldots$$

If we stop the decimal expansion of any number at a certain place, we get an approximation to the number. For instance, we can write

$$\pi \approx 3.14159265$$

where the symbol $\approx$ is read "is approximately equal to." The more decimal places we retain, the better the approximation we get.

The real numbers can be represented by points on a line as in Figure 1. The positive direction (to the right) is indicated by an arrow. We choose an arbitrary reference point O, called the **origin**, which corresponds to the real number 0. Given any convenient unit of measurement, each positive number x is represented by the point on the line a distance of x units to the right of the origin, and each negative number $-x$ is represented by the point x units to the left of the origin. Thus every real number is represented by a point on the line, and every point P on the line corresponds to exactly one real number. The number associated with the point P is called the **coordinate** of P and the line is then called a

coordinate line, or a **real number line**, or simply a **real line**. Often we identify the point with its coordinate and think of a number as being a point on the real line.

FIGURE 1

The real numbers are ordered. We say *a is less than b* and write $a < b$ if $b - a$ is a positive number. Geometrically this means that a lies to the left of b on the number line. (Equivalently, we say *b is greater than a* and write $b > a$.) The symbol $a \leq b$ (or $b \geq a$) means that either $a < b$ or $a = b$ and is read "a is less than or equal to b." For instance, the following are true inequalities:

$$7 < 7.4 < 7.5 \qquad -3 > -\pi \qquad \sqrt{2} < 2 \qquad \sqrt{2} \leq 2 \qquad 2 \leq 2$$

In what follows we need to use *set notation*. A **set** is a collection of objects, and these objects are called the **elements** of the set. If S is a set, the notation $a \in S$ means that a is an element of S, and $a \notin S$ means that a is not an element of S. For example, if Z represents the set of integers, then $-3 \in Z$ but $\pi \notin Z$. If S and T are sets, then their **union** $S \cup T$ is the set consisting of all elements that are in S or T (or in both S and T). The **intersection** of S and T is the set $S \cap T$ consisting of all elements that are in both S and T. In other words, $S \cap T$ is the common part of S and T. The empty set, denoted by $\varnothing$, is the set that contains no element.

Some sets can be described by listing their elements between braces. For instance, the set A consisting of all positive integers less than 7 can be written as

$$A = \{1, 2, 3, 4, 5, 6\}$$

We could also write A in *set-builder notation* as

$$A = \{x \mid x \text{ is an integer and } 0 < x < 7\}$$

which is read "A is the set of x such that x is an integer and $0 < x < 7$."

■ Intervals

Certain sets of real numbers, called **intervals**, occur frequently in calculus and correspond geometrically to line segments. For example, if $a < b$, the **open interval** from a to b consists of all numbers between a and b and is denoted by the symbol (a, b). Using set-builder notation, we can write

$$(a, b) = \{x \mid a < x < b\}$$

Notice that the endpoints of the interval—namely, a and b—are excluded. This is indicated by the round brackets $(\ )$ and by the open dots in Figure 2. The **closed interval** from a to b is the set

$$[a, b] = \{x \mid a \leq x \leq b\}$$

Here the endpoints of the interval are included. This is indicated by the square brackets $[\ ]$ and by the solid dots in Figure 3. It is also possible to include only one endpoint in an interval, as shown in Table 1.

FIGURE 2
Open interval (a, b)

FIGURE 3
Closed interval $[a, b]$

We also need to consider infinite intervals such as

$$(a, \infty) = \{x \mid x > a\}$$

This does not mean that ∞ ("infinity") is a number. The notation (a, ∞) stands for the set of all numbers that are greater than a, so the symbol ∞ simply indicates that the interval extends indefinitely far in the positive direction.

1 **Table of Intervals**

Table 1 lists the nine possible types of intervals. When these intervals are discussed, it is always assumed that $a < b$.

Notation	Set description	Picture
(a, b)	$\{x \mid a < x < b\}$	
$[a, b]$	$\{x \mid a \leqslant x \leqslant b\}$	
$[a, b)$	$\{x \mid a \leqslant x < b\}$	
$(a, b]$	$\{x \mid a < x \leqslant b\}$	
(a, ∞)	$\{x \mid x > a\}$	
$[a, \infty)$	$\{x \mid x \geqslant a\}$	
$(-\infty, b)$	$\{x \mid x < b\}$	
$(-\infty, b]$	$\{x \mid x \leqslant b\}$	
$(-\infty, \infty)$	$\mathbb{R}$ (set of all real numbers)	

■ Inequalities

When working with inequalities, note the following rules.

---

2 **Rules for Inequalities**

1. If $a < b$, then $a + c < b + c$.
2. If $a < b$ and $c < d$, then $a + c < b + d$.
3. If $a < b$ and $c > 0$, then $ac < bc$.
4. If $a < b$ and $c < 0$, then $ac > bc$.
5. If $0 < a < b$, then $1/a > 1/b$.

Rule 1 says that we can add any number to both sides of an inequality, and Rule 2 says that two inequalities can be added. However, we have to be careful with multiplication. Rule 3 says that we can multiply both sides of an inequality by a *positive* number, but Rule 4 says that if we multiply both sides of an inequality by a negative number, then we reverse the direction of the inequality. For example, if we take the inequality $3 < 5$ and multiply by 2, we get $6 < 10$, but if we multiply by -2, we get $-6 > -10$. Finally, Rule 5 says that if we take reciprocals, then we reverse the direction of an inequality (provided the numbers are positive).

EXAMPLE 1 Solve the inequality $1 + x < 7x + 5$.

SOLUTION The given inequality is satisfied by some values of x but not by others. To *solve* an inequality means to determine the set of numbers x for which the inequality is true. This is called the *solution set*.

First we subtract 1 from each side of the inequality (using Rule 1 with $c = -1$):

$$x < 7x + 4$$

Then we subtract $7x$ from both sides (Rule 1 with $c = -7x$):

$$-6x < 4$$

Now we divide both sides by -6 $\left(\text{Rule 4 with } c = -\frac{1}{6}\right)$:

$$x > -\tfrac{4}{6} = -\tfrac{2}{3}$$

These steps can all be reversed, so the solution set consists of all numbers greater than $-\frac{2}{3}$. In other words, the solution of the inequality is the interval $\left(-\frac{2}{3}, \infty\right)$. ∎

EXAMPLE 2 Solve the inequalities $4 \leqslant 3x - 2 < 13$.

SOLUTION Here the solution set consists of all values of x that satisfy both inequalities. Using the rules given in (2), we see that the following inequalities are equivalent:

$$4 \leqslant 3x - 2 < 13$$

$$6 \leqslant 3x < 15 \qquad \text{(add 2)}$$

$$2 \leqslant x < 5 \qquad \text{(divide by 3)}$$

Therefore the solution set is $[2, 5)$. ∎

EXAMPLE 3 Solve the inequality $x^2 - 5x + 6 \leqslant 0$.

SOLUTION First we factor the left side:

$$(x - 2)(x - 3) \leqslant 0$$

We know that the corresponding equation $(x - 2)(x - 3) = 0$ has the solutions 2 and 3. The numbers 2 and 3 divide the real line into three intervals:

$$(-\infty, 2) \qquad (2, 3) \qquad (3, \infty)$$

A visual method for solving Example 3 is to use a graphing device to graph the parabola $y = x^2 - 5x + 6$ (as in Figure 4) and observe that the curve lies on or below the x-axis when $2 \leqslant x \leqslant 3$.

On each of these intervals we determine the signs of the factors. For instance,

$$x \in (-\infty, 2) \quad \Rightarrow \quad x < 2 \quad \Rightarrow \quad x - 2 < 0$$

Then we record these signs in the following chart:

Interval	$x - 2$	$x - 3$	$(x - 2)(x - 3)$
$x < 2$	$-$	$-$	$+$
$2 < x < 3$	$+$	$-$	$-$
$x > 3$	$+$	$+$	$+$

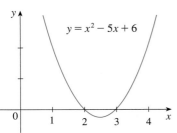

FIGURE 4

Another method for obtaining the information in the chart is to use *test values*. For instance, if we use the test value $x = 1$ for the interval $(-\infty, 2)$, then substitution in $x^2 - 5x + 6$ gives

$$1^2 - 5(1) + 6 = 2$$

The polynomial $x^2 - 5x + 6$ doesn't change sign inside any of the three intervals, so we conclude that it is positive on $(-\infty, 2)$.

Then we read from the chart that $(x - 2)(x - 3)$ is negative when $2 < x < 3$. Thus the solution of the inequality $(x - 2)(x - 3) \leq 0$ is

$$\{x \mid 2 \leq x \leq 3\} = [2, 3]$$

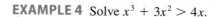

FIGURE 5

Notice that we have included the endpoints 2 and 3 because we are looking for values of x such that the product is either negative or zero. The solution is illustrated in Figure 5.   ∎

EXAMPLE 4 Solve $x^3 + 3x^2 > 4x$.

SOLUTION First we take all nonzero terms to one side of the inequality sign and factor the resulting expression:

$$x^3 + 3x^2 - 4x > 0 \qquad \text{or} \qquad x(x - 1)(x + 4) > 0$$

As in Example 3 we solve the corresponding equation $x(x - 1)(x + 4) = 0$ and use the solutions $x = -4$, $x = 0$, and $x = 1$ to divide the real line into four intervals $(-\infty, -4)$, $(-4, 0)$, $(0, 1)$, and $(1, \infty)$. On each interval the product keeps a constant sign, which we list in the following chart:

Interval	x	$x - 1$	$x + 4$	$x(x - 1)(x + 4)$
$x < -4$	$-$	$-$	$-$	$-$
$-4 < x < 0$	$-$	$-$	$+$	$+$
$0 < x < 1$	$+$	$-$	$+$	$-$
$x > 1$	$+$	$+$	$+$	$+$

Then we read from the chart that the solution set is

$$\{x \mid -4 < x < 0 \text{ or } x > 1\} = (-4, 0) \cup (1, \infty)$$

FIGURE 6

The solution is illustrated in Figure 6.   ∎

■ Absolute Value

The **absolute value** of a number a, denoted by $|a|$, is the distance from a to 0 on the real number line. Distances are always positive or 0, so we have

$$|a| \geq 0 \qquad \text{for every number } a$$

For example,

$$|3| = 3 \qquad |-3| = 3 \qquad |0| = 0 \qquad |\sqrt{2} - 1| = \sqrt{2} - 1 \qquad |3 - \pi| = \pi - 3$$

In general, we have

Remember that if a is negative, then $-a$ is positive.

3

$$|a| = a \qquad \text{if } a \geq 0$$
$$|a| = -a \qquad \text{if } a < 0$$

EXAMPLE 5 Express $|3x - 2|$ without using the absolute-value symbol.

SOLUTION

$$|3x - 2| = \begin{cases} 3x - 2 & \text{if } 3x - 2 \geq 0 \\ -(3x - 2) & \text{if } 3x - 2 < 0 \end{cases}$$

$$= \begin{cases} 3x - 2 & \text{if } x \geq \frac{2}{3} \\ 2 - 3x & \text{if } x < \frac{2}{3} \end{cases}$$

Recall that the symbol $\sqrt{}$ means "the positive square root of." Thus $\sqrt{r} = s$ means $s^2 = r$ and $s \geq 0$. Therefore the equation $\sqrt{a^2} = a$ is not always true. It is true only when $a \geq 0$. If $a < 0$, then $-a > 0$, so we have $\sqrt{a^2} = -a$. In view of (3), we then have the equation

4

$$\sqrt{a^2} = |a|$$

which is true for all values of a.

Hints for the proofs of the following properties are given in the exercises.

> **5** **Properties of Absolute Values** Suppose a and b are any real numbers and n is an integer. Then
>
> **1.** $|ab| = |a||b|$ **2.** $\left|\dfrac{a}{b}\right| = \dfrac{|a|}{|b|}$ $(b \neq 0)$ **3.** $|a^n| = |a|^n$

For solving equations or inequalities involving absolute values, it's often very helpful to use the following statements.

> **6** Suppose $a > 0$. Then
>
> **4.** $|x| = a$ if and only if $x = \pm a$
>
> **5.** $|x| < a$ if and only if $-a < x < a$
>
> **6.** $|x| > a$ if and only if $x > a$ or $x < -a$

FIGURE 7

For instance, the inequality $|x| < a$ says that the distance from x to the origin is less than a, and you can see from Figure 7 that this is true if and only if x lies between $-a$ and a.

If a and b are any real numbers, then the distance between a and b is the absolute value of the difference, namely, $|a - b|$, which is also equal to $|b - a|$. (See Figure 8.)

FIGURE 8
Length of a line segment $= |a - b|$

EXAMPLE 6 Solve $|2x - 5| = 3$.

SOLUTION By Property 4 of (6), $|2x - 5| = 3$ is equivalent to

$$2x - 5 = 3 \quad \text{or} \quad 2x - 5 = -3$$

So $2x = 8$ or $2x = 2$. Thus $x = 4$ or $x = 1$.

EXAMPLE 7 Solve $|x - 5| < 2$.

SOLUTION 1 By Property 5 of (6), $|x - 5| < 2$ is equivalent to

$$-2 < x - 5 < 2$$

Therefore, adding 5 to each side, we have

$$3 < x < 7$$

and the solution set is the open interval $(3, 7)$.

SOLUTION 2 Geometrically the solution set consists of all numbers x whose distance from 5 is less than 2. From Figure 9 we see that this is the interval $(3, 7)$. ∎

FIGURE 9

EXAMPLE 8 Solve $|3x + 2| \geqslant 4$.

SOLUTION By Properties 4 and 6 of (6), $|3x + 2| \geqslant 4$ is equivalent to

$$3x + 2 \geqslant 4 \qquad \text{or} \qquad 3x + 2 \leqslant -4$$

In the first case $3x \geqslant 2$, which gives $x \geqslant \frac{2}{3}$. In the second case $3x \leqslant -6$, which gives $x \leqslant -2$. So the solution set is

$$\{x \mid x \leqslant -2 \text{ or } x \geqslant \tfrac{2}{3}\} = (-\infty, -2] \cup \left[\tfrac{2}{3}, \infty\right)$$ ∎

Another important property of absolute value, called the Triangle Inequality, is used frequently not only in calculus but throughout mathematics in general.

---

7 **The Triangle Inequality** If a and b are any real numbers, then

$$|a + b| \leqslant |a| + |b|$$

Observe that if the numbers a and b are both positive or both negative, then the two sides in the Triangle Inequality are actually equal. But if a and b have opposite signs, the left side involves a subtraction and the right side does not. This makes the Triangle Inequality seem reasonable, but we can prove it as follows.

Notice that

$$-|a| \leqslant a \leqslant |a|$$

is always true because a equals either $|a|$ or $-|a|$. The corresponding statement for b is

$$-|b| \leqslant b \leqslant |b|$$

Adding these inequalities, we get

$$-(|a| + |b|) \leqslant a + b \leqslant |a| + |b|$$

If we now apply Properties 4 and 5 $\bigl($with x replaced by $a + b$ and a by $|a| + |b|\bigr)$, we obtain

$$|a + b| \leqslant |a| + |b|$$

which is what we wanted to show.

EXAMPLE 9 If $|x - 4| < 0.1$ and $|y - 7| < 0.2$, use the Triangle Inequality to estimate $|(x + y) - 11|$.

SOLUTION In order to use the given information, we use the Triangle Inequality with $a = x - 4$ and $b = y - 7$:

$$|(x + y) - 11| = |(x - 4) + (y - 7)|$$
$$\leq |x - 4| + |y - 7|$$
$$< 0.1 + 0.2 = 0.3$$

Thus $|(x + y) - 11| < 0.3$ ■

A   EXERCISES

1–12 Rewrite the expression without using the absolute-value symbol.

1. $|5 - 23|$

2. $|5| - |-23|$

3. $|-\pi|$

4. $|\pi - 2|$

5. $|\sqrt{5} - 5|$

6. $||-2| - |-3||$

7. $|x - 2|$ if $x < 2$

8. $|x - 2|$ if $x > 2$

9. $|x + 1|$

10. $|2x - 1|$

11. $|x^2 + 1|$

12. $|1 - 2x^2|$

13–38 Solve the inequality in terms of intervals and illustrate the solution set on the real number line.

13. $2x + 7 > 3$

14. $3x - 11 < 4$

15. $1 - x \leq 2$

16. $4 - 3x \geq 6$

17. $2x + 1 < 5x - 8$

18. $1 + 5x > 5 - 3x$

19. $-1 < 2x - 5 < 7$

20. $1 < 3x + 4 \leq 16$

21. $0 \leq 1 - x < 1$

22. $-5 \leq 3 - 2x \leq 9$

23. $4x < 2x + 1 \leq 3x + 2$

24. $2x - 3 < x + 4 < 3x - 2$

25. $(x - 1)(x - 2) > 0$

26. $(2x + 3)(x - 1) \geq 0$

27. $2x^2 + x \leq 1$

28. $x^2 < 2x + 8$

29. $x^2 + x + 1 > 0$

30. $x^2 + x > 1$

31. $x^2 < 3$

32. $x^2 \geq 5$

33. $x^3 - x^2 \leq 0$

34. $(x + 1)(x - 2)(x + 3) \geq 0$

35. $x^3 > x$

36. $x^3 + 3x < 4x^2$

37. $\dfrac{1}{x} < 4$

38. $-3 < \dfrac{1}{x} \leq 1$

39. The relationship between the Celsius and Fahrenheit temperature scales is given by $C = \frac{5}{9}(F - 32)$, where C is the temperature in degrees Celsius and F is the temperature in degrees Fahrenheit. What interval on the Celsius scale corresponds to the temperature range $50 \leq F \leq 95$?

40. Use the relationship between C and F given in Exercise 39 to find the interval on the Fahrenheit scale corresponding to the temperature range $20 \leq C \leq 30$.

41. As dry air moves upward, it expands and in so doing cools at a rate of about 1°C for each 100-m rise, up to about 12 km.
   (a) If the ground temperature is 20°C, write a formula for the temperature at height h.
   (b) What range of temperature can be expected if a plane takes off and reaches a maximum height of 5 km?

42. If a ball is thrown upward from the top of a building 128 ft high with an initial velocity of 16 ft/s, then the height h above the ground t seconds later will be

$$h = 128 + 16t - 16t^2$$

During what time interval will the ball be at least 32 ft above the ground?

43–46 Solve the equation for x.

43. $|2x| = 3$

44. $|3x + 5| = 1$

45. $|x + 3| = |2x + 1|$

46. $\left|\dfrac{2x - 1}{x + 1}\right| = 3$

47–56 Solve the inequality.

47. $|x| < 3$

48. $|x| \geq 3$

49. $|x - 4| < 1$

50. $|x - 6| < 0.1$

51. $|x + 5| \geq 2$

52. $|x + 1| \geq 3$

53. $|2x - 3| \leq 0.4$

54. $|5x - 2| < 6$

55. $1 \leq |x| \leq 4$

56. $0 < |x - 5| < \frac{1}{2}$

57–58 Solve for x, assuming a, b, and c are positive constants.

57. $a(bx - c) \geq bc$

58. $a \leq bx + c < 2a$

59–60 Solve for x, assuming a, b, and c are negative constants.

59. $ax + b < c$

60. $\dfrac{ax + b}{c} \leq b$

61. Suppose that $|x - 2| < 0.01$ and $|y - 3| < 0.04$. Use the Triangle Inequality to show that $|(x + y) - 5| < 0.05$.

62. Show that if $|x + 3| < \frac{1}{2}$, then $|4x + 13| < 3$.

63. Show that if $a < b$, then $a < \dfrac{a + b}{2} < b$.

64. Use Rule 3 to prove Rule 5 of (2).

65. Prove that $|ab| = |a||b|$. [*Hint:* Use Equation 4.]

66. Prove that $\left|\dfrac{a}{b}\right| = \dfrac{|a|}{|b|}$.

67. Show that if $0 < a < b$, then $a^2 < b^2$.

68. Prove that $|x - y| \geq |x| - |y|$. [*Hint:* Use the Triangle Inequality with $a = x - y$ and $b = y$.]

69. Show that the sum, difference, and product of rational numbers are rational numbers.

70. (a) Is the sum of two irrational numbers always an irrational number?
(b) Is the product of two irrational numbers always an irrational number?

B   Coordinate Geometry and Lines

Just as the points on a line can be identified with real numbers by assigning them coordinates, as described in Appendix A, so the points in a plane can be identified with ordered pairs of real numbers. We start by drawing two perpendicular coordinate lines that intersect at the origin O on each line. Usually one line is horizontal with positive direction to the right and is called the *x*-axis; the other line is vertical with positive direction upward and is called the *y*-axis.

Any point P in the plane can be located by a unique ordered pair of numbers as follows. Draw lines through P perpendicular to the *x*- and *y*-axes. These lines intersect the axes in points with coordinates a and b as shown in Figure 1. Then the point P is assigned the ordered pair (a, b). The first number a is called the **x-coordinate** of P; the second number b is called the **y-coordinate** of P. We say that P is the point with coordinates (a, b), and we denote the point by the symbol $P(a, b)$. Several points are labeled with their coordinates in Figure 2.

FIGURE 1

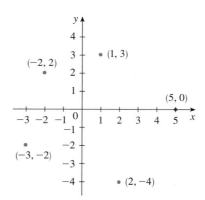

FIGURE 2

By reversing the preceding process we can start with an ordered pair (a, b) and arrive at the corresponding point P. Often we identify the point P with the ordered pair (a, b) and refer to "the point (a, b)." [Although the notation used for an open interval (a, b) is the same as the notation used for a point (a, b), you will be able to tell from the context which meaning is intended.]

This coordinate system is called the **rectangular coordinate system** or the **Cartesian coordinate system** in honor of the French mathematician René Descartes (1596–1650), even though another Frenchman, Pierre Fermat (1601–1665), invented the principles of analytic geometry at about the same time as Descartes. The plane supplied with this coordinate system is called the **coordinate plane** or the **Cartesian plane** and is denoted by $\mathbb{R}^2$.

The x- and y-axes are called the **coordinate axes** and divide the Cartesian plane into four quadrants, which are labeled I, II, III, and IV in Figure 1. Notice that the first quadrant consists of those points whose x- and y-coordinates are both positive.

EXAMPLE 1 Describe and sketch the regions given by the following sets.
(a) $\{(x, y) \mid x \geq 0\}$ (b) $\{(x, y) \mid y = 1\}$ (c) $\{(x, y) \mid |y| < 1\}$

SOLUTION
(a) The points whose x-coordinates are 0 or positive lie on the y-axis or to the right of it as indicated by the shaded region in Figure 3(a).

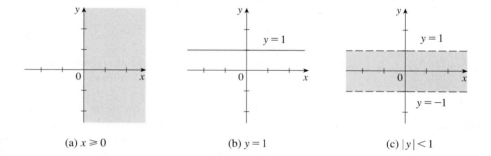

FIGURE 3   (a) $x \geq 0$ (b) $y = 1$ (c) $|y| < 1$

(b) The set of all points with y-coordinate 1 is a horizontal line one unit above the x-axis [see Figure 3(b)].

(c) Recall from Appendix A that

$$|y| < 1 \qquad \text{if and only if} \qquad -1 < y < 1$$

The given region consists of those points in the plane whose y-coordinates lie between -1 and 1. Thus the region consists of all points that lie between (but not on) the horizontal lines $y = 1$ and $y = -1$. [These lines are shown as dashed lines in Figure 3(c) to indicate that the points on these lines don't lie in the set.] ∎

Recall from Appendix A that the distance between points a and b on a number line is $|a - b| = |b - a|$. Thus the distance between points $P_1(x_1, y_1)$ and $P_3(x_2, y_1)$ on a horizontal line must be $|x_2 - x_1|$ and the distance between $P_2(x_2, y_2)$ and $P_3(x_2, y_1)$ on a vertical line must be $|y_2 - y_1|$. (See Figure 4.)

To find the distance $|P_1P_2|$ between any two points $P_1(x_1, y_1)$ and $P_2(x_2, y_2)$, we note that triangle $P_1P_2P_3$ in Figure 4 is a right triangle, and so by the Pythagorean Theorem we have

$$|P_1P_2| = \sqrt{|P_1P_3|^2 + |P_2P_3|^2} = \sqrt{|x_2 - x_1|^2 + |y_2 - y_1|^2}$$

$$= \sqrt{(x_2 - x_1)^2 + (y_2 - y_1)^2}$$

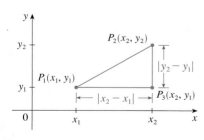

FIGURE 4

> **1** **Distance Formula** The distance between the points $P_1(x_1, y_1)$ and $P_2(x_2, y_2)$ is
>
> $$|P_1P_2| = \sqrt{(x_2 - x_1)^2 + (y_2 - y_1)^2}$$

EXAMPLE 2 The distance between $(1, -2)$ and $(5, 3)$ is

$$\sqrt{(5 - 1)^2 + [3 - (-2)]^2} = \sqrt{4^2 + 5^2} = \sqrt{41}$$ ∎

■ Lines

We want to find an equation of a given line L; such an equation is satisfied by the coordinates of the points on L and by no other point. To find the equation of L we use its *slope,* which is a measure of the steepness of the line.

FIGURE 5

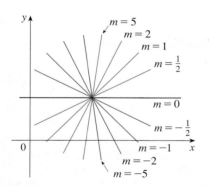

FIGURE 6

> **2** **Definition** The **slope** of a nonvertical line that passes through the points $P_1(x_1, y_1)$ and $P_2(x_2, y_2)$ is
>
> $$m = \frac{\Delta y}{\Delta x} = \frac{y_2 - y_1}{x_2 - x_1}$$
>
> The slope of a vertical line is not defined.

Thus the slope of a line is the ratio of the change in y, Δy, to the change in x, Δx. (See Figure 5.) The slope is therefore the rate of change of y with respect to x. The fact that the line is straight means that the rate of change is constant.

Figure 6 shows several lines labeled with their slopes. Notice that lines with positive slope slant upward to the right, whereas lines with negative slope slant downward to the right. Notice also that the steepest lines are the ones for which the absolute value of the slope is largest, and a horizontal line has slope 0.

Now let's find an equation of the line that passes through a given point $P_1(x_1, y_1)$ and has slope m. A point $P(x, y)$ with $x \neq x_1$ lies on this line if and only if the slope of the line through P_1 and P is equal to m; that is,

$$\frac{y - y_1}{x - x_1} = m$$

This equation can be rewritten in the form

$$y - y_1 = m(x - x_1)$$

and we observe that this equation is also satisfied when $x = x_1$ and $y = y_1$. Therefore it is an equation of the given line.

> **3** **Point-Slope Form of the Equation of a Line** An equation of the line passing through the point $P_1(x_1, y_1)$ and having slope m is
>
> $$y - y_1 = m(x - x_1)$$

EXAMPLE 3 Find an equation of the line through $(1, -7)$ with slope $-\frac{1}{2}$.

SOLUTION Using (3) with $m = -\frac{1}{2}$, $x_1 = 1$, and $y_1 = -7$, we obtain an equation of the line as

$$y + 7 = -\tfrac{1}{2}(x - 1)$$

which we can rewrite as

$$2y + 14 = -x + 1 \qquad \text{or} \qquad x + 2y + 13 = 0 \qquad\blacksquare$$

EXAMPLE 4 Find an equation of the line through the points $(-1, 2)$ and $(3, -4)$.

SOLUTION By Definition 2 the slope of the line is

$$m = \frac{-4 - 2}{3 - (-1)} = -\frac{3}{2}$$

Using the point-slope form with $x_1 = -1$ and $y_1 = 2$, we obtain

$$y - 2 = -\tfrac{3}{2}(x + 1)$$

which simplifies to $\qquad\qquad 3x + 2y = 1 \qquad\blacksquare$

FIGURE 7

Suppose a nonvertical line has slope m and y-intercept b. (See Figure 7.) This means it intersects the y-axis at the point $(0, b)$, so the point-slope form of the equation of the line, with $x_1 = 0$ and $y_1 = b$, becomes

$$y - b = m(x - 0)$$

This simplifies as follows.

> **4** **Slope-Intercept Form of the Equation of a Line** An equation of the line with slope m and y-intercept b is
>
> $$y = mx + b$$

In particular, if a line is horizontal, its slope is $m = 0$, so its equation is $y = b$, where b is the y-intercept (see Figure 8). A vertical line does not have a slope, but we can write its equation as $x = a$, where a is the x-intercept, because the x-coordinate of every point on the line is a.

Observe that the equation of every line can be written in the form

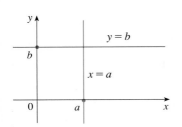

FIGURE 8

$$\boxed{5} \qquad\qquad Ax + By + C = 0$$

because a vertical line has the equation $x = a$ or $x - a = 0$ ($A = 1$, $B = 0$, $C = -a$) and a nonvertical line has the equation $y = mx + b$ or $-mx + y - b = 0$ ($A = -m$, $B = 1$, $C = -b$). Conversely, if we start with a general first-degree equation, that is, an equation of the form (5), where A, B, and C are constants and A and B are not both 0, then we can show that it is the equation of a line. If $B = 0$, the equation becomes $Ax + C = 0$ or $x = -C/A$, which represents a vertical line with x-intercept $-C/A$. If $B \neq 0$, the equation can be rewritten by solving for y:

$$y = -\frac{A}{B}x - \frac{C}{B}$$

and we recognize this as being the slope-intercept form of the equation of a line ($m = -A/B$, $b = -C/B$). Therefore an equation of the form (5) is called a **linear equation** or the **general equation of a line**. For brevity, we often refer to "the line $Ax + By + C = 0$" instead of "the line whose equation is $Ax + By + C = 0$."

EXAMPLE 5 Sketch the graph of the equation $3x - 5y = 15$.

SOLUTION Since the equation is linear, its graph is a line. To draw the graph, we can simply find two points on the line. It's easiest to find the intercepts. Substituting $y = 0$ (the equation of the x-axis) in the given equation, we get $3x = 15$, so $x = 5$ is the x-intercept. Substituting $x = 0$ in the equation, we see that the y-intercept is -3. This allows us to sketch the graph as in Figure 9.   ∎

FIGURE 9

EXAMPLE 6 Graph the inequality $x + 2y > 5$.

SOLUTION We are asked to sketch the graph of the set $\{(x, y) \mid x + 2y > 5\}$ and we begin by solving the inequality for y:

$$x + 2y > 5$$

$$2y > -x + 5$$

$$y > -\tfrac{1}{2}x + \tfrac{5}{2}$$

Compare this inequality with the equation $y = -\tfrac{1}{2}x + \tfrac{5}{2}$, which represents a line with slope $-\tfrac{1}{2}$ and y-intercept $\tfrac{5}{2}$. We see that the given graph consists of points whose y-coordinates are *larger* than those on the line $y = -\tfrac{1}{2}x + \tfrac{5}{2}$. Thus the graph is the region that lies *above* the line, as illustrated in Figure 10.   ∎

FIGURE 10

■ Parallel and Perpendicular Lines

Slopes can be used to show that lines are parallel or perpendicular. The following facts are proved, for instance, in *Precalculus: Mathematics for Calculus,* Seventh Edition, by Stewart, Redlin, and Watson (Belmont, CA, 2016).

---

6 **Parallel and Perpendicular Lines**

1. Two nonvertical lines are parallel if and only if they have the same slope.

2. Two lines with slopes m_1 and m_2 are perpendicular if and only if $m_1 m_2 = -1$; that is, their slopes are negative reciprocals:

$$m_2 = -\frac{1}{m_1}$$

EXAMPLE 7 Find an equation of the line through the point $(5, 2)$ that is parallel to the line $4x + 6y + 5 = 0$.

SOLUTION The given line can be written in the form

$$y = -\tfrac{2}{3}x - \tfrac{5}{6}$$

which is in slope-intercept form with $m = -\frac{2}{3}$. Parallel lines have the same slope, so the required line has slope $-\frac{2}{3}$ and its equation in point-slope form is

$$y - 2 = -\tfrac{2}{3}(x - 5)$$

We can write this equation as $2x + 3y = 16$.   ■

EXAMPLE 8 Show that the lines $2x + 3y = 1$ and $6x - 4y - 1 = 0$ are perpendicular.

SOLUTION The equations can be written as

$$y = -\tfrac{2}{3}x + \tfrac{1}{3} \qquad \text{and} \qquad y = \tfrac{3}{2}x - \tfrac{1}{4}$$

from which we see that the slopes are

$$m_1 = -\tfrac{2}{3} \qquad \text{and} \qquad m_2 = \tfrac{3}{2}$$

Since $m_1 m_2 = -1$, the lines are perpendicular.   ■

B   EXERCISES

1–6 Find the distance between the points.

1. $(1, 1),\quad (4, 5)$

2. $(1, -3),\quad (5, 7)$

3. $(6, -2),\quad (-1, 3)$

4. $(1, -6),\quad (-1, -3)$

5. $(2, 5),\quad (4, -7)$

6. $(a, b),\quad (b, a)$

7–10 Find the slope of the line through P and Q.

7. $P(1, 5),\quad Q(4, 11)$

8. $P(-1, 6),\quad Q(4, -3)$

9. $P(-3, 3),\quad Q(-1, -6)$

10. $P(-1, -4),\quad Q(6, 0)$

11. Show that the triangle with vertices $A(0, 2)$, $B(-3, -1)$, and $C(-4, 3)$ is isosceles.

12. (a) Show that the triangle with vertices $A(6, -7)$, $B(11, -3)$, and $C(2, -2)$ is a right triangle using the converse of the Pythagorean Theorem.
 (b) Use slopes to show that ABC is a right triangle.
 (c) Find the area of the triangle.

13. Show that the points $(-2, 9)$, $(4, 6)$, $(1, 0)$, and $(-5, 3)$ are the vertices of a square.

14. (a) Show that the points $A(-1, 3)$, $B(3, 11)$, and $C(5, 15)$ are collinear (lie on the same line) by showing that $|AB| + |BC| = |AC|$.
 (b) Use slopes to show that A, B, and C are collinear.

15. Show that $A(1, 1)$, $B(7, 4)$, $C(5, 10)$, and $D(-1, 7)$ are vertices of a parallelogram.

16. Show that $A(1, 1)$, $B(11, 3)$, $C(10, 8)$, and $D(0, 6)$ are vertices of a rectangle.

17–20 Sketch the graph of the equation.

17. $x = 3$

18. $y = -2$

19. $xy = 0$

20. $|y| = 1$

21–36 Find an equation of the line that satisfies the given conditions.

21. Through $(2, -3)$,   slope 6

22. Through $(-1, 4)$,   slope -3

23. Through $(1, 7)$,   slope $\frac{2}{3}$

24. Through $(-3, -5)$,   slope $-\frac{7}{2}$

25. Through $(2, 1)$ and $(1, 6)$

26. Through $(-1, -2)$ and $(4, 3)$

27. Slope 3, y-intercept -2

28. Slope $\frac{2}{5}$, y-intercept 4

29. x-intercept 1, y-intercept -3

30. x-intercept -8, y-intercept 6

31. Through $(4, 5)$,   parallel to the x-axis

32. Through $(4, 5)$,   parallel to the y-axis

33. Through $(1, -6)$,   parallel to the line $x + 2y = 6$

34. y-intercept 6,   parallel to the line $2x + 3y + 4 = 0$

35. Through $(-1, -2)$,   perpendicular to the line $2x + 5y + 8 = 0$

36. Through $\left(\frac{1}{2}, -\frac{2}{3}\right)$,   perpendicular to the line $4x - 8y = 1$

37–42 Find the slope and y-intercept of the line and draw its graph.

37. $x + 3y = 0$

38. $2x - 5y = 0$

39. $y = -2$

40. $2x - 3y + 6 = 0$

41. $3x - 4y = 12$

42. $4x + 5y = 10$

43–52 Sketch the region in the xy-plane.

43. $\{(x, y) \mid x < 0\}$

44. $\{(x, y) \mid y > 0\}$

45. $\{(x, y) \mid xy < 0\}$

46. $\{(x, y) \mid x \geqslant 1 \text{ and } y < 3\}$

47. $\{(x, y) \mid |x| \leqslant 2\}$

48. $\{(x, y) \mid |x| < 3 \text{ and } |y| < 2\}$

49. $\{(x, y) \mid 0 \leqslant y \leqslant 4 \text{ and } x \leqslant 2\}$

50. $\{(x, y) \mid y > 2x - 1\}$

51. $\{(x, y) \mid 1 + x \leqslant y \leqslant 1 - 2x\}$

52. $\{(x, y) \mid -x \leqslant y < \frac{1}{2}(x + 3)\}$

---

53. Find a point on the y-axis that is equidistant from $(5, -5)$ and $(1, 1)$.

54. Show that the midpoint of the line segment from $P_1(x_1, y_1)$ to $P_2(x_2, y_2)$ is

$$\left(\frac{x_1 + x_2}{2}, \frac{y_1 + y_2}{2} \right)$$

55. Find the midpoint of the line segment joining the given points.
(a) $(1, 3)$ and $(7, 15)$
(b) $(-1, 6)$ and $(8, -12)$

56. Find the lengths of the medians of the triangle with vertices $A(1, 0)$, $B(3, 6)$, and $C(8, 2)$. (A median is a line segment from a vertex to the midpoint of the opposite side.)

57. Show that the lines $2x - y = 4$ and $6x - 2y = 10$ are not parallel and find their point of intersection.

58. Show that the lines $3x - 5y + 19 = 0$ and $10x + 6y - 50 = 0$ are perpendicular and find their point of intersection.

59. Find an equation of the perpendicular bisector of the line segment joining the points $A(1, 4)$ and $B(7, -2)$.

60. (a) Find equations for the sides of the triangle with vertices $P(1, 0)$, $Q(3, 4)$, and $R(-1, 6)$.
(b) Find equations for the medians of this triangle. Where do they intersect?

61. (a) Show that if the x- and y-intercepts of a line are nonzero numbers a and b, then the equation of the line can be put in the form

$$\frac{x}{a} + \frac{y}{b} = 1$$

This equation is called the **two-intercept form** of an equation of a line.
(b) Use part (a) to find an equation of the line whose x-intercept is 6 and whose y-intercept is -8.

62. A car leaves Detroit at 2:00 PM, traveling at a constant speed west along I-96. It passes Ann Arbor, 40 mi from Detroit, at 2:50 PM.
(a) Express the distance traveled in terms of the time elapsed.
(b) Draw the graph of the equation in part (a).
(c) What is the slope of this line? What does it represent?

C   Graphs of Second-Degree Equations

In Appendix B we saw that a first-degree, or linear, equation $Ax + By + C = 0$ represents a line. In this section we discuss second-degree equations such as

$$x^2 + y^2 = 1 \qquad y = x^2 + 1 \qquad \frac{x^2}{9} + \frac{y^2}{4} = 1 \qquad x^2 - y^2 = 1$$

which represent a circle, a parabola, an ellipse, and a hyperbola, respectively.

The graph of such an equation in x and y is the set of all points (x, y) that satisfy the equation; it gives a visual representation of the equation. Conversely, given a curve in the xy-plane, we may have to find an equation that represents it, that is, an equation satisfied by the coordinates of the points on the curve and by no other point. This is the other half of the basic principle of analytic geometry as formulated by Descartes and Fermat. The idea is that if a geometric curve can be represented by an algebraic equation, then the rules of algebra can be used to analyze the geometric problem.

▮ Circles

As an example of this type of problem, let's find an equation of the circle with radius r and center (h, k). By definition, the circle is the set of all points $P(x, y)$ whose distance

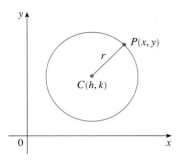

FIGURE 1

from the center $C(h, k)$ is r. (See Figure 1.) Thus P is on the circle if and only if $|PC| = r$. From the distance formula, we have

$$\sqrt{(x - h)^2 + (y - k)^2} = r$$

or equivalently, squaring both sides, we get

$$(x - h)^2 + (y - k)^2 = r^2$$

This is the desired equation.

---

1   Equation of a Circle An equation of the circle with center (h, k) and radius r is

$$(x - h)^2 + (y - k)^2 = r^2$$

In particular, if the center is the origin $(0, 0)$, the equation is

$$x^2 + y^2 = r^2$$

EXAMPLE 1 Find an equation of the circle with radius 3 and center $(2, -5)$.

SOLUTION From Equation 1 with $r = 3$, $h = 2$, and $k = -5$, we obtain

$$(x - 2)^2 + (y + 5)^2 = 9$$ ∎

EXAMPLE 2 Sketch the graph of the equation $x^2 + y^2 + 2x - 6y + 7 = 0$ by first showing that it represents a circle and then finding its center and radius.

SOLUTION We first group the x-terms and y-terms as follows:

$$(x^2 + 2x) + (y^2 - 6y) = -7$$

Then we complete the square within each grouping, adding the appropriate constants (the squares of half the coefficients of x and y) to both sides of the equation:

$$(x^2 + 2x + 1) + (y^2 - 6y + 9) = -7 + 1 + 9$$

or $$(x + 1)^2 + (y - 3)^2 = 3$$

Comparing this equation with the standard equation of a circle (1), we see that $h = -1$, $k = 3$, and $r = \sqrt{3}$, so the given equation represents a circle with center $(-1, 3)$ and radius $\sqrt{3}$. It is sketched in Figure 2.

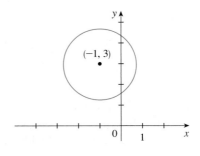

FIGURE 2
$x^2 + y^2 + 2x - 6y + 7 = 0$

Parabolas

The geometric properties of parabolas are reviewed in Section 10.5. Here we regard a parabola as a graph of an equation of the form $y = ax^2 + bx + c$.

EXAMPLE 3 Draw the graph of the parabola $y = x^2$.

SOLUTION We set up a table of values, plot points, and join them by a smooth curve to obtain the graph in Figure 3.

x	$y = x^2$
0	0
$\pm\frac{1}{2}$	$\frac{1}{4}$
± 1	1
± 2	4
± 3	9

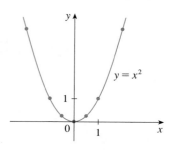

FIGURE 3

Figure 4 shows the graphs of several parabolas with equations of the form $y = ax^2$ for various values of the number a. In each case the *vertex*, the point where the parabola changes direction, is the origin. We see that the parabola $y = ax^2$ opens upward if $a > 0$ and downward if $a < 0$ (as in Figure 5).

FIGURE 4

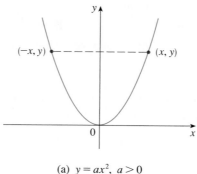

(a) $y = ax^2$, $a > 0$

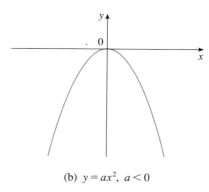

(b) $y = ax^2$, $a < 0$

FIGURE 5

Notice that if (x, y) satisfies $y = ax^2$, then so does $(-x, y)$. This corresponds to the geometric fact that if the right half of the graph is reflected about the y-axis, then the left half of the graph is obtained. We say that the graph is **symmetric with respect to the y-axis**.

> The graph of an equation is symmetric with respect to the y-axis if the equation is unchanged when x is replaced by $-x$.

If we interchange x and y in the equation $y = ax^2$, the result is $x = ay^2$, which also represents a parabola. (Interchanging x and y amounts to reflecting about the diagonal line $y = x$.) The parabola $x = ay^2$ opens to the right if $a > 0$ and to the left if $a < 0$. (See

Figure 6.) This time the parabola is symmetric with respect to the x-axis because if (x, y) satisfies $x = ay^2$, then so does $(x, -y)$.

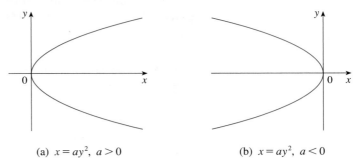

FIGURE 6

(a) $x = ay^2$, $a > 0$ (b) $x = ay^2$, $a < 0$

> The graph of an equation is symmetric with respect to the x-axis if the equation is unchanged when y is replaced by $-y$.

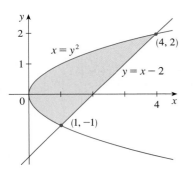

FIGURE 7

EXAMPLE 4   Sketch the region bounded by the parabola $x = y^2$ and the line $y = x - 2$.

SOLUTION   First we find the points of intersection by solving the two equations. Substituting $x = y + 2$ into the equation $x = y^2$, we get $y + 2 = y^2$, which gives

$$0 = y^2 - y - 2 = (y - 2)(y + 1)$$

so $y = 2$ or -1. Thus the points of intersection are $(4, 2)$ and $(1, -1)$, and we draw the line $y = x - 2$ passing through these points. We then sketch the parabola $x = y^2$ by referring to Figure 6(a) and having the parabola pass through $(4, 2)$ and $(1, -1)$. The region bounded by $x = y^2$ and $y = x - 2$ means the finite region whose boundaries are these curves. It is sketched in Figure 7.   ∎

◼ Ellipses

The curve with equation

2
$$\frac{x^2}{a^2} + \frac{y^2}{b^2} = 1$$

where a and b are positive numbers, is called an **ellipse** in standard position. (Geometric properties of ellipses are discussed in Section 10.5.) Observe that Equation 2 is unchanged if x is replaced by $-x$ or y is replaced by $-y$, so the ellipse is symmetric with respect to both axes. As a further aid to sketching the ellipse, we find its intercepts.

> The **x-intercepts** of a graph are the x-coordinates of the points where the graph intersects the x-axis. They are found by setting $y = 0$ in the equation of the graph.
>
> The **y-intercepts** are the y-coordinates of the points where the graph intersects the y-axis. They are found by setting $x = 0$ in its equation.

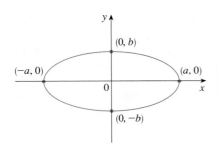

FIGURE 8
$$\frac{x^2}{a^2} + \frac{y^2}{b^2} = 1$$

If we set $y = 0$ in Equation 2, we get $x^2 = a^2$ and so the x-intercepts are $\pm a$. Setting $x = 0$, we get $y^2 = b^2$, so the y-intercepts are $\pm b$. Using this information, together with symmetry, we sketch the ellipse in Figure 8. If $a = b$, the ellipse is a circle with radius a.

EXAMPLE 5 Sketch the graph of $9x^2 + 16y^2 = 144$.

SOLUTION We divide both sides of the equation by 144:

$$\frac{x^2}{16} + \frac{y^2}{9} = 1$$

The equation is now in the standard form for an ellipse (2), so we have $a^2 = 16$, $b^2 = 9$, $a = 4$, and $b = 3$. The x-intercepts are ± 4; the y-intercepts are ± 3. The graph is sketched in Figure 9.

FIGURE 9
$9x^2 + 16y^2 = 144$

Hyperbolas

The curve with equation

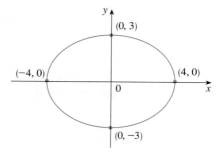

$$\boxed{3} \qquad \frac{x^2}{a^2} - \frac{y^2}{b^2} = 1$$

is called a **hyperbola** in standard position. Again, Equation 3 is unchanged when x is replaced by $-x$ or y is replaced by $-y$, so the hyperbola is symmetric with respect to both axes. To find the x-intercepts we set $y = 0$ and obtain $x^2 = a^2$ and $x = \pm a$. However, if we put $x = 0$ in Equation 3, we get $y^2 = -b^2$, which is impossible, so there is no y-intercept. In fact, from Equation 3 we obtain

$$\frac{x^2}{a^2} = 1 + \frac{y^2}{b^2} \geq 1$$

which shows that $x^2 \geq a^2$ and so $|x| = \sqrt{x^2} \geq a$. Therefore we have $x \geq a$ or $x \leq -a$. This means that the hyperbola consists of two parts, called its *branches*. It is sketched in Figure 10.

In drawing a hyperbola it is useful to draw first its *asymptotes*, which are the lines $y = (b/a)x$ and $y = -(b/a)x$ shown in Figure 10. Both branches of the hyperbola approach the asymptotes; that is, they come arbitrarily close to the asymptotes. This involves the idea of a limit, which is discussed in Chapter 2. (See also Exercise 4.5.73.)

By interchanging the roles of x and y we get an equation of the form

$$\frac{y^2}{a^2} - \frac{x^2}{b^2} = 1$$

which also represents a hyperbola and is sketched in Figure 11.

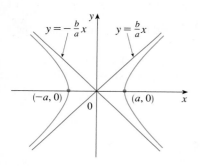

FIGURE 10
The hyperbola $\dfrac{x^2}{a^2} - \dfrac{y^2}{b^2} = 1$

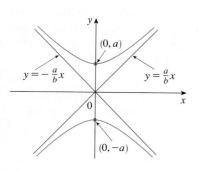

FIGURE 11
The hyperbola $\dfrac{y^2}{a^2} - \dfrac{x^2}{b^2} = 1$

EXAMPLE 6 Sketch the curve $9x^2 - 4y^2 = 36$.

SOLUTION Dividing both sides by 36, we obtain

$$\frac{x^2}{4} - \frac{y^2}{9} = 1$$

which is the standard form of the equation of a hyperbola (Equation 3). Since $a^2 = 4$, the x-intercepts are ± 2. Since $b^2 = 9$, we have $b = 3$ and the asymptotes are $y = \pm\frac{3}{2}x$. The hyperbola is sketched in Figure 12.

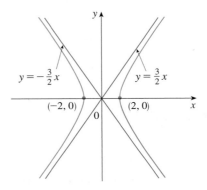

FIGURE 12
The hyperbola $9x^2 - 4y^2 = 36$

If $b = a$, a hyperbola has the equation $x^2 - y^2 = a^2$ (or $y^2 - x^2 = a^2$) and is called an *equilateral hyperbola* [see Figure 13(a)]. Its asymptotes are $y = \pm x$, which are perpendicular. If an equilateral hyperbola is rotated by $45°$, the asymptotes become the x- and y-axes, and it can be shown that the new equation of the hyperbola is $xy = k$, where k is a constant [see Figure 13(b)].

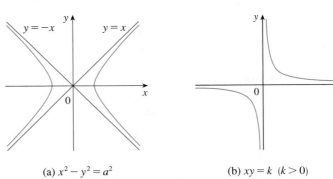

FIGURE 13
Equilateral hyperbolas

(a) $x^2 - y^2 = a^2$

(b) $xy = k$ $(k > 0)$

▪ Shifted Conics

Recall that an equation of the circle with center the origin and radius r is $x^2 + y^2 = r^2$, but if the center is the point (h, k), then the equation of the circle becomes

$$(x - h)^2 + (y - k)^2 = r^2$$

Similarly, if we take the ellipse with equation

4

$$\frac{x^2}{a^2} + \frac{y^2}{b^2} = 1$$

and translate it (shift it) so that its center is the point (h, k), then its equation becomes

5

$$\frac{(x - h)^2}{a^2} + \frac{(y - k)^2}{b^2} = 1$$

(See Figure 14.)

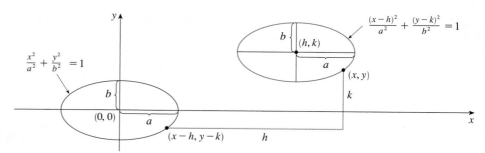

FIGURE 14

Notice that in shifting the ellipse, we replaced x by $x - h$ and y by $y - k$ in Equation 4 to obtain Equation 5. We use the same procedure to shift the parabola $y = ax^2$ so that its vertex (the origin) becomes the point (h, k) as in Figure 15. Replacing x by $x - h$ and y by $y - k$, we see that the new equation is

$$y - k = a(x - h)^2 \qquad \text{or} \qquad y = a(x - h)^2 + k$$

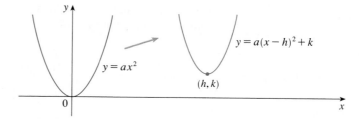

FIGURE 15

EXAMPLE 7 Sketch the graph of the equation $y = 2x^2 - 4x + 1$.

SOLUTION First we complete the square:

$$y = 2(x^2 - 2x) + 1 = 2(x - 1)^2 - 1$$

In this form we see that the equation represents the parabola obtained by shifting $y = 2x^2$ so that its vertex is at the point $(1, -1)$. The graph is sketched in Figure 16.

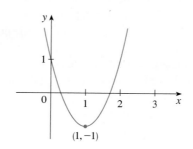

FIGURE 16
$y = 2x^2 - 4x + 1$

EXAMPLE 8 Sketch the curve $x = 1 - y^2$.

SOLUTION This time we start with the parabola $x = -y^2$ (as in Figure 6 with $a = -1$) and shift one unit to the right to get the graph of $x = 1 - y^2$. (See Figure 17.)

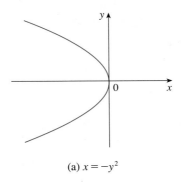

FIGURE 17

(a) $x = -y^2$ (b) $x = 1 - y^2$ ■

C   EXERCISES

1–4 Find an equation of a circle that satisfies the given conditions.

1. Center $(3, -1)$, radius 5

2. Center $(-2, -8)$, radius 10

3. Center at the origin, passes through $(4, 7)$

4. Center $(-1, 5)$, passes through $(-4, -6)$

5–9 Show that the equation represents a circle and find the center and radius.

5. $x^2 + y^2 - 4x + 10y + 13 = 0$

6. $x^2 + y^2 + 6y + 2 = 0$

7. $x^2 + y^2 + x = 0$

8. $16x^2 + 16y^2 + 8x + 32y + 1 = 0$

9. $2x^2 + 2y^2 - x + y = 1$

---

10. Under what condition on the coefficients a, b, and c does the equation $x^2 + y^2 + ax + by + c = 0$ represent a circle? When that condition is satisfied, find the center and radius of the circle.

11–32 Identify the type of curve and sketch the graph. Do not plot points. Just use the standard graphs given in Figures 5, 6, 8, 10, and 11 and shift if necessary.

11. $y = -x^2$

12. $y^2 - x^2 = 1$

13. $x^2 + 4y^2 = 16$

14. $x = -2y^2$

15. $16x^2 - 25y^2 = 400$

16. $25x^2 + 4y^2 = 100$

17. $4x^2 + y^2 = 1$

18. $y = x^2 + 2$

19. $x = y^2 - 1$

20. $9x^2 - 25y^2 = 225$

21. $9y^2 - x^2 = 9$

22. $2x^2 + 5y^2 = 10$

23. $xy = 4$

24. $y = x^2 + 2x$

25. $9(x - 1)^2 + 4(y - 2)^2 = 36$

26. $16x^2 + 9y^2 - 36y = 108$

27. $y = x^2 - 6x + 13$

28. $x^2 - y^2 - 4x + 3 = 0$

29. $x = 4 - y^2$

30. $y^2 - 2x + 6y + 5 = 0$

31. $x^2 + 4y^2 - 6x + 5 = 0$

32. $4x^2 + 9y^2 - 16x + 54y + 61 = 0$

33–34 Sketch the region bounded by the curves.

33. $y = 3x$, $y = x^2$

34. $y = 4 - x^2$, $x - 2y = 2$

---

35. Find an equation of the parabola with vertex $(1, -1)$ that passes through the points $(-1, 3)$ and $(3, 3)$.

36. Find an equation of the ellipse with center at the origin that passes through the points $\left(1, -10\sqrt{2}/3\right)$ and $\left(-2, 5\sqrt{5}/3\right)$.

37–40 Sketch the graph of the set.

37. $\{(x, y) \mid x^2 + y^2 \leqslant 1\}$

38. $\{(x, y) \mid x^2 + y^2 > 4\}$

39. $\{(x, y) \mid y \geqslant x^2 - 1\}$

40. $\{(x, y) \mid x^2 + 4y^2 \leqslant 4\}$

D Trigonometry

◼ Angles

Angles can be measured in degrees or in radians (abbreviated as rad). The angle given by a complete revolution contains 360°, which is the same as 2π rad. Therefore

$$\boxed{1} \qquad \boxed{\pi \text{ rad} = 180°}$$

and

$$\boxed{2} \qquad 1 \text{ rad} = \left(\frac{180}{\pi}\right)^{\circ} \approx 57.3° \qquad 1° = \frac{\pi}{180} \text{ rad} \approx 0.017 \text{ rad}$$

EXAMPLE 1

(a) Find the radian measure of 60°.   (b) Express $5\pi/4$ rad in degrees.

SOLUTION

(a) From Equation 1 or 2 we see that to convert from degrees to radians we multiply by $\pi/180$. Therefore

$$60° = 60\left(\frac{\pi}{180}\right) = \frac{\pi}{3} \text{ rad}$$

(b) To convert from radians to degrees we multiply by $180/\pi$. Thus

$$\frac{5\pi}{4} \text{ rad} = \frac{5\pi}{4}\left(\frac{180}{\pi}\right) = 225°$$

◼

In calculus we use radians to measure angles except when otherwise indicated. The following table gives the correspondence between degree and radian measures of some common angles.

Degrees	0°	30°	45°	60°	90°	120°	135°	150°	180°	270°	360°
Radians	0	$\dfrac{\pi}{6}$	$\dfrac{\pi}{4}$	$\dfrac{\pi}{3}$	$\dfrac{\pi}{2}$	$\dfrac{2\pi}{3}$	$\dfrac{3\pi}{4}$	$\dfrac{5\pi}{6}$	π	$\dfrac{3\pi}{2}$	2π

Figure 1 shows a sector of a circle with central angle θ and radius r subtending an arc with length a. Since the length of the arc is proportional to the size of the angle, and since the entire circle has circumference $2\pi r$ and central angle 2π, we have

$$\frac{\theta}{2\pi} = \frac{a}{2\pi r}$$

Solving this equation for θ and for a, we obtain

FIGURE 1

$$\boxed{3} \qquad \boxed{\theta = \frac{a}{r}} \qquad\qquad \boxed{a = r\theta}$$

Remember that Equations 3 are valid only when θ is measured in radians.

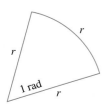

FIGURE 2

In particular, putting $a = r$ in Equation 3, we see that an angle of 1 rad is the angle subtended at the center of a circle by an arc equal in length to the radius of the circle (see Figure 2).

EXAMPLE 2

(a) If the radius of a circle is 5 cm, what angle is subtended by an arc of 6 cm?
(b) If a circle has radius 3 cm, what is the length of an arc subtended by a central angle of $3\pi/8$ rad?

SOLUTION
(a) Using Equation 3 with $a = 6$ and $r = 5$, we see that the angle is

$$\theta = \tfrac{6}{5} = 1.2 \text{ rad}$$

(b) With $r = 3$ cm and $\theta = 3\pi/8$ rad, the arc length is

$$a = r\theta = 3\left(\frac{3\pi}{8}\right) = \frac{9\pi}{8} \text{ cm} \qquad \blacksquare$$

The **standard position** of an angle occurs when we place its vertex at the origin of a coordinate system and its initial side on the positive x-axis as in Figure 3. A **positive** angle is obtained by rotating the initial side counterclockwise until it coincides with the terminal side. Likewise, **negative** angles are obtained by clockwise rotation as in Figure 4.

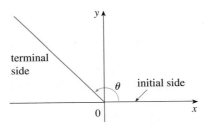

FIGURE 3 $\theta \geqslant 0$

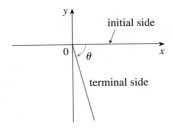

FIGURE 4 $\theta < 0$

Figure 5 shows several examples of angles in standard position. Notice that different angles can have the same terminal side. For instance, the angles $3\pi/4$, $-5\pi/4$, and $11\pi/4$ have the same initial and terminal sides because

$$\frac{3\pi}{4} - 2\pi = -\frac{5\pi}{4} \qquad \frac{3\pi}{4} + 2\pi = \frac{11\pi}{4}$$

and 2π rad represents a complete revolution.

FIGURE 5
Angles in standard position

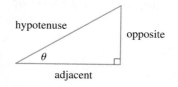

FIGURE 6

The Trigonometric Functions

For an acute angle θ the six trigonometric functions are defined as ratios of lengths of sides of a right triangle as follows (see Figure 6).

[4]

$$\sin \theta = \frac{\text{opp}}{\text{hyp}} \qquad \csc \theta = \frac{\text{hyp}}{\text{opp}}$$

$$\cos \theta = \frac{\text{adj}}{\text{hyp}} \qquad \sec \theta = \frac{\text{hyp}}{\text{adj}}$$

$$\tan \theta = \frac{\text{opp}}{\text{adj}} \qquad \cot \theta = \frac{\text{adj}}{\text{opp}}$$

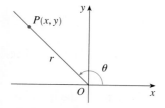

FIGURE 7

This definition doesn't apply to obtuse or negative angles, so for a general angle θ in standard position we let $P(x, y)$ be any point on the terminal side of θ and we let r be the distance $|OP|$ as in Figure 7. Then we define

[5]

$$\sin \theta = \frac{y}{r} \qquad \csc \theta = \frac{r}{y}$$

$$\cos \theta = \frac{x}{r} \qquad \sec \theta = \frac{r}{x}$$

$$\tan \theta = \frac{y}{x} \qquad \cot \theta = \frac{x}{y}$$

If we put $r = 1$ in Definition 5 and draw a unit circle with center the origin and label θ as in Figure 8, then the coordinates of P are $(\cos \theta, \sin \theta)$.

Since division by 0 is not defined, $\tan \theta$ and $\sec \theta$ are undefined when $x = 0$ and $\csc \theta$ and $\cot \theta$ are undefined when $y = 0$. Notice that the definitions in (4) and (5) are consistent when θ is an acute angle.

If θ is a number, the convention is that $\sin \theta$ means the sine of the angle whose *radian* measure is θ. For example, the expression $\sin 3$ implies that we are dealing with an angle of 3 rad. When finding a calculator approximation to this number, we must remember to set our calculator in radian mode, and then we obtain

$$\sin 3 \approx 0.14112$$

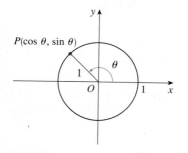

FIGURE 8

If we want to know the sine of the angle 3° we would write $\sin 3°$ and, with our calculator in degree mode, we find that

$$\sin 3° \approx 0.05234$$

The exact trigonometric ratios for certain angles can be read from the triangles in Figure 9. For instance,

FIGURE 9

$$\sin \frac{\pi}{4} = \frac{1}{\sqrt{2}} \qquad \sin \frac{\pi}{6} = \frac{1}{2} \qquad \sin \frac{\pi}{3} = \frac{\sqrt{3}}{2}$$

$$\cos \frac{\pi}{4} = \frac{1}{\sqrt{2}} \qquad \cos \frac{\pi}{6} = \frac{\sqrt{3}}{2} \qquad \cos \frac{\pi}{3} = \frac{1}{2}$$

$$\tan \frac{\pi}{4} = 1 \qquad \tan \frac{\pi}{6} = \frac{1}{\sqrt{3}} \qquad \tan \frac{\pi}{3} = \sqrt{3}$$

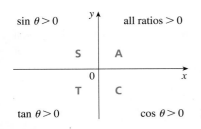

$\sin \theta > 0$ \qquad all ratios > 0

S \qquad A

T \qquad C

$\tan \theta > 0$ \qquad $\cos \theta > 0$

FIGURE 10

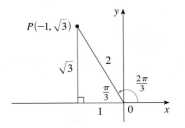

$P(-1, \sqrt{3})$

FIGURE 11

The signs of the trigonometric functions for angles in each of the four quadrants can be remembered by means of the rule "All Students Take Calculus" shown in Figure 10.

EXAMPLE 3 Find the exact trigonometric ratios for $\theta = 2\pi/3$.

SOLUTION From Figure 11 we see that a point on the terminal line for $\theta = 2\pi/3$ is $P(-1, \sqrt{3})$. Therefore, taking

$$x = -1 \qquad y = \sqrt{3} \qquad r = 2$$

in the definitions of the trigonometric ratios, we have

$$\sin \frac{2\pi}{3} = \frac{\sqrt{3}}{2} \qquad \cos \frac{2\pi}{3} = -\frac{1}{2} \qquad \tan \frac{2\pi}{3} = -\sqrt{3}$$

$$\csc \frac{2\pi}{3} = \frac{2}{\sqrt{3}} \qquad \sec \frac{2\pi}{3} = -2 \qquad \cot \frac{2\pi}{3} = -\frac{1}{\sqrt{3}} \qquad \blacksquare$$

The following table gives some values of $\sin \theta$ and $\cos \theta$ found by the method of Example 3.

θ	0	$\dfrac{\pi}{6}$	$\dfrac{\pi}{4}$	$\dfrac{\pi}{3}$	$\dfrac{\pi}{2}$	$\dfrac{2\pi}{3}$	$\dfrac{3\pi}{4}$	$\dfrac{5\pi}{6}$	π	$\dfrac{3\pi}{2}$	2π
$\sin \theta$	0	$\dfrac{1}{2}$	$\dfrac{1}{\sqrt{2}}$	$\dfrac{\sqrt{3}}{2}$	1	$\dfrac{\sqrt{3}}{2}$	$\dfrac{1}{\sqrt{2}}$	$\dfrac{1}{2}$	0	-1	0
$\cos \theta$	1	$\dfrac{\sqrt{3}}{2}$	$\dfrac{1}{\sqrt{2}}$	$\dfrac{1}{2}$	0	$-\dfrac{1}{2}$	$-\dfrac{1}{\sqrt{2}}$	$-\dfrac{\sqrt{3}}{2}$	-1	0	1

EXAMPLE 4 If $\cos \theta = \frac{2}{5}$ and $0 < \theta < \pi/2$, find the other five trigonometric functions of θ.

SOLUTION Since $\cos \theta = \frac{2}{5}$, we can label the hypotenuse as having length 5 and the adjacent side as having length 2 in Figure 12. If the opposite side has length x, then the Pythagorean Theorem gives $x^2 + 4 = 25$ and so $x^2 = 21$, $x = \sqrt{21}$. We can now use the diagram to write the other five trigonometric functions:

$$\sin \theta = \frac{\sqrt{21}}{5} \qquad \tan \theta = \frac{\sqrt{21}}{2}$$

$$\csc \theta = \frac{5}{\sqrt{21}} \qquad \sec \theta = \frac{5}{2} \qquad \cot \theta = \frac{2}{\sqrt{21}} \qquad \blacksquare$$

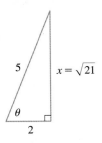

5 \qquad $x = \sqrt{21}$

θ

2

FIGURE 12

16

x

$40°$

FIGURE 13

EXAMPLE 5 Use a calculator to approximate the value of x in Figure 13.

SOLUTION From the diagram we see that

$$\tan 40° = \frac{16}{x}$$

Therefore \qquad $$x = \frac{16}{\tan 40°} \approx 19.07 \qquad \blacksquare$$

■ Trigonometric Identities

A trigonometric identity is a relationship among the trigonometric functions. The most elementary are the following, which are immediate consequences of the definitions of the trigonometric functions.

$$
\boxed{6} \qquad \csc\theta = \frac{1}{\sin\theta} \qquad \sec\theta = \frac{1}{\cos\theta} \qquad \cot\theta = \frac{1}{\tan\theta}
$$

$$
\tan\theta = \frac{\sin\theta}{\cos\theta} \qquad \cot\theta = \frac{\cos\theta}{\sin\theta}
$$

For the next identity we refer back to Figure 7. The distance formula (or, equivalently, the Pythagorean Theorem) tells us that $x^2 + y^2 = r^2$. Therefore

$$
\sin^2\theta + \cos^2\theta = \frac{y^2}{r^2} + \frac{x^2}{r^2} = \frac{x^2 + y^2}{r^2} = \frac{r^2}{r^2} = 1
$$

We have therefore proved one of the most useful of all trigonometric identities:

$$
\boxed{7} \qquad \sin^2\theta + \cos^2\theta = 1
$$

If we now divide both sides of Equation 7 by $\cos^2\theta$ and use Equations 6, we get

$$
\boxed{8} \qquad \tan^2\theta + 1 = \sec^2\theta
$$

Similarly, if we divide both sides of Equation 7 by $\sin^2\theta$, we get

$$
\boxed{9} \qquad 1 + \cot^2\theta = \csc^2\theta
$$

The identities

$$
\boxed{10a} \qquad \sin(-\theta) = -\sin\theta
$$

$$
\boxed{10b} \qquad \cos(-\theta) = \cos\theta
$$

Odd functions and even functions are discussed in Section 1.1.

show that sine is an odd function and cosine is an even function. They are easily proved by drawing a diagram showing θ and $-\theta$ in standard position (see Exercise 39).

Since the angles θ and $\theta + 2\pi$ have the same terminal side, we have

$$
\boxed{11} \qquad \sin(\theta + 2\pi) = \sin\theta \qquad \cos(\theta + 2\pi) = \cos\theta
$$

These identities show that the sine and cosine functions are periodic with period 2π.

The remaining trigonometric identities are all consequences of two basic identities called the **addition formulas**:

12a

$$\sin(x + y) = \sin x \cos y + \cos x \sin y$$

12b

$$\cos(x + y) = \cos x \cos y - \sin x \sin y$$

The proofs of these addition formulas are outlined in Exercises 85, 86, and 87.

By substituting $-y$ for y in Equations 12a and 12b and using Equations 10a and 10b, we obtain the following **subtraction formulas**:

13a

$$\sin(x - y) = \sin x \cos y - \cos x \sin y$$

13b

$$\cos(x - y) = \cos x \cos y + \sin x \sin y$$

Then, by dividing the formulas in Equations 12 or Equations 13, we obtain the corresponding formulas for $\tan(x \pm y)$:

14a

$$\tan(x + y) = \frac{\tan x + \tan y}{1 - \tan x \tan y}$$

14b

$$\tan(x - y) = \frac{\tan x - \tan y}{1 + \tan x \tan y}$$

If we put $y = x$ in the addition formulas (12), we get the **double-angle formulas**:

15a

$$\sin 2x = 2 \sin x \cos x$$

15b

$$\cos 2x = \cos^2 x - \sin^2 x$$

Then, by using the identity $\sin^2 x + \cos^2 x = 1$, we obtain the following alternate forms of the double-angle formulas for $\cos 2x$:

16a

$$\cos 2x = 2 \cos^2 x - 1$$

16b

$$\cos 2x = 1 - 2 \sin^2 x$$

If we now solve these equations for $\cos^2 x$ and $\sin^2 x$, we get the following **half-angle formulas**, which are useful in integral calculus:

17a

$$\cos^2 x = \frac{1 + \cos 2x}{2}$$

17b

$$\sin^2 x = \frac{1 - \cos 2x}{2}$$

Finally, we state the **product formulas**, which can be deduced from Equations 12 and 13:

18a	$\sin x \cos y = \frac{1}{2}[\sin(x + y) + \sin(x - y)]$
18b	$\cos x \cos y = \frac{1}{2}[\cos(x + y) + \cos(x - y)]$
18c	$\sin x \sin y = \frac{1}{2}[\cos(x - y) - \cos(x + y)]$

There are many other trigonometric identities, but those we have stated are the ones used most often in calculus. If you forget any of the identities 13–18, remember that they can all be deduced from Equations 12a and 12b.

EXAMPLE 6 Find all values of x in the interval $[0, 2\pi]$ such that $\sin x = \sin 2x$.

SOLUTION Using the double-angle formula (15a), we rewrite the given equation as

$$\sin x = 2 \sin x \cos x \qquad \text{or} \qquad \sin x(1 - 2 \cos x) = 0$$

Therefore there are two possibilities:

$$\sin x = 0 \qquad \text{or} \qquad 1 - 2 \cos x = 0$$

$$x = 0, \pi, 2\pi \qquad\qquad \cos x = \frac{1}{2}$$

$$x = \frac{\pi}{3}, \frac{5\pi}{3}$$

The given equation has five solutions: 0, $\pi/3$, π, $5\pi/3$, and 2π. ∎

■ Graphs of the Trigonometric Functions

The graph of the function $f(x) = \sin x$, shown in Figure 14(a), is obtained by plotting points for $0 \le x \le 2\pi$ and then using the periodic nature of the function (from Equation 11) to complete the graph. Notice that the zeros of the sine function occur at the

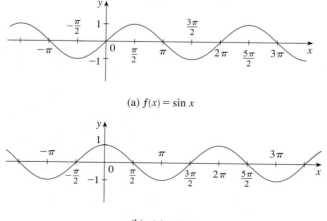

(a) $f(x) = \sin x$

(b) $g(x) = \cos x$

FIGURE 14

integer multiples of π, that is,

$$\sin x = 0 \qquad \text{whenever } x = n\pi, \quad n \text{ an integer}$$

Because of the identity

$$\cos x = \sin\left(x + \frac{\pi}{2}\right)$$

(which can be verified using Equation 12a), the graph of cosine is obtained by shifting the graph of sine by an amount $\pi/2$ to the left [see Figure 14(b)]. Note that for both the sine and cosine functions the domain is $(-\infty, \infty)$ and the range is the closed interval $[-1, 1]$. Thus, for all values of x, we have

$$-1 \leqslant \sin x \leqslant 1 \qquad -1 \leqslant \cos x \leqslant 1$$

The graphs of the remaining four trigonometric functions are shown in Figure 15 and their domains are indicated there. Notice that tangent and cotangent have range $(-\infty, \infty)$, whereas cosecant and secant have range $(-\infty, -1] \cup [1, \infty)$. All four functions are periodic: tangent and cotangent have period π, whereas cosecant and secant have period 2π.

(a) $y = \tan x$

(b) $y = \cot x$

(c) $y = \csc x$

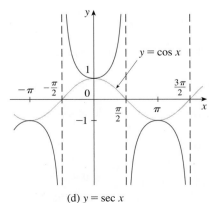

(d) $y = \sec x$

FIGURE 15

D   EXERCISES

1–6 Convert from degrees to radians.

1. $210°$

2. $300°$

3. $9°$

4. $-315°$

5. $900°$

6. $36°$

7–12 Convert from radians to degrees.

7. 4π

8. $-\dfrac{7\pi}{2}$

9. $\dfrac{5\pi}{12}$

10. $\dfrac{8\pi}{3}$

11. $-\dfrac{3\pi}{8}$

12. 5

---

13. Find the length of a circular arc subtended by an angle of $\pi/12$ rad if the radius of the circle is 36 cm.

14. If a circle has radius 10 cm, find the length of the arc subtended by a central angle of $72°$.

15. A circle has radius 1.5 m. What angle is subtended at the center of the circle by an arc 1 m long?

16. Find the radius of a circular sector with angle $3\pi/4$ and arc length 6 cm.

17–22 Draw, in standard position, the angle whose measure is given.

17. $315°$

18. $-150°$

19. $-\dfrac{3\pi}{4}$ rad

20. $\dfrac{7\pi}{3}$ rad

21. 2 rad

22. -3 rad

23–28 Find the exact trigonometric ratios for the angle whose radian measure is given.

23. $\dfrac{3\pi}{4}$

24. $\dfrac{4\pi}{3}$

25. $\dfrac{9\pi}{2}$

26. -5π

27. $\dfrac{5\pi}{6}$

28. $\dfrac{11\pi}{4}$

29–34 Find the remaining trigonometric ratios.

29. $\sin\theta = \dfrac{3}{5}, \quad 0 < \theta < \dfrac{\pi}{2}$

30. $\tan\alpha = 2, \quad 0 < \alpha < \dfrac{\pi}{2}$

31. $\sec\phi = -1.5, \quad \dfrac{\pi}{2} < \phi < \pi$

32. $\cos x = -\dfrac{1}{3}, \quad \pi < x < \dfrac{3\pi}{2}$

33. $\cot\beta = 3, \quad \pi < \beta < 2\pi$

34. $\csc\theta = -\dfrac{4}{3}, \quad \dfrac{3\pi}{2} < \theta < 2\pi$

35–38 Find, correct to five decimal places, the length of the side labeled x.

35.

36.

37.

38.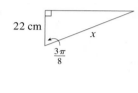

39–41 Prove each equation.

39. (a) Equation 10a   (b) Equation 10b

40. (a) Equation 14a   (b) Equation 14b

41. (a) Equation 18a   (b) Equation 18b
      (c) Equation 18c

42–58 Prove the identity.

42. $\cos\left(\dfrac{\pi}{2} - x\right) = \sin x$

43. $\sin\left(\dfrac{\pi}{2} + x\right) = \cos x$ **44.** $\sin(\pi - x) = \sin x$

45. $\sin\theta \cot\theta = \cos\theta$ **46.** $(\sin x + \cos x)^2 = 1 + \sin 2x$

47. $\sec y - \cos y = \tan y \sin y$

48. $\tan^2\alpha - \sin^2\alpha = \tan^2\alpha \sin^2\alpha$

49. $\cot^2\theta + \sec^2\theta = \tan^2\theta + \csc^2\theta$

50. $2\csc 2t = \sec t \csc t$

51. $\tan 2\theta = \dfrac{2\tan\theta}{1 - \tan^2\theta}$

52. $\dfrac{1}{1 - \sin\theta} + \dfrac{1}{1 + \sin\theta} = 2\sec^2\theta$

53. $\sin x \sin 2x + \cos x \cos 2x = \cos x$

54. $\sin^2 x - \sin^2 y = \sin(x + y)\sin(x - y)$

55. $\dfrac{\sin\phi}{1 - \cos\phi} = \csc\phi + \cot\phi$

56. $\tan x + \tan y = \dfrac{\sin(x + y)}{\cos x \cos y}$

57. $\sin 3\theta + \sin \theta = 2 \sin 2\theta \cos \theta$

58. $\cos 3\theta = 4 \cos^3\theta - 3 \cos \theta$

59–64 If $\sin x = \frac{1}{3}$ and $\sec y = \frac{5}{4}$, where x and y lie between 0 and $\pi/2$, evaluate the expression.

59. $\sin(x + y)$ **60.** $\cos(x + y)$

61. $\cos(x - y)$ **62.** $\sin(x - y)$

63. $\sin 2y$ **64.** $\cos 2y$

65–72 Find all values of x in the interval $[0, 2\pi]$ that satisfy the equation.

65. $2 \cos x - 1 = 0$ **66.** $3 \cot^2x = 1$

67. $2 \sin^2x = 1$ **68.** $|\tan x| = 1$

69. $\sin 2x = \cos x$ **70.** $2 \cos x + \sin 2x = 0$

71. $\sin x = \tan x$ **72.** $2 + \cos 2x = 3 \cos x$

73–76 Find all values of x in the interval $[0, 2\pi]$ that satisfy the inequality.

73. $\sin x \leq \frac{1}{2}$ **74.** $2 \cos x + 1 > 0$

75. $-1 < \tan x < 1$ **76.** $\sin x > \cos x$

77–82 Graph the function by starting with the graphs in Figures 14 and 15 and applying the transformations of Section 1.3 where appropriate.

77. $y = \cos\left(x - \dfrac{\pi}{3}\right)$ **78.** $y = \tan 2x$

79. $y = \dfrac{1}{3} \tan\left(x - \dfrac{\pi}{2}\right)$ **80.** $y = 1 + \sec x$

81. $y = |\sin x|$ **82.** $y = 2 + \sin\left(x + \dfrac{\pi}{4}\right)$

83. Prove the **Law of Cosines**: If a triangle has sides with lengths a, b, and c, and θ is the angle between the sides with lengths a and b, then
$$c^2 = a^2 + b^2 - 2ab \cos\theta$$

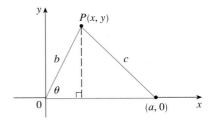

[*Hint:* Introduce a coordinate system so that θ is in standard

position, as in the figure. Express x and y in terms of θ and then use the distance formula to compute c.]

84. In order to find the distance $|AB|$ across a small inlet, a point C was located as in the figure and the following measurements were recorded:
$$\angle C = 103° \qquad |AC| = 820 \text{ m} \qquad |BC| = 910 \text{ m}$$
Use the Law of Cosines from Exercise 83 to find the required distance.

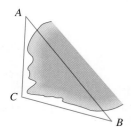

85. Use the figure to prove the subtraction formula
$$\cos(\alpha - \beta) = \cos \alpha \cos \beta + \sin \alpha \sin \beta$$
[*Hint:* Compute c^2 in two ways (using the Law of Cosines from Exercise 83 and also using the distance formula) and compare the two expressions.]

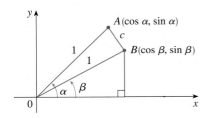

86. Use the formula in Exercise 85 to prove the addition formula for cosine (12b).

87. Use the addition formula for cosine and the identities
$$\cos\left(\dfrac{\pi}{2} - \theta\right) = \sin \theta \qquad \sin\left(\dfrac{\pi}{2} - \theta\right) = \cos \theta$$
to prove the subtraction formula (13a) for the sine function.

88. Show that the area of a triangle with sides of lengths a and b and with included angle θ is
$$A = \tfrac{1}{2}ab \sin \theta$$

89. Find the area of triangle ABC, correct to five decimal places, if
$$|AB| = 10 \text{ cm} \qquad |BC| = 3 \text{ cm} \qquad \angle ABC = 107°$$

E   Sigma Notation

A convenient way of writing sums uses the Greek letter Σ (capital sigma, corresponding to our letter S) and is called **sigma notation**.

This tells us to end with $i = n$.

This tells us to add.

This tells us to start with $i = m$.

$$\sum_{i=m}^{n} a_i$$

> **1** **Definition** If $a_m, a_{m+1}, \ldots, a_n$ are real numbers and m and n are integers such that $m \leqslant n$, then
>
> $$\sum_{i=m}^{n} a_i = a_m + a_{m+1} + a_{m+2} + \cdots + a_{n-1} + a_n$$

With function notation, Definition 1 can be written as

$$\sum_{i=m}^{n} f(i) = f(m) + f(m + 1) + f(m + 2) + \cdots + f(n - 1) + f(n)$$

Thus the symbol $\sum_{i=m}^{n}$ indicates a summation in which the letter i (called the **index of summation**) takes on consecutive integer values beginning with m and ending with n, that is, $m, m + 1, \ldots, n$. Other letters can also be used as the index of summation.

EXAMPLE 1

(a) $\displaystyle\sum_{i=1}^{4} i^2 = 1^2 + 2^2 + 3^2 + 4^2 = 30$

(b) $\displaystyle\sum_{i=3}^{n} i = 3 + 4 + 5 + \cdots + (n - 1) + n$

(c) $\displaystyle\sum_{j=0}^{5} 2^j = 2^0 + 2^1 + 2^2 + 2^3 + 2^4 + 2^5 = 63$

(d) $\displaystyle\sum_{k=1}^{n} \frac{1}{k} = 1 + \frac{1}{2} + \frac{1}{3} + \cdots + \frac{1}{n}$

(e) $\displaystyle\sum_{i=1}^{3} \frac{i - 1}{i^2 + 3} = \frac{1 - 1}{1^2 + 3} + \frac{2 - 1}{2^2 + 3} + \frac{3 - 1}{3^2 + 3} = 0 + \frac{1}{7} + \frac{1}{6} = \frac{13}{42}$

(f) $\displaystyle\sum_{i=1}^{4} 2 = 2 + 2 + 2 + 2 = 8$ ∎

EXAMPLE 2 Write the sum $2^3 + 3^3 + \cdots + n^3$ in sigma notation.

SOLUTION There is no unique way of writing a sum in sigma notation. We could write

$$2^3 + 3^3 + \cdots + n^3 = \sum_{i=2}^{n} i^3$$

or

$$2^3 + 3^3 + \cdots + n^3 = \sum_{j=1}^{n-1} (j + 1)^3$$

or

$$2^3 + 3^3 + \cdots + n^3 = \sum_{k=0}^{n-2} (k + 2)^3$$ ∎

The following theorem gives three simple rules for working with sigma notation.

> **2 Theorem** If c is any constant (that is, it does not depend on i), then
>
> (a) $\displaystyle\sum_{i=m}^{n} ca_i = c \sum_{i=m}^{n} a_i$ (b) $\displaystyle\sum_{i=m}^{n} (a_i + b_i) = \sum_{i=m}^{n} a_i + \sum_{i=m}^{n} b_i$
>
> (c) $\displaystyle\sum_{i=m}^{n} (a_i - b_i) = \sum_{i=m}^{n} a_i - \sum_{i=m}^{n} b_i$

PROOF To see why these rules are true, all we have to do is write both sides in expanded form. Rule (a) is just the distributive property of real numbers:

$$ca_m + ca_{m+1} + \cdots + ca_n = c(a_m + a_{m+1} + \cdots + a_n)$$

Rule (b) follows from the associative and commutative properties:

$$(a_m + b_m) + (a_{m+1} + b_{m+1}) + \cdots + (a_n + b_n)$$
$$= (a_m + a_{m+1} + \cdots + a_n) + (b_m + b_{m+1} + \cdots + b_n)$$

Rule (c) is proved similarly. ∎

EXAMPLE 3 Find $\displaystyle\sum_{i=1}^{n} 1$.

SOLUTION
$$\sum_{i=1}^{n} 1 = \underbrace{1 + 1 + \cdots + 1}_{n \text{ terms}} = n$$

■

EXAMPLE 4 Prove the formula for the sum of the first n positive integers:

$$\sum_{i=1}^{n} i = 1 + 2 + 3 + \cdots + n = \frac{n(n + 1)}{2}$$

SOLUTION This formula can be proved by mathematical induction (see page 72) or by the following method used by the German mathematician Karl Friedrich Gauss (1777–1855) when he was ten years old.

Write the sum S twice, once in the usual order and once in reverse order:

$$S = 1 + \quad 2 \quad + \quad 3 \quad + \cdots + (n - 1) + n$$
$$S = n + (n - 1) + (n - 2) + \cdots + \quad 2 \quad + 1$$

Adding all columns vertically, we get

$$2S = (n + 1) + (n + 1) + (n + 1) + \cdots + (n + 1) + (n + 1)$$

On the right side there are n terms, each of which is $n + 1$, so

$$2S = n(n + 1) \qquad \text{or} \qquad S = \frac{n(n + 1)}{2}$$

■

EXAMPLE 5 Prove the formula for the sum of the squares of the first n positive integers:

$$\sum_{i=1}^{n} i^2 = 1^2 + 2^2 + 3^2 + \cdots + n^2 = \frac{n(n + 1)(2n + 1)}{6}$$

Most terms cancel in pairs.

SOLUTION 1 Let S be the desired sum. We start with the *telescoping sum* (or collapsing sum):

$$\sum_{i=1}^{n} [(1 + i)^3 - i^3] = (2^3 - 1^3) + (3^3 - 2^3) + (4^3 - 3^3) + \cdots + [(n + 1)^3 - n^3]$$

$$= (n + 1)^3 - 1^3 = n^3 + 3n^2 + 3n$$

On the other hand, using Theorem 2 and Examples 3 and 4, we have

$$\sum_{i=1}^{n} [(1 + i)^3 - i^3] = \sum_{i=1}^{n} [3i^2 + 3i + 1] = 3 \sum_{i=1}^{n} i^2 + 3 \sum_{i=1}^{n} i + \sum_{i=1}^{n} 1$$

$$= 3S + 3 \frac{n(n + 1)}{2} + n = 3S + \tfrac{3}{2}n^2 + \tfrac{5}{2}n$$

Thus we have

$$n^3 + 3n^2 + 3n = 3S + \tfrac{3}{2}n^2 + \tfrac{5}{2}n$$

Solving this equation for S, we obtain

$$3S = n^3 + \tfrac{3}{2}n^2 + \tfrac{1}{2}n$$

or

$$S = \frac{2n^3 + 3n^2 + n}{6} = \frac{n(n + 1)(2n + 1)}{6}$$

Principle of Mathematical Induction
Let S_n be a statement involving the positive integer n. Suppose that
1. S_1 is true.
2. If S_k is true, then S_{k+1} is true.
Then S_n is true for all positive integers n.

See pages 72 and 74 for a more thorough discussion of mathematical induction.

SOLUTION 2 Let S_n be the given formula.

1. S_1 is true because $1^2 = \dfrac{1(1 + 1)(2 \cdot 1 + 1)}{6}$

2. Assume that S_k is true; that is,

$$1^2 + 2^2 + 3^2 + \cdots + k^2 = \frac{k(k + 1)(2k + 1)}{6}$$

Then

$$1^2 + 2^2 + 3^2 + \cdots + (k + 1)^2 = (1^2 + 2^2 + 3^2 + \cdots + k^2) + (k + 1)^2$$

$$= \frac{k(k + 1)(2k + 1)}{6} + (k + 1)^2$$

$$= (k + 1) \frac{k(2k + 1) + 6(k + 1)}{6}$$

$$= (k + 1) \frac{2k^2 + 7k + 6}{6}$$

$$= \frac{(k + 1)(k + 2)(2k + 3)}{6}$$

$$= \frac{(k + 1)[(k + 1) + 1][2(k + 1) + 1]}{6}$$

So S_{k+1} is true.

By the Principle of Mathematical Induction, S_n is true for all n. ■

We list the results of Examples 3, 4, and 5 together with a similar result for cubes (see Exercises 37–40) as Theorem 3. These formulas are needed for finding areas and evaluating integrals in Chapter 5.

---

3 Theorem Let c be a constant and n a positive integer. Then

(a) $\displaystyle\sum_{i=1}^{n} 1 = n$

(b) $\displaystyle\sum_{i=1}^{n} c = nc$

(c) $\displaystyle\sum_{i=1}^{n} i = \frac{n(n+1)}{2}$

(d) $\displaystyle\sum_{i=1}^{n} i^2 = \frac{n(n+1)(2n+1)}{6}$

(e) $\displaystyle\sum_{i=1}^{n} i^3 = \left[\frac{n(n+1)}{2}\right]^2$

EXAMPLE 6 Evaluate $\displaystyle\sum_{i=1}^{n} i(4i^2 - 3)$.

SOLUTION Using Theorems 2 and 3, we have

$$\sum_{i=1}^{n} i(4i^2 - 3) = \sum_{i=1}^{n} (4i^3 - 3i) = 4\sum_{i=1}^{n} i^3 - 3\sum_{i=1}^{n} i$$

$$= 4\left[\frac{n(n+1)}{2}\right]^2 - 3\frac{n(n+1)}{2}$$

$$= \frac{n(n+1)[2n(n+1) - 3]}{2}$$

$$= \frac{n(n+1)(2n^2 + 2n - 3)}{2}$$ ■

EXAMPLE 7 Find $\displaystyle\lim_{n\to\infty} \sum_{i=1}^{n} \frac{3}{n}\left[\left(\frac{i}{n}\right)^2 + 1\right]$.

The type of calculation in Example 7 arises in Chapter 5 when we compute areas.

SOLUTION

$$\lim_{n\to\infty} \sum_{i=1}^{n} \frac{3}{n}\left[\left(\frac{i}{n}\right)^2 + 1\right] = \lim_{n\to\infty} \sum_{i=1}^{n} \left[\frac{3}{n^3} i^2 + \frac{3}{n}\right]$$

$$= \lim_{n\to\infty} \left[\frac{3}{n^3} \sum_{i=1}^{n} i^2 + \frac{3}{n} \sum_{i=1}^{n} 1\right]$$

$$= \lim_{n\to\infty} \left[\frac{3}{n^3} \frac{n(n+1)(2n+1)}{6} + \frac{3}{n} \cdot n\right]$$

$$= \lim_{n\to\infty} \left[\frac{1}{2} \cdot \frac{n}{n} \cdot \left(\frac{n+1}{n}\right)\left(\frac{2n+1}{n}\right) + 3\right]$$

$$= \lim_{n\to\infty} \left[\frac{1}{2} \cdot 1\left(1 + \frac{1}{n}\right)\left(2 + \frac{1}{n}\right) + 3\right]$$

$$= \tfrac{1}{2} \cdot 1 \cdot 1 \cdot 2 + 3 = 4$$ ■

E   EXERCISES

1–10 Write the sum in expanded form.

1. $\displaystyle\sum_{i=1}^{5} \sqrt{i}$

2. $\displaystyle\sum_{i=1}^{6} \frac{1}{i+1}$

3. $\displaystyle\sum_{i=4}^{6} 3^i$

4. $\displaystyle\sum_{i=4}^{6} i^3$

5. $\displaystyle\sum_{k=0}^{4} \frac{2k-1}{2k+1}$

6. $\displaystyle\sum_{k=5}^{8} x^k$

7. $\displaystyle\sum_{i=1}^{n} i^{10}$

8. $\displaystyle\sum_{j=n}^{n+3} j^2$

9. $\displaystyle\sum_{j=0}^{n-1} (-1)^j$

10. $\displaystyle\sum_{i=1}^{n} f(x_i)\,\Delta x_i$

11–20 Write the sum in sigma notation.

11. $1 + 2 + 3 + 4 + \cdots + 10$

12. $\sqrt{3} + \sqrt{4} + \sqrt{5} + \sqrt{6} + \sqrt{7}$

13. $\frac{1}{2} + \frac{2}{3} + \frac{3}{4} + \frac{4}{5} + \cdots + \frac{19}{20}$

14. $\frac{3}{7} + \frac{4}{8} + \frac{5}{9} + \frac{6}{10} + \cdots + \frac{23}{27}$

15. $2 + 4 + 6 + 8 + \cdots + 2n$

16. $1 + 3 + 5 + 7 + \cdots + (2n-1)$

17. $1 + 2 + 4 + 8 + 16 + 32$

18. $\frac{1}{1} + \frac{1}{4} + \frac{1}{9} + \frac{1}{16} + \frac{1}{25} + \frac{1}{36}$

19. $x + x^2 + x^3 + \cdots + x^n$

20. $1 - x + x^2 - x^3 + \cdots + (-1)^n x^n$

21–35 Find the value of the sum.

21. $\displaystyle\sum_{i=4}^{8} (3i-2)$

22. $\displaystyle\sum_{i=3}^{6} i(i+2)$

23. $\displaystyle\sum_{j=1}^{6} 3^{j+1}$

24. $\displaystyle\sum_{k=0}^{8} \cos k\pi$

25. $\displaystyle\sum_{n=1}^{20} (-1)^n$

26. $\displaystyle\sum_{i=1}^{100} 4$

27. $\displaystyle\sum_{i=0}^{4} (2^i + i^2)$

28. $\displaystyle\sum_{i=-2}^{4} 2^{3-i}$

29. $\displaystyle\sum_{i=1}^{n} 2i$

30. $\displaystyle\sum_{i=1}^{n} (2-5i)$

31. $\displaystyle\sum_{i=1}^{n} (i^2 + 3i + 4)$

32. $\displaystyle\sum_{i=1}^{n} (3+2i)^2$

33. $\displaystyle\sum_{i=1}^{n} (i+1)(i+2)$

34. $\displaystyle\sum_{i=1}^{n} i(i+1)(i+2)$

35. $\displaystyle\sum_{i=1}^{n} (i^3 - i - 2)$

---

36. Find the number n such that $\displaystyle\sum_{i=1}^{n} i = 78$.

37. Prove formula (b) of Theorem 3.

38. Prove formula (e) of Theorem 3 using mathematical induction.

39. Prove formula (e) of Theorem 3 using a method similar to that of Example 5, Solution 1 [start with $(1+i)^4 - i^4$].

40. Prove formula (e) of Theorem 3 using the following method published by Abu Bekr Mohammed ibn Alhusain Alkarchi in about AD 1010. The figure shows a square $ABCD$ in which sides AB and AD have been divided into segments of lengths $1, 2, 3, \ldots, n$. Thus the side of the square has length $n(n+1)/2$ so the area is $[n(n+1)/2]^2$. But the area is also the sum of the areas of the n "gnomons" $G_1, G_2, \ldots, G_n$ shown in the figure. Show that the area of G_i is i^3 and conclude that formula (e) is true.

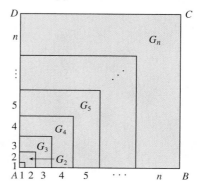

41. Evaluate each telescoping sum.

(a) $\displaystyle\sum_{i=1}^{n} [i^4 - (i-1)^4]$

(b) $\displaystyle\sum_{i=1}^{100} (5^i - 5^{i-1})$

(c) $\displaystyle\sum_{i=3}^{99} \left(\frac{1}{i} - \frac{1}{i+1}\right)$

(d) $\displaystyle\sum_{i=1}^{n} (a_i - a_{i-1})$

42. Prove the generalized triangle inequality:

$$\left|\sum_{i=1}^{n} a_i\right| \leq \sum_{i=1}^{n} |a_i|$$

43–46 Find the limit.

43. $\displaystyle\lim_{n\to\infty} \sum_{i=1}^{n} \frac{1}{n}\left(\frac{i}{n}\right)^2$

44. $\displaystyle\lim_{n\to\infty} \sum_{i=1}^{n} \frac{1}{n}\left[\left(\frac{i}{n}\right)^3 + 1\right]$

45. $\displaystyle\lim_{n\to\infty} \sum_{i=1}^{n} \frac{2}{n}\left[\left(\frac{2i}{n}\right)^3 + 5\left(\frac{2i}{n}\right)\right]$

46. $\displaystyle\lim_{n\to\infty} \sum_{i=1}^{n} \frac{3}{n}\left[\left(1 + \frac{3i}{n}\right)^3 - 2\left(1 + \frac{3i}{n}\right)\right]$

47. Prove the formula for the sum of a finite geometric series with first term a and common ratio $r \neq 1$:

$$\sum_{i=1}^{n} ar^{i-1} = a + ar + ar^2 + \cdots + ar^{n-1} = \frac{a(r^n - 1)}{r - 1}$$

48. Evaluate $\displaystyle\sum_{i=1}^{n} \frac{3}{2^{i-1}}$.

49. Evaluate $\displaystyle\sum_{i=1}^{n} (2i + 2^i)$.

50. Evaluate $\displaystyle\sum_{i=1}^{m}\left[\sum_{j=1}^{n} (i + j)\right]$.

F   Proofs of Theorems

In this appendix we present proofs of several theorems that are stated in the main body of the text. The sections in which they occur are indicated in the margin.

Section 2.3

> **Limit Laws** Suppose that c is a constant and the limits
>
> $$\lim_{x\to a} f(x) = L \qquad \text{and} \qquad \lim_{x\to a} g(x) = M$$
>
> exist. Then
>
> **1.** $\displaystyle\lim_{x\to a} [f(x) + g(x)] = L + M$
> **2.** $\displaystyle\lim_{x\to a} [f(x) - g(x)] = L - M$
>
> **3.** $\displaystyle\lim_{x\to a} [cf(x)] = cL$
> **4.** $\displaystyle\lim_{x\to a} [f(x)g(x)] = LM$
>
> **5.** $\displaystyle\lim_{x\to a} \frac{f(x)}{g(x)} = \frac{L}{M}$ if $M \neq 0$

PROOF OF LAW 4 Let $\varepsilon > 0$ be given. We want to find $\delta > 0$ such that

$$\text{if} \qquad 0 < |x - a| < \delta \qquad \text{then} \qquad |f(x)g(x) - LM| < \varepsilon$$

In order to get terms that contain $|f(x) - L|$ and $|g(x) - M|$, we add and subtract $Lg(x)$ as follows:

$$
\begin{aligned}
|f(x)g(x) - LM| &= |f(x)g(x) - Lg(x) + Lg(x) - LM| \\
&= |[f(x) - L]g(x) + L[g(x) - M]| \\
&\leq |[f(x) - L]g(x)| + |L[g(x) - M]| \qquad \text{(Triangle Inequality)} \\
&= |f(x) - L||g(x)| + |L||g(x) - M|
\end{aligned}
$$

We want to make each of these terms less than $\varepsilon/2$.

Since $\lim_{x\to a} g(x) = M$, there is a number $\delta_1 > 0$ such that

$$\text{if} \qquad 0 < |x - a| < \delta_1 \qquad \text{then} \qquad |g(x) - M| < \frac{\varepsilon}{2(1 + |L|)}$$

Also, there is a number $\delta_2 > 0$ such that if $0 < |x - a| < \delta_2$, then

$$|g(x) - M| < 1$$

and therefore

$$|g(x)| = |g(x) - M + M| \leq |g(x) - M| + |M| < 1 + |M|$$

Since $\lim_{x \to a} f(x) = L$, there is a number $\delta_3 > 0$ such that

$$\text{if} \quad 0 < |x - a| < \delta_3 \quad \text{then} \quad |f(x) - L| < \frac{\varepsilon}{2(1 + |M|)}$$

Let $\delta = \min\{\delta_1, \delta_2, \delta_3\}$. If $0 < |x - a| < \delta$, then we have $0 < |x - a| < \delta_1$, $0 < |x - a| < \delta_2$, and $0 < |x - a| < \delta_3$, so we can combine the inequalities to obtain

$$|f(x)g(x) - LM| \leq |f(x) - L||g(x)| + |L||g(x) - M|$$

$$< \frac{\varepsilon}{2(1 + |M|)}(1 + |M|) + |L|\frac{\varepsilon}{2(1 + |L|)}$$

$$< \frac{\varepsilon}{2} + \frac{\varepsilon}{2} = \varepsilon$$

This shows that $\lim_{x \to a}[f(x)g(x)] = LM$.   ∎

PROOF OF LAW 3   If we take $g(x) = c$ in Law 4, we get

$$\lim_{x \to a}[cf(x)] = \lim_{x \to a}[g(x)f(x)] = \lim_{x \to a} g(x) \cdot \lim_{x \to a} f(x)$$

$$= \lim_{x \to a} c \cdot \lim_{x \to a} f(x)$$

$$= c \lim_{x \to a} f(x) \qquad \text{(by Law 7)}$$ ∎

PROOF OF LAW 2   Using Law 1 and Law 3 with $c = -1$, we have

$$\lim_{x \to a}[f(x) - g(x)] = \lim_{x \to a}[f(x) + (-1)g(x)] = \lim_{x \to a} f(x) + \lim_{x \to a}(-1)g(x)$$

$$= \lim_{x \to a} f(x) + (-1)\lim_{x \to a} g(x) = \lim_{x \to a} f(x) - \lim_{x \to a} g(x)$$ ∎

PROOF OF LAW 5   First let us show that

$$\lim_{x \to a} \frac{1}{g(x)} = \frac{1}{M}$$

To do this we must show that, given $\varepsilon > 0$, there exists $\delta > 0$ such that

$$\text{if} \quad 0 < |x - a| < \delta \quad \text{then} \quad \left|\frac{1}{g(x)} - \frac{1}{M}\right| < \varepsilon$$

Observe that
$$\left|\frac{1}{g(x)} - \frac{1}{M}\right| = \frac{|M - g(x)|}{|Mg(x)|}$$

We know that we can make the numerator small. But we also need to know that the denominator is not small when x is near a. Since $\lim_{x \to a} g(x) = M$, there is a number $\delta_1 > 0$ such that, whenever $0 < |x - a| < \delta_1$, we have

$$|g(x) - M| < \frac{|M|}{2}$$

and therefore
$$|M| = |M - g(x) + g(x)| \leq |M - g(x)| + |g(x)|$$

$$< \frac{|M|}{2} + |g(x)|$$

This shows that

$$\text{if} \quad 0 < |x - a| < \delta_1 \quad \text{then} \quad |g(x)| > \frac{|M|}{2}$$

and so, for these values of x,

$$\frac{1}{|Mg(x)|} = \frac{1}{|M||g(x)|} < \frac{1}{|M|} \cdot \frac{2}{|M|} = \frac{2}{M^2}$$

Also, there exists $\delta_2 > 0$ such that

$$\text{if} \quad 0 < |x - a| < \delta_2 \quad \text{then} \quad |g(x) - M| < \frac{M^2}{2}\varepsilon$$

Let $\delta = \min\{\delta_1, \delta_2\}$. Then, for $0 < |x - a| < \delta$, we have

$$\left| \frac{1}{g(x)} - \frac{1}{M} \right| = \frac{|M - g(x)|}{|Mg(x)|} < \frac{2}{M^2}\frac{M^2}{2}\varepsilon = \varepsilon$$

It follows that $\lim_{x \to a} 1/g(x) = 1/M$. Finally, using Law 4, we obtain

$$\lim_{x \to a} \frac{f(x)}{g(x)} = \lim_{x \to a} \left(f(x) \cdot \frac{1}{g(x)} \right) = \lim_{x \to a} f(x) \lim_{x \to a} \frac{1}{g(x)} = L \cdot \frac{1}{M} = \frac{L}{M} \qquad \blacksquare$$

---

2 **Theorem** If $f(x) \leqslant g(x)$ for all x in an open interval that contains a (except possibly at a) and

$$\lim_{x \to a} f(x) = L \quad \text{and} \quad \lim_{x \to a} g(x) = M$$

then $L \leqslant M$.

PROOF We use the method of proof by contradiction. Suppose, if possible, that $L > M$. Law 2 of limits says that

$$\lim_{x \to a} [g(x) - f(x)] = M - L$$

Therefore, for any $\varepsilon > 0$, there exists $\delta > 0$ such that

$$\text{if} \quad 0 < |x - a| < \delta \quad \text{then} \quad |[g(x) - f(x)] - (M - L)| < \varepsilon$$

In particular, taking $\varepsilon = L - M$ (noting that $L - M > 0$ by hypothesis), we have a number $\delta > 0$ such that

$$\text{if} \quad 0 < |x - a| < \delta \quad \text{then} \quad |[g(x) - f(x)] - (M - L)| < L - M$$

Since $b \leqslant |b|$ for any number b, we have

$$\text{if} \quad 0 < |x - a| < \delta \quad \text{then} \quad [g(x) - f(x)] - (M - L) < L - M$$

which simplifies to

$$\text{if} \quad 0 < |x - a| < \delta \quad \text{then} \quad g(x) < f(x)$$

But this contradicts $f(x) \leqslant g(x)$. Thus the inequality $L > M$ must be false. Therefore $L \leqslant M$. $\qquad \blacksquare$

> **3** **The Squeeze Theorem** If $f(x) \leqslant g(x) \leqslant h(x)$ for all x in an open interval that contains a (except possibly at a) and
>
> $$\lim_{x \to a} f(x) = \lim_{x \to a} h(x) = L$$
>
> then $$\lim_{x \to a} g(x) = L$$

PROOF Let $\varepsilon > 0$ be given. Since $\lim_{x \to a} f(x) = L$, there is a number $\delta_1 > 0$ such that

$$\text{if} \qquad 0 < |x - a| < \delta_1 \qquad \text{then} \qquad |f(x) - L| < \varepsilon$$

that is,

$$\text{if} \qquad 0 < |x - a| < \delta_1 \qquad \text{then} \qquad L - \varepsilon < f(x) < L + \varepsilon$$

Since $\lim_{x \to a} h(x) = L$, there is a number $\delta_2 > 0$ such that

$$\text{if} \qquad 0 < |x - a| < \delta_2 \qquad \text{then} \qquad |h(x) - L| < \varepsilon$$

that is,

$$\text{if} \qquad 0 < |x - a| < \delta_2 \qquad \text{then} \qquad L - \varepsilon < h(x) < L + \varepsilon$$

Let $\delta = \min\{\delta_1, \delta_2\}$. If $0 < |x - a| < \delta$, then $0 < |x - a| < \delta_1$ and $0 < |x - a| < \delta_2$, so

$$L - \varepsilon < f(x) \leqslant g(x) \leqslant h(x) < L + \varepsilon$$

In particular, $$L - \varepsilon < g(x) < L + \varepsilon$$

and so $|g(x) - L| < \varepsilon$. Therefore $\lim_{x \to a} g(x) = L$. ∎

Section 2.5

> **Theorem** If f is a one-to-one continuous function defined on an interval (a, b), then its inverse function f^{-1} is also continuous.

PROOF First we show that if f is both one-to-one and continuous on (a, b), then it must be either increasing or decreasing on (a, b). If it were neither increasing nor decreasing, then there would exist numbers x_1, x_2, and x_3 in (a, b) with $x_1 < x_2 < x_3$ such that $f(x_2)$ does not lie between $f(x_1)$ and $f(x_3)$. There are two possibilities: either (1) $f(x_3)$ lies between $f(x_1)$ and $f(x_2)$ or (2) $f(x_1)$ lies between $f(x_2)$ and $f(x_3)$. (Draw a picture.) In case (1) we apply the Intermediate Value Theorem to the continuous function f to get a number c between x_1 and x_2 such that $f(c) = f(x_3)$. In case (2) the Intermediate Value Theorem gives a number c between x_2 and x_3 such that $f(c) = f(x_1)$. In either case we have contradicted the fact that f is one-to-one.

Let us assume, for the sake of definiteness, that f is increasing on (a, b). We take any number y_0 in the domain of f^{-1} and we let $f^{-1}(y_0) = x_0$; that is, x_0 is the number in (a, b) such that $f(x_0) = y_0$. To show that f^{-1} is continuous at y_0 we take any $\varepsilon > 0$ such that the interval $(x_0 - \varepsilon, x_0 + \varepsilon)$ is contained in the interval (a, b). Since f is increasing, it maps the numbers in the interval $(x_0 - \varepsilon, x_0 + \varepsilon)$ onto the numbers in the interval $(f(x_0 - \varepsilon), f(x_0 + \varepsilon))$ and f^{-1} reverses the correspondence. If we let δ denote the smaller of the numbers $\delta_1 = y_0 - f(x_0 - \varepsilon)$ and $\delta_2 = f(x_0 + \varepsilon) - y_0$, then the interval $(y_0 - \delta, y_0 + \delta)$ is contained in the interval $(f(x_0 - \varepsilon), f(x_0 + \varepsilon))$ and so is mapped into the interval $(x_0 - \varepsilon, x_0 + \varepsilon)$ by f^{-1}. (See the arrow diagram in Figure 1.) We have therefore found a number $\delta > 0$ such that

$$\text{if} \quad |y - y_0| < \delta \quad \text{then} \quad |f^{-1}(y) - f^{-1}(y_0)| < \varepsilon$$

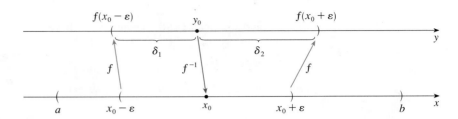

FIGURE 1

This shows that $\lim_{y \to y_0} f^{-1}(y) = f^{-1}(y_0)$ and so f^{-1} is continuous at any number y_0 in its domain. ∎

---

8 **Theorem** If f is continuous at b and $\lim_{x \to a} g(x) = b$, then

$$\lim_{x \to a} f(g(x)) = f(b)$$

PROOF Let $\varepsilon > 0$ be given. We want to find a number $\delta > 0$ such that

$$\text{if} \quad 0 < |x - a| < \delta \quad \text{then} \quad |f(g(x)) - f(b)| < \varepsilon$$

Since f is continuous at b, we have

$$\lim_{y \to b} f(y) = f(b)$$

and so there exists $\delta_1 > 0$ such that

$$\text{if} \quad 0 < |y - b| < \delta_1 \quad \text{then} \quad |f(y) - f(b)| < \varepsilon$$

Since $\lim_{x \to a} g(x) = b$, there exists $\delta > 0$ such that

$$\text{if} \quad 0 < |x - a| < \delta \quad \text{then} \quad |g(x) - b| < \delta_1$$

Combining these two statements, we see that whenever $0 < |x - a| < \delta$ we have $|g(x) - b| < \delta_1$, which implies that $|f(g(x)) - f(b)| < \varepsilon$. Therefore we have proved that $\lim_{x \to a} f(g(x)) = f(b)$. ∎

Section 3.3

The proof of the following result was promised when we proved that $\lim_{\theta \to 0} \dfrac{\sin \theta}{\theta} = 1$.

Theorem If $0 < \theta < \pi/2$, then $\theta \leqslant \tan \theta$.

PROOF Figure 2 shows a sector of a circle with center O, central angle θ, and radius 1. Then

$$|AD| = |OA| \tan \theta = \tan \theta$$

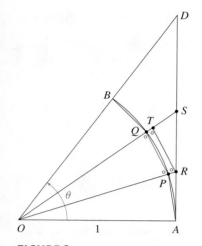

FIGURE 2

We approximate the arc AB by an inscribed polygon consisting of n equal line segments and we look at a typical segment PQ. We extend the lines OP and OQ to meet AD in the points R and S. Then we draw $RT \parallel PQ$ as in Figure 2. Observe that

$$\angle RTO = \angle PQO < 90°$$

and so $\angle RTS > 90°$. Therefore we have

$$|PQ| < |RT| < |RS|$$

If we add n such inequalities, we get

$$L_n < |AD| = \tan \theta$$

where L_n is the length of the inscribed polygon. Thus, by Theorem 2.3.2, we have

$$\lim_{n \to \infty} L_n \leqslant \tan \theta$$

But the arc length is defined in Equation 8.1.1 as the limit of the lengths of inscribed polygons, so

$$\theta = \lim_{n \to \infty} L_n \leqslant \tan \theta \qquad \blacksquare$$

Section 4.3

> **Concavity Test**
> (a) If $f''(x) > 0$ for all x in I, then the graph of f is concave upward on I.
> (b) If $f''(x) < 0$ for all x in I, then the graph of f is concave downward on I.

PROOF OF (a) Let a be any number in I. We must show that the curve $y = f(x)$ lies above the tangent line at the point $(a, f(a))$. The equation of this tangent is

$$y = f(a) + f'(a)(x - a)$$

So we must show that

$$f(x) > f(a) + f'(a)(x - a)$$

whenever $x \in I$ ($x \neq a$). (See Figure 3.)

    First let us take the case where $x > a$. Applying the Mean Value Theorem to f on the interval $[a, x]$, we get a number c, with $a < c < x$, such that

1 $$f(x) - f(a) = f'(c)(x - a)$$

Since $f'' > 0$ on I, we know from the Increasing/Decreasing Test that f' is increasing on I. Thus, since $a < c$, we have

$$f'(a) < f'(c)$$

and so, multiplying this inequality by the positive number $x - a$, we get

2 $$f'(a)(x - a) < f'(c)(x - a)$$

FIGURE 3

Now we add $f(a)$ to both sides of this inequality:

$$f(a) + f'(a)(x - a) < f(a) + f'(c)(x - a)$$

But from Equation 1 we have $f(x) = f(a) + f'(c)(x - a)$. So this inequality becomes

$$\boxed{3} \qquad\qquad f(x) > f(a) + f'(a)(x - a)$$

which is what we wanted to prove.

For the case where $x < a$ we have $f'(c) < f'(a)$, but multiplication by the negative number $x - a$ reverses the inequality, so we get (2) and (3) as before. ∎

Section 4.4

In order to give the promised proof of l'Hospital's Rule, we first need a generalization of the Mean Value Theorem. The following theorem is named after another French mathematician, Augustin-Louis Cauchy (1789–1857).

See the biographical sketch of Cauchy on page 109.

$\boxed{1}$ **Cauchy's Mean Value Theorem** Suppose that the functions f and g are continuous on $[a, b]$ and differentiable on (a, b), and $g'(x) \neq 0$ for all x in (a, b). Then there is a number c in (a, b) such that

$$\frac{f'(c)}{g'(c)} = \frac{f(b) - f(a)}{g(b) - g(a)}$$

Notice that if we take the special case in which $g(x) = x$, then $g'(c) = 1$ and Theorem 1 is just the ordinary Mean Value Theorem. Furthermore, Theorem 1 can be proved in a similar manner. You can verify that all we have to do is change the function h given by Equation 4.2.4 to the function

$$h(x) = f(x) - f(a) - \frac{f(b) - f(a)}{g(b) - g(a)}[g(x) - g(a)]$$

and apply Rolle's Theorem as before.

L'Hospital's Rule Suppose f and g are differentiable and $g'(x) \neq 0$ on an open interval I that contains a (except possibly at a). Suppose that

$$\lim_{x \to a} f(x) = 0 \qquad \text{and} \qquad \lim_{x \to a} g(x) = 0$$

or that

$$\lim_{x \to a} f(x) = \pm\infty \qquad \text{and} \qquad \lim_{x \to a} g(x) = \pm\infty$$

$\left(\text{In other words, we have an indeterminate form of type } \dfrac{0}{0} \text{ or } \infty/\infty.\right)$ Then

$$\lim_{x \to a} \frac{f(x)}{g(x)} = \lim_{x \to a} \frac{f'(x)}{g'(x)}$$

if the limit on the right side exists (or is ∞ or $-\infty$).

PROOF OF L'HOSPITAL'S RULE We are assuming that $\lim_{x \to a} f(x) = 0$ and $\lim_{x \to a} g(x) = 0$. Let

$$L = \lim_{x \to a} \frac{f'(x)}{g'(x)}$$

We must show that $\lim_{x \to a} f(x)/g(x) = L$. Define

$$F(x) = \begin{cases} f(x) & \text{if } x \neq a \\ 0 & \text{if } x = a \end{cases} \qquad G(x) = \begin{cases} g(x) & \text{if } x \neq a \\ 0 & \text{if } x = a \end{cases}$$

Then F is continuous on I since f is continuous on $\{x \in I \mid x \neq a\}$ and

$$\lim_{x \to a} F(x) = \lim_{x \to a} f(x) = 0 = F(a)$$

Likewise, G is continuous on I. Let $x \in I$ and $x > a$. Then F and G are continuous on $[a, x]$ and differentiable on (a, x) and $G' \neq 0$ there (since $F' = f'$ and $G' = g'$). Therefore, by Cauchy's Mean Value Theorem, there is a number y such that $a < y < x$ and

$$\frac{F'(y)}{G'(y)} = \frac{F(x) - F(a)}{G(x) - G(a)} = \frac{F(x)}{G(x)}$$

Here we have used the fact that, by definition, $F(a) = 0$ and $G(a) = 0$. Now, if we let $x \to a^+$, then $y \to a^+$ (since $a < y < x$), so

$$\lim_{x \to a^+} \frac{f(x)}{g(x)} = \lim_{x \to a^+} \frac{F(x)}{G(x)} = \lim_{y \to a^+} \frac{F'(y)}{G'(y)} = \lim_{y \to a^+} \frac{f'(y)}{g'(y)} = L$$

A similar argument shows that the left-hand limit is also L. Therefore

$$\lim_{x \to a} \frac{f(x)}{g(x)} = L$$

This proves l'Hospital's Rule for the case where a is finite.

If a is infinite, we let $t = 1/x$. Then $t \to 0^+$ as $x \to \infty$, so we have

$$\lim_{x \to \infty} \frac{f(x)}{g(x)} = \lim_{t \to 0^+} \frac{f(1/t)}{g(1/t)}$$

$$= \lim_{t \to 0^+} \frac{f'(1/t)(-1/t^2)}{g'(1/t)(-1/t^2)} \qquad \text{(by l'Hospital's Rule for finite } a)$$

$$= \lim_{t \to 0^+} \frac{f'(1/t)}{g'(1/t)} = \lim_{x \to \infty} \frac{f'(x)}{g'(x)} \qquad \blacksquare$$

Section 11.8

In order to prove Theorem 11.8.4, we first need the following results.

Theorem

1. If a power series $\sum c_n x^n$ converges when $x = b$ (where $b \neq 0$), then it converges whenever $|x| < |b|$.

2. If a power series $\sum c_n x^n$ diverges when $x = d$ (where $d \neq 0$), then it diverges whenever $|x| > |d|$.

PROOF OF 1 Suppose that $\Sigma\, c_n b^n$ converges. Then, by Theorem 11.2.6, we have $\lim_{n\to\infty} c_n b^n = 0$. According to Definition 11.1.2 with $\varepsilon = 1$, there is a positive integer N such that $|c_n b^n| < 1$ whenever $n \geq N$. Thus, for $n \geq N$, we have

$$\left| c_n x^n \right| = \left| \frac{c_n b^n x^n}{b^n} \right| = \left| c_n b^n \right| \left| \frac{x}{b} \right|^n < \left| \frac{x}{b} \right|^n$$

If $|x| < |b|$, then $|x/b| < 1$, so $\Sigma\, |x/b|^n$ is a convergent geometric series. Therefore, by the Comparison Test, the series $\Sigma_{n=N}^{\infty} |c_n x^n|$ is convergent. Thus the series $\Sigma\, c_n x^n$ is absolutely convergent and therefore convergent. ∎

PROOF OF 2 Suppose that $\Sigma\, c_n d^n$ diverges. If x is any number such that $|x| > |d|$, then $\Sigma\, c_n x^n$ cannot converge because, by part 1, the convergence of $\Sigma\, c_n x^n$ would imply the convergence of $\Sigma\, c_n d^n$. Therefore $\Sigma\, c_n x^n$ diverges whenever $|x| > |d|$. ∎

Theorem For a power series $\Sigma\, c_n x^n$ there are only three possibilities:

1. The series converges only when $x = 0$.
2. The series converges for all x.
3. There is a positive number R such that the series converges if $|x| < R$ and diverges if $|x| > R$.

PROOF Suppose that neither case 1 nor case 2 is true. Then there are nonzero numbers b and d such that $\Sigma\, c_n x^n$ converges for $x = b$ and diverges for $x = d$. Therefore the set $S = \{x \mid \Sigma\, c_n x^n \text{ converges}\}$ is not empty. By the preceding theorem, the series diverges if $|x| > |d|$, so $|x| \leq |d|$ for all $x \in S$. This says that $|d|$ is an upper bound for the set S. Thus, by the Completeness Axiom (see Section 11.1), S has a least upper bound R. If $|x| > R$, then $x \notin S$, so $\Sigma\, c_n x^n$ diverges. If $|x| < R$, then $|x|$ is not an upper bound for S and so there exists $b \in S$ such that $b > |x|$. Since $b \in S$, $\Sigma\, c_n x^n$ converges, so by the preceding theorem $\Sigma\, c_n x^n$ converges. ∎

---

4 **Theorem** For a power series $\Sigma\, c_n(x - a)^n$ there are only three possibilities:

1. The series converges only when $x = a$.
2. The series converges for all x.
3. There is a positive number R such that the series converges if $|x - a| < R$ and diverges if $|x - a| > R$.

PROOF If we make the change of variable $u = x - a$, then the power series becomes $\Sigma\, c_n u^n$ and we can apply the preceding theorem to this series. In case 3 we have convergence for $|u| < R$ and divergence for $|u| > R$. Thus we have convergence for $|x - a| < R$ and divergence for $|x - a| > R$. ∎

Section 14.3

> **Clairaut's Theorem** Suppose f is defined on a disk D that contains the point (a, b). If the functions f_{xy} and f_{yx} are both continuous on D, then $f_{xy}(a, b) = f_{yx}(a, b)$.

PROOF For small values of h, $h \neq 0$, consider the difference

$$\Delta(h) = [f(a + h, b + h) - f(a + h, b)] - [f(a, b + h) - f(a, b)]$$

Notice that if we let $g(x) = f(x, b + h) - f(x, b)$, then

$$\Delta(h) = g(a + h) - g(a)$$

By the Mean Value Theorem, there is a number c between a and $a + h$ such that

$$g(a + h) - g(a) = g'(c)h = h[f_x(c, b + h) - f_x(c, b)]$$

Applying the Mean Value Theorem again, this time to f_x, we get a number d between b and $b + h$ such that

$$f_x(c, b + h) - f_x(c, b) = f_{xy}(c, d)h$$

Combining these equations, we obtain

$$\Delta(h) = h^2 f_{xy}(c, d)$$

If $h \to 0$, then $(c, d) \to (a, b)$, so the continuity of f_{xy} at (a, b) gives

$$\lim_{h \to 0} \frac{\Delta(h)}{h^2} = \lim_{(c, d) \to (a, b)} f_{xy}(c, d) = f_{xy}(a, b)$$

Similarly, by writing

$$\Delta(h) = [f(a + h, b + h) - f(a, b + h)] - [f(a + h, b) - f(a, b)]$$

and using the Mean Value Theorem twice and the continuity of f_{yx} at (a, b), we obtain

$$\lim_{h \to 0} \frac{\Delta(h)}{h^2} = f_{yx}(a, b)$$

It follows that $f_{xy}(a, b) = f_{yx}(a, b)$. ∎

Section 14.4

> **8** **Theorem** If the partial derivatives f_x and f_y exist near (a, b) and are continuous at (a, b), then f is differentiable at (a, b).

PROOF Let

$$\Delta z = f(a + \Delta x, b + \Delta y) - f(a, b)$$

According to (14.4.7), to prove that f is differentiable at (a, b) we have to show that we can write Δz in the form

$$\Delta z = f_x(a, b)\,\Delta x + f_y(a, b)\,\Delta y + \varepsilon_1\,\Delta x + \varepsilon_2\,\Delta y$$

where ε_1 and $\varepsilon_2 \to 0$ as $(\Delta x, \Delta y) \to (0, 0)$.

Referring to Figure 4, we write

$$\boxed{1} \quad \Delta z = [f(a + \Delta x, b + \Delta y) - f(a, b + \Delta y)] + [f(a, b + \Delta y) - f(a, b)]$$

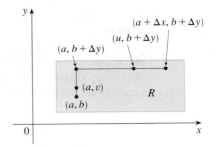

FIGURE 4

Observe that the function of a single variable

$$g(x) = f(x, b + \Delta y)$$

is defined on the interval $[a, a + \Delta x]$ and $g'(x) = f_x(x, b + \Delta y)$. If we apply the Mean Value Theorem to g, we get

$$g(a + \Delta x) - g(a) = g'(u) \, \Delta x$$

where u is some number between a and $a + \Delta x$. In terms of f, this equation becomes

$$f(a + \Delta x, b + \Delta y) - f(a, b + \Delta y) = f_x(u, b + \Delta y) \, \Delta x$$

This gives us an expression for the first part of the right side of Equation 1. For the second part we let $h(y) = f(a, y)$. Then h is a function of a single variable defined on the interval $[b, b + \Delta y]$ and $h'(y) = f_y(a, y)$. A second application of the Mean Value Theorem then gives

$$h(b + \Delta y) - h(b) = h'(v) \, \Delta y$$

where v is some number between b and $b + \Delta y$. In terms of f, this becomes

$$f(a, b + \Delta y) - f(a, b) = f_y(a, v) \, \Delta y$$

We now substitute these expressions into Equation 1 and obtain

$$\Delta z = f_x(u, b + \Delta y) \, \Delta x + f_y(a, v) \, \Delta y$$

$$= f_x(a, b) \, \Delta x + [f_x(u, b + \Delta y) - f_x(a, b)] \, \Delta x + f_y(a, b) \, \Delta y$$

$$+ [f_y(a, v) - f_y(a, b)] \, \Delta y$$

$$= f_x(a, b) \, \Delta x + f_y(a, b) \, \Delta y + \varepsilon_1 \, \Delta x + \varepsilon_2 \, \Delta y$$

where
$$\varepsilon_1 = f_x(u, b + \Delta y) - f_x(a, b)$$
$$\varepsilon_2 = f_y(a, v) - f_y(a, b)$$

Since $(u, b + \Delta y) \to (a, b)$ and $(a, v) \to (a, b)$ as $(\Delta x, \Delta y) \to (0, 0)$ and since f_x and f_y are continuous at (a, b), we see that $\varepsilon_1 \to 0$ and $\varepsilon_2 \to 0$ as $(\Delta x, \Delta y) \to (0, 0)$.

Therefore f is differentiable at (a, b). ∎

G The Logarithm Defined as an Integral

Our treatment of exponential and logarithmic functions until now has relied on our intuition, which is based on numerical and visual evidence. (See Sections 1.4, 1.5, and 3.1.) Here we use the Fundamental Theorem of Calculus to give an alternative treatment that provides a surer footing for these functions.

Instead of starting with b^x and defining $\log_b x$ as its inverse, this time we start by defining $\ln x$ as an integral and then define the exponential function as its inverse. You should bear in mind that we do not use any of our previous definitions and results concerning exponential and logarithmic functions.

■ The Natural Logarithm

We first define $\ln x$ as an integral.

> **1 Definition** The **natural logarithmic function** is the function defined by
> $$\ln x = \int_1^x \frac{1}{t}\, dt \qquad x > 0$$

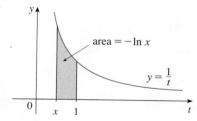

FIGURE 1

The existence of this function depends on the fact that the integral of a continuous function always exists. If $x > 1$, then $\ln x$ can be interpreted geometrically as the area under the hyperbola $y = 1/t$ from $t = 1$ to $t = x$. (See Figure 1.) For $x = 1$, we have

$$\ln 1 = \int_1^1 \frac{1}{t}\, dt = 0$$

For $0 < x < 1$,
$$\ln x = \int_1^x \frac{1}{t}\, dt = -\int_x^1 \frac{1}{t}\, dt < 0$$

and so $\ln x$ is the negative of the area shown in Figure 2.

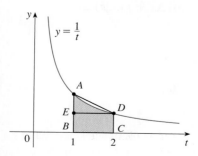

FIGURE 2

EXAMPLE 1

(a) By comparing areas, show that $\frac{1}{2} < \ln 2 < \frac{3}{4}$.

(b) Use the Midpoint Rule with $n = 10$ to estimate the value of $\ln 2$.

SOLUTION

(a) We can interpret $\ln 2$ as the area under the curve $y = 1/t$ from 1 to 2. From Figure 3 we see that this area is larger than the area of rectangle $BCDE$ and smaller than the area of trapezoid $ABCD$. Thus we have

$$\tfrac{1}{2} \cdot 1 < \ln 2 < 1 \cdot \tfrac{1}{2}\left(1 + \tfrac{1}{2}\right)$$

$$\tfrac{1}{2} < \ln 2 < \tfrac{3}{4}$$

(b) If we use the Midpoint Rule with $f(t) = 1/t$, $n = 10$, and $\Delta t = 0.1$, we get

$$\ln 2 = \int_1^2 \frac{1}{t}\, dt \approx (0.1)[f(1.05) + f(1.15) + \cdots + f(1.95)]$$

$$= (0.1)\left(\frac{1}{1.05} + \frac{1}{1.15} + \cdots + \frac{1}{1.95}\right) \approx 0.693$$

FIGURE 3

Notice that the integral that defines ln x is exactly the type of integral discussed in Part 1 of the Fundamental Theorem of Calculus (see Section 5.3). In fact, using that theorem, we have

$$\frac{d}{dx}\int_1^x \frac{1}{t}\,dt = \frac{1}{x}$$

and so

2
$$\frac{d}{dx}(\ln x) = \frac{1}{x}$$

We now use this differentiation rule to prove the following properties of the logarithm function.

3 **Laws of Logarithms** If x and y are positive numbers and r is a rational number, then

1. $\ln(xy) = \ln x + \ln y$ **2.** $\ln\left(\dfrac{x}{y}\right) = \ln x - \ln y$ **3.** $\ln(x^r) = r\ln x$

PROOF

1. Let $f(x) = \ln(ax)$, where a is a positive constant. Then, using Equation 2 and the Chain Rule, we have

$$f'(x) = \frac{1}{ax}\frac{d}{dx}(ax) = \frac{1}{ax}\cdot a = \frac{1}{x}$$

Therefore $f(x)$ and ln x have the same derivative and so they must differ by a constant:

$$\ln(ax) = \ln x + C$$

Putting $x = 1$ in this equation, we get ln $a = \ln 1 + C = 0 + C = C$. Thus

$$\ln(ax) = \ln x + \ln a$$

If we now replace the constant a by any number y, we have

$$\ln(xy) = \ln x + \ln y$$

2. Using Law 1 with $x = 1/y$, we have

$$\ln\frac{1}{y} + \ln y = \ln\left(\frac{1}{y}\cdot y\right) = \ln 1 = 0$$

and so
$$\ln\frac{1}{y} = -\ln y$$

Using Law 1 again, we have

$$\ln\left(\frac{x}{y}\right) = \ln\left(x\cdot\frac{1}{y}\right) = \ln x + \ln\frac{1}{y} = \ln x - \ln y$$

The proof of Law 3 is left as an exercise. ∎

In order to graph $y = \ln x$, we first determine its limits:

$\boxed{4}$

$$\text{(a) } \lim_{x \to \infty} \ln x = \infty \qquad \text{(b) } \lim_{x \to 0^+} \ln x = -\infty$$

PROOF

(a) Using Law 3 with $x = 2$ and $r = n$ (where n is any positive integer), we have $\ln(2^n) = n \ln 2$. Now $\ln 2 > 0$, so this shows that $\ln(2^n) \to \infty$ as $n \to \infty$. But $\ln x$ is an increasing function since its derivative $1/x > 0$. Therefore $\ln x \to \infty$ as $x \to \infty$.

(b) If we let $t = 1/x$, then $t \to \infty$ as $x \to 0^+$. Thus, using (a), we have

$$\lim_{x \to 0^+} \ln x = \lim_{t \to \infty} \ln\left(\frac{1}{t}\right) = \lim_{t \to \infty} (-\ln t) = -\infty$$

■

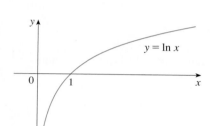

FIGURE 4

If $y = \ln x$, $x > 0$, then

$$\frac{dy}{dx} = \frac{1}{x} > 0 \qquad \text{and} \qquad \frac{d^2y}{dx^2} = -\frac{1}{x^2} < 0$$

which shows that $\ln x$ is increasing and concave downward on $(0, \infty)$. Putting this information together with (4), we draw the graph of $y = \ln x$ in Figure 4.

Since $\ln 1 = 0$ and $\ln x$ is an increasing continuous function that takes on arbitrarily large values, the Intermediate Value Theorem shows that there is a number where $\ln x$ takes on the value 1. (See Figure 5.) This important number is denoted by e.

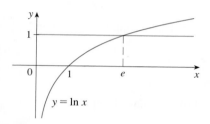

FIGURE 5

$\boxed{5}$ **Definition** e is the number such that $\ln e = 1$.

We will show (in Theorem 19) that this definition is consistent with our previous definition of e.

■ The Natural Exponential Function

Since $\ln$ is an increasing function, it is one-to-one and therefore has an inverse function, which we denote by exp. Thus, according to the definition of an inverse function,

$f^{-1}(x) = y \iff f(y) = x$

$\boxed{6}$

$$\exp(x) = y \iff \ln y = x$$

and the cancellation equations are

$f^{-1}(f(x)) = x$
$f(f^{-1}(x)) = x$

$\boxed{7}$

$$\exp(\ln x) = x \qquad \text{and} \qquad \ln(\exp x) = x$$

In particular, we have

$$\exp(0) = 1 \quad \text{since} \quad \ln 1 = 0$$

$$\exp(1) = e \quad \text{since} \quad \ln e = 1$$

We obtain the graph of $y = \exp x$ by reflecting the graph of $y = \ln x$ about the line

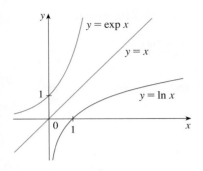

FIGURE 6

$y = x$. (See Figure 6.) The domain of exp is the range of ln, that is, $(-\infty, \infty)$; the range of exp is the domain of ln, that is, $(0, \infty)$.

If r is any rational number, then the third law of logarithms gives

$$\ln(e^r) = r \ln e = r$$

Therefore, by (6), $\exp(r) = e^r$

Thus $\exp(x) = e^x$ whenever x is a rational number. This leads us to define e^x, even for irrational values of x, by the equation

$$e^x = \exp(x)$$

In other words, for the reasons given, we define e^x to be the inverse of the function $\ln x$. In this notation (6) becomes

8
$$e^x = y \iff \ln y = x$$

and the cancellation equations (7) become

9
$$e^{\ln x} = x \qquad x > 0$$

10
$$\ln(e^x) = x \qquad \text{for all } x$$

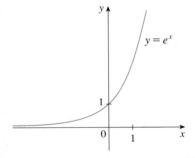

FIGURE 7
The natural exponential function

The natural exponential function $f(x) = e^x$ is one of the most frequently occurring functions in calculus and its applications, so it is important to be familiar with its graph (Figure 7) and its properties (which follow from the fact that it is the inverse of the natural logarithmic function).

> **Properties of the Exponential Function** The exponential function $f(x) = e^x$ is an increasing continuous function with domain $\mathbb{R}$ and range $(0, \infty)$. Thus $e^x > 0$ for all x. Also
> $$\lim_{x \to -\infty} e^x = 0 \qquad \lim_{x \to \infty} e^x = \infty$$
> So the x-axis is a horizontal asymptote of $f(x) = e^x$.

We now verify that f has the other properties expected of an exponential function.

> **11 Laws of Exponents** If x and y are real numbers and r is rational, then
>
> **1.** $e^{x+y} = e^x e^y$ **2.** $e^{x-y} = \dfrac{e^x}{e^y}$ **3.** $(e^x)^r = e^{rx}$

PROOF OF LAW 1 Using the first law of logarithms and Equation 10, we have

$$\ln(e^x e^y) = \ln(e^x) + \ln(e^y) = x + y = \ln(e^{x+y})$$

Since ln is a one-to-one function, it follows that $e^x e^y = e^{x+y}$.

Laws 2 and 3 are proved similarly (see Exercises 6 and 7). As we will soon see, Law 3 actually holds when r is any real number.   ∎

We now prove the differentiation formula for e^x.

12
$$\frac{d}{dx}(e^x) = e^x$$

PROOF The function $y = e^x$ is differentiable because it is the inverse function of $y = \ln x$, which we know is differentiable with nonzero derivative. To find its derivative, we use the inverse function method. Let $y = e^x$. Then $\ln y = x$ and, differentiating this latter equation implicitly with respect to x, we get

$$\frac{1}{y}\frac{dy}{dx} = 1$$

$$\frac{dy}{dx} = y = e^x$$ ∎

■ General Exponential Functions

If $b > 0$ and r is any rational number, then by (9) and (11),

$$b^r = (e^{\ln b})^r = e^{r \ln b}$$

Therefore, even for irrational numbers x, we *define*

13
$$b^x = e^{x \ln b}$$

Thus, for instance,

$$2^{\sqrt{3}} = e^{\sqrt{3} \ln 2} \approx e^{1.20} \approx 3.32$$

The function $f(x) = b^x$ is called the **exponential function with base b**. Notice that b^x is positive for all x because e^x is positive for all x.

Definition 13 allows us to extend one of the laws of logarithms. We already know that $\ln(b^r) = r \ln b$ when r is rational. But if we now let r be *any* real number we have, from Definition 13,

$$\ln b^r = \ln(e^{r \ln b}) = r \ln b$$

Thus

14
$$\ln b^r = r \ln b \qquad \text{for any real number } r$$

The general laws of exponents follow from Definition 13 together with the laws of exponents for e^x.

---

15 **Laws of Exponents** If x and y are real numbers and $a, b > 0$, then

1. $b^{x+y} = b^x b^y$ **2.** $b^{x-y} = b^x / b^y$ **3.** $(b^x)^y = b^{xy}$ **4.** $(ab)^x = a^x b^x$

PROOF

1. Using Definition 13 and the laws of exponents for e^x, we have

$$b^{x+y} = e^{(x+y)\ln b} = e^{x\ln b + y\ln b}$$

$$= e^{x\ln b} e^{y\ln b} = b^x b^y$$

3. Using Equation 14 we obtain

$$(b^x)^y = e^{y\ln(b^x)} = e^{yx\ln b} = e^{xy\ln b} = b^{xy}$$

The remaining proofs are left as exercises. ∎

The differentiation formula for exponential functions is also a consequence of Definition 13:

16
$$\frac{d}{dx}(b^x) = b^x \ln b$$

PROOF

$$\frac{d}{dx}(b^x) = \frac{d}{dx}(e^{x\ln b}) = e^{x\ln b}\frac{d}{dx}(x\ln b) = b^x \ln b \qquad ∎$$

If $b > 1$, then $\ln b > 0$, so $(d/dx)\, b^x = b^x \ln b > 0$, which shows that $y = b^x$ is increasing (see Figure 8). If $0 < b < 1$, then $\ln b < 0$ and so $y = b^x$ is decreasing (see Figure 9).

▪ General Logarithmic Functions

If $b > 0$ and $b \neq 1$, then $f(x) = b^x$ is a one-to-one function. Its inverse function is called the **logarithmic function with base b** and is denoted by $\log_b$. Thus

17
$$\log_b x = y \iff b^y = x$$

In particular, we see that

$$\log_e x = \ln x$$

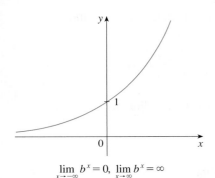

$$\lim_{x \to -\infty} b^x = 0, \ \lim_{x \to \infty} b^x = \infty$$

FIGURE 8 $y = b^x$, $b > 1$

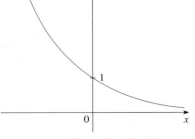

$$\lim_{x \to -\infty} b^x = \infty, \ \lim_{x \to \infty} b^x = 0$$

FIGURE 9 $y = b^x$, $0 < b < 1$

The laws of logarithms are similar to those for the natural logarithm and can be deduced from the laws of exponents (see Exercise 10).

To differentiate $y = \log_b x$, we write the equation as $b^y = x$. From Equation 14 we have $y \ln b = \ln x$, so

$$\log_b x = y = \frac{\ln x}{\ln b}$$

Since $\ln b$ is a constant, we can differentiate as follows:

$$\frac{d}{dx}(\log_b x) = \frac{d}{dx}\frac{\ln x}{\ln b} = \frac{1}{\ln b}\frac{d}{dx}(\ln x) = \frac{1}{x \ln b}$$

[18]
$$\frac{d}{dx}(\log_b x) = \frac{1}{x \ln b}$$

■ The Number e Expressed as a Limit

In this section we defined e as the number such that $\ln e = 1$. The next theorem shows that this is the same as the number e defined in Section 3.1 (see Equation 3.6.5).

[19]
$$e = \lim_{x \to 0}(1 + x)^{1/x}$$

PROOF Let $f(x) = \ln x$. Then $f'(x) = 1/x$, so $f'(1) = 1$. But, by the definition of derivative,

$$f'(1) = \lim_{h \to 0}\frac{f(1 + h) - f(1)}{h} = \lim_{x \to 0}\frac{f(1 + x) - f(1)}{x}$$

$$= \lim_{x \to 0}\frac{\ln(1 + x) - \ln 1}{x} = \lim_{x \to 0}\frac{1}{x}\ln(1 + x) = \lim_{x \to 0}\ln(1 + x)^{1/x}$$

Because $f'(1) = 1$, we have

$$\lim_{x \to 0}\ln(1 + x)^{1/x} = 1$$

Then, by Theorem 2.5.8 and the continuity of the exponential function, we have

$$e = e^1 = e^{\lim_{x \to 0}\ln(1+x)^{1/x}} = \lim_{x \to 0}e^{\ln(1+x)^{1/x}} = \lim_{x \to 0}(1 + x)^{1/x} \qquad ■$$

G   EXERCISES

1. (a) By comparing areas, show that

$$\tfrac{1}{3} < \ln 1.5 < \tfrac{5}{12}$$

(b) Use the Midpoint Rule with $n = 10$ to estimate $\ln 1.5$.

2. Refer to Example 1.

(a) Find the equation of the tangent line to the curve $y = 1/t$ that is parallel to the secant line AD.

(b) Use part (a) to show that $\ln 2 > 0.66$.

3. By comparing areas, show that

$$\frac{1}{2} + \frac{1}{3} + \cdots + \frac{1}{n} < \ln n < 1 + \frac{1}{2} + \frac{1}{3} + \cdots + \frac{1}{n-1}$$

4. (a) By comparing areas, show that $\ln 2 < 1 < \ln 3$.

(b) Deduce that $2 < e < 3$.

5. Prove the third law of logarithms. [*Hint:* Start by showing that both sides of the equation have the same derivative.]

6. Prove the second law of exponents for e^x [see (11)].

7. Prove the third law of exponents for e^x [see (11)].

8. Prove the second law of exponents [see (15)].

9. Prove the fourth law of exponents [see (15)].

10. Deduce the following laws of logarithms from (15):
(a) $\log_b(xy) = \log_b x + \log_b y$
(b) $\log_b(x/y) = \log_b x - \log_b y$
(c) $\log_b(x^y) = y \log_b x$

H   Complex Numbers

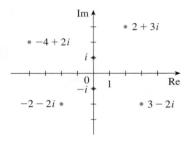

FIGURE 1
Complex numbers as points in the Argand plane

A **complex number** can be represented by an expression of the form $a + bi$, where a and b are real numbers and i is a symbol with the property that $i^2 = -1$. The complex number $a + bi$ can also be represented by the ordered pair (a, b) and plotted as a point in a plane (called the Argand plane) as in Figure 1. Thus the complex number $i = 0 + 1 \cdot i$ is identified with the point $(0, 1)$.

The **real part** of the complex number $a + bi$ is the real number a and the **imaginary part** is the real number b. Thus the real part of $4 - 3i$ is 4 and the imaginary part is -3. Two complex numbers $a + bi$ and $c + di$ are **equal** if $a = c$ and $b = d$, that is, their real parts are equal and their imaginary parts are equal. In the Argand plane the horizontal axis is called the real axis and the vertical axis is called the imaginary axis.

The sum and difference of two complex numbers are defined by adding or subtracting their real parts and their imaginary parts:

$$(a + bi) + (c + di) = (a + c) + (b + d)i$$

$$(a + bi) - (c + di) = (a - c) + (b - d)i$$

For instance,

$$(1 - i) + (4 + 7i) = (1 + 4) + (-1 + 7)i = 5 + 6i$$

The product of complex numbers is defined so that the usual commutative and distributive laws hold:

$$(a + bi)(c + di) = a(c + di) + (bi)(c + di)$$

$$= ac + adi + bci + bdi^2$$

Since $i^2 = -1$, this becomes

$$(a + bi)(c + di) = (ac - bd) + (ad + bc)i$$

EXAMPLE 1

$$(-1 + 3i)(2 - 5i) = (-1)(2 - 5i) + 3i(2 - 5i)$$

$$= -2 + 5i + 6i - 15(-1) = 13 + 11i \qquad \blacksquare$$

Division of complex numbers is much like rationalizing the denominator of a rational expression. For the complex number $z = a + bi$, we define its **complex conjugate** to be $\bar{z} = a - bi$. To find the quotient of two complex numbers we multiply numerator and denominator by the complex conjugate of the denominator.

EXAMPLE 2 Express the number $\dfrac{-1 + 3i}{2 + 5i}$ in the form $a + bi$.

SOLUTION   We multiply numerator and denominator by the complex conjugate of $2 + 5i$, namely, $2 - 5i$, and we take advantage of the result of Example 1:

$$\frac{-1 + 3i}{2 + 5i} = \frac{-1 + 3i}{2 + 5i} \cdot \frac{2 - 5i}{2 - 5i} = \frac{13 + 11i}{2^2 + 5^2} = \frac{13}{29} + \frac{11}{29} i$$ ■

The geometric interpretation of the complex conjugate is shown in Figure 2: $\bar{z}$ is the reflection of z in the real axis. We list some of the properties of the complex conjugate in the following box. The proofs follow from the definition and are requested in Exercise 18.

FIGURE 2

> **Properties of Conjugates**
>
> $$\overline{z + w} = \bar{z} + \bar{w} \qquad \overline{zw} = \bar{z}\,\bar{w} \qquad \overline{z^n} = \bar{z}^n$$

The **modulus**, or **absolute value**, $|z|$ of a complex number $z = a + bi$ is its distance from the origin. From Figure 3 we see that if $z = a + bi$, then

$$|z| = \sqrt{a^2 + b^2}$$

FIGURE 3

Notice that

$$z\bar{z} = (a + bi)(a - bi) = a^2 + abi - abi - b^2 i^2 = a^2 + b^2$$

and so

$$z\bar{z} = |z|^2$$

This explains why the division procedure in Example 2 works in general:

$$\frac{z}{w} = \frac{z\bar{w}}{w\bar{w}} = \frac{z\bar{w}}{|w|^2}$$

Since $i^2 = -1$, we can think of i as a square root of -1. But notice that we also have $(-i)^2 = i^2 = -1$ and so $-i$ is also a square root of -1. We say that i is the **principal square root** of -1 and write $\sqrt{-1} = i$. In general, if c is any positive number, we write

$$\sqrt{-c} = \sqrt{c}\, i$$

With this convention, the usual derivation and formula for the roots of the quadratic equation $ax^2 + bx + c = 0$ are valid even when $b^2 - 4ac < 0$:

$$x = \frac{-b \pm \sqrt{b^2 - 4ac}}{2a}$$

EXAMPLE 3   Find the roots of the equation $x^2 + x + 1 = 0$.

SOLUTION   Using the quadratic formula, we have

$$x = \frac{-1 \pm \sqrt{1^2 - 4 \cdot 1}}{2} = \frac{-1 \pm \sqrt{-3}}{2} = \frac{-1 \pm \sqrt{3}\, i}{2}$$ ■

We observe that the solutions of the equation in Example 3 are complex conjugates of each other. In general, the solutions of any quadratic equation $ax^2 + bx + c = 0$ with real coefficients a, b, and c are always complex conjugates. (If z is real, $\bar{z} = z$, so z is its own conjugate.)

We have seen that if we allow complex numbers as solutions, then every quadratic equation has a solution. More generally, it is true that every polynomial equation

$$a_n x^n + a_{n-1} x^{n-1} + \cdots + a_1 x + a_0 = 0$$

of degree at least one has a solution among the complex numbers. This fact is known as the Fundamental Theorem of Algebra and was proved by Gauss.

Polar Form

We know that any complex number $z = a + bi$ can be considered as a point (a, b) and that any such point can be represented by polar coordinates (r, θ) with $r \geqslant 0$. In fact,

$$a = r \cos \theta \qquad b = r \sin \theta$$

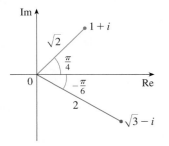

FIGURE 4

as in Figure 4. Therefore we have

$$z = a + bi = (r \cos \theta) + (r \sin \theta)i$$

Thus we can write any complex number z in the form

$$\boxed{z = r(\cos \theta + i \sin \theta)}$$

where $\qquad r = |z| = \sqrt{a^2 + b^2} \qquad$ and $\qquad \tan \theta = \dfrac{b}{a}$

The angle θ is called the **argument** of z and we write $\theta = \arg(z)$. Note that $\arg(z)$ is not unique; any two arguments of z differ by an integer multiple of 2π.

EXAMPLE 4 Write the following numbers in polar form.

(a) $z = 1 + i$ 　　　　　　　　　　　　　(b) $w = \sqrt{3} - i$

SOLUTION

(a) We have $r = |z| = \sqrt{1^2 + 1^2} = \sqrt{2}$ and $\tan \theta = 1$, so we can take $\theta = \pi/4$. Therefore the polar form is

$$z = \sqrt{2}\left(\cos \frac{\pi}{4} + i \sin \frac{\pi}{4}\right)$$

(b) Here we have $r = |w| = \sqrt{3 + 1} = 2$ and $\tan \theta = -1/\sqrt{3}$. Since w lies in the fourth quadrant, we take $\theta = -\pi/6$ and

$$w = 2\left[\cos\left(-\frac{\pi}{6}\right) + i \sin\left(-\frac{\pi}{6}\right)\right]$$

FIGURE 5

The numbers z and w are shown in Figure 5.

The polar form of complex numbers gives insight into multiplication and division. Let

$$z_1 = r_1(\cos\theta_1 + i\sin\theta_1) \qquad z_2 = r_2(\cos\theta_2 + i\sin\theta_2)$$

be two complex numbers written in polar form. Then

$$z_1z_2 = r_1r_2(\cos\theta_1 + i\sin\theta_1)(\cos\theta_2 + i\sin\theta_2)$$
$$= r_1r_2[(\cos\theta_1\cos\theta_2 - \sin\theta_1\sin\theta_2) + i(\sin\theta_1\cos\theta_2 + \cos\theta_1\sin\theta_2)]$$

Therefore, using the addition formulas for cosine and sine, we have

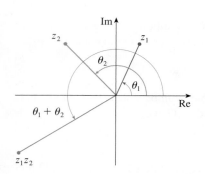

FIGURE 6

$$\boxed{1} \qquad z_1z_2 = r_1r_2[\cos(\theta_1 + \theta_2) + i\sin(\theta_1 + \theta_2)]$$

This formula says that *to multiply two complex numbers we multiply the moduli and add the arguments*. (See Figure 6.)

A similar argument using the subtraction formulas for sine and cosine shows that *to divide two complex numbers we divide the moduli and subtract the arguments*.

$$\frac{z_1}{z_2} = \frac{r_1}{r_2}[\cos(\theta_1 - \theta_2) + i\sin(\theta_1 - \theta_2)] \qquad z_2 \ne 0$$

In particular, taking $z_1 = 1$ and $z_2 = z$ (and therefore $\theta_1 = 0$ and $\theta_2 = \theta$), we have the following, which is illustrated in Figure 7.

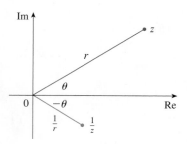

FIGURE 7

$$\text{If} \quad z = r(\cos\theta + i\sin\theta), \quad \text{then} \quad \frac{1}{z} = \frac{1}{r}(\cos\theta - i\sin\theta).$$

EXAMPLE 5 Find the product of the complex numbers $1 + i$ and $\sqrt{3} - i$ in polar form.

SOLUTION From Example 4 we have

$$1 + i = \sqrt{2}\left(\cos\frac{\pi}{4} + i\sin\frac{\pi}{4}\right)$$

and

$$\sqrt{3} - i = 2\left[\cos\left(-\frac{\pi}{6}\right) + i\sin\left(-\frac{\pi}{6}\right)\right]$$

So, by Equation 1,

$$(1 + i)(\sqrt{3} - i) = 2\sqrt{2}\left[\cos\left(\frac{\pi}{4} - \frac{\pi}{6}\right) + i\sin\left(\frac{\pi}{4} - \frac{\pi}{6}\right)\right]$$

$$= 2\sqrt{2}\left(\cos\frac{\pi}{12} + i\sin\frac{\pi}{12}\right)$$

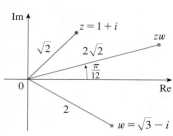

FIGURE 8

This is illustrated in Figure 8. ■

Repeated use of Formula 1 shows how to compute powers of a complex number. If

$$z = r(\cos\theta + i\sin\theta)$$

then

$$z^2 = r^2(\cos 2\theta + i\sin 2\theta)$$

and

$$z^3 = zz^2 = r^3(\cos 3\theta + i\sin 3\theta)$$

In general, we obtain the following result, which is named after the French mathematician Abraham De Moivre (1667–1754).

> **2** **De Moivre's Theorem** If $z = r(\cos\theta + i\sin\theta)$ and n is a positive integer, then
>
> $$z^n = [r(\cos\theta + i\sin\theta)]^n = r^n(\cos n\theta + i\sin n\theta)$$

This says that *to take the nth power of a complex number we take the nth power of the modulus and multiply the argument by n.*

EXAMPLE 6 Find $\left(\frac{1}{2} + \frac{1}{2}i\right)^{10}$.

SOLUTION Since $\frac{1}{2} + \frac{1}{2}i = \frac{1}{2}(1 + i)$, it follows from Example 4(a) that $\frac{1}{2} + \frac{1}{2}i$ has the polar form

$$\frac{1}{2} + \frac{1}{2}i = \frac{\sqrt{2}}{2}\left(\cos\frac{\pi}{4} + i\sin\frac{\pi}{4}\right)$$

So by De Moivre's Theorem,

$$\left(\frac{1}{2} + \frac{1}{2}i\right)^{10} = \left(\frac{\sqrt{2}}{2}\right)^{10}\left(\cos\frac{10\pi}{4} + i\sin\frac{10\pi}{4}\right)$$

$$= \frac{2^5}{2^{10}}\left(\cos\frac{5\pi}{2} + i\sin\frac{5\pi}{2}\right) = \frac{1}{32}i \qquad\blacksquare$$

De Moivre's Theorem can also be used to find the nth roots of complex numbers. An nth root of the complex number z is a complex number w such that

$$w^n = z$$

Writing these two numbers in trigonometric form as

$$w = s(\cos\phi + i\sin\phi) \qquad\text{and}\qquad z = r(\cos\theta + i\sin\theta)$$

and using De Moivre's Theorem, we get

$$s^n(\cos n\phi + i\sin n\phi) = r(\cos\theta + i\sin\theta)$$

The equality of these two complex numbers shows that

$$s^n = r \qquad\text{or}\qquad s = r^{1/n}$$

and

$$\cos n\phi = \cos\theta \qquad\text{and}\qquad \sin n\phi = \sin\theta$$

From the fact that sine and cosine have period 2π, it follows that

$$n\phi = \theta + 2k\pi \qquad \text{or} \qquad \phi = \frac{\theta + 2k\pi}{n}$$

Thus

$$w = r^{1/n}\left[\cos\left(\frac{\theta + 2k\pi}{n}\right) + i\sin\left(\frac{\theta + 2k\pi}{n}\right)\right]$$

Since this expression gives a different value of w for $k = 0, 1, 2, \ldots, n - 1$, we have the following.

---

3 **Roots of a Complex Number** Let $z = r(\cos\theta + i\sin\theta)$ and let n be a positive integer. Then z has the n distinct nth roots

$$w_k = r^{1/n}\left[\cos\left(\frac{\theta + 2k\pi}{n}\right) + i\sin\left(\frac{\theta + 2k\pi}{n}\right)\right]$$

where $k = 0, 1, 2, \ldots, n - 1$.

Notice that each of the nth roots of z has modulus $|w_k| = r^{1/n}$. Thus all the nth roots of z lie on the circle of radius $r^{1/n}$ in the complex plane. Also, since the argument of each successive nth root exceeds the argument of the previous root by $2\pi/n$, we see that the nth roots of z are equally spaced on this circle.

EXAMPLE 7 Find the six sixth roots of $z = -8$ and graph these roots in the complex plane.

SOLUTION In trigonometric form, $z = 8(\cos\pi + i\sin\pi)$. Applying Equation 3 with $n = 6$, we get

$$w_k = 8^{1/6}\left(\cos\frac{\pi + 2k\pi}{6} + i\sin\frac{\pi + 2k\pi}{6}\right)$$

We get the six sixth roots of -8 by taking $k = 0, 1, 2, 3, 4, 5$ in this formula:

$$w_0 = 8^{1/6}\left(\cos\frac{\pi}{6} + i\sin\frac{\pi}{6}\right) = \sqrt{2}\left(\frac{\sqrt{3}}{2} + \frac{1}{2}i\right)$$

$$w_1 = 8^{1/6}\left(\cos\frac{\pi}{2} + i\sin\frac{\pi}{2}\right) = \sqrt{2}\,i$$

$$w_2 = 8^{1/6}\left(\cos\frac{5\pi}{6} + i\sin\frac{5\pi}{6}\right) = \sqrt{2}\left(-\frac{\sqrt{3}}{2} + \frac{1}{2}i\right)$$

$$w_3 = 8^{1/6}\left(\cos\frac{7\pi}{6} + i\sin\frac{7\pi}{6}\right) = \sqrt{2}\left(-\frac{\sqrt{3}}{2} - \frac{1}{2}i\right)$$

$$w_4 = 8^{1/6}\left(\cos\frac{3\pi}{2} + i\sin\frac{3\pi}{2}\right) = -\sqrt{2}\,i$$

$$w_5 = 8^{1/6}\left(\cos\frac{11\pi}{6} + i\sin\frac{11\pi}{6}\right) = \sqrt{2}\left(\frac{\sqrt{3}}{2} - \frac{1}{2}i\right)$$

All these points lie on the circle of radius $\sqrt{2}$ as shown in Figure 9. ∎

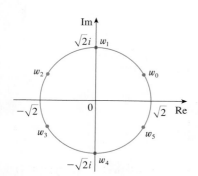

FIGURE 9
The six sixth roots of $z = -8$

■ Complex Exponentials

We also need to give a meaning to the expression e^z when $z = x + iy$ is a complex number. The theory of infinite series as developed in Chapter 11 can be extended to the case where the terms are complex numbers. Using the Taylor series for e^x (11.10.11) as our guide, we define

4    $$e^z = \sum_{n=0}^{\infty} \frac{z^n}{n!} = 1 + z + \frac{z^2}{2!} + \frac{z^3}{3!} + \cdots$$

and it turns out that this complex exponential function has the same properties as the real exponential function. In particular, it is true that

5    $$e^{z_1 + z_2} = e^{z_1} e^{z_2}$$

If we put $z = iy$, where y is a real number, in Equation 4, and use the facts that

$$i^2 = -1, \quad i^3 = i^2 i = -i, \quad i^4 = 1, \quad i^5 = i, \quad \ldots$$

we get $e^{iy} = 1 + iy + \frac{(iy)^2}{2!} + \frac{(iy)^3}{3!} + \frac{(iy)^4}{4!} + \frac{(iy)^5}{5!} + \cdots$

$$= 1 + iy - \frac{y^2}{2!} - i\frac{y^3}{3!} + \frac{y^4}{4!} + i\frac{y^5}{5!} + \cdots$$

$$= \left(1 - \frac{y^2}{2!} + \frac{y^4}{4!} - \frac{y^6}{6!} + \cdots\right) + i\left(y - \frac{y^3}{3!} + \frac{y^5}{5!} - \cdots\right)$$

$$= \cos y + i \sin y$$

Here we have used the Taylor series for $\cos y$ and $\sin y$ (Equations 11.10.16 and 11.10.15). The result is a famous formula called **Euler's formula**:

6    $$e^{iy} = \cos y + i \sin y$$

Combining Euler's formula with Equation 5, we get

7    $$e^{x+iy} = e^x e^{iy} = e^x(\cos y + i \sin y)$$

EXAMPLE 8 Evaluate: (a) $e^{i\pi}$ (b) $e^{-1+i\pi/2}$

SOLUTION

(a) From Euler's equation (6) we have

$$e^{i\pi} = \cos \pi + i \sin \pi = -1 + i(0) = -1$$

We could write the result of Example 8(a) as

$$e^{i\pi} + 1 = 0$$

This equation relates the five most famous numbers in all of mathematics: 0, 1, e, i, and π.

(b) Using Equation 7 we get

$$e^{-1+i\pi/2} = e^{-1}\left(\cos \frac{\pi}{2} + i \sin \frac{\pi}{2}\right) = \frac{1}{e}[0 + i(1)] = \frac{i}{e}$$

■

Finally, we note that Euler's equation provides us with an easier method of proving De Moivre's Theorem:

$$[r(\cos \theta + i \sin \theta)]^n = (re^{i\theta})^n = r^n e^{in\theta} = r^n(\cos n\theta + i \sin n\theta)$$

H   EXERCISES

1–14 Evaluate the expression and write your answer in the form $a + bi$.

1. $(5 - 6i) + (3 + 2i)$

2. $\left(4 - \frac{1}{2}i\right) - \left(9 + \frac{5}{2}i\right)$

3. $(2 + 5i)(4 - i)$

4. $(1 - 2i)(8 - 3i)$

5. $\overline{12 + 7i}$

6. $\overline{2i\left(\frac{1}{2} - i\right)}$

7. $\dfrac{1 + 4i}{3 + 2i}$

8. $\dfrac{3 + 2i}{1 - 4i}$

9. $\dfrac{1}{1 + i}$

10. $\dfrac{3}{4 - 3i}$

11. i^3

12. i^{100}

13. $\sqrt{-25}$

14. $\sqrt{-3}\,\sqrt{-12}$

15–17 Find the complex conjugate and the modulus of the number.

15. $12 - 5i$

16. $-1 + 2\sqrt{2}\,i$

17. $-4i$

18. Prove the following properties of complex numbers.
   (a) $\overline{z + w} = \bar{z} + \bar{w}$ (b) $\overline{zw} = \bar{z}\,\bar{w}$
   (c) $\overline{z^n} = \bar{z}^n$, where n is a positive integer
   [*Hint:* Write $z = a + bi$, $w = c + di$.]

19–24 Find all solutions of the equation.

19. $4x^2 + 9 = 0$

20. $x^4 = 1$

21. $x^2 + 2x + 5 = 0$

22. $2x^2 - 2x + 1 = 0$

23. $z^2 + z + 2 = 0$

24. $z^2 + \frac{1}{2}z + \frac{1}{4} = 0$

25–28 Write the number in polar form with argument between 0 and 2π.

25. $-3 + 3i$

26. $1 - \sqrt{3}\,i$

27. $3 + 4i$

28. $8i$

29–32 Find polar forms for zw, z/w, and $1/z$ by first putting z and w into polar form.

29. $z = \sqrt{3} + i$, $w = 1 + \sqrt{3}\,i$

30. $z = 4\sqrt{3} - 4i$, $w = 8i$

31. $z = 2\sqrt{3} - 2i$, $w = -1 + i$

32. $z = 4\left(\sqrt{3} + i\right)$, $w = -3 - 3i$

33–36 Find the indicated power using De Moivre's Theorem.

33. $(1 + i)^{20}$

34. $\left(1 - \sqrt{3}\,i\right)^5$

35. $\left(2\sqrt{3} + 2i\right)^5$

36. $(1 - i)^8$

37–40 Find the indicated roots. Sketch the roots in the complex plane.

37. The eighth roots of 1

38. The fifth roots of 32

39. The cube roots of i

40. The cube roots of $1 + i$

41–46 Write the number in the form $a + bi$.

41. $e^{i\pi/2}$

42. $e^{2\pi i}$

43. $e^{i\pi/3}$

44. $e^{-i\pi}$

45. $e^{2 + i\pi}$

46. $e^{\pi + i}$

47. Use De Moivre's Theorem with $n = 3$ to express $\cos 3\theta$ and $\sin 3\theta$ in terms of $\cos\theta$ and $\sin\theta$.

48. Use Euler's formula to prove the following formulas for $\cos x$ and $\sin x$:
$$\cos x = \frac{e^{ix} + e^{-ix}}{2} \qquad \sin x = \frac{e^{ix} - e^{-ix}}{2i}$$

49. If $u(x) = f(x) + ig(x)$ is a complex-valued function of a real variable x and the real and imaginary parts $f(x)$ and $g(x)$ are differentiable functions of x, then the derivative of u is defined to be $u'(x) = f'(x) + ig'(x)$. Use this together with Equation 7 to prove that if $F(x) = e^{rx}$, then $F'(x) = re^{rx}$ when $r = a + bi$ is a complex number.

50. (a) If u is a complex-valued function of a real variable, its indefinite integral $\int u(x)\, dx$ is an antiderivative of u. Evaluate
$$\int e^{(1+i)x}\, dx$$

   (b) By considering the real and imaginary parts of the integral in part (a), evaluate the real integrals
$$\int e^x \cos x\, dx \qquad \text{and} \qquad \int e^x \sin x\, dx$$

   (c) Compare with the method used in Example 7.1.4.

| Answers to Odd-Numbered Exercises

CHAPTER 1

EXERCISES 1.1 ■ PAGE 19

1. Yes
3. (a) 3   (b) -0.2 (c) $0, 3$ (d) -0.8
(e) $[-2, 4], [-1, 3]$ (f) $[-2, 1]$
5. $[-85, 115]$ **7.** No
9. Yes, $[-3, 2], [-3, -2) \cup [-1, 3]$
11. (a) $13.8°C$ (b) 1990   (c) 1910, 2005   (d) $[13.5, 14.5]$
13.

15. (a) 500 MW; 730 MW   (b) 4 AM; noon; yes
17.

19.

21.

23. (a)   (b) $74°F$

25. $12, 16, 3a^2 - a + 2, 3a^2 + a + 2, 3a^2 + 5a + 4,$
$6a^2 - 2a + 4, 12a^2 - 2a + 2, 3a^4 - a^2 + 2,$
$9a^4 - 6a^3 + 13a^2 - 4a + 4, 3a^2 + 6ah + 3h^2 - a - h + 2$
27. $-3 - h$ **29.** $-1/(ax)$
31. $(-\infty, -3) \cup (-3, 3) \cup (3, \infty)$ **33.** $(-\infty, \infty)$
35. $(-\infty, 0) \cup (5, \infty)$ **37.** $[0, 4]$

39. $(-\infty, \infty)$

41. $-1, 1, -1$

43. $-2, 0, 4$

45.

47.

49.

51. $f(x) = \frac{5}{2}x - \frac{11}{2}, 1 \le x \le 5$ **53.** $f(x) = 1 - \sqrt{-x}$
55. $f(x) = \begin{cases} -x + 3 & \text{if } 0 \le x \le 3 \\ 2x - 6 & \text{if } 3 < x \le 5 \end{cases}$
57. $A(L) = 10L - L^2, 0 < L < 10$
59. $A(x) = \sqrt{3}x^2/4, x > 0$ **61.** $S(x) = x^2 + (8/x), x > 0$
63. $V(x) = 4x^3 - 64x^2 + 240x, 0 < x < 6$
65. $F(x) = \begin{cases} 15(40 - x) & \text{if } 0 \le x < 40 \\ 0 & \text{if } 40 \le x \le 65 \\ 15(x - 65) & \text{if } x > 65 \end{cases}$

67. (a)   (b) $400, $1900

(c)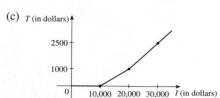

69. f is odd, g is even   **71.** (a) $(-5, 3)$ (b) $(-5, -3)$
73. Odd   **75.** Neither   **77.** Even
79. Even; odd; neither (unless $f = 0$ or $g = 0$)

EXERCISES 1.2 ■ **PAGE 33**

1. (a) Logarithmic   (b) Root   (c) Rational
(d) Polynomial, degree 2   (e) Exponential   (f) Trigonometric
3. (a) h (b) f (c) g
5. $\{x \mid x \neq \pi/2 + 2n\pi\}$, n an integer

7. (a) $y = 2x + b$,
where b is the y-intercept.

(b) $y = mx + 1 - 2m$,
where m is the slope.

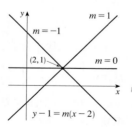

(c) $y = 2x - 3$

9. Their graphs have slope -1.

11. $f(x) = -3x(x + 1)(x - 2)$
13. (a) 8.34, change in mg for every 1 year change
(b) 8.34 mg
15. (a)

(b) $\frac{9}{5}$, change in °F for every 1°C change; 32, Fahrenheit temperature corresponding to 0°C
17. (a) $T = \frac{1}{6}N + \frac{307}{6}$ (b) $\frac{1}{6}$, change in °F for every chirp per minute change   (c) 76°F
19. (a) $P = 0.434d + 15$ (b) 196 ft
21. (a) Cosine   (b) Linear
23. (a)

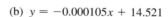

A linear model is appropriate.

(b) $y = -0.000105x + 14.521$

(c) $y = -0.00009979x + 13.951$
(d) About 11.5 per 100 population
(e) About 6%   (f) No

25. (a) See graph in part (b).
(b) $y = 1.88074x + 82.64974$

(c) 182.3 cm

27. (a) A linear model is appropriate. See graph in part (b).
(b) $y = 1116.64x + 60,188.33$

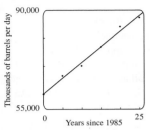

(c) In thousands of barrels per day: 79,171 and 90,338
29. Four times as bright
31. (a) $N = 3.1046A^{0.308}$ (b) 18

EXERCISES 1.3 ■ PAGE 42

1. (a) $y = f(x) + 3$ (b) $y = f(x) - 3$ (c) $y = f(x - 3)$
(d) $y = f(x + 3)$ (e) $y = -f(x)$ (f) $y = f(-x)$
(g) $y = 3f(x)$ (h) $y = \frac{1}{3}f(x)$
3. (a) 3 (b) 1 (c) 4 (d) 5 (e) 2
5. (a)   (b)

(c)   (d)

7. $y = -\sqrt{-x^2 - 5x - 4} - 1$
9.

11.

13.

15.

17.

19.

21.

23.

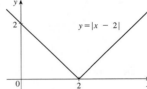

25. $L(t) = 12 + 2 \sin\left[\dfrac{2\pi}{365}(t - 80)\right]$
27. $V(t) = 250 \sin(\pi t/2) + 2250$
29. (a) The portion of the graph of $y = f(x)$ to the right of the y-axis is reflected about the y-axis.
(b) (c)

31. (a) $(f + g)(x) = x^3 + 5x^2 - 1, (-\infty, \infty)$
(b) $(f - g)(x) = x^3 - x^2 + 1, (-\infty, \infty)$
(c) $(fg)(x) = 3x^5 + 6x^4 - x^3 - 2x^2, (-\infty, \infty)$

(d) $(f/g)(x) = \dfrac{x^3 + 2x^2}{3x^2 - 1}, \left\{x \mid x \neq \pm\dfrac{1}{\sqrt{3}}\right\}$

33. (a) $(f \circ g)(x) = 3x^2 + 3x + 5, (-\infty, \infty)$
(b) $(g \circ f)(x) = 9x^2 + 33x + 30, (-\infty, \infty)$
(c) $(f \circ f)(x) = 9x + 20, (-\infty, \infty)$
(d) $(g \circ g)(x) = x^4 + 2x^3 + 2x^2 + x, (-\infty, \infty)$

35. (a) $(f \circ g)(x) = \sqrt{4x - 2}, \left[\frac{1}{2}, \infty\right)$
(b) $(g \circ f)(x) = 4\sqrt{x + 1} - 3, [-1, \infty)$
(c) $(f \circ f)(x) = \sqrt{\sqrt{x + 1} + 1}, [-1, \infty)$
(d) $(g \circ g)(x) = 16x - 15, (-\infty, \infty)$

37. (a) $(f \circ g)(x) = \dfrac{2x^2 + 6x + 5}{(x + 2)(x + 1)}, \{x \mid x \neq -2, -1\}$

(b) $(g \circ f)(x) = \dfrac{x^2 + x + 1}{(x + 1)^2}, \{x \mid x \neq -1, 0\}$

(c) $(f \circ f)(x) = \dfrac{x^4 + 3x^2 + 1}{x(x^2 + 1)}, \{x \mid x \neq 0\}$

(d) $(g \circ g)(x) = \dfrac{2x + 3}{3x + 5}, \left\{x \mid x \neq -2, -\frac{5}{3}\right\}$

39. $(f \circ g \circ h)(x) = 3\sin(x^2) - 2$
41. $(f \circ g \circ h)(x) = \sqrt{x^6 + 4x^3 + 1}$
43. $g(x) = 2x + x^2, f(x) = x^4$
45. $g(x) = \sqrt[3]{x}, f(x) = x/(1 + x)$
47. $g(t) = t^2, f(t) = \sec t \tan t$
49. $h(x) = \sqrt{x}, g(x) = x - 1, f(x) = \sqrt{x}$
51. $h(t) = \cos t, g(t) = \sin t, f(t) = t^2$
53. (a) 4   (b) 3   (c) 0   (d) Does not exist; $f(6) = 6$ is not in the domain of g.   (e) 4   (f) -2
55. (a) $r(t) = 60t$ (b) $(A \circ r)(t) = 3600\pi t^2$; the area of the circle as a function of time
57. (a) $s = \sqrt{d^2 + 36}$ (b) $d = 30t$
(c) $(f \circ g)(t) = \sqrt{900t^2 + 36}$; the distance between the lighthouse and the ship as a function of the time elapsed since noon
59. (a)   (b)

$$V(t) = 120H(t)$$

(c)

$$V(t) = 240H(t - 5)$$

61. Yes; m_1m_2
63. (a) $f(x) = x^2 + 6$ (b) $g(x) = x^2 + x - 1$
65. Yes

EXERCISES 1.4   ▪   PAGE 53

1. (a) 4   (b) $x^{-4/3}$
3. (a) $16b^{12}$ (b) $648y^7$
5. (a) $f(x) = b^x, b > 0$ (b) $\mathbb{R}$ (c) $(0, \infty)$
(d) See Figures 4(c), 4(b), and 4(a), respectively.
7.

All approach 0 as $x \to -\infty$, all pass through $(0, 1)$, and all are increasing. The larger the base, the faster the rate of increase.

9.

The functions with base greater than 1 are increasing and those with base less than 1 are decreasing. The latter are reflections of the former about the y-axis.

11.

13.

15.

17. (a) $y = e^x - 2$ (b) $y = e^{x-2}$ (c) $y = -e^x$
(d) $y = e^{-x}$ (e) $y = -e^{-x}$
19. (a) $(-\infty, -1) \cup (-1, 1) \cup (1, \infty)$ (b) $(-\infty, \infty)$
21. $f(x) = 3 \cdot 2^x$ **27.** At $x \approx 35.8$
29. (a) See graph in part (c).
(b) $f(t) = 36.89301(1.06614)^t$
(c)   About 10.87 h

31. (a) 25 mg   (b) $200 \cdot 2^{-t/5}$ mg
(c) 10.9 mg   (d) 38.2 days

33. 3.5 days

35. $P = 2614.086(1.01693)^t$; 5381 million; 8466 million

EXERCISES 1.5 ▪ PAGE 66

1. (a) See Definition 1.
(b) It must pass the Horizontal Line Test.
3. No   **5.** No   **7.** Yes   **9.** Yes   **11.** No   **13.** No
15. (a) 6   (b) 3   **17.** 0
19. $F = \frac{9}{5}C + 32$; the Fahrenheit temperature as a function of the Celsius temperature; $[-273.15, \infty)$
21. $f^{-1}(x) = \frac{1}{3}(x-1)^2 - \frac{2}{3}, x \geqslant 1$
23. $f^{-1}(x) = \frac{1}{2}(1 + \ln x)$ **25.** $y = e^x - 3$
27. $f^{-1}(x) = \frac{1}{4}(x^2 - 3), x \geqslant 0$

29.

31. (a) $f^{-1}(x) = \sqrt{1 - x^2}, 0 \leqslant x \leqslant 1$; f^{-1} and f are the same function.   (b) Quarter-circle in the first quadrant
33. (a) It's defined as the inverse of the exponential function with base b, that is, $\log_b x = y \iff b^y = x$.
(b) $(0, \infty)$ (c) $\mathbb{R}$ (d) See Figure 11.
35. (a) 5   (b) $\frac{1}{3}$ **37.** (a) 2   (b) $\frac{2}{3}$
39. $\ln 250$ **41.** $\ln \dfrac{\sqrt{x}}{x+1}$

43.

All graphs approach $-\infty$ as $x \to 0^+$, all pass through $(1, 0)$, and all are increasing. The larger the base, the slower the rate of increase.

45. About 1,084,588 mi

47. (a)   (b)

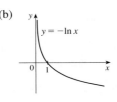

49. (a) $(0, \infty)$; $(-\infty, \infty)$ (b) e^{-2}
(c)

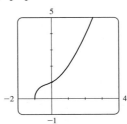

51. (a) $\frac{1}{4}(7 - \ln 6)$ (b) $\frac{1}{3}(e^2 + 10)$
53. (a) $5 + \log_2 3$ or $5 + (\ln 3)/\ln 2$ (b) $\frac{1}{2}(1 + \sqrt{1 + 4e})$
55. (a) $0 < x < 1$ (b) $x > \ln 5$
57. (a) $(\ln 3, \infty)$ (b) $f^{-1}(x) = \ln(e^x + 3)$; $\mathbb{R}$
59. The graph passes the Horizontal Line Test.

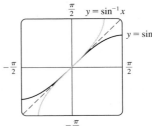

$f^{-1}(x) = -\frac{1}{6}\sqrt[3]{4}\left(\sqrt[3]{D - 27x^2 + 20} - \sqrt[3]{D + 27x^2 - 20} + \sqrt[3]{2}\right)$, where $D = 3\sqrt{3}\sqrt{27x^4 - 40x^2 + 16}$; two of the expressions are complex.

61. (a) $f^{-1}(n) = (3/\ln 2) \ln(n/100)$; the time elapsed when there are n bacteria   (b) After about 26.9 hours
63. (a) π (b) $\pi/6$
65. (a) $\pi/4$ (b) $\pi/2$
67. (a) $5\pi/6$ (b) $\pi/3$
71. $x/\sqrt{1 + x^2}$
73.

The second graph is the reflection of the first graph about the line $y = x$

75. $\left[-\frac{2}{3}, 0\right], [-\pi/2, \pi/2]$
77. (a) $g^{-1}(x) = f^{-1}(x) - c$ (b) $h^{-1}(x) = (1/c)f^{-1}(x)$

CHAPTER 1 REVIEW ▪ PAGE 69

True-False Quiz
1. False   **3.** False   **5.** True   **7.** False   **9.** True
11. False   **13.** False

Exercises
1. (a) 2.7   (b) 2.3, 5.6   (c) $[-6, 6]$ (d) $[-4, 4]$
(e) $[-4, 4]$ (f) No; it fails the Horizontal Line Test.
(g) Odd; its graph is symmetric about the origin.
3. $2a + h - 2$ **5.** $\left(-\infty, \frac{1}{3}\right) \cup \left(\frac{1}{3}, \infty\right)$, $(-\infty, 0) \cup (0, \infty)$
7. $(-6, \infty)$, $\mathbb{R}$

9. (a) Shift the graph 8 units upward.
(b) Shift the graph 8 units to the left.
(c) Stretch the graph vertically by a factor of 2, then shift it 1 unit upward.
(d) Shift the graph 2 units to the right and 2 units downward.
(e) Reflect the graph about the x-axis.
(f) Reflect the graph about the line $y = x$ (assuming f is one-to-one).

11.

$y = (x - 2)^3$

13.

$y = (x - 1)^2 + 1$

15.

$y = -\cos 2x$

17. (a) Neither   (b) Odd   (c) Even   (d) Neither
19. (a) $(f \circ g)(x) = \ln(x^2 - 9), (-\infty, -3) \cup (3, \infty)$
(b) $(g \circ f)(x) = (\ln x)^2 - 9, (0, \infty)$
(c) $(f \circ f)(x) = \ln \ln x, (1, \infty)$
(d) $(g \circ g)(x) = (x^2 - 9)^2 - 9, (-\infty, \infty)$

21. $y = 0.2493x - 423.4818$; about 77.6 years
23. 1   **25.** (a) 9   (b) 2   (c) $1/\sqrt{3}$ (d) $\frac{3}{5}$
27. (a) $\frac{1}{16}$ g   (b) $m(t) = 2^{-t/4}$
(c) $t(m) = -4 \log_2 m$; the time elapsed when there are m grams of ^{100}Pd
(d) About 26.6 days

PRINCIPLES OF PROBLEM SOLVING ▪ PAGE 76

1. $a = 4\sqrt{h^2 - 16}/h$, where a is the length of the altitude and h is the length of the hypotenuse
3. $-\frac{7}{3}, 9$

5.

7.

9. (a)

$f(x) = \max\{x, 1/x\}$

(b)

$f(x) = \max\{\sin x, \cos x\}$

(c)

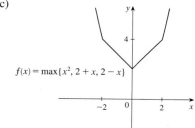

$f(x) = \max\{x^2, 2 + x, 2 - x\}$

11. 5   **13.** $x \in \left[-1, 1 - \sqrt{3}\right) \cup \left(1 + \sqrt{3}, 3\right]$
15. 40 mi/h   **19.** $f_n(x) = x^{2^{n+1}}$

CHAPTER 2

EXERCISES 2.1 ▪ PAGE 82

1. (a) $-44.4, -38.8, -27.8, -22.2, -16.\overline{6}$
(b) -33.3 (c) $-33\frac{1}{3}$
3. (a) (i) 2   (ii) 1.111111   (iii) 1.010101   (iv) 1.001001
(v) 0.666667   (vi) 0.909091   (vii) 0.990099
(viii) 0.999001   (b) 1   (c) $y = x - 3$
5. (a) (i) -32 ft/s   (ii) -25.6 ft/s   (iii) -24.8 ft/s
(iv) -24.16 ft/s   (b) -24 ft/s
7. (a) (i) 29.3 ft/s   (ii) 32.7 ft/s   (iii) 45.6 ft/s
(iv) 48.75 ft/s   (b) 29.7 ft/s
9. (a) 0, 1.7321, $-1.0847, -2.7433, 4.3301, -2.8173, 0,$
$-2.1651, -2.6061, -5, 3.4202$; no   (c) -31.4

EXERCISES 2.2 ▪ PAGE 92

1. Yes
3. (a) $\lim_{x \to -3} f(x) = \infty$ means that the values of $f(x)$ can be made arbitrarily large (as large as we please) by taking x sufficiently close to -3 (but not equal to -3).
(b) $\lim_{x \to 4^+} f(x) = -\infty$ means that the values of $f(x)$ can be made arbitrarily large negative by taking x sufficiently close to 4 through values larger than 4.

5. (a) 2   (b) 1   (c) 4   (d) Does not exist   (e) 3
7. (a) -1 (b) -2 (c) Does not exist   (d) 2   (e) 0
(f) Does not exist   (g) 1   (h) 3
9. (a) $-\infty$ (b) ∞ (c) ∞ (d) $-\infty$ (e) ∞
(f) $x = -7, x = -3, x = 0, x = 6$
11. $\lim\limits_{x \to a} f(x)$ exists for all a except $a = -1$.
13. (a) 1   (b) 0   (c) Does not exist
15.   **17.**

   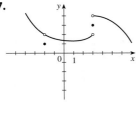

19. $\frac{1}{2}$ **21.** 5   **23.** 0.25   **25.** 1.5   **27.** 1
29. (a) -1.5 **31.** ∞ **33.** ∞ **35.** $-\infty$ **37.** $-\infty$
39. $-\infty$ **41.** ∞ **43.** $-\infty$ **45.** $-\infty; \infty$
47. (a) 2.71828   (b)

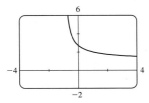

49. (a) 0.998000, 0.638259, 0.358484, 0.158680, 0.038851,
0.008928, 0.001465; 0
(b) 0.000572, -0.000614, -0.000907, -0.000978, -0.000993,
-0.001000; -0.001

51. No matter how many times we zoom in toward the origin, the
graph appears to consist of almost-vertical lines. This indicates
more and more frequent oscillations as $x \to 0$.
53. $x \approx \pm 0.90, \pm 2.24; x = \pm \sin^{-1}(\pi/4), \pm(\pi - \sin^{-1}(\pi/4))$
55. (a) 6   (b) Within 0.0649 of 1

EXERCISES 2.3 ■ **PAGE 102**
1. (a) -6 (b) -8 (c) 2   (d) -6
(e) Does not exist   (f) 0
3. 105   **5.** $\frac{7}{8}$ **7.** 390   **9.** $\frac{3}{2}$ **11.** 4
13. Does not exist   **15.** $\frac{6}{5}$ **17.** -10 **19.** $\frac{1}{12}$
21. $\frac{1}{6}$ **23.** $-\frac{1}{9}$ **25.** 1   **27.** $\frac{1}{128}$ **29.** $-\frac{1}{2}$
31. $3x^2$ **33.** (a), (b) $\frac{2}{3}$ **37.** 7   **41.** 6   **43.** -4
45. Does not exist

47. (a)   (b) (i) 1
                              (ii) -1
                                                (iii) Does not exist
                                                (iv) 1

49. (a) (i) 5   (ii) -5 (b) Does not exist
(c)

51. 7
53. (a) (i) -2 (ii) Does not exist   (iii) -3
(b) (i) $n - 1$ (ii) n (c) a is not an integer.
59. 8   **65.** 15; -1

EXERCISES 2.4 ■ **PAGE 113**
1. 0.1 (or any smaller positive number)
3. 1.44 (or any smaller positive number)
5. 0.0906 (or any smaller positive number)
7. 0.011 (or any smaller positive number)
9. (a) 0.01 (or any smaller positive number)

(b) $\lim\limits_{x \to 2^+} \dfrac{1}{\ln(x-1)} = \infty$

11. (a) $\sqrt{1000/\pi}$ cm   (b) Within approximately 0.0445 cm
(c) Radius; area; $\sqrt{1000/\pi}$; 1000; 5; ≈ 0.0445
13. (a) 0.025   (b) 0.0025
35. (a) 0.093   (b) $\delta = (B^{2/3} - 12)/(6B^{1/3}) - 1$, where
$B = 216 + 108\varepsilon + 12\sqrt{336 + 324\varepsilon + 81\varepsilon^2}$
41. Within 0.1

EXERCISES 2.5 ■ **PAGE 124**
1. $\lim_{x \to 4} f(x) = f(4)$
3. (a) $-4, -2, 2, 4$; $f(-4)$ is not defined and $\lim\limits_{x \to a} f(x)$ does not
exist for $a = -2, 2$, and 4
(b) -4, neither; -2, left; 2, right; 4, right

5.   **7.**

9. (a)

17. $f(-2)$ is undefined.

$y = \dfrac{1}{x+2}$

$x = -2$

19. $\lim\limits_{x \to -1} f(x)$ does not exist.   **21.** $\lim\limits_{x \to 0} f(x) \neq f(0)$

$y = x + 3$

$y = 2^x$

23. Define $f(2) = 3$.   **25.** $(-\infty, \infty)$
27. $\left(-\infty, \sqrt[3]{2}\right) \cup \left(\sqrt[3]{2}, \infty\right)$ **29.** $[-1, 0]$
31. $(-\infty, -1] \cup (0, \infty)$
33. $x = 0$

35. 8   **37.** $\ln 2$
41. -1, right   **43.** 0, right; 1, left

$(-1, 1)$

$(1, 1)$

$(-1, -1)$

$(0, 2)$ $(1, e)$
$(0, 1)$ $(1, 1)$

45. $\frac{2}{3}$ **47.** 4
49. (a) $g(x) = x^3 + x^2 + x + 1$ (b) $g(x) = x^2 + x$
57. (b) $(0.86, 0.87)$ **59.** (b) 70.347 **67.** None
69. Yes

EXERCISES 2.6 ■ PAGE 137

1. (a) As x becomes large, $f(x)$ approaches 5.
(b) As x becomes large negative, $f(x)$ approaches 3.
3. (a) -2 (b) 2   (c) ∞ (d) $-\infty$
(e) $x = 1, x = 3, y = -2, y = 2$

5.

$y = 5$

$y = -5$

7.

$x = 2$

9.

$y = 3$

$x = 4$

11. 0   **13.** $\frac{2}{5}$ **15.** $\frac{3}{2}$ **17.** 0   **19.** -1 **21.** 4
23. -2 **25.** $\frac{\sqrt{3}}{4}$ **27.** $\frac{1}{6}$ **29.** $\frac{1}{2}(a-b)$ **31.** ∞
33. $-\infty$ **35.** $\pi/2$ **37.** $-\frac{1}{2}$ **39.** 0   **41.** ∞
43. (a) (i) 0   (ii) $-\infty$ (iii) ∞ (b) ∞
(c)

$y = \dfrac{x}{\ln x}$

$x = 1$

45. (a), (b) $-\frac{1}{2}$ **47.** $y = 4, x = -3$
49. $y = 2; x = -2, x = 1$ **51.** $x = 5$ **53.** $y = 3$
55. (a) 0   (b) $\pm\infty$
57. $f(x) = \dfrac{2-x}{x^2(x-3)}$ **59.** (a) $\frac{5}{4}$ (b) 5
61. $-\infty, -\infty$ **63.** $-\infty, \infty$

65. (a) 0   (b) An infinite number of times

67. 5

69. (a) v^* (b)   1.2 ≈ 0.47 s

71. $N \geqslant 15$ **73.** $N \leqslant -9, N \leqslant -19$
75. (a) $x > 100$

EXERCISES 2.7 ■ PAGE 148

1. (a) $\dfrac{f(x) - f(3)}{x - 3}$ (b) $\displaystyle\lim_{x \to 3} \dfrac{f(x) - f(3)}{x - 3}$

3. (a) 2   (b) $y = 2x + 1$ (c)

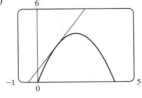

5. $y = -8x + 12$ **7.** $y = \frac{1}{2}x + \frac{1}{2}$
9. (a) $8a - 6a^2$ (b) $y = 2x + 3, y = -8x + 19$
(c)

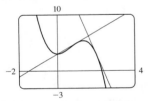

11. (a) Right: $0 < t < 1$ and $4 < t < 6$; left: $2 < t < 3$;
standing still: $1 < t < 2$ and $3 < t < 4$
(b)

13. -24 ft/s
15. $-2/a^3$ m/s; -2 m/s; $-\frac{1}{4}$ m/s; $-\frac{2}{27}$ m/s
17. $g'(0), 0, g'(4), g'(2), g'(-2)$
19. (a) 26   (b) No   (c) Yes
21. $f(2) = 3; f'(2) = 4$
23.

25.

27. $y = 3x - 1$

29. (a) $-\frac{3}{5}; y = -\frac{3}{5}x + \frac{16}{5}$ (b)

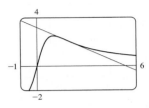

31. $6a - 4$ **33.** $\dfrac{5}{(a + 3)^2}$ **35.** $-\dfrac{1}{\sqrt{1 - 2a}}$
37. $f(x) = \sqrt{x}, a = 9$ **39.** $f(x) = x^6, a = 2$
41. $f(x) = \cos x, a = \pi$ or $f(x) = \cos(\pi + x), a = 0$
43. 32 m/s; 32 m/s

45.   Greater (in magnitude)

47. (a) In (mg/mL)/h: (i) -0.15 (ii) -0.12
(iii) -0.12 (iv) -0.11 (b) -0.12 (mg/mL)/h; After
2 hours, the BAC is decreasing at a rate of 0.12 (mg/mL)/h.

49. (a) 1169.6 thousands of barrels of oil per day per year; oil
consumption rose by an average of 1169.6 thousands of barrels of
oil per day each year from 1990 to 2005.
(b) 1397.8 thousands of barrels of oil per day per year

51. (a) (i) $20.25/unit   (ii) $20.05/unit   (b) $20/unit

53. (a) The rate at which the cost is changing per ounce of gold
produced; dollars per ounce
(b) When the 800th ounce of gold is produced, the cost of
production is $17/oz.
(c) Decrease in the short term; increase in the long term

55. (a) The rate at which daily heating costs change with respect
to temperature when the temperature is 58°F; dollars/°F
(b) Negative; If the outside temperature increases, the building
should require less heating.

57. (a) The rate at which the oxygen solubility changes with
respect to the water temperature; (mg/L)/°C
(b) $S'(16) \approx -0.25$; as the temperature increases past 16°C,
the oxygen solubility is decreasing at a rate of 0.25 (mg/L)/°C.

59. Does not exist

61. (a)

Slope appears to be 1.

(b)

Yes

(c)

Yes; 0

EXERCISES 2.8 ■ PAGE 160

1. (a) -0.2 (b) 0 (c) 1 (d) 2
(e) 1 (f) 0 (g) -0.2

3. (a) II   (b) IV   (c) I   (d) III

5.   **7.**

9.   **11.**

13. (a) The instantaneous rate of change of percentage of full capacity with respect to elapsed time in hours

(b)

The rate of change of percentage of full capacity is decreasing and approaching 0.

15.

1963 to 1971

17.

$f'(x) = e^x$

19. (a) $0, 1, 2, 4$ (b) $-1, -2, -4$ (c) $f'(x) = 2x$
21. $f'(x) = 3$, $\mathbb{R}$, $\mathbb{R}$ **23.** $f'(t) = 5t + 6$, $\mathbb{R}$, $\mathbb{R}$
25. $f'(x) = 2x - 6x^2$, $\mathbb{R}$, $\mathbb{R}$
27. $g'(x) = -\dfrac{1}{2\sqrt{9-x}}$, $(-\infty, 9]$, $(-\infty, 9)$
29. $G'(t) = \dfrac{-7}{(3+t)^2}$, $(-\infty, -3) \cup (-3, \infty)$, $(-\infty, -3) \cup (-3, \infty)$
31. $f'(x) = 4x^3$, $\mathbb{R}$, $\mathbb{R}$ **33.** (a) $f'(x) = 4x^3 + 2$
35. (a) The rate at which the unemployment rate is changing, in percent unemployed per year
(b)

t	$U'(t)$	t	$U'(t)$
2003	-0.50	2008	2.35
2004	-0.45	2009	1.90
2005	-0.45	2010	-0.20
2006	-0.25	2011	-0.75
2007	0.60	2012	-0.80

37.

t	14	21	28	35	42	49
$H'(t)$	$\frac{13}{7}$	$\frac{23}{14}$	$\frac{9}{7}$	1	$\frac{11}{14}$	$\frac{5}{7}$

39. (a) The rate at which the percentage of electrical power produced by solar panels is changing, in percentage points per year. **(b)** On January 1, 2002, the percentage of electrical power produced by solar panels was increasing at a rate of 3.5 percentage points per year.
41. -4 (corner); 0 (discontinuity)
43. 1 (not defined); 5 (vertical tangent)
45.

Differentiable at -1; not differentiable at 0

47. $f''(1)$ **49.** $a = f, b = f', c = f''$
51. a = acceleration, b = velocity, c = position
53. $6x + 2$; 6

55.

$f'(x) = 4x - 3x^2$,
$f''(x) = 4 - 6x$,
$f'''(x) = -6$,
$f^{(4)}(x) = 0$

57. (a) $\frac{1}{3}a^{-2/3}$

59. $f'(x) = \begin{cases} -1 & \text{if } x < 6 \\ 1 & \text{if } x > 6 \end{cases}$

or $f'(x) = \dfrac{x - 6}{|x - 6|}$

61. (a)

(b) All x
(c) $f'(x) = 2|x|$

65. (a)

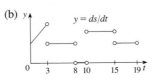

(b)

67. $63°$

CHAPTER 2 REVIEW ■ **PAGE 166**

True-False Quiz
1. False   **3.** True   **5.** True   **7.** False   **9.** True
11. True   **13.** True   **15.** False   **17.** True
19. True   **21.** False   **23.** False   **25.** True

Exercises
1. (a) (i) 3 **(ii)** 0 **(iii)** Does not exist   **(iv)** 2
(v) ∞ **(vi)** $-\infty$ **(vii)** 4 **(viii)** -1
(b) $y = 4$, $y = -1$ **(c)** $x = 0$, $x = 2$ **(d)** $-3, 0, 2, 4$
3. 1 **5.** $\frac{3}{2}$ **7.** 3 **9.** ∞ **11.** $\frac{4}{7}$ **13.** $\frac{1}{2}$
15. $-\infty$ **17.** 2 **19.** $\pi/2$ **21.** $x = 0, y = 0$ **23.** 1
29. (a) (i) 3 **(ii)** 0 **(iii)** Does not exist
(iv) 0 **(v)** 0 **(vi)** 0
(b) At 0 and 3 **(c)**

31. $\mathbb{R}$ **35. (a)** -8 **(b)** $y = -8x + 17$
37. (a) (i) 3 m/s   **(ii)** 2.75 m/s   **(iii)** 2.625 m/s
(iv) 2.525 m/s   **(b)** 2.5 m/s
39. (a) 10 **(b)** $y = 10x - 16$
(c)

41. (a) The rate at which the cost changes with respect to the interest rate; dollars/(percent per year)
(b) As the interest rate increases past 10%, the cost is increasing at a rate of $\$1200$/(percent per year).
(c) Always positive
43.

45. (a) $f'(x) = -\frac{5}{2}(3 - 5x)^{-1/2}$ **(b)** $\left(-\infty, \frac{3}{5}\right], \left(-\infty, \frac{3}{5}\right)$
(c)

47. -4 (discontinuity), -1 (corner), 2 (discontinuity), 5 (vertical tangent)

49.

51. The rate at which the number of US $20 bills in circulation is changing with respect to time; 0.156 billion bills per year
53. 0

PROBLEMS PLUS ▪ PAGE 169

1. $\frac{2}{3}$ **3.** -4 **5.** (a) Does not exist   (b) 1
7. $a = \frac{1}{2} \pm \frac{1}{2}\sqrt{5}$ **9.** -5 **11.** (b) Yes   (c) Yes; no
13. (a) 0   (b) 1   (c) $f'(x) = x^2 + 1$

CHAPTER 3

EXERCISES 3.1 ▪ PAGE 180

1. (a) e is the number such that $\lim\limits_{h \to 0} \dfrac{e^h - 1}{h} = 1$.
(b) 0.99, 1.03; $2.7 < e < 2.8$
3. $f'(x) = 0$ **5.** $f'(x) = 5.2$ **7.** $f'(t) = 6t^2 - 6t - 4$
9. $g'(x) = 2x - 6x^2$ **11.** $g'(t) = -\frac{3}{2}t^{-7/4}$
13. $F'(r) = -15/r^4$ **15.** $R'(a) = 18a + 6$
17. $S'(p) = \frac{1}{2}p^{-1/2} - 1$ **19.** $y' = 3e^x - \frac{4}{3}x^{-4/3}$
21. $h'(u) = 3Au^2 + 2Bu + C$
23. $y' = \frac{3}{2}\sqrt{x} + \dfrac{2}{\sqrt{x}} - \dfrac{3}{2x\sqrt{x}}$ **25.** $j'(x) = 2.4x^{1.4}$
27. $G'(q) = -2q^{-2} - 2q^{-3}$ **29.** $f'(v) = -\frac{2}{3}v^{-5/3} - 2e^v$
31. $z' = -10A/y^{11} + Be^y$ **33.** $y = 4x - 1$
35. $y = \frac{1}{2}x + 2$
37. Tangent: $y = 2x + 2$; normal: $y = -\frac{1}{2}x + 2$
39. $y = 3x - 1$ **41.** $f'(x) = 4x^3 - 6x^2 + 2x$
43. (a)   (c) $4x^3 - 9x^2 - 12x + 7$

45. $f'(x) = 0.005x^4 - 0.06x^2$, $f''(x) = 0.02x^3 - 0.12x$
47. $f'(x) = 2 - \frac{15}{4}x^{-1/4}$, $f''(x) = \frac{15}{16}x^{-5/4}$
49. (a) $v(t) = 3t^2 - 3$, $a(t) = 6t$ (b) 12 m/s²
(c) $a(1) = 6$ m/s²
51. 1.718; instantaneous rate of change of the length with respect to the age at 12 yr
53. (a) $V = 5.3/P$
(b) -0.00212; instantaneous rate of change of the volume with respect to the pressure at 25°C; m³/kPa
55. $(-2, 21)$, $(1, -6)$ **59.** $y = 3x - 3$, $y = 3x - 7$
61. $y = -2x + 3$
63. $(\pm 2, 4)$ **67.** $P(x) = x^2 - x + 3$

69. $y = \frac{3}{16}x^3 - \frac{9}{4}x + 3$
71. No

73. (a) Not differentiable at 3 or -3
$$f'(x) = \begin{cases} 2x & \text{if } |x| > 3 \\ -2x & \text{if } |x| < 3 \end{cases}$$
(b)

75. $y = 2x^2 - x$ **77.** $a = -\frac{1}{2}$, $b = 2$ **79.** $-\frac{1}{3}$
81. $m = 4$, $b = -4$ **83.** 1000   **85.** 3; 1

EXERCISES 3.2 ▪ PAGE 188

1. $1 - 2x + 6x^2 - 8x^3$ **3.** $f'(x) = e^x(3x^2 + x - 5)$
5. $y' = \dfrac{1 - x}{e^x}$ **7.** $g'(x) = \dfrac{10}{(3 - 4x)^2}$ **9.** $H'(u) = 2u - 1$
11. $F'(y) = 5 + \dfrac{14}{y^2} + \dfrac{9}{y^4}$ **13.** $y' = \dfrac{x(-x^3 - 3x - 2)}{(x^3 - 1)^2}$
15. $y' = \dfrac{t^4 - 8t^3 + 6t^2 + 9}{(t^2 - 4t + 3)^2}$
17. $y' = e^p\left(1 + \frac{3}{2}\sqrt{p} + p + p\sqrt{p}\right)$ **19.** $y' = \dfrac{3 - 2\sqrt{s}}{2s^{5/2}}$
21. $f'(t) = \dfrac{-2t - 3}{3t^{2/3}(t - 3)^2}$ **23.** $f'(x) = \dfrac{xe^x(x^3 + 2e^x)}{(x^2 + e^x)^2}$
25. $f'(x) = \dfrac{2cx}{(x^2 + c)^2}$
27. $(x^3 + 3x^2 + 1)e^x$; $(x^3 + 6x^2 + 6x + 1)e^x$
29. $\dfrac{x(2 + 2e^x - xe^x)}{(1 + e^x)^2}$;
$$\dfrac{2 + 4e^x - 4xe^x - x^2e^x + 2e^{2x} - 4xe^{2x} + x^2e^{2x}}{(1 + e^x)^3}$$
31. $y = \frac{2}{3}x - \frac{2}{3}$ **33.** $y = 2x$; $y = -\frac{1}{2}x$
35. (a) $y = \frac{1}{2}x + 1$ (b)

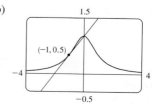

37. (a) $e^x(x^3 + 3x^2 - x - 1)$
(b)

39. (a) $f'(x) = \dfrac{4x}{(x^2 + 1)^2}$; $f''(x) = \dfrac{4(1 - 3x^2)}{(x^2 + 1)^3}$
(b)

41. $\frac{1}{4}$ **43.** (a) -16 (b) $-\frac{20}{9}$ (c) 20 **45.** 7
47. $y = -2x + 18$ **49.** (a) 0 (b) $-\frac{2}{3}$
51. (a) $y' = xg'(x) + g(x)$ (b) $y' = \dfrac{g(x) - xg'(x)}{[g(x)]^2}$
(c) $y' = \dfrac{xg'(x) - g(x)}{x^2}$
53. Two, $\left(-2 \pm \sqrt{3}, \frac{1}{2}(1 \mp \sqrt{3})\right)$ **55.** 1
57. \$1.627 billion/year
59. $\dfrac{0.0021}{(0.015 + [S])^2}$;
The rate of change of the rate of an enzymatic reaction with respect to the concentration of a substrate S.
61. (c) $3e^{3x}$
63. $f'(x) = (x^2 + 2x)e^x$, $f''(x) = (x^2 + 4x + 2)e^x$,
$f'''(x) = (x^2 + 6x + 6)e^x$, $f^{(4)}(x) = (x^2 + 8x + 12)e^x$,
$f^{(5)}(x) = (x^2 + 10x + 20)e^x$; $f^{(n)}(x) = [x^2 + 2nx + n(n - 1)]e^x$

EXERCISES 3.3 ■ PAGE 196

1. $f'(x) = x^2 \cos x + 2x \sin x$ **3.** $f'(x) = e^x(\cos x - \sin x)$
5. $y' = \sec\theta\,(\sec^2\theta + \tan^2\theta)$
7. $y' = -c \sin t + t(t \cos t + 2 \sin t)$
9. $y' = \dfrac{2 - \tan x + x \sec^2 x}{(2 - \tan x)^2}$ **11.** $f'(\theta) = \dfrac{1}{1 + \cos\theta}$
13. $y' = \dfrac{(t^2 + t)\cos t + \sin t}{(1 + t)^2}$
15. $f'(\theta) = \frac{1}{2} \sin 2\theta + \theta \cos 2\theta$
21. $y = x + 1$ **23.** $y = x - \pi - 1$
25. (a) $y = 2x$ (b) $\frac{3\pi}{2}$

27. (a) $\sec x \tan x - 1$
29. $\theta \cos\theta + \sin\theta$; $2 \cos\theta - \theta \sin\theta$

31. (a) $f'(x) = (1 + \tan x)/\sec x$ (b) $f'(x) = \cos x + \sin x$
33. $(2n + 1)\pi \pm \frac{1}{3}\pi$, n an integer
35. (a) $v(t) = 8 \cos t$, $a(t) = -8 \sin t$
(b) $4\sqrt{3}$, -4, $-4\sqrt{3}$; to the left
37. 5 ft/rad   **39.** $\frac{5}{3}$ **41.** 3 **43.** $-\frac{3}{4}$
45. $\frac{1}{2}$ **47.** $-\frac{1}{4}$ **49.** $-\sqrt{2}$ **51.** $-\cos x$
53. $A = -\frac{3}{10}$, $B = -\frac{1}{10}$
55. (a) $\sec^2 x = \dfrac{1}{\cos^2 x}$ (b) $\sec x \tan x = \dfrac{\sin x}{\cos^2 x}$
(c) $\cos x - \sin x = \dfrac{\cot x - 1}{\csc x}$
57. 1

EXERCISES 3.4 ■ PAGE 204

1. $\dfrac{4}{3\sqrt[3]{(1 + 4x)^2}}$ **3.** $\pi \sec^2 \pi x$ **5.** $\dfrac{e^{\sqrt{x}}}{2\sqrt{x}}$
7. $F'(x) = 24x^{11}(5x^3 + 2)^3(5x^3 + 1)$
9. $f'(x) = \dfrac{5}{2\sqrt{5x + 1}}$ **11.** $f'(\theta) = -2\theta \sin(\theta^2)$
13. $y' = xe^{-3x}(2 - 3x)$ **15.** $f'(t) = e^{at}(b \cos bt + a \sin bt)$
17. $f'(x) = (2x - 3)^3(x^2 + x + 1)^4(28x^2 - 12x - 7)$
19. $h'(t) = \frac{2}{3}(t + 1)^{-1/3}(2t^2 - 1)^2(20t^2 + 18t - 1)$
21. $y' = \dfrac{1}{2\sqrt{x}\,(x + 1)^{3/2}}$ **23.** $y' = (\sec^2\theta)\,e^{\tan\theta}$
25. $g'(u) = \dfrac{48u^2(u^3 - 1)^7}{(u^3 + 1)^9}$ **27.** $r'(t) = \dfrac{(\ln 10)10^{2\sqrt{t}}}{\sqrt{t}}$
29. $H'(r) = \dfrac{2(r^2 - 1)^2(r^2 + 3r + 5)}{(2r + 1)^6}$
31. $F'(t) = e^{t \sin 2t}(2t \cos 2t + \sin 2t)$
33. $G'(x) = -C(\ln 4)\dfrac{4^{C/x}}{x^2}$
35. $y' = \dfrac{4e^{2x}}{(1 + e^{2x})^2} \sin \dfrac{1 - e^{2x}}{1 + e^{2x}}$
37. $y' = -2 \cos\theta \cot(\sin\theta) \csc^2(\sin\theta)$
39. $f'(t) = -\sec^2(\sec(\cos t)) \sec(\cos t) \tan(\cos t) \sin t$
41. $f'(t) = 4 \sin(e^{\sin^2 t}) \cos(e^{\sin^2 t}) e^{\sin^2 t} \sin t \cos t$
43. $g'(x) = 2r^2 p(\ln a)(2ra^{rx} + n)^{p-1} a^{rx}$
45. $y' = \dfrac{-\pi \cos(\tan \pi x) \sec^2(\pi x) \sin\sqrt{\sin(\tan \pi x)}}{2\sqrt{\sin(\tan \pi x)}}$
47. $y' = -3 \cos 3\theta \sin(\sin 3\theta)$;
$y'' = -9 \cos^2(3\theta) \cos(\sin 3\theta) + 9(\sin 3\theta) \sin(\sin 3\theta)$
49. $y' = \dfrac{-\sec t \tan t}{2\sqrt{1 - \sec t}}$;
$y'' = \dfrac{\sec t (3 \sec^3 t - 4 \sec^2 t - \sec t + 2)}{4(1 - \sec t)^{3/2}}$
51. $y = (\ln 2)x + 1$ **53.** $y = -x + \pi$
55. (a) $y = \frac{1}{2}x + 1$ (b)

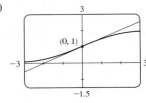

57. (a) $f'(x) = \dfrac{2 - 2x^2}{\sqrt{2 - x^2}}$

59. $((\pi/2) + 2n\pi, 3), ((3\pi/2) + 2n\pi, -1), n$ an integer

61. 24   **63.** (a) 30   (b) 36

65. (a) $\frac{3}{4}$ (b) Does not exist   (c) -2 **67.** $-\frac{1}{6}\sqrt{2}$

69. (a) $F'(x) = e^x f'(e^x)$ (b) $G'(x) = e^{f(x)} f'(x)$

71. 120   **73.** 96

77. $-2^{50} \cos 2x$ **79.** $v(t) = \frac{5}{2}\pi \cos(10\pi t)$ cm/s

81. (a) $\dfrac{dB}{dt} = \dfrac{7\pi}{54} \cos \dfrac{2\pi t}{5.4}$ (b) 0.16

83. $v(t) = 2e^{-1.5t}(2\pi \cos 2\pi t - 1.5 \sin 2\pi t)$

85. (a) 0.0075 (mg/mL)/min   (b) 0.0030 (mg/mL)/min

87. dv/dt is the rate of change of velocity with respect to time; dv/ds is the rate of change of velocity with respect to displacement

89. (a) $Q = ab^t$ where $a \approx 100.01244$ and $b \approx 0.000045146$

(b) -670.63 µA

91. (b) The factored form   **95.** (b) $-n \cos^{n-1}x \sin[(n + 1)x]$

EXERCISES 3.5 ■ PAGE 215

1. (a) $y' = 9x/y$ (b) $y = \pm\sqrt{9x^2 - 1}, y' = \pm 9x/\sqrt{9x^2 - 1}$

3. (a) $y' = -\sqrt{y}/\sqrt{x}$ (b) $y = (1 - \sqrt{x})^2, y' = 1 - 1/\sqrt{x}$

5. $y' = \dfrac{2y - x}{y - 2x}$ **7.** $y' = -\dfrac{2x(2x^2 + y^2)}{y(2x^2 + 3y)}$

9. $y' = \dfrac{x(x + 2y)}{2x^2y + 4xy^2 + 2y^3 + x^2}$ **11.** $y' = \dfrac{2x + y \sin x}{\cos x - 2y}$

13. $y' = \dfrac{1 - 8x^3\sqrt{x + y}}{8y^3\sqrt{x + y} - 1}$ **15.** $y' = \dfrac{y(y - e^{x/y})}{y^2 - xe^{x/y}}$

17. $y' = \dfrac{1 + x^4y^2 + y^2 + x^4y^4 - 2xy}{x^2 - 2xy - 2x^5y^3}$

19. $y' = -\dfrac{y \cos(xy) + \sin(x + y)}{x \cos(xy) + \sin(x + y)}$ **21.** $-\frac{16}{13}$

23. $x' = \dfrac{-2x^4y + x^3 - 6xy^2}{4x^3y^2 - 3x^2y + 2y^3}$ **25.** $y = \frac{1}{2}x$

27. $y = \frac{3}{4}x - \frac{1}{2}$ **29.** $y = x + \frac{1}{2}$ **31.** $y = -\frac{9}{13}x + \frac{40}{13}$

33. (a) $y = \frac{9}{2}x - \frac{5}{2}$ (b)

35. $-1/(4y^3)$ **37.** $\dfrac{\cos^2 y \cos x + \sin^2 x \sin y}{\cos^3 y}$ **39.** $1/e^2$

41. (a)

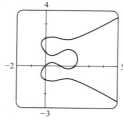

Eight; $x \approx 0.42, 1.58$

(b) $y = -x + 1, y = \frac{1}{3}x + 2$ (c) $1 \mp \frac{1}{3}\sqrt{3}$

43. $(\pm\frac{5}{4}\sqrt{3}, \pm\frac{5}{4})$ **45.** $(x_0x/a^2) - (y_0y/b^2) = 1$

49. $y' = \dfrac{2 \tan^{-1}x}{1 + x^2}$ **51.** $y' = \dfrac{1}{\sqrt{-x^2 - x}}$

53. $F'(x) = \dfrac{3}{\sqrt{x^6 - 1}} + \sec^{-1}(x^3)$ **55.** $h'(t) = 0$

57. $y' = \sin^{-1}x$ **59.** $y' = \dfrac{\sqrt{a^2 - b^2}}{a + b \cos x}$

61. $1 - \dfrac{x \arcsin x}{\sqrt{1 - x^2}}$

65.

67.

71. (a) $\dfrac{V^3(nb - V)}{PV^3 - n^2aV + 2n^3ab}$ (b) -4.04 L/atm

73. $(\pm\sqrt{3}, 0)$ **75.** $(-1, -1), (1, 1)$ **77.** (b) $\frac{3}{2}$

79. (a) 0   (b) $-\frac{1}{2}$

EXERCISES 3.6 ■ PAGE 223

1. The differentiation formula is simplest.

3. $f'(x) = \dfrac{\cos(\ln x)}{x}$ **5.** $f'(x) = -\dfrac{1}{x}$

7. $f'(x) = \dfrac{-\sin x}{(1 + \cos x) \ln 10}$ **9.** $g'(x) = \dfrac{1}{x} - 2$

11. $F'(t) = \ln t \left(\ln t \cos t + \dfrac{2 \sin t}{t} \right)$

13. $G'(y) = \dfrac{10}{2y + 1} - \dfrac{y}{y^2 + 1}$ **15.** $F'(s) = \dfrac{1}{s \ln s}$

17. $T'(z) = 2^z \left(\dfrac{1}{z \ln 2} + \ln z \right)$

19. $y' = \dfrac{-x}{1 + x}$ **21.** $y' = \sec^2[\ln(ax + b)]\dfrac{a}{ax + b}$

23. $y' = (2 + \ln x)/(2\sqrt{x}); y'' = -\ln x/(4x\sqrt{x})$

25. $y' = \tan x; y'' = \sec^2 x$

27. $f'(x) = \dfrac{2x - 1 - (x - 1) \ln(x - 1)}{(x - 1)[1 - \ln(x - 1)]^2};$

$(1, 1 + e) \cup (1 + e, \infty)$

29. $f'(x) = \dfrac{2(x - 1)}{x(x - 2)}; (-\infty, 0) \cup (2, \infty)$ **31.** 2

33. $y = 3x - 9$ **35.** $\cos x + 1/x$ **37.** 7

39. $y' = (x^2 + 2)^2 (x^4 + 4)^4 \left(\dfrac{4x}{x^2 + 2} + \dfrac{16x^3}{x^4 + 4} \right)$

41. $y' = \sqrt{\dfrac{x-1}{x^4+1}} \left(\dfrac{1}{2x-2} - \dfrac{2x^3}{x^4+1} \right)$

43. $y' = x^x (1 + \ln x)$

45. $y' = x^{\sin x} \left(\dfrac{\sin x}{x} + \cos x \ln x \right)$

47. $y' = (\cos x)^x (-x \tan x + \ln \cos x)$

49. $y' = (\tan x)^{1/x} \left(\dfrac{\sec^2 x}{x \tan x} - \dfrac{\ln \tan x}{x^2} \right)$

51. $y' = \dfrac{2x}{x^2 + y^2 - 2y}$ **53.** $f^{(n)}(x) = \dfrac{(-1)^{n-1}(n-1)!}{(x-1)^n}$

EXERCISES 3.7 ▪ PAGE 233

1. (a) $3t^2 - 16t + 24$ (b) 11 ft/s   (c) Never   (d) Always
(e) 72 ft
(f)
$t = 0$
$s = 0$
$t = 6$
$s = 72$
s
(g) $6t - 16$; -10 ft/s^2

(h)   80
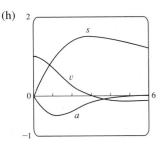
0
s
v
a
6
-20
(i) Speeding up when $t > \frac{8}{3}$; slowing down when $0 \le t < \frac{8}{3}$

3. (a) $(\pi/2) \cos(\pi t/2)$ (b) 0 ft/s
(c) $t = 2n + 1$, t a nonnegative integer
(d) $0 < t < 1$, $3 < t < 5$, $7 < t < 9$, and so on   (e) 6 ft
(f)
$t = 6$
$s = 1.2$
$t = 0$
$s = 0$
$t = 3$
$s = 1.5$
s
(g) $(-\pi^2/4) \sin(\pi t/2)$; $-\pi^2/4$ ft/s^2

(h)   2
s
v
0
6
a
-1
(i) Speeding up when $1 < t < 2$, $3 < t < 4$, and $5 < t < 6$; slowing down when $0 < t < 1$, $2 < t < 3$, and $4 < t < 5$
5. (a) Speeding up when $0 < t < 1$ or $2 < t < 3$; slowing down when $1 < t < 2$

(b) Speeding up when $1 < t < 2$ or $3 < t < 4$; slowing down when $0 < t < 1$ or $2 < t < 3$
7. (a) 4.9 m/s; -14.7 m/s   (b) After 2.5 s   (c) $32\frac{5}{8}$ m
(d) ≈ 5.08 s   (e) ≈ -25.3 m/s
9. (a) 7.56 m/s   (b) 6.24 m/s; -6.24 m/s
11. (a) 30 mm^2/mm; the rate at which the area is increasing with respect to side length as x reaches 15 mm
(b) $\Delta A \approx 2x\, \Delta x$
13. (a) (i) 5π (ii) 4.5π (iii) 4.1π
(b) 4π (c) $\Delta A \approx 2\pi r\, \Delta r$
15. (a) 8π ft^2/ft   (b) 16π ft^2/ft   (c) 24π ft^2/ft
The rate increases as the radius increases.
17. (a) 6 kg/m   (b) 12 kg/m   (c) 18 kg/m
At the right end; at the left end
19. (a) 4.75 A   (b) 5 A; $t = \frac{2}{3}$ s
23. (a) $dV/dP = -C/P^2$ (b) At the beginning
25. $400(3^t) \ln 3$; ≈ 6850 bacteria/h
27. (a) 16 million/year; 78.5 million/year
(b) $P(t) = at^3 + bt^2 + ct + d$, where $a \approx -0.0002849$,
$b \approx 0.5224331$, $c \approx -6.395641$, $d \approx 1720.586$
(c) $P'(t) = 3at^2 + 2bt + c$
(d) 14.16 million/year (smaller); 71.72 million/year (smaller)
(e) $f'(t) = (1.43653 \times 10^9) \cdot (1.01395)^t \ln 1.01395$
(f) 26.25 million/year (larger); 60.28 million/year (smaller)
(g) $P'(85) \approx 76.24$ million/year, $f'(85) = 64.61$ million/year
29. (a) 0.926 cm/s; 0.694 cm/s; 0
(b) 0; -92.6 (cm/s)/cm; -185.2 (cm/s)/cm
(c) At the center; at the edge
31. (a) $C'(x) = 3 + 0.02x + 0.0006x^2$
(b) \$11/pair; the rate at which the cost is changing as the 100th pair of jeans is being produced; the cost of the 101st pair
(c) \$11.07
33. (a) $[xp'(x) - p(x)]/x^2$; the average productivity increases as new workers are added.
35. $\dfrac{dt}{dc} = \dfrac{3\sqrt{9c^2 - 8c} + 9c - 4}{\sqrt{9c^2 - 8c}\,(3c + \sqrt{9c^2 - 8c}\,)}$; the rate of change of duration of dialysis required with respect to the initial urea concentration
37. -0.2436 K/min
39. (a) 0 and 0   (b) $C = 0$
(c) $(0, 0)$, $(500, 50)$; it is possible for the species to coexist.

EXERCISES 3.8 ▪ PAGE 242

1. About 235
3. (a) $100(4.2)^t$ (b) ≈ 7409 (c) $\approx 10{,}632$ bacteria/h
(d) $(\ln 100)/(\ln 4.2) \approx 3.2$ h
5. (a) 1508 million, 1871 million   (b) 2161 million
(c) 3972 million; wars in the first half of century, increased life expectancy in second half
7. (a) $Ce^{-0.0005t}$ (b) $-2000 \ln 0.9 \approx 211$ s
9. (a) $100 \times 2^{-t/30}$ mg   (b) ≈ 9.92 mg   (c) ≈ 199.3 years
11. ≈ 2500 years   **13.** Yes; 12.5 billion years
15. (a) $\approx 137°$F   (b) ≈ 116 min
17. (a) $13.\overline{3}°$C   (b) ≈ 67.74 min
19. (a) ≈ 64.5 kPa   (b) ≈ 39.9 kPa

21. (a) (i) \$3828.84 (ii) \$3840.25 (iii) \$3850.08
(iv) \$3851.61 (v) \$3852.01 (vi) \$3852.08
(b) $dA/dt = 0.05A$, $A(0) = 3000$

EXERCISES 3.9 ■ PAGE 249

1. $dV/dt = 3x^2\, dx/dt$ **3.** 48 cm²/s **5.** $3/(25\pi)$ m/min
7. 128π cm²/min **9.** (a) 1 (b) 25 **11.** -18
13. (a) The plane's altitude is 1 mi and its speed is 500 mi/h.
(b) The rate at which the distance from the plane to the station is
increasing when the plane is 2 mi from the station
(c) (d) $y^2 = x^2 + 1$
(e) $250\sqrt{3}$ mi/h

15. (a) The height of the pole (15 ft), the height of the man (6 ft),
and the speed of the man (5 ft/s)
(b) The rate at which the tip of the man's shadow is moving when
he is 40 ft from the pole
(c)

(d) $\dfrac{15}{6} = \dfrac{x + y}{y}$ (e) $\dfrac{25}{3}$ ft/s

17. 65 mi/h **19.** $837/\sqrt{8674} \approx 8.99$ ft/s
21. -1.6 cm/min **23.** $\frac{720}{13} \approx 55.4$ km/h
25. $(10{,}000 + 800{,}000\pi/9) \approx 2.89 \times 10^5$ cm³/min
27. $\frac{10}{3}$ cm/min **29.** $6/(5\pi) \approx 0.38$ ft/min
31. $150\sqrt{3}$ cm²/min **33.** 5 m **35.** $\pi r^2/60$ cm²/min
37. 80 cm³/min **39.** $\frac{107}{810} \approx 0.132$ Ω/s
41. $\sqrt{7}\,\pi/21 \approx 0.396$ m/min
43. (a) 360 ft/s (b) 0.096 rad/s
45. $\frac{10}{9}\pi$ km/min **47.** $1650/\sqrt{31} \approx 296$ km/h
49. $\frac{7}{4}\sqrt{15} \approx 6.78$ m/s

EXERCISES 3.10 ■ PAGE 256

1. $L(x) = 16x + 23$ **3.** $L(x) = \frac{1}{4}x + 1$
5. $\sqrt{1 - x} \approx 1 - \frac{1}{2}x$;
$\sqrt{0.9} \approx 0.95$,
$\sqrt{0.99} \approx 0.995$

7. $-0.383 < x < 0.516$ **9.** $-0.368 < x < 0.677$
11. (a) $dy = (1 - 4x)e^{-4x}\, dx$ (b) $dy = -\dfrac{2t^3}{\sqrt{1 - t^4}}\, dt$
13. (a) $dy = \dfrac{\sec^2\sqrt{t}}{2\sqrt{t}}\, dt$ (b) $dy = \dfrac{-4v}{(1 + v^2)^2}\, dv$
15. (a) $dy = \frac{1}{10}e^{x/10}\, dx$ (b) 0.01
17. (a) $dy = \dfrac{x}{\sqrt{3 + x^2}}\, dx$ (b) -0.05

19. $\Delta y = 1.25$, $dy = 1$

21. $\Delta y \approx 0.34$, $dy = 0.4$

23. 15.968 **25.** $10.00\overline{3}$ **27.** 1.1
33. (a) 270 cm³, 0.01, 1% (b) 36 cm², 0.00$\overline{6}$, 0.$\overline{6}$%
35. (a) $84/\pi \approx 27$ cm²; $\frac{1}{84} \approx 0.012 = 1.2\%$
(b) $1764/\pi^2 \approx 179$ cm³; $\frac{1}{56} \approx 0.018 = 1.8\%$
37. (a) $2\pi r h\, \Delta r$ (b) $\pi(\Delta r)^2 h$
43. (a) 4.8, 5.2 (b) Too large

EXERCISES 3.11 ■ PAGE 264

1. (a) 0 (b) 1 **3.** (a) $\frac{13}{5}$ (b) $\frac{1}{2}(e^5 + e^{-5}) \approx 74.20995$
5. (a) 1 (b) 0
21. $\operatorname{sech} x = \frac{3}{5}$, $\sinh x = \frac{4}{3}$, $\operatorname{csch} x = \frac{3}{4}$, $\tanh x = \frac{4}{5}$, $\coth x = \frac{5}{4}$
23. (a) 1 (b) -1 (c) ∞ (d) $-\infty$ (e) 0 (f) 1
(g) ∞ (h) $-\infty$ (i) 0 (j) $\frac{1}{2}$
31. $f'(x) = \dfrac{\operatorname{sech}^2\sqrt{x}}{2\sqrt{x}}$ **33.** $h'(x) = 2x \cosh(x^2)$
35. $G'(t) = \dfrac{t^2 + 1}{2t^2}$
37. $y' = 3e^{\cosh 3x} \sinh 3x$
39. $g'(t) = \coth\sqrt{t^2 + 1} - \dfrac{t^2}{\sqrt{t^2 + 1}}\operatorname{csch}^2\sqrt{t^2 + 1}$
41. $y' = \dfrac{1}{2\sqrt{x(x - 1)}}$
43. $y' = \sinh^{-1}(x/3)$ **45.** $y' = -\csc x$
51. (a) 0.3572 (b) 70.34°
53. (a) 164.50 m (b) 120 m; 164.13 m
55. (b) $y = 2\sinh 3x - 4\cosh 3x$
57. $\left(\ln\left(1 + \sqrt{2}\right), \sqrt{2}\right)$

CHAPTER 3 REVIEW ■ PAGE 266

True-False Quiz
1. True **3.** True **5.** False **7.** False **9.** True
11. True **13.** True **15.** True

Exercises

1. $4x^7(x+1)^3(3x+2)$ **3.** $\frac{3}{2}\sqrt{x} - \dfrac{1}{2\sqrt{x}} - \dfrac{1}{\sqrt{x^3}}$

5. $x(\pi x \cos \pi x + 2 \sin \pi x)$

7. $\dfrac{8t^3}{(t^4+1)^2}$ **9.** $\dfrac{1+\ln x}{x \ln x}$ **11.** $\dfrac{\cos\sqrt{x} - \sqrt{x}\sin\sqrt{x}}{2\sqrt{x}}$

13. $-\dfrac{e^{1/x}(1+2x)}{x^4}$ **15.** $\dfrac{2xy - \cos y}{1 - x \sin y - x^2}$

17. $\dfrac{1}{2\sqrt{\arctan x}\,(1+x^2)}$ **19.** $\dfrac{1-t^2}{(1+t^2)^2}\sec^2\!\left(\dfrac{t}{1+t^2}\right)$

21. $3^{x\ln x}(\ln 3)(1+\ln x)$ **23.** $-(x-1)^{-2}$

25. $\dfrac{2x - y\cos(xy)}{x\cos(xy)+1}$ **27.** $\dfrac{2}{(1+2x)\ln 5}$

29. $\cot x - \sin x \cos x$ **31.** $\dfrac{4x}{1+16x^2} + \tan^{-1}(4x)$

33. $5 \sec 5x$ **35.** $-6x \csc^2(3x^2+5)$

37. $\cos\!\left(\tan\sqrt{1+x^3}\right)\!\left(\sec^2\sqrt{1+x^3}\right)\dfrac{3x^2}{2\sqrt{1+x^3}}$

39. $2 \cos\theta \tan(\sin\theta) \sec^2(\sin\theta)$

41. $\dfrac{(2-x)^4(3x^2 - 55x - 52)}{2\sqrt{x+1}\,(x+3)^8}$ **43.** $2x^2 \cosh(x^2) + \sinh(x^2)$

45. $3 \tanh 3x$ **47.** $\dfrac{\cosh x}{\sqrt{\sinh^2 x - 1}}$

49. $\dfrac{-3\sin\!\left(e^{\sqrt{\tan 3x}}\right)e^{\sqrt{\tan 3x}}\sec^2(3x)}{2\sqrt{\tan 3x}}$ **51.** $-\frac{4}{27}$

53. $-5x^4/y^{11}$ **57.** $y = 2\sqrt{3}x + 1 - \pi\sqrt{3}/3$

59. $y = 2x + 1$ **61.** $y = -x + 2;\ y = x + 2$

63. (a) $\dfrac{10 - 3x}{2\sqrt{5-x}}$ (b) $y = \frac{7}{4}x + \frac{1}{4},\ y = -x + 8$

(c)

65. $\left(\pi/4, \sqrt{2}\right), \left(5\pi/4, -\sqrt{2}\right)$

69. (a) 4   (b) 6   (c) $\frac{7}{9}$ (d) 12

71. $2xg(x) + x^2 g'(x)$ **73.** $2g(x)g'(x)$

75. $g'(e^x)e^x$ **77.** $g'(x)/g(x)$

79. $\dfrac{f'(x)[g(x)]^2 + g'(x)\,[f(x)]^2}{[f(x)+g(x)]^2}$

81. $f'(g(\sin 4x))g'(\sin 4x)(\cos 4x)(4)$

83. $(-3, 0)$ **85.** $y = -\frac{2}{3}x^2 + \frac{14}{3}x$

87. $v(t) = -Ae^{-ct}[c \cos(\omega t + \delta) + \omega \sin(\omega t + \delta)],$
$a(t) = Ae^{-ct}[(c^2 - \omega^2)\cos(\omega t + \delta) + 2c\omega \sin(\omega t + \delta)]$

89. (a) $v(t) = 3t^2 - 12;\ a(t) = 6t$ (b) $t > 2;\ 0 \le t < 2$
(c) 23   (d)

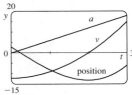

(e) $t > 2;\ 0 < t < 2$
91. 4 kg/m
93. (a) $200(3.24)^t$ (b) $\approx 22{,}040$
(c) $\approx 25{,}910$ bacteria/h   (d) $(\ln 50)/(\ln 3.24) \approx 3.33$ h
95. (a) $C_0 e^{-kt}$ (b) ≈ 100 h   **97.** $\frac{4}{3}$ cm^2/min
99. 13 ft/s   **101.** 400 ft/h
103. (a) $L(x) = 1 + x; \sqrt[3]{1+3x} \approx 1 + x; \sqrt[3]{1.03} \approx 1.01$
(b) $-0.23 < x < 0.40$
105. $12 + \frac{3}{2}\pi \approx 16.7$ cm^2 **107.** $\frac{1}{32}$ **109.** $\frac{1}{4}$ **111.** $\frac{1}{8}x^2$

PROBLEMS PLUS   ■   PAGE 271

1. $\left(\pm\sqrt{3}/2, \frac{1}{4}\right)$ **5.** $3\sqrt{2}$ **11.** $\left(0, \frac{5}{4}\right)$

13. 3 lines; $(0, 2), \left(\frac{4}{3}\sqrt{2}, \frac{2}{3}\right)$ and $\left(\frac{2}{3}\sqrt{2}, \frac{10}{3}\right), \left(-\frac{4}{3}\sqrt{2}, \frac{2}{3}\right)$ and
$\left(-\frac{2}{3}\sqrt{2}, \frac{10}{3}\right)$

15. (a) $4\pi\sqrt{3}/\sqrt{11}$ rad/s   (b) $40\!\left(\cos\theta + \sqrt{8 + \cos^2\theta}\right)$ cm
(c) $-480\pi \sin\theta \left(1 + \cos\theta/\sqrt{8 + \cos^2\theta}\right)$ cm/s

19. $x_T \in (3, \infty),\ y_T \in (2, \infty),\ x_N \in \left(0, \frac{5}{3}\right),\ y_N \in \left(-\frac{5}{2}, 0\right)$
21. (b) (i) $53°$ (or $127°$)   (ii) $63°$ (or $117°$)
23. R approaches the midpoint of the radius AO.
25. $-\sin a$ **27.** $2\sqrt{e}$ **31.** $(1, -2), (-1, 0)$
33. $\sqrt{29}/58$ **35.** $2 + \frac{375}{128}\pi \approx 11.204$ cm^3/min

CHAPTER 4

EXERCISES 4.1   ■   PAGE 283

Abbreviations: abs, absolute; loc, local; max, maximum;
min, minimum
1. Abs min: smallest function value on the entire domain of the
function; loc min at c: smallest function value when x is near c
3. Abs max at s, abs min at r, loc max at c, loc min at b and r,
neither a max nor a min at a and d
5. Abs max $f(4) = 5$, loc max $f(4) = 5$ and $f(6) = 4$,
loc min $f(2) = 2$ and $f(1) = f(5) = 3$

7.

9.

11. (a) (b)

(c)

13. (a) (b)

15. Abs max $f(3) = 4$ **17.** Abs max $f(1) = 1$
19. Abs min $f(0) = 0$
21. Abs max $f(\pi/2) = 1$; abs min $f(-\pi/2) = -1$
23. Abs max $f(2) = \ln 2$ **25.** Abs max $f(0) = 1$
27. Abs min $f(1) = -1$ **29.** $\frac{1}{3}$ **31.** $-2, 3$ **33.** 0
35. $0, 2$ **37.** $0, \frac{4}{9}$ **39.** $0, \frac{8}{7}, 4$ **41.** $n\pi$ (n an integer)
43. $0, \frac{2}{3}$ **45.** 10 **47.** $f(2) = 16, f(5) = 7$
49. $f(-1) = 8, f(2) = -19$ **51.** $f(-2) = 33, f(2) = -31$
53. $f(0.2) = 5.2, f(1) = 2$
55. $f(4) = 4 - \sqrt[3]{4}, f(\sqrt{3}/9) = -2\sqrt{3}/9$
57. $f(\pi/6) = \frac{3}{2}\sqrt{3}, f(\pi/2) = 0$
59. $f(e^{1/2}) = 1/(2e), f(\frac{1}{2}) = -4\ln 2$
61. $f(1) = \ln 3, f(-\frac{1}{2}) = \ln\frac{3}{4}$

63. $f\left(\dfrac{a}{a+b}\right) = \dfrac{a^a b^b}{(a+b)^{a+b}}$

65. (a) $2.19, 1.81$ (b) $\frac{6}{25}\sqrt{\frac{3}{5}} + 2, -\frac{6}{25}\sqrt{\frac{3}{5}} + 2$
67. (a) $0.32, 0.00$ (b) $\frac{3}{16}\sqrt{3}, 0$
69. 0.177 mg/mL; 21.4 min   **71.** $\approx 3.9665°C$
73. About 4.1 months after Jan. 1
75. (a) $r = \frac{2}{3}r_0$ (b) $v = \frac{4}{27}kr_0^3$

(c)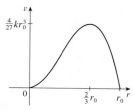

EXERCISES 4.2 ■ **PAGE 291**

1. $1, 5$
3. (a) g is continuous on $[0, 8]$ and differentiable on $(0, 8)$.
(b) $2.1, 6.3$ (c) $3.6, 5.6$
5. 1 **7.** π
9. f is not differentiable on $(-1, 1)$ **11.** 1
13. $3/\ln 4$ **15.** 1; yes   **17.** f is not continous at 3
25. 16 **27.** No   **33.** No

EXERCISES 4.3 ■ **PAGE 300**

1. (a) $(1, 3), (4, 6)$ (b) $(0, 1), (3, 4)$ (c) $(0, 2)$
(d) $(2, 4), (4, 6)$ (e) $(2, 3)$
3. (a) I/D Test   (b) Concavity Test
(c) Find points at which the concavity changes.
5. (a) Inc on $(1, 5)$; dec on $(0, 1)$ and $(5, 6)$
(b) Loc max at $x = 5$, loc min at $x = 1$
7. (a) $3, 5$ (b) $2, 4, 6$ (c) $1, 7$
9. (a) Inc on $(-\infty, -1), (3, \infty)$; dec on $(-1, 3)$
(b) Loc max $f(-1) = 9$; loc min $f(3) = -23$
(c) CU on $(1, \infty)$, CD on $(-\infty, 1)$; IP $(1, -7)$
11. (a) Inc on $(-1, 0), (1, \infty)$; dec on $(-\infty, -1), (0, 1)$
(b) Loc max $f(0) = 3$; loc min $f(\pm 1) = 2$
(c) CU on $\left(-\infty, -\sqrt{3}/3\right), \left(\sqrt{3}/3, \infty\right)$;
CD on $\left(-\sqrt{3}/3, \sqrt{3}/3\right)$; IP $\left(\pm\sqrt{3}/3, \frac{22}{9}\right)$
13. (a) Inc on $(0, \pi/4), (5\pi/4, 2\pi)$; dec on $(\pi/4, 5\pi/4)$
(b) Loc max $f(\pi/4) = \sqrt{2}$; loc min $f(5\pi/4) = -\sqrt{2}$
(c) CU on $(3\pi/4, 7\pi/4)$; CD on $(0, 3\pi/4), (7\pi/4, 2\pi)$;
IP $(3\pi/4, 0), (7\pi/4, 0)$
15. (a) Inc on $\left(-\frac{1}{3}\ln 2, \infty\right)$; dec on $\left(-\infty, -\frac{1}{3}\ln 2\right)$
(b) Loc min $f\left(-\frac{1}{3}\ln 2\right) = 2^{-2/3} + 2^{1/3}$ (c) CU on $(-\infty, \infty)$
17. (a) Inc on $(1, \infty)$; dec on $(0, 1)$ (b) Loc min $f(1) = 0$
(c) CU on $(0, \infty)$; No IP
19. Loc max $f(1) = 2$; loc min $f(0) = 1$
21. Loc min $f\left(\frac{1}{16}\right) = -\frac{1}{4}$
23. (a) f has a local maximum at 2.
(b) f has a horizontal tangent at 6.
25. (a) (b)

27.   **29.**

31.

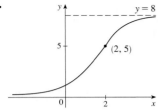

33. (a) No
(b) Yes

(c)

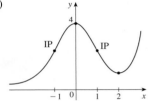

35. (a) Inc on $(0, 2)$, $(4, 6)$, $(8, \infty)$;
dec on $(2, 4)$, $(6, 8)$
(b) Loc max at $x = 2, 6$;
loc min at $x = 4, 8$
(c) CU on $(3, 6)$, $(6, \infty)$;
CD on $(0, 3)$ (d) 3
(e) See graph at right.

37. (a) Inc on $(-\infty, -2)$, $(2, \infty)$; dec on $(-2, 2)$
(b) Loc max $f(-2) = 18$; loc min $f(2) = -14$
(c) CU on $(0, \infty)$, CD on $(-\infty, 0)$; IP $(0, 2)$
(d)

39. (a) Inc on $(-2, 0)$, $(2, \infty)$; dec on $(-\infty, -2)$, $(0, 2)$
(b) Loc max $f(0) = 3$; loc min $f(\pm 2) = -5$

(c) CU on $\left(-\infty, -\dfrac{2}{\sqrt{3}}\right)$, $\left(\dfrac{2}{\sqrt{3}}, \infty\right)$; CD on $\left(-\dfrac{2}{\sqrt{3}}, \dfrac{2}{\sqrt{3}}\right)$;

IPs $\left(\pm\dfrac{2}{\sqrt{3}}, -\dfrac{13}{9}\right)$

(d)

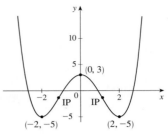

41. (a) Inc on $(-\infty, -2)$, $(0, \infty)$;
dec on $(-2, 0)$
(b) Loc max $h(-2) = 7$;
loc min $h(0) = -1$
(c) CU on $(-1, \infty)$;
CD on $(-\infty, -1)$; IP $(-1, 3)$
(d) See graph at right.

43. (a) Inc on $(-\infty, 4)$;
dec on $(4, 6)$
(b) Loc max $F(4) = 4\sqrt{2}$
(c) CD on $(-\infty, 6)$; No IP
(d) See graph at right.

45. (a) Inc on $(-1, \infty)$;
dec on $(-\infty, -1)$
(b) Loc min $C(-1) = -3$
(c) CU on $(-\infty, 0)$, $(2, \infty)$;
CD on $(0, 2)$;
IPs $(0, 0)$, $\left(2, 6\sqrt[3]{2}\right)$
(d) See graph at right.

47. (a) Inc on $(\pi, 2\pi)$;
dec on $(0, \pi)$
(b) Loc min $f(\pi) = -1$
(c) CU on $(\pi/3, 5\pi/3)$;
CD on $(0, \pi/3)$, $(5\pi/3, 2\pi)$;
IPs $\left(\pi/3, \frac{5}{4}\right)$, $\left(5\pi/3, \frac{5}{4}\right)$
(d) See graph at right.

49. (a) VA $x = 0$; HA $y = 1$
(b) Inc on $(0, 2)$;
dec on $(-\infty, 0)$, $(2, \infty)$
(c) Loc max $f(2) = \frac{5}{4}$
(d) CU on $(3, \infty)$;
CD on $(-\infty, 0)$, $(0, 3)$; IP $\left(3, \frac{11}{9}\right)$
(e) See graph at right.

51. (a) HA $y = 0$
(b) Dec on $(-\infty, \infty)$
(c) None
(d) CU on $(-\infty, \infty)$
(e) See graph at right.

(d)
$(0, 3)$
IP IP
$(-2, -5)$ $(2, -5)$

53. (a) HA $y = 0$
(b) Inc on $(-\infty, 0)$,
dec on $(0, \infty)$
(c) Loc max $f(0) = 1$
(d) CU on $\left(-\infty, -1\sqrt{2}\right)$,
$\left(1/\sqrt{2}, \infty\right)$; CD on $\left(-1/\sqrt{2}, 1/\sqrt{2}\right)$; IPs $\left(\pm 1/\sqrt{2}, e^{-1/2}\right)$
(e) See graph at right.

55. (a) VA $x = 0$, $x = e$
(b) Dec on $(0, e)$
(c) None
(d) CU on $(0, 1)$; CD on $(1, e)$;
IP $(1, 0)$
(e) See graph at right.

57. $(3, \infty)$
59. (a) Loc and abs max $f(1) = \sqrt{2}$, no min   (b) $\frac{1}{4}\left(3 - \sqrt{17}\right)$
61. (b) CU on $(0, 0.85)$, $(1.57, 2.29)$; CD on $(0.85, 1.57)$,
$(2.29, \pi)$; IP $(0.85, 0.74)$, $(1.57, 0)$, $(2.29, -0.74)$
63. CU on $(-\infty, -0.6)$, $(0.0, \infty)$; CD on $(-0.6, 0.0)$
65. (a) The rate of increase is initially very small, increases to a
maximum at $t \approx 8$ h, then decreases toward 0.
(b) When $t = 8$ (c) CU on $(0, 8)$; CD on $(8, 18)$
(d) $(8, 350)$
67. If $D(t)$ is the size of the deficit as a function of time, then at
the time of the speech $D'(t) > 0$, but $D''(t) < 0$.
69. $K(3) - K(2)$; CD
71. 28.57 min, when the rate of increase of drug level in the
bloodstream is greatest; 85.71 min, when rate of decrease is
greatest
73. $f(x) = \frac{1}{9}(2x^3 + 3x^2 - 12x + 7)$
75. (a) $a = 0$, $b = -1$ (b) $y = -x$ at $(0, 0)$

EXERCISES 4.4 ▪ PAGE 311

1. (a) Indeterminate   (b) 0   (c) 0
(d) ∞, $-\infty$, or does not exist   (e) Indeterminate
3. (a) $-\infty$ (b) Indeterminate   (c) ∞
5. $\frac{9}{4}$ **7.** 1   **9.** 6   **11.** $-\frac{1}{3}$
13. $-\infty$ **15.** 2   **17.** $\frac{1}{4}$ **19.** 0   **21.** $-\infty$
23. $\frac{8}{5}$ **25.** 3   **27.** $\frac{1}{2}$ **29.** 1   **31.** 1
33. $1/\ln 3$ **35.** 0   **37.** 0   **39.** a/b
41. $\frac{1}{24}$ **43.** π **45.** $\frac{5}{3}$ **47.** 0   **49.** $-2/\pi$
51. $\frac{1}{2}$ **53.** $\frac{1}{2}$ **55.** ∞ **57.** 1   **59.** e^{-2}
61. $1/e$ **63.** 1   **65.** e^4 **67.** e^3 **69.** e^2
71. $\frac{1}{4}$ **75.** 1
77. f has an absolute minimum for $c > 0$. As c increases, the
minimum points get farther away from the origin.
81. (a) M; the population should approach its maximum size as
time increases   (b) $P_0 e^{kt}$; exponential
83. $\frac{16}{9}a$ **85.** $\frac{1}{2}$ **87.** 56   **91.** (a) 0

EXERCISES 4.5 ▪ PAGE 321

Abbreviation: int, intercept; SA, slant asymptote
1. A. $\mathbb{R}$ B. y-int 0; x-int $-3, 0$
C. None   D. None
E. Inc on $(-\infty, -2)$, $(0, \infty)$;
dec on $(-2, 0)$
F. Loc max $f(-2) = 4$;
loc min $f(0) = 0$
G. CU on $(-1, \infty)$; CD on $(-\infty, -1)$;
IP $(-1, 2)$
H. See graph at right.

3. A. $\mathbb{R}$ B. y-int 0; x-int 0, $\sqrt[3]{4}$
C. None   D. None
E. Inc on $(1, \infty)$; dec on $(-\infty, 1)$
F. Loc min $f(1) = -3$
G. CU on $(-\infty, \infty)$
H. See graph at right.

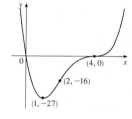

5. A. $\mathbb{R}$ B. y-int 0; x-int 0, 4
C. None   D. None
E. Inc on $(1, \infty)$; dec on $(-\infty, 1)$
F. Loc min $f(1) = -27$
G. CU on $(-\infty, 2)$, $(4, \infty)$;
CD on $(2, 4)$;
IPs $(2, -16)$, $(4, 0)$
H. See graph at right.

7. A. $\mathbb{R}$ B. y-int 0; x-int 0
C. About $(0, 0)$ D. None
E. Inc on $(-\infty, \infty)$
F. None
G. CU on $(-2, 0)$, $(2, \infty)$;
CD on $(-\infty, -2)$, $(0, 2)$;
IPs $\left(-2, -\frac{256}{15}\right)$, $(0, 0)$, $\left(2, \frac{256}{15}\right)$
H. See graph at right.

9. A. $\{x \mid x \neq 1\}$ B. y-int 0; x-int 0
C. None   D. VA $x = 1$, HA $y = 1$
E. Dec on $(-\infty, 1)$, $(1, \infty)$
F. None
G. CU on $(1, \infty)$; CD on $(-\infty, 1)$
H. See graph at right.

11. A. $(-\infty, 1) \cup (1, 2) \cup (2, \infty)$
B. y-int 0; x-int 0   C. None
D. HA $y = -1$; VA $x = 2$
E. Inc on $(-\infty, 1)$, $(1, 2)$, $(2, \infty)$
F. None
G. CU on $(-\infty, 1)$, $(1, 2)$;
CD on $(2, \infty)$
H. See graph at right.

13. A. $(-\infty, -2) \cup (-2, 2) \cup (2, \infty)$ **B.** y-int 0; x-int 0
C. About $(0, 0)$ **D.** VA $x = \pm 2$; HA $y = 0$
E. Dec on $(-\infty, -2)$, $(-2, 2)$, $(2, \infty)$
F. No local extrema
G. CU on $(-2, 0)$, $(2, \infty)$;
CD on $(-\infty, -2)$, $(0, 2)$; IP $(0, 0)$
H. See graph at right.

15. A. $\mathbb{R}$ **B.** y-int 0; x-int 0
C. About y-axis   **D.** HA $y = 1$
E. Inc on $(0, \infty)$; dec on $(-\infty, 0)$
F. Loc min $f(0) = 0$
G. CU on $(-1, 1)$;
CD on $(-\infty, -1)$, $(1, \infty)$; IPs $\left(\pm 1, \frac{1}{4}\right)$
H. See graph at right.

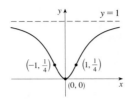

17. A. $(-\infty, 0) \cup (0, \infty)$ **B.** x-int 1
C. None   **D.** HA $y = 0$; VA $x = 0$
E. Inc on $(0, 2)$;
dec on $(-\infty, 0)$, $(2, \infty)$
F. Loc max $f(2) = \frac{1}{4}$
G. CU on $(3, \infty)$;
CD on $(-\infty, 0)$, $(0, 3)$; IP $\left(3, \frac{2}{9}\right)$
H. See graph at right.

19. A. $(-\infty, -1) \cup (-1, \infty)$
B. y-int 0; x-int 0   **C.** None
D. VA $x = -1$; HA $y = 1$
E. Inc on $(-\infty, -1)$, $(-1, \infty)$;
F. None
G. CU on $(-\infty, -1)$, $\left(0, \sqrt[3]{\frac{1}{2}}\right)$;
CD on $(-1, 0)$, $\left(\sqrt[3]{\frac{1}{2}}, \infty\right)$;
IPs $(0, 0)$, $\left(\sqrt[3]{\frac{1}{2}}, \frac{1}{3}\right)$
H. See graph at right.

21. A. $[0, \infty)$ **B.** y-int 0; x-int 0, 3
C. None   **D.** None
E. Inc on $(1, \infty)$; dec on $(0, 1)$
F. Loc min $f(1) = -2$
G. CU on $(0, \infty)$
H. See graph at right.

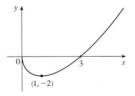

23. A. $(-\infty, -2] \cup [1, \infty)$
B. x-int $-2, 1$ **C.** None
D. None
E. Inc on $(1, \infty)$; dec on $(-\infty, -2)$
F. None
G. CD on $(-\infty, -2)$, $(1, \infty)$
H. See graph at right.

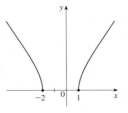

25. A. $\mathbb{R}$ **B.** y-int 0; x-int 0
C. About $(0, 0)$
D. HA $y = \pm 1$
E. Inc on $(-\infty, \infty)$ **F.** None
G. CU on $(-\infty, 0)$;
CD on $(0, \infty)$; IP $(0, 0)$
H. See graph at right.

27. A. $[-1, 0) \cup (0, 1]$ **B.** x-int ± 1 **C.** About $(0, 0)$
D. VA $x = 0$
E. Dec on $(-1, 0)$, $(0, 1)$
F. None
G. CU on $\left(-1, -\sqrt{2/3}\right)$,
$\left(0, \sqrt{2/3}\right)$;
CD on $\left(-\sqrt{2/3}, 0\right)$, $\left(\sqrt{2/3}, 1\right)$;
IPs $\left(\pm\sqrt{2/3}, \pm 1/\sqrt{2}\right)$
H. See graph at right.

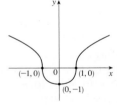

29. A. $\mathbb{R}$ **B.** y-int 0; x-int $\pm 3\sqrt{3}, 0$ **C.** About $(0, 0)$
D. None   **E.** Inc on $(-\infty, -1)$, $(1, \infty)$; dec on $(-1, 1)$
F. Loc max $f(-1) = 2$;
loc min $f(1) = -2$
G. CU on $(0, \infty)$;
CD on $(-\infty, 0)$; IP $(0, 0)$
H. See graph at right.

31. A. $\mathbb{R}$ **B.** y-int -1; x-int ± 1
C. About the y-axis   **D.** None
E. Inc on $(0, \infty)$; dec on $(-\infty, 0)$
F. Loc min $f(0) = -1$
G. CU on $(-1, 1)$;
CD on $(-\infty, -1)$, $(1, \infty)$; IPs $(\pm 1, 0)$
H. See graph at right.

33. A. $\mathbb{R}$ **B.** y-int 0; x-int $n\pi$ (n an integer)
C. About $(0, 0)$, period 2π **D.** None
E–G answers for $0 \le x \le \pi$:
E. Inc on $(0, \pi/2)$; dec on $(\pi/2, \pi)$ **F.** Loc max $f(\pi/2) = 1$
G. Let $\alpha = \sin^{-1}\sqrt{2/3}$; CU on $(0, \alpha)$, $(\pi - \alpha, \pi)$;
CD on $(\alpha, \pi - \alpha)$; IPs at $x = 0, \pi, \alpha, \pi - \alpha$
H.

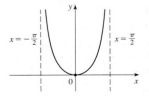

35. A. $(-\pi/2, \pi/2)$ **B.** y-int 0; x-int 0   **C.** About y-axis
D. VA $x = \pm\pi/2$
E. Inc on $(0, \pi/2)$;
dec on $(-\pi/2, 0)$
F. Loc min $f(0) = 0$
G. CU on $(-\pi/2, \pi/2)$
H. See graph at right.

37. A. $[-2\pi, 2\pi]$ B. y-int $\sqrt{3}$; x-int $-4\pi/3, -\pi/3, 2\pi/3$,
$5\pi/3$ C. Period 2π D. None
E. Inc on $(-2\pi, -11\pi/6), (-5\pi/6, \pi/6), (7\pi/6, 2\pi)$;
dec on $(-11\pi/6, -5\pi/6), (\pi/6, 7\pi/6)$
F. Loc max $f(-11\pi/6) = f(\pi/6) = 2$; loc min
$f(-5\pi/6) = f(7\pi/6) = -2$
G. CU on $(-4\pi/3, -\pi/3)$,
$(2\pi/3, 5\pi/3)$;
CD on $(-2\pi, -4\pi/3)$,
$(-\pi/3, 2\pi/3), (5\pi/3, 2\pi)$;
IPs $(-4\pi/3, 0), (-\pi/3, 0)$,
$(2\pi/3, 0), (5\pi/3, 0)$
H. See graph at right.

39. A. All reals except $(2n + 1)\pi$ (n an integer)
B. y-int 0; x-int $2n\pi$ C. About the origin, period 2π
D. VA $x = (2n + 1)\pi$ E. Inc on $((2n - 1)\pi, (2n + 1)\pi)$
F. None   G. CU on $(2n\pi, (2n + 1)\pi)$;
CD on $((2n - 1)\pi, 2n\pi)$; IPs $(2n\pi, 0)$
H.

41. A. $\mathbb{R}$ B. y-int $\pi/4$
C. None
D. HA $y = 0, y = \pi/2$
E. Inc on $(-\infty, \infty)$ F. None
G. CU on $(-\infty, 0)$;
CD on $(0, \infty)$; IP $(0, \pi/4)$
H. See graph at right.

43. A. $\mathbb{R}$ B. y-int $\frac{1}{2}$ C. None
D. HA $y = 0, y = 1$
E. Inc on $\mathbb{R}$ F. None
G. CU on $(-\infty, 0)$;
CD on $(0, \infty)$;   IP $(0, \frac{1}{2})$
H. See graph at right.

45. A. $(0, \infty)$ B. None
C. None   D. VA $x = 0$
E. Inc on $(1, \infty)$; dec on $(0, 1)$
F. Loc min $f(1) = 1$
G. CU on $(0, 2)$; CD on $(2, \infty)$;
IP $(2, \frac{1}{2} + \ln 2)$
H. See graph at right.

47. A. $\mathbb{R}$ B. y-int $\frac{1}{4}$
C. None
D. HA $y = 0, y = 1$
E. Dec on $\mathbb{R}$ F. None
G. CU on $\left(\ln \frac{1}{2}, \infty\right)$;
CD on $\left(-\infty, \ln \frac{1}{2}\right)$; IP $\left(\ln \frac{1}{2}, \frac{4}{9}\right)$
H. See graph at right.

49. A. All x in $(2n\pi, (2n + 1)\pi)$ (n an integer)
B. x-int $\pi/2 + 2n\pi$ C. Period 2π D. VA $x = n\pi$
E. Inc on $(2n\pi, \pi/2 + 2n\pi)$; dec on $(\pi/2 + 2n\pi, (2n + 1)\pi)$
F. Loc max $f(\pi/2 + 2n\pi) = 0$ G. CD on $(2n\pi, (2n + 1)\pi)$
H.

51. A. $(-\infty, 0) \cup (0, \infty)$
B. None   C. None
D. VA $x = 0$
E. Inc on $(-\infty, -1), (0, \infty)$;
dec on $(-1, 0)$
F. Loc max $f(-1) = -e$
G. CU on $(0, \infty)$; CD on $(-\infty, 0)$
H. See graph at right.

53. A. $\mathbb{R}$ B. y-int 1
C. None   D. HA $y = e^{\pm\pi/2}$
E. Inc on $\mathbb{R}$ F. None
G. CU on $\left(-\infty, \frac{1}{2}\right)$; CD on $\left(\frac{1}{2}, \infty\right)$;
IP $\left(\frac{1}{2}, e^{\arctan(1/2)}\right)$
H. See graph at right.

55.

57. (a) When $t = (\ln a)/k$ (b) When $t = (\ln a)/k$
(c)

59.

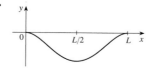

61. $y = x - 1$ **63.** $y = 2x - 3$

65. A. $(-\infty, 1) \cup (1, \infty)$
B. y-int 0; x-int 0   C. None
D. VA $x = 1$; SA $y = x + 1$
E. Inc on $(-\infty, 0)$, $(2, \infty)$;
dec on $(0, 1)$, $(1, 2)$
F. Loc max $f(0) = 0$;
loc min $f(2) = 4$
G. CU on $(1, \infty)$; CD on $(-\infty, 1)$
H. See graph at right.

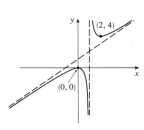

67. A. $(-\infty, 0) \cup (0, \infty)$
B. x-int $-\sqrt[3]{4}$ C. None
D. VA $x = 0$; SA $y = x$
E. Inc on $(-\infty, 0)$, $(2, \infty)$;
dec on $(0, 2)$
F. Loc min $f(2) = 3$
G. CU on $(-\infty, 0)$, $(0, \infty)$
H. See graph at right.

69. A. $\mathbb{R}$ B. y-int 2
C. None
D. SA $y = 1 + \frac{1}{2}x$
E. Inc on $(\ln 2, \infty)$;
dec on $(-\infty, \ln 2)$
F. Loc min $f(\ln 2) = \frac{3}{2} + \frac{1}{2}\ln 2$
G. CU on $(-\infty, \infty)$
H. See graph at right.

71.

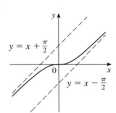

75. VA $x = 0$, asymptotic to $y = x^3$

EXERCISES 4.6 ■ PAGE 329

1. Inc on $(-\infty, -1.50)$, $(0.04, 2.62)$, $(2.84, \infty)$; dec on
$(-1.50, 0.04)$, $(2.62, 2.84)$; loc max $f(-1.50) \approx 36.47$,
$f(2.62) \approx 56.83$; loc min $f(0.04) \approx -0.04$, $f(2.84) \approx 56.73$;
CU on $(-0.89, 1.15)$, $(2.74, \infty)$; CD on $(-\infty, -0.89)$, $(1.15, 2.74)$;
IPs $(-0.89, 20.90)$, $(1.15, 26.57)$, $(2.74, 56.78)$

3. Inc on $(-1.31, -0.84)$, $(1.06, 2.50)$, $(2.75, \infty)$; dec on
$(-\infty, -1.31)$, $(-0.84, 1.06)$, $(2.50, 2.75)$; loc max
$f(-0.84) \approx 23.71, f(2.50) \approx -11.02$; loc min
$f(-1.31) \approx 20.72, f(1.06) \approx -33.12, f(2.75) \approx -11.33$; CU
on $(-\infty. -1.10)$, $(0.08, 1.72)$, $(2.64, \infty)$; CD on $(-1.10, 0.08)$,
$(1.72, 2.64)$; IPs $(-1.10, 22.09)$, $(0.08, -3.88)$, $(1.72, -22.53)$,
$(2.64, -11.18)$

5. Inc on $(-\infty, -1.47)$, $(-1.47, 0.66)$; dec on $(0.66, \infty)$;
loc max $f(0.66) \approx 0.38$; CU on $(-\infty, -1.47)$, $(-0.49, 0)$,
$(1.10, \infty)$; CD on $(-1.47, -0.49)$, $(0, 1.10)$;
IPs $(-0.49, -0.44)$, $(1.10, 0.31)$, $(0, 0)$

7. Inc on $(-1.40, -0.44)$, $(0.44, 1.40)$; dec on $(-\pi, -1.40)$,
$(-0.44, 0)$, $(0, 0.44)$, $(1.40, \pi)$; loc max $f(-0.44) \approx -4.68$,
$f(1.40) \approx 6.09$; loc min $f(-1.40) \approx -6.09$, $f(0.44) \approx 4.68$;
CU on $(-\pi, -0.77)$, $(0, 0.77)$; CD on $(-0.77, 0)$, $(0.77, \pi)$;
IPs $(-0.77, -5.22)$, $(0.77, 5.22)$

9. Inc on $\left(-8 - \sqrt{61}, -8 + \sqrt{61}\right)$; dec on $\left(-\infty, -8 - \sqrt{61}\right)$,
$\left(-8 + \sqrt{61}, 0\right)$, $(0, \infty)$; CU on $\left(-12 - \sqrt{138}, -12 + \sqrt{138}\right)$,
$(0, \infty)$; CD on $\left(-\infty, -12 - \sqrt{138}\right)$, $\left(-12 + \sqrt{138}, 0\right)$

11. (a)

(b) $\lim_{x \to 0^+} f(x) = 0$ (c) Loc min $f\left(1/\sqrt{e}\right) = -1/(2e)$;
CD on $(0, e^{-3/2})$; CU on $(e^{-3/2}, \infty)$
13. Loc max $f(-5.6) \approx 0.018$, $f(0.82) \approx -281.5$,
$f(5.2) \approx 0.0145$; loc min $f(3) = 0$

15. $f'(x) = -\dfrac{x(x+1)^2(x^3 + 18x^2 - 44x - 16)}{(x-2)^3(x-4)^5}$

$f''(x) = 2\dfrac{(x+1)(x^6 + 36x^5 + 6x^4 - 628x^3 + 684x^2 + 672x + 64)}{(x-2)^4(x-4)^6}$

CU on $(-35.3, -5.0)$, $(-1, -0.5)$, $(-0.1, 2)$, $(2, 4)$, $(4, \infty)$;
CD on $(-\infty, -35.3)$, $(-5.0, -1)$, $(-0.5, -0.1)$;
IPs $(-35.3, -0.015)$, $(-5.0, -0.005)$, $(-1, 0)$, $(-0.5, 0.00001)$,
$(-0.1, 0.0000066)$
17. Inc on $(-9.41, -1.29)$, $(0, 1.05)$;
dec on $(-\infty, -9.41)$, $(-1.29, 0)$, $(1.05, \infty)$;
loc max $f(-1.29) \approx 7.49$, $f(1.05) \approx 2.35$;
loc min $f(-9.41) \approx -0.056$, $f(0) = 0.5$;
CU on $(-13.81, -1.55)$, $(-1.03, 0.60)$, $(1.48, \infty)$;
CD on $(-\infty, -13.81)$, $(-1.55, -1.03)$, $(0.60, 1.48)$;
IPs $(-13.81, -0.05)$, $(-1.55, 5.64)$, $(-1.03, 5.39)$, $(0.60, 1.52)$,
$(1.48, 1.93)$

19. Inc on $(-4.91, -4.51)$, $(0, 1.77)$, $(4.91, 8.06)$, $(10.79, 14.34)$,
$(17.08, 20)$;
dec on $(-4.51, -4.10)$, $(1.77, 4.10)$, $(8.06, 10.79)$, $(14.34, 17.08)$;
loc max $f(-4.51) \approx 0.62$, $f(1.77) \approx 2.58$, $f(8.06) \approx 3.60$,
$f(14.34) \approx 4.39$;
loc min $f(10.79) \approx 2.43$, $f(17.08) \approx 3.49$;

CU on $(9.60, 12.25)$, $(15.81, 18.65)$;
CD on $(-4.91, -4.10)$, $(0, 4.10)$, $(4.91, 9.60)$, $(12.25, 15.81)$,
$(18.65, 20)$;
IPs $(9.60, 2.95)$, $(12.25, 3.27)$, $(15.81, 3.91)$, $(18.65, 4.20)$

21. Inc on $(-\infty, 0)$, $(0, \infty)$;
CU on $(-\infty, -0.42)$, $(0, 0.42)$;
CD on $(-0.42, 0)$, $(0.42, \infty)$;
IPs $(\mp0.42, \pm0.83)$

23.

25. (a)

(b) $\lim_{x \to 0^+} x^{1/x} = 0$, $\lim_{x \to \infty} x^{1/x} = 1$
(c) Loc max $f(e) = e^{1/e}$ (d) IPs at $x \approx 0.58, 4.37$
27. Max $f(0.59) \approx 1$, $f(0.68) \approx 1$, $f(1.96) \approx 1$;
min $f(0.64) \approx 0.99996$, $f(1.46) \approx 0.49$, $f(2.73) \approx -0.51$;
IPs $(0.61, 0.99998)$, $(0.66, 0.99998)$, $(1.17, 0.72)$,
$(1.75, 0.77)$, $(2.28, 0.34)$

29. For $c < 0$, there is a loc min that moves toward $(-3, -9)$ as c increases. For $0 < c < 8$, there is a loc min that moves toward $(-3, -9)$ and a loc max that moves toward the origin as c decreases. For all $c > 0$, there is a first quadrant loc min that moves toward the origin as c decreases. $c = 0$ is a transitional value that gives the graph of a parabola. For all nonzero c, the y-axis is a VA and there is an IP that moves toward the origin as $|c| \to 0$.
$c \le 0$:

$c \ge 0$:

31. For $c < 0$, there is no extreme point and one IP, which decreases along the x-axis. For $c > 0$, there is no IP, and one minimum point.

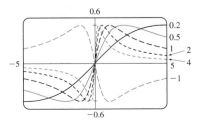

33. For $c > 0$, the maximum and minimum values are always $\pm\frac{1}{2}$, but the extreme points and IPs move closer to the y-axis as c increases. $c = 0$ is a transitional value: when c is replaced by $-c$, the curve is reflected in the x-axis.

35. For $|c| < 1$, the graph has loc max and min values; for $|c| \ge 1$ it does not. The function increases for $c \ge 1$ and decreases for $c \le -1$. As c changes, the IPs move vertically but not horizontally.

37.

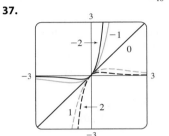

For $c > 0$, $\lim_{x \to \infty} f(x) = 0$ and $\lim_{x \to -\infty} f(x) = -\infty$.
For $c < 0$, $\lim_{x \to \infty} f(x) = \infty$ and $\lim_{x \to -\infty} f(x) = 0$.
As $|c|$ increases, the max and min points and the IPs get closer to the origin.

39. (a) Positive   (b)

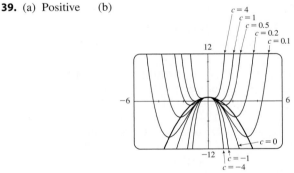

EXERCISES 4.7 ■ PAGE 336

1. (a) 11, 12   (b) 11.5, 11.5   **3.** 10, 10   **5.** $\frac{9}{4}$
7. 25 m by 25 m   **9.** $N = 1$
11. (a)

(b)

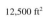

(c) $A = xy$ (d) $5x + 2y = 750$ (e) $A(x) = 375x - \frac{5}{2}x^2$
(f) $14{,}062.5$ ft^2
13. 1000 ft by 1500 ft   **15.** 4000 cm^3 **17.** $\$191.28$
19. $20\sqrt{30}$ ft by $\frac{40}{3}\sqrt{30}$ ft   **21.** $\left(-\frac{6}{5}, \frac{3}{5}\right)$
23. $\left(-\frac{1}{3}, \pm\frac{4}{3}\sqrt{2}\right)$ **25.** Square, side $\sqrt{2}\,r$
27. $L/2,\ \sqrt{3}\,L/4$ **29.** Base $\sqrt{3}\,r$, height $3r/2$
31. $4\pi r^3/(3\sqrt{3})$ **33.** $\pi r^2(1 + \sqrt{5})$ **35.** 24 cm by 36 cm
37. (a) Use all of the wire for the square
(b) $40\sqrt{3}/(9 + 4\sqrt{3})$ m for the square
39. 16 in.   **41.** $V = 2\pi R^3/(9\sqrt{3})$ **45.** $E^2/(4r)$
47. (a) $\frac{3}{2}s^2\csc\theta\,(\csc\theta - \sqrt{3}\cot\theta)$ (b) $\cos^{-1}(1/\sqrt{3}) \approx 55°$
(c) $6s\left[h + s/(2\sqrt{2})\right]$
49. Row directly to B **51.** ≈ 4.85 km east of the refinery
53. $10\sqrt[3]{3}/(1 + \sqrt[3]{3})$ ft from the stronger source
55. $(a^{2/3} + b^{2/3})^{3/2}$ **57.** $2\sqrt{6}$
59. (b) (i) $\$342{,}491$; $\$342/$unit; $\$390/$unit   (ii) 400
(iii) $\$320/$unit
61. (a) $p(x) = 19 - \frac{1}{3000}x$ (b) $\$9.50$
63. (a) $p(x) = 500 - \frac{1}{8}x$ (b) $\$250$ (c) $\$310$
69. 9.35 m   **73.** $x = 6$ in.   **75.** $\pi/6$
77. At a distance $5 - 2\sqrt{5}$ from A **79.** $\frac{1}{2}(L + W)^2$
81. (a) About 5.1 km from B (b) C is close to B; C is close
to D; $W/L = \sqrt{25 + x^2}/x$, where $x = |BC|$
(c) ≈ 1.07; no such value   (d) $\sqrt{41}/4 \approx 1.6$

EXERCISES 4.8 ■ PAGE 348

1. (a) $x_2 \approx 7.3,\ x_3 \approx 6.8$ (b) Yes
3. $\frac{9}{2}$ **5.** a, b, c **7.** 1.5215 **9.** -1.25
11. 2.94283096 **13.** (b) 2.630020 **15.** -1.964636
17. $-3.637958, -1.862365, 0.889470$
19. $-1.257691, 0.653483$ **21.** $0, \pm 0.902025$
23. $-1.69312029, -0.74466668, 1.26587094$
25. 0.76682579 **27.** $0.21916368, 1.08422462$
29. (b) 31.622777
35. (a) $-1.293227, -0.441731, 0.507854$ (b) -2.0212
37. $(1.519855, 2.306964)$ **39.** $(0.410245, 0.347810)$
41. 0.76286%

EXERCISES 4.9 ■ PAGE 355

1. $F(x) = 2x^2 + 7x + C$ **3.** $F(x) = \frac{1}{2}x^4 - \frac{2}{9}x^3 + \frac{5}{2}x^2 + C$
5. $F(x) = 4x^3 + 4x^2 + C$ **7.** $F(x) = 5x^{7/5} + 40x^{1/5} + C$
9. $F(x) = \sqrt{2}x + C$ **11.** $F(x) = 2x^{3/2} - \frac{3}{2}x^{4/3} + C$
13. $F(x) = \begin{cases} \frac{1}{5}x - 2\ln|x| + C_1 & \text{if } x < 0 \\ \frac{1}{5}x - 2\ln|x| + C_2 & \text{if } x > 0 \end{cases}$
15. $G(t) = 2t^{1/2} + \frac{2}{3}t^{3/2} + \frac{2}{5}t^{5/2} + C$
17. $H(\theta) = -2\cos\theta - \tan\theta + C_n$ on $(n\pi - \pi/2,\ n\pi + \pi/2)$,
n an integer
19. $F(x) = 2^x/\ln 2 + 4\cosh x + C$
21. $F(x) = x^2 + 4x + 1/x + C,\ x > 0$
23. $F(x) = x^5 - \frac{1}{3}x^6 + 4$
25. $f(x) = x^5 - x^4 + x^3 + Cx + D$
27. $f(x) = \frac{1}{3}x^3 + 3e^x + Cx + D$

29. $f(t) = 2t^3 + \cos t + Ct^2 + Dt + E$
31. $f(x) = x + 2x^{3/2} + 5$ **33.** $f(t) = 4\arctan t - \pi$
35. $f(x) = 3x^{5/3} - 75$
37. $f(t) = \tan t + \sec t - 2 - \sqrt{2}$
39. $f(x) = -x^2 + 2x^3 - x^4 + 12x + 4$
41. $f(\theta) = -\sin\theta - \cos\theta + 5\theta + 4$
43. $f(x) = 2x^2 + x^3 + 2x^4 + 2x + 3$
45. $f(x) = e^x + 2\sin x - \dfrac{2}{\pi}(e^{\pi/2} + 4)x + 2$
47. $f(x) = -\ln x + (\ln 2)x - \ln 2$ **49.** 8 **51.** b
53.

55.

57.

59. $s(t) = 1 - \cos t - \sin t$ **61.** $s(t) = \frac{1}{3}t^3 + \frac{1}{2}t^2 - 2t + 3$
63. $s(t) = -10\sin t - 3\cos t + (6/\pi)t + 3$
65. (a) $s(t) = 450 - 4.9t^2$ (b) $\sqrt{450/4.9} \approx 9.58$ s
(c) $-9.8\sqrt{450/4.9} \approx -93.9$ m/s   (d) About 9.09 s
69. 225 ft   **71.** $\$742.08$ **73.** $\frac{130}{11} \approx 11.8$ s
75. $\frac{88}{15} \approx 5.87$ ft/s^2 **77.** $62{,}500$ km/h$^2 \approx 4.82$ m/s^2
79. (a) 22.9125 mi   (b) 21.675 mi   (c) 30 min 33 s
(d) 55.425 mi

CHAPTER 4 REVIEW ■ PAGE 358

True-False Quiz

1. False   **3.** False   **5.** True   **7.** False   **9.** True
11. True   **13.** False   **15.** True   **17.** True
19. True   **21.** False

Exercises

1. Abs max $f(2) = f(5) = 18$, abs min $f(0) = -2$,
loc max $f(2) = 18$, loc min $f(4) = 14$
3. Abs max $f(2) = \frac{2}{5}$, abs and loc min $f(-\frac{1}{3}) = -\frac{9}{2}$
5. Abs and loc max $f(\pi/6) = \pi/6 + \sqrt{3}$,
abs min $f(-\pi) = -\pi - 2$, loc min $f(5\pi/6) = 5\pi/6 - \sqrt{3}$
7. 1 **9.** 4 **11.** 0 **13.** $\frac{1}{2}$
15.

17.

19. A. $\mathbb{R}$ **B.** y-int 2
C. None   **D.** None
E. Dec on $(-\infty, \infty)$ **F.** None
G. CU on $(-\infty, 0)$;
CD on $(0, \infty)$; IP $(0, 2)$
H. See graph at right.

21. A. $\mathbb{R}$ **B.** y-int 2
C. None   **D.** None
E. Inc on $(1, \infty)$; dec on $(-\infty, 1)$
F. Loc min $f(1) = 1$
G. CU on $(-\infty, 0)$, $\left(\frac{2}{3}, \infty\right)$;
CD on $\left(0, \frac{2}{3}\right)$; IPs $(0, 2)$, $\left(\frac{2}{3}, \frac{38}{27}\right)$
H. See graph at right.

23. A. $\{x \mid x \neq 0, 3\}$
B. None   **C.** None
D. HA $y = 0$; VA $x = 0$, $x = 3$
E. Inc on $(1, 3)$;
dec on $(-\infty, 0)$, $(0, 1)$, $(3, \infty)$
F. Loc min $f(1) = \frac{1}{4}$
G. CU on $(0, 3)$, $(3, \infty)$;
CD on $(-\infty, 0)$
H. See graph at right.

25. A. $(-\infty, 0) \cup (0, \infty)$
B. x-int 1   **C.** None
D. VA $x = 0$; SA $y = x - 3$
E. Inc on $(-\infty, -2)$, $(0, \infty)$;
dec on $(-2, 0)$
F. Loc max $f(-2) = -\frac{27}{4}$
G. CU on $(1, \infty)$; CD on $(-\infty, 0)$,
$(0, 1)$; IP $(1, 0)$
H. See graph at right.

27. A. $[-2, \infty)$
B. y-int 0; x-int $-2, 0$
C. None   **D.** None
E. Inc on $\left(-\frac{4}{3}, \infty\right)$, dec on $\left(-2, -\frac{4}{3}\right)$
F. Loc min $f\left(-\frac{4}{3}\right) = -\frac{4}{9}\sqrt{6}$
G. CU on $(-2, \infty)$
H. See graph at right.

29. A. $[-\pi, \pi]$ **B.** y-int 0; x-int $-\pi, 0, \pi$
C. None   **D.** None
E. Inc on $(-\pi/4, 3\pi/4)$; dec on $(-\pi, -\pi/4)$, $(3\pi/4, \pi)$
F. Loc max $f(3\pi/4) = \frac{1}{2}\sqrt{2}\,e^{3\pi/4}$,
loc min $f(-\pi/4) = -\frac{1}{2}\sqrt{2}\,e^{-\pi/4}$

G. CU on $(-\pi/2, \pi/2)$; CD on $(-\pi, -\pi/2)$, $(\pi/2, \pi)$;
IPs $(-\pi/2, -e^{-\pi/2})$, $(\pi/2, e^{\pi/2})$
H.

31. A. $\{x \mid |x| \geq 1\}$
B. None   **C.** About $(0, 0)$
D. HA $y = 0$
E. Dec on $(-\infty, -1)$, $(1, \infty)$
F. None
G. CU on $(1, \infty)$;
CD on $(-\infty, -1)$
H. See graph at right.

33. A. $\mathbb{R}$
B. y-int -2; x-int 2
C. None   **D.** HA $y = 0$
E. Inc on $(-\infty, 3)$; dec on $(3, \infty)$
F. Loc max $f(3) = e^{-3}$
G. CU on $(4, \infty)$;
CD on $(-\infty, 4)$;
IP $(4, 2e^{-4})$
H. See graph at right.

35. Inc on $\left(-\sqrt{3}, 0\right)$, $\left(0, \sqrt{3}\right)$;
dec on $\left(-\infty, -\sqrt{3}\right)$, $\left(\sqrt{3}, \infty\right)$;
loc max $f\left(\sqrt{3}\right) = \frac{2}{9}\sqrt{3}$,
loc min $f\left(-\sqrt{3}\right) = -\frac{2}{9}\sqrt{3}$;
CU on $\left(-\sqrt{6}, 0\right)$, $\left(\sqrt{6}, \infty\right)$;
CD on $\left(-\infty, -\sqrt{6}\right)$, $\left(0, \sqrt{6}\right)$;
IPs $\left(\sqrt{6}, \frac{5}{36}\sqrt{6}\right)$, $\left(-\sqrt{6}, -\frac{5}{36}\sqrt{6}\right)$

37. Inc on $(-0.23, 0)$, $(1.62, \infty)$; dec on $(-\infty, -0.23)$, $(0, 1.62)$;
loc max $f(0) = 2$; loc min $f(-0.23) \approx 1.96$, $f(1.62) \approx -19.2$;
CU on $(-\infty, -0.12)$, $(1.24, \infty)$;
CD on $(-0.12, 1.24)$; IPs $(-0.12, 1.98)$, $(1.24, -12.1)$

39.

$(\pm 0.82, 0.22)$; $\left(\pm\sqrt{2/3}, e^{-3/2}\right)$

41. $-2.96, -0.18, 3.01; -1.57, 1.57; -2.16, -0.75, 0.46, 2.21$
43. For $C > -1$, f is periodic with period 2π and has local maxima at $2n\pi + \pi/2$, n an integer. For $C \le -1$, f has no graph. For $-1 < C \le 1$, f has vertical asymptotes. For $C > 1$, f is continuous on $\mathbb{R}$. As C increases, f moves upward and its oscillations become less pronounced.
49. (a) 0   (b) CU on $\mathbb{R}$ **53.** $3\sqrt{3}\,r^2$
55. $4/\sqrt{3}$ cm from D **57.** $L = C$ **59.** \$11.50
61. 1.297383   **63.** 1.16718557
65. $F(x) = \frac{8}{3}x^{3/2} - 2x^3 + 3x + C$
67. $F(t) = -2\cos t - 3e^t + C$
69. $f(t) = t^2 + 3\cos t + 2$
71. $f(x) = \frac{1}{2}x^2 - x^3 + 4x^4 + 2x + 1$
73. $s(t) = t^2 - \tan^{-1} t + 1$
75. (b) $0.1e^x - \cos x + 0.9$ (c)

77. No
79. (b) About 8.5 in. by 2 in.   (c) $20/\sqrt{3}$ in. by $20\sqrt{2/3}$ in.
85. (a) $20\sqrt{2} \approx 28$ ft
(b) $\dfrac{dI}{dt} = \dfrac{-480k(h-4)}{[(h-4)^2 + 1600]^{5/2}}$, where k is the constant of proportionality

PROBLEMS PLUS ■ PAGE 363

3. Abs max $f(-5) = e^{45}$, no abs min   **7.** 24
9. $(-2, 4), (2, -4)$ **13.** $(1 + \sqrt{5})/2$ **15.** $(m/2, m^2/4)$
17. $a \le e^{1/e}$
21. (a) $T_1 = D/c_1$, $T_2 = (2h\sec\theta)/c_1 + (D - 2h\tan\theta)/c_2$, $T_3 = \sqrt{4h^2 + D^2}/c_1$
(c) $c_1 \approx 3.85$ km/s, $c_2 \approx 7.66$ km/s, $h \approx 0.42$ km
25. $3/(\sqrt[3]{2} - 1) \approx 11\frac{1}{2}$ h

CHAPTER 5

EXERCISES 5.1 ■ PAGE 375

1. (a) $R_5 \approx 12$, $L_5 \approx 22$

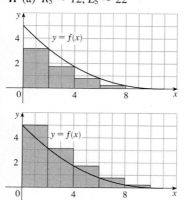

(b) $R_{10} \approx 14.4$, $L_{10} \approx 19.4$

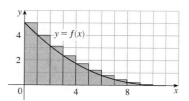

3. (a) 0.6345, underestimate   (b) 0.7595, overestimate

5. (a) 8, 6.875   (b) 5, 5.375

(c) 5.75, 5.9375

(d) M_6

7. $n = 2$: upper $= 3\pi \approx 9.42$, lower $= 2\pi \approx 6.28$

$n = 4$: upper $= \left(10 + \sqrt{2}\,\right)(\pi/4) \approx 8.96$,

lower $= \left(8 + \sqrt{2}\,\right)(\pi/4) \approx 7.39$

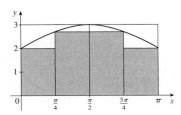

$n = 8$: upper ≈ 8.65, lower ≈ 7.86

9. 0.2533, 0.2170, 0.2101, 0.2050; 0.2
11. (a) Left: 0.8100, 0.7937, 0.7904;
right: 0.7600, 0.7770, 0.7804
13. 34.7 ft, 44.8 ft   **15.** 63.2 L, 70 L   **17.** 155 ft
19. 7840   **21.** $\displaystyle\lim_{n\to\infty} \sum_{i=1}^{n} \frac{2(1 + 2i/n)}{(1 + 2i/n)^2 + 1} \cdot \frac{2}{n}$
23. $\displaystyle\lim_{n\to\infty} \sum_{i=1}^{n} \sqrt{\sin(\pi i/n)} \cdot \frac{\pi}{n}$
25. The region under the graph of $y = \tan x$ from 0 to $\pi/4$
27. (a) $L_n < A < R_n$
29. (a) $\displaystyle\lim_{n\to\infty} \frac{64}{n^6} \sum_{i=1}^{n} i^5$ (b) $\dfrac{n^2(n + 1)^2(2n^2 + 2n - 1)}{12}$
(c) $\frac{32}{3}$
31. $\sin b$, 1

EXERCISES 5.2 ■ PAGE 388

1. -10
The Riemann sum represents
the sum of the areas of the two
rectangles above the x-axis minus
the sum of the areas of the three
rectangles below the x-axis; that is,
the *net area* of the rectangles with
respect to the x-axis.

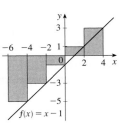

3. $-\frac{49}{16}$
The Riemann sum represents the
sum of the areas of the two rectangles above the x-axis minus the
sum of the areas of the four
rectangles below the x-axis.

5. (a) 6   (b) 4   (c) 2
7. Lower, $L_5 = -64$; upper, $R_5 = 16$
9. 6.1820   **11.** 0.9071   **13.** 0.9029, 0.9018
15.

n	R_n
5	1.933766
10	1.983524
50	1.999342
100	1.999836

The values of R_n appear to be approaching 2.

17. $\displaystyle\int_0^1 \frac{e^x}{1 + x}\, dx$ **19.** $\displaystyle\int_2^7 (5x^3 - 4x)\, dx$
21. -9 **23.** $\frac{2}{3}$ **25.** $-\frac{3}{4}$
29. $\displaystyle\lim_{n\to\infty} \sum_{i=1}^{n} \sqrt{4 + (1 + 2i/n)} \cdot \frac{2}{n}$
31. $\displaystyle\lim_{n\to\infty} \sum_{i=1}^{n} \left(\sin \frac{5\pi i}{n}\right) \frac{\pi}{n} = \frac{2}{5}$
33. (a) 4   (b) 10   (c) -3 (d) 2
35. $\frac{3}{2}$ **37.** $3 + \frac{9}{4}\pi$ **39.** $\frac{25}{4}$ **41.** 0   **43.** 3
45. $e^5 - e^3$ **47.** $\int_{-1}^5 f(x)\, dx$ **49.** 122
51. $B < E < A < D < C$ **53.** 15
59. $0 \le \int_0^1 x^3\, dx \le 1$ **61.** $\dfrac{\pi}{12} \le \displaystyle\int_{\pi/4}^{\pi/3} \tan x\, dx \le \dfrac{\pi}{12}\sqrt{3}$
63. $0 \le \int_0^2 xe^{-x}\, dx \le 2/e$ **67.** $\int_1^2 \arctan x\, dx$
73. $\int_0^1 x^4\, dx$ **75.** $\frac{1}{2}$

EXERCISES 5.3 ■ PAGE 399

1. One process undoes what the other one does. See the Fundamental Theorem of Calculus, page 398.
3. (a) 0, 2, 5, 7, 3   (d)
(b) (0, 3)
(c) $x = 3$

5.   (a), (b) x^2

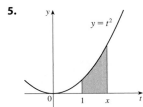

7. $g'(x) = \sqrt{x + x^3}$ **9.** $g'(s) = (s - s^2)^8$

11. $F'(x) = -\sqrt{1 + \sec x}$ **13.** $h'(x) = xe^x$

15. $y' = \dfrac{3(3x + 2)}{1 + (3x + 2)^3}$

17. $y' = -\frac{1}{2}\tan\sqrt{x}$ **19.** $\frac{26}{3}$ **21.** 2   **23.** $\frac{52}{3}$

25. $1 + \sqrt{3}/2$ **27.** $-\frac{37}{6}$ **29.** $\frac{82}{5}$ **31.** 1   **33.** $\frac{15}{4}$

35. $\ln 2 + 7$ **37.** $\dfrac{1}{e + 1} + e - 1$ **39.** $4\pi/3$

41. $\dfrac{15}{\ln 2}$ **43.** 0   **45.** $\frac{16}{3}$ **47.** $\frac{32}{3}$

49. $\frac{243}{4}$ **51.** 2

53. 3.75

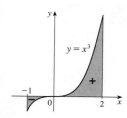

55. The function $f(x) = x^{-4}$ is not continuous on the interval $[-2, 1]$, so FTC2 cannot be applied.

57. The function $f(\theta) = \sec\theta\,\tan\theta$ is not continuous on the interval $[\pi/3, \pi]$, so FTC2 cannot be applied.

59. $g'(x) = \dfrac{-2(4x^2 - 1)}{4x^2 + 1} + \dfrac{3(9x^2 - 1)}{9x^2 + 1}$

61. $F'(x) = 2xe^{x^4} - e^{x^2}$

63. $y' = \sin x\,\ln(1 + 2\cos x) + \cos x\,\ln(1 + 2\sin x)$

65. $(-4, 0)$ **67.** $y = e^4 x - 2e^4$ **69.** 29

71. (a) $-2\sqrt{n},\ \sqrt{4n - 2}$, n an integer > 0

(b) $(0, 1),\ (-\sqrt{4n - 1},\ -\sqrt{4n - 3})$, and $(\sqrt{4n - 1},\ \sqrt{4n + 1})$, n an integer > 0 (c) 0.74

73. (a) Loc max at 1 and 5; loc min at 3 and 7

(b) $x = 9$

(c) $(\frac{1}{2}, 2),\ (4, 6),\ (8, 9)$

(d) See graph at right.

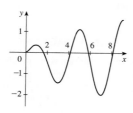

75. $\frac{7}{10}$ **83.** $f(x) = x^{3/2},\ a = 9$

85. (b) Average expenditure over $[0, t]$; minimize average expenditure

EXERCISES 5.4 ■ **PAGE 408**

5. $\frac{1}{2.3}x^{2.3} + 2x^{3.5} + C$ **7.** $5x + \frac{2}{9}x^3 + \frac{3}{16}x^4 + C$

9. $\frac{2}{3}u^3 + \frac{9}{2}u^2 + 4u + C$ **11.** $\ln|x| + 2\sqrt{x} + x + C$

13. $-\cos x + \cosh x + C$ **15.** $\theta + \tan\theta + C$

17. $\dfrac{2^t}{\ln 2} + \dfrac{10^t}{\ln 10} + C$

19. $\sin x + \frac{1}{4}x^2 + C$

21. $-\frac{10}{3}$ **23.** $\frac{21}{5}$ **25.** -2 **27.** $5e^\pi + 1$ **29.** 36

31. $\frac{55}{63}$ **33.** $\frac{3}{4} - 2\ln 2$ **35.** $\dfrac{1}{11} + \dfrac{9}{\ln 10}$

37. $1 + \pi/4$ **39.** $\frac{69}{4}$ **41.** $\pi/3$ **43.** $\pi/6$ **45.** -3.5

47. ≈ 1.36 **49.** $\frac{4}{3}$

51. The increase in the child's weight (in pounds) between the ages of 5 and 10

53. Number of gallons of oil leaked in the first 2 hours

55. Increase in revenue when production is increased from 1000 to 5000 units

57. Newton-meters (or joules)   **59.** (a) $-\frac{3}{2}$ m   (b) $\frac{41}{6}$ m

61. (a) $v(t) = \frac{1}{2}t^2 + 4t + 5$ m/s   (b) $416\frac{2}{3}$ m

63. $46\frac{2}{3}$ kg   **65.** 1.4 mi   **67.** \$58,000   **69.** 39.8 ft/s

71. 5443 bacteria   **73.** 4.75×10^5 megawatt-hours

EXERCISES 5.5 ■ **PAGE 418**

1. $\frac{1}{2}\sin 2x + C$ **3.** $\frac{2}{9}(x^3 + 1)^{3/2} + C$

5. $\frac{1}{4}\ln|x^4 - 5| + C$ **7.** $-\frac{1}{3}(1 - x^2)^{3/2} + C$

9. $-\frac{1}{20}(1 - 2x)^{10} + C$ **11.** $(2/\pi)\sin(\pi t/2) + C$

13. $-\frac{1}{3}\ln|5 - 3x| + C$

15. $-\frac{1}{4}\cos^4\theta + C$ **17.** $\dfrac{1}{1 - e^u} + C$

19. $\frac{2}{3}\sqrt{3ax + bx^3} + C$ **21.** $\frac{1}{3}(\ln x)^3 + C$ **23.** $\frac{1}{4}\tan^4\theta + C$

25. $\frac{2}{3}(1 + e^x)^{3/2} + C$ **27.** $\frac{1}{15}(x^3 + 3x)^5 + C$

29. $-\dfrac{1}{\ln 5}\cos(5^t) + C$ **31.** $\frac{1}{3}(\arctan x)^3 + C$

33. $\frac{1}{5}\sin(1 + 5t) + C$ **35.** $-\frac{2}{3}(\cot x)^{3/2} + C$

37. $\frac{1}{3}\sinh^3 x + C$ **39.** $-\ln(1 + \cos^2 x) + C$

41. $\ln|\sin x| + C$ **43.** $\ln|\sin^{-1}x| + C$

45. $\tan^{-1}x + \frac{1}{2}\ln(1 + x^2) + C$

47. $\frac{1}{40}(2x + 5)^{10} - \frac{5}{36}(2x + 5)^9 + C$

49. $\frac{1}{8}(x^2 - 1)^4 + C$ **51.** $-e^{\cos x} + C$

53. $2/\pi$ **55.** $\frac{45}{28}$ **57.** $2/\sqrt{3} - 1$ **59.** $e - \sqrt{e}$

61. 0   **63.** 3   **65.** $\frac{1}{3}(2\sqrt{2} - 1)a^3$ **67.** $\frac{16}{15}$ **69.** 2

71. $\ln(e + 1)$ **73.** $\frac{1}{6}$ **75.** $\sqrt{3} - \frac{1}{3}$ **77.** 6π

79. All three areas are equal.   **81.** ≈ 4512 L

83. $\dfrac{5}{4\pi}\left(1 - \cos\dfrac{2\pi t}{5}\right)$ L

85. $C_0(1 - e^{-30r/V})$; the total amount of urea removed from the blood in the first 30 minutes of dialysis treatment
87. 5   **93.** $\pi^2/4$

CHAPTER 5 REVIEW ■ PAGE 421

True-False Quiz
1. True   **3.** True   **5.** False   **7.** True   **9.** True
11. False   **13.** True   **15.** False   **17.** False

Exercises
1. (a) 8   (b) 5.7

   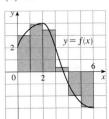

3. $\frac{1}{2} + \pi/4$ **5.** 3   **7.** f is c, f' is b, $\int_0^x f(t)\,dt$ is a
9. 3, 0   **11.** 37   **13.** $\frac{9}{10}$ **15.** -76 **17.** $\frac{21}{4}$
19. Does not exist   **21.** $\frac{1}{3}\sin 1$ **23.** 0
25. $-(1/x) - 2\ln|x| + x + C$
27. $\sqrt{x^2 + 4x} + C$ **29.** $[1/(2\pi)]\sin^2 \pi t + C$
31. $2e^{\sqrt{x}} + C$ **33.** $-\frac{1}{2}[\ln(\cos x)]^2 + C$
35. $\frac{1}{4}\ln(1 + x^4) + C$ **37.** $\ln|1 + \sec\theta| + C$ **39.** $\frac{23}{3}$
41. $2\sqrt{1 + \sin x} + C$ **43.** $\frac{64}{5}$ **45.** $F'(x) = x^2/(1 + x^3)$
47. $g'(x) = 4x^3\cos(x^8)$ **49.** $y' = (2e^x - e^{\sqrt{x}})/(2x)$
51. $4 \le \int_1^3 \sqrt{x^2 + 3}\,dx \le 4\sqrt{3}$ **57.** 0.280981
59. Number of barrels of oil consumed from Jan. 1, 2000, through Jan. 1, 2008
61. 72,400   **63.** 3   **65.** $c \approx 1.62$
67. $f(x) = e^{2x}(2x - 1)/(1 - e^{-x})$ **73.** $\frac{2}{3}$

PROBLEMS PLUS ■ PAGE 425

1. $\pi/2$ **3.** $2k$ **5.** -1 **7.** e^{-2} **9.** $[-1, 2]$
11. (a) $\frac{1}{2}(n - 1)n$
(b) $\frac{1}{2}\llbracket b \rrbracket(2b - \llbracket b \rrbracket - 1) - \frac{1}{2}\llbracket a \rrbracket(2a - \llbracket a \rrbracket - 1)$
17. $y = -\dfrac{2b}{a^2}x^2 + \dfrac{3b}{a}x$ **19.** $2(\sqrt{2} - 1)$

CHAPTER 6

EXERCISES 6.1 ■ PAGE 434

1. $\frac{45}{4} - \ln 8$ **3.** $e - (1/e) + \frac{10}{3}$ **5.** $e - (1/e) + \frac{4}{3}$
7. $\frac{9}{2}$ **9.** $\ln 2 - \frac{1}{2}$ **11.** $\frac{8}{3}$ **13.** 72   **15.** $6\sqrt{3}$
17. $\frac{32}{3}$ **19.** $2/\pi + \frac{2}{3}$ **21.** $2 - 2\ln 2$
23. $\frac{47}{3} - \frac{9}{2}\sqrt[3]{12}$ **25.** $\frac{13}{5}$ **27.** $\ln 2$
29. (a) 39   (b) 15   **31.** $\frac{1}{6}\ln 2$ **33.** $\frac{5}{2}$ **35.** $\frac{3}{2}\sqrt{3} - 1$
37. 0, 0.90; 0.04   **39.** $-1.11, 1.25, 2.86$; 8.38
41. 2.80123   **43.** 0.25142   **45.** $12\sqrt{6} - 9$
47. $117\frac{1}{3}$ ft   **49.** 4232 cm^2

51. (a) Twelfth ($t \approx 11.26$)   (b) Eighteenth ($t \approx 17.18$)
(c) 706
53. (a) Car A   (b) The distance by which A is ahead of B after 1 minute   (c) Car A   (d) $t \approx 2.2$ min
55. $\frac{24}{5}\sqrt{3}$ **57.** $4^{2/3}$ **59.** ± 6
61. $0 < m < 1$; $m - \ln m - 1$

EXERCISES 6.2 ■ PAGE 446

1. $26\pi/3$

3. 8π

5. 162π

7. $4\pi/21$

9. $64\pi/15$

11. $11\pi/30$

13. $2\pi\left(\frac{4}{3}\pi - \sqrt{3}\right)$

15. $3\pi/5$

17. $10\sqrt{2}\,\pi/3$

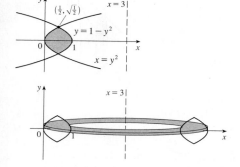

19. $\pi/3$ **21.** $\pi/3$ **23.** $\pi/3$
25. $13\pi/45$ **27.** $\pi/3$ **29.** $17\pi/45$

31. (a) $2\pi\int_0^1 e^{-2x^2}\,dx \approx 3.75825$
(b) $2\pi\int_0^1\left(e^{-2x^2} + 2e^{-x^2}\right)dx \approx 13.14312$
33. (a) $2\pi\int_0^2 8\sqrt{1 - x^2/4}\ dx \approx 78.95684$
(b) $2\pi\int_0^1 8\sqrt{4 - 4y^2}\ dy \approx 78.95684$
35. $-4.091, -1.467, 1.091; 89.023$ **37.** $\frac{11}{8}\pi^2$
39. Solid obtained by rotating the region $0 \le x \le \pi$,
$0 \le y \le \sqrt{\sin x}$ about the x-axis
41. Solid obtained by rotating the region above the x-axis
bounded by $x = y^2$ and $x = y^4$ about the y-axis
43. 1110 cm^3 **45.** (a) 196 (b) 838
47. $\frac{1}{3}\pi r^2 h$ **49.** $\pi h^2\left(r - \frac{1}{3}h\right)$ **51.** $\frac{2}{3}b^2h$
53. 10 cm^3 **55.** 24 **57.** $\frac{1}{3}$ **59.** $\frac{8}{15}$ **61.** $4\pi/15$
63. (a) $8\pi R\int_0^r \sqrt{r^2 - y^2}\ dy$ (b) $2\pi^2 r^2 R$
65. (b) $\pi r^2 h$ **67.** $\frac{5}{12}\pi r^3$ **69.** $8\int_0^r \sqrt{R^2 - y^2}\,\sqrt{r^2 - y^2}\ dy$

EXERCISES 6.3   ▪   PAGE 453

1. Circumference $= 2\pi x$, height $= x(x - 1)^2$; $\pi/15$

3. $6\pi/7$ **5.** $\pi(1 - 1/e)$ **7.** 8π
9. 4π **11.** 192π **13.** $16\pi/3$
15. $264\pi/5$ **17.** $8\pi/3$ **19.** $13\pi/3$
21. (a) $2\pi\int_0^2 x^2 e^{-x}\,dx$ (b) 4.06300
23. (a) $4\pi\int_{-\pi/2}^{\pi/2}(\pi - x)\cos^4 x\,dx$ (b) 46.50942
25. (a) $\int_0^\pi 2\pi(4 - y)\sqrt{\sin y}\ dy$ (b) 36.57476
27. 3.68
29. Solid obtained by rotating the region $0 \le y \le x^4, 0 \le x \le 3$
about the y-axis
31. Solid obtained (using shells) by rotating the region
$0 \le x \le 1/y^2, 1 \le y \le 4$ about the line $y = -2$
33. $0, 2.175; 14.450$ **35.** $\frac{1}{32}\pi^3$ **37.** 8π
39. $4\sqrt{3}\,\pi$ **41.** $4\pi/3$
43. $117\pi/5$ **45.** $\frac{4}{3}\pi r^3$ **47.** $\frac{1}{3}\pi r^2 h$

EXERCISES 6.4   ▪   PAGE 458

1. (a) 7200 ft-lb   (b) 7200 ft-lb
3. 4.5 ft-lb   **5.** 180 J   **7.** $\frac{15}{4}$ ft-lb
9. (a) $\frac{25}{24} \approx 1.04$ J   (b) 10.8 cm   **11.** $W_2 = 3W_1$
13. (a) 625 ft-lb   (b) $\frac{1875}{4}$ ft-lb   **15.** $650{,}000$ ft-lb
17. 3857 J   **19.** 62.5 ft-lb   **21.** 2450 J
23. $\approx 1.06 \times 10^6$ J   **25.** $\approx 1.04 \times 10^5$ ft-lb   **27.** 2.0 m
31. (b) $161.\overline{3}$ ft-lb
33. (a) $Gm_1 m_2\left(\dfrac{1}{a} - \dfrac{1}{b}\right)$ (b) $\approx 8.50 \times 10^9$ J

EXERCISES 6.5 ■ PAGE 463

1. 7   **3.** $6/\pi$ **5.** $(2/\pi)(e-1)$ **7.** $2/(5\pi)$

9. (a) 1   (b) 2, 4   (c)

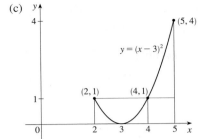

11. (a) $4/\pi$ (b) $\approx 1.24, 2.81$
(c)

15. $\frac{9}{8}$ **17.** $(50 + 28/\pi)°F \approx 59°F$ **19.** 6 kg/m
21. About 4056 million (or 4 billion) people
23. $5/(4\pi) \approx 0.4$ L

CHAPTER 6 REVIEW ■ PAGE 466

Exercises

1. $\frac{8}{3}$ **3.** $\frac{7}{12}$ **5.** $\frac{4}{3} + 4/\pi$ **7.** $64\pi/15$ **9.** $1656\pi/5$
11. $\frac{4}{3}\pi(2ah + h^2)^{3/2}$ **13.** $\int_{-\pi/3}^{\pi/3} 2\pi\left(\pi/2 - x\right)\left(\cos^2 x - \frac{1}{4}\right) dx$
15. (a) $2\pi/15$ (b) $\pi/6$ (c) $8\pi/15$
17. (a) 0.38   (b) 0.87
19. Solid obtained by rotating the region $0 \le y \le \cos x$,
$0 \le x \le \pi/2$ about the y-axis
21. Solid obtained by rotating the region $0 \le x \le \pi$,
$0 \le y \le 2 - \sin x$ about the x-axis
23. 36   **25.** $\frac{125}{3}\sqrt{3}$ m³   **27.** 3.2 J
29. (a) $8000\pi/3 \approx 8378$ ft-lb   (b) 2.1 ft
31. $4/\pi$ **33.** $f(x)$

PROBLEMS PLUS ■ PAGE 468

1. (a) $f(t) = 3t^2$ (b) $f(x) = \sqrt{2x/\pi}$ **3.** $\frac{32}{27}$
5. (b) 0.2261   (c) 0.6736 m
(d) (i) $1/(105\pi) \approx 0.003$ in/s   (ii) $370\pi/3$ s ≈ 6.5 min
9. $y = \frac{32}{9}x^2$
11. (a) $V = \int_0^h \pi[f(y)]^2 dy$
(c) $f(y) = \sqrt{kA/(\pi C)}\, y^{1/4}$. Advantage: the markings on the
container are equally spaced.
13. $b = 2a$ **15.** $B = 16A$

CHAPTER 7

EXERCISES 7.1 ■ PAGE 476

1. $\frac{1}{2}xe^{2x} - \frac{1}{4}e^{2x} + C$ **3.** $\frac{1}{5}x \sin 5x + \frac{1}{25} \cos 5x + C$
5. $-\frac{1}{3}te^{-3t} - \frac{1}{9}e^{-3t} + C$
7. $(x^2 + 2x) \sin x + (2x + 2) \cos x - 2 \sin x + C$
9. $x \cos^{-1}x - \sqrt{1 - x^2} + C$ **11.** $\frac{1}{5}t^5 \ln t - \frac{1}{25}t^5 + C$
13. $-t \cot t + \ln|\sin t| + C$
15. $x(\ln x)^2 - 2x \ln x + 2x + C$
17. $\frac{1}{13}e^{2\theta}(2 \sin 3\theta - 3 \cos 3\theta) + C$
19. $z^3 e^z - 3z^2 e^z + 6ze^z - 6e^z + C$
21. $\dfrac{e^{2x}}{4(2x + 1)} + C$ **23.** $\dfrac{\pi - 2}{2\pi^2}$
25. $2 \cosh 2 - \sinh 2$ **27.** $\frac{4}{5} - \frac{1}{5} \ln 5$ **29.** $-\pi/4$
31. $2e^{-1} - 6e^{-5}$ **33.** $\frac{1}{2} \ln 2 - \frac{1}{2}$
35. $\frac{32}{5}(\ln 2)^2 - \frac{64}{25} \ln 2 + \frac{62}{125}$
37. $2\sqrt{x}\, e^{\sqrt{x}} - 2e^{\sqrt{x}} + C$ **39.** $-\frac{1}{2} - \pi/4$
41. $\frac{1}{2}(x^2 - 1) \ln(1 + x) - \frac{1}{4}x^2 + \frac{1}{2}x + \frac{3}{4} + C$
43. $-\frac{1}{2}xe^{-2x} - \frac{1}{4}e^{-2x} + C$

45. $\frac{1}{3}x^2(1 + x^2)^{3/2} - \frac{2}{15}(1 + x^2)^{5/2} + C$

47. (b) $-\frac{1}{4} \cos x \sin^3 x + \frac{3}{8}x - \frac{3}{16} \sin 2x + C$
49. (b) $\frac{2}{3}, \frac{8}{15}$
55. $x[(\ln x)^3 - 3(\ln x)^2 + 6 \ln x - 6] + C$
57. $\frac{16}{9} \ln 2 - \frac{29}{9}$ **59.** $-1.75119, 1.17210; 3.99926$
61. $4 - 8/\pi$ **63.** $2\pi e$
65. (a) $2\pi(2 \ln 2 - \frac{3}{4})$ (b) $2\pi[(\ln 2)^2 - 2 \ln 2 + 1]$
67. $x S(x) + \frac{1}{\pi} \cos(\frac{1}{2}\pi x^2) + C$
69. $2 - e^{-t}(t^2 + 2t + 2)$ m   **71.** 2

EXERCISES 7.2 ■ PAGE 484

1. $\frac{1}{3} \sin^3 x - \frac{1}{5} \sin^5 x + C$ **3.** $\frac{1}{120}$
5. $-\frac{1}{14} \cos^7(2t) + \frac{1}{5} \cos^5(2t) - \frac{1}{6} \cos^3(2t) + C$
7. $\pi/4$ **9.** $3\pi/8$ **11.** $\pi/16$
13. $\frac{2}{7}(\cos \theta)^{7/2} - \frac{2}{3}(\cos \theta)^{3/2} + C$
15. $\ln|\sin x| - \frac{1}{2}\sin^2 x + C$ **17.** $\frac{1}{2}\sin^4 x + C$
19. $\frac{1}{4}t^2 - \frac{1}{4}t \sin 2t - \frac{1}{8} \cos 2t + C$ **21.** $\frac{1}{3} \sec^3 x + C$
23. $\tan x - x + C$ **25.** $\frac{1}{9} \tan^9 x + \frac{2}{7} \tan^7 x + \frac{1}{5} \tan^5 x + C$
27. $\frac{1}{3} \sec^3 x - \sec x + C$ **29.** $\frac{1}{8} \tan^8 x + \frac{1}{3} \tan^6 x + \frac{1}{4} \tan^4 x + C$
31. $\frac{1}{4} \sec^4 x - \tan^2 x + \ln|\sec x| + C$

33. $x \sec x - \ln|\sec x + \tan x| + C$ **35.** $\sqrt{3} - \frac{1}{3}\pi$

37. $\frac{22}{105}\sqrt{2} - \frac{8}{105}$ **39.** $\ln|\csc x - \cot x| + C$

41. $-\frac{1}{6}\cos 3x - \frac{1}{26}\cos 13x + C$ **43.** $\frac{1}{15}$

45. $\frac{1}{2}\sqrt{2}$ **47.** $\frac{1}{2}\sin 2x + C$

49. $x \tan x - \ln|\sec x| - \frac{1}{2}x^2 + C$

51. $\frac{1}{4}x^2 - \frac{1}{4}\sin(x^2)\cos(x^2) + C$

53. $\frac{1}{6}\sin 3x - \frac{1}{18}\sin 9x + C$

55. 0   **57.** $\frac{1}{2}\pi - \frac{4}{3}$ **59.** 0   **61.** $\pi^2/4$

63. $\pi\left(2\sqrt{2} - \frac{5}{2}\right)$ **65.** $s = (1 - \cos^3\omega t)/(3\omega)$

EXERCISES 7.3 ■ PAGE 491

1. $-\dfrac{\sqrt{4 - x^2}}{4x} + C$ **3.** $\sqrt{x^2 - 4} - 2\sec^{-1}\left(\dfrac{x}{2}\right) + C$

5. $\frac{1}{3}\dfrac{(x^2 - 1)^{3/2}}{x^3} + C$ **7.** $\dfrac{1}{\sqrt{2}\,a^2}$

9. $\frac{2}{3}\sqrt{3} - \frac{3}{4}\sqrt{2}$ **11.** $\frac{1}{12}$

13. $\frac{1}{6}\sec^{-1}(x/3) - \sqrt{x^2 - 9}/(2x^2) + C$

15. $\frac{1}{16}\pi a^4$ **17.** $\sqrt{x^2 - 7} + C$

19. $\ln\left|\left(\sqrt{1 + x^2} - 1\right)/x\right| + \sqrt{1 + x^2} + C$ **21.** $\frac{9}{500}\pi$

23. $\ln\left|\sqrt{x^2 + 2x + 5} + x + 1\right| + C$

25. $4\sin^{-1}\left(\dfrac{x - 1}{2}\right) + \frac{1}{4}(x - 1)^3\sqrt{3 + 2x - x^2}$
$$- \frac{2}{3}(3 + 2x - x^2)^{3/2} + C$$

27. $\frac{1}{2}(x + 1)\sqrt{x^2 + 2x} - \frac{1}{2}\ln\left|x + 1 + \sqrt{x^2 + 2x}\right| + C$

29. $\frac{1}{4}\sin^{-1}(x^2) + \frac{1}{4}x^2\sqrt{1 - x^4} + C$

33. $\frac{1}{6}\left(\sqrt{48} - \sec^{-1}7\right)$ **37.** $\frac{3}{8}\pi^2 + \frac{3}{4}\pi$

41. $2\pi^2 R r^2$ **43.** $r\sqrt{R^2 - r^2} + \pi r^2/2 - R^2\arcsin(r/R)$

EXERCISES 7.4 ■ PAGE 501

1. (a) $\dfrac{A}{1 + 2x} + \dfrac{B}{3 - x}$ (b) $\dfrac{A}{x} + \dfrac{B}{x^2} + \dfrac{C}{x^3} + \dfrac{D}{1 + x}$

3. (a) $\dfrac{A}{x} + \dfrac{B}{x^2} + \dfrac{Cx + D}{1 + x^2}$ (b) $1 + \dfrac{A}{x} + \dfrac{B}{x - 1} + \dfrac{C}{x - 2}$

5. (a) $x^4 + 4x^2 + 16 + \dfrac{A}{x + 2} + \dfrac{B}{x - 2}$

(b) $\dfrac{Ax + B}{x^2 - x + 1} + \dfrac{Cx + D}{x^2 + 2} + \dfrac{Ex + F}{(x^2 + 2)^2}$

7. $\frac{1}{4}x^4 + \frac{1}{3}x^3 + \frac{1}{2}x^2 + x + \ln|x - 1| + C$

9. $\frac{1}{2}\ln|2x + 1| + 2\ln|x - 1| + C$ **11.** $2\ln\frac{3}{2}$

13. $a\ln|x - b| + C$ **15.** $\frac{5}{2} - \ln 2 - \ln 3$ $\left(\text{or } \frac{5}{2} - \ln 6\right)$

17. $\frac{27}{5}\ln 2 - \frac{9}{5}\ln 3$ $\left(\text{or } \frac{9}{5}\ln\frac{8}{3}\right)$

19. $\frac{1}{2} - 5\ln 2 + 3\ln 3$ $\left(\text{or } \frac{1}{2} + \ln\frac{27}{32}\right)$

21. $\dfrac{1}{4}\left[\ln|t + 1| - \dfrac{1}{t + 1} - \ln|t - 1| - \dfrac{1}{t - 1}\right] + C$

23. $\ln|x - 1| - \frac{1}{2}\ln(x^2 + 9) - \frac{1}{3}\tan^{-1}(x/3) + C$

25. $-2\ln|x + 1| + \ln(x^2 + 1) + 2\tan^{-1}x + C$

27. $\frac{1}{2}\ln(x^2 + 1) + \tan^{-1}x - \frac{1}{2}\tan^{-1}\left(\dfrac{x}{2}\right) + C$

29. $\frac{1}{2}\ln(x^2 + 2x + 5) + \frac{3}{2}\tan^{-1}\left(\dfrac{x + 1}{2}\right) + C$

31. $\frac{1}{3}\ln|x - 1| - \frac{1}{6}\ln(x^2 + x + 1) - \dfrac{1}{\sqrt{3}}\tan^{-1}\dfrac{2x + 1}{\sqrt{3}} + C$

33. $\frac{1}{4}\ln\frac{8}{3}$

35. $2\ln|x| + \frac{3}{2}\ln(x^2 + 1) + \frac{1}{2}\tan^{-1}x + \dfrac{x}{2(x^2 + 1)} + C$

37. $\frac{7}{8}\sqrt{2}\,\tan^{-1}\left(\dfrac{x - 2}{\sqrt{2}}\right) + \dfrac{3x - 8}{4(x^2 - 4x + 6)} + C$

39. $2\tan^{-1}\sqrt{x - 1} + C$

41. $-2\ln\sqrt{x} - \dfrac{2}{\sqrt{x}} + 2\ln(\sqrt{x} + 1) + C$

43. $\frac{3}{10}(x^2 + 1)^{5/3} - \frac{3}{4}(x^2 + 1)^{2/3} + C$

45. $2\sqrt{x} + 3\sqrt[3]{x} + 6\sqrt[6]{x} + 6\ln\left|\sqrt[6]{x} - 1\right| + C$

47. $\ln\left[\dfrac{(e^x + 2)^2}{(e^x + 1)}\right] + C$

49. $\ln|\tan t + 1| - \ln|\tan t + 2| + C$

51. $x - \ln(e^x + 1) + C$

53. $\left(x - \frac{1}{2}\right)\ln(x^2 - x + 2) - 2x + \sqrt{7}\,\tan^{-1}\left(\dfrac{2x - 1}{\sqrt{7}}\right) + C$

55. $-\frac{1}{2}\ln 3 \approx -0.55$

57. $\frac{1}{2}\ln\left|\dfrac{x - 2}{x}\right| + C$ **61.** $\frac{1}{5}\ln\left|\dfrac{2\tan(x/2) - 1}{\tan(x/2) + 2}\right| + C$

63. $4\ln\frac{2}{3} + 2$ **65.** $-1 + \frac{11}{3}\ln 2$

67. $t = \ln\dfrac{10,000}{P} + 11\ln\dfrac{P - 9000}{1000}$

69. (a) $\dfrac{24,110}{4879}\dfrac{1}{5x + 2} - \dfrac{668}{323}\dfrac{1}{2x + 1} - \dfrac{9438}{80,155}\dfrac{1}{3x - 7}$
$$+ \dfrac{1}{260,015}\dfrac{22,098x + 48,935}{x^2 + x + 5}$$

(b) $\dfrac{4822}{4879}\ln|5x + 2| - \dfrac{334}{323}\ln|2x + 1|$
$$- \dfrac{3146}{80,155}\ln|3x - 7|$$
$$+ \dfrac{11,049}{260,015}\ln(x^2 + x + 5) + \dfrac{75,772}{260,015\sqrt{19}}\tan^{-1}\dfrac{2x + 1}{\sqrt{19}} + C$$

The CAS omits the absolute value signs and the constant of integration.

75. $\dfrac{1}{a^n(x - a)} - \dfrac{1}{a^n x} - \dfrac{1}{a^{n-1}x^2} - \cdots - \dfrac{1}{ax^n}$

EXERCISES 7.5 ■ PAGE 507

1. $-\ln(1 - \sin x) + C$ **3.** $\frac{32}{3}\ln 2 - \frac{28}{9}$

5. $\frac{1}{2\sqrt{2}}\tan^{-1}\left(\frac{t^2}{\sqrt{2}}\right) + C$ **7.** $e^{\pi/4} - e^{-\pi/4}$

9. $\frac{4}{5}\ln 2 + \frac{1}{5}\ln 3$ (or $\frac{1}{5}\ln 48$) **11.** $\frac{1}{2}\sec^{-1}x + \dfrac{\sqrt{x^2 - 1}}{2x^2} + C$

13. $-\frac{1}{5}\cos^5 t + \frac{2}{7}\cos^7 t - \frac{1}{9}\cos^9 t + C$

15. $x \sec x - \ln|\sec x + \tan x| + C$

17. $\frac{1}{4}\pi^2$ **19.** $e^{e^x} + C$ **21.** $(x + 1)\arctan\sqrt{x} - \sqrt{x} + C$

23. $\frac{4097}{45}$ **25.** $4 - \ln 4$ **27.** $x - \ln(1 + e^x) + C$

29. $x\ln\left(x + \sqrt{x^2 - 1}\right) - \sqrt{x^2 - 1} + C$

31. $\sin^{-1}x - \sqrt{1 - x^2} + C$

33. $2\sin^{-1}\left(\dfrac{x + 1}{2}\right) + \dfrac{x + 1}{2}\sqrt{3 - 2x - x^2} + C$

35. 0 **37.** $\frac{1}{4}$ **39.** $\ln|\sec\theta - 1| - \ln|\sec\theta| + C$

41. $\theta\tan\theta - \frac{1}{2}\theta^2 - \ln|\sec\theta| + C$ **43.** $\frac{2}{3}\tan^{-1}(x^{3/2}) + C$

45. $-\frac{1}{3}(x^3 + 1)e^{-x^3} + C$

47. $\ln|x - 1| - 3(x - 1)^{-1} - \frac{3}{2}(x - 1)^{-2} - \frac{1}{3}(x - 1)^{-3} + C$

49. $\ln\left|\dfrac{\sqrt{4x + 1} - 1}{\sqrt{4x + 1} + 1}\right| + C$ **51.** $-\ln\left|\dfrac{\sqrt{4x^2 + 1} + 1}{2x}\right| + C$

53. $\dfrac{1}{m}x^2\cosh mx - \dfrac{2}{m^2}x\sinh mx + \dfrac{2}{m^3}\cosh mx + C$

55. $2\ln\sqrt{x} - 2\ln\left(1 + \sqrt{x}\right) + C$

57. $\frac{3}{7}(x + c)^{7/3} - \frac{3}{4}c(x + c)^{4/3} + C$

59. $\dfrac{1}{32}\ln\left|\dfrac{x - 2}{x + 2}\right| - \dfrac{1}{16}\tan^{-1}\left(\dfrac{x}{2}\right) + C$

61. $\csc\theta - \cot\theta + C$ or $\tan(\theta/2) + C$

63. $2(x - 2\sqrt{x} + 2)e^{\sqrt{x}} + C$

65. $-\tan^{-1}(\cos^2 x) + C$ **67.** $\frac{2}{3}[(x + 1)^{3/2} - x^{3/2}] + C$

69. $\sqrt{2} - 2/\sqrt{3} + \ln\left(2 + \sqrt{3}\right) - \ln\left(1 + \sqrt{2}\right)$

71. $e^x - \ln(1 + e^x) + C$

73. $-\sqrt{1 - x^2} + \frac{1}{2}(\arcsin x)^2 + C$ **75.** $\ln|\ln x - 1| + C$

77. $2(x - 2)\sqrt{1 + e^x} + 2\ln\dfrac{\sqrt{1 + e^x} + 1}{\sqrt{1 + e^x} - 1} + C$

79. $\frac{1}{3}x\sin^3 x + \frac{1}{3}\cos x - \frac{1}{9}\cos^3 x + C$

81. $2\sqrt{1 + \sin x} + C$ **83.** $xe^{x^2} + C$

EXERCISES 7.6 ■ PAGE 512

1. $-\frac{5}{21}$ **3.** $\sqrt{13} - \frac{3}{4}\ln\left(4 + \sqrt{13}\right) - \frac{1}{2} + \frac{3}{4}\ln 3$

5. $\dfrac{\pi}{8}\arctan\dfrac{\pi}{4} - \frac{1}{4}\ln\left(1 + \frac{1}{16}\pi^2\right)$ **7.** $\frac{1}{6}\ln\left|\dfrac{\sin x - 3}{\sin x + 3}\right| + C$

9. $-\dfrac{\sqrt{9x^2 + 4}}{x} + 3\ln\left(3x + \sqrt{9x^2 + 4}\right) + C$

11. $5\pi/16$ **13.** $2\sqrt{x}\arctan\sqrt{x} - \ln(1 + x) + C$

15. $-\ln|\sinh(1/y)| + C$

17. $\dfrac{2y - 1}{8}\sqrt{6 + 4y - 4y^2} + \dfrac{7}{8}\sin^{-1}\left(\dfrac{2y - 1}{\sqrt{7}}\right)$
$\qquad - \frac{1}{12}(6 + 4y - 4y^2)^{3/2} + C$

19. $\frac{1}{9}\sin^3 x\,[3\ln(\sin x) - 1] + C$

21. $\dfrac{1}{2\sqrt{3}}\ln\left|\dfrac{e^x + \sqrt{3}}{e^x - \sqrt{3}}\right| + C$

23. $\frac{1}{4}\tan x\sec^3 x + \frac{3}{8}\tan x\sec x + \frac{3}{8}\ln|\sec x + \tan x| + C$

25. $\frac{1}{2}(\ln x)\sqrt{4 + (\ln x)^2} + 2\ln\left[\ln x + \sqrt{4 + (\ln x)^2}\right] + C$

27. $-\frac{1}{2}x^{-2}\cos^{-1}(x^{-2}) + \frac{1}{2}\sqrt{1 - x^{-4}} + C$

29. $\sqrt{e^{2x} - 1} - \cos^{-1}(e^{-x}) + C$

31. $\frac{1}{5}\ln|x^5 + \sqrt{x^{10} - 2}| + C$ **33.** $\frac{3}{8}\pi^2$

37. $\frac{1}{3}\tan x\sec^2 x + \frac{2}{3}\tan x + C$

39. $\frac{1}{4}x(x^2 + 2)\sqrt{x^2 + 4} - 2\ln\left(\sqrt{x^2 + 4} + x\right) + C$

41. $\frac{1}{4}\cos^3 x\sin x + \frac{3}{8}x + \frac{3}{8}\sin x\cos x + C$

43. $-\ln|\cos x| - \frac{1}{2}\tan^2 x + \frac{1}{4}\tan^4 x + C$

45. (a) $-\ln\left|\dfrac{1 + \sqrt{1 - x^2}}{x}\right| + C$;

both have domain $(-1, 0) \cup (0, 1)$

EXERCISES 7.7 ■ PAGE 524

1. (a) $L_2 = 6, R_2 = 12, M_2 \approx 9.6$
(b) L_2 is an underestimate, R_2 and M_2 are overestimates.
(c) $T_2 = 9 < I$ (d) $L_n < T_n < I < M_n < R_n$
3. (a) $T_4 \approx 0.895759$ (underestimate)
(b) $M_4 \approx 0.908907$ (overestimate); $T_4 < I < M_4$
5. (a) $M_{10} \approx 0.806598, E_M \approx -0.001879$
(b) $S_{10} \approx 0.804779, E_S \approx -0.000060$
7. (a) 1.506361 (b) 1.518362 (c) 1.511519
9. (a) 2.660833 (b) 2.664377 (c) 2.663244
11. (a) -7.276910 (b) -4.818251 (c) -5.605350
13. (a) -2.364034 (b) -2.310690 (c) -2.346520
15. (a) 0.243747 (b) 0.243748 (c) 0.243751
17. (a) 8.814278 (b) 8.799212 (c) 8.804229
19. (a) $T_8 \approx 0.902333, M_8 \approx 0.905620$
(b) $|E_T| \leq 0.0078, |E_M| \leq 0.0039$
(c) $n = 71$ for T_n, $n = 50$ for M_n
21. (a) $T_{10} \approx 1.983524, E_T \approx 0.016476$;
$M_{10} \approx 2.008248, E_M \approx -0.008248$;
$S_{10} \approx 2.000110, E_S \approx -0.000110$
(b) $|E_T| \leq 0.025839, |E_M| \leq 0.012919, |E_S| \leq 0.000170$
(c) $n = 509$ for T_n, $n = 360$ for M_n, $n = 22$ for S_n
23. (a) 2.8 (b) 7.954926518 (c) 0.2894
(d) 7.954926521 (e) The actual error is much smaller.
(f) 10.9 (g) 7.953789422 (h) 0.0593
(i) The actual error is smaller. (j) $n \geq 50$
25.

n	L_n	R_n	T_n	M_n
5	0.742943	1.286599	1.014771	0.992621
10	0.867782	1.139610	1.003696	0.998152
20	0.932967	1.068881	1.000924	0.999538

n	E_L	E_R	E_T	E_M
5	0.257057	-0.286599	-0.014771	0.007379
10	0.132218	-0.139610	-0.003696	0.001848
20	0.067033	-0.068881	-0.000924	0.000462

Observations are the same as after Example 1.

27.

n	T_n	M_n	S_n
6	6.695473	6.252572	6.403292
12	6.474023	6.363008	6.400206

n	E_T	E_M	E_S
6	-0.295473	0.147428	-0.003292
12	-0.074023	0.036992	-0.000206

Observations are the same as after Example 1.
29. (a) 19   (b) 18.6   (c) $18.\overline{6}$
31. (a) 14.4   (b) $\frac{1}{2}$
33. 70.8°F   **35.** $37.7\overline{3}$ ft/s   **37.** 10,177 megawatt-hours
39. (a) 190   (b) 828
41. 28   **43.** 59.4
45.

EXERCISES 7.8 ■ PAGE 534

Abbreviations: C, convergent; D, divergent
1. (a), (d) Infinite discontinuity   (b), (c) Infinite interval
3. $\frac{1}{2} - 1/(2t^2)$; 0.495, 0.49995, 0.4999995; 0.5
5. 2   **7.** D   **9.** $\frac{1}{5}e^{-10}$ **11.** D   **13.** 0   **15.** D
17. ln 2   **19.** $-\frac{1}{4}$ **21.** D   **23.** $-\pi/8$ **25.** 2
27. D   **29.** $\frac{32}{3}$ **31.** D   **33.** $\frac{9}{2}$ **35.** D   **37.** $-\frac{1}{4}$
39. $-2/e$
41. $1/e$
43. $\frac{1}{2}\ln 2$

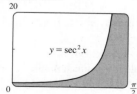

45. Infinite area

47. (a)

t	$\int_1^t [(\sin^2 x)/x^2]\, dx$
2	0.447453
5	0.577101
10	0.621306
100	0.668479
1,000	0.672957
10,000	0.673407

It appears that the integral is convergent.

(c)

49. C   **51.** D   **53.** D   **55.** π **57.** $p < 1, 1/(1-p)$
59. $p > -1, -1/(p+1)^2$ **63.** π **65.** $\sqrt{2GM/R}$
67. (a)

(b) The rate at which the fraction $F(t)$ increases as t increases
(c) 1; all bulbs burn out eventually

69. $\gamma = \dfrac{cN}{\lambda(k+\lambda)}$ **71.** 1000
73. (a) $F(s) = 1/s$, $s > 0$ (b) $F(s) = 1/(s-1)$, $s > 1$
(c) $F(s) = 1/s^2$, $s > 0$
79. $C = 1$; ln 2   **81.** No

CHAPTER 7 REVIEW ■ PAGE 537

True-False Quiz
1. False   **3.** False   **5.** False   **7.** False
9. (a) True   (b) False   **11.** False   **13.** False

Exercises
1. $\frac{7}{2} + \ln 2$ **3.** $e^{\sin x} + C$ **5.** $\ln|2t+1| - \ln|t+1| + C$
7. $\frac{2}{15}$ **9.** $-\cos(\ln t) + C$ **11.** $\sqrt{3} - \frac{1}{3}\pi$
13. $3e^{\sqrt[3]{x}}(x^{2/3} - 2x^{1/3} + 2) + C$
15. $-\frac{1}{2}\ln|x| + \frac{3}{2}\ln|x+2| + C$
17. $x \sinh x - \cosh x + C$
19. $\frac{1}{18}\ln(9x^2 + 6x + 5) + \frac{1}{9}\tan^{-1}\left[\frac{1}{2}(3x+1)\right] + C$
21. $\ln|x - 2 + \sqrt{x^2 - 4x}| + C$
23. $\ln\left|\dfrac{\sqrt{x^2+1} - 1}{x}\right| + C$
25. $\frac{3}{2}\ln(x^2 + 1) - 3\tan^{-1}x + \sqrt{2}\tan^{-1}(x/\sqrt{2}) + C$
27. $\frac{2}{5}$ **29.** 0   **31.** $6 - \frac{3}{2}\pi$
33. $\dfrac{x}{\sqrt{4 - x^2}} - \sin^{-1}\left(\dfrac{x}{2}\right) + C$
35. $4\sqrt{1 + \sqrt{x}} + C$ **37.** $\frac{1}{4}\sin 2x - \frac{1}{8}\cos 4x + C$
39. $\frac{1}{8}e - \frac{1}{4}$ **41.** $\frac{1}{36}$ **43.** D
45. $4\ln 4 - 8$ **47.** $-\frac{4}{3}$ **49.** $\pi/4$
51. $(x+1)\ln(x^2 + 2x + 2) + 2\arctan(x+1) - 2x + C$
53. 0
55. $\frac{1}{4}(2x - 1)\sqrt{4x^2 - 4x - 3}$
$\qquad\qquad - \ln|2x - 1 + \sqrt{4x^2 - 4x - 3}| + C$
57. $\frac{1}{2}\sin x\sqrt{4 + \sin^2 x} + 2\ln\left(\sin x + \sqrt{4 + \sin^2 x}\right) + C$
61. No
63. (a) 1.925444   (b) 1.920915   (c) 1.922470
65. (a) 0.01348, $n \geq 368$ (b) 0.00674, $n \geq 260$

67. 8.6 mi

69. (a) 3.8   (b) 1.7867, 0.000646   (c) $n \geqslant 30$

71. (a) D   (b) C

73. 2   **75.** $\frac{3}{16}\pi^2$

PROBLEMS PLUS ■ PAGE 541

1. About 1.85 inches from the center   **3.** 0

7. $f(\pi) = -\pi/2$ **11.** $(b^b a^{-a})^{1/(b-a)} e^{-1}$ **13.** $\frac{1}{8}\pi - \frac{1}{12}$

15. $2 - \sin^{-1}(2/\sqrt{5})$

CHAPTER 8

EXERCISES 8.1 ■ PAGE 548

1. $4\sqrt{5}$ **3.** 3.8202   **5.** 3.4467   **7.** 3.6095

9. $\frac{2}{243}(82\sqrt{82} - 1)$ **11.** $\frac{59}{24}$ **13.** $\frac{32}{3}$

15. $\ln(\sqrt{2} + 1)$ **17.** $\frac{3}{4} + \frac{1}{2}\ln 2$ **19.** $\ln 3 - \frac{1}{2}$

21. $\sqrt{2} + \ln(1 + \sqrt{2})$ **23.** 10.0556

25. 15.498085; 15.374568   **27.** 7.094570; 7.118819

29. (a), (b)   3

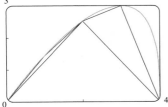

$L_1 = 4,$
$L_2 \approx 6.43,$
$L_4 \approx 7.50$

(c) $\int_0^4 \sqrt{1 + [4(3 - x)/(3(4 - x)^{2/3})]^2}\, dx$ (d) 7.7988

31. $\sqrt{1 + e^4} - \ln(1 + \sqrt{1 + e^4}) + 2 - \sqrt{2} + \ln(1 + \sqrt{2})$

33. 6

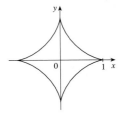

35. $s(x) = \frac{2}{27}\left[(1 + 9x)^{3/2} - 10\sqrt{10}\right]$ **37.** $2\sqrt{2}(\sqrt{1 + x} - 1)$

41. 209.1 m   **43.** 29.36 in.   **45.** 12.4

EXERCISES 8.2 ■ PAGE 555

1. (a) (i) $\int_0^{\pi/3} 2\pi \tan x \sqrt{1 + \sec^4 x}\, dx$

(ii) $\int_0^{\pi/3} 2\pi x \sqrt{1 + \sec^4 x}\, dx$ (b) (i) 10.5017   (ii) 7.9353

3. (a) (i) $\int_{-1}^{1} 2\pi e^{-x^2} \sqrt{1 + 4x^2 e^{-2x^2}}\, dx$

(ii) $\int_0^1 2\pi x \sqrt{1 + 4x^2 e^{-2x^2}}\, dx$ (b) (i) 11.0753   (ii) 3.9603

5. (a) (i) $\int_0^1 2\pi y \sqrt{1 + (1 + 3y^2)^2}\, dy$

(ii) $\int_0^1 2\pi (y + y^3)\sqrt{1 + (1 + 3y^2)^2}\, dy$

(b) (i) 8.5302   (ii) 13.5134

7. $\frac{1}{27}\pi(145\sqrt{145} - 1)$ **9.** $\frac{1}{6}\pi(27\sqrt{27} - 5\sqrt{5})$

11. $\pi\sqrt{5} + 4\pi \ln\left(\dfrac{1 + \sqrt{5}}{2}\right)$ **13.** $\frac{21}{2}\pi$ **15.** $\frac{3712}{15}\pi$

17. πa^2 **19.** 1,230,507   **21.** 24.145807

23. $\frac{1}{4}\pi\left[4\ln(\sqrt{17} + 4) - 4\ln(\sqrt{2} + 1) - \sqrt{17} + 4\sqrt{2}\right]$

25. $\frac{1}{6}\pi\left[\ln(\sqrt{10} + 3) + 3\sqrt{10}\right]$

29. (a) $\frac{1}{3}\pi a^2$ (b) $\frac{56}{45}\pi\sqrt{3}\, a^2$

31. (a) $2\pi\left[b^2 + \dfrac{a^2 b \sin^{-1}(\sqrt{a^2 - b^2}/a)}{\sqrt{a^2 - b^2}}\right]$

(b) $2\pi a^2 + \dfrac{2\pi ab^2}{\sqrt{a^2 - b^2}} \ln \dfrac{a + \sqrt{a^2 - b^2}}{b}$

33. $\int_a^b 2\pi[c - f(x)]\sqrt{1 + [f'(x)]^2}\, dx$ **35.** $4\pi^2 r^2$

37. Both equal $\pi \int_a^b (e^{x/2} + e^{-x/2})^2\, dx$.

EXERCISES 8.3 ■ PAGE 565

1. (a) 187.5 lb/ft^2 (b) 1875 lb   (c) 562.5 lb

3. 7000 lb   **5.** 2.36×10^7 N   **7.** 9.8×10^3 N

9. 889 lb   **11.** $\frac{2}{3}\delta ah^2$ **13.** 5.27×10^5 N

15. (a) 314 N   (b) 353 N

17. (a) 5.63×10^3 lb   (b) 5.06×10^4 lb

(c) 4.88×10^4 lb   (d) 3.03×10^5 lb

19. 4148 lb   **21.** 330; 22

23. 10; 14; (1.4, 1)   **25.** $(\frac{2}{3}, \frac{2}{3})$

27. $\left(\dfrac{1}{e - 1}, \dfrac{e + 1}{4}\right)$ **29.** $(\frac{9}{20}, \frac{9}{20})$

31. $\left(\dfrac{\pi\sqrt{2} - 4}{4(\sqrt{2} - 1)}, \dfrac{1}{4(\sqrt{2} - 1)}\right)$ **33.** $(\frac{8}{5}, -\frac{1}{2})$

35. $\left(\dfrac{28}{3(\pi + 2)}, \dfrac{10}{3(\pi + 2)}\right)$ **37.** $(-\frac{1}{5}, -\frac{12}{35})$

41. $(0, \frac{1}{12})$ **45.** $\frac{1}{3}\pi r^2 h$ **47.** $\left(\dfrac{8}{\pi}, \dfrac{8}{\pi}\right)$

49. $4\pi^2 rR$

EXERCISES 8.4 ■ PAGE 572

1. $21,104   **3.** $140,000; $60,000   **5.** $407.25

7. $166,666.67   **9.** (a) 3800   (b) $324,900

11. 3727; $37,753   **13.** $\frac{2}{3}(16\sqrt{2} - 8) \approx 9.75$ million

15. $65,230.48   **17.** $\dfrac{(1 - k)(b^{2-k} - a^{2-k})}{(2 - k)(b^{1-k} - a^{1-k})}$

19. 1.19×10^{-4} cm^3/s   **21.** 6.59 L/min   **23.** 5.77 L/min

EXERCISES 8.5 ■ PAGE 579

1. (a) The probability that a randomly chosen tire will have a lifetime between 30,000 and 40,000 miles

(b) The probability that a randomly chosen tire will have a lifetime of at least 25,000 miles

3. (a) $f(x) \geqslant 0$ for all x and $\int_{-\infty}^{\infty} f(x)\, dx = 1$ (b) $\frac{17}{81}$

5. (a) $1/\pi$ (b) $\frac{1}{2}$

7. (a) $f(x) \geqslant 0$ for all x and $\int_{-\infty}^{\infty} f(x)\, dx = 1$ (b) 5

11. (a) $\approx 46.5\%$ (b) $\approx 15.3\%$ (c) About 4.8 s

13. $\approx 59.4\%$ (b) 40 min   **15.** $\approx 44\%$

17. (a) 0.0668   (b) $\approx 5.21\%$ **19.** ≈ 0.9545

21. (b) $0; a_0$

(c) 1×10^{10}

(d) $1 - 41e^{-8} \approx 0.986$ (e) $\frac{3}{2}a_0$

CHAPTER 8 REVIEW ▪ PAGE 581

Exercises

1. $\frac{1}{54}\left(109\sqrt{109} - 1\right)$ **3.** $\frac{53}{6}$

5. (a) 3.5121 (b) 22.1391 (c) 29.8522

7. 3.8202 **9.** $\frac{124}{5}$ **11.** ≈ 458 lb **13.** $\left(\frac{8}{5}, 1\right)$

15. $\left(\frac{4}{3}, \frac{4}{3}\right)$ **17.** $2\pi^2$ **19.** $7166.67

21. (a) $f(x) \geq 0$ for all x and $\int_{-\infty}^{\infty} f(x)\,dx = 1$

(b) ≈ 0.3455 (c) 5; yes

23. (a) $1 - e^{-3/8} \approx 0.31$ (b) $e^{-5/4} \approx 0.29$

(c) $8 \ln 2 \approx 5.55$ min

PROBLEMS PLUS ▪ PAGE 583

1. $\frac{2}{3}\pi - \frac{1}{2}\sqrt{3}$

3. (a) $2\pi r(r \pm d)$ (b) $\approx 3.36 \times 10^6$ mi^2

(d) $\approx 7.84 \times 10^7$ mi^2

5. (a) $P(z) = P_0 + g\int_0^z \rho(x)\,dx$

(b) $(P_0 - \rho_0 gH)(\pi r^2) + \rho_0 gHe^{L/H}\int_{-r}^{r} e^{x/H} \cdot 2\sqrt{r^2 - x^2}\,dx$

7. Height $\sqrt{2}\,b$, volume $\left(\frac{28}{27}\sqrt{6} - 2\right)\pi b^3$ **9.** 0.14 m

11. $2/\pi; 1/\pi$ **13.** $(0, -1)$

CHAPTER 9

EXERCISES 9.1 ▪ PAGE 590

3. (a) $\frac{1}{2}, -1$ **5.** (d)

7. (a) It must be either 0 or decreasing

(c) $y = 0$ (d) $y = 1/(x + 2)$

9. (a) $0 < P < 4200$ (b) $P > 4200$

(c) $P = 0, P = 4200$

13. (a) III (b) I (c) IV (d) II

15. (a) At the beginning; stays positive, but decreases

(c)

17. It approaches 0 as c approaches c_s.

EXERCISES 9.2 ▪ PAGE 597

1. (a)

(b) $y = 0.5, y = 1.5$

3. III **5.** IV

7.

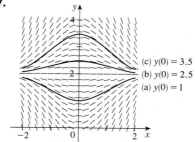

(c) $y(0) = 3.5$
(b) $y(0) = 2.5$
(a) $y(0) = 1$

9.

11.

13.

15.

17.

$-2 \leqslant c \leqslant 2; -2, 0, 2$

19. (a) (i) 1.4 (ii) 1.44 (iii) 1.4641
(b)

Underestimates

(c) (i) 0.0918 (ii) 0.0518 (iii) 0.0277
It appears that the error is also halved (approximately).
21. $-1, -3, -6.5, -12.25$ **23.** 1.7616
25. (a) (i) 3 (ii) 2.3928 (iii) 2.3701 (iv) 2.3681
(c) (i) -0.6321 (ii) -0.0249 (iii) -0.0022 (iv) -0.0002
It appears that the error is also divided by 10 (approximately).
27. (a), (d)   (b) 3

(c) Yes, $Q = 3$
(e) 2.77 C

EXERCISES 9.3 ■ **PAGE 605**
1. $y = -1/(x^3 + C), \ y = 0$
3. $y = \pm\sqrt{x^2 + 2\ln|x| + C}$
5. $e^y - y = 2x + \sin x + C$
7. $\theta \sin \theta + \cos \theta = -\frac{1}{2}e^{-t^2} + C$ **9.** $p = Ke^{t^3/3 - t} - 1$
11. $y = -\ln\left(1 - \frac{1}{2}x^2\right)$ **13.** $u = -\sqrt{t^2 + \tan t + 25}$
15. $\frac{1}{2}y^2 + \frac{1}{3}(3 + y^2)^{3/2} = \frac{1}{2}x^2 \ln x - \frac{1}{4}x^2 + \frac{41}{12}$
17. $y = \dfrac{4a}{\sqrt{3}} \sin x - a$
19. $y = \sqrt{x^2 + 4}$ **21.** $y = Ke^x - x - 1$

23. (a) $\sin^{-1} y = x^2 + C$
(b) $y = \sin(x^2), \ -\sqrt{\pi/2} \leqslant x \leqslant \sqrt{\pi/2}$ (c) No
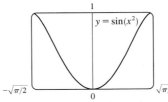
25. $\cos y = \cos x - 1$
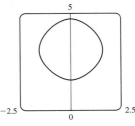

27. (a)   (b) $y = \dfrac{1}{K - x}$

29. $y = Cx^2$

31. $x^2 - y^2 = C$
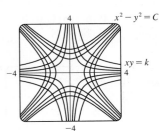

33. $y = 1 + e^{2 - x^2/2}$ **35.** $y = \left(\frac{1}{2}x^2 + 2\right)^2$
37. $Q(t) = 3 - 3e^{-4t}; \ 3$ **39.** $P(t) = M - Me^{-kt}; \ M$
41. (a) $x = a - \dfrac{4}{(kt + 2/\sqrt{a})^2}$

(b) $t = \dfrac{2}{k\sqrt{a - b}}\left(\tan^{-1}\sqrt{\dfrac{b}{a - b}} - \tan^{-1}\sqrt{\dfrac{b - x}{a - b}}\right)$

43. (a) $C(t) = (C_0 - r/k)e^{-kt} + r/k$ (b) r/k; the concentration approaches r/k regardless of the value of C_0

45. (a) $15e^{-t/100}$ kg   (b) $15e^{-0.2} \approx 12.3$ kg

47. About 4.9%   **49.** g/k

51. (a) $L_1 = KL_2{}^k$ (b) $B = KV^{0.0794}$

53. (a) $dA/dt = k\sqrt{A}\,(M - A)$ (b) $A(t) = M\left(\dfrac{Ce^{\sqrt{M}\,kt} - 1}{Ce^{\sqrt{M}\,kt} + 1}\right)^2$,

where $C = \dfrac{\sqrt{M} + \sqrt{A_0}}{\sqrt{M} - \sqrt{A_0}}$ and $A_0 = A(0)$

EXERCISES 9.4 ■ PAGE 617

1. (a) 1200; 0.04 (b) $P(t) = \dfrac{1200}{1 + 19e^{-0.04t}}$ (c) 87

3. (a) 100; 0.05 (b) Where P is close to 0 or 100; on the line $P = 50$; $0 < P_0 < 100$; $P_0 > 100$

(c)

Solutions approach 100; some increase and some decrease, some have an inflection point but others don't; solutions with $P_0 = 20$ and $P_0 = 40$ have inflection points at $P = 50$

(d) $P = 0$, $P = 100$; other solutions move away from $P = 0$ and toward $P = 100$

5. (a) 3.23×10^7 kg   (b) ≈ 1.55 years   **7.** 9000

9. (a) $\dfrac{dP}{dt} = \dfrac{1}{305}P\left(1 - \dfrac{P}{20}\right)$

(b) 6.24 billion   (c) 7.57 billion; 13.87 billion

11. (a) $dy/dt = ky(1 - y)$ (b) $y = \dfrac{y_0}{y_0 + (1 - y_0)e^{-kt}}$

(c) 3:36 PM

15. $P_E(t) = 1909.7761(1.0796)^t + 94{,}000$;

$P_L(t) = \dfrac{33{,}086.4394}{1 + 12.3428\,e^{-0.1657t}} + 94{,}000$

17. (a) $P(t) = \dfrac{m}{k} + \left(P_0 - \dfrac{m}{k}\right)e^{kt}$ (b) $m < kP_0$

(c) $m = kP_0$, $m > kP_0$ (d) Declining

19. (a) Fish are caught at a rate of 15 per week.

(b) See part (d).   (c) $P = 250$, $P = 750$

(d)

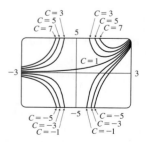

$0 < P_0 < 250$: $P \to 0$;
$P_0 = 250$: $P \to 250$;
$P_0 > 250$: $P \to 750$

(e) $P(t) = \dfrac{250 - 750ke^{t/25}}{1 - ke^{t/25}}$

where $k = \frac{1}{11}, -\frac{1}{9}$

21. (b)

$0 < P_0 < 200$: $P \to 0$;
$P_0 = 200$: $P \to 200$;
$P_0 > 200$: $P \to 1000$

(c) $P(t) = \dfrac{m(M - P_0) + M(P_0 - m)e^{(M-m)(k/M)t}}{M - P_0 + (P_0 - m)e^{(M-m)(k/M)t}}$

23. (a) $P(t) = P_0 e^{(k/r)[\sin(rt - \phi) + \sin \phi]}$ (b) Does not exist

EXERCISES 9.5 ■ PAGE 625

1. No   **3.** Yes   **5.** $y = 1 + Ce^{-x}$

7. $y = x - 1 + Ce^{-x}$ **9.** $y = \frac{2}{3}\sqrt{x} + C/x$

11. $y = x^2(\ln x + C)$ **13.** $y = \frac{1}{3}t^{-3}(1 + t^2)^{3/2} + Ct^{-3}$

15. $y = \dfrac{1}{x}\ln x - \dfrac{1}{x} + \dfrac{3}{x^2}$ **17.** $u = -t^2 + t^3$

19. $y = -x \cos x - x$

21. $y = \dfrac{(x - 1)e^x + C}{x^2}$

25. $y = \pm\left(Cx^4 + \dfrac{2}{5x}\right)^{-1/2}$

27. (a) $I(t) = 4 - 4e^{-5t}$ (b) $4 - 4e^{-1/2} \approx 1.57$ A

29. $Q(t) = 3(1 - e^{-4t})$, $I(t) = 12e^{-4t}$

31. $P(t) = M + Ce^{-kt}$

33. $y = \frac{2}{5}(100 + 2t) - 40{,}000(100 + 2t)^{-3/2}$; 0.2275 kg/L

35. (b) mg/c (c) $(mg/c)[t + (m/c)e^{-ct/m}] - m^2g/c^2$

37. (b) $P(t) = \dfrac{M}{1 + MCe^{-kt}}$

EXERCISES 9.6 ▪ PAGE 631

1. (a) x = predators, y = prey; growth is restricted only by predators, which feed only on prey.
(b) x = prey, y = predators; growth is restricted by carrying capacity and by predators, which feed only on prey.

3. (a) Competition
(b) (i) $x = 0$, $y = 0$: zero populations
(ii) $x = 0$, $y = 400$: In the absence of an x-population, the y-population stabilizes at 400.
(iii) $x = 125$, $y = 0$: In the absence of a y-population, the x-population stabilizes at 125.
(iv) $x = 50$, $y = 300$: Both populations are stable.

5. (a) The rabbit population starts at about 300, increases to 2400, then decreases back to 300. The fox population starts at 100, decreases to about 20, increases to about 315, decreases to 100, and the cycle starts again.
(b)

7.

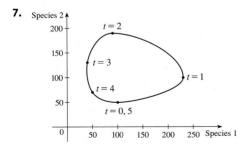

11. (a) Population stabilizes at 5000.
(b) (i) $W = 0$, $R = 0$: Zero populations
(ii) $W = 0$, $R = 5000$: In the absence of wolves, the rabbit population is always 5000.
(iii) $W = 64$, $R = 1000$: Both populations are stable.
(c) The populations stabilize at 1000 rabbits and 64 wolves.

(d)

CHAPTER 9 REVIEW ▪ PAGE 634

True-False Quiz

1. True   **3.** False   **5.** True   **7.** True

Exercises

1. (a)

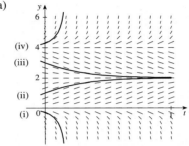

(b) $0 \le c \le 4$; $y = 0$, $y = 2$, $y = 4$

3. (a)

$y(0.3) \approx 0.8$

(b) 0.75676
(c) $y = x$ and $y = -x$; there is a loc max or loc min

5. $y = \left(\frac{1}{2}x^2 + C\right)e^{-\sin x}$

7. $y = \pm\sqrt{\ln(x^2 + 2x^{3/2} + C)}$

9. $r(t) = 5e^{t - t^2}$ **11.** $y = \frac{1}{2}x(\ln x)^2 + 2x$ **13.** $x = C - \frac{1}{2}y^2$

15. (a) $P(t) = \dfrac{2000}{1 + 19e^{-0.1t}}$; ≈ 560 (b) $t = -10 \ln \frac{2}{57} \approx 33.5$

17. (a) $L(t) = L_\infty - [L_\infty - L(0)]e^{-kt}$ (b) $L(t) = 53 - 43e^{-0.2t}$

19. 15 days   **21.** $k \ln h + h = (-R/V)t + C$

23. (a) Stabilizes at 200,000
(b) (i) $x = 0$, $y = 0$: Zero populations
(ii) $x = 200{,}000$, $y = 0$: In the absence of birds, the insect population is always 200,000.
(iii) $x = 25{,}000$, $y = 175$: Both populations are stable.
(c) The populations stabilize at 25,000 insects and 175 birds.
(d)

PROBLEMS PLUS ■ **PAGE 637**

1. $f(x) = \pm 10e^x$ **5.** $y = x^{1/n}$ **7.** 20°C

9. (b) $f(x) = \dfrac{x^2 - L^2}{4L} - \dfrac{L}{2} \ln\left(\dfrac{x}{L}\right)$ (c) No

11. (a) 9.8 h   (b) $31{,}900\pi$ ft²; 2000π ft²/h   (c) 5.1 h
13. $x^2 + (y - 6)^2 = 25$ **15.** $y = K/x, \quad K \neq 0$

CHAPTER 10

EXERCISES 10.1 ■ **PAGE 645**

1.

3.

5. (a)

(b) $y = \frac{1}{4}x + \frac{5}{4}$

7. (a)
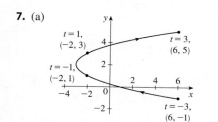
(b) $x = y^2 - 4y + 1$, $-1 \leqslant y \leqslant 5$

9. (a)

(b) $y = 1 - x^2, x \geqslant 0$

11. (a) $x^2 + y^2 = 1, y \geqslant 0$ (b)

13. (a) $y = 1/x, y > 1$ (b)

15. (a) $x = e^{2y}$ (b)

17. (a) $y^2 - x^2 = 1, y \geqslant 1$ (b)

19. Moves counterclockwise along the circle
$$\left(\dfrac{x - 5}{2}\right)^2 + \left(\dfrac{y - 3}{2}\right)^2 = 1 \text{ from } (3, 3) \text{ to } (7, 3)$$
21. Moves 3 times clockwise around the ellipse
$(x^2/25) + (y^2/4) = 1$, starting and ending at $(0, -2)$
23. It is contained in the rectangle described by $1 \leqslant x \leqslant 4$
and $2 \leqslant y \leqslant 3$.

25.

27.

29.
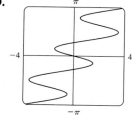

31. (b) $x = -2 + 5t, y = 7 - 8t, 0 \leqslant t \leqslant 1$
33. (a) $x = 2 \cos t, y = 1 - 2 \sin t, 0 \leqslant t \leqslant 2\pi$
(b) $x = 2 \cos t, y = 1 + 2 \sin t, 0 \leqslant t \leqslant 6\pi$
(c) $x = 2 \cos t, y = 1 + 2 \sin t, \pi/2 \leqslant t \leqslant 3\pi/2$
37. The curve $y = x^{2/3}$ is generated in (a). In (b), only the portion
with $x \geqslant 0$ is generated, and in (c) we get only the portion with
$x > 0$.

41. $x = a \cos \theta, y = b \sin \theta; (x^2/a^2) + (y^2/b^2) = 1$, ellipse

43.

45. (a) Two points of intersection

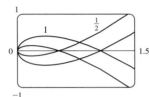

(b) One collision point at $(-3, 0)$ when $t = 3\pi/2$
(c) There are still two intersection points, but no collision point.

47. For $c = 0$, there is a cusp; for $c > 0$, there is a loop whose size increases as c increases.

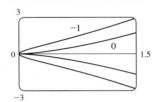

49. The curves roughly follow the line $y = x$, and they start having loops when a is between 1.4 and 1.6 (more precisely, when $a > \sqrt{2}$). The loops increase in size as a increases.
51. As n increases, the number of oscillations increases; a and b determine the width and height.

EXERCISES 10.2 ▪ **PAGE 655**

1. $\frac{1}{2}(1 + t)^{3/2}$ **3.** $y = -x$ **5.** $y = \pi x + \pi^2$
7. $y = 2x + 1$
9. $y = 3x + 3$

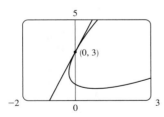

11. $\dfrac{2t + 1}{2t}, -\dfrac{1}{4t^3}, t < 0$ **13.** $e^{-2t}(1 - t), e^{-3t}(2t - 3), t > \frac{3}{2}$
15. $\dfrac{t + 1}{t - 1}, \dfrac{-2t}{(t - 1)^3}, 0 < t < 1$

17. Horizontal at $(0, -3)$, vertical at $(\pm 2, -2)$
19. Horizontal at $\left(\frac{1}{2}, -1\right)$ and $\left(-\frac{1}{2}, 1\right)$, no vertical
21. $(0.6, 2); \left(5 \cdot 6^{-6/5}, e^{6^{-1/5}}\right)$
23.

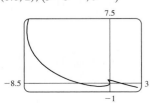

25. $y = x, y = -x$

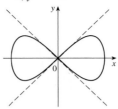

27. (a) $d \sin \theta/(r - d \cos \theta)$ **29.** $(4, 0)$ **31.** πab
33. $\frac{24}{5}$ **35.** $2\pi r^2 + \pi d^2$ **37.** $\int_0^2 \sqrt{2 + 2e^{-2t}}\, dt \approx 3.1416$
39. $\int_0^{4\pi} \sqrt{5 - 4 \cos t}\, dt \approx 26.7298$ **41.** $4\sqrt{2} - 2$
43. $\frac{1}{2}\sqrt{2} + \frac{1}{2}\ln\left(1 + \sqrt{2}\right)$
45. $\sqrt{2}\,(e^\pi - 1)$

47. 16.7102

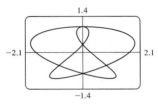

49. 612.3053   **51.** $6\sqrt{2}, \sqrt{2}$
55. (a)

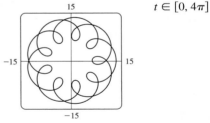

$t \in [0, 4\pi]$

(b) 294
57. $\int_0^{\pi/2} 2\pi t \cos t \sqrt{t^2 + 1}\, dt \approx 4.7394$
59. $\int_0^1 2\pi e^{-t}\sqrt{1 + 2e^t + e^{2t} + e^{-2t}}\, dt \approx 10.6705$
61. $\frac{2}{1215}\pi\left(247\sqrt{3} + 64\right)$ **63.** $\frac{6}{5}\pi a^2$
65. $\frac{24}{5}\pi\left(949\sqrt{26} + 1\right)$ **71.** $\frac{1}{4}$

EXERCISES 10.3 ▪ **PAGE 666**

1. (a)

(b)

$(1, 9\pi/4), (-1, 5\pi/4)$ $(2, \pi/2), (-2, 7\pi/2)$

(c)

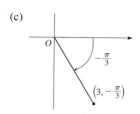

$(3, 5\pi/3), (-3, 2\pi/3)$

3. (a)

$(0, -2)$

(b)

$(1, 1)$

(c)

$(-\sqrt{3}/2, 1/2)$

5. (a) (i) $(4\sqrt{2}, 3\pi/4)$ (ii) $(-4\sqrt{2}, 7\pi/4)$
(b) (i) $(6, \pi/3)$ (ii) $(-6, 4\pi/3)$

7.

9.

11.

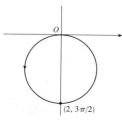

13. $2\sqrt{7}$ **15.** Circle, center O, radius $\sqrt{5}$
17. Circle, center $(5/2, 0)$, radius $5/2$
19. Hyperbola, center O, foci on x-axis
21. $r = 2\csc\theta$ **23.** $r = 1/(\sin\theta - 3\cos\theta)$
25. $r = 2c\cos\theta$ **27.** (a) $\theta = \pi/6$ (b) $x = 3$
29.

31.

33.

35.

37.

39.

41.

43.

45.

47.

49.

51.

53. (a) For $c < -1$, the inner loop begins at $\theta = \sin^{-1}(-1/c)$
and ends at $\theta = \pi - \sin^{-1}(-1/c)$; for $c > 1$, it begins at
$\theta = \pi + \sin^{-1}(1/c)$ and ends at $\theta = 2\pi - \sin^{-1}(1/c)$.
55. $1/\sqrt{3}$ **57.** $-\pi$ **59.** 1
61. Horizontal at $(3/\sqrt{2}, \pi/4), (-3/\sqrt{2}, 3\pi/4)$;
vertical at $(3, 0), (0, \pi/2)$
63. Horizontal at $(\frac{3}{2}, \pi/3), (0, \pi)$ [the pole], and $(\frac{3}{2}, 5\pi/3)$;
vertical at $(2, 0), (\frac{1}{2}, 2\pi/3), (\frac{1}{2}, 4\pi/3)$
65. Center $(b/2, a/2)$, radius $\sqrt{a^2 + b^2}/2$

67.

69.

71.

73. By counterclockwise rotation through angle $\pi/6$, $\pi/3$, or α about the origin

75. For $c = 0$, the curve is a circle. As c increases, the left side gets flatter, then has a dimple for $0.5 < c < 1$, a cusp for $c = 1$, and a loop for $c > 1$.

EXERCISES 10.4 ■ **PAGE 672**

1. $e^{-\pi/4} - e^{-\pi/2}$ **3.** $\pi/2$ **5.** $\frac{1}{2}$ **7.** $\frac{41}{4}\pi$
9. π **11.** 11π

13. $\frac{9}{2}\pi$

15. $\frac{3}{2}\pi$

17. $\frac{4}{3}\pi$ **19.** $\frac{1}{16}\pi$ **21.** $\pi - \frac{3}{2}\sqrt{3}$ **23.** $\frac{4}{3}\pi + 2\sqrt{3}$
25. $4\sqrt{3} - \frac{4}{3}\pi$ **27.** π **29.** $\frac{9}{8}\pi - \frac{9}{4}$ **31.** $\frac{1}{2}\pi - 1$
33. $-\sqrt{3} + 2 + \frac{1}{3}\pi$ **35.** $\frac{1}{4}(\pi + 3\sqrt{3})$
37. $\left(\frac{1}{2}, \pi/6\right), \left(\frac{1}{2}, 5\pi/6\right)$, and the pole
39. $(1, \theta)$ where $\theta = \pi/12, 5\pi/12, 13\pi/12, 17\pi/12$
and $(-1, \theta)$ where $\theta = 7\pi/12, 11\pi/12, 19\pi/12, 23\pi/12$
41. $\left(\frac{1}{2}\sqrt{3}, \pi/3\right), \left(\frac{1}{2}\sqrt{3}, 2\pi/3\right)$, and the pole
43. Intersection at $\theta \approx 0.89, 2.25$; area ≈ 3.46

45. 2π **47.** $\frac{8}{3}[(\pi^2 + 1)^{3/2} - 1]$ **49.** $\frac{16}{3}$
51. 2.4221   **53.** 8.0091
55. (b) $2\pi(2 - \sqrt{2})$

EXERCISES 10.5 ■ **PAGE 680**
1. $(0, 0), \left(0, \frac{3}{2}\right), y = -\frac{3}{2}$ **3.** $(0, 0), \left(-\frac{1}{2}, 0\right), x = \frac{1}{2}$

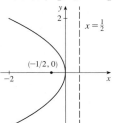

5. $(-2, 3)$ $(-2, 5), y = 1$ **7.** $(4, -3), \left(\frac{7}{2}, -3\right), x = \frac{9}{2}$

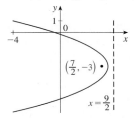

9. $x = -y^2$, focus $\left(\frac{1}{4}, 0\right)$, directrix $x = \frac{1}{4}$
11. $(0, \pm 2), \left(0, \pm\sqrt{2}\right)$ **13.** $(\pm 3, 0), \left(\pm 2\sqrt{2}, 0\right)$

15. $(1, \pm 3), \left(1, \pm\sqrt{5}\right)$ **17.** $\dfrac{x^2}{4} + \dfrac{y^2}{9} = 1$, foci $\left(0, \pm\sqrt{5}\right)$

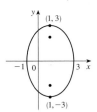

19. $(0, \pm 5); \left(0, \pm\sqrt{34}\right); y = \pm\frac{5}{3}x$

21. $(\pm10, 0)$, $(\pm10\sqrt{2}, 0)$, $y = \pm x$

23. $(\pm1, 1)$, $(\pm\sqrt{2}, 1)$, $y - 1 = \pm x$

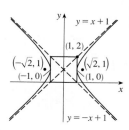

25. Hyperbola, $(\pm1, 0)$, $(\pm\sqrt{5}, 0)$
27. Ellipse, $(\pm\sqrt{2}, 1)$, $(\pm1, 1)$
29. Parabola, $(1, -2)$, $\left(1, -\frac{11}{6}\right)$
31. $y^2 = 4x$ **33.** $y^2 = -12(x + 1)$
35. $(y + 1)^2 = -\frac{1}{2}(x - 3)$
37. $\dfrac{x^2}{25} + \dfrac{y^2}{21} = 1$ **39.** $\dfrac{x^2}{12} + \dfrac{(y - 4)^2}{16} = 1$
41. $\dfrac{(x + 1)^2}{12} + \dfrac{(y - 4)^2}{16} = 1$ **43.** $\dfrac{x^2}{9} - \dfrac{y^2}{16} = 1$
45. $\dfrac{(y - 1)^2}{25} - \dfrac{(x + 3)^2}{39} = 1$ **47.** $\dfrac{x^2}{9} - \dfrac{y^2}{36} = 1$
49. $\dfrac{x^2}{3,763,600} + \dfrac{y^2}{3,753,196} = 1$
51. (a) $\dfrac{121x^2}{1,500,625} - \dfrac{121y^2}{3,339,375} = 1$ (b) ≈248 mi
55. (a) Ellipse   (b) Hyperbola   (c) No curve
59. 15.9
61. $\dfrac{b^2 c}{a} + ab \ln\left(\dfrac{a}{b + c}\right)$ where $c^2 = a^2 + b^2$
63. $(0, 4/\pi)$

EXERCISES 10.6 ■ PAGE 688

1. $r = \dfrac{4}{2 + \cos\theta}$ **3.** $r = \dfrac{6}{2 + 3\sin\theta}$
5. $r = \dfrac{10}{3 - 2\cos\theta}$ **7.** $r = \dfrac{6}{1 + \sin\theta}$
9. (a) $\frac{4}{5}$ (b) Ellipse   (c) $y = -1$
(d)

11. (a) 1   (b) Parabola   (c) $y = \frac{2}{3}$
(d)

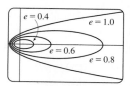

13. (a) $\frac{1}{3}$ (b) Ellipse   (c) $x = \frac{9}{2}$
(d)

15. (a) 2   (b) Hyperbola   (c) $x = -\frac{3}{8}$
(d)

17. (a) 2, $y = -\frac{1}{2}$

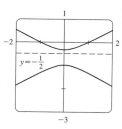

(b) $r = \dfrac{1}{1 - 2\sin(\theta - 3\pi/4)}$

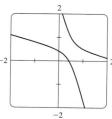

19. The ellipse is nearly circular when e is close to 0 and becomes more elongated as $e \rightarrow 1^-$. At $e = 1$, the curve becomes a parabola.

25. $r = \dfrac{2.26 \times 10^8}{1 + 0.093\cos\theta}$ **27.** $r = \dfrac{1.07}{1 + 0.97\cos\theta}$; 35.64 AU
29. 7.0×10^7 km   **31.** 3.6×10^8 km

CHAPTER 10 REVIEW ■ PAGE 689

True-False Quiz

1. False   **3.** False   **5.** True   **7.** False   **9.** True

Exercises

1. $x = y^2 - 8y + 12$

3. $y = 1/x$

5. $x = t,\ y = \sqrt{t}$; $x = t^4,\ y = t^2$; $x = \tan^2 t,\ y = \tan t,\ 0 \le t < \pi/2$

7. (a)

$(-2, 2\sqrt{3})$

(b) $(3\sqrt{2}, 3\pi/4)$, $(-3\sqrt{2}, 7\pi/4)$

9.

11.

13.

15.

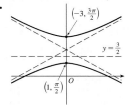

17. $r = \dfrac{2}{\cos\theta + \sin\theta}$

19.

21. 2   **23.** -1

25. $\dfrac{1 + \sin t}{1 + \cos t}$, $\dfrac{1 + \cos t + \sin t}{(1 + \cos t)^3}$

27. $\left(\tfrac{11}{8}, \tfrac{3}{4}\right)$

29. Vertical tangent at $\left(\tfrac{3}{2}a, \pm\tfrac{1}{2}\sqrt{3}\,a\right)$, $(-3a, 0)$; horizontal tangent at $(a, 0)$, $\left(-\tfrac{1}{2}a, \pm\tfrac{3}{2}\sqrt{3}\,a\right)$

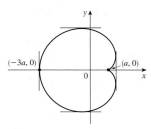

31. 18   **33.** $(2, \pm\pi/3)$ **35.** $\tfrac{1}{2}(\pi - 1)$

37. $2(5\sqrt{5} - 1)$

39. $\dfrac{2\sqrt{\pi^2 + 1} - \sqrt{4\pi^2 + 1}}{2\pi} + \ln\left(\dfrac{2\pi + \sqrt{4\pi^2 + 1}}{\pi + \sqrt{\pi^2 + 1}}\right)$

41. $471{,}295\pi/1024$

43. All curves have the vertical asymptote $x = 1$. For $c < -1$, the curve bulges to the right. At $c = -1$, the curve is the line $x = 1$. For $-1 < c < 0$, it bulges to the left. At $c = 0$ there is a cusp at $(0, 0)$. For $c > 0$, there is a loop.

45. $(\pm 1, 0)$, $(\pm 3, 0)$ **47.** $\left(-\tfrac{25}{24}, 3\right)$, $(-1, 3)$

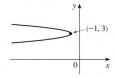

49. $\dfrac{x^2}{25} + \dfrac{y^2}{9} = 1$ **51.** $\dfrac{y^2}{72/5} - \dfrac{x^2}{8/5} = 1$

53. $\dfrac{x^2}{25} + \dfrac{(8y - 399)^2}{160{,}801} = 1$ **55.** $r = \dfrac{4}{3 + \cos\theta}$

57. $x = a(\cot\theta + \sin\theta\cos\theta)$, $y = a(1 + \sin^2\theta)$

PROBLEMS PLUS ■ PAGE 692

1. $\ln(\pi/2)$ **3.** $\left[-\tfrac{3}{4}\sqrt{3}, \tfrac{3}{4}\sqrt{3}\right] \times [-1, 2]$

CHAPTER 11

EXERCISES 11.1 ■ PAGE 704

Abbreviations: C, convergent; D, divergent

1. (a) A sequence is an ordered list of numbers. It can also be defined as a function whose domain is the set of positive integers.
(b) The terms a_n approach 8 as n becomes large.
(c) The terms a_n become large as n becomes large.

3. $\tfrac{2}{3}, \tfrac{4}{5}, \tfrac{8}{7}, \tfrac{16}{9}, \tfrac{32}{11}$ **5.** $\tfrac{1}{5}, -\tfrac{1}{25}, \tfrac{1}{125}, -\tfrac{1}{625}, \tfrac{1}{3125}$ **7.** $\tfrac{1}{2}, \tfrac{1}{6}, \tfrac{1}{24}, \tfrac{1}{120}, \tfrac{1}{720}$

9. 1, 2, 7, 32, 157   **11.** $2, \tfrac{2}{3}, \tfrac{2}{5}, \tfrac{2}{7}, \tfrac{2}{9}$ **13.** $a_n = 1/(2n)$

15. $a_n = -3\left(-\tfrac{2}{3}\right)^{n-1}$ **17.** $a_n = (-1)^{n+1}\dfrac{n^2}{n + 1}$

19. 0.4286, 0.4615, 0.4737, 0.4800, 0.4839, 0.4865, 0.4884, 0.4898, 0.4909, 0.4918; yes; $\tfrac{1}{2}$

21. 0.5000, 1.2500, 0.8750, 1.0625, 0.9688, 1.0156, 0.9922, 1.0039, 0.9980, 1.0010; yes; 1

23. 5   **25.** D   **27.** 0   **29.** 1   **31.** 2

33. D   **35.** 0   **37.** 0   **39.** D   **41.** 0   **43.** 0

45. 1   **47.** e^2 **49.** $\ln 2$ **51.** $\pi/2$ **53.** D   **55.** D

57. D   **59.** $\pi/4$ **61.** D   **63.** 0

65. (a) 1060, 1123.60, 1191.02, 1262.48, 1338.23   (b) D
67. (b) 5734   **69.** $-1 < r < 1$
71. Convergent by the Monotonic Sequence Theorem; $5 \leq L < 8$
73. Decreasing; yes   **75.** Not monotonic; no
77. Increasing; yes
79. 2   **81.** $\frac{1}{2}(3 + \sqrt{5})$ **83.** (b) $\frac{1}{2}(1 + \sqrt{5})$
85. (a) 0   (b) 9, 11

EXERCISES 11.2 ■ PAGE 715

1. (a) A sequence is an ordered list of numbers whereas a series is the *sum* of a list of numbers.
(b) A series is convergent if the sequence of partial sums is a convergent sequence. A series is divergent if it is not convergent.
3. 2
5. 0.5, 0.55, 0.5611, 0.5648, 0.5663, 0.5671, 0.5675, 0.5677; C
7. 1, 1.7937, 2.4871, 3.1170, 3.7018, 4.2521, 4.7749, 5.2749; D
9. $-2.40000, -1.92000,$
$-2.01600, -1.99680,$
$-2.00064, -1.99987,$
$-2.00003, -1.99999,$
$-2.00000, -2.00000;$
convergent, sum $= -2$

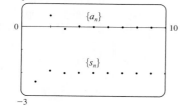

11. 0.44721, 1.15432,
1.98637, 2.88080,
3.80927, 4.75796,
5.71948, 6.68962,
7.66581, 8.64639;
divergent

13. 1.00000, 1.33333,
1.50000, 1.60000,
1.66667, 1.71429,
1.75000, 1.77778,
1.80000, 1.81818;
convergent, sum $= 2$

15. (a) Yes   (b) No   **17.** D   **19.** $\frac{25}{3}$ **21.** $\frac{400}{9}$
23. $\frac{1}{7}$ **25.** D   **27.** D   **29.** D   **31.** 9   **33.** D
35. $\dfrac{\sin 100}{1 - \sin 100}$
37. D   **39.** D   **41.** $e/(e - 1)$ **43.** $\frac{3}{2}$ **45.** $\frac{11}{6}$
47. $e - 1$
49. (b) 1   (c) 2   (d) All rational numbers with a terminating decimal representation, except 0
51. $\frac{8}{9}$ **53.** $\frac{838}{333}$ **55.** 45,679/37,000
57. $-\dfrac{1}{5} < x < \dfrac{1}{5}; \dfrac{-5x}{1 + 5x}$

59. $-1 < x < 5; \dfrac{3}{5 - x}$
61. $x > 2$ or $x < -2; \dfrac{x}{x - 2}$ **63.** $x < 0; \dfrac{1}{1 - e^x}$
65. 1   **67.** $a_1 = 0, a_n = \dfrac{2}{n(n + 1)}$ for $n > 1$, sum $= 1$
69. (a) 120 mg; 124 mg
(b) $Q_{n+1} = 100 + 0.20Q_n$ (c) 125 mg
71. (a) 157.875 mg; $\frac{3000}{19}(1 - 0.05^n)$ (b) 157.895 mg
73. (a) $S_n = \dfrac{D(1 - c^n)}{1 - c}$ (b) 5   **75.** $\frac{1}{2}(\sqrt{3} - 1)$
79. $\dfrac{1}{n(n + 1)}$ **81.** The series is divergent.
87. $\{s_n\}$ is bounded and increasing.
89. (a) $0, \frac{1}{9}, \frac{2}{9}, \frac{1}{3}, \frac{2}{3}, \frac{7}{9}, \frac{8}{9}, 1$
91. (a) $\frac{1}{2}, \frac{5}{6}, \frac{23}{24}, \frac{119}{120}; \dfrac{(n + 1)! - 1}{(n + 1)!}$ (c) 1

EXERCISES 11.3 ■ PAGE 725

1. C

3. C   **5.** D   **7.** D   **9.** C   **11.** C   **13.** D
15. C   **17.** C   **19.** D   **21.** D   **23.** C   **25.** C
27. f is neither positive nor decreasing.
29. $p > 1$ **31.** $p < -1$ **33.** $(1, \infty)$
35. (a) $\frac{9}{10}\pi^4$ (b) $\frac{1}{90}\pi^4 - \frac{17}{16}$
37. (a) 1.54977, error ≤ 0.1 (b) 1.64522, error ≤ 0.005
(c) 1.64522 compared to 1.64493   (d) $n > 1000$
39. 0.00145   **45.** $b < 1/e$

EXERCISES 11.4 ■ PAGE 731

1. (a) Nothing   (b) C   **3.** C   **5.** D   **7.** C   **9.** D
11. C   **13.** C   **15.** D   **17.** D   **19.** C   **21.** D
23. C   **25.** D   **27.** C   **29.** C   **31.** D
33. 0.1993, error $< 2.5 \times 10^{-5}$
35. 0.0739, error $< 6.4 \times 10^{-8}$
45. Yes

EXERCISES 11.5 ■ PAGE 736

1. (a) A series whose terms are alternately positive and negative   (b) $0 < b_{n+1} \leq b_n$ and $\lim_{n \to \infty} b_n = 0$, where $b_n = |a_n|$ (c) $|R_n| \leq b_{n+1}$
3. D   **5.** C   **7.** D   **9.** C   **11.** C   **13.** D
15. C   **17.** C   **19.** D   **21.** -0.5507 **23.** 5
25. 5   **27.** -0.4597 **29.** -0.1050
31. An underestimate
33. p is not a negative integer.   **35.** $\{b_n\}$ is not decreasing.

EXERCISES 11.6 ▪ PAGE 742

Abbreviations: AC, absolutely convergent;
CC, conditionally convergent
1. (a) D   (b) C   (c) May converge or diverge
3. CC   **5.** AC   **7.** AC   **9.** D   **11.** AC
13. AC   **15.** D   **17.** AC   **19.** AC   **21.** AC
23. D   **25.** AC   **27.** AC   **29.** D   **31.** CC
33. AC   **35.** D   **37.** AC   **39.** D   **41.** AC
43. (a) and (d)
47. (a) $\frac{661}{960} \approx 0.68854$, error < 0.00521
(b) $n \geq 11$, 0.693109
53. (b) $\sum_{n=2}^{\infty} \frac{(-1)^n}{n \ln n}$; $\sum_{n=1}^{\infty} \frac{(-1)^{n-1}}{n}$

EXERCISES 11.7 ▪ PAGE 746

1. D   **3.** CC   **5.** D   **7.** D   **9.** C   **11.** C
13. C   **15.** C   **17.** C   **19.** C   **21.** D   **23.** D
25. C   **27.** C   **29.** C   **31.** D
33. C   **35.** D   **37.** C

EXERCISES 11.8 ▪ PAGE 751

1. A series of the form $\sum_{n=0}^{\infty} c_n(x - a)^n$, where x is a variable
and a and the c_n's are constants
3. $1, (-1, 1)$ **5.** $1, [-1, 1)$
7. $\infty, (-\infty, \infty)$ **9.** $4, [-4, 4]$
11. $\frac{1}{4}, \left(-\frac{1}{4}, \frac{1}{4}\right]$ **13.** $2, [-2, 2)$
15. $1, [1, 3]$ **17.** $2, [-4, 0)$
19. $\infty, (-\infty, \infty)$ **21.** $b, (a - b, a + b)$ **23.** $0, \left\{\frac{1}{2}\right\}$
25. $\frac{1}{5}, \left[\frac{3}{5}, 1\right]$ **27.** $\infty, (-\infty, \infty)$
29. (a) Yes   (b) No
31. k^k **33.** No
35. (a) $(-\infty, \infty)$
(b), (c)

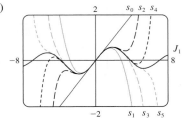

37. $(-1, 1)$, $f(x) = (1 + 2x)/(1 - x^2)$ **41.** 2

EXERCISES 11.9 ▪ PAGE 757

1. 10   **3.** $\sum_{n=0}^{\infty} (-1)^n x^n, (-1, 1)$ **5.** $2\sum_{n=0}^{\infty} \frac{1}{3^{n+1}} x^n, (-3, 3)$
7. $\sum_{n=0}^{\infty} \frac{(-1)^n x^{4n+2}}{2^{4n+4}}, (-2, 2)$ **9.** $-\frac{1}{2} - \sum_{n=1}^{\infty} \frac{(-1)^n 3 x^n}{2^{n+1}}, (-2, 2)$
11. $\sum_{n=0}^{\infty} \left(-1 - \frac{1}{3^{n+1}}\right) x^n, (-1, 1)$

13. (a) $\sum_{n=0}^{\infty} (-1)^n(n + 1)x^n, R = 1$

(b) $\frac{1}{2} \sum_{n=0}^{\infty} (-1)^n(n + 2)(n + 1)x^n, R = 1$

(c) $\frac{1}{2} \sum_{n=2}^{\infty} (-1)^n n(n - 1)x^n, R = 1$

15. $\ln 5 - \sum_{n=1}^{\infty} \frac{x^n}{n5^n}, R = 5$

17. $\sum_{n=0}^{\infty} (-1)^n 4^n(n + 1)x^{n+1}, R = \frac{1}{4}$

19. $\sum_{n=0}^{\infty} (2n + 1)x^n, R = 1$

21. $\sum_{n=0}^{\infty} (-1)^n x^{2n+2}, R = 1$

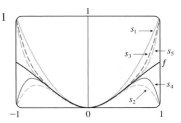

23. $\sum_{n=0}^{\infty} \frac{2x^{2n+1}}{2n + 1}, R = 1$

25. $C + \sum_{n=0}^{\infty} \frac{t^{8n+2}}{8n + 2}, R = 1$

27. $C + \sum_{n=1}^{\infty} (-1)^n \frac{x^{n+3}}{n(n + 3)}, R = 1$

29. 0.044522 **31.** 0.000395
33. 0.19740
35. (b) 0.920 **39.** $[-1, 1], [-1, 1), (-1, 1)$

EXERCISES 11.10 ▪ PAGE 771

1. $b_8 = f^{(8)}(5)/8!$ **3.** $\sum_{n=0}^{\infty} (n + 1)x^n, R = 1$
5. $x + x^2 + \frac{1}{2}x^3 + \frac{1}{6}x^4$
7. $2 + \frac{1}{12}(x - 8) - \frac{1}{288}(x - 8)^2 + \frac{5}{20,736}(x - 8)^3$
9. $\frac{1}{2} + \frac{\sqrt{3}}{2}\left(x - \frac{\pi}{6}\right) - \frac{1}{4}\left(x - \frac{\pi}{6}\right)^2 - \frac{\sqrt{3}}{12}\left(x - \frac{\pi}{6}\right)^3$
11. $\sum_{n=0}^{\infty} (n + 1)x^n, R = 1$ **13.** $\sum_{n=0}^{\infty} (-1)^n \frac{x^{2n}}{(2n)!}, R = \infty$

15. $\displaystyle\sum_{n=0}^{\infty} \frac{(\ln 2)^n}{n!} x^n, R = \infty$ **17.** $\displaystyle\sum_{n=0}^{\infty} \frac{x^{2n+1}}{(2n+1)!}, R = \infty$

19. $50 + 105(x-2) + 92(x-2)^2 + 42(x-2)^3 + 10(x-2)^4$
$+ (x-2)^5, R = \infty$

21. $\ln 2 + \displaystyle\sum_{n=1}^{\infty} (-1)^{n+1} \frac{1}{n 2^n} (x-2)^n, R = 2$

23. $\displaystyle\sum_{n=0}^{\infty} \frac{2^n e^6}{n!} (x-3)^n, R = \infty$

25. $\displaystyle\sum_{n=0}^{\infty} \frac{(-1)^{n+1}}{(2n+1)!} (x-\pi)^{2n+1}, R = \infty$

31. $1 - \dfrac{1}{4}x - \displaystyle\sum_{n=2}^{\infty} \frac{3 \cdot 7 \cdot \ \cdots \ (4n-5)}{4^n \cdot n!} x^n, R = 1$

33. $\displaystyle\sum_{n=0}^{\infty} (-1)^n \frac{(n+1)(n+2)}{2^{n+4}} x^n, R = 2$

35. $\displaystyle\sum_{n=0}^{\infty} (-1)^n \frac{1}{2n+1} x^{4n+2}, R = 1$

37. $\displaystyle\sum_{n=0}^{\infty} (-1)^n \frac{2^{2n}}{(2n)!} x^{2n+1}, R = \infty$

39. $\displaystyle\sum_{n=0}^{\infty} (-1)^n \frac{1}{2^{2n}(2n)!} x^{4n+1}, R = \infty$

41. $\dfrac{1}{2}x + \displaystyle\sum_{n=1}^{\infty} (-1)^n \frac{1 \cdot 3 \cdot 5 \cdot \ \cdots \ (2n-1)}{n! \, 2^{3n+1}} x^{2n+1}, R = 2$

43. $\displaystyle\sum_{n=1}^{\infty} (-1)^{n+1} \frac{2^{2n-1}}{(2n)!} x^{2n}, R = \infty$

45. $\displaystyle\sum_{n=0}^{\infty} (-1)^n \frac{1}{(2n)!} x^{4n}, R = \infty$

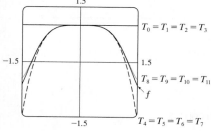

47. $\displaystyle\sum_{n=1}^{\infty} \frac{(-1)^{n-1}}{(n-1)!} x^n, R = \infty$

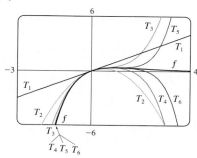

49. 0.99619

51. (a) $1 + \displaystyle\sum_{n=1}^{\infty} \frac{1 \cdot 3 \cdot 5 \cdot \ \cdots \ (2n-1)}{2^n n!} x^{2n}$

(b) $x + \displaystyle\sum_{n=1}^{\infty} \frac{1 \cdot 3 \cdot 5 \cdot \ \cdots \ (2n-1)}{(2n+1) 2^n n!} x^{2n+1}$

53. $C + \displaystyle\sum_{n=0}^{\infty} \binom{\frac{1}{2}}{n} \frac{x^{3n+1}}{3n+1}, R = 1$

55. $C + \displaystyle\sum_{n=1}^{\infty} (-1)^n \frac{1}{2n(2n)!} x^{2n}, R = \infty$

57. 0.0059   **59.** 0.40102   **61.** $\frac{1}{2}$ **63.** $\frac{1}{120}$ **65.** $\frac{3}{5}$

67. $1 - \frac{3}{2}x^2 + \frac{25}{24}x^4$ **69.** $1 + \frac{1}{6}x^2 + \frac{7}{360}x^4$

71. $x - \frac{2}{3}x^4 + \frac{23}{45}x^6$

73. e^{-x^4} **75.** $\ln \frac{8}{5}$

77. $1/\sqrt{2}$ **79.** $e^3 - 1$

EXERCISES 11.11 ■ PAGE 780

1. (a) $T_0(x) = 0$, $T_1(x) = T_2(x) = x$, $T_3(x) = T_4(x) = x - \frac{1}{6}x^3$,
$T_5(x) = x - \frac{1}{6}x^3 + \frac{1}{120}x^5$

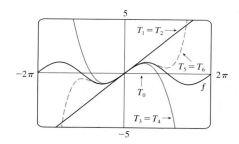

(b)

x	f	T_0	$T_1 = T_2$	$T_3 = T_4$	T_5
$\pi/4$	0.7071	0	0.7854	0.7047	0.7071
$\pi/2$	1	0	1.5708	0.9248	1.0045
π	0	0	3.1416	-2.0261	0.5240

(c) As n increases, $T_n(x)$ is a good approximation to $f(x)$ on a larger and larger interval.

3. $e + e(x-1) + \frac{1}{2}e(x-1)^2 + \frac{1}{6}e(x-1)^3$

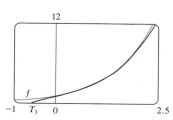

5. $-\left(x - \dfrac{\pi}{2}\right) + \dfrac{1}{6}\left(x - \dfrac{\pi}{2}\right)^3$

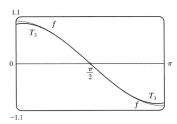

7. $(x - 1) - \dfrac{1}{2}(x - 1)^2 + \dfrac{1}{3}(x - 1)^3$

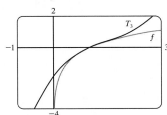

9. $x - 2x^2 + 2x^3$

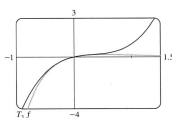

11. $T_5(x) = 1 - 2\left(x - \dfrac{\pi}{4}\right) + 2\left(x - \dfrac{\pi}{4}\right)^2 - \dfrac{8}{3}\left(x - \dfrac{\pi}{4}\right)^3$
$+ \dfrac{10}{3}\left(x - \dfrac{\pi}{4}\right)^4 - \dfrac{64}{15}\left(x - \dfrac{\pi}{4}\right)^5$

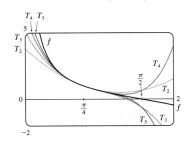

13. (a) $1 - (x - 1) + (x - 1)^2$ (b) 0.006 482 7
15. (a) $1 + \dfrac{2}{3}(x - 1) - \dfrac{1}{9}(x - 1)^2 + \dfrac{4}{81}(x - 1)^3$ (b) 0.000 097
17. (a) $1 + \dfrac{1}{2}x^2$ (b) 0.0015
19. (a) $1 + x^2$ (b) 0.000 06   **21.** (a) $x^2 - \dfrac{1}{6}x^4$ (b) 0.042

23. 0.17365   **25.** Four   **27.** $-1.037 < x < 1.037$
29. $-0.86 < x < 0.86$ **31.** 21 m, no
37. (c) They differ by about 8×10^{-9} km.

CHAPTER 11 REVIEW ■ PAGE 784

True-False Quiz
1. False   **3.** True   **5.** False   **7.** False   **9.** False
11. True   **13.** True   **15.** False   **17.** True
19. True   **21.** True

Exercises
1. $\dfrac{1}{2}$ **3.** D   **5.** 0   **7.** e^{12} **9.** 2   **11.** C
13. C   **15.** D   **17.** C   **19.** C   **21.** C   **23.** CC
25. AC   **27.** $\dfrac{1}{11}$ **29.** $\pi/4$ **31.** e^{-e} **35.** 0.9721
37. 0.189 762 24, error $< 6.4 \times 10^{-7}$
41. $4, [-6, 2)$ **43.** 0.5, [2.5, 3.5)
45. $\dfrac{1}{2} \displaystyle\sum_{n=0}^{\infty} (-1)^n \left[\dfrac{1}{(2n)!}\left(x - \dfrac{\pi}{6}\right)^{2n} + \dfrac{\sqrt{3}}{(2n + 1)!}\left(x - \dfrac{\pi}{6}\right)^{2n+1} \right]$
47. $\displaystyle\sum_{n=0}^{\infty} (-1)^n x^{n+2}, R = 1$ **49.** $\ln 4 - \displaystyle\sum_{n=1}^{\infty} \dfrac{x^n}{n4^n}, R = 4$
51. $\displaystyle\sum_{n=0}^{\infty} (-1)^n \dfrac{x^{8n+4}}{(2n + 1)!}, R = \infty$
53. $\dfrac{1}{2} + \displaystyle\sum_{n=1}^{\infty} \dfrac{1 \cdot 5 \cdot 9 \cdot \cdots \cdot (4n - 3)}{n!\,2^{6n+1}} x^n, R = 16$
55. $C + \ln |x| + \displaystyle\sum_{n=1}^{\infty} \dfrac{x^n}{n \cdot n!}$
57. (a) $1 + \dfrac{1}{2}(x - 1) - \dfrac{1}{8}(x - 1)^2 + \dfrac{1}{16}(x - 1)^3$
(b)   (c) 0.000 006

59. $-\dfrac{1}{6}$

PROBLEMS PLUS ■ PAGE 787

1. $15!/5! = 10{,}897{,}286{,}400$
3. (b) 0 if $x = 0$, $(1/x) - \cot x$ if $x \neq k\pi$, k an integer
5. (a) $s_n = 3 \cdot 4^n, l_n = 1/3^n, p_n = 4^n/3^{n-1}$ (c) $\dfrac{2}{5}\sqrt{3}$
9. $\dfrac{3\pi}{4}$ **11.** $(-1, 1)$, $\dfrac{x^3 + 4x^2 + x}{(1 - x)^4}$ **13.** $\ln \dfrac{1}{2}$
17. (a) $\dfrac{250}{101}\pi \left(e^{-(n-1)\pi/5} - e^{-n\pi/5}\right)$ (b) $\dfrac{250}{101}\pi$
19. $\dfrac{\pi}{2\sqrt{3}} - 1$
21. $-\left(\dfrac{\pi}{2} - \pi k\right)^2$, where k is a positive integer

CHAPTER 12

EXERCISES 12.1 ■ **PAGE 796**

1. $(4, 0, -3)$ **3.** $C; A$
5. A line parallel to the y-axis and 4 units to the right of it; a vertical plane parallel to the yz-plane and 4 units in front of it.

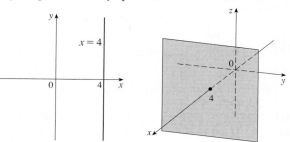

7. A vertical plane that intersects the xy-plane in the line $y = 2 - x, z = 0$

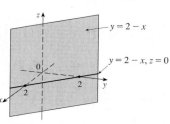

9. (a) $|PQ| = 6, |QR| = 2\sqrt{10}, |RP| = 6$; isosceles triangle
11. (a) No (b) Yes
13. $(x + 3)^2 + (y - 2)^2 + (z - 5)^2 = 16$;
$(y - 2)^2 + (z - 5)^2 = 7, x = 0$ (a circle)
15. $(x - 3)^2 + (y - 8)^2 + (z - 1)^2 = 30$
17. $(1, 2, -4), 6$ **19.** $(2, 0, -6), 9/\sqrt{2}$
21. (b) $\frac{5}{2}, \frac{1}{2}\sqrt{94}, \frac{1}{2}\sqrt{85}$
23. (a) $(x - 2)^2 + (y + 3)^2 + (z - 6)^2 = 36$
(b) $(x - 2)^2 + (y + 3)^2 + (z - 6)^2 = 4$
(c) $(x - 2)^2 + (y + 3)^2 + (z - 6)^2 = 9$
25. A plane parallel to the yz-plane and 5 units in front of it
27. A half-space consisting of all points to the left of the plane $y = 8$
29. All points on or between the horizontal planes $z = 0$ and $z = 6$
31. All points on a circle with radius 2 with center on the z-axis that is contained in the plane $z = -1$
33. All point on a sphere with radius 2 and center $(0, 0, 0)$
35. All points on or between spheres with radii 1 and $\sqrt{5}$ and centers $(0, 0, 0)$
37. All points on or inside a circular cylinder of radius 3 with axis the y-axis
39. $0 < x < 5$ **41.** $r^2 < x^2 + y^2 + z^2 < R^2$
43. (a) $(2, 1, 4)$ (b)

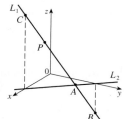

45. $14x - 6y - 10z = 9$, a plane perpendicular to AB
47. $2\sqrt{3} - 3$

EXERCISES 12.2 ■ **PAGE 805**

1. (a) Scalar (b) Vector (c) Vector (d) Scalar
3. $\overrightarrow{AB} = \overrightarrow{DC}, \overrightarrow{DA} = \overrightarrow{CB}, \overrightarrow{DE} = \overrightarrow{EB}, \overrightarrow{EA} = \overrightarrow{CE}$
5. (a) (b)

(c) (d)

(e) (f)

7. $\mathbf{c} = \frac{1}{2}\mathbf{a} + \frac{1}{2}\mathbf{b}, \mathbf{d} = \frac{1}{2}\mathbf{b} - \frac{1}{2}\mathbf{a}$

9. $\mathbf{a} = \langle 3, 1 \rangle$ **11.** $\mathbf{a} = \langle -1, 4 \rangle$

13. $\mathbf{a} = \langle 2, 0, -2 \rangle$ **15.** $\langle 5, 2 \rangle$

17. $\langle 3, 8, 1 \rangle$

19. $\langle 6, 3 \rangle, \langle 6, 14 \rangle, 5, 13$
21. $6\mathbf{i} - 3\mathbf{j} - 2\mathbf{k}, 20\mathbf{i} - 12\mathbf{j}, \sqrt{29}, 7$
23. $\left\langle \dfrac{3}{\sqrt{10}}, -\dfrac{1}{\sqrt{10}} \right\rangle$ **25.** $\frac{8}{9}\mathbf{i} - \frac{1}{9}\mathbf{j} + \frac{4}{9}\mathbf{k}$ **27.** $60°$
29. $\langle 2, 2\sqrt{3} \rangle$ **31.** ≈ 45.96 ft/s, ≈ 38.57 ft/s
33. $100\sqrt{7} \approx 264.6$ N, $\approx 139.1°$
35. $\sqrt{493} \approx 22.2$ mi/h, N8°W

9. $y = c_1 e^{3x} + c_2 e^{-2x} - \frac{1}{6} - \frac{1}{5}xe^{-2x}$

11. $y = 5 - 2e^{-6(x-1)}$ **13.** $y = (e^{4x} - e^x)/3$

15. No solution   **17.** $\displaystyle\sum_{n=0}^{\infty} \frac{(-2)^n n!}{(2n+1)!} x^{2n+1}$

19. $Q(t) = -0.02e^{-10t}(\cos 10t + \sin 10t) + 0.03$

21. (c) $2\pi/k \approx 85$ min   (d) $\approx 17{,}600$ mi/h

APPENDIXES

EXERCISES A ■ PAGE A9

1. 18   **3.** π **5.** $5 - \sqrt{5}$ **7.** $2 - x$

9. $|x + 1| = \begin{cases} x + 1 & \text{for } x \geq -1 \\ -x - 1 & \text{for } x < -1 \end{cases}$ **11.** $x^2 + 1$

13. $(-2, \infty)$

15. $[-1, \infty)$

17. $(3, \infty)$

19. $(2, 6)$

21. $(0, 1]$

23. $\left[-1, \frac{1}{2}\right)$

25. $(-\infty, 1) \cup (2, \infty)$

27. $\left[-1, \frac{1}{2}\right]$

29. $(-\infty, \infty)$

31. $\left(-\sqrt{3}, \sqrt{3}\right)$

33. $(-\infty, 1]$

35. $(-1, 0) \cup (1, \infty)$

37. $(-\infty, 0) \cup \left(\frac{1}{4}, \infty\right)$

39. $10 \leq C \leq 35$ **41.** (a) $T = 20 - 10h, 0 \leq h \leq 12$
(b) $-30°C \leq T \leq 20°C$ **43.** $\pm\frac{3}{2}$ **45.** $2, -\frac{4}{3}$
47. $(-3, 3)$ **49.** $(3, 5)$ **51.** $(-\infty, -7] \cup [-3, \infty)$
53. $[1.3, 1.7]$ **55.** $[-4, -1] \cup [1, 4]$
57. $x \geq (a + b)c/(ab)$ **59.** $x > (c - b)/a$

EXERCISES B ■ PAGE A15

1. 5   **3.** $\sqrt{74}$ **5.** $2\sqrt{37}$ **7.** 2   **9.** $-\frac{9}{2}$
17.

19.

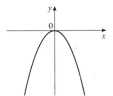

21. $y = 6x - 15$ **23.** $2x - 3y + 19 = 0$

25. $5x + y = 11$ **27.** $y = 3x - 2$ **29.** $y = 3x - 3$

31. $y = 5$ **33.** $x + 2y + 11 = 0$ **35.** $5x - 2y + 1 = 0$

37. $m = -\frac{1}{3},$ **39.** $m = 0,$ **41.** $m = \frac{3}{4},$
  $b = 0$ $b = -2$ $b = -3$

43.   **45.**

47.   **49.**

51.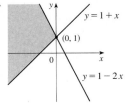

53. $(0, -4)$ **55.** (a) $(4, 9)$ (b) $(3.5, -3)$ **57.** $(1, -2)$
59. $y = x - 3$ **61.** (b) $4x - 3y - 24 = 0$

EXERCISES C ■ PAGE A23

1. $(x - 3)^2 + (y + 1)^2 = 25$ **3.** $x^2 + y^2 = 65$

5. $(2, -5), 4$ **7.** $\left(-\frac{1}{2}, 0\right), \frac{1}{2}$ **9.** $\left(\frac{1}{4}, -\frac{1}{4}\right), \sqrt{10}/4$

11. Parabola   **13.** Ellipse

15. Hyperbola

17. Ellipse

19. Parabola

21. Hyperbola

23. Hyperbola

25. Ellipse

27. Parabola

29. Parabola

31. Ellipse

33.

35. $y = x^2 - 2x$

37.

39.

EXERCISES D ■ **PAGE A32**

1. $7\pi/6$ **3.** $\pi/20$ **5.** 5π **7.** $720°$ **9.** $75°$

11. $-67.5°$ **13.** 3π cm   **15.** $\frac{2}{3}$ rad $= (120/\pi)°$

17.

19.

21.

23. $\sin(3\pi/4) = 1/\sqrt{2}$, $\cos(3\pi/4) = -1/\sqrt{2}$, $\tan(3\pi/4) = -1$,
$\csc(3\pi/4) = \sqrt{2}$, $\sec(3\pi/4) = -\sqrt{2}$, $\cot(3\pi/4) = -1$
25. $\sin(9\pi/2) = 1$, $\cos(9\pi/2) = 0$, $\csc(9\pi/2) = 1$, $\cot(9\pi/2) = 0$
, $\tan(9\pi/2)$ and $\sec(9\pi/2)$ undefined
27. $\sin(5\pi/6) = \frac{1}{2}$, $\cos(5\pi/6) = -\sqrt{3}/2$, $\tan(5\pi/6) = -1/\sqrt{3}$,
$\csc(5\pi/6) = 2$, $\sec(5\pi/6) = -2/\sqrt{3}$, $\cot(5\pi/6) = -\sqrt{3}$
29. $\cos\theta = \frac{4}{5}$, $\tan\theta = \frac{3}{4}$, $\csc\theta = \frac{5}{3}$, $\sec\theta = \frac{5}{4}$, $\cot\theta = \frac{4}{3}$
31. $\sin\phi = \sqrt{5}/3$, $\cos\phi = -\frac{2}{3}$, $\tan\phi = -\sqrt{5}/2$, $\csc\phi = 3/\sqrt{5}$,
$\cot\phi = -2/\sqrt{5}$
33. $\sin\beta = -1/\sqrt{10}$, $\cos\beta = -3/\sqrt{10}$, $\tan\beta = \frac{1}{3}$,
$\csc\beta = -\sqrt{10}$, $\sec\beta = -\sqrt{10}/3$
35. 5.73576 cm   **37.** 24.62147 cm   **59.** $\frac{1}{15}(4 + 6\sqrt{2})$
61. $\frac{1}{15}(3 + 8\sqrt{2})$ **63.** $\frac{24}{25}$ **65.** $\pi/3, 5\pi/3$
67. $\pi/4, 3\pi/4, 5\pi/4, 7\pi/4$ **69.** $\pi/6, \pi/2, 5\pi/6, 3\pi/2$
71. $0, \pi, 2\pi$ **73.** $0 \leqslant x \leqslant \pi/6$ and $5\pi/6 \leqslant x \leqslant 2\pi$
75. $0 \leqslant x < \pi/4, 3\pi/4 < x < 5\pi/4, 7\pi/4 < x \leqslant 2\pi$
77.

79.

81.

89. 14.34457 cm^2

EXERCISES E ■ PAGE A38

1. $\sqrt{1} + \sqrt{2} + \sqrt{3} + \sqrt{4} + \sqrt{5}$ **3.** $3^4 + 3^5 + 3^6$

5. $-1 + \frac{1}{3} + \frac{3}{5} + \frac{5}{7} + \frac{7}{9}$ **7.** $1^{10} + 2^{10} + 3^{10} + \cdots + n^{10}$

9. $1 - 1 + 1 - 1 + \cdots + (-1)^{n-1}$ **11.** $\sum_{i=1}^{10} i$

13. $\sum_{i=1}^{19} \frac{i}{i+1}$ **15.** $\sum_{i=1}^{n} 2i$ **17.** $\sum_{i=0}^{5} 2^i$ **19.** $\sum_{i=1}^{n} x^i$

21. 80 **23.** 3276 **25.** 0 **27.** 61 **29.** $n(n+1)$

31. $n(n^2 + 6n + 17)/3$ **33.** $n(n^2 + 6n + 11)/3$

35. $n(n^3 + 2n^2 - n - 10)/4$

41. (a) n^4 (b) $5^{100} - 1$ (c) $\frac{97}{300}$ (d) $a_n - a_0$

43. $\frac{1}{3}$ **45.** 14 **49.** $2^{n+1} + n^2 + n - 2$

EXERCISES G ■ PAGE A56

1. (b) 0.405

EXERCISES H ■ PAGE A64

1. $8 - 4i$ **3.** $13 + 18i$ **5.** $12 - 7i$ **7.** $\frac{11}{13} + \frac{10}{13}i$

9. $\frac{1}{2} - \frac{1}{2}i$ **11.** $-i$ **13.** $5i$ **15.** $12 + 5i, 13$

17. $4i, 4$ **19.** $\pm\frac{3}{2}i$ **21.** $-1 \pm 2i$

23. $-\frac{1}{2} \pm (\sqrt{7}/2)i$ **25.** $3\sqrt{2}\,[\cos(3\pi/4) + i\sin(3\pi/4)]$

27. $5\{\cos[\tan^{-1}(\frac{4}{3})] + i\sin[\tan^{-1}(\frac{4}{3})]\}$

29. $4[\cos(\pi/2) + i\sin(\pi/2)], \cos(-\pi/6) + i\sin(-\pi/6),$
$\frac{1}{2}[\cos(-\pi/6) + i\sin(-\pi/6)]$

31. $4\sqrt{2}\,[\cos(7\pi/12) + i\sin(7\pi/12)],$
$(2\sqrt{2})[\cos(13\pi/12) + i\sin(13\pi/12)], \frac{1}{4}[\cos(\pi/6) + i\sin(\pi/6)]$

33. -1024 **35.** $-512\sqrt{3} + 512i$

37. $\pm 1, \pm i, (1/\sqrt{2})(\pm 1 \pm i)$ **39.** $\pm(\sqrt{3}/2) + \frac{1}{2}i, -i$

41. i **43.** $\frac{1}{2} + (\sqrt{3}/2)i$ **45.** $-e^2$

47. $\cos 3\theta = \cos^3\theta - 3\cos\theta\sin^2\theta,$
$\sin 3\theta = 3\cos^2\theta\sin\theta - \sin^3\theta$

Index

Cut here and keep for reference

ALGEBRA

Arithmetic Operations

$$a(b + c) = ab + ac$$

$$\frac{a}{b} + \frac{c}{d} = \frac{ad + bc}{bd}$$

$$\frac{a + c}{b} = \frac{a}{b} + \frac{c}{b}$$

$$\frac{\frac{a}{b}}{\frac{c}{d}} = \frac{a}{b} \times \frac{d}{c} = \frac{ad}{bc}$$

Exponents and Radicals

$$x^m x^n = x^{m+n}$$

$$\frac{x^m}{x^n} = x^{m-n}$$

$$(x^m)^n = x^{mn}$$

$$x^{-n} = \frac{1}{x^n}$$

$$(xy)^n = x^n y^n$$

$$\left(\frac{x}{y}\right)^n = \frac{x^n}{y^n}$$

$$x^{1/n} = \sqrt[n]{x}$$

$$x^{m/n} = \sqrt[n]{x^m} = \left(\sqrt[n]{x}\right)^m$$

$$\sqrt[n]{xy} = \sqrt[n]{x}\,\sqrt[n]{y}$$

$$\sqrt[n]{\frac{x}{y}} = \frac{\sqrt[n]{x}}{\sqrt[n]{y}}$$

Factoring Special Polynomials

$$x^2 - y^2 = (x + y)(x - y)$$

$$x^3 + y^3 = (x + y)(x^2 - xy + y^2)$$

$$x^3 - y^3 = (x - y)(x^2 + xy + y^2)$$

Binomial Theorem

$$(x + y)^2 = x^2 + 2xy + y^2 \qquad (x - y)^2 = x^2 - 2xy + y^2$$

$$(x + y)^3 = x^3 + 3x^2y + 3xy^2 + y^3$$

$$(x - y)^3 = x^3 - 3x^2y + 3xy^2 - y^3$$

$$(x + y)^n = x^n + nx^{n-1}y + \frac{n(n - 1)}{2}x^{n-2}y^2$$

$$+ \cdots + \binom{n}{k}x^{n-k}y^k + \cdots + nxy^{n-1} + y^n$$

where $\binom{n}{k} = \dfrac{n(n - 1) \cdots (n - k + 1)}{1 \cdot 2 \cdot 3 \cdot \cdots \cdot k}$

Quadratic Formula

If $ax^2 + bx + c = 0$, then $x = \dfrac{-b \pm \sqrt{b^2 - 4ac}}{2a}$.

Inequalities and Absolute Value

If $a < b$ and $b < c$, then $a < c$.

If $a < b$, then $a + c < b + c$.

If $a < b$ and $c > 0$, then $ca < cb$.

If $a < b$ and $c < 0$, then $ca > cb$.

If $a > 0$, then

$|x| = a$ means $x = a$ or $x = -a$

$|x| < a$ means $-a < x < a$

$|x| > a$ means $x > a$ or $x < -a$

GEOMETRY

Geometric Formulas

Formulas for area A, circumference C, and volume V:

Triangle
$$A = \tfrac{1}{2} bh$$
$$= \tfrac{1}{2} ab \sin \theta$$

Circle
$$A = \pi r^2$$
$$C = 2\pi r$$

Sector of Circle
$$A = \tfrac{1}{2} r^2 \theta$$
$$s = r\theta \quad (\theta \text{ in radians})$$

Sphere
$$V = \tfrac{4}{3} \pi r^3$$
$$A = 4\pi r^2$$

Cylinder
$$V = \pi r^2 h$$

Cone
$$V = \tfrac{1}{3} \pi r^2 h$$
$$A = \pi r \sqrt{r^2 + h^2}$$

Distance and Midpoint Formulas

Distance between $P_1(x_1, y_1)$ and $P_2(x_2, y_2)$:

$$d = \sqrt{(x_2 - x_1)^2 + (y_2 - y_1)^2}$$

Midpoint of $\overline{P_1P_2}$: $\left(\dfrac{x_1 + x_2}{2}, \dfrac{y_1 + y_2}{2}\right)$

Lines

Slope of line through $P_1(x_1, y_1)$ and $P_2(x_2, y_2)$:

$$m = \frac{y_2 - y_1}{x_2 - x_1}$$

Point-slope equation of line through $P_1(x_1, y_1)$ with slope m:

$$y - y_1 = m(x - x_1)$$

Slope-intercept equation of line with slope m and y-intercept b:

$$y = mx + b$$

Circles

Equation of the circle with center (h, k) and radius r:

$$(x - h)^2 + (y - k)^2 = r^2$$

TRIGONOMETRY

Angle Measurement

π radians $= 180°$

$$1° = \frac{\pi}{180} \text{ rad} \qquad 1 \text{ rad} = \frac{180°}{\pi}$$

$$s = r\theta$$

(θ in radians)

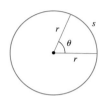

Right Angle Trigonometry

$$\sin\theta = \frac{\text{opp}}{\text{hyp}} \qquad \csc\theta = \frac{\text{hyp}}{\text{opp}}$$

$$\cos\theta = \frac{\text{adj}}{\text{hyp}} \qquad \sec\theta = \frac{\text{hyp}}{\text{adj}}$$

$$\tan\theta = \frac{\text{opp}}{\text{adj}} \qquad \cot\theta = \frac{\text{adj}}{\text{opp}}$$

Trigonometric Functions

$$\sin\theta = \frac{y}{r} \qquad \csc\theta = \frac{r}{y}$$

$$\cos\theta = \frac{x}{r} \qquad \sec\theta = \frac{r}{x}$$

$$\tan\theta = \frac{y}{x} \qquad \cot\theta = \frac{x}{y}$$

Graphs of Trigonometric Functions

Trigonometric Functions of Important Angles

θ	radians	$\sin\theta$	$\cos\theta$	$\tan\theta$
$0°$	0	0	1	0
$30°$	$\pi/6$	$1/2$	$\sqrt{3}/2$	$\sqrt{3}/3$
$45°$	$\pi/4$	$\sqrt{2}/2$	$\sqrt{2}/2$	1
$60°$	$\pi/3$	$\sqrt{3}/2$	$1/2$	$\sqrt{3}$
$90°$	$\pi/2$	1	0	—

Fundamental Identities

$$\csc\theta = \frac{1}{\sin\theta} \qquad\qquad \sec\theta = \frac{1}{\cos\theta}$$

$$\tan\theta = \frac{\sin\theta}{\cos\theta} \qquad\qquad \cot\theta = \frac{\cos\theta}{\sin\theta}$$

$$\cot\theta = \frac{1}{\tan\theta} \qquad\qquad \sin^2\theta + \cos^2\theta = 1$$

$$1 + \tan^2\theta = \sec^2\theta \qquad 1 + \cot^2\theta = \csc^2\theta$$

$$\sin(-\theta) = -\sin\theta \qquad \cos(-\theta) = \cos\theta$$

$$\tan(-\theta) = -\tan\theta \qquad \sin\left(\frac{\pi}{2} - \theta\right) = \cos\theta$$

$$\cos\left(\frac{\pi}{2} - \theta\right) = \sin\theta \qquad \tan\left(\frac{\pi}{2} - \theta\right) = \cot\theta$$

The Law of Sines

$$\frac{\sin A}{a} = \frac{\sin B}{b} = \frac{\sin C}{c}$$

The Law of Cosines

$$a^2 = b^2 + c^2 - 2bc\cos A$$

$$b^2 = a^2 + c^2 - 2ac\cos B$$

$$c^2 = a^2 + b^2 - 2ab\cos C$$

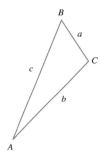

Addition and Subtraction Formulas

$$\sin(x + y) = \sin x\cos y + \cos x\sin y$$

$$\sin(x - y) = \sin x\cos y - \cos x\sin y$$

$$\cos(x + y) = \cos x\cos y - \sin x\sin y$$

$$\cos(x - y) = \cos x\cos y + \sin x\sin y$$

$$\tan(x + y) = \frac{\tan x + \tan y}{1 - \tan x\tan y}$$

$$\tan(x - y) = \frac{\tan x - \tan y}{1 + \tan x\tan y}$$

Double-Angle Formulas

$$\sin 2x = 2\sin x\cos x$$

$$\cos 2x = \cos^2 x - \sin^2 x = 2\cos^2 x - 1 = 1 - 2\sin^2 x$$

$$\tan 2x = \frac{2\tan x}{1 - \tan^2 x}$$

Half-Angle Formulas

$$\sin^2 x = \frac{1 - \cos 2x}{2} \qquad \cos^2 x = \frac{1 + \cos 2x}{2}$$

SPECIAL FUNCTIONS

Power Functions $f(x) = x^a$

(i) $f(x) = x^n$, n a positive integer

n even

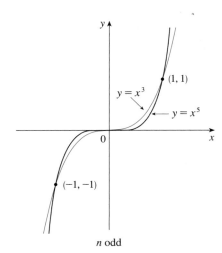

n odd

(ii) $f(x) = x^{1/n} = \sqrt[n]{x}$, n a positive integer

$f(x) = \sqrt{x}$

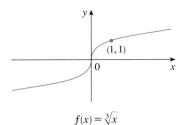

$f(x) = \sqrt[3]{x}$

(iii) $f(x) = x^{-1} = \dfrac{1}{x}$

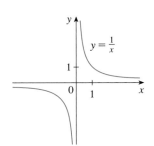

Inverse Trigonometric Functions

$\arcsin x = \sin^{-1}x = y \iff \sin y = x \quad \text{and} \quad -\dfrac{\pi}{2} \leqslant y \leqslant \dfrac{\pi}{2}$

$\arccos x = \cos^{-1}x = y \iff \cos y = x \quad \text{and} \quad 0 \leqslant y \leqslant \pi$

$\arctan x = \tan^{-1}x = y \iff \tan y = x \quad \text{and} \quad -\dfrac{\pi}{2} < y < \dfrac{\pi}{2}$

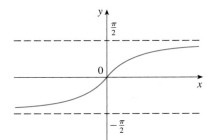

$y = \tan^{-1}x = \arctan x$

$\lim\limits_{x \to -\infty} \tan^{-1}x = -\dfrac{\pi}{2}$

$\lim\limits_{x \to \infty} \tan^{-1}x = \dfrac{\pi}{2}$

SPECIAL FUNCTIONS

Exponential and Logarithmic Functions

$\log_b x = y \iff b^y = x$

$\ln x = \log_e x, \quad \text{where} \quad \ln e = 1$

$\ln x = y \iff e^y = x$

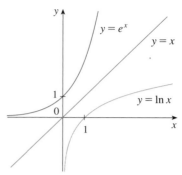

Cancellation Equations	Laws of Logarithms
$\log_b(b^x) = x \quad b^{\log_b x} = x$	**1.** $\log_b(xy) = \log_b x + \log_b y$
$\ln(e^x) = x \qquad e^{\ln x} = x$	**2.** $\log_b\left(\dfrac{x}{y}\right) = \log_b x - \log_b y$
	3. $\log_b(x^r) = r \log_b x$

$\lim_{x \to -\infty} e^x = 0 \qquad \lim_{x \to \infty} e^x = \infty$

$\lim_{x \to 0^+} \ln x = -\infty \qquad \lim_{x \to \infty} \ln x = \infty$

Exponential functions

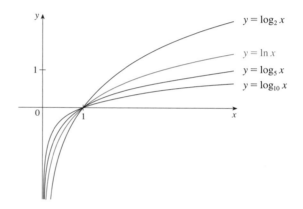

Logarithmic functions

Hyperbolic Functions

$\sinh x = \dfrac{e^x - e^{-x}}{2}$ 　　　 $\operatorname{csch} x = \dfrac{1}{\sinh x}$

$\cosh x = \dfrac{e^x + e^{-x}}{2}$ 　　　 $\operatorname{sech} x = \dfrac{1}{\cosh x}$

$\tanh x = \dfrac{\sinh x}{\cosh x}$ 　　　 $\coth x = \dfrac{\cosh x}{\sinh x}$

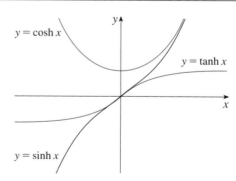

Inverse Hyperbolic Functions

$y = \sinh^{-1} x \iff \sinh y = x$ 　　　 $\sinh^{-1} x = \ln\left(x + \sqrt{x^2 + 1}\right)$

$y = \cosh^{-1} x \iff \cosh y = x \quad \text{and} \quad y \geq 0$ 　　　 $\cosh^{-1} x = \ln\left(x + \sqrt{x^2 - 1}\right)$

$y = \tanh^{-1} x \iff \tanh y = x$ 　　　 $\tanh^{-1} x = \tfrac{1}{2}\ln\left(\dfrac{1 + x}{1 - x}\right)$

DIFFERENTIATION RULES

General Formulas

1. $\dfrac{d}{dx}(c) = 0$

2. $\dfrac{d}{dx}[cf(x)] = cf'(x)$

3. $\dfrac{d}{dx}[f(x) + g(x)] = f'(x) + g'(x)$

4. $\dfrac{d}{dx}[f(x) - g(x)] = f'(x) - g'(x)$

5. $\dfrac{d}{dx}[f(x)g(x)] = f(x)g'(x) + g(x)f'(x)$ (Product Rule)

6. $\dfrac{d}{dx}\left[\dfrac{f(x)}{g(x)}\right] = \dfrac{g(x)f'(x) - f(x)g'(x)}{[g(x)]^2}$ (Quotient Rule)

7. $\dfrac{d}{dx}f(g(x)) = f'(g(x))g'(x)$ (Chain Rule)

8. $\dfrac{d}{dx}(x^n) = nx^{n-1}$ (Power Rule)

Exponential and Logarithmic Functions

9. $\dfrac{d}{dx}(e^x) = e^x$

10. $\dfrac{d}{dx}(b^x) = b^x \ln b$

11. $\dfrac{d}{dx}\ln|x| = \dfrac{1}{x}$

12. $\dfrac{d}{dx}(\log_b x) = \dfrac{1}{x \ln b}$

Trigonometric Functions

13. $\dfrac{d}{dx}(\sin x) = \cos x$

14. $\dfrac{d}{dx}(\cos x) = -\sin x$

15. $\dfrac{d}{dx}(\tan x) = \sec^2 x$

16. $\dfrac{d}{dx}(\csc x) = -\csc x \cot x$

17. $\dfrac{d}{dx}(\sec x) = \sec x \tan x$

18. $\dfrac{d}{dx}(\cot x) = -\csc^2 x$

Inverse Trigonometric Functions

19. $\dfrac{d}{dx}(\sin^{-1}x) = \dfrac{1}{\sqrt{1 - x^2}}$

20. $\dfrac{d}{dx}(\cos^{-1}x) = -\dfrac{1}{\sqrt{1 - x^2}}$

21. $\dfrac{d}{dx}(\tan^{-1}x) = \dfrac{1}{1 + x^2}$

22. $\dfrac{d}{dx}(\csc^{-1}x) = -\dfrac{1}{x\sqrt{x^2 - 1}}$

23. $\dfrac{d}{dx}(\sec^{-1}x) = \dfrac{1}{x\sqrt{x^2 - 1}}$

24. $\dfrac{d}{dx}(\cot^{-1}x) = -\dfrac{1}{1 + x^2}$

Hyperbolic Functions

25. $\dfrac{d}{dx}(\sinh x) = \cosh x$

26. $\dfrac{d}{dx}(\cosh x) = \sinh x$

27. $\dfrac{d}{dx}(\tanh x) = \operatorname{sech}^2 x$

28. $\dfrac{d}{dx}(\operatorname{csch} x) = -\operatorname{csch} x \coth x$

29. $\dfrac{d}{dx}(\operatorname{sech} x) = -\operatorname{sech} x \tanh x$

30. $\dfrac{d}{dx}(\coth x) = -\operatorname{csch}^2 x$

Inverse Hyperbolic Functions

31. $\dfrac{d}{dx}(\sinh^{-1}x) = \dfrac{1}{\sqrt{1 + x^2}}$

32. $\dfrac{d}{dx}(\cosh^{-1}x) = \dfrac{1}{\sqrt{x^2 - 1}}$

33. $\dfrac{d}{dx}(\tanh^{-1}x) = \dfrac{1}{1 - x^2}$

34. $\dfrac{d}{dx}(\operatorname{csch}^{-1}x) = -\dfrac{1}{|x|\sqrt{x^2 + 1}}$

35. $\dfrac{d}{dx}(\operatorname{sech}^{-1}x) = -\dfrac{1}{x\sqrt{1 - x^2}}$

36. $\dfrac{d}{dx}(\coth^{-1}x) = \dfrac{1}{1 - x^2}$

TABLE OF INTEGRALS

Basic Forms

1. $\displaystyle\int u \, dv = uv - \int v \, du$

2. $\displaystyle\int u^n \, du = \frac{u^{n+1}}{n+1} + C, \quad n \neq -1$

3. $\displaystyle\int \frac{du}{u} = \ln |u| + C$

4. $\displaystyle\int e^u \, du = e^u + C$

5. $\displaystyle\int b^u \, du = \frac{b^u}{\ln b} + C$

6. $\displaystyle\int \sin u \, du = -\cos u + C$

7. $\displaystyle\int \cos u \, du = \sin u + C$

8. $\displaystyle\int \sec^2 u \, du = \tan u + C$

9. $\displaystyle\int \csc^2 u \, du = -\cot u + C$

10. $\displaystyle\int \sec u \tan u \, du = \sec u + C$

11. $\displaystyle\int \csc u \cot u \, du = -\csc u + C$

12. $\displaystyle\int \tan u \, du = \ln |\sec u| + C$

13. $\displaystyle\int \cot u \, du = \ln |\sin u| + C$

14. $\displaystyle\int \sec u \, du = \ln |\sec u + \tan u| + C$

15. $\displaystyle\int \csc u \, du = \ln |\csc u - \cot u| + C$

16. $\displaystyle\int \frac{du}{\sqrt{a^2 - u^2}} = \sin^{-1} \frac{u}{a} + C, \quad a > 0$

17. $\displaystyle\int \frac{du}{a^2 + u^2} = \frac{1}{a} \tan^{-1} \frac{u}{a} + C$

18. $\displaystyle\int \frac{du}{u\sqrt{u^2 - a^2}} = \frac{1}{a} \sec^{-1} \frac{u}{a} + C$

19. $\displaystyle\int \frac{du}{a^2 - u^2} = \frac{1}{2a} \ln \left| \frac{u + a}{u - a} \right| + C$

20. $\displaystyle\int \frac{du}{u^2 - a^2} = \frac{1}{2a} \ln \left| \frac{u - a}{u + a} \right| + C$

Forms Involving $\sqrt{a^2 + u^2}$, $a > 0$

21. $\displaystyle\int \sqrt{a^2 + u^2} \, du = \frac{u}{2} \sqrt{a^2 + u^2} + \frac{a^2}{2} \ln\left(u + \sqrt{a^2 + u^2}\right) + C$

22. $\displaystyle\int u^2 \sqrt{a^2 + u^2} \, du = \frac{u}{8}\left(a^2 + 2u^2\right) \sqrt{a^2 + u^2} - \frac{a^4}{8} \ln\left(u + \sqrt{a^2 + u^2}\right) + C$

23. $\displaystyle\int \frac{\sqrt{a^2 + u^2}}{u} \, du = \sqrt{a^2 + u^2} - a \ln \left| \frac{a + \sqrt{a^2 + u^2}}{u} \right| + C$

24. $\displaystyle\int \frac{\sqrt{a^2 + u^2}}{u^2} \, du = -\frac{\sqrt{a^2 + u^2}}{u} + \ln\left(u + \sqrt{a^2 + u^2}\right) + C$

25. $\displaystyle\int \frac{du}{\sqrt{a^2 + u^2}} = \ln\left(u + \sqrt{a^2 + u^2}\right) + C$

26. $\displaystyle\int \frac{u^2 \, du}{\sqrt{a^2 + u^2}} = \frac{u}{2} \sqrt{a^2 + u^2} - \frac{a^2}{2} \ln\left(u + \sqrt{a^2 + u^2}\right) + C$

27. $\displaystyle\int \frac{du}{u\sqrt{a^2 + u^2}} = -\frac{1}{a} \ln \left| \frac{\sqrt{a^2 + u^2} + a}{u} \right| + C$

28. $\displaystyle\int \frac{du}{u^2\sqrt{a^2 + u^2}} = -\frac{\sqrt{a^2 + u^2}}{a^2 u} + C$

29. $\displaystyle\int \frac{du}{(a^2 + u^2)^{3/2}} = \frac{u}{a^2\sqrt{a^2 + u^2}} + C$

TABLE OF INTEGRALS

Forms Involving $\sqrt{a^2 - u^2}$, $a > 0$

30. $\displaystyle \int \sqrt{a^2 - u^2}\, du = \frac{u}{2} \sqrt{a^2 - u^2} + \frac{a^2}{2} \sin^{-1} \frac{u}{a} + C$

31. $\displaystyle \int u^2 \sqrt{a^2 - u^2}\, du = \frac{u}{8} (2u^2 - a^2) \sqrt{a^2 - u^2} + \frac{a^4}{8} \sin^{-1} \frac{u}{a} + C$

32. $\displaystyle \int \frac{\sqrt{a^2 - u^2}}{u}\, du = \sqrt{a^2 - u^2} - a \ln \left| \frac{a + \sqrt{a^2 - u^2}}{u} \right| + C$

33. $\displaystyle \int \frac{\sqrt{a^2 - u^2}}{u^2}\, du = -\frac{1}{u} \sqrt{a^2 - u^2} - \sin^{-1} \frac{u}{a} + C$

34. $\displaystyle \int \frac{u^2\, du}{\sqrt{a^2 - u^2}} = -\frac{u}{2} \sqrt{a^2 - u^2} + \frac{a^2}{2} \sin^{-1} \frac{u}{a} + C$

35. $\displaystyle \int \frac{du}{u \sqrt{a^2 - u^2}} = -\frac{1}{a} \ln \left| \frac{a + \sqrt{a^2 - u^2}}{u} \right| + C$

36. $\displaystyle \int \frac{du}{u^2 \sqrt{a^2 - u^2}} = -\frac{1}{a^2 u} \sqrt{a^2 - u^2} + C$

37. $\displaystyle \int (a^2 - u^2)^{3/2}\, du = -\frac{u}{8} (2u^2 - 5a^2) \sqrt{a^2 - u^2} + \frac{3a^4}{8} \sin^{-1} \frac{u}{a} + C$

38. $\displaystyle \int \frac{du}{(a^2 - u^2)^{3/2}} = \frac{u}{a^2 \sqrt{a^2 - u^2}} + C$

Forms Involving $\sqrt{u^2 - a^2}$, $a > 0$

39. $\displaystyle \int \sqrt{u^2 - a^2}\, du = \frac{u}{2} \sqrt{u^2 - a^2} - \frac{a^2}{2} \ln \left| u + \sqrt{u^2 - a^2} \right| + C$

40. $\displaystyle \int u^2 \sqrt{u^2 - a^2}\, du = \frac{u}{8} (2u^2 - a^2) \sqrt{u^2 - a^2} - \frac{a^4}{8} \ln \left| u + \sqrt{u^2 - a^2} \right| + C$

41. $\displaystyle \int \frac{\sqrt{u^2 - a^2}}{u}\, du = \sqrt{u^2 - a^2} - a \cos^{-1} \frac{a}{|u|} + C$

42. $\displaystyle \int \frac{\sqrt{u^2 - a^2}}{u^2}\, du = -\frac{\sqrt{u^2 - a^2}}{u} + \ln \left| u + \sqrt{u^2 - a^2} \right| + C$

43. $\displaystyle \int \frac{du}{\sqrt{u^2 - a^2}} = \ln \left| u + \sqrt{u^2 - a^2} \right| + C$

44. $\displaystyle \int \frac{u^2\, du}{\sqrt{u^2 - a^2}} = \frac{u}{2} \sqrt{u^2 - a^2} + \frac{a^2}{2} \ln \left| u + \sqrt{u^2 - a^2} \right| + C$

45. $\displaystyle \int \frac{du}{u^2 \sqrt{u^2 - a^2}} = \frac{\sqrt{u^2 - a^2}}{a^2 u} + C$

46. $\displaystyle \int \frac{du}{(u^2 - a^2)^{3/2}} = -\frac{u}{a^2 \sqrt{u^2 - a^2}} + C$

(continued)

TABLE OF INTEGRALS

Forms Involving $a + bu$

47. $\displaystyle\int \frac{u\,du}{a + bu} = \frac{1}{b^2}\big(a + bu - a \ln|a + bu|\big) + C$

48. $\displaystyle\int \frac{u^2\,du}{a + bu} = \frac{1}{2b^3}\Big[(a + bu)^2 - 4a(a + bu) + 2a^2 \ln|a + bu|\Big] + C$

49. $\displaystyle\int \frac{du}{u(a + bu)} = \frac{1}{a}\ln\left|\frac{u}{a + bu}\right| + C$

50. $\displaystyle\int \frac{du}{u^2(a + bu)} = -\frac{1}{au} + \frac{b}{a^2}\ln\left|\frac{a + bu}{u}\right| + C$

51. $\displaystyle\int \frac{u\,du}{(a + bu)^2} = \frac{a}{b^2(a + bu)} + \frac{1}{b^2}\ln|a + bu| + C$

52. $\displaystyle\int \frac{du}{u(a + bu)^2} = \frac{1}{a(a + bu)} - \frac{1}{a^2}\ln\left|\frac{a + bu}{u}\right| + C$

53. $\displaystyle\int \frac{u^2\,du}{(a + bu)^2} = \frac{1}{b^3}\left(a + bu - \frac{a^2}{a + bu} - 2a \ln|a + bu|\right) + C$

54. $\displaystyle\int u\sqrt{a + bu}\,du = \frac{2}{15b^2}(3bu - 2a)(a + bu)^{3/2} + C$

55. $\displaystyle\int \frac{u\,du}{\sqrt{a + bu}} = \frac{2}{3b^2}(bu - 2a)\sqrt{a + bu} + C$

56. $\displaystyle\int \frac{u^2\,du}{\sqrt{a + bu}} = \frac{2}{15b^3}(8a^2 + 3b^2u^2 - 4abu)\sqrt{a + bu} + C$

57. $\displaystyle\int \frac{du}{u\sqrt{a + bu}} = \frac{1}{\sqrt{a}}\ln\left|\frac{\sqrt{a + bu} - \sqrt{a}}{\sqrt{a + bu} + \sqrt{a}}\right| + C,\quad \text{if } a > 0$

$\displaystyle\qquad\qquad = \frac{2}{\sqrt{-a}}\tan^{-1}\sqrt{\frac{a + bu}{-a}} + C,\qquad \text{if } a < 0$

58. $\displaystyle\int \frac{\sqrt{a + bu}}{u}\,du = 2\sqrt{a + bu} + a\int \frac{du}{u\sqrt{a + bu}}$

59. $\displaystyle\int \frac{\sqrt{a + bu}}{u^2}\,du = -\frac{\sqrt{a + bu}}{u} + \frac{b}{2}\int \frac{du}{u\sqrt{a + bu}}$

60. $\displaystyle\int u^n\sqrt{a + bu}\,du = \frac{2}{b(2n + 3)}\left[u^n(a + bu)^{3/2} - na\int u^{n-1}\sqrt{a + bu}\,du\right]$

61. $\displaystyle\int \frac{u^n\,du}{\sqrt{a + bu}} = \frac{2u^n\sqrt{a + bu}}{b(2n + 1)} - \frac{2na}{b(2n + 1)}\int \frac{u^{n-1}\,du}{\sqrt{a + bu}}$

62. $\displaystyle\int \frac{du}{u^n\sqrt{a + bu}} = -\frac{\sqrt{a + bu}}{a(n - 1)u^{n-1}} - \frac{b(2n - 3)}{2a(n - 1)}\int \frac{du}{u^{n-1}\sqrt{a + bu}}$

Cut here and keep for reference

TABLE OF INTEGRALS

Trigonometric Forms

63. $\int \sin^2 u \, du = \frac{1}{2}u - \frac{1}{4}\sin 2u + C$

64. $\int \cos^2 u \, du = \frac{1}{2}u + \frac{1}{4}\sin 2u + C$

65. $\int \tan^2 u \, du = \tan u - u + C$

66. $\int \cot^2 u \, du = -\cot u - u + C$

67. $\int \sin^3 u \, du = -\frac{1}{3}(2 + \sin^2 u)\cos u + C$

68. $\int \cos^3 u \, du = \frac{1}{3}(2 + \cos^2 u)\sin u + C$

69. $\int \tan^3 u \, du = \frac{1}{2}\tan^2 u + \ln|\cos u| + C$

70. $\int \cot^3 u \, du = -\frac{1}{2}\cot^2 u - \ln|\sin u| + C$

71. $\int \sec^3 u \, du = \frac{1}{2}\sec u \tan u + \frac{1}{2}\ln|\sec u + \tan u| + C$

72. $\int \csc^3 u \, du = -\frac{1}{2}\csc u \cot u + \frac{1}{2}\ln|\csc u - \cot u| + C$

73. $\int \sin^n u \, du = -\frac{1}{n}\sin^{n-1}u \cos u + \frac{n-1}{n}\int \sin^{n-2}u \, du$

74. $\int \cos^n u \, du = \frac{1}{n}\cos^{n-1}u \sin u + \frac{n-1}{n}\int \cos^{n-2}u \, du$

75. $\int \tan^n u \, du = \frac{1}{n-1}\tan^{n-1}u - \int \tan^{n-2}u \, du$

76. $\int \cot^n u \, du = \frac{-1}{n-1}\cot^{n-1}u - \int \cot^{n-2}u \, du$

77. $\int \sec^n u \, du = \frac{1}{n-1}\tan u \sec^{n-2}u + \frac{n-2}{n-1}\int \sec^{n-2}u \, du$

78. $\int \csc^n u \, du = \frac{-1}{n-1}\cot u \csc^{n-2}u + \frac{n-2}{n-1}\int \csc^{n-2}u \, du$

79. $\int \sin au \, \sin bu \, du = \frac{\sin(a-b)u}{2(a-b)} - \frac{\sin(a+b)u}{2(a+b)} + C$

80. $\int \cos au \, \cos bu \, du = \frac{\sin(a-b)u}{2(a-b)} + \frac{\sin(a+b)u}{2(a+b)} + C$

81. $\int \sin au \, \cos bu \, du = -\frac{\cos(a-b)u}{2(a-b)} - \frac{\cos(a+b)u}{2(a+b)} + C$

82. $\int u \sin u \, du = \sin u - u \cos u + C$

83. $\int u \cos u \, du = \cos u + u \sin u + C$

84. $\int u^n \sin u \, du = -u^n \cos u + n \int u^{n-1}\cos u \, du$

85. $\int u^n \cos u \, du = u^n \sin u - n \int u^{n-1}\sin u \, du$

86. $\int \sin^n u \, \cos^m u \, du = -\frac{\sin^{n-1}u \, \cos^{m+1}u}{n+m} + \frac{n-1}{n+m}\int \sin^{n-2}u \, \cos^m u \, du$

$$= \frac{\sin^{n+1}u \, \cos^{m-1}u}{n+m} + \frac{m-1}{n+m}\int \sin^n u \, \cos^{m-2}u \, du$$

Inverse Trigonometric Forms

87. $\int \sin^{-1}u \, du = u \sin^{-1}u + \sqrt{1-u^2} + C$

88. $\int \cos^{-1}u \, du = u \cos^{-1}u - \sqrt{1-u^2} + C$

89. $\int \tan^{-1}u \, du = u \tan^{-1}u - \frac{1}{2}\ln(1+u^2) + C$

90. $\int u \sin^{-1}u \, du = \frac{2u^2-1}{4}\sin^{-1}u + \frac{u\sqrt{1-u^2}}{4} + C$

91. $\int u \cos^{-1}u \, du = \frac{2u^2-1}{4}\cos^{-1}u - \frac{u\sqrt{1-u^2}}{4} + C$

92. $\int u \tan^{-1}u \, du = \frac{u^2+1}{2}\tan^{-1}u - \frac{u}{2} + C$

93. $\int u^n \sin^{-1}u \, du = \frac{1}{n+1}\left[u^{n+1}\sin^{-1}u - \int \frac{u^{n+1}\,du}{\sqrt{1-u^2}}\right], \quad n \neq -1$

94. $\int u^n \cos^{-1}u \, du = \frac{1}{n+1}\left[u^{n+1}\cos^{-1}u + \int \frac{u^{n+1}\,du}{\sqrt{1-u^2}}\right], \quad n \neq -1$

95. $\int u^n \tan^{-1}u \, du = \frac{1}{n+1}\left[u^{n+1}\tan^{-1}u - \int \frac{u^{n+1}\,du}{1+u^2}\right], \quad n \neq -1$

(continued)

TABLE OF INTEGRALS

Exponential and Logarithmic Forms

96. $\displaystyle\int ue^{au}\,du = \frac{1}{a^2}(au - 1)e^{au} + C$

97. $\displaystyle\int u^n e^{au}\,du = \frac{1}{a}u^n e^{au} - \frac{n}{a}\int u^{n-1}e^{au}\,du$

98. $\displaystyle\int e^{au}\sin bu\,du = \frac{e^{au}}{a^2 + b^2}(a\sin bu - b\cos bu) + C$

99. $\displaystyle\int e^{au}\cos bu\,du = \frac{e^{au}}{a^2 + b^2}(a\cos bu + b\sin bu) + C$

100. $\displaystyle\int \ln u\,du = u\ln u - u + C$

101. $\displaystyle\int u^n \ln u\,du = \frac{u^{n+1}}{(n+1)^2}[(n+1)\ln u - 1] + C$

102. $\displaystyle\int \frac{1}{u\ln u}\,du = \ln|\ln u| + C$

Hyperbolic Forms

103. $\displaystyle\int \sinh u\,du = \cosh u + C$

104. $\displaystyle\int \cosh u\,du = \sinh u + C$

105. $\displaystyle\int \tanh u\,du = \ln\cosh u + C$

106. $\displaystyle\int \coth u\,du = \ln|\sinh u| + C$

107. $\displaystyle\int \operatorname{sech} u\,du = \tan^{-1}|\sinh u| + C$

108. $\displaystyle\int \operatorname{csch} u\,du = \ln\left|\tanh \tfrac{1}{2}u\right| + C$

109. $\displaystyle\int \operatorname{sech}^2 u\,du = \tanh u + C$

110. $\displaystyle\int \operatorname{csch}^2 u\,du = -\coth u + C$

111. $\displaystyle\int \operatorname{sech} u\tanh u\,du = -\operatorname{sech} u + C$

112. $\displaystyle\int \operatorname{csch} u\coth u\,du = -\operatorname{csch} u + C$

Forms Involving $\sqrt{2au - u^2}$, $a > 0$

113. $\displaystyle\int \sqrt{2au - u^2}\,du = \frac{u-a}{2}\sqrt{2au - u^2} + \frac{a^2}{2}\cos^{-1}\left(\frac{a-u}{a}\right) + C$

114. $\displaystyle\int u\sqrt{2au - u^2}\,du = \frac{2u^2 - au - 3a^2}{6}\sqrt{2au - u^2} + \frac{a^3}{2}\cos^{-1}\left(\frac{a-u}{a}\right) + C$

115. $\displaystyle\int \frac{\sqrt{2au - u^2}}{u}\,du = \sqrt{2au - u^2} + a\cos^{-1}\left(\frac{a-u}{a}\right) + C$

116. $\displaystyle\int \frac{\sqrt{2au - u^2}}{u^2}\,du = -\frac{2\sqrt{2au - u^2}}{u} - \cos^{-1}\left(\frac{a-u}{a}\right) + C$

117. $\displaystyle\int \frac{du}{\sqrt{2au - u^2}} = \cos^{-1}\left(\frac{a-u}{a}\right) + C$

118. $\displaystyle\int \frac{u\,du}{\sqrt{2au - u^2}} = -\sqrt{2au - u^2} + a\cos^{-1}\left(\frac{a-u}{a}\right) + C$

119. $\displaystyle\int \frac{u^2\,du}{\sqrt{2au - u^2}} = -\frac{(u+3a)}{2}\sqrt{2au - u^2} + \frac{3a^2}{2}\cos^{-1}\left(\frac{a-u}{a}\right) + C$

120. $\displaystyle\int \frac{du}{u\sqrt{2au - u^2}} = -\frac{\sqrt{2au - u^2}}{au} + C$

1. (a) What is a function? What are its domain and range?

A function f is a rule that assigns to each element x in a set D exactly one element, called $f(x)$, in a set E. The domain is the set D and the range is the set of all possible values of $f(x)$ as x varies throughout the domain.

(b) What is the graph of a function?

The graph of a function f consists of all points (x, y) such that $y = f(x)$ and x is in the domain of f.

(c) How can you tell whether a given curve is the graph of a function?

Use the Vertical Line Test: a curve in the xy-plane is the graph of a function of x if and only if no vertical line intersects the curve more than once.

2. Discuss four ways of representing a function. Illustrate your discussion with examples.

A function can be represented verbally, numerically, visually, or algebraically. An example of each is given below.

Verbally: An assignment of students to chairs in a classroom (a description in words)

Numerically: A tax table that assigns an amount of tax to an income (a table of values)

Visually: A graphical history of the Dow Jones average (a graph)

Algebraically: A relationship between the area A and side length s of a square: $A = s^2$ (an explicit formula)

3. (a) What is an even function? How can you tell if a function is even by looking at its graph? Give three examples of an even function.

A function f is even if it satisfies $f(-x) = f(x)$ for every number x in its domain. If the graph of a function is symmetric with respect to the y-axis, then f is even. Examples are $f(x) = x^2$, $f(x) = \cos x$, $f(x) = |x|$.

(b) What is an odd function? How can you tell if a function is odd by looking at its graph? Give three examples of an odd function.

A function f is odd if it satisfies $f(-x) = -f(x)$ for every number x in its domain. If the graph of a function is symmetric with respect to the origin, then f is odd. Examples are $f(x) = x^3$, $f(x) = \sin x$, $f(x) = 1/x$.

4. What is an increasing function?

A function f is increasing on an interval I if $f(x_1) < f(x_2)$ whenever $x_1 < x_2$ in I.

5. What is a mathematical model?

A mathematical model is a mathematical description (often by means of a function or an equation) of a real-world phenomenon. (See the discussion on pages 23–24.)

6. Give an example of each type of function.

(a) Linear function: $f(x) = 2x + 1$, $f(x) = ax + b$

(b) Power function: $f(x) = x^2$, $f(x) = x^n$

(c) Exponential function: $f(x) = 2^x$, $f(x) = b^x$

(d) Quadratic function: $f(x) = x^2 + x + 1$, $f(x) = ax^2 + bx + c$

(e) Polynomial of degree 5: $f(x) = x^5 + 2x^4 - 3x^2 + 7$

(f) Rational function: $f(x) = \dfrac{x}{x + 2}$, $f(x) = \dfrac{P(x)}{Q(x)}$

where $P(x)$ and $Q(x)$ are polynomials

7. Sketch by hand, on the same axes, the graphs of the following functions.

(a) $f(x) = x$ **(b)** $g(x) = x^2$

(c) $h(x) = x^3$ **(d)** $j(x) = x^4$

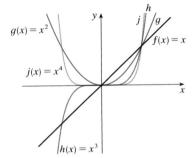

8. Draw, by hand, a rough sketch of the graph of each function.

(a) $y = \sin x$

(b) $y = \tan x$

(c) $y = e^x$ **(d)** $y = \ln x$

(continued)

(e) $y = 1/x$

(f) $y = |x|$

(g) $y = \sqrt{x}$

(h) $y = \tan^{-1}x$

9. Suppose that f has domain A and g has domain B.

(a) What is the domain of $f + g$?

The domain of $f + g$ is the intersection of the domain of f and the domain of g; that is, $A \cap B$.

(b) What is the domain of fg?

The domain of fg is also $A \cap B$.

(c) What is the domain of f/g?

The domain of f/g must exclude values of x that make g equal to 0; that is, $\{x \in A \cap B \mid g(x) \neq 0\}$.

10. How is the composite function $f \circ g$ defined? What is its domain?

The composition of f and g is defined by $(f \circ g)(x) = f(g(x))$. The domain is the set of all x in the domain of g such that $g(x)$ is in the domain of f.

11. Suppose the graph of f is given. Write an equation for each of the graphs that are obtained from the graph of f as follows.

(a) Shift 2 units upward: $y = f(x) + 2$

(b) Shift 2 units downward: $y = f(x) - 2$

(c) Shift 2 units to the right: $y = f(x - 2)$

(d) Shift 2 units to the left: $y = f(x + 2)$

(e) Reflect about the x-axis: $y = -f(x)$

(f) Reflect about the y-axis: $y = f(-x)$

(g) Stretch vertically by a factor of 2: $y = 2f(x)$

(h) Shrink vertically by a factor of 2: $y = \frac{1}{2}f(x)$

(i) Stretch horizontally by a factor of 2: $y = f\left(\frac{1}{2}x\right)$

(j) Shrink horizontally by a factor of 2: $y = f(2x)$

12. (a) What is a one-to-one function? How can you tell if a function is one-to-one by looking at its graph?

A function f is one-to-one if it never takes on the same value twice; that is $f(x_1) \neq f(x_2)$ whenever $x_1 \neq x_2$. When looking at a graph, use the Horizontal Line Test: a function is one-to-one if and only if no horizontal line intersects its graph more than once.

(b) If f is a one-to-one function, how is its inverse function f^{-1} defined? How do you obtain the graph of f^{-1} from the graph of f?

If f is a one-to-one function with domain A and range B, then its inverse function f^{-1} has domain B and range A and is defined by

$$f^{-1}(y) = x \iff f(x) = y$$

for any y in B.

The graph of f^{-1} is obtained by reflecting the graph of f about the line $y = x$.

13. (a) How is the inverse sine function $f(x) = \sin^{-1}x$ defined? What are its domain and range?

The inverse sine function $f(x) = \sin^{-1}x$ is defined as

$$\sin^{-1}x = y \iff \sin y = x$$

and $\qquad -\pi/2 \leq y \leq \pi/2$

Its domain is $-1 \leq x \leq 1$ and its range is $-\pi/2 \leq y \leq \pi/2$.

(b) How is the inverse cosine function $f(x) = \cos^{-1}x$ defined? What are its domain and range?

The inverse cosine function $f(x) = \cos^{-1}x$ is defined as

$$\cos^{-1}x = y \iff \cos y = x$$

and $\qquad 0 \leq y \leq \pi$

Its domain is $-1 \leq x \leq 1$ and its range is $0 \leq y \leq \pi$.

(c) How is the inverse tangent function $f(x) = \tan^{-1}x$ defined? What are its domain and range?

The inverse tangent function $f(x) = \tan^{-1}x$ is defined as

$$\tan^{-1}x = y \iff \tan y = x$$

and $\qquad -\pi/2 < y < \pi/2$

Its domain is $\mathbb{R}$ and its range is $-\pi/2 < y < \pi/2$.

1. Explain what each of the following means and illustrate with a sketch.

(a) $\lim\limits_{x \to a} f(x) = L$ means that the values of $f(x)$ approach L as the values of x approach a (but $x \neq a$).

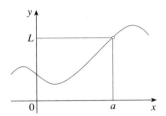

(b) $\lim\limits_{x \to a^+} f(x) = L$ means that the values of $f(x)$ approach L as the values of x approach a through values greater than a.

(c) $\lim\limits_{x \to a^-} f(x) = L$ means that the values of $f(x)$ approach L as the values of x approach a through values less than a.

(d) $\lim\limits_{x \to a} f(x) = \infty$ means that the values of $f(x)$ can be made arbitrarily large by taking x sufficiently close to a (but not equal to a).

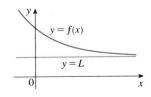

(e) $\lim\limits_{x \to \infty} f(x) = L$ means that the values of $f(x)$ can be made arbitrarily close to L by requiring x to be sufficiently large.

2. Describe several ways in which a limit can fail to exist. Illustrate with sketches.

In general, the limit of a function fails to exist when the function values do not approach a fixed number. For each of the following functions, the limit fails to exist at $x = 2$.

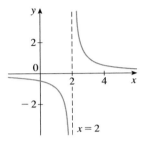

The left and right limits are not equal.

There is an infinite discontinuity.

The function values oscillate between 1 and -1 infinitely often.

3. State the following Limit Laws.

(a) Sum Law

The limit of a sum is the sum of the limits:
$$\lim_{x \to a} [f(x) + g(x)] = \lim_{x \to a} f(x) + \lim_{x \to a} g(x)$$

(b) Difference Law

The limit of a difference is the difference of the limits:
$$\lim_{x \to a} [f(x) - g(x)] = \lim_{x \to a} f(x) - \lim_{x \to a} g(x)$$

(c) Constant Multiple Law

The limit of a constant times a function is the constant times the limit of the function: $\lim\limits_{x \to a} [cf(x)] = c \lim\limits_{x \to a} f(x)$

(d) Product Law

The limit of a product is the product of the limits:
$$\lim_{x \to a} [f(x)g(x)] = \lim_{x \to a} f(x) \cdot \lim_{x \to a} g(x)$$

(e) Quotient Law

The limit of a quotient is the quotient of the limits, provided that the limit of the denominator is not 0:

$$\lim_{x \to a} \frac{f(x)}{g(x)} = \frac{\lim\limits_{x \to a} f(x)}{\lim\limits_{x \to a} g(x)} \quad \text{if } \lim_{x \to a} g(x) \neq 0$$

(continued)

(f) Power Law

The limit of a power is the power of the limit:

$$\lim_{x \to a} [f(x)]^n = \left[\lim_{x \to a} f(x)\right]^n \quad \text{(for } n \text{ a positive integer)}$$

(g) Root Law

The limit of a root is the root of the limit:

$$\lim_{x \to a} \sqrt[n]{f(x)} = \sqrt[n]{\lim_{x \to a} f(x)} \quad \text{(for } n \text{ a positive integer)}$$

4. What does the Squeeze Theorem say?

If $f(x) \leqslant g(x) \leqslant h(x)$ when x is near a (except possibly at a) and $\lim_{x \to a} f(x) = \lim_{x \to a} h(x) = L$, then $\lim_{x \to a} g(x) = L$. In other words, if $g(x)$ is squeezed between $f(x)$ and $h(x)$ near a, and if f and h have the same limit L at a, then g is forced to have the same limit L at a.

5. (a) What does it mean to say that the line $x = a$ is a vertical asymptote of the curve $y = f(x)$? Draw curves to illustrate the various possibilities.

It means that the limit of $f(x)$ as x approaches a from one or both sides is positive or negative infinity.

(b) What does it mean to say that the line $y = L$ is a horizontal asymptote of the curve $y = f(x)$? Draw curves to illustrate the various possibilities.

It means that $\lim_{x \to \infty} f(x) = L$ or $\lim_{x \to -\infty} f(x) = L$.

6. Which of the following curves have vertical asymptotes? Which have horizontal asymptotes?

(a) $y = x^4$: No asymptotes

(b) $y = \sin x$: No asymptotes

(c) $y = \tan x$: Vertical asymptotes $x = \pi/2 + \pi n$, n an integer

(d) $y = \tan^{-1} x$: Horizontal asymptotes $y = \pm\pi/2$

(e) $y = e^x$: Horizontal asymptote $y = 0$ $\left(\lim_{x \to -\infty} e^x = 0\right)$

(f) $y = \ln x$: Vertical asymptote $x = 0$ $\left(\lim_{x \to 0^+} \ln x = -\infty\right)$

(g) $y = 1/x$: Vertical asymptote $x = 0$, horizontal asymptote $y = 0$

(h) $y = \sqrt{x}$: No asymptotes

7. (a) What does it mean for f to be continuous at a?

A function f is continuous at a number a if the value of the function at $x = a$ is the same as the limit when x approaches a; that is, $\lim_{x \to a} f(x) = f(a)$.

(b) What does it mean for f to be continuous on the interval $(-\infty, \infty)$? What can you say about the graph of such a function?

A function f is continuous on the interval $(-\infty, \infty)$ if it is continuous at every real number a.

The graph of such a function has no hole or break in it.

8. (a) Give examples of functions that are continuous on $[-1, 1]$.

$f(x) = x^3 - x$, $g(x) = 2^x$, $y = \sin x$, $y = \tan x$, $y = 1/(x - 3)$, and $h(x) = |x|$ are all continuous on $[-1, 1]$.

(b) Give an example of a function that is not continuous on $[0, 1]$.

$$f(x) = \frac{1}{x - \frac{1}{2}} \quad \left(f(x) \text{ is not defined at } x = \frac{1}{2}\right)$$

9. What does the Intermediate Value Theorem say?

If f is continuous on $[a, b]$ and N is any number between $f(a)$ and $f(b)$ $[f(a) \neq f(b)]$, Then there exists a number c in (a, b) such that $f(c) = N$. In other words, a continuous function takes on every intermediate value between the function values $f(a)$ and $f(b)$.

10. Write an expression for the slope of the tangent line to the curve $y = f(x)$ at the point $(a, f(a))$.

The slope of the tangent line is given by

$$\lim_{x \to a} \frac{f(x) - f(a)}{x - a} \quad \text{or} \quad \lim_{h \to 0} \frac{f(a + h) - f(a)}{h}$$

(continued)

11. Suppose an object moves along a straight line with position $f(t)$ at time t. Write an expression for the instantaneous velocity of the object at time $t = a$. How can you interpret this velocity in terms of the graph of f?

The instantaneous velocity at time $t = a$ is

$$v(a) = \lim_{h \to 0} \frac{f(a + h) - f(a)}{h}$$

It is equal to the slope of the tangent line to the graph of f at the point $P(a, f(a))$.

12. If $y = f(x)$ and x changes from x_1 to x_2, write expressions for the following.

(a) The average rate of change of y with respect to x over the interval $[x_1, x_2]$:

$$\frac{\Delta y}{\Delta x} = \frac{f(x_2) - f(x_1)}{x_2 - x_1}$$

(b) The instantaneous rate of change of y with respect to x at $x = x_1$:

$$\lim_{\Delta x \to 0} \frac{\Delta y}{\Delta x} = \lim_{x_2 \to x_1} \frac{f(x_2) - f(x_1)}{x_2 - x_1}$$

13. Define the derivative $f'(a)$. Discuss two ways of interpreting this number.

$$f'(a) = \lim_{h \to 0} \frac{f(a + h) - f(a)}{h}$$

or, equivalently,

$$f'(a) = \lim_{x \to a} \frac{f(x) - f(a)}{x - a}$$

The derivative $f'(a)$ is the instantaneous rate of change of $y = f(x)$ (with respect to x) when $x = a$ and also represents the slope of the tangent line to the graph of f at the point $P(a, f(a))$.

14. Define the second derivative of f. If $f(t)$ is the position function of a particle, how can you interpret the second derivative?

The second derivative of f, f'', is the derivative of f'.

If $f(t)$ is the position function of a particle, then the first derivative is velocity and the second derivative is the instantaneous rate of change of velocity, namely, acceleration.

15. (a) What does it mean for f to be differentiable at a?

f is differentiable at a if the derivative $f'(a)$ exists.

(b) What is the relation between the differentiability and continuity of a function?

If f is differentiable at a, then f is continuous at a.

(c) Sketch the graph of a function that is continuous but not differentiable at $a = 2$.

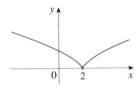

The graph of f changes direction abruptly at $x = 2$, so f has no tangent line there.

16. Describe several ways in which a function can fail to be differentiable. Illustrate with sketches.

A function is not differentiable at any value where the graph has a "corner," where the graph has a discontinuity, or where it has a vertical tangent line.

A corner   A discontinuity

A vertical tangent

CHAPTER 3 CONCEPT CHECK ANSWERS

1. State each differentiation rule both in symbols and in words.

(a) The Power Rule

If n is any real number, then $\dfrac{d}{dx}(x^n) = nx^{n-1}$.

To find the derivative of a variable raised to a constant power, we multiply the expression by the exponent and then subtract one from the exponent.

(b) The Constant Multiple Rule

If c is a constant and f is a differentiable function, then

$$\frac{d}{dx}[cf(x)] = c\,\frac{d}{dx}f(x)$$

The derivative of a constant times a function is the constant times the derivative of the function.

(c) The Sum Rule

If f and g are both differentiable, then

$$\frac{d}{dx}[f(x) + g(x)] = \frac{d}{dx}f(x) + \frac{d}{dx}g(x)$$

The derivative of a sum of functions is the sum of the derivatives.

(d) The Difference Rule

If f and g are both differentiable, then

$$\frac{d}{dx}[f(x) - g(x)] = \frac{d}{dx}f(x) - \frac{d}{dx}g(x)$$

The derivative of a difference of functions is the difference of the derivatives.

(e) The Product Rule

If f and g are both differentiable, then

$$\frac{d}{dx}[f(x)g(x)] = f(x)\,\frac{d}{dx}[g(x)] + g(x)\,\frac{d}{dx}[f(x)]$$

The derivative of a product of two functions is the first function times the derivative of the second function plus the second function times the derivative of the first function.

(f) The Quotient Rule

If f and g are both differentiable, then

$$\frac{d}{dx}\left[\frac{f(x)}{g(x)}\right] = \frac{g(x)\,\dfrac{d}{dx}[f(x)] - f(x)\,\dfrac{d}{dx}[g(x)]}{[g(x)]^2}$$

The derivative of a quotient of functions is the denominator times the derivative of the numerator minus the numerator times the derivative of the denominator, all divided by the square of the denominator.

(g) The Chain Rule

If g is differentiable at x and f is differentiable at $g(x)$, then the composite function defined by $F(x) = f(g(x))$ is differentiable at x and F' is given by the product

$$F'(x) = f'(g(x))g'(x)$$

The derivative of a composite function is the derivative of the outer function evaluated at the inner function times the derivative of the inner function.

2. State the derivative of each function.

(a) $y = x^n$: $y' = nx^{n-1}$

(b) $y = e^x$: $y' = e^x$

(c) $y = b^x$: $y' = b^x \ln b$

(d) $y = \ln x$: $y' = 1/x$

(e) $y = \log_b x$: $y' = 1/(x \ln b)$

(f) $y = \sin x$: $y' = \cos x$

(g) $y = \cos x$: $y' = -\sin x$

(h) $y = \tan x$: $y' = \sec^2 x$

(i) $y = \csc x$: $y' = -\csc x \cot x$

(j) $y = \sec x$: $y' = \sec x \tan x$

(k) $y = \cot x$: $y' = -\csc^2 x$

(l) $y = \sin^{-1} x$: $y' = 1/\sqrt{1 - x^2}$

(m) $y = \cos^{-1} x$: $y' = -1/\sqrt{1 - x^2}$

(n) $y = \tan^{-1} x$: $y' = 1/(1 + x^2)$

(o) $y = \sinh x$: $y' = \cosh x$

(p) $y = \cosh x$: $y' = \sinh x$

(q) $y = \tanh x$: $y' = \text{sech}^2 x$

(r) $y = \sinh^{-1} x$: $y' = 1/\sqrt{1 + x^2}$

(s) $y = \cosh^{-1} x$: $y' = 1/\sqrt{x^2 - 1}$

(t) $y = \tanh^{-1} x$: $y' = 1/(1 - x^2)$

3. (a) How is the number e defined?

e is the number such that $\displaystyle\lim_{h \to 0} \frac{e^h - 1}{h} = 1$.

(b) Express e as a limit.

$$e = \lim_{x \to 0} (1 + x)^{1/x}$$

(c) Why is the natural exponential function $y = e^x$ used more often in calculus than the other exponential functions $y = b^x$?

The differentiation formula for $y = b^x$ $[dy/dx = b^x \ln b]$ is simplest when $b = e$ because $\ln e = 1$.

(continued)

(d) Why is the natural logarithmic function $y = \ln x$ used more often in calculus than the other logarithmic functions $y = \log_b x$?

The differentiation formula for $y = \log_b x$ $[dy/dx = 1/(x \ln b)]$ is simplest when $b = e$ because $\ln e = 1$.

4. (a) Explain how implicit differentiation works.

Implicit differentiation consists of differentiating both sides of an equation with respect to x, treating y as a function of x. Then we solve the resulting equation for y'.

(b) Explain how logarithmic differentiation works.

Logarithmic differentiation consists of taking natural logarithms of both sides of an equation $y = f(x)$, simplifying using the Laws of Logarithms, differentiating implicitly with respect to x, and then solving the resulting equation for y'.

5. Give several examples of how the derivative can be interpreted as a rate of change in physics, chemistry, biology, economics, or other sciences.

In physics, interpretations of the derivative include velocity, linear density, electrical current, power (the rate of change of work), and the rate of radioactive decay. Chemists can use derivatives to measure reaction rates and the compressibility of a substance under pressure. In biology the derivative measures rates of population growth and blood flow. In economics, the derivative measures marginal cost (the rate of change of cost as more items are produced) and marginal profit. Other examples include the rate of heat flow in geology, the rate of performance improvement in psychology, and the rate at which a rumor spreads in sociology.

6. (a) Write a differential equation that expresses the law of natural growth.

If $y(t)$ is the value of a quantity y at time t, then

$$\frac{dy}{dt} = ky \qquad \text{where } k > 0 \text{ is a constant}$$

(b) Under what circumstances is this an appropriate model for population growth?

The equation in part (a) is an appropriate model when there is enough room and nutrition to support growth.

(c) What are the solutions of this equation?

If $y(0) = y_0$, then the solutions are $y(t) = y_0 e^{kt}$.

7. (a) Write an expression for the linearization of f at a.

$$L(x) = f(a) + f'(a)(x - a)$$

(b) If $y = f(x)$, write an expression for the differential dy.

$$dy = f'(x)\, dx$$

(c) If $dx = \Delta x$, draw a picture showing the geometric meanings of Δy and dy.

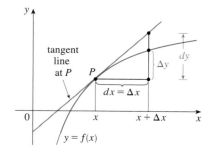

1. Explain the difference between an absolute maximum and a local maximum. Illustrate with a sketch.

The function value $f(c)$ is the absolute maximum value of f if $f(c)$ is the largest function value on the entire domain of f, whereas $f(c)$ is a local maximum value if it is the largest function value when x is near c.

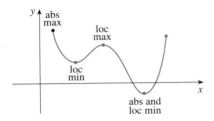

2. (a) What does the Extreme Value Theorem say?

If f is a continuous function on a closed interval $[a, b]$, then it always attains an absolute maximum and an absolute minimum value on that interval.

(b) Explain how the Closed Interval Method works.

To find the absolute maximum and minimum values of a continuous function f on a closed interval $[a, b]$, we follow these three steps:

- Find the critical numbers of f in the interval (a, b) and compute the values of f at these numbers.
- Find the values of f at the endpoints of the interval.
- The largest of the values from the previous two steps is the absolute maximum value; the smallest of these values is the absolute minimum value.

3. (a) State Fermat's Theorem.

If f has a local maximum or minimum at c, and if $f'(c)$ exists, then $f'(c) = 0$.

(b) Define a critical number of f.

A critical number of a function f is a number c in the domain of f such that either $f'(c) = 0$ or $f'(c)$ does not exist.

4. (a) State Rolle's Theorem.

Let f be a function that satisfies the following three hypotheses:

- f is continuous on the closed interval $[a, b]$.
- f is differentiable on the open interval (a, b).
- $f(a) = f(b)$

Then there is a number c in (a, b) such that $f'(c) = 0$.

(b) State the Mean Value Theorem and give a geometric interpretation.

If f is continuous on the interval $[a, b]$ and differentiable on (a, b), then there exists a number c between a and b such that

$$f'(c) = \frac{f(b) - f(a)}{b - a}$$

Geometrically, the theorem says that there is a point $P(c, f(c))$, where $a < c < b$, on the graph of f where the tangent line is parallel to the secant line that connects $(a, f(a))$ and $(b, f(b))$.

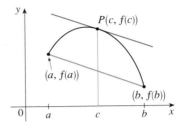

5. (a) State the Increasing/Decreasing Test.

If $f'(x) > 0$ on an interval, then f is increasing on that interval.

If $f'(x) < 0$ on an interval, then f is decreasing on that interval.

(b) What does it mean to say that f is concave upward on an interval I?

f is concave upward on an interval if the graph of f lies above all of its tangents on that interval.

(c) State the Concavity Test.

If $f''(x) > 0$ on an interval, then the graph of f is concave upward on that interval.

If $f''(x) < 0$ on an interval, then the graph of f is concave downward on that interval.

(d) What are inflection points? How do you find them?

Inflection points on the graph of a continuous function f are points where the curve changes from concave upward to concave downward or from concave downward to concave upward. They can be found by determining the values at which the second derivative changes sign.

6. (a) State the First Derivative Test.

Suppose that c is a critical number of a continuous function f.

- If f' changes from positive to negative at c, then f has a local maximum at c.
- If f' changes from negative to positive at c, then f has a local minimum at c.
- If f' is positive to the left and right of c, or negative to the left and right of c, then f has no local maximum or minimum at c.

(b) State the Second Derivative Test.

Suppose f'' is continuous near c.

- If $f'(c) = 0$ and $f''(c) > 0$, then f has a local minimum at c.
- If $f'(c) = 0$ and $f''(c) < 0$, then f has a local maximum at c.

(continued)

(c) What are the relative advantages and disadvantages of these tests?

The Second Derivative Test is sometimes easier to use, but it is inconclusive when $f''(c) = 0$ and fails if $f''(c)$ does not exist. In either case the First Derivative Test must be used.

7. (a) What does l'Hospital's Rule say?

L'Hospital's Rule says that *if* the limit of a quotient of functions is an indeterminate form of type $0/0$ or ∞/∞, then the limit is equal to the limit of the quotient of their derivatives:

$$\lim_{x \to a} \frac{f(x)}{g(x)} = \lim_{x \to a} \frac{f'(x)}{g'(x)}$$

(b) How can you use l'Hospital's Rule if you have a product $f(x)\,g(x)$ where $f(x) \to 0$ and $g(x) \to \infty$ as $x \to a$?

Write fg as $\dfrac{f}{1/g}$ or $\dfrac{g}{1/f}$ so that the limit becomes an indeterminate form of type $0/0$ or ∞/∞.

(c) How can you use l'Hospital's Rule if you have a difference $f(x) - g(x)$ where $f(x) \to \infty$ and $g(x) \to \infty$ as $x \to a$?

Convert the difference into a quotient by using a common denominator, rationalizing, factoring, or by some other method.

(d) How can you use l'Hospital's Rule if you have a power $[f(x)]^{g(x)}$ where $f(x) \to 0$ and $g(x) \to 0$ as $x \to a$?

By taking the natural logarithm of both sides of $y = [f(x)]^{g(x)}$, we can convert the power to a product by writing $\ln y = g(x) \ln f(x)$. Alternatively, we can write the function as an exponential: $[f(x)]^{g(x)} = e^{g(x) \ln f(x)}$.

8. State whether each of the following limit forms is indeterminate. Where possible, state the limit.

(a) $\dfrac{0}{0}$: indeterminate. L'Hospital's Rule can be applied to this form. Note that every derivative is a limit of this form.

(b) $\dfrac{\infty}{\infty}$: indeterminate. L'Hospital's Rule can be applied to this form.

(c) $\dfrac{0}{\infty}$: not indeterminate. A limit of this form has value 0.

(d) $\dfrac{\infty}{0}$: not indeterminate. A limit of this form could equal ∞, $-\infty$, or may not exist (but it cannot equal a finite value).

(e) $\infty + \infty$: not indeterminate. A limit of this form is equal to ∞.

(f) $\infty - \infty$: indeterminate

(g) $\infty \cdot \infty$: not indeterminate. A limit of this form is equal to ∞.

(h) $\infty \cdot 0$: indeterminate

(i) 0^0: indeterminate

(j) 0^∞: not indeterminate. A limit of this form has value 0.

(k) ∞^0: indeterminate

(l) 1^∞: indeterminate

9. If you have a graphing calculator or computer, why do you need calculus to graph a function?

Calculus reveals all the important aspects of a graph, such as local extreme values and inflection points, that can be missed when relying solely on technology. In many cases we can find exact locations of these key points rather than approximations. Using derivatives to identify the behavior of the graph also helps us choose an appropriate viewing window and alerts us to where we may wish to zoom in on a graph.

10. (a) Given an initial approximation x_1 to a root of the equation $f(x) = 0$, explain geometrically, with a diagram, how the second approximation x_2 in Newton's method is obtained.

We find the tangent line L to the graph of $y = f(x)$ at the point $(x_1, f(x_1))$. Then x_2 is the x-intercept of L.

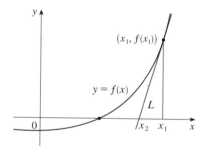

(b) Write an expression for x_2 in terms of x_1, $f(x_1)$, and $f'(x_1)$.

$$x_2 = x_1 - \frac{f(x_1)}{f'(x_1)}$$

(c) Write an expression for x_{n+1} in terms of x_n, $f(x_n)$, and $f'(x_n)$.

$$x_{n+1} = x_n - \frac{f(x_n)}{f'(x_n)}$$

(d) Under what circumstances is Newton's method likely to fail or to work very slowly?

Newton's method is likely to fail or to work very slowly when $f'(x_1)$ is close to 0. It also fails when $f'(x_i)$ is undefined.

11. (a) What is an antiderivative of a function f?

A function F is an antiderivative of f if $F'(x) = f(x)$.

(b) Suppose F_1 and F_2 are both antiderivatives of f on an interval I. How are F_1 and F_2 related?

They are identical or they differ by a constant.

1. (a) Write an expression for a Riemann sum of a function f. Explain the meaning of the notation that you use.

A Riemann sum of f is

$$\sum_{i=1}^{n} f(x_i^*)\,\Delta x$$

where x_i^* is a point in the ith subinterval $[x_{i-1}, x_i]$ and Δx is the length of the subintervals.

(b) If $f(x) \geqslant 0$, what is the geometric interpretation of a Riemann sum? Illustrate with a diagram.

If f is positive, then a Riemann sum can be interpreted as the sum of areas of approximating rectangles, as shown in the figure.

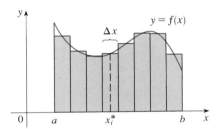

(c) If $f(x)$ takes on both positive and negative values, what is the geometric interpretation of a Riemann sum? Illustrate with a diagram.

If f takes on both positive and negative values then the Riemann sum is the sum of the areas of the rectangles that lie above the x-axis and the negatives of the areas of the rectangles that lie below the x-axis (the areas of the blue rectangles minus the areas of the gray rectangles).

2. (a) Write the definition of the definite integral of a continuous function from a to b.

If f is a continuous function on the interval $[a, b]$, then we divide $[a, b]$ into n subintervals of equal width $\Delta x = (b - a)/n$. We let $x_0\ (= a),\ x_1, x_2, \ldots,$ $x_n\ (= b)$ be the endpoints of these subintervals. Then

$$\int_a^b f(x)\,dx = \lim_{n\to\infty} \sum_{i=1}^{n} f(x_i^*)\,\Delta x$$

where x_i^* is any sample point in the ith subinterval $[x_{i-1}, x_i]$.

(b) What is the geometric interpretation of $\int_a^b f(x)\,dx$ if $f(x) \geqslant 0$?

If f is positive, then $\int_a^b f(x)\,dx$ can be interpreted as the area under the graph of $y = f(x)$ and above the x-axis for $a \leqslant x \leqslant b$.

(c) What is the geometric interpretation of $\int_a^b f(x)\,dx$ if $f(x)$ takes on both positive and negative values? Illustrate with a diagram.

In this case $\int_a^b f(x)\,dx$ can be interpreted as a "net area," that is, the area of the region above the x-axis and below the graph of f (labeled "+" in the figure) minus the area of the region below the x-axis and above the graph of f (labeled "−").

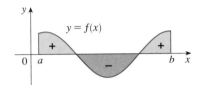

3. State the Midpoint Rule.

If f is a continuous function on the interval $[a, b]$ and we divide $[a, b]$ into n subintervals of equal width $\Delta x = (b - a)/n$, then

$$\int_a^b f(x)\,dx \approx \sum_{i=1}^{n} f(\bar{x}_i)\,\Delta x$$

where $\bar{x}_i = $ midpoint of $[x_{i-1}, x_i] = \frac{1}{2}(x_{i-1} + x_i)$.

4. State both parts of the Fundamental Theorem of Calculus.

Suppose f is continuous on $[a, b]$.

Part 1. If $g(x) = \int_a^x f(t)\,dt$, then $g'(x) = f(x)$.

Part 2. $\int_a^b f(x)\,dx = F(b) - F(a)$, where F is any antiderivative of f, that is, $F' = f$.

5. (a) State the Net Change Theorem.

The integral of a rate of change is the net change:

$$\int_a^b F'(x)\,dx = F(b) - F(a)$$

(b) If $r(t)$ is the rate at which water flows into a reservoir, what does $\int_{t_1}^{t_2} r(t)\,dt$ represent?

$\int_{t_1}^{t_2} r(t)\,dt$ represents the change in the amount of water in the reservoir between time t_1 and time t_2.

6. Suppose a particle moves back and forth along a straight line with velocity $v(t)$, measured in feet per second, and acceleration $a(t)$.

(a) What is the meaning of $\int_{60}^{120} v(t)\,dt$?

$\int_{60}^{120} v(t)\,dt$ represents the net change in position (the displacement) of the particle from $t = 60$ to $t = 120$ seconds, in other words, in the second minute.

(b) What is the meaning of $\int_{60}^{120} |v(t)|\,dt$?

$\int_{60}^{120} |v(t)|\,dt$ represents the total distance traveled by the particle in the second minute.

(continued)

(c) What is the meaning of $\int_{60}^{120} a(t)\, dt$?

$\int_{60}^{120} a(t)\, dt$ represents the change in velocity of the particle in the second minute.

7. (a) Explain the meaning of the indefinite integral $\int f(x)\, dx$.

The indefinite integral $\int f(x)\, dx$ is another name for an antiderivative of f, so $\int f(x)\, dx = F(x)$ means that $F'(x) = f(x)$.

(b) What is the connection between the definite integral $\int_a^b f(x)\, dx$ and the indefinite integral $\int f(x)\, dx$?

The connection is given by Part 2 of the Fundamental Theorem:

$$\int_a^b f(x)\, dx = \Big[\int f(x)\, dx \Big]_a^b$$

if f is continuous on $[a, b]$.

8. Explain exactly what is meant by the statement that "differentiation and integration are inverse processes."

Part 1 of the Fundamental Theorem of Calculus can be rewritten as

$$\frac{d}{dx} \int_a^x f(t)\, dt = f(x)$$

which says that if f is integrated and then the result is differentiated, we arrive back at the original function f.

Since $F'(x) = f(x)$, Part 2 of the theorem (or, equivalently, the Net Change Theorem) states that

$$\int_a^b F'(x)\, dx = F(b) - F(a)$$

This says that if we take a function F, first differentiate it, and then integrate the result, we arrive back at the original function, but in the form $F(b) - F(a)$.

Also, the indefinite integral $\int f(x)\, dx$ represents an antiderivative of f, so

$$\frac{d}{dx} \int f(x)\, dx = f(x)$$

9. State the Substitution Rule. In practice, how do you use it?

If $u = g(x)$ is a differentiable function and f is continuous on the range of g, then

$$\int f(g(x))\, g'(x)\, dx = \int f(u)\, du$$

In practice, we make the substitutions $u = g(x)$ and $du = g'(x)\, dx$ in the integrand to make the integral simpler to evaluate.

CHAPTER 6 CONCEPT CHECK ANSWERS

1. (a) Draw two typical curves $y = f(x)$ and $y = g(x)$, where $f(x) \geq g(x)$ for $a \leq x \leq b$. Show how to approximate the area between these curves by a Riemann sum and sketch the corresponding approximating rectangles. Then write an expression for the exact area.

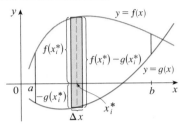

A Riemann sum that approximates the area between these curves is $\sum_{i=1}^{n} [f(x_i^*) - g(x_i^*)] \Delta x$. A sketch of the corresponding approximating rectangles:

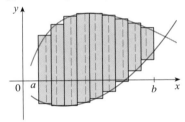

An expression for the exact area is

$$\lim_{n \to \infty} \sum_{i=1}^{n} [f(x_i^*) - g(x_i^*)] \Delta x = \int_a^b [f(x) - g(x)] dx$$

(b) Explain how the situation changes if the curves have equations $x = f(y)$ and $x = g(y)$, where $f(y) \geq g(y)$ for $c \leq y \leq d$.

Instead of using "top minus bottom" and integrating from left to right, we use "right minus left" and integrate from bottom to top: $A = \int_c^d [f(y) - g(y)] dy$

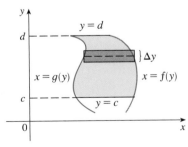

2. Suppose that Sue runs faster than Kathy throughout a 1500-meter race. What is the physical meaning of the area between their velocity curves for the first minute of the race?

It represents the number of meters by which Sue is ahead of Kathy after 1 minute.

3. (a) Suppose S is a solid with known cross-sectional areas. Explain how to approximate the volume of S by a

Riemann sum. Then write an expression for the exact volume.

We slice S into n "slabs" of equal width Δx. The volume of the ith slab is approximately $A(x_i^*) \Delta x$, where x_i^* is a sample point in the ith slab and $A(x_i^*)$ is the cross-sectional area of S at x_i^*. Then the volume of S is approximately $\sum_{i=1}^{n} A(x_i^*) \Delta x$ and the exact volume is

$$V = \lim_{n \to \infty} \sum_{i=1}^{n} A(x_i^*) \Delta x = \int_a^b A(x) dx$$

(b) If S is a solid of revolution, how do you find the cross-sectional areas?

If the cross-section is a disk, find the radius in terms of x or y and use $A = \pi(\text{radius})^2$. If the cross-section is a washer, find the inner radius r_{in} and outer radius r_{out} and use $A = \pi(r_{out}^2) - \pi(r_{in}^2)$.

4. (a) What is the volume of a cylindrical shell?

$$V = 2\pi r h \, \Delta r = (\text{circumference})(\text{height})(\text{thickness})$$

(b) Explain how to use cylindrical shells to find the volume of a solid of revolution.

We approximate the region to be revolved by rectangles, oriented so that revolution forms cylindrical shells rather than disks or washers. For a typical shell, find the circumference and height in terms of x or y and calculate

$$V = \int_a^b (\text{circumference})(\text{height})(dx \text{ or } dy)$$

(c) Why might you want to use the shell method instead of slicing?

Sometimes slicing produces washers or disks whose radii are difficult (or impossible) to find explicitly. On other occasions, the cylindrical shell method leads to an easier integral than slicing does.

5. Suppose that you push a book across a 6-meter-long table by exerting a force $f(x)$ at each point from $x = 0$ to $x = 6$. What does $\int_0^6 f(x) dx$ represent? If $f(x)$ is measured in newtons, what are the units for the integral?

$\int_0^6 f(x) dx$ represents the amount of work done. Its units are newton-meters, or joules.

6. (a) What is the average value of a function f on an interval $[a, b]$?

$$f_{ave} = \frac{1}{b - a} \int_a^b f(x) dx$$

(b) What does the Mean Value Theorem for Integrals say? What is its geometric interpretation?

If f is continuous on $[a, b]$, then there is a number c in $[a, b]$ at which the value of f is exactly equal to the average value of the function, that is, $f(c) = f_{ave}$. This means that for positive functions f, there is a number c such that the rectangle with base $[a, b]$ and height $f(c)$ has the same area as the region under the graph of f from a to b.

1. State the rule for integration by parts. In practice, how do you use it?

To integrate $\int f(x)g'(x)\,dx$, let $u = f(x)$ and $v = g(x)$. Then $\int u\,dv = uv - \int v\,du$.

In practice, try to choose $u = f(x)$ to be a function that becomes simpler when differentiated (or at least not more complicated) at long as $dv = g'(x)\,dx$ can be readily integrated to give v.

2. How do you evaluate $\int \sin^m x \cos^n x\,dx$ if m is odd? What if n is odd? What if m and n are both even?

If m is odd, use $\sin^2 x = 1 - \cos^2 x$ to write all sine factors except one in terms of cosine. Then substitute $u = \cos x$.

If n is odd, use $\cos^2 x = 1 - \sin^2 x$ to write all cosine factors except one in terms of sine. Then substitute $u = \sin x$.

If m and n are even, use the half-angle identities

$$\sin^2 x = \tfrac{1}{2}(1 - \cos 2x) \qquad \cos^2 x = \tfrac{1}{2}(1 + \cos 2x)$$

3. If the expression $\sqrt{a^2 - x^2}$ occurs in an integral, what substitution might you try? What if $\sqrt{a^2 + x^2}$ occurs? What if $\sqrt{x^2 - a^2}$ occurs?

If $\sqrt{a^2 - x^2}$ occurs, try $x = a \sin\theta$; if $\sqrt{a^2 + x^2}$ occurs, try $x = a \tan\theta$, and if $\sqrt{x^2 - a^2}$ occurs, try $x = a \sec\theta$.

4. What is the form of the partial fraction decomposition of a rational function $P(x)/Q(x)$ if the degree of P is less than the degree of Q and $Q(x)$ has only distinct linear factors? What if a linear factor is repeated? What if $Q(x)$ has an irreducible quadratic factor (not repeated)? What if the quadratic factor is repeated?

For distinct linear factors,

$$\frac{P(x)}{Q(x)} = \frac{A_1}{a_1 x + b_1} + \frac{A_2}{a_2 x + b_2} + \cdots + \frac{A_k}{a_k x + b_k}$$

If the linear factor $a_1 x + b_1$ is repeated r times, then we must include all the terms

$$\frac{B_1}{a_1 x + b_1} + \frac{B_2}{(a_1 x + b_1)^2} + \cdots + \frac{B_r}{(a_1 x + b_1)^r}$$

If $Q(x)$ has an irreducible quadratic factor (not repeated), then we include a term of the form

$$\frac{Ax + B}{ax^2 + bx + c}$$

If the irreducible quadratic factor is repeated r times, then we include all the terms

$$\frac{A_1 x + B_1}{ax^2 + bx + c} + \frac{A_2 x + B_2}{(ax^2 + bx + c)^2} + \cdots + \frac{A_r x + B_r}{(ax^2 + bx + c)^r}$$

5. State the rules for approximating the definite integral $\int_a^b f(x)\,dx$ with the Midpoint Rule, the Trapezoidal Rule, and Simpson's Rule. Which would you expect to give the best estimate? How do you approximate the error for each rule?

Let $a \leqslant x \leqslant b$, $I = \int_a^b f(x)\,dx$, and $\Delta x = (b - a)/n$.

Midpoint Rule:

$$I \approx M_n = \Delta x\,[f(\bar{x}_1) + f(\bar{x}_2) + \cdots + f(\bar{x}_n)]$$

where $\bar{x}_i$ is the midpoint of $[x_{i-1}, x_i]$.

Trapezoidal Rule:

$$I \approx T_n$$
$$= \frac{\Delta x}{2}[f(x_0) + 2f(x_1) + 2f(x_2) + \cdots + 2f(x_{n-1}) + f(x_n)]$$

where $x_i = a + i\,\Delta x$.

Simpson's Rule:

$$I \approx S_n$$
$$= \frac{\Delta x}{3}[f(x_0) + 4f(x_1) + 2f(x_2) + 4f(x_3) + \cdots$$
$$+ 2f(x_{n-2}) + 4f(x_{n-1}) + f(x_n)]$$

where n is even.

We would expect the best estimate to be given by Simpson's Rule.

Suppose $|f''(x)| \leqslant K$ and $|f^{(4)}(x)| \leqslant L$ for $a \leqslant x \leqslant b$. The errors in the Midpoint, Trapezoidal, and Simpson's Rules are given by, respectively,

$$|E_M| \leqslant \frac{K(b - a)^3}{24n^2} \qquad |E_T| \leqslant \frac{K(b - a)^3}{12n^2}$$

$$|E_S| \leqslant \frac{L(b - a)^5}{180n^4}$$

6. Define the following improper integrals.

(a) $\displaystyle\int_a^\infty f(x)\,dx = \lim_{t \to \infty} \int_a^t f(x)\,dx$

(b) $\displaystyle\int_{-\infty}^b f(x)\,dx = \lim_{t \to -\infty} \int_t^b f(x)\,dx$

(c) $\displaystyle\int_{-\infty}^\infty f(x)\,dx = \int_{-\infty}^a f(x)\,dx + \int_a^\infty f(x)\,dx$, where a is any real number (assuming that both integrals are convergent)

7. Define the improper integral $\int_a^b f(x)\,dx$ for each of the following cases.

(a) f has an infinite discontinuity at a.

If f is continuous on $(a, b]$, then

$$\int_a^b f(x)\,dx = \lim_{t \to a^+} \int_t^b f(x)\,dx$$

if this limit exists (as a finite number).

(b) f has an infinite discontinuity at b.

If f is continuous on $[a, b)$, then

$$\int_a^b f(x)\,dx = \lim_{t \to b^-} \int_a^t f(x)\,dx$$

if this limit exists (as a finite number).

(c) f has an infinite discontinuity at c, where $a < c < b$.

If both $\int_a^c f(x)\,dx$ and $\int_c^b f(x)\,dx$ are convergent, then

$$\int_a^b f(x)\,dx = \int_a^c f(x)\,dx + \int_c^b f(x)\,dx$$

8. State the Comparison Theorem for improper integrals.

Suppose that f and g are continuous functions with $f(x) \geqslant g(x) \geqslant 0$ for $x \geqslant a$.

(a) If $\int_a^\infty f(x)\,dx$ is convergent, then $\int_a^\infty g(x)\,dx$ is convergent.

(b) If $\int_a^\infty g(x)\,dx$ is divergent, then $\int_a^\infty f(x)\,dx$ is divergent.

1. (a) How is the length of a curve defined?

We can approximate a curve C by a polygon with vertices P_i along C. The length L of C is defined to be the limit of the lengths of these inscribed polygons:

$$L = \lim_{n \to \infty} \sum_{i=1}^{n} |P_{i-1}P_i|$$

(b) Write an expression for the length of a smooth curve given by $y = f(x)$, $a \le x \le b$.

$$L = \int_a^b \sqrt{1 + [f'(x)]^2}\, dx$$

(c) What if x is given as a function of y?

If $x = g(y)$, $c \le y \le d$, then $L = \int_c^d \sqrt{1 + [g'(y)]^2}\, dy$.

2. (a) Write an expression for the surface area of the surface obtained by rotating the curve $y = f(x)$, $a \le x \le b$, about the x-axis.

$$S = \int_a^b 2\pi f(x) \sqrt{1 + [f'(x)]^2}\, dx$$

(b) What if x is given as a function of y?

If $x = g(y)$, $c \le y \le d$, then $S = \int_c^d 2\pi y \sqrt{1 + [g'(y)]^2}\, dy$.

(c) What if the curve is rotated about the y-axis?

$$S = \int_a^b 2\pi x \sqrt{1 + [f'(x)]^2}\, dx$$

or

$$S = \int_c^d 2\pi g(y) \sqrt{1 + [g'(y)]^2}\, dy$$

3. Describe how we can find the hydrostatic force against a vertical wall submersed in a fluid.

We divide the wall into horizontal strips of equal height Δx and approximate each by a rectangle with horizontal length $f(x_i)$ at depth x_i. If δ is the weight density of the fluid, then the hydrostatic force is

$$F = \lim_{n \to \infty} \sum_{i=1}^{n} \delta x_i f(x_i)\, \Delta x = \int_a^b \delta x f(x)\, dx$$

4. (a) What is the physical significance of the center of mass of a thin plate?

The center of mass is the point at which the plate balances horizontally.

(b) If the plate lies between $y = f(x)$ and $y = 0$, where $a \le x \le b$, write expressions for the coordinates of the center of mass.

$$\bar{x} = \frac{1}{A} \int_a^b x f(x)\, dx \qquad \text{and} \qquad \bar{y} = \frac{1}{A} \int_a^b \tfrac{1}{2}[f(x)]^2\, dx$$

where $A = \int_a^b f(x)\, dx$.

5. What does the Theorem of Pappus say?

If a plane region $\mathcal{R}$ that lies entirely on one side of a line ℓ in its plane is rotated about ℓ, then the volume of the resulting solid is the product of the area of $\mathcal{R}$ and the distance traveled by the centroid of $\mathcal{R}$.

6. Given a demand function $p(x)$, explain what is meant by the consumer surplus when the amount of a commodity currently available is X and the current selling price is P. Illustrate with a sketch.

The consumer surplus represents the amount of money saved by consumers in purchasing the commodity at price P [when they were willing to purchase it at price $p(x)$], corresponding to an amount demanded of X.

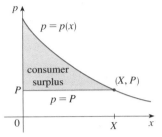

7. (a) What is the cardiac output of the heart?

It is the volume of blood pumped by the heart per unit time, that is, the rate of flow into the aorta.

(b) Explain how the cardiac output can be measured by the dye dilution method.

An amount A of dye is injected into part of the heart and its concentration $c(t)$ leaving the heart is measured over a time interval $[0, T]$ until the dye has cleared. The cardiac output is given by $A / \int_0^T c(t)\, dt$.

8. What is a probability density function? What properties does such a function have?

Given a random variable X, its probability density function f is a function such that $\int_a^b f(x)\, dx$ gives the probability that X lies between a and b. The function f has the properties that $f(x) \ge 0$ for all x, and $\int_{-\infty}^{\infty} f(x)\, dx = 1$.

9. Suppose $f(x)$ is the probability density function for the weight of a female college student, where x is measured in pounds.

(a) What is the meaning of the integral $\int_0^{130} f(x)\, dx$?

It represents the probability that a randomly chosen female college student weighs less than 130 pounds.

(b) Write an expression for the mean of this density function.

$$\mu = \int_{-\infty}^{\infty} x f(x)\, dx = \int_0^{\infty} x f(x)\, dx$$

[since $f(x) = 0$ for $x < 0$]

(c) How can we find the median of this density function?

The median of f is the number m such that

$$\int_m^{\infty} f(x)\, dx = \frac{1}{2}$$

10. What is a normal distribution? What is the significance of the standard deviation?

A normal distribution corresponds to a random variable X that has a probability density function with a bell-shaped graph and equation given by

$$f(x) = \frac{1}{\sigma \sqrt{2\pi}} e^{-(x-\mu)^2/(2\sigma^2)}$$

where μ is the mean and the positive constant σ is the standard deviation. σ measures how spread out the values of X are.

1. (a) What is a differential equation?

It is an equation that contains an unknown function and one or more of its derivatives.

(b) What is the order of a differential equation?

It is the order of the highest derivative that occurs in the equation.

(c) What is an initial condition?

It is a condition of the form $y(t_0) = y_0$.

2. What can you say about the solutions of the equation $y' = x^2 + y^2$ just by looking at the differential equation?

The equation tells us that the slope of a solution curve at any point (x, y) is $x^2 + y^2$. Note that $x^2 + y^2$ is always positive except at the origin, where $y' = x^2 + y^2 = 0$. Thus there is a horizontal tangent at $(0, 0)$ but nowhere else and the solution curves are increasing everywhere.

3. What is a direction field for the differential equation $y' = F(x, y)$?

A direction field (or slope field) for the differential equation $y' = F(x, y)$ is a two-dimensional graph consisting of short line segments with slope $F(x, y)$ at point (x, y).

4. Explain how Euler's method works.

Euler's method says to start at the point given by the initial value and proceed in the direction indicated by the direction field. Stop after a short time, look at the slope at the new location, and proceed in that direction. Keep stopping and changing direction according to the direction field until the approximation is complete.

5. What is a separable differential equation? How do you solve it?

It is a differential equation in which the expression for dy/dx can be factored as a function of x times a function of y, that is, $dy/dx = g(x) f(y)$. We can solve the equation by rewriting it as $[1/f(y)] \, dy = g(x) \, dx$, integrating both sides, and solving for y.

6. What is a first-order linear differential equation? How do you solve it?

A first-order linear differential equation is a differential equation that can be put in the form

$$\frac{dy}{dx} + P(x)y = Q(x)$$

where P and Q are continuous functions on a given interval. To solve such an equation, we multiply both sides by the integrating factor $I(x) = e^{\int P(x) \, dx}$ to put it in the form

$(I(x)y)' = I(x) Q(x)$. We then integrate both sides and solve for y.

7. (a) Write a differential equation that expresses the law of natural growth. What does it say in terms of relative growth rate?

If $P(t)$ is the value of a quantity y at time t and if the rate of change of P with respect to t is proportional to its size $P(t)$ at any time, then $\dfrac{dP}{dt} = kP$.

In this case the relative growth rate, $\dfrac{1}{P} \dfrac{dP}{dt}$, is constant.

(b) Under what circumstances is this an appropriate model for population growth?

It is an appropriate model under ideal conditions: unlimited environment, adequate nutrition, absence of predators and disease.

(c) What are the solutions of this equation?

If $P(0) = P_0$, the initial value, then the solutions are $P(t) = P_0 e^{kt}$.

8. (a) Write the logistic differential equation.

The logistic differential equation is

$$\frac{dP}{dt} = kP\left(1 - \frac{P}{M}\right)$$

where M is the carrying capacity.

(b) Under what circumstances is this an appropriate model for population growth?

It is an appropriate model for population growth if the population grows at a rate proportional to the size of the population in the beginning, but eventually levels off and approaches its carrying capacity because of limited resources.

9. (a) Write Lotka-Volterra equations to model populations of food-fish (F) and sharks (S).

$$\frac{dF}{dt} = kF - aFS \quad \text{and} \quad \frac{dS}{dt} = -rS + bFS$$

(b) What do these equations say about each population in the absence of the other?

In the absence of sharks, an ample food supply would support exponential growth of the fish population, that is, $dF/dt = kF$, where k is a positive constant. In the absence of fish, we assume that the shark population would decline at a rate proportional to itself, that is $dS/dt = -rS$, where r is a positive constant.

1. (a) What is a parametric curve?

A parametric curve is a set of points of the form $(x, y) = (f(t), g(t))$, where f and g are functions of a variable t, the parameter.

(b) How do you sketch a parametric curve?

Sketching a parametric curve, like sketching the graph of a function, is difficult to do in general. We can plot points on the curve by finding $f(t)$ and $g(t)$ for various values of t, either by hand or with a calculator or computer. Sometimes, when f and g are given by formulas, we can eliminate t from the equations $x = f(t)$ and $y = g(t)$ to get a Cartesian equation relating x and y. It may be easier to graph that equation than to work with the original formulas for x and y in terms of t.

2. (a) How do you find the slope of a tangent to a parametric curve?

You can find dy/dx as a function of t by calculating

$$\frac{dy}{dx} = \frac{dy/dt}{dx/dt} \qquad \text{if } dx/dt \neq 0$$

(b) How do you find the area under a parametric curve?

If the curve is traced out once by the parametric equations $x = f(t)$, $y = g(t)$, $\alpha \leqslant t \leqslant \beta$, then the area is

$$A = \int_a^b y\, dx = \int_\alpha^\beta g(t) f'(t)\, dt$$

[or $\int_\beta^\alpha g(t) f'(t)\, dt$ if the leftmost point is $(f(\beta), g(\beta))$ rather than $(f(\alpha), g(\alpha))$].

3. Write an expression for each of the following:

(a) The length of a parametric curve

If the curve is traced out once by the parametric equations $x = f(t)$, $y = g(t)$, $\alpha \leqslant t \leqslant \beta$, then the length is

$$L = \int_\alpha^\beta \sqrt{(dx/dt)^2 + (dy/dt)^2}\, dt$$

$$= \int_\alpha^\beta \sqrt{[f'(t)]^2 + [g'(t)]^2}\, dt$$

(b) The area of the surface obtained by rotating a parametric curve about the x-axis

$$S = \int_\alpha^\beta 2\pi y \sqrt{(dx/dt)^2 + (dy/dt)^2}\, dt$$

$$= \int_\alpha^\beta 2\pi g(t) \sqrt{[f'(t)]^2 + [g'(t)]^2}\, dt$$

4. (a) Use a diagram to explain the meaning of the polar coordinates (r, θ) of a point.

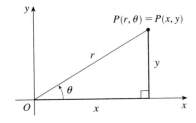

(b) Write equations that express the Cartesian coordinates (x, y) of a point in terms of the polar coordinates.

$$x = r \cos \theta \qquad y = r \sin \theta$$

(c) What equations would you use to find the polar coordinates of a point if you knew the Cartesian coordinates?

To find a polar representation (r, θ) with $r \geqslant 0$ and $0 \leqslant \theta < 2\pi$, first calculate $r = \sqrt{x^2 + y^2}$. Then θ is specified by $\tan \theta = y/x$. Be sure to choose θ so that (r, θ) lies in the correct quadrant.

5. (a) How do you find the slope of a tangent line to a polar curve?

$$\frac{dy}{dx} = \frac{\dfrac{dy}{d\theta}}{\dfrac{dx}{d\theta}} = \frac{\dfrac{d}{d\theta}(y)}{\dfrac{d}{d\theta}(x)} = \frac{\dfrac{d}{d\theta}(r \sin \theta)}{\dfrac{d}{d\theta}(r \cos \theta)}$$

$$= \frac{\left(\dfrac{dr}{d\theta}\right) \sin \theta + r \cos \theta}{\left(\dfrac{dr}{d\theta}\right) \cos \theta - r \sin \theta} \qquad \text{where } r = f(\theta)$$

(b) How do you find the area of a region bounded by a polar curve?

$$A = \int_a^b \tfrac{1}{2} r^2\, d\theta = \int_a^b \tfrac{1}{2} [f(\theta)]^2\, d\theta$$

(c) How do you find the length of a polar curve?

$$L = \int_a^b \sqrt{(dx/d\theta)^2 + (dy/d\theta)^2}\, d\theta$$

$$= \int_a^b \sqrt{r^2 + (dr/d\theta)^2}\, d\theta$$

$$= \int_a^b \sqrt{[f(\theta)]^2 + [f'(\theta)]^2}\, d\theta$$

6. (a) Give a geometric definition of a parabola.

A parabola is a set of points in a plane whose distances from a fixed point F (the focus) and a fixed line l (the directrix) are equal.

(b) Write an equation of a parabola with focus $(0, p)$ and directrix $y = -p$. What if the focus is $(p, 0)$ and the directrix is $x = -p$?

In the first case an equation is $x^2 = 4py$ and in the second case, $y^2 = 4px$.

7. (a) Give a definition of an ellipse in terms of foci.

An ellipse is a set of points in a plane the sum of whose distances from two fixed points (the foci) is a constant.

(b) Write an equation for the ellipse with foci $(\pm c, 0)$ and vertices $(\pm a, 0)$.

$$\frac{x^2}{a^2} + \frac{y^2}{b^2} = 1$$

where $a \geqslant b > 0$ and $c^2 = a^2 - b^2$.

(continued)

8. (a) Give a definition of a hyperbola in terms of foci.

A hyperbola is a set of points in a plane the difference of whose distances from two fixed points (the foci) is a constant. This difference should be interpreted as the larger distance minus the smaller distance.

(b) Write an equation for the hyperbola with foci $(\pm c, 0)$ and vertices $(\pm a, 0)$.

$$\frac{x^2}{a^2} - \frac{y^2}{b^2} = 1$$

where $c^2 = a^2 + b^2$.

(c) Write equations for the asymptotes of the hyperbola in part (b).

$$y = \pm \frac{b}{a} x$$

9. (a) What is the eccentricity of a conic section?

If a conic section has focus F and corresponding directrix l, then the eccentricity e is the fixed ratio $|PF|/|Pl|$ for points P of the conic section.

(b) What can you say about the eccentricity if the conic section is an ellipse? A hyperbola? A parabola?

$e < 1$ for an ellipse; $e > 1$ for a hyperbola; $e = 1$ for a parabola

(c) Write a polar equation for a conic section with eccentricity e and directrix $x = d$. What if the directrix is $x = -d$? $y = d$? $y = -d$?

$$\text{directrix } x = d: \quad r = \frac{ed}{1 + e \cos \theta}$$

$$x = -d: \quad r = \frac{ed}{1 - e \cos \theta}$$

$$y = d: \quad r = \frac{ed}{1 + e \sin \theta}$$

$$y = -d: \quad r = \frac{ed}{1 - e \sin \theta}$$

1. (a) What is a convergent sequence?

A convergent sequence $\{a_n\}$ is an ordered list of numbers where $\lim_{n \to \infty} a_n$ exists.

(b) What is a convergent series?

A series $\Sigma\, a_n$ is the *sum* of a sequence of numbers. It is convergent if the partial sums $s_n = \sum_{i=1}^{n} a_n$ approach a finite value, that is, $\lim_{n \to \infty} s_n$ exists as a real number.

(c) What does $\lim_{n \to \infty} a_n = 3$ mean?

The terms of the sequence $\{a_n\}$ approach 3 as n becomes large.

(d) What does $\Sigma_{n=1}^{\infty} a_n = 3$ mean?

By adding sufficiently many terms of the series, we can make the partial sums as close to 3 as we like.

2. (a) What is a bounded sequence?

A sequence $\{a_n\}$ is bounded if there are numbers m and M such that $m \leq a_n \leq M$ for all $n \geq 1$.

(b) What is a monotonic sequence?

A sequence is monotonic if it is either increasing or decreasing for all $n \geq 1$.

(c) What can you say about a bounded monotonic sequence?

Every bounded, monotonic sequence is convergent.

3. (a) What is a geometric series? Under what circumstances is it convergent? What is its sum?

A geometric series is of the form

$$\sum_{n=1}^{\infty} ar^{n-1} = a + ar + ar^2 + \cdots$$

It is convergent if $|r| < 1$ and its sum is $\dfrac{a}{1-r}$.

(b) What is a p-series? Under what circumstances is it convergent?

A p-series is of the form $\sum_{n=1}^{\infty} \dfrac{1}{n^p}$. It is convergent if $p > 1$.

4. Suppose $\Sigma\, a_n = 3$ and s_n is the nth partial sum of the series. What is $\lim_{n \to \infty} a_n$? What is $\lim_{n \to \infty} s_n$?

If $\Sigma\, a_n = 3$, then $\lim_{n \to \infty} a_n = 0$ and $\lim_{n \to \infty} s_n = 3$.

5. State the following.

(a) The Test for Divergence

If $\lim_{n \to \infty} a_n$ does not exist or if $\lim_{n \to \infty} a_n \neq 0$, then the series $\sum_{n=1}^{\infty} a_n$ is divergent.

(b) The Integral Test

Suppose f is a continuous, positive, decreasing function on $[1, \infty)$ and let $a_n = f(n)$.

- If $\int_1^{\infty} f(x)\,dx$ is convergent, then $\sum_{n=1}^{\infty} a_n$ is convergent.

- If $\int_1^{\infty} f(x)\,dx$ is divergent, then $\sum_{n=1}^{\infty} a_n$ is divergent.

(c) The Comparison Test

Suppose that $\Sigma\, a_n$ and $\Sigma\, b_n$ are series with positive terms.

- If $\Sigma\, b_n$ is convergent and $a_n \leq b_n$ for all n, then $\Sigma\, a_n$ is also convergent.

- If $\Sigma\, b_n$ is divergent and $a_n \geq b_n$ for all n, then $\Sigma\, a_n$ is also divergent.

(d) The Limit Comparison Test

Suppose that $\Sigma\, a_n$ and $\Sigma\, b_n$ are series with positive terms. If $\lim_{n \to \infty} a_n/b_n = c$, where c is a finite number and $c > 0$, then either both series converge or both diverge.

(e) The Alternating Series Test

If the alternating series

$$\sum_{n=1}^{\infty} (-1)^{n-1} b_n = b_1 - b_2 + b_3 - b_4 + b_5 - b_6 + \cdots$$

where $b_n > 0$ satisfies (i) $b_{n+1} \leq b_n$ for all n and (ii) $\lim_{n \to \infty} b_n = 0$, then the series is convergent.

(f) The Ratio Test

- If $\lim_{n \to \infty} \left| \dfrac{a_{n+1}}{a_n} \right| = L < 1$, then the series $\sum_{n=1}^{\infty} a_n$ is absolutely convergent (and therefore convergent).

- If $\lim_{n \to \infty} \left| \dfrac{a_{n+1}}{a_n} \right| = L > 1$ or $\lim_{n \to \infty} \left| \dfrac{a_{n+1}}{a_n} \right| = \infty$, then the series $\sum_{n=1}^{\infty} a_n$ is divergent.

- If $\lim_{n \to \infty} \left| \dfrac{a_{n+1}}{a_n} \right| = 1$, the Ratio Test is inconclusive.

(g) The Root Test

- If $\lim_{n \to \infty} \sqrt[n]{|a_n|} = L < 1$, then the series $\sum_{n=1}^{\infty} a_n$ is absolutely convergent (and therefore convergent).

- If $\lim_{n \to \infty} \sqrt[n]{|a_n|} = L > 1$ or $\lim_{n \to \infty} \sqrt[n]{|a_n|} = \infty$, then the series $\sum_{n=1}^{\infty} a_n$ is divergent.

- If $\lim_{n \to \infty} \sqrt[n]{|a_n|} = 1$, the Root Test is inconclusive.

6. (a) What is an absolutely convergent series?

A series $\Sigma\, a_n$ is called absolutely convergent if the series of absolute values $\Sigma\, |a_n|$ is convergent.

(b) What can you say about such a series?

If a series $\Sigma\, a_n$ is absolutely convergent, then it is convergent.

(c) What is a conditionally convergent series?

A series $\Sigma\, a_n$ is called conditionally convergent if it is convergent but not absolutely convergent.

(continued)

7. (a) If a series is convergent by the Integral Test, how do you estimate its sum?

The sum s can be estimated by the inequality

$$s_n + \int_{n+1}^{\infty} f(x)\, dx \leq s \leq s_n + \int_{n}^{\infty} f(x)\, dx$$

where s_n is the nth partial sum.

(b) If a series is convergent by the Comparison Test, how do you estimate its sum?

We first estimate the remainder for the comparison series. This gives an upper bound for the remainder of the original series (as in Example 11.4.5).

(c) If a series is convergent by the Alternating Series Test, how do you estimate its sum?

We can use a partial sum s_n of an alternating series as an approximation to the total sum. The size of the error is guaranteed to be no more than $|a_{n+1}|$, the absolute value of the first neglected term.

8. (a) Write the general form of a power series.

A power series centered at a is

$$\sum_{n=0}^{\infty} c_n(x - a)^n$$

(b) What is the radius of convergence of a power series?

Given the power series $\sum_{n=0}^{\infty} c_n(x - a)^n$, the radius of convergence is:

(i) 0 if the series converges only when $x = a$,

(ii) ∞ if the series converges for all x, or

(iii) a positive number R such that the series converges if $|x - a| < R$ and diverges if $|x - a| > R$.

(c) What is the interval of convergence of a power series?

The interval of convergence of a power series is the interval that consists of all values of x for which the series converges. Corresponding to the cases in part (b), the interval of convergence is (i) the single point $\{a\}$, (ii) $(-\infty, \infty)$, or (iii) an interval with endpoints $a - R$ and $a + R$ that can contain neither, either, or both of the endpoints.

9. Suppose $f(x)$ is the sum of a power series with radius of convergence R.

(a) How do you differentiate f? What is the radius of convergence of the series for f'?

If $f(x) = \sum_{n=0}^{\infty} c_n(x - a)^n$, then $f'(x) = \sum_{n=1}^{\infty} n c_n(x - a)^{n-1}$ with radius of convergence R.

(b) How do you integrate f? What is the radius of convergence of the series for $\int f(x)\, dx$?

$$\int f(x)\, dx = C + \sum_{n=0}^{\infty} c_n \frac{(x - a)^{n+1}}{n + 1} \text{ with radius of}$$

convergence R.

10. (a) Write an expression for the nth-degree Taylor polynomial of f centered at a.

$$T_n(x) = \sum_{i=0}^{n} \frac{f^{(i)}(a)}{i!}(x - a)^i$$

(b) Write an expression for the Taylor series of f centered at a.

$$\sum_{n=0}^{\infty} \frac{f^{(n)}(a)}{n!}(x - a)^n$$

(c) Write an expression for the Maclaurin series of f.

$$\sum_{n=0}^{\infty} \frac{f^{(n)}(0)}{n!} x^n \qquad [a = 0 \text{ in part (b)}]$$

(d) How do you show that $f(x)$ is equal to the sum of its Taylor series?

If $f(x) = T_n(x) + R_n(x)$, where $T_n(x)$ is the nth-degree Taylor polynomial of f and $R_n(x)$ is the remainder of the Taylor series, then we must show that

$$\lim_{n \to \infty} R_n(x) = 0$$

(e) State Taylor's Inequality.

If $|f^{(n+1)}(x)| \leq M$ for $|x - a| \leq d$, then the remainder $R_n(x)$ of the Taylor series satisfies the inequality

$$|R_n(x)| \leq \frac{M}{(n + 1)!} |x - a|^{n+1} \qquad \text{for } |x - a| \leq d$$

11. Write the Maclaurin series and the interval of convergence for each of the following functions.

(a) $\dfrac{1}{1 - x} = \sum_{n=0}^{\infty} x^n, \quad R = 1$

(b) $e^x = \sum_{n=0}^{\infty} \dfrac{x^n}{n!}, \quad R = \infty$

(c) $\sin x = \sum_{n=0}^{\infty} (-1)^n \dfrac{x^{2n+1}}{(2n + 1)!}, \quad R = \infty$

(d) $\cos x = \sum_{n=0}^{\infty} (-1)^n \dfrac{x^{2n}}{(2n)!}, \quad R = \infty$

(e) $\tan^{-1} x = \sum_{n=0}^{\infty} (-1)^n \dfrac{x^{2n+1}}{2n + 1}, \quad R = 1$

(f) $\ln(1 + x) = \sum_{n=1}^{\infty} (-1)^{n-1} \dfrac{x^n}{n}, \quad R = 1$

12. Write the binomial series expansion of $(1 + x)^k$. What is the radius of convergence of this series?

If k is any real number and $|x| < 1$, then

$$(1 + x)^k = \sum_{n=0}^{\infty} \binom{k}{n} x^n$$

$$= 1 + kx + \frac{k(k - 1)}{2!} x^2 + \frac{k(k - 1)(k - 2)}{3!} x^3 + \cdots$$

The radius of convergence for the binomial series is 1.

1. What is the difference between a vector and a scalar?

A scalar is a real number, whereas a vector is a quantity that has both a real-valued magnitude and a direction.

2. How do you add two vectors geometrically? How do you add them algebraically?

To add two vectors geometrically, we can use either the Triangle Law or the Parallelogram Law:

 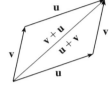

Triangle Law   Parallelogram Law

Algebraically, we add the corresponding components of the vectors.

3. If **a** is a vector and c is a scalar, how is c**a** related to **a** geometrically? How do you find c**a** algebraically?

For $c > 0$, c**a** is a vector with the same direction as **a** and length c times the length of **a**. If $c < 0$, c**a** points in the direction opposite to **a** and has length $|c|$ times the length of **a**. Algebraically, to find c**a** we multiply each component of **a** by c.

4. How do you find the vector from one point to another?

The vector from point $A(x_1, y_1, z_1)$ to point $B(x_2, y_2, z_2)$ is given by
$$\langle x_2 - x_1, y_2 - y_1, z_2 - z_1 \rangle$$

5. How do you find the dot product **a** · **b** of two vectors if you know their lengths and the angle between them? What if you know their components?

If θ is the angle between the vectors **a** and **b**, then
$$\mathbf{a} \cdot \mathbf{b} = |\mathbf{a}||\mathbf{b}| \cos \theta$$

If $\mathbf{a} = \langle a_1, a_2, a_3 \rangle$ and $\mathbf{b} = \langle b_1, b_2, b_3 \rangle$, then
$$\mathbf{a} \cdot \mathbf{b} = a_1 b_1 + a_2 b_2 + a_3 b_3$$

6. How are dot products useful?

The dot product can be used to find the angle between two vectors. In particular, it can be used to determine whether two vectors are orthogonal. We can also use the dot product to find the scalar projection of one vector onto another. Additionally, if a constant force moves an object, the work done is the dot product of the force and displacement vectors.

7. Write expressions for the scalar and vector projections of **b** onto **a**. Illustrate with diagrams.

Scalar projection of **b** onto **a**: $\text{comp}_{\mathbf{a}}\, \mathbf{b} = \dfrac{\mathbf{a} \cdot \mathbf{b}}{|\mathbf{a}|}$

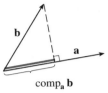

$\text{comp}_{\mathbf{a}}\, \mathbf{b}$

Vector projection of **b** onto **a**:

$$\text{proj}_{\mathbf{a}}\, \mathbf{b} = \left(\frac{\mathbf{a} \cdot \mathbf{b}}{|\mathbf{a}|} \right) \frac{\mathbf{a}}{|\mathbf{a}|} = \frac{\mathbf{a} \cdot \mathbf{b}}{|\mathbf{a}|^2}\, \mathbf{a}$$

$\text{proj}_{\mathbf{a}}\, \mathbf{b}$

8. How do you find the cross product **a** × **b** of two vectors if you know their lengths and the angle between them? What if you know their components?

If θ is the angle between **a** and **b** ($0 \le \theta \le \pi$), then **a** × **b** is the vector with length $|\mathbf{a} \times \mathbf{b}| = |\mathbf{a}||\mathbf{b}| \sin \theta$ and direction orthogonal to both **a** and **b**, as given by the right-hand rule. If
$$\mathbf{a} = \langle a_1, a_2, a_3 \rangle \qquad \text{and} \qquad \mathbf{b} = \langle b_1, b_2, b_3 \rangle$$
then
$$\mathbf{a} \times \mathbf{b} = \begin{vmatrix} \mathbf{i} & \mathbf{j} & \mathbf{k} \\ a_1 & a_2 & a_3 \\ b_1 & b_2 & b_3 \end{vmatrix}$$
$$= \langle a_2 b_3 - a_3 b_2,\ a_3 b_1 - a_1 b_3,\ a_1 b_2 - a_2 b_1 \rangle$$

9. How are cross products useful?

The cross product can be used to create a vector orthogonal to two given vectors and it can be used to compute the area of a parallelogram determined by two vectors. Two nonzero vectors are parallel if and only if their cross product is **0**. In addition, if a force acts on a rigid body, then the torque vector is the cross product of the position and force vectors.

(continued)

Cut here and keep for reference

10. (a) How do you find the area of the parallelogram determined by **a** and **b**?

The area of the parallelogram determined by **a** and **b** is the length of the cross product: $|\mathbf{a} \times \mathbf{b}|$.

(b) How do you find the volume of the parallelepiped determined by **a**, **b**, and **c**?

The volume of the parallelepiped determined by **a**, **b**, and **c** is the magnitude of their scalar triple product: $|\mathbf{a} \cdot (\mathbf{b} \times \mathbf{c})|$.

11. How do you find a vector perpendicular to a plane?

If an equation of the plane is known, it can be written in the form $ax + by + cz + d = 0$. A normal vector, which is perpendicular to the plane, is $\langle a, b, c \rangle$ (or any nonzero scalar multiple of $\langle a, b, c \rangle$). If an equation is not known, we can use points on the plane to find two nonparallel vectors that lie in the plane. The cross product of these vectors is a vector perpendicular to the plane.

12. How do you find the angle between two intersecting planes?

The angle between two intersecting planes is defined as the acute angle θ between their normal vectors. If $\mathbf{n}_1$ and $\mathbf{n}_2$ are the normal vectors, then

$$\cos \theta = \frac{\mathbf{n}_1 \cdot \mathbf{n}_2}{|\mathbf{n}_1||\mathbf{n}_2|}$$

13. Write a vector equation, parametric equations, and symmetric equations for a line.

A vector equation for a line that is parallel to a vector **v** and that passes through a point with position vector $\mathbf{r}_0$ is $\mathbf{r} = \mathbf{r}_0 + t\mathbf{v}$. Parametric equations for a line through the point (x_0, y_0, z_0) and parallel to the vector $\langle a, b, c \rangle$ are

$$x = x_0 + at \quad y = y_0 + bt \quad z = z_0 + ct$$

while symmetric equations are

$$\frac{x - x_0}{a} = \frac{y - y_0}{b} = \frac{z - z_0}{c}$$

14. Write a vector equation and a scalar equation for a plane.

A vector equation of a plane that passes through a point with position vector $\mathbf{r}_0$ and that has normal vector **n** (meaning **n** is orthogonal to the plane) is $\mathbf{n} \cdot (\mathbf{r} - \mathbf{r}_0) = 0$ or, equivalently, $\mathbf{n} \cdot \mathbf{r} = \mathbf{n} \cdot \mathbf{r}_0$.

A scalar equation of a plane through a point (x_0, y_0, z_0) with normal vector $\mathbf{n} = \langle a, b, c \rangle$ is

$$a(x - x_0) + b(y - y_0) + c(z - z_0) = 0$$

15. (a) How do you tell if two vectors are parallel?

Two (nonzero) vectors are parallel if and only if one is a scalar multiple of the other. In addition, two nonzero vectors are parallel if and only if their cross product is **0**.

(b) How do you tell if two vectors are perpendicular?

Two vectors are perpendicular if and only if their dot product is 0.

(c) How do you tell if two planes are parallel?

Two planes are parallel if and only if their normal vectors are parallel.

16. (a) Describe a method for determining whether three points P, Q, and R lie on the same line.

Determine the vectors $\overrightarrow{PQ} = \mathbf{a}$ and $\overrightarrow{PR} = \mathbf{b}$. If there is a scalar t such that $\mathbf{a} = t\mathbf{b}$, then the vectors are parallel and the points must all lie on the same line.

Alternatively, if $\overrightarrow{PQ} \times \overrightarrow{PR} = \mathbf{0}$, then $\overrightarrow{PQ}$ and $\overrightarrow{PR}$ are parallel, so P, Q, and R are collinear.

An algebraic method is to determine an equation of the line joining two of the points, and then check whether or not the third point satisfies this equation.

(b) Describe a method for determining whether four points P, Q, R, and S lie in the same plane.

Find the vectors $\overrightarrow{PQ} = \mathbf{a}$, $\overrightarrow{PR} = \mathbf{b}$, $\overrightarrow{PS} = \mathbf{c}$. Then $\mathbf{a} \times \mathbf{b}$ is normal to the plane formed by P, Q, and R, and so S lies on this plane if $\mathbf{a} \times \mathbf{b}$ and $\mathbf{c}$ are orthogonal, that is, if $(\mathbf{a} \times \mathbf{b}) \cdot \mathbf{c} = 0$.

Alternatively, we can check if the volume of the parallelepiped determined by **a**, **b**, and **c** is 0 (see Example 12.4.5).

An algebraic method is to find an equation of the plane determined by three of the points, and then check whether or not the fourth point satisfies this equation.

17. (a) How do you find the distance from a point to a line?

Let P be a point not on the line L that passes through the points Q and R and let $\mathbf{a} = \overrightarrow{QR}$, $\mathbf{b} = \overrightarrow{QP}$. The distance from the point P to the line L is

$$d = \frac{|\mathbf{a} \times \mathbf{b}|}{|\mathbf{a}|}$$

(b) How do you find the distance from a point to a plane?

Let $P_0(x_0, y_0, z_0)$ be any point in the plane $ax + by + cz + d = 0$ and let $P_1(x_1, y_1, z_1)$ be a point not in the plane. If $\mathbf{b} = \overrightarrow{P_0P_1} = \langle x_1 - x_0, y_1 - y_0, z_1 - z_0 \rangle$, then the distance D from P_1 to the plane is equal to the absolute value of the scalar projection of **b** onto the plane's normal vector $\mathbf{n} = \langle a, b, c \rangle$:

$$D = |\text{comp}_\mathbf{n} \mathbf{b}| = \frac{|\mathbf{n} \cdot \mathbf{b}|}{|\mathbf{n}|} = \frac{|ax_1 + by_1 + cz_1 + d|}{\sqrt{a^2 + b^2 + c^2}}$$

(c) How do you find the distance between two lines?

Two skew lines L_1 and L_2 can be viewed as lying on two parallel planes, each with normal vector $\mathbf{n} = \mathbf{v}_1 \times \mathbf{v}_2$, where $\mathbf{v}_1$ and $\mathbf{v}_2$ are the direction vectors of L_1 and L_2. After choosing one point on L_1 and determining the equation of the plane containing L_2, we can proceed as in part (b). (See Example 12.5.10.)

(continued)

18. What are the traces of a surface? How do you find them?

The traces of a surface are the curves of intersection of the surface with planes parallel to the coordinate planes. We can find the trace in the plane $x = k$ (parallel to the yz-plane) by setting $x = k$ and determining the curve represented by the resulting equation. Traces in the planes $y = k$ (parallel to the xz-plane) and $z = k$ (parallel to the xy-plane) are found similarly.

19. Write equations in standard form of the six types of quadric surfaces.

Equations for the quadric surfaces symmetric with respect to the z-axis are as follows.

Ellipsoid: $\dfrac{x^2}{a^2} + \dfrac{y^2}{b^2} + \dfrac{z^2}{c^2} = 1$

Cone: $\dfrac{z^2}{c^2} = \dfrac{x^2}{a^2} + \dfrac{y^2}{b^2}$

Elliptic paraboloid: $\dfrac{z}{c} = \dfrac{x^2}{a^2} + \dfrac{y^2}{b^2}$

Hyperboloid of one sheet: $\dfrac{x^2}{a^2} + \dfrac{y^2}{b^2} - \dfrac{z^2}{c^2} = 1$

Hyperboloid of two sheets: $-\dfrac{x^2}{a^2} - \dfrac{y^2}{b^2} + \dfrac{z^2}{c^2} = 1$

Hyperbolic paraboloid: $\dfrac{z}{c} = \dfrac{x^2}{a^2} - \dfrac{y^2}{b^2}$

1. What is a vector function? How do you find its derivative and its integral?

A vector function is a function whose domain is a set of real numbers and whose range is a set of vectors. To find the derivative or integral, we can differentiate or integrate each component function of the vector function.

2. What is the connection between vector functions and space curves?

A continuous vector function $\mathbf{r}$ defines a space curve that is traced out by the tip of the moving position vector $\mathbf{r}(t)$.

3. How do you find the tangent vector to a smooth curve at a point? How do you find the tangent line? The unit tangent vector?

The tangent vector to a smooth curve at a point P with position vector $\mathbf{r}(t)$ is the vector $\mathbf{r}'(t)$. The tangent line at P is the line through P parallel to the tangent vector $\mathbf{r}'(t)$. The unit tangent vector is $\mathbf{T}(t) = \dfrac{\mathbf{r}'(t)}{|\mathbf{r}'(t)|}$.

4. If $\mathbf{u}$ and $\mathbf{v}$ are differentiable vector functions, c is a scalar, and f is a real-valued function, write the rules for differentiating the following vector functions.

(a) $\mathbf{u}(t) + \mathbf{v}(t)$

$$\frac{d}{dt}[\mathbf{u}(t) + \mathbf{v}(t)] = \mathbf{u}'(t) + \mathbf{v}'(t)$$

(b) $c\mathbf{u}(t)$

$$\frac{d}{dt}[c\mathbf{u}(t)] = c\mathbf{u}'(t)$$

(c) $f(t)\mathbf{u}(t)$

$$\frac{d}{dt}[f(t)\mathbf{u}(t)] = f'(t)\mathbf{u}(t) + f(t)\mathbf{u}'(t)$$

(d) $\mathbf{u}(t) \cdot \mathbf{v}(t)$

$$\frac{d}{dt}[\mathbf{u}(t) \cdot \mathbf{v}(t)] = \mathbf{u}'(t) \cdot \mathbf{v}(t) + \mathbf{u}(t) \cdot \mathbf{v}'(t)$$

(e) $\mathbf{u}(t) \times \mathbf{v}(t)$

$$\frac{d}{dt}[\mathbf{u}(t) \times \mathbf{v}(t)] = \mathbf{u}'(t) \times \mathbf{v}(t) + \mathbf{u}(t) \times \mathbf{v}'(t)$$

(f) $\mathbf{u}(f(t))$

$$\frac{d}{dt}[\mathbf{u}(f(t))] = f'(t)\mathbf{u}'(f(t))$$

5. How do you find the length of a space curve given by a vector function $\mathbf{r}(t)$?

If $\mathbf{r}(t) = \langle f(t), g(t), h(t) \rangle$, $a \le t \le b$, and the curve is traversed exactly once as t increases from a to b, then the length is

$$L = \int_a^b |\mathbf{r}'(t)|\,dt = \int_a^b \sqrt{[f'(t)]^2 + [g'(t)]^2 + [h'(t)]^2}\,dt$$

6. (a) What is the definition of curvature?

The curvature of a curve is $\kappa = \left|\dfrac{d\mathbf{T}}{ds}\right|$ where $\mathbf{T}$ is the unit tangent vector.

(b) Write a formula for curvature in terms of $\mathbf{r}'(t)$ and $\mathbf{T}'(t)$.

$$\kappa(t) = \frac{|\mathbf{T}'(t)|}{|\mathbf{r}'(t)|}$$

(c) Write a formula for curvature in terms of $\mathbf{r}'(t)$ and $\mathbf{r}''(t)$.

$$\kappa(t) = \frac{|\mathbf{r}'(t) \times \mathbf{r}''(t)|}{|\mathbf{r}'(t)|^3}$$

(d) Write a formula for the curvature of a plane curve with equation $y = f(x)$.

$$\kappa(x) = \frac{|f''(x)|}{[1 + (f'(x))^2]^{3/2}}$$

7. (a) Write formulas for the unit normal and binormal vectors of a smooth space curve $\mathbf{r}(t)$.

Unit normal vector: $\mathbf{N}(t) = \dfrac{\mathbf{T}'(t)}{|\mathbf{T}'(t)|}$

Binormal vector: $\mathbf{B}(t) = \mathbf{T}(t) \times \mathbf{N}(t)$

(b) What is the normal plane of a curve at a point? What is the osculating plane? What is the osculating circle?

The normal plane of a curve at a point P is the plane determined by the normal and binormal vectors $\mathbf{N}$ and $\mathbf{B}$ at P. The tangent vector $\mathbf{T}$ is orthogonal to the normal plane.

The osculating plane at P is the plane determined by the vectors $\mathbf{T}$ and $\mathbf{N}$. It is the plane that comes closest to containing the part of the curve near P.

The osculating circle at P is the circle that lies in the osculating plane of C at P, has the same tangent as C at P, lies on the concave side of C (toward which $\mathbf{N}$ points), and has radius $\rho = 1/\kappa$ (the reciprocal of the curvature). It is the circle that best describes how C behaves near P; it shares the same tangent, normal, and curvature at P.

(continued)

8. (a) How do you find the velocity, speed, and acceleration of a particle that moves along a space curve?

If $\mathbf{r}(t)$ is the position vector of the particle on the space curve, the velocity vector is $\mathbf{v}(t) = \mathbf{r}'(t)$, the speed is given by $|\mathbf{v}(t)|$, and the acceleration is $\mathbf{a}(t) = \mathbf{v}'(t) = \mathbf{r}''(t)$.

(b) Write the acceleration in terms of its tangential and normal components.

$\mathbf{a} = a_T\mathbf{T} + a_N\mathbf{N}$, where $a_T = v'$ and $a_N = \kappa v^2$ ($v = |\mathbf{v}|$ is speed and κ is the curvature).

9. State Kepler's Laws.

- A planet revolves around the sun in an elliptical orbit with the sun at one focus.

- The line joining the sun to a planet sweeps out equal areas in equal times.

- The square of the period of revolution of a planet is proportional to the cube of the length of the major axis of its orbit.

1. (a) What is a function of two variables?

A function f of two variables is a rule that assigns to each ordered pair (x, y) of real numbers in its domain a unique real number denoted by $f(x, y)$.

(b) Describe three methods for visualizing a function of two variables.

One way to visualize a function of two variables is by graphing it, resulting in the surface $z = f(x, y)$. Another method is a contour map, consisting of level curves $f(x, y) = k$ (k a constant), which are horizontal traces of the graph of the function projected onto the xy-plane. Also, we can use an arrow diagram such as the one below.

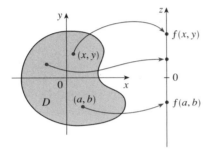

2. What is a function of three variables? How can you visualize such a function?

A function f of three variables is a rule that assigns to each ordered triple (x, y, z) in its domain a unique real number $f(x, y, z)$. We can visualize a function of three variables by examining its level surfaces $f(x, y, z) = k$, where k is a constant.

3. What does

$$\lim_{(x, y) \to (a, b)} f(x, y) = L$$

mean? How can you show that such a limit does not exist?

$\lim_{(x, y) \to (a, b)} f(x, y) = L$ means that the values of $f(x, y)$ approach the number L as the point (x, y) approaches the point (a, b) along any path that is within the domain of f. We can show that a limit at a point does not exist by finding two different paths approaching the point along which $f(x, y)$ has different limits.

4. (a) What does it mean to say that f is continuous at (a, b)?

A function f of two variables is continuous at (a, b) if

$$\lim_{(x, y) \to (a, b)} f(x, y) = f(a, b)$$

(b) If f is continuous on $\mathbb{R}^2$, what can you say about its graph?

If f is continuous on $\mathbb{R}^2$, its graph will appear as a surface without holes or breaks.

5. (a) Write expressions for the partial derivatives $f_x(a, b)$ and $f_y(a, b)$ as limits.

$$f_x(a, b) = \lim_{h \to 0} \frac{f(a + h, b) - f(a, b)}{h}$$

$$f_y(a, b) = \lim_{h \to 0} \frac{f(a, b + h) - f(a, b)}{h}$$

(b) How do you interpret $f_x(a, b)$ and $f_y(a, b)$ geometrically? How do you interpret them as rates of change?

If $f(a, b) = c$, then the point $P(a, b, c)$ lies on the surface S given by $z = f(x, y)$. We can interpret $f_x(a, b)$ as the slope of the tangent line at P to the curve of intersection of the vertical plane $y = b$ and S. In other words, if we restrict ourselves to the path along S through P that is parallel to the xz-plane, then $f_x(a, b)$ is the slope at P looking in the positive x-direction. Similarly, $f_y(a, b)$ is the slope of the tangent line at P to the curve of intersection of the vertical plane $x = a$ and S.

If $z = f(x, y)$, then $f_x(x, y)$ can be interpreted as the rate of change of z with respect to x when y is fixed. Thus $f_x(a, b)$ is the rate of change of z (with respect to x) when y is fixed at b and x is allowed to vary from a. Similarly, $f_y(a, b)$ is the rate of change of z (with respect to y) when x is fixed at a and y is allowed to vary from b.

(c) If $f(x, y)$ is given by a formula, how do you calculate f_x and f_y?

To find f_x, regard y as a constant and differentiate $f(x, y)$ with respect to x. To find f_y, regard x as a constant and differentiate $f(x, y)$ with respect to y.

6. What does Clairaut's Theorem say?

If f is a function of two variables that is defined on a disk D containing the point (a, b) and the functions f_{xy} and f_{yx} are both continuous on D, then Clairaut's Theorem states that $f_{xy}(a, b) = f_{yx}(a, b)$.

7. How do you find a tangent plane to each of the following types of surfaces?

(a) A graph of a function of two variables, $z = f(x, y)$

If f has continuous partial derivatives, an equation of the tangent plane to the surface $z = f(x, y)$ at the point (x_0, y_0, z_0) is

$$z - z_0 = f_x(x_0, y_0)(x - x_0) + f_y(x_0, y_0)(y - y_0)$$

(b) A level surface of a function of three variables, $F(x, y, z) = k$

The tangent plane to the level surface $F(x, y, z) = k$ at $P(x_0, y_0, z_0)$ is the plane that passes through P and has normal vector $\nabla F(x_0, y_0, z_0)$:

$$F_x(x_0, y_0, z_0)(x - x_0) + F_y(x_0, y_0, z_0)(y - y_0)$$
$$+ F_z(x_0, y_0, z_0)(z - z_0) = 0$$

(continued)

8. Define the linearization of f at (a, b). What is the corresponding linear approximation? What is the geometric interpretation of the linear approximation?

The linearization of f at (a, b) is the linear function whose graph is the tangent plane to the surface $z = f(x, y)$ at the point $(a, b, f(a, b))$:

$$L(x, y) = f(a, b) + f_x(a, b)(x - a) + f_y(a, b)(y - b)$$

The linear approximation of f at (a, b) is

$$f(x, y) \approx f(a, b) + f_x(a, b)(x - a) + f_y(a, b)(y - b)$$

Geometrically, the linear approximation says that function values $f(x, y)$ can be approximated by values $L(x, y)$ from the tangent plane to the graph of f at $(a, b, f(a, b))$ when (x, y) is near (a, b).

9. (a) What does it mean to say that f is differentiable at (a, b)?

If $z = f(x, y)$, then f is differentiable at (a, b) if Δz can be expressed in the form

$$\Delta z = f_x(a, b)\,\Delta x + f_y(a, b)\,\Delta y + \varepsilon_1\,\Delta x + \varepsilon_2\,\Delta y$$

where ε_1 and $\varepsilon_2 \to 0$ as $(\Delta x, \Delta y) \to (0, 0)$. In other words, a differentiable function is one for which the linear approximation as stated above is a good approximation when (x, y) is near (a, b).

(b) How do you usually verify that f is differentiable?

If the partial derivatives f_x and f_y exist near (a, b) and are continuous at (a, b), then f is differentiable at (a, b).

10. If $z = f(x, y)$, what are the differentials dx, dy, and dz?

The differentials dx and dy are independent variables that can be given any values. If f is differentiable, the differential dz is then defined by

$$dz = f_x(x, y)\,dx + f_y(x, y)\,dy$$

11. State the Chain Rule for the case where $z = f(x, y)$ and x and y are functions of one variable. What if x and y are functions of two variables?

Suppose that $z = f(x, y)$ is a differentiable function of x and y, where $x = g(t)$ and $y = h(t)$ are both differentiable functions of t. Then z is a differentiable function of t and

$$\frac{dz}{dt} = \frac{\partial f}{\partial x}\frac{dx}{dt} + \frac{\partial f}{\partial y}\frac{dy}{dt}$$

If $z = f(x, y)$ is a differentiable function of x and y, where $x = g(s, t)$ and $y = h(s, t)$ are differentiable functions of s and t, then

$$\frac{\partial z}{\partial s} = \frac{\partial z}{\partial x}\frac{\partial x}{\partial s} + \frac{\partial z}{\partial y}\frac{\partial y}{\partial s} \qquad \frac{\partial z}{\partial t} = \frac{\partial z}{\partial x}\frac{\partial x}{\partial t} + \frac{\partial z}{\partial y}\frac{\partial y}{\partial t}$$

12. If z is defined implicitly as a function of x and y by an equation of the form $F(x, y, z) = 0$, how do you find $\partial z/\partial x$ and $\partial z/\partial y$?

If F is differentiable and $\partial F/\partial z \neq 0$, then

$$\frac{\partial z}{\partial x} = -\frac{\dfrac{\partial F}{\partial x}}{\dfrac{\partial F}{\partial z}} \qquad \frac{\partial z}{\partial y} = -\frac{\dfrac{\partial F}{\partial y}}{\dfrac{\partial F}{\partial z}}$$

13. (a) Write an expression as a limit for the directional derivative of f at (x_0, y_0) in the direction of a unit vector $\mathbf{u} = \langle a, b \rangle$. How do you interpret it as a rate? How do you interpret it geometrically?

The directional derivative of f at (x_0, y_0) in the direction of $\mathbf{u}$ is

$$D_\mathbf{u} f(x_0, y_0) = \lim_{h \to 0} \frac{f(x_0 + ha, y_0 + hb) - f(x_0, y_0)}{h}$$

if this limit exists.

We can interpret it as the rate of change of f (with respect to distance) at (x_0, y_0) in the direction of $\mathbf{u}$.

Geometrically, if P is the point $(x_0, y_0, f(x_0, y_0))$ on the graph of f and C is the curve of intersection of the graph of f with the vertical plane that passes through P in the direction of $\mathbf{u}$, then $D_\mathbf{u} f(x_0, y_0)$ is the slope of the tangent line to C at P.

(b) If f is differentiable, write an expression for $D_\mathbf{u} f(x_0, y_0)$ in terms of f_x and f_y.

$$D_\mathbf{u} f(x_0, y_0) = f_x(x_0, y_0)\,a + f_y(x_0, y_0)\,b$$

14. (a) Define the gradient vector ∇f for a function f of two or three variables.

If f is a function of two variables, then

$$\nabla f(x, y) = \langle f_x(x, y), f_y(x, y) \rangle = \frac{\partial f}{\partial x}\mathbf{i} + \frac{\partial f}{\partial y}\mathbf{j}$$

For a function f of three variables,

$$\nabla f(x, y, z) = \langle f_x(x, y, z), f_y(x, y, z), f_z(x, y, z) \rangle$$

$$= \frac{\partial f}{\partial x}\mathbf{i} + \frac{\partial f}{\partial y}\mathbf{j} + \frac{\partial f}{\partial z}\mathbf{k}$$

(b) Express $D_\mathbf{u} f$ in terms of ∇f.

$$D_\mathbf{u} f(x, y) = \nabla f(x, y) \cdot \mathbf{u}$$

or $\qquad D_\mathbf{u} f(x, y, z) = \nabla f(x, y, z) \cdot \mathbf{u}$

(continued)

(c) Explain the geometric significance of the gradient.

The gradient vector of f gives the direction of maximum rate of increase of f. On the graph of $z = f(x, y)$, ∇f points in the direction of steepest ascent. Also, the gradient vector is perpendicular to the level curves or level surfaces of a function.

15. What do the following statements mean?

(a) f has a local maximum at (a, b).

f has a local maximum at (a, b) if $f(x, y) \leqslant f(a, b)$ when (x, y) is near (a, b).

(b) f has an absolute maximum at (a, b).

f has an absolute maximum at (a, b) if $f(x, y) \leqslant f(a, b)$ for all points (x, y) in the domain of f.

(c) f has a local minimum at (a, b).

f has a local minimum at (a, b) if $f(x, y) \geqslant f(a, b)$ when (x, y) is near (a, b).

(d) f has an absolute minimum at (a, b).

f has an absolute minimum at (a, b) if $f(x, y) \geqslant f(a, b)$ for all points (x, y) in the domain of f.

(e) f has a saddle point at (a, b).

f has a saddle point at (a, b) if $f(a, b)$ is a local maximum in one direction but a local minimum in another.

16. (a) If f has a local maximum at (a, b), what can you say about its partial derivatives at (a, b)?

If f has a local maximum at (a, b) and the first-order partial derivatives of f exist there, then $f_x(a, b) = 0$ and $f_y(a, b) = 0$.

(b) What is a critical point of f?

A critical point of f is a point (a, b) such that $f_x(a, b) = 0$ and $f_y(a, b) = 0$ or one of these partial derivatives does not exist.

17. State the Second Derivatives Test.

Suppose the second partial derivatives of f are continuous on a disk with center (a, b), and suppose that $f_x(a, b) = 0$ and $f_y(a, b) = 0$ [that is, (a, b) is a critical point of f]. Let

$$D = D(a, b) = f_{xx}(a, b) f_{yy}(a, b) - [f_{xy}(a, b)]^2$$

- If $D > 0$ and $f_{xx}(a, b) > 0$, then $f(a, b)$ is a local minimum.
- If $D > 0$ and $f_{xx}(a, b) < 0$, then $f(a, b)$ is a local maximum.
- If $D < 0$, then $f(a, b)$ is not a local maximum or minimum. The point (a, b) is a saddle point of f.

18. (a) What is a closed set in $\mathbb{R}^2$? What is a bounded set?

A closed set in $\mathbb{R}^2$ is one that contains all its boundary points. If one or more points on the boundary curve are omitted, the set is not closed.

A bounded set is one that is contained within some disk. In other words, it is finite in extent.

(b) State the Extreme Value Theorem for functions of two variables.

If f is continuous on a closed, bounded set D in $\mathbb{R}^2$, then f attains an absolute maximum value $f(x_1, y_1)$ and an absolute minimum value $f(x_2, y_2)$ at some points (x_1, y_1) and (x_2, y_2) in D.

(c) How do you find the values that the Extreme Value Theorem guarantees?

- Find the values of f at the critical points of f in D.
- Find the extreme values of f on the boundary of D.
- The largest of the values from the above steps is the absolute maximum value; the smallest of these values is the absolute minimum value.

19. Explain how the method of Lagrange multipliers works in finding the extreme values of $f(x, y, z)$ subject to the constraint $g(x, y, z) = k$. What if there is a second constraint $h(x, y, z) = c$?

To find the maximum and minimum values of $f(x, y, z)$ subject to the constraint $g(x, y, z) = k$ [assuming that these extreme values exist and $\nabla g \neq \mathbf{0}$ on the surface $g(x, y, z) = k$], we first find all values of x, y, z, and λ where $\nabla f(x, y, z) = \lambda \nabla g(x, y, z)$ and $g(x, y, z) = k$. (Thus we are finding the points from the constraint where the gradient vectors ∇f and ∇g are parallel.) Evaluate f at all the resulting points (x, y, z); the largest of these values is the maximum value of f, and the smallest is the minimum value of f.

If there is a second constraint $h(x, y, z) = c$, then we find all values of x, y, z, λ, and μ such that

$$\nabla f(x, y, z) = \lambda \nabla g(x, y, z) + \mu \nabla h(x, y, z)$$

Again we find the extreme values of f by evaluating f at the resulting points (x, y, z).

1. Suppose f is a continuous function defined on a rectangle $R = [a, b] \times [c, d]$.

   (a) Write an expression for a double Riemann sum of f. If $f(x, y) \geqslant 0$, what does the sum represent?

   A double Riemann sum of f is

   $$\sum_{i=1}^{m} \sum_{j=1}^{n} f(x_{ij}^*, y_{ij}^*) \, \Delta A$$

   where ΔA is the area of each subrectangle and (x_{ij}^*, y_{ij}^*) is a sample point in each subrectangle. If $f(x, y) \geqslant 0$, this sum represents an approximation to the volume of the solid that lies above the rectangle R and below the graph of f.

   (b) Write the definition of $\iint_R f(x, y) \, dA$ as a limit.

   $$\iint_R f(x, y) \, dA = \lim_{m,\, n \to \infty} \sum_{i=1}^{m} \sum_{j=1}^{n} f(x_{ij}^*, y_{ij}^*) \, \Delta A$$

   (c) What is the geometric interpretation of $\iint_R f(x, y) \, dA$ if $f(x, y) \geqslant 0$? What if f takes on both positive and negative values?

   If $f(x, y) \geqslant 0$, $\iint_R f(x, y) \, dA$ represents the volume of the solid that lies above the rectangle R and below the surface $z = f(x, y)$. If f takes on both positive and negative values, then $\iint_R f(x, y) \, dA$ is $V_1 - V_2$, where V_1 is the volume above R and below the surface $z = f(x, y)$, and V_2 is the volume below R and above the surface.

   (d) How do you evaluate $\iint_R f(x, y) \, dA$?

   We usually evaluate $\iint_R f(x, y) \, dA$ as an iterated integral according to Fubini's Theorem:

   $$\iint_R f(x, y) \, dA = \int_a^b \int_c^d f(x, y) \, dy \, dx = \int_c^d \int_a^b f(x, y) \, dx \, dy$$

   (e) What does the Midpoint Rule for double integrals say?

   The Midpoint Rule for double integrals says that we approximate the double integral $\iint_R f(x, y) \, dA$ by the double Riemann sum $\sum_{i=1}^{m} \sum_{j=1}^{n} f(\bar{x}_i, \bar{y}_j) \, \Delta A$, where the sample points $(\bar{x}_i, \bar{y}_j)$ are the centers of the subrectangles.

   (f) Write an expression for the average value of f.

   $$f_{\text{ave}} = \frac{1}{A(R)} \iint_R f(x, y) \, dA$$

   where $A(R)$ is the area of R.

2. (a) How do you define $\iint_D f(x, y) \, dA$ if D is a bounded region that is not a rectangle?

   Since D is bounded, it can be enclosed in a rectangular region R. We define a new function F with domain R by

   $$F(x, y) = \begin{cases} f(x, y) & \text{if } (x, y) \text{ is in } D \\ 0 & \text{if } (x, y) \text{ is in } R \text{ but not in } D \end{cases}$$

Then we define

$$\iint_D f(x, y) \, dA = \iint_R F(x, y) \, dA$$

(b) What is a type I region? How do you evaluate $\iint_D f(x, y) \, dA$ if D is a type I region?

A region D is of type I if it lies between the graphs of two continuous functions of x, that is,

$$D = \{(x, y) \mid a \leqslant x \leqslant b, \ g_1(x) \leqslant y \leqslant g_2(x)\}$$

where g_1 and g_2 are continuous on $[a, b]$. Then

$$\iint_D f(x, y) \, dA = \int_a^b \int_{g_1(x)}^{g_2(x)} f(x, y) \, dy \, dx$$

(c) What is a type II region? How do you evaluate $\iint_D f(x, y) \, dA$ if D is a type II region?

A region D is of type II if it lies between the graphs of two continuous functions of y, that is,

$$D = \{(x, y) \mid c \leqslant y \leqslant d, \ h_1(y) \leqslant x \leqslant h_2(y)\}$$

where h_1 and h_2 are continuous on $[c, d]$. Then

$$\iint_D f(x, y) \, dA = \int_c^d \int_{h_1(y)}^{h_2(y)} f(x, y) \, dx \, dy$$

(d) What properties do double integrals have?

- $\displaystyle \iint_D [f(x, y) + g(x, y)] \, dA$

  $$= \iint_D f(x, y) \, dA + \iint_D g(x, y) \, dA$$

- $\displaystyle \iint_D c f(x, y) \, dA = c \iint_D f(x, y) \, dA$

  where c is a constant

- If $f(x, y) \geqslant g(x, y)$ for all (x, y) in D, then

  $$\iint_D f(x, y) \, dA \geqslant \iint_D g(x, y) \, dA$$

- If $D = D_1 \cup D_2$, where D_1 and D_2 don't overlap except perhaps on their boundaries, then

  $$\iint_D f(x, y) \, dA = \iint_{D_1} f(x, y) \, dA + \iint_{D_2} f(x, y) \, dA$$

- $\displaystyle \iint_D 1 \, dA = A(D)$, the area of D.

- If $m \leqslant f(x, y) \leqslant M$ for all (x, y) in D, then

  $$mA(D) \leqslant \iint_D f(x, y) \, dA \leqslant MA(D)$$

(continued)

3. How do you change from rectangular coordinates to polar coordinates in a double integral? Why would you want to make the change?

We may want to change from rectangular to polar coordinates in a double integral if the region D of integration is more easily described in polar coordinates:

$$D = \{(r, \theta) \mid \alpha \leqslant \theta \leqslant \beta, \ h_1(\theta) \leqslant r \leqslant h_2(\theta)\}$$

To evaluate $\iint_R f(x, y) \, dA$, we replace x by $r \cos \theta$, y by $r \sin \theta$, and dA by $r \, dr \, d\theta$ (and use appropriate limits of integration):

$$\iint_D f(x, y) \, dA = \int_\alpha^\beta \int_{h_1(\theta)}^{h_2(\theta)} f(r \cos \theta, r \sin \theta) \, r \, dr \, d\theta$$

4. If a lamina occupies a plane region D and has density function $\rho(x, y)$, write expressions for each of the following in terms of double integrals.

(a) The mass: $m = \displaystyle\iint_D \rho(x, y) \, dA$

(b) The moments about the axes:

$$M_x = \iint_D y \rho(x, y) \, dA \qquad M_y = \iint_D x \rho(x, y) \, dA$$

(c) The center of mass:

$$(\bar{x}, \bar{y}), \qquad \text{where } \bar{x} = \frac{M_y}{m} \quad \text{and} \quad \bar{y} = \frac{M_x}{m}$$

(d) The moments of inertia about the axes and the origin:

$$I_x = \iint_D y^2 \rho(x, y) \, dA$$

$$I_y = \iint_D x^2 \rho(x, y) \, dA$$

$$I_0 = \iint_D (x^2 + y^2) \rho(x, y) \, dA$$

5. Let f be a joint density function of a pair of continuous random variables X and Y.

(a) Write a double integral for the probability that X lies between a and b and Y lies between c and d.

$$P(a \leqslant X \leqslant b, \ c \leqslant Y \leqslant d) = \int_a^b \int_c^d f(x, y) \, dy \, dx$$

(b) What properties does f possess?

$$f(x, y) \geqslant 0 \qquad \iint_{\mathbb{R}^2} f(x, y) \, dA = 1$$

(c) What are the expected values of X and Y?

The expected value of X is $\mu_1 = \displaystyle\iint_{\mathbb{R}^2} x f(x, y) \, dA$

The expected value of Y is $\mu_2 = \displaystyle\iint_{\mathbb{R}^2} y f(x, y) \, dA$

6. Write an expression for the area of a surface with equation $z = f(x, y)$, $(x, y) \in D$.

$$A(S) = \iint_D \sqrt{[f_x(x, y)]^2 + [f_y(x, y)]^2 + 1} \ dA$$

(assuming that f_x and f_y are continuous).

7. (a) Write the definition of the triple integral of f over a rectangular box B.

$$\iiint_B f(x, y, z) \, dV = \lim_{l, m, n \to \infty} \sum_{i=1}^l \sum_{j=1}^m \sum_{k=1}^n f(x_{ijk}^*, y_{ijk}^*, z_{ijk}^*) \, \Delta V$$

where ΔV is the volume of each sub-box and $(x_{ijk}^*, y_{ijk}^*, z_{ijk}^*)$ is a sample point in each sub-box.

(b) How do you evaluate $\iiint_B f(x, y, z) \, dV$?

We usually evaluate $\iiint_B f(x, y, z) \, dV$ as an iterated integral according to Fubini's Theorem for Triple Integrals: If f is continuous on $B = [a, b] \times [c, d] \times [r, s]$, then

$$\iiint_B f(x, y, z) \, dV = \int_r^s \int_c^d \int_a^b f(x, y, z) \, dx \, dy \, dz$$

Note that there are five other orders of integration that we can use.

(c) How do you define $\iiint_E f(x, y, z) \, dV$ if E is a bounded solid region that is not a box?

Since E is bounded, it can be enclosed in a box B as described in part (b). We define a new function F with domain B by

$$F(x, y, z) = \begin{cases} f(x, y, z) & \text{if } (x, y, z) \text{ is in } E \\ 0 & \text{if } (x, y, z) \text{ is in } B \text{ but not in } E \end{cases}$$

Then we define

$$\iiint_E f(x, y, z) \, dV = \iiint_B F(x, y, z) \, dV$$

(continued)

(d) What is a type 1 solid region? How do you evaluate $\iiint_E f(x, y, z)\, dV$ if E is such a region?

A region E is of type 1 if it lies between the graphs of two continuous functions of x and y, that is,

$$E = \{(x, y, z) \mid (x, y) \in D,\ u_1(x, y) \leq z \leq u_2(x, y)\}$$

where D is the projection of E onto the xy-plane. Then

$$\iiint_E f(x, y, z)\, dV = \iint_D \left[\int_{u_1(x, y)}^{u_2(x, y)} f(x, y, z)\, dz\right] dA$$

(e) What is a type 2 solid region? How do you evaluate $\iiint_E f(x, y, z)\, dV$ if E is such a region?

A type 2 region is of the form

$$E = \{(x, y, z) \mid (y, z) \in D,\ u_1(y, z) \leq x \leq u_2(y, z)\}$$

where D is the projection of E onto the yz-plane. Then

$$\iiint_E f(x, y, z)\, dV = \iint_D \left[\int_{u_1(y, z)}^{u_2(y, z)} f(x, y, z)\, dx\right] dA$$

(f) What is a type 3 solid region? How do you evaluate $\iiint_E f(x, y, z)\, dV$ if E is such a region?

A type 3 region is of the form

$$E = \{(x, y, z) \mid (x, z) \in D,\ u_1(x, z) \leq y \leq u_2(x, z)\}$$

where D is the projection of E onto the xz-plane. Then

$$\iiint_E f(x, y, z)\, dV = \iint_D \left[\int_{u_1(x, z)}^{u_2(x, z)} f(x, y, z)\, dy\right] dA$$

8. Suppose a solid object occupies the region E and has density function $\rho(x, y, z)$. Write expressions for each of the following.

(a) The mass:

$$m = \iiint_E \rho(x, y, z)\, dV$$

(b) The moments about the coordinate planes:

$$M_{yz} = \iiint_E x\rho(x, y, z)\, dV$$

$$M_{xz} = \iiint_E y\rho(x, y, z)\, dV$$

$$M_{xy} = \iiint_E z\rho(x, y, z)\, dV$$

(c) The coordinates of the center of mass:

$$(\bar{x}, \bar{y}, \bar{z}),\ \text{where}\ \bar{x} = \frac{M_{yz}}{m},\ \bar{y} = \frac{M_{xz}}{m},\ \bar{z} = \frac{M_{xy}}{m}$$

(d) The moments of inertia about the axes:

$$I_x = \iiint_E (y^2 + z^2)\rho(x, y, z)\, dV$$

$$I_y = \iiint_E (x^2 + z^2)\rho(x, y, z)\, dV$$

$$I_z = \iiint_E (x^2 + y^2)\rho(x, y, z)\, dV$$

(continued)

9. (a) How do you change from rectangular coordinates to cylindrical coordinates in a triple integral?

$$\iiint_E f(x, y, z)\, dV = \int_\alpha^\beta \int_{h_1(\theta)}^{h_2(\theta)} \int_{u_1(r\cos\theta,\, r\sin\theta)}^{u_2(r\cos\theta,\, r\sin\theta)} f(r\cos\theta, r\sin\theta, z)\, r\, dz\, dr\, d\theta$$

where

$$E = \left\{(r, \theta, z) \mid \alpha \le \theta \le \beta, h_1(\theta) \le r \le h_2(\theta), u_1(r\cos\theta, r\sin\theta) \le z \le u_2(r\cos\theta, r\sin\theta)\right\}$$

Thus we replace x by $r\cos\theta$, y by $r\sin\theta$, dV by $r\, dz\, dr\, d\theta$, and use appropriate limits of integration.

(b) How do you change from rectangular coordinates to spherical coordinates in a triple integral?

$$\iiint_E f(x, y, z)\, dV = \int_c^d \int_\alpha^\beta \int_{g_1(\theta,\, \phi)}^{g_2(\theta,\, \phi)} f(\rho\sin\phi\cos\theta, \rho\sin\phi\sin\theta, \rho\cos\phi)\, \rho^2 \sin\phi\, d\rho\, d\theta\, d\phi$$

where

$$E = \left\{(\rho, \theta, \phi) \mid \alpha \le \theta \le \beta,\ c \le \phi \le d,\ g_1(\theta, \phi) \le \rho \le g_2(\theta, \phi)\right\}$$

Thus we replace x by $\rho\sin\phi\cos\theta$, y by $\rho\sin\phi\sin\theta$, z by $\rho\cos\phi$, dV by $\rho^2 \sin\phi\, d\rho\, d\theta\, d\phi$, and use appropriate limits of integration.

(c) In what situations would you change to cylindrical or spherical coordinates?

We may want to change from rectangular to cylindrical or spherical coordinates in a triple integral if the region E of integration is more easily described in cylindrical or spherical coordinates. Regions that involve symmetry about the z-axis are often simpler to describe using cylindrical coordinates, and regions that are symmetrical about the origin are often simpler in spherical coordinates. Also, sometimes the integrand is easier to integrate using cylindrical or spherical coordinates.

10. (a) If a transformation T is given by $x = g(u, v)$, $y = h(u, v)$, what is the Jacobian of T?

$$\frac{\partial(x, y)}{\partial(u, v)} = \begin{vmatrix} \dfrac{\partial x}{\partial u} & \dfrac{\partial x}{\partial v} \\[2mm] \dfrac{\partial y}{\partial u} & \dfrac{\partial y}{\partial v} \end{vmatrix} = \frac{\partial x}{\partial u}\frac{\partial y}{\partial v} - \frac{\partial x}{\partial v}\frac{\partial y}{\partial u}$$

(b) How do you change variables in a double integral?

We change from an integral in x and y to an integral in u and v by expressing x and y in terms of u and v and writing

$$dA = \left| \frac{\partial(x, y)}{\partial(u, v)} \right| du\, dv$$

Thus, under the appropriate conditions,

$$\iint_R f(x, y)\, dA = \iint_S f(x(u, v), y(u, v)) \left| \frac{\partial(x, y)}{\partial(u, v)} \right| du\, dv$$

where R is the image of S under the transformation.

(c) How do you change variables in a triple integral?

Similarly to the case of two variables in part (b),

$$\iiint_R f(x, y, z)\, dV = \iiint_S f(x(u, v, w), y(u, v, w), z(u, v, w)) \left| \frac{\partial(x, y, z)}{\partial(u, v, w)} \right| du\, dv\, dw$$

where

$$\frac{\partial(x, y, z)}{\partial(u, v, w)} = \begin{vmatrix} \dfrac{\partial x}{\partial u} & \dfrac{\partial x}{\partial v} & \dfrac{\partial x}{\partial w} \\[2mm] \dfrac{\partial y}{\partial u} & \dfrac{\partial y}{\partial v} & \dfrac{\partial y}{\partial w} \\[2mm] \dfrac{\partial z}{\partial u} & \dfrac{\partial z}{\partial v} & \dfrac{\partial z}{\partial w} \end{vmatrix}$$

is the Jacobian.

CHAPTER 16 · CONCEPT CHECK ANSWERS

1. What is a vector field? Give three examples that have physical meaning.

A vector field is a function that assigns a vector to each point in its domain.

A vector field can represent, for example, the wind velocity at any location in space, the speed and direction of the ocean current at any location, or the force vector of the earth's gravitational field at a location in space.

2. (a) What is a conservative vector field?

A conservative vector field $\mathbf{F}$ is a vector field that is the gradient of some scalar function f, that is, $\mathbf{F} = \nabla f$.

(b) What is a potential function?

The function f in part (a) is called a potential function for $\mathbf{F}$.

3. (a) Write the definition of the line integral of a scalar function f along a smooth curve C with respect to arc length.

If C is given by the parametric equations $x = x(t)$, $y = y(t)$, $a \leqslant t \leqslant b$, we divide the parameter interval $[a, b]$ into n subintervals $[t_{i-1}, t_i]$ of equal width. The ith subinterval corresponds to a subarc of C with length Δs_i. Then

$$\int_C f(x, y)\, ds = \lim_{n \to \infty} \sum_{i=1}^{n} f(x_i^*, y_i^*)\, \Delta s_i$$

where (x_i^*, y_i^*) is any sample point in the ith subarc.

(b) How do you evaluate such a line integral?

$$\int_C f(x, y)\, ds = \int_a^b f(x(t), y(t)) \sqrt{\left(\frac{dx}{dt}\right)^2 + \left(\frac{dy}{dt}\right)^2}\, dt$$

Similarly, if C is a smooth space curve, then

$$\int_C f(x, y, z)\, ds$$

$$= \int_a^b f(x(t), y(t), z(t)) \sqrt{\left(\frac{dx}{dt}\right)^2 + \left(\frac{dy}{dt}\right)^2 + \left(\frac{dz}{dt}\right)^2}\, dt$$

(c) Write expressions for the mass and center of mass of a thin wire shaped like a curve C if the wire has linear density function $\rho(x, y)$.

The mass is $m = \int_C \rho(x, y)\, ds$.

The center of mass is $(\bar{x}, \bar{y})$, where

$$\bar{x} = \frac{1}{m} \int_C x\rho(x, y)\, ds$$

$$\bar{y} = \frac{1}{m} \int_C y\rho(x, y)\, ds$$

(d) Write the definitions of the line integrals along C of a scalar function f with respect to x, y, and z.

$$\int_C f(x, y, z)\, dx = \lim_{n \to \infty} \sum_{i=1}^{n} f(x_i^*, y_i^*, z_i^*)\, \Delta x_i$$

$$\int_C f(x, y, z)\, dy = \lim_{n \to \infty} \sum_{i=1}^{n} f(x_i^*, y_i^*, z_i^*)\, \Delta y_i$$

$$\int_C f(x, y, z)\, dz = \lim_{n \to \infty} \sum_{i=1}^{n} f(x_i^*, y_i^*, z_i^*)\, \Delta z_i$$

(We have similar results when f is a function of two variables.)

(e) How do you evaluate these line integrals?

$$\int_C f(x, y, z)\, dx = \int_a^b f(x(t), y(t), z(t))\, x'(t)\, dt$$

$$\int_C f(x, y, z)\, dy = \int_a^b f(x(t), y(t), z(t))\, y'(t)\, dt$$

$$\int_C f(x, y, z)\, dz = \int_a^b f(x(t), y(t), z(t))\, z'(t)\, dt$$

4. (a) Define the line integral of a vector field $\mathbf{F}$ along a smooth curve C given by a vector function $\mathbf{r}(t)$.

If $\mathbf{F}$ is a continuous vector field and C is given by a vector function $\mathbf{r}(t)$, $a \leqslant t \leqslant b$, then

$$\int_C \mathbf{F} \cdot d\mathbf{r} = \int_a^b \mathbf{F}(\mathbf{r}(t)) \cdot \mathbf{r}'(t)\, dt = \int_C \mathbf{F} \cdot \mathbf{T}\, ds$$

(b) If $\mathbf{F}$ is a force field, what does this line integral represent?

It represents the work done by $\mathbf{F}$ in moving a particle along the curve C.

(c) If $\mathbf{F} = \langle P, Q, R \rangle$, what is the connection between the line integral of $\mathbf{F}$ and the line integrals of the component functions P, Q, and R?

$$\int_C \mathbf{F} \cdot d\mathbf{r} = \int_C P\, dx + Q\, dy + R\, dz$$

5. State the Fundamental Theorem for Line Integrals.

If C is a smooth curve given by $\mathbf{r}(t)$, $a \leqslant t \leqslant b$, and f is a differentiable function whose gradient vector ∇f is continuous on C, then

$$\int_C \nabla f \cdot d\mathbf{r} = f(\mathbf{r}(b)) - f(\mathbf{r}(a))$$

6. (a) What does it mean to say that $\int_C \mathbf{F} \cdot d\mathbf{r}$ is independent of path?

$\int_C \mathbf{F} \cdot d\mathbf{r}$ is independent of path if the line integral has the same value for any two curves that have the same initial points and the same terminal points.

(b) If you know that $\int_C \mathbf{F} \cdot d\mathbf{r}$ is independent of path, what can you say about $\mathbf{F}$?

We know that $\mathbf{F}$ is a conservative vector field, that is, there exists a function f such that $\nabla f = \mathbf{F}$.

(continued)

7. State Green's Theorem.

Let C be a positively oriented, piecewise-smooth, simple closed curve in the plane and let D be the region bounded by C. If P and Q have continuous partial derivatives on an open region that contains D, then

$$\int_C P\,dx + Q\,dy = \iint_D \left(\frac{\partial Q}{\partial x} - \frac{\partial P}{\partial y}\right) dA$$

8. Write expressions for the area enclosed by a curve C in terms of line integrals around C.

$$A = \oint_C x\,dy = -\oint_C y\,dx = \tfrac{1}{2}\oint_C x\,dy - y\,dx$$

9. Suppose $\mathbf{F}$ is a vector field on $\mathbb{R}^3$.

(a) Define curl $\mathbf{F}$.

$$\text{curl }\mathbf{F} = \left(\frac{\partial R}{\partial y} - \frac{\partial Q}{\partial z}\right)\mathbf{i} + \left(\frac{\partial P}{\partial z} - \frac{\partial R}{\partial x}\right)\mathbf{j} + \left(\frac{\partial Q}{\partial x} - \frac{\partial P}{\partial y}\right)\mathbf{k}$$
$$= \nabla \times \mathbf{F}$$

(b) Define div $\mathbf{F}$.

$$\text{div }\mathbf{F} = \frac{\partial P}{\partial x} + \frac{\partial Q}{\partial y} + \frac{\partial R}{\partial z} = \nabla \cdot \mathbf{F}$$

(c) If $\mathbf{F}$ is a velocity field in fluid flow, what are the physical interpretations of curl $\mathbf{F}$ and div $\mathbf{F}$?

At a point in the fluid, the vector curl $\mathbf{F}$ aligns with the axis about which the fluid tends to rotate, and its length measures the speed of rotation; div $\mathbf{F}$ at a point measures the tendency of the fluid to flow away (diverge) from that point.

10. If $\mathbf{F} = P\,\mathbf{i} + Q\,\mathbf{j}$, how do you determine whether $\mathbf{F}$ is conservative? What if $\mathbf{F}$ is a vector field on $\mathbb{R}^3$?

If P and Q have continuous first-order derivatives and $\frac{\partial P}{\partial y} = \frac{\partial Q}{\partial x}$, then $\mathbf{F}$ is conservative.

If $\mathbf{F}$ is a vector field on $\mathbb{R}^3$ whose component functions have continuous partial derivatives and curl $\mathbf{F} = \mathbf{0}$, then $\mathbf{F}$ is conservative.

11. (a) What is a parametric surface? What are its grid curves?

A parametric surface S is a surface in $\mathbb{R}^3$ described by a vector function

$$\mathbf{r}(u, v) = x(u, v)\,\mathbf{i} + y(u, v)\,\mathbf{j} + z(u, v)\,\mathbf{k}$$

of two parameters u and v. Equivalent parametric equations are

$$x = x(u, v) \qquad y = y(u, v) \qquad z = z(u, v)$$

The grid curves of S are the curves that correspond to holding either u or v constant.

(b) Write an expression for the area of a parametric surface.

If S is a smooth parametric surface given by

$$\mathbf{r}(u, v) = x(u, v)\,\mathbf{i} + y(u, v)\,\mathbf{j} + z(u, v)\,\mathbf{k}$$

where $(u, v) \in D$ and S is covered just once as (u, v) ranges throughout D, then the surface area of S is

$$A(S) = \iint_D |\mathbf{r}_u \times \mathbf{r}_v|\,dA$$

(c) What is the area of a surface given by an equation $z = g(x, y)$?

$$A(S) = \iint_D \sqrt{1 + \left(\frac{\partial z}{\partial x}\right)^2 + \left(\frac{\partial z}{\partial y}\right)^2}\,dA$$

12. (a) Write the definition of the surface integral of a scalar function f over a surface S.

We divide S into "patches" S_{ij}. Then

$$\iint_S f(x, y, z)\,dS = \lim_{m,\,n \to \infty} \sum_{i=1}^{m} \sum_{j=1}^{n} f(P_{ij}^*)\,\Delta S_{ij}$$

where ΔS_{ij} is the area of the patch S_{ij} and P_{ij}^* is a sample point from the patch. (S is divided into patches in such a way that ensures that $\Delta S_{ij} \to 0$ as $m, n \to \infty$.)

(b) How do you evaluate such an integral if S is a parametric surface given by a vector function $\mathbf{r}(u, v)$?

$$\iint_S f(x, y, z)\,dS = \iint_D f(\mathbf{r}(u, v))\,|\mathbf{r}_u \times \mathbf{r}_v|\,dA$$

where D is the parameter domain of S.

(c) What if S is given by an equation $z = g(x, y)$?

$$\iint_S f(x, y, z)\,dS$$

$$= \iint_D f(x, y, g(x, y))\sqrt{\left(\frac{\partial z}{\partial x}\right)^2 + \left(\frac{\partial z}{\partial y}\right)^2 + 1}\,dA$$

(d) If a thin sheet has the shape of a surface S, and the density at (x, y, z) is $\rho(x, y, z)$, write expressions for the mass and center of mass of the sheet.

The mass is

$$m = \iint_S \rho(x, y, z)\,dS$$

The center of mass is $(\bar{x}, \bar{y}, \bar{z})$, where

$$\bar{x} = \frac{1}{m}\iint_S x\,\rho(x, y, z)\,dS$$

$$\bar{y} = \frac{1}{m}\iint_S y\,\rho(x, y, z)\,dS$$

$$\bar{z} = \frac{1}{m}\iint_S z\,\rho(x, y, z)\,dS$$

(continued)

13. (a) What is an oriented surface? Give an example of a non-orientable surface.

An oriented surface S is one for which we can choose a unit normal vector $\mathbf{n}$ at every point so that $\mathbf{n}$ varies continuously over S. The choice of $\mathbf{n}$ provides S with an orientation.

A Möbius strip is a nonorientable surface. (It has only one side.)

(b) Define the surface integral (or flux) of a vector field $\mathbf{F}$ over an oriented surface S with unit normal vector $\mathbf{n}$.

$$\iint_S \mathbf{F} \cdot d\mathbf{S} = \iint_S \mathbf{F} \cdot \mathbf{n}\, dS$$

(c) How do you evaluate such an integral if S is a parametric surface given by a vector function $\mathbf{r}(u, v)$?

$$\iint_S \mathbf{F} \cdot d\mathbf{S} = \iint_D \mathbf{F} \cdot (\mathbf{r}_u \times \mathbf{r}_v)\, dA$$

We multiply by -1 if the opposite orientation of S is desired.

(d) What if S is given by an equation $z = g(x, y)$?

If $\mathbf{F} = \langle P, Q, R \rangle$,

$$\iint_S \mathbf{F} \cdot d\mathbf{S} = \iint_D \left(-P\frac{\partial g}{\partial x} - Q\frac{\partial g}{\partial y} + R \right) dA$$

for the upward orientation of S; we multiply by -1 for the downward orientation.

14. State Stokes' Theorem.

Let S be an oriented piecewise-smooth surface that is bounded by a simple, closed, piecewise-smooth boundary curve C with positive orientation. Let $\mathbf{F}$ be a vector field whose components have continuous partial derivatives on an open region in $\mathbb{R}^3$ that contains S. Then

$$\int_C \mathbf{F} \cdot d\mathbf{r} = \iint_S \operatorname{curl} \mathbf{F} \cdot d\mathbf{S}$$

15. State the Divergence Theorem.

Let E be a simple solid region and let S be the boundary surface of E, given with positive (outward) orientation. Let $\mathbf{F}$ be a vector field whose component functions have continuous partial derivatives on an open region that contains E. Then

$$\iint_S \mathbf{F} \cdot d\mathbf{S} = \iiint_E \operatorname{div} \mathbf{F}\, dV$$

16. In what ways are the Fundamental Theorem for Line Integrals, Green's Theorem, Stokes' Theorem, and the Divergence Theorem similar?

In each theorem, we integrate a "derivative" over a region, and this integral is equal to an expression involving the values of the original function only on the *boundary* of the region.

1. (a) Write the general form of a second-order homogeneous linear differential equation with constant coefficients.

$$ay'' + by' + cy = 0$$

where a, b, and c are constants and $a \neq 0$.

(b) Write the auxiliary equation.

$$ar^2 + br + c = 0$$

(c) How do you use the roots of the auxiliary equation to solve the differential equation? Write the form of the solution for each of the three cases that can occur.

If the auxiliary equation has two distinct real roots r_1 and r_2, the general solution of the differential equation is

$$y = c_1 e^{r_1 x} + c_2 e^{r_2 x}$$

If the roots are real and equal, the solution is

$$y = c_1 e^{rx} + c_2 x e^{rx}$$

where r is the common root.

If the roots are complex, we can write $r_1 = \alpha + i\beta$ and $r_2 = \alpha - i\beta$, and the solution is

$$y = e^{\alpha x}(c_1 \cos \beta x + c_2 \sin \beta x)$$

2. (a) What is an initial-value problem for a second-order differential equation?

An initial-value problem consists of finding a solution y of the differential equation that also satisfies given conditions $y(x_0) = y_0$ and $y'(x_0) = y_1$, where y_0 and y_1 are constants.

(b) What is a boundary-value problem for such an equation?

A boundary-value problem consists of finding a solution y of the differential equation that also satisfies given boundary conditions $y(x_0) = y_0$ and $y(x_1) = y_1$.

3. (a) Write the general form of a second-order nonhomogeneous linear differential equation with constant coefficients.

$ay'' + by' + cy = G(x)$, where a, b, and c are constants and G is a continuous function.

(b) What is the complementary equation? How does it help solve the original differential equation?

The complementary equation is the related homogeneous equation $ay'' + by' + cy = 0$. If we find the general solution y_c of the complementary equation and y_p is any particular solution of the nonhomogeneous differential equation, then the general solution of the original differential equation is $y(x) = y_p(x) + y_c(x)$.

(c) Explain how the method of undetermined coefficients works.

To determine a particular solution y_p of $ay'' + by' + cy = G(x)$, we make an initial guess that y_p is a general function of the same type as G. If $G(x)$

is a polynomial, choose y_p to be a general polynomial of the same degree. If $G(x)$ is of the form Ce^{kx}, choose $y_p(x) = Ae^{kx}$. If $G(x)$ is $C \cos kx$ or $C \sin kx$, choose $y_p(x) = A \cos kx + B \sin kx$. If $G(x)$ is a product of functions, choose y_p to be a product of functions of the same type. Some examples are:

$G(x)$	$y_p(x)$
x^2	$Ax^2 + Bx + C$
e^{2x}	Ae^{2x}
$\sin 2x$	$A \cos 2x + B \sin 2x$
xe^{-x}	$(Ax + B)e^{-x}$

We then substitute y_p, y_p', and y_p'' into the differential equation and determine the coefficients.

If y_p happens to be a solution of the complementary equation, then multiply the initial trial solution by x (or x^2 if necessary).

If $G(x)$ is a sum of functions, we find a particular solution for each function and then y_p is the sum of these.

The general solution of the differential equation is

$$y(x) = y_p(x) + y_c(x)$$

(d) Explain how the method of variation of parameters works.

We write the solution of the complementary equation $ay'' + by' + cy = 0$ as $y_c(x) = c_1 y_1(x) + c_2 y_2(x)$, where y_1 and y_2 are linearly independent solutions. We then take $y_p(x) = u_1(x) y_1(x) + u_2(x) y_2(x)$ as a particular solution, where $u_1(x)$ and $u_2(x)$ are arbitrary functions. After computing y_p', we impose the condition that

$$u_1' y_1 + u_2' y_2 = 0 \qquad \textbf{(1)}$$

and then compute y_p''. Substituting y_p, y_p', and y_p'' into the original differential equation gives

$$a(u_1' y_1' + u_2' y_2') = G \qquad \textbf{(2)}$$

We then solve equations **(1)** and **(2)** for the unknown functions u_1' and u_2'. If we are able to integrate these functions, then a particular solution is $y_p(x) = u_1(x) y_1(x) + u_2(x) y_2(x)$ and the general solution is $y(x) = y_p(x) + y_c(x)$.

4. Discuss two applications of second-order linear differential equations.

The motion of an object with mass m at the end of a spring is an example of simple harmonic motion and is described by the second-order linear differential equation

$$m\frac{d^2x}{dt^2} + kx = 0$$

(continued)

where k is the spring constant and x is the distance the spring is stretched (or compressed) from its natural length. If there are external forces acting on the spring, then the differential equation is modified.

Second-order linear differential equations are also used to analyze electrical circuits involving an electromotive force, a resistor, an inductor, and a capacitor in series.

See the discussion in Section 17.3 for additional details.

5. How do you use power series to solve a differential equation?

We first assume that the differential equation has a power series solution of the form

$$y = \sum_{n=0}^{\infty} c_n x^n = c_0 + c_1 x + c_2 x^2 + c_3 x^3 + \cdots$$

Differentiating gives

$$y' = \sum_{n=1}^{\infty} n c_n x^{n-1} = \sum_{n=0}^{\infty} (n+1)c_{n+1}x^n$$

and

$$y'' = \sum_{n=2}^{\infty} n(n-1)c_n x^{n-2} = \sum_{n=0}^{\infty} (n+2)(n+1)c_{n+2}x^n$$

We substitute these expressions into the differential equation and equate the coefficients of x^n to find a recursion relation involving the constants c_n. Solving the recursion relation gives a formula for c_n and then

$$y = \sum_{n=0}^{\infty} c_n x^n$$

is the solution of the differential equation.